CURA ANIMARUM • ERUDITIO ARTIS MEDICINAE
MDCCCXLVIII
HAHNEMANN UNIVERSITY
PHILADELPHIA
HAHNEMANN UNIVERSITY
LIBRARY

DATE DUE
JUL 0 6 2004
JUN 2 7 2006
NOV 16 2007
D1212843
GAYLORD
#3523PI
Printed in USA

FACIAL GROWTH

3rd Edition

Donald H. Enlow, Ph.D.

Thomas Hill Distinguished Professor Emeritus
Department of Orthodontics
School of Dentistry
Case Western Reserve University
Cleveland, Ohio

Illustrations by

William Roger Poston, II

Director of Biomedical Communications
University of Louisville

1990

W.B. SAUNDERS COMPANY
Harcourt Brace Jovanovich, Inc.

Philadelphia London Toronto Montreal Sydney Tokyo

W. B. SAUNDERS COMPANY
Harcourt Brace Jovanovich, Inc.

The Curtis Center
Independence Square West
Philadelphia, PA 19106

Library of Congress Cataloging-in-Publication Data

Enlow, Donald H.

Facial growth / Donald H. Enlow.—3rd ed.

p. cm.

Includes bibliographical references.

1. Face—Growth. 2. Facial bones—Growth. 3. Skull—Growth. I. Title.

QM535.E47 1990 621′.92—dc20

ISBN 0–7216–2843–5 89–37208

Listed here is the latest translated edition of this book together with the language of the translation and the publisher.

Japanese (1st Edition)—Ishiyaku Publishers, Inc., Tokyo, Japan

Italian (2nd Edition)—Editrice Cides Odonto Edizioni Internazionali, Turin, Italy

German (2nd Edition)—Quintessenz Verlag, Berlin, W. Germany

Spanish (1st Edition)—Inter-Medica, Buenos Aires, Argentina

Spanish (2nd Edition)—Nueva Editorial Interamericana S.A., Mexico City, Mexico

Editor: John Dyson
Designer: Joan Owen
Production Manager: Linda R. Turner
Manuscript Editor: Jessie Raymond
Illustration Coordinator: Ceil Kunkle
Indexer: Linda Van Pelt

Facial Growth, Third Edition ISBN 0–7216–2843–5

Printed in the United States of America.

Last digit is the print number: 9 8 7 6 5 4 3 2 1

Contributors

Rolf G. Behrents, D.D.S., M.S., Ph.D.
Professor and Chairman, Department of Orthodontics, College of Dentistry, University of Tennessee, Memphis, Tennessee

B. Holly Broadbent, Jr., D.D.S.
Clinical Professor of Orthodontics, School of Dentistry, and Director, Bolton-Brush Growth Study Center, Case Western Reserve University, Cleveland, Ohio

David S. Carlson, Ph.D.
Professor of Dentistry, School of Dentistry, and Center for Human Growth and Development, The University of Michigan, Ann Arbor, Michigan

Robert Cederquist, L.D.S., D.D.S., M.S., M.A., Ph.D.
Chairman, Department of Orthodontics, School of Dental and Oral Surgery, Columbia University, New York, New York

M. Michael Cohen, Jr., D.M.D., Ph.D.
Professor of Oral Pathology, Faculty of Dentistry, Dalhousie University, Halifax, Nova Scotia, Canada

J. M. H. Dibbets, D.D.S., Ph.D.
Professor and Chairman, Orthodontic Department, School of Dentistry, University of Groningen, Groningen, The Netherlands

W. Stuart Hunter, D.D.S., Ph.D.
Professor and Chairman, Department of Pediatric and Community Dentistry, The University of Western Ontario, London, Ontario, Canada

William W. Merow, D.D.S. (Deceased)

Robert E. Moyers, D.D.S., Ph.D.
Center for Human Growth and Development, The University of Michigan, Ann Arbor, Michigan

Ordean J. Oyen, Ph.D.
Departments of Oral Biology and Orthodontics, School of Dentistry, Case Western Reserve University, Cleveland, Ohio

Preface

Material in the former edition that I find I have not been including in my own graduate and undergraduate facial growth courses has been eliminated, and much of the rest of the book also has been tightened up. The reason is to make room for other information that I *do* include but which has not, until now, been part of the book. Thus, additions have been patched in throughout, but as concisely as possible, in order to update or to complement. Importantly, new sections are added by persons known for their expertise in particular subjects. Thus, now included are textbook-type introductions to TMJ, birth defects, facial heredity, adult facial growth, growth control, biomechanics, and general body growth. The purpose is to offer more complete coverage of the range of subjects most teachers feel is needed for a balanced course.

As before, a few of the chapters have a Part 1 section that presents a "short form" of that chapter's subject. It gives a less-detailed overview appropriate for undergraduate courses, "short courses," subject review, preparation for special seminars and lectures, study clubs, etc. The Part 2 that follows gives much more in-depth information. At the graduate level, many teachers will wish to give a follow-up course in which the classic literature is presented, evaluated, and discussed by the graduate students and residents.

In the development and preparation of this third edition, I am pleased to acknowledge the skills and the considerable and tolerant efforts of Marion Hurwitz. I continue to be most grateful for the expertise, cheerful cooperation, and competent collaboration provided by the editors and staff of W. B. Saunders Company.

DONALD H. ENLOW

Contents

CHAPTER ONE

Faces

In your lifetime, you have seen the faces of thousands of people, and each face is recognizable to you as distinctively individual. No two are quite alike, even those of identical twins. Every person's face is a custom-made original; there has never been another face exactly the same before, and there never will be again. Yet consider how relatively few parts compose a face: a lower jaw and chin, cheekbones, a mouth and upper jaw, a nose, and two orbits. Add a forehead and supraorbital ridges for the neurocranial parts of the face. How is it possible that so few components can underlie such great variation in facial form?

The answer is that we have the ability to perceive exceedingly subtle differences in the relative shape, spread, and proportions of both hard and soft tissue parts and minute variations in the topographic contours among all of them. A very slight alteration in the configuration of the nose, for example, makes a substantial difference in the appearance and the character of one's face as a whole. (See Figure 1–1, which shows a sketch from photographs of the same person before and after rhinoplasty.) Furthermore, there is the particular "set" to a person's mouth, the personal sparkle in the eyes, and the tone in the muscles of facial expression that are quite individualized. Often we ask, "Who does that person remind you of?" because there is some unique combination of nasal contour, lip configuration, jaw shape, and so on, that resembles some other face known to us.

Anthropologists can "reconstruct" the face from a dry skull by use of normative population data that provide integument thicknesses in the different areas of the face. However, the results can provide only a general approximation, because population "averages" can never match the delicate features of a given individual in all, or even most, regards. Everybody is familiar with the method by which a police department artist attempts to draw a suspected felon's face from the recollections of eyewitnesses. Sometimes the artist's "composite" picture can be close enough to give a more or less recognizable likeness, but sometimes it is not. It depends on how thoroughly the witness can recall and visualize key facial features. Also, the effectiveness of the artist's rendering depends on how accurately the witness can select the proper features from the police department's "catalogue," picturing different noses, cheekbones, hairlines, eyebrows, chins, and so on. As pointed out before, relatively subtle

FIGURE 1–1

differences in a given feature can produce a quite noticeably different overall facial "character."

In the next few pages, the biologic rationale underlying common variations in facial features is described. Three general considerations are taken into account: (1) different facial types as they relate to variations in the overall form and shape of the whole head, (2) male and female facial differences, and (3) child and adult facial differences. As you study these variations, you will begin to realize that some of the same characteristics relate to all three categories, essentially for similar physiologic and morphologic reasons.

HEAD FORM

Two general extremes exist for the shape of the head: the long, narrow (dolichocephalic) head form and the wide, short, globular (brachycephalic) head form. The facial complex attaches to the basicranium, and the cranial floor is the template that establishes many of the dimensional, angular, and topographic characteristics of the face. The dolichocephalic head form, therefore, sets up a face that is correspondingly narrow, long, and protrusive. This facial type is termed leptoprosopic. Conversely, the brachycephalic head form establishes a face that is broad but somewhat less protrusive, and this is called a euryprosopic facial type.

In Figure 1–2, observe what happens when a skull is dolicho- or brachycephalized. If faces are cast onto rubber balloons and the balloons are then either squeezed or stretched, as shown, distinctively divergent facial patterns occur with regard to the forehead, the shape of the nose, the set of the eyes, the prominence of the cheekbones, the contour of the facial profile, the degree of flatness (depth) of the face, and the position of the mandible. Note that the dolichocephalic nose is vertically longer and much more protrusive. The pug-like brachycephalic nose is shorter, and it has a more rounded tip. Even though quite different in configuration, the design is such that approximately equivalent airway capacity exists, since the proportionately wider nasal chambers and nasopharynx in the brachycephalic type tend to be vertically shorter. This sets

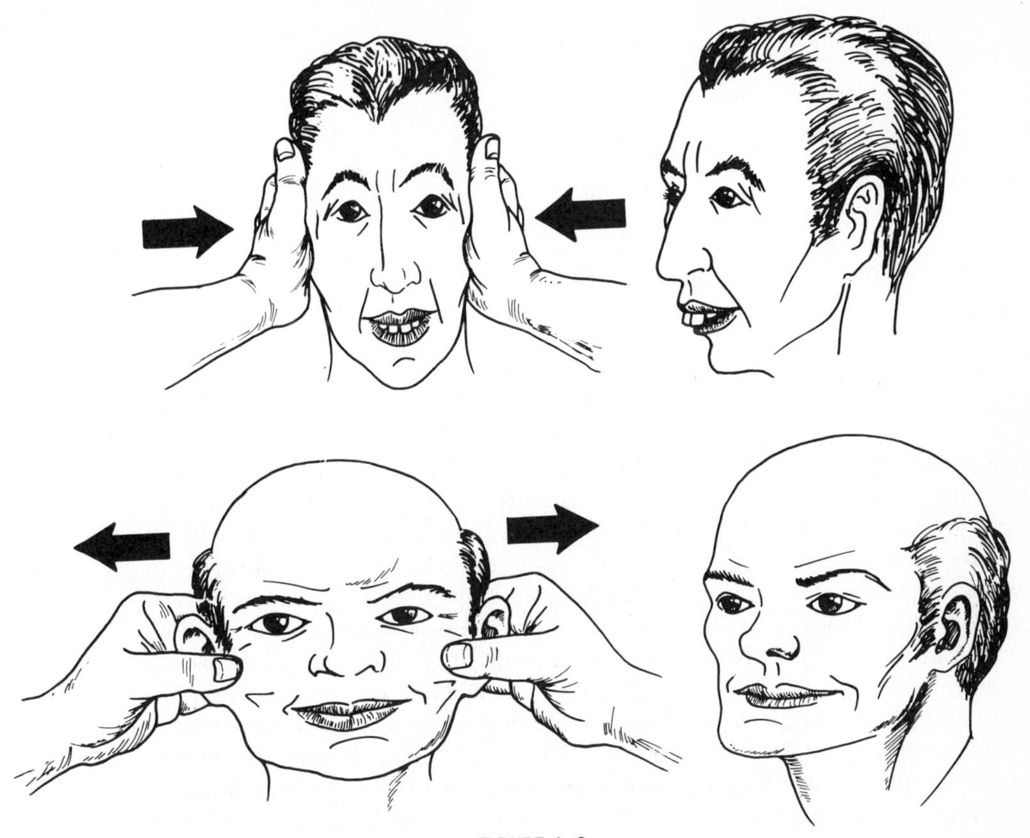

FIGURE 1–2

up the vertically shorter midfacial feature of the wide-face type, which in turn establishes a number of other facial features distinguishing it from the longer and more narrow midface of the leptoprosopic (including differing malocclusion tendencies, as described in a later chapter). Because the proboscis in the long and narrow type of face is also much more protrusive, the bridge and root of the nose tend to be much higher. In the dolichocephalic, also, the slope of the nasal profile tends to follow the same slope of the forehead, in contrast to the brachycephalic nose as it breaks from a more bulbous forehead. Because the **upper** part of the dolichocephalic nose is also quite protrusive, the nose sometimes "bends" to produce an aquiline ("Roman," "Dick Tracy") type of convex nasal contour; and the end of the more pointed nose frequently tips down. The degree of the bending and down-turning increases with increasing height of the nose. Thus, aquiline convexity becomes much more marked in persons having vertically longer noses. In contrast, the more stubby brachycephalic nose tends to be straighter or often concave, and it frequently tips up with the external nares showing in a face-on view. (Note: A "third" nasal configuration also exists among some long-nosed dolichocephalics in which the **middle** part of the external nose is protrusive relative to an upper part that is much less so. In this type, the nose displays a graceful, recurved, S-shaped configuration. The interorbital region of the nasal bridge is lower, and supraorbital protrusion is less marked.)

Because the nasal part of the narrow (leptoprosopic) type of face is more protrusive, the external bony table of the contiguous forehead is correspondingly more sloping, and the glabella and upper orbital rims tend to be much

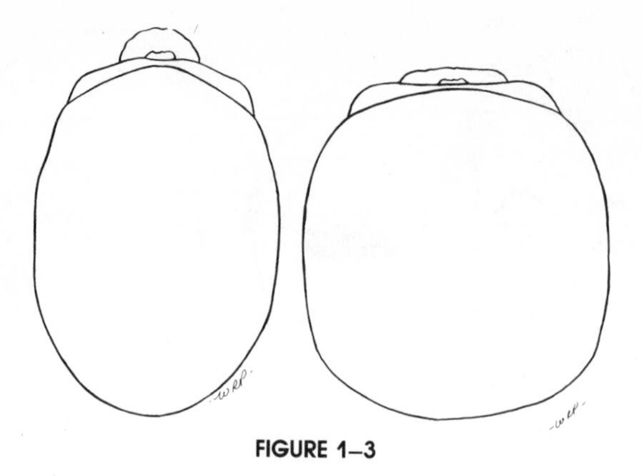

FIGURE 1–3

more prominent. The forehead of the wide (euryprosopic) facial type is more bulbous and upright, and the frontal sinus tends to be thinner because of the lesser degree of separation between the inner and outer tables of the forehead. The more protrusive nature of the nasal region and the supraorbital ridges in the dolichocephalic type of head form gives the cheekbones a much less prominent appearance, and the eyes look more deep-set for the same reason. As seen from above (Fig. 1–3) as well as laterally, the dolichocephalic face is more angular and less flat. In the brachycephalic head form, the wider, flatter, and less protrusive face gives the cheekbones a noticeably squared configuration and a more prominent-looking character. The brachycephalic eyeballs tend to be more exophthalmic (bulging, proptotic) because of the shorter anterior cranial fossa (the floor of which serves as the roof for each orbit). The broad brachycephalic face appears quite shallow in comparison with the deep and topographically more bold contours of the dolichocephalic face. The vertically long nature of the midface and the "open" (obtuse) form of the cranial base flexure in dolichocephalics (Chapter 6) relate to a downward-backward rotational alignment of the mandible. This results in a tendency for a retrusively placed mandible and lower lip and a retrognathic (convex) facial profile. The brachycephalic face relates to a more "closed" basicranial flexure; and as a result, the lower jaw tends to be variably more protrusive, with a greater tendency for a straighter or even concave facial profile and a more prominent-appearing chin. The vertically shorter midface tends to highlight a more prominent appearance of the lower jaw. The more upright (closed) nature of the brachycephalic basicranium produces a tendency for more erect head posture, in contrast to a tendency for a more slumped stance in many individuals with a dolichocephalic head form. The narrow but longer anterior cranial fossa in the dolichocephalic head form results in a correspondingly longer but narrower, deeper maxillary arch and palate. The broad but anteroposteriorly shorter brachycephalic type of anterior cranial fossa sets up a wider but shorter palate and maxillary arch. The palate, therefore, is a proportionate projection of the anterior cranial fossa, and the apical base of the maxillary dental arch, in turn, is established by the perimeter of the palate. A link thereby exists between brain and basicranium down to arch configuration. These same long-narrow and short-wide cranial and facial relationships are also seen in other mammalian species, e.g., the Doberman pinscher or collie versus the bulldog or boxer.

Among most of the world's different population groups, either the brachycephalic or the dolichocephalic type of head form tends to predominate in any given group. However, a distribution **range** from one extreme to the other also usually exists within a group, even though one particular side of the range is more common. An intermediate head form type (mesocephalic) can occur, and the facial features tend to be correspondingly intermediate. In the northern and southern edges of some parts of continental Europe, as well as in England, Scotland, Scandinavia, northern Africa, and some Near and Middle Eastern countries (e.g., Iran, Afghanistan, India, Iraq, and Arabia), the dolichocephalic head form tends to predominate. In central Europe (the Alpine head form) and the Far East (Oriental), the tendency is toward brachycephaly. Interestingly, at the geographic interface between the dolicho- and brachycephalic regions of the world, a "third" and quite distinctive type of head form commonly occurs, the **dinaric** (after the Dinaric Alps, in Yugoslavia). Such interface areas include regions located between middle and northern Europe, between central and southern Europe, and between Europe and the Near East. Thus, geographically separate lines of this head form have independently appeared and have become fairly common. Admixtures of head form types do not, however, necessarily produce a consistent "mesocephalic" result, although this as well as pure "dolicho/brachy" offspring (mendelian-type) ratios can occur in an individual family. In the dinaric, although technically brachycephalic because it is antero-posteriorly short, it is primarily the **posterior** part of the skull that has been brachycephalized (Fig. 1–4). The occipital or lambdoidal regions have become widened and/or markedly **flattened.** Lateral peaking (bossing) of the parietal regions can occur, and the skull often has a distinctively triangular configuration when viewed from the top. A common variation also occurs in which bossing is directed more upward than bilaterally, thereby forming an elevated hump in the posterosuperior part of the skull dome. Cranial configuration is less triangular when viewed from the top. Whatever the manner of bossing, it accommodates the volumetric mass of a brain that has undergone alteration in overall shape. The old practice of "cradling"* intensifies the degree of occipital flattening. Indeed, the sleeping posture of an infant, whether or not cradled as such, may be a factor in "dinarizing" a growing head to some greater or lesser

*Immobilizing an infant on its back by swaddling clothes.

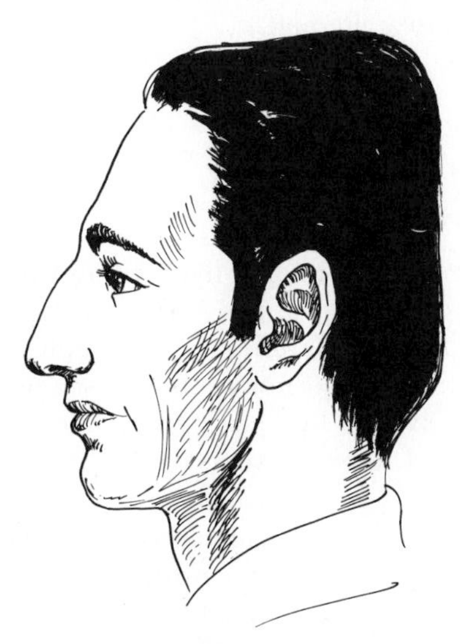

FIGURE 1–4

extent. It has been argued that such influences during infancy represent the dominant reason for the dinaric form, since it has been shown that American descendants of Old World dinaric grandparents can lose the dinaric features when not cradled, and cradling is not a common custom in the United States. It is also argued, however, that a "genetic" dinaric form also exists, but studies have not to date entirely clarified the issues involved.

The ears of the dinaric characteristically appear much closer to the back of the head because of the occipital flattening. The anterior part of the skull, however, has retained the relative narrowness that characterizes the dolichocephalic pattern. The narrow face from the dolichocephalic side of the ancestral heritage has perhaps "constrained" the anterior cranial fossa, thereby sustaining a narrow dimension in this part of the basicranium. Even though the head form is technically brachycephalic, the form of the face itself is distinctively leptoprosopic, although the posterior facial parts (such as the mandibular ramus) tend to flare laterally in many individuals because they grade farther back on a widening cranial triangle. The forehead is usually quite sloping, the supraorbital ridges are prominent, and the face is long and topographically protrusive. The nose tends to be very large and often aquiline (this occurs in many females as well as males), and the nasal bridge is high. The mandible tends to be less retrusive and the face less retrognathic. This is because the basicranial flexure is compressed and more closed (see page 209). The midface, however, tends to be vertically long in proportion, even more so than among dolichocephalic leptoprosopics. These various leptoprosopic features often appear exaggerated in character in the dinaric head form, almost as though the brachycephalized, flattened posterior part of the cranium "pushed" the face even more protrusively than in a routine dolichocephalic head form. Any malocclusion in a dinaric will probably have a combination of structural features different from that in a dolichocephalic. Both, in turn, will be different in malocclusion anatomy from a brachycephalic, and treatment responses and rebound tendencies will also probably be different.

While the dinaric head form has been historically perceived as a "brachycephalized dolichocephalic," it is probable that **any** headform type, including the brachycephalic, is susceptible to this modification, whether genetically or ontogenetically. Various degrees of modification, further, appear to exist but have yet to be studied and catalogued.

FACIAL FEATURES

Male Versus Female

A reasonably talented artist can effectively render male versus female faces, and the viewer has no problem recognizing either sex from sketches or portraits of adults. (Children are another matter, as will be seen.) However, many artists, as well as the average citizen, are not really conscious of the actual, specific anatomic differences involved. They just "know." In our mind's eye, we have all subconsciously associated, over the years, the topographic characteristics that relate to facial dimorphism.

The overall body size of the male tends to be larger than that of the female, and the male lungs are correspondingly more sizable to provide for the relatively more massive muscles and body organs. This calls for a larger airway, beginning with the nose and nasopharynx. A principal sexual dimorphic

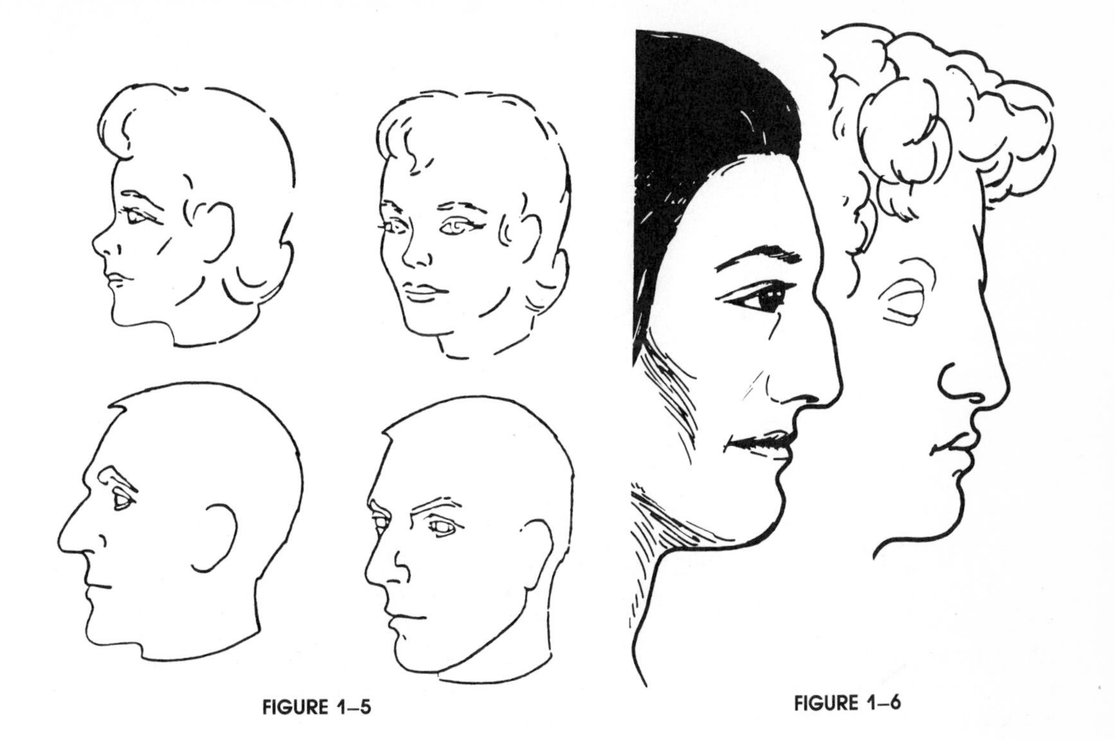

FIGURE 1–5

FIGURE 1–6

difference, therefore, is the size (and configuration) of the nose, and this in turn leads to collateral differences in other topographic structures of the face.

The male nose is proportionately larger than the female nose (Fig. 1–5). This is a "population" feature based on general comparisons among large numbers of people; any given female or male, of course, can display a smaller or larger nose. The male nose, in general, tends to be more protrusive, longer, wider, and more fleshy and to have larger and more flaring nostrils. The interorbital part of the nasal bridge in the male tends to be much higher. All of this is in contrast to a relatively thin and less protrusive female nose. The male nose usually ranges from a straight to a convex (aquiline) profile, whereas the female nose tends to range from a straight to a somewhat concave profile. The tip of the male nose is often more pointed and has a greater tendency to turn downward, and the somewhat more rounded female nose often tips upward. A variation of the aquiline (Roman) type of nose, which is also much more prevalent among males than females, is the classic "Greek" nose, in which the nasal profile drops almost straight downward from a protruding forehead (Fig. 1–6). The reason for these male variations lies in the more protuberant nature of the whole nasal region. Both the upper and lower parts of the whole external nose are protrusive, but the lower part can be constrained to a degree by the palate and maxillary arch. The contour of the nose can thus "rotate"; it either "bends" to give an aquiline configuration or rotates into a straight but more vertical alignment.

Because of the larger, more protuberant character of the male nose, the part of the forehead contiguous with it also necessarily grows into a more protrusive position. Therefore, the male forehead tends to be more sloping, in contrast to a more bulbous, upright female forehead. The supraorbital and glabellar parts of the male forehead tend to be **quite** protrusive, as compared with the much **less** Neanderthal-like character of the female forehead. This,

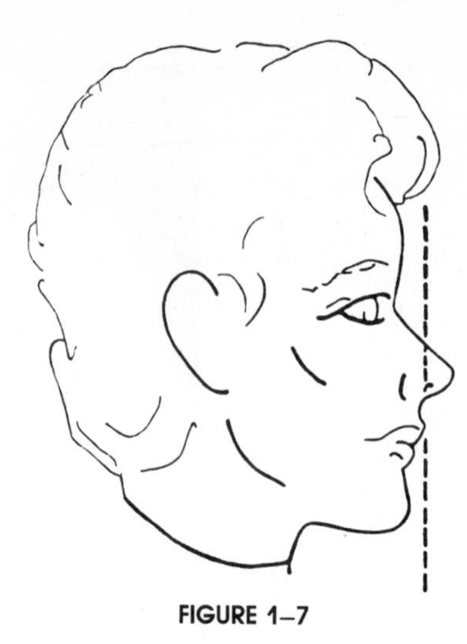

FIGURE 1–7

together with the relative size and vertical alignment of the nose, provides two of the features most readily recognizable in our subconscious perceptions of male and female facial differences. In Figure 1–7, note the dashed line passing vertically along the surface of the upper lip perpendicular to the neutral orbital axis. In the female, this line usually crosses about midway along the slope of the upper nasal margin, and the forehead lies well behind the line. Conversely, in the male, the nose and forehead are often so protrusive that the forehead bulges out as far as this line or sometimes even beyond it, and the greater part of the nose often lies ahead of it.

Because of the greater extent of protrusiveness of the male forehead and nose, the eyes appear more deep-set. In the female the eyes appear more proptotic and "closer to the front" of the face. Female cheekbones also "look" much more prominent for the same reason; that is, the malar protuberances seem more apparent because the nose and forehead are less prominent. Indeed, high cheekbones are a classic feature of femininity, much emphasized by beauty analysts. Of course, the malar protuberances are not actually "higher," they are just more conspicuous. This topographic feature of the female face is better seen in a 45-degree view of the face (see Fig. 1–5). In addition, the temporal region along the side of the forehead tends to be less bulgy in the female.

The protuberant supraorbital part of the more sloping male forehead (because of the larger nose) is produced by a greater extent of separation between the inner and outer tables of the frontal bone. In both sexes, the growth of the inner table stops when the enlargement of the frontal lobes of the cerebrum ceases around 5 to 6 years of age. The outer table, however, continues to remodel forward, until contiguous nasal growth ceases some years later. The inner and outer tables thereby separate, and the cancellous bone between them is hollowed into the frontal sinus. Because the nasal part of the male face continues to grow for several years more than that of the female, the frontal sinus is therefore much larger in the male face than in the more juvenile-like female face. Also, because of the smaller frontal sinuses in a female, the temporal regions of the lateral forehead often appear less full.

Because the forehead and nose are less protrusive in the female face, the upper jaw "looks" more prominent and muzzle-like. For the same reason (less protrusion of the forehead and nose), together with the more prominent and squared-appearing cheekbones, the female face appears flatter than the topographically more coarse, irregular, and deep male face.

Now remember some of the facial features that distinguish the dolichocephalic from the brachycephalic head form (degree of supraorbital protrusion, forehead slope, cheekbone prominence, nasal configuration, and depth or flatness of the whole face). Some of these head form characteristics are the same as those that distinguish the male from the female face, and for essentially the same reasons. The long, narrow **dolichocephalic** nose leads to facial features that essentially parallel those of the **male** face; the short, wide **brachycephalic** nose sets up facial features that also characterize the **female** face. It is the nasal part of the face that thereby underlies much of the overall character of a person's face. In the "brachy/dolicho" situation, it is the width and length of the basicranium that establish the basis for nasal form and size. In the male/female comparison, it is relative body size that leads to corresponding nasal characteristics, which in turn establish the other facial features that are analogous to those associated with head form type.

Then what about a female dolichocephalic? Or a male brachycephalic? Or a female dinaric? Most people feel a comforting confidence that they can nevertheless usually distinguish sex by a person's face. However, recognition tests have shown that it is actually not all that easy. If ancillary clues (e.g., hairstyle, cosmetics) are removed from facial photographs and other kinds of clues are not available (e.g., voice, clothing, gait, presence or absence of a protruding larynx, neck circumference, breadth of shoulders), recognition tests in which facial photographs are presented can lead to dismay and despair. The chances of a correct identification for many individual faces are often not much better than a toss-up.

In a female **brachycephalic,** the characteristics of a wider and flatter face, smaller nose, squared cheekbones, and an upright forehead tend to augment and emphasize the same dimorphic features that are also related to sex. Conversely, in the female **dolichocephalic,** the narrow, more protrusive facial characteristics associated with this head form type tend to give a more "male-like" cast to the face, although, of course, these features are **not** really masculine at all, but, rather, head form–linked. Thus, a narrow-faced woman can have a more sloping forehead, greater supraorbital protrusion, a higher nasal bridge, a longer nose, an aquiline or more vertically aligned nasal contour, and a downturned and more pointed nasal tip. The average citizen has subconsciously learned the difference between brachycephalic and dolichocephalic females, even though few have any idea what a brachycephalic or a dolichocephalic is. In a male brachycephalic, the reverse situation exists. The "female-like" characteristics of the brachycephalic present a flatter, wider face with more prominent cheekbones; a more bulbous forehead; a smaller, less protrusive nose with a lower nasal bridge; and a tendency for a concave nasal profile with a more rounded and upturned tip. One frequently sees female impersonators on TV (for example, a detective disguised as a lady of the street); the realism of the disguise is favored if the male is a wide-faced, small-nosed brachycephalic.

Child Versus Adult

The faces of prepubertal boys and girls are essentially comparable. How many times have you been embarrassed by calling a little boy a girl? In the female, facial development begins to slow markedly after about 13 or so years of age. At about the time of puberty for the male, however, the sex-related dimorphic facial features described earlier begin to be manifested, and this maturation process of the facial superstructures continues actively throughout the adolescent period and into early adulthood. This is a factor to be taken into account by the orthodontist and orthognathic surgeon in treatment planning for girls and boys.

Whether a young child's head form is dolichocephalic or brachycephalic, the youthful face itself looks more brachycephalic because it is still relatively wide and vertically short. It is wide because the brain, and therefore the basicranium, is precocious relative to facial development. The neurocranium grows earlier, faster, and to a much greater extent than the contiguous facial complex. The wider basicranium, because it establishes the positions of the glenoid fossae for the mandible and the facial-cranial sutures for the nasomaxillary complex, is thereby a template that also paces the early **width** of the growing face. The face is **vertically** short, however, because (1) the nasal part of the face is still small (overall body and lung size is still correspondingly small); (2) the primary and secondary dentition has not yet become fully established; and (3) the jaw bones have not yet grown to the vertical extent that will later support the full dentition and the enlarging masticatory muscles and airway.

Note the features of the child's face, regardless of sex or head form type: the nose is short, rounded, and pug-like; the nasal bridge is low; the nasal profile is concave; the forehead is bulbous and upright; the cheekbones are prominent; the face is flat; and the eyes are rather wide-set and bulging.

The same situation, noted twice before, thus exists; some of the **same** facial features that characterize **both** head form and sexual dimorphism **also** relate to the difference between the facial features of the child and the adult. In all three categories, the character of the nasal part of the face is a key factor that relates to the other facial features (forehead slope, nasal configuration, height of nasal bridge, cheekbone prominence, flatness of the face, and the general extent of facial protrusiveness).

How does one recognize advancing age in an **older adult** by external facial appearance? There was a time when edentulism caused major changes in facial structure and topography. In many parts of the world, modern dentistry has effectively precluded much of this. Other facial changes, however, still occur. The velvety, soft, pink, tightly bound, resilient, and firm skin of the child becomes replaced over the years by the more leathery, crinkled, open-pored, limp, blemished skin that characterizes more aged persons. As one advances through middle age, the integument begins to droop and sag noticeably. Several physical and biochemical changes are occurring in the connective tissue of the dermis and hypodermis that cause the skin to become less firmly anchored to underlying bone or facial muscles. First, if general loss of body weight occurs for whatever reason as one ages, resorption of subcutaneous adipose results in a "surplus" of skin, which leads to sagging, wrinkling, and creasing. Loss of adipose thus exaggerates age appearance. After dieting, for example, a face often looks older. The effect can occur in children as well; the sight of a severely malnourished child with a wrinkled, lined, hollow face is not easily forgotten. Second, the distribution and character of the collagenous matrix change with advancing age. Fibers increase in massiveness, and the whole skin decreases in resilience. Third, fibroblasts decline in number as well as cellular activities. The latter includes a marked decrease in the secretion and overall amount of water-bound protein mucopolysaccharides (proteoglycans). Because of this, a widespread subcutaneous dehydration occurs that contributes significantly to shrunken facial volume and skin surplus, with consequent skin wrinkling. In advanced old age, a person's face can become an expansive carpet of noble ripples and lines. Resorption of adipose in the orbit can lead to a sunken appearance of the eyes, and the more visible venous plexus in the thinned suborbital hypodermis produces a darkening of the skin below the eyes. The suborbital integument can also begin to sag perceptibly to form "bags." Something also happens to the youthful "sparkle" in many persons' eyes as they age. By waxing artificial wrinkles onto the face and blue-tinting the suborbital region, a good Hollywood make-up technician can "age" a face in minutes. Observe closely, however, and you will note that, unlike real skin, the artificial furrows are not as motile during attempts at expressive facial movements.

Facial lines and wrinkles develop in specific and characteristic locations, particularly during the middle-age period (Fig. 1–8). One of the initial lines to appear is the prominent **nasolabial furrow.** This "smile line" is seen at any age when one grins, but it becomes a fixed feature of the face sometime during the late 30s or 40s in many people. It extends from the lateral side of each nasal ala down to the corners of the mouth. This is a particular facial feature that we have subconsciously learned to associate with the onset of middle age. Stopping smiling does not seem to help.

Other wrinkles and creases begin to develop as crow's feet at the lateral corners of the eyes, horizontal lines on the forehead, vertical corrugations overlying the glabella, vertical furrows along the upper lip, lines extending down from the corner of the mouth lateral to the chin, a horizontal crease just above the chin, suborbital lines, drooping jowls over the sides of the mandible, and a "turkey gobbler" bag of skin sagging down over the neck below the chin. The placement, alignment, prominence, and number of such lines and creases, as well as many other topographic facial features, are utilized as presumptive clues by the physiognomist (a practitioner of the ancient Chinese art of face reading) to judge a person's character, temperament, and ultimate fate. However, no real functional, preprogrammed, cause-and-effect relationships are likely to exist in most such correlations; there are just too many physiologic, anatomic, environmental, social, developmental, and ethnic variables involved in the biology of the human face.

How about the person who "looks younger than his or her years"? Or

FIGURE 1–8. (From Enlow, D. H.: Faces. Dent. Dimensions, 1:4, 1977.)

older? For reasons only partially understood, the onset of the smile line and some other facial wrinkles is delayed, or at least such lines look less marked, in youthful-appearing individuals. Conversely, in others, lines can appear more harsh and begin to develop at an earlier age. Intrinsic physiologic as well as environmental factors can contribute to this. For example, sun damage to the facial integument, particularly in lighter-complexioned individuals, is known to accelerate the aging process of skin. Furthermore, chronic alcoholism causes the muscles of facial expression within the skin to sag because of the long-term anesthetized-like and limp state of their tone. Alcohol, further, is a dehydrator. Smoking tends to intensify wrinkling because tobacco is a peripheral vasoconstrictor. Major loss of adipose can also accelerate the onset of facial wrinkles, as previously explained. In addition, a euryprosopic (brachycephalic) type of face appears more juvenile-like because it resembles the wide-face configuration characterizing a child. A dolichocephalic adult face "looks" more mature because the nasal region is vertically longer. A fat face looks younger (1) because the subcutaneous adipose tends to smooth out wrinkles that would otherwise be much more prominent and (2) because it resembles the buccal fat-padded face of a child. Thus, a wide-faced, more chunky, sober, nonsmoking, darker-skinned individual, particularly one protected from undue sun exposure, tends to retain a more youthful appearance somewhat longer.

TOPOGRAPHIC FACIAL VARIATIONS

The word "face" is one of those much used terms that have multiple definitions and almost limitless shades of meaning. Indeed, the unabridged dictionary gives, literally, several dozen different connotations. The primary derivation is "to make" or "to form" (as to sur**face** a piece of wood in furniture making). Our meaning, of course, has to do with what we face the world with, and as a professional student of the face, you can take every opportunity to study people's faces whenever and wherever possible—watching them on TV, studying noses while waiting in lines, analyzing profiles seen in magazine photographs, and so on.

Subtle variations in topographic contours and proportions have a major impact on the character of a person's face. As a career professional dealing with the faces of individual human beings, from this day forward you will never again be content with just eyeballing a face without analyzing it. You will take note of head form and facial type, whether the facial features fully conform with classic descriptions for dolichocephalics and brachycephalics, dinarics, and leptoprosopics or euryprosopics, male and female, child or adult, or whether some combination of virtually limitless facial variations occurs.

As a start, take note of the relative height of the nasomaxillary complex, compared with the mandible, and the marked effect this proportion has on overall facial pattern. Mentally divide the front of the head in a face-on view into thirds: the forehead; the midface, from the eyebrows to the bottom of the nose; and the lower face, from the inferior edge of the nose to the bottom of the chin. Are they approximately equal, giving "ideal" facial proportions, or does one or another appear longer or shorter? Observe the degree of facial profile convexity or concavity versus a straight line (orthognathic) profile. Is the mandible massive, full, and heavily built or relatively gracile? Is there a single frontal boss or two? Does the hairline come to a midline peak with two lateral recesses, or is it uniformly rounded? Observe the degree of roundness of the orbits and any obliquity, and note whether the eyebrows follow the supraorbital rims. Where does a vertical line from the pupil lie with respect to the corner of the mouth? Which tooth lies at the mouth's corner? Is the external nose broad or narrow? If broad-appearing, is this only because the nasal bridge is low and the nose relatively flat? Note whether the overall prominence of the maxillary "muzzle" is a principal profile feature. Note the effect of incisor and canine dentoalveolar form on the lips. Is the upper lip vertically short, long, or proportionately balanced with the face? Lip thickness? Are the upper incisors tipped outward by the lowers to produce bimaxillary protrusion? Extent of overjet and overbite? Open bite? How prominent is the "Cupid's bow" of the upper lip? Are the lips closed? Mouth breathing? Does one lip protrude beyond the other? Is the upper lip concave or flat? Which teeth show when the mouth is open, the uppers or the lowers? Is there any noticeable bilateral asymmetry among any of the orbital, nasal, or maxillomandibular parts? Note whether the transverse biorbital width is greater or less than frontal width, with either the cheekbones or the temporal region protruding laterally beyond the other. Are the eyes more wide-set or tending toward hypotelorism? Do the eyes sparkle, or do they lack luster? Are the facial muscles expressionless, or are they alive with vibrant tone?

Figures 1–9 to 1–12 catalogue a number of such topographic facial variations. A good artist can render a "composite" facial representation of a given person by combining individualized feature variations, adding character-

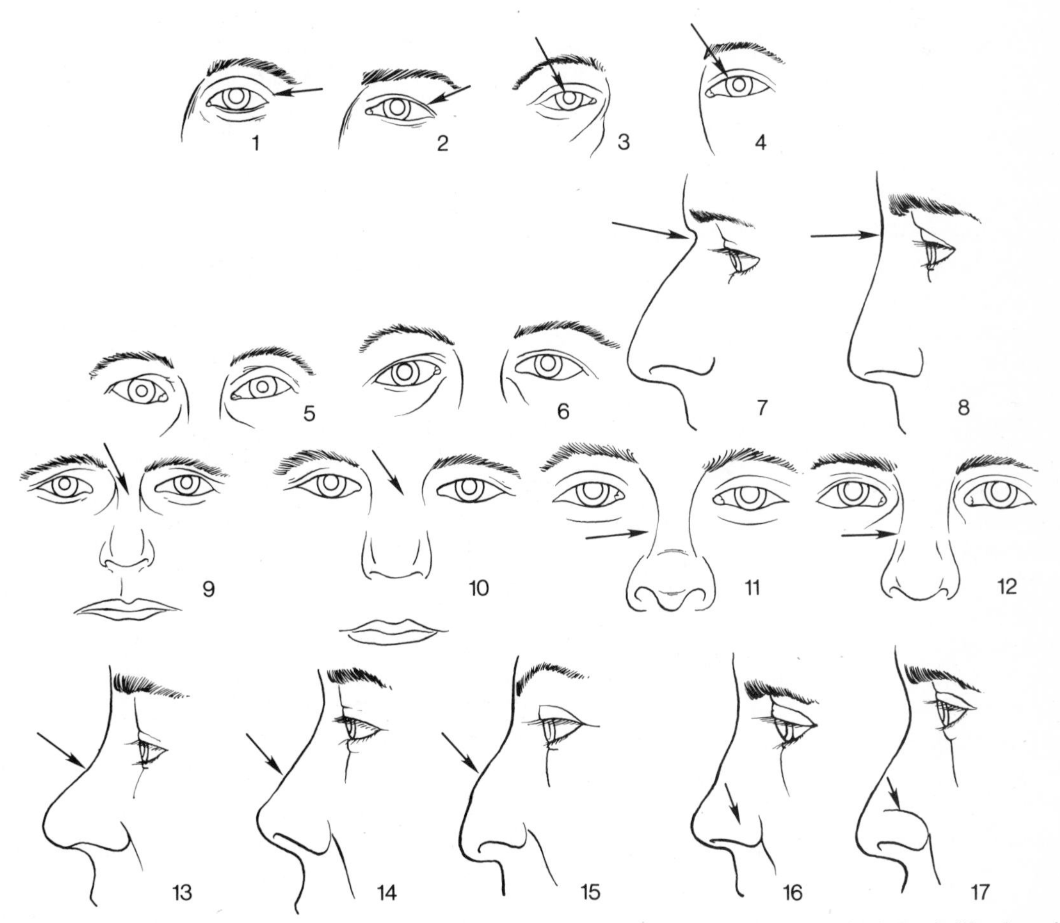

FIGURE 1–9. (Modified from Hulanicka, B.: Nadbitka Z Nru 86, Materialow 1 Prac antropologicznych Wroclaw, 115, 1973.)

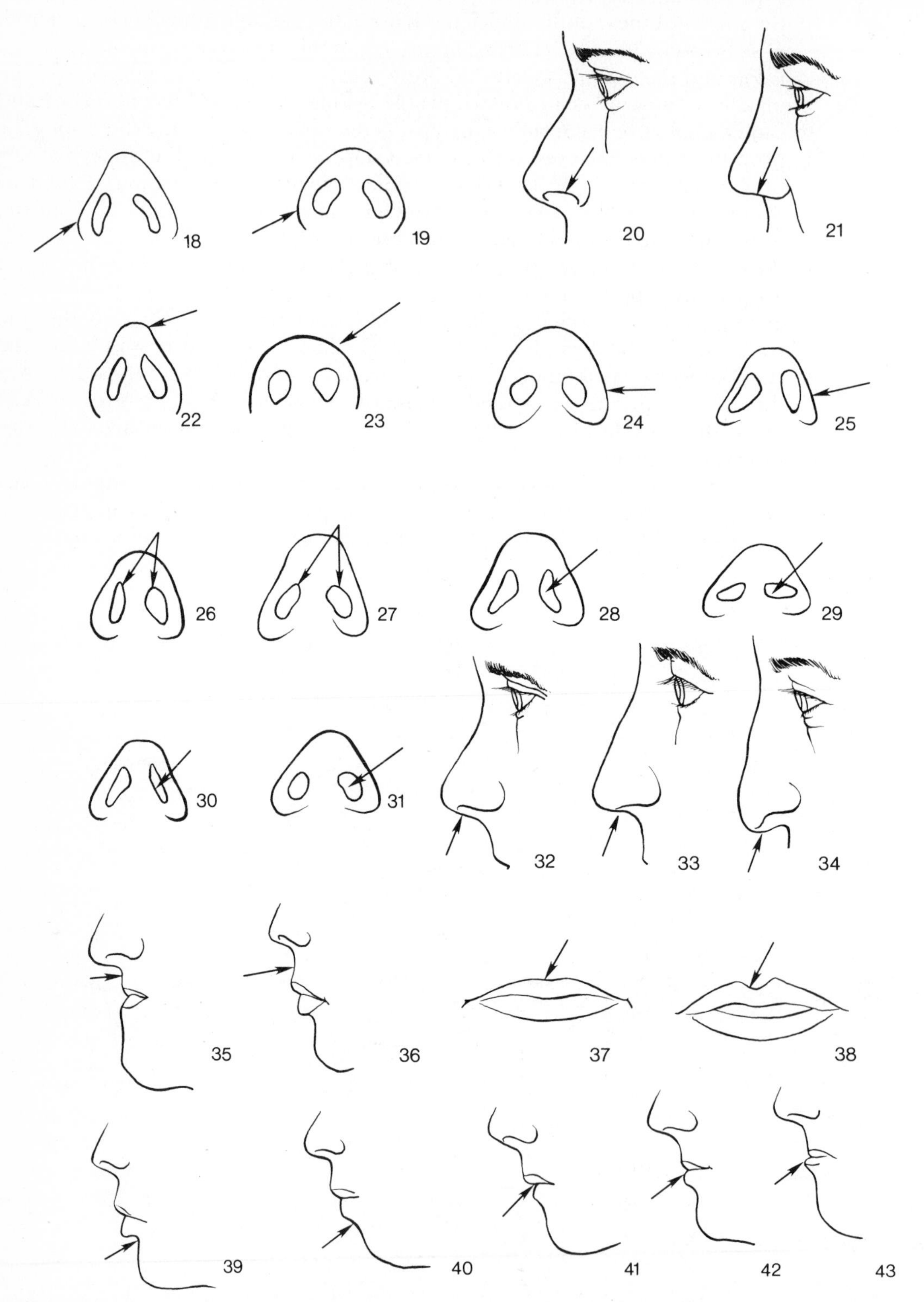

FIGURE 1–10. (Modified from Hulanicka, B.: Nadbitka Z Nru 86, Materialow 1 Prac antropologicznych Wroclaw, 115, 1973.)

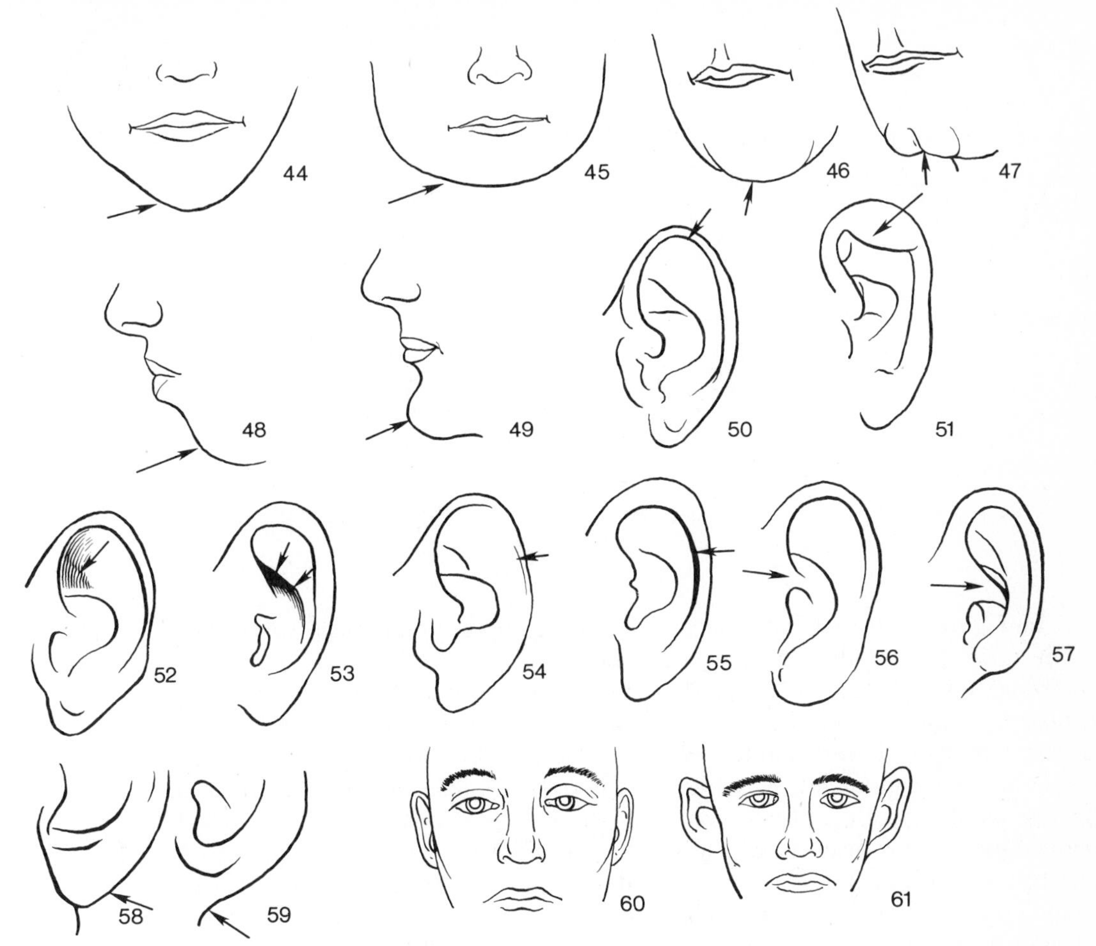

FIGURE 1–11. (Modified from Hulanicka, B.: Nadbitka Z Nru 86, Materialow 1 Prac antropologicznych Wroclaw, 115, 1973.)

istics such as hairstyle and color, eyebrow form, age wrinkles, skin complexion, and so on. Police departments have need to engage in this exercise frequently utilizing eyewitness descriptions (admittedly much more scanty than outlined above, and thus much less effective in recognizing some suspected felon). Indeed, many law enforcement agencies now use commercial "kits" containing examples of common features, feature by feature each separately shown on clear film, that can be compiled into a composite, individualized face.

For a face-responsible clinician, notations of the kind of facial features outlined here are useful as *clues* in properly sizing up a clinical situation.

If you have access to a darkroom, try this test for the nature of the symmetry of your own face. Look at the two sides of your face in the mirror. They look about the same, right? Probably not. Prepare two frontal photographic prints of the face, but reverse the negative for one. Then cut the prints into equal right and left halves, and reassemble them so that the two right halves and the two left halves each form a full face. Compare the two reassemblies. Two different faces? Which side, right or left, is the more "masculine," and which the more "feminine"? Which side of your own face would you want to favor in facing the camera to show your best profile? Are you right- or left-faced (similar to being right- or left-handed)?

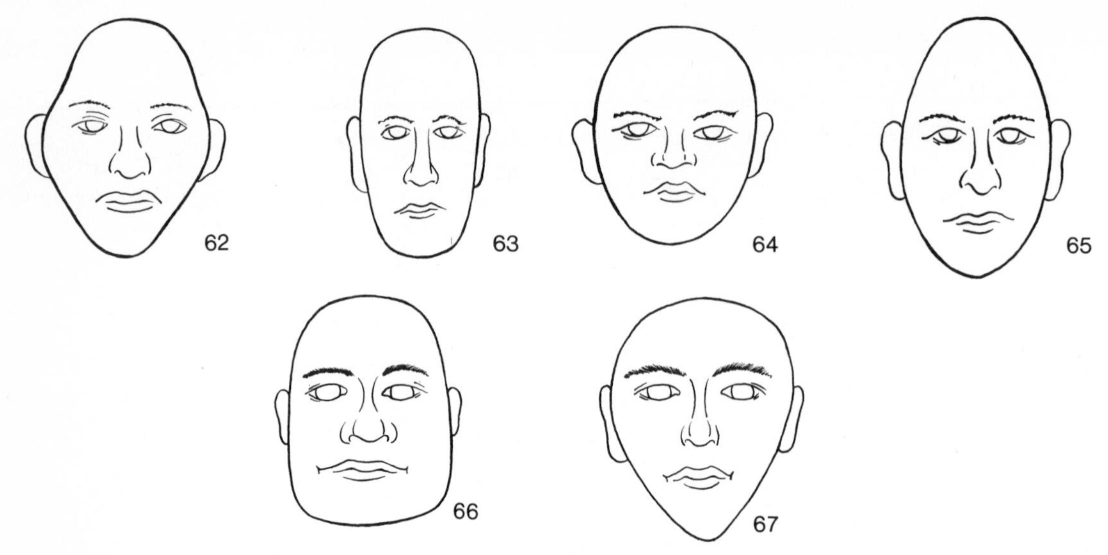

FIGURE 1–12

In Figures 1–9 to 1–12, note the following topographic facial variations: 1, the tarsal part of the upper eyelid exposed; 2, the eye laterally covered by an eyelid fold; 3, the iris covered by the upper eyelid; 4, most of the iris exposed; 5, the lateral corner of eye higher than the medial corner; 6, the lateral eye corner lower than the medial corner; 7, the top of the nasal bridge (root) markedly indented; 8, a high nasal root (so-called "Greek nose"); 9, a narrow nasal root; 10, a broad nasal root; 11, a narrow nasal slope; 12, a broad nasal slope; 13, a concave nasal profile; 14, a straight nasal profile; 15, a convex nasal profile; 16, inconspicuous nasal wings; 17, prominent nasal wings; 18, V-shaped nasal wings; 19, rounded nasal wings; 20, arched nasal wings; 21, straight nasal wings; 22, a narrow nasal tip; 23, a broad, flattened nasal tip; 24, a thick, fleshy nasal wing; 25, a thin nasal wing; 26, asymmetrical nasal openings; 27, symmetrical openings; 28, posterolaterally directed openings; 29, laterally directed openings; 30, narrow, elongate openings; 31, rounded nasal openings; 32, an upward nasal inclination; 33, a straight lower nasal border; 34, a downward inclined nasal border; 35, a vertically short upper lip; 36, a long upper lip (check also to see whether the upper lip profile is straight or concave); 37, the upper lip without a midline "Cupid's bow"; 38, a deep midline notch in the upper lip (look also for a more conspicuous philtrum above the upper lip, and check for thinness or thickness of the red part of both the upper and lower lips); 39, an acutely curved lower border (concavity) below the lower lip; 40, lesser concavity between lower lip and chin and a greater distance between the lip and mentolabial sulcus; 41, the lower lip retrusive; 42, the lips equally protruding; 43, the lower lip protrusive; 44, a pointed mandible; 45, a squared mandible; 46, no chin cleft; 47, a bifid chin; 48, a retrusive mandible (and chin); 49, a prominent chin; 50, slight rolling of the upper border of ear helix; 51, pronounced helix rolling; 52, a flat, shallow ear scapha; 53, a pronounced, deep groove below the scapha; 54, slight rolling of the middle part of the helix; 55, pronounced middle helix rolling; 56, a short, low crus; 57, a prominent, long crus; 58, a dangling ear lobe; 59, an ear lobe fused with facial skin; 60, slight ear protrusion; 61, marked ear protrusion; 62, a diamond-shaped face; 63, a long, narrow face; 64, a round, short face; 65, an oval face; 66, a square face; 67, an egg-shaped face.

THE CHANGING FEATURES OF THE GROWING FACE

The "baby face" has large-appearing eyes, dainty jaws, a small pug nose, puffy cheeks with buccal fat pads, a high intellectual-like forehead without coarse eyebrow ridges, a low nasal bridge, a small mouth, velvety skin, and overall wide and short proportions. It is a cute face. It warms the cockles of parental hearts. A parent can worry, though, because the otherwise great little face "has no chin," or "the jaw is much too small," or "the eyes are too far apart." However, these and many of the other features of the baby's face gradually undergo marked changes as the face grows and develops through the years. The chin develops, jaw size catches up, and the eyes appear less wide-set. From the many possible variations that can exist among different individuals, a person's own facial characteristics take on, month by month, definitive adult form. The general features of any fully grown face are quite different from those of the same individual as an infant and young child. Trying to decide which parent the infant "looks like" or which uncle it "takes after" is fun but usually more or less futile. There is little in the general shape and proportions of the infantile face, at least topographically, to give a hint as to what form it will take in later years. Unless, of course, the adult happens to have a euryprosopic face with pudgy cheeks, wide-set eyes, pug nose, etc., all childlike features.

Outlined below are some major features that contrast the characteristics of the child and adult faces. Study them carefully. Later chapters will explain the actual growth processes that underlie them.

The word combination "growth and development" is frequently heard. Why both terms? Growth, as will be seen, is not merely a process of size increases. Rather, progressive facial enlargement is a "differential" growth process in which each of the many component parts matures earlier or later than the others, to different extents in different facial regions, in a multitude of different directions, and at different rates. It is a gradual maturational process involving a complex of different but functionally interrelated organs and tissues. The growth process also involves a bewildering succession of regional changes in proportions and requires countless localized "adjustments" to achieve proper fitting and function among all of the parts. The phrase "growth and development" is appropriate and descriptive. The child's face is not merely a miniature of the adult, as dramatically illustrated in Figure 1–13, which shows a neonatal skull enlarged to the same size as a fully grown one.

The baby's face appears diminutive relative to the larger, more precocious cranium above and behind it (Fig. 1–14). The respective proportions change significantly, however. The growth of the brain slows considerably after about the third or fourth year of childhood, but the facial bones continue to enlarge markedly for many more years.

The eyes appear large in the young child. As facial growth continues, however, the nasal and jaw regions grow much faster and to a much greater extent than the earlier growing orbit and its soft tissues. As a result, the eyes of the adult appear smaller in proportion.

The ears of the infant and child appear to be low; in the adult they are much higher with respect to the face. Do the ears actually rise? No; they in fact move downward during continued growth. However, the face enlarges inferiorly even farther, so that the **relative** position of the ears seems to rise. In an infant, the body of the mandible is in near alignment with the auditory meatus, thus reflecting their commonality of embryonic origin. Later, the corpus

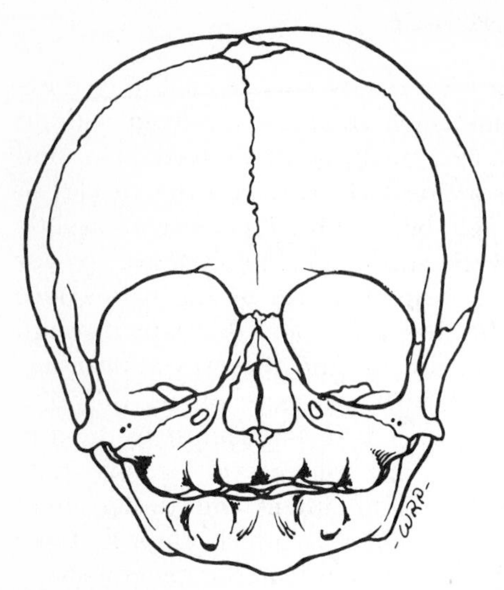

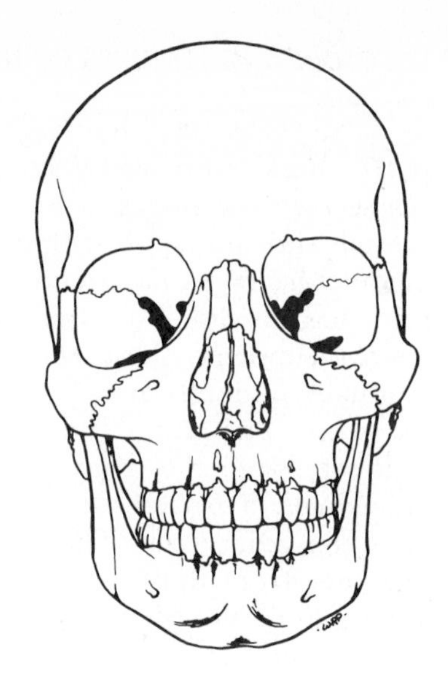

FIGURE 1–13

descends as the midface and ramus lengthen vertically, and this relationship becomes more obscure.

The young child's forehead is upright and bulbous. The forehead of the adult is much more sloping (the amount of slope is related to sex and ethnic origin, as already explained). The forehead region of the child seems very large and high because the face beneath it is still relatively small. The child's forehead continues to enlarge during the early years, but the face enlarges much more, so that the proportionate size of the forehead becomes reduced.

The child's face (Fig. 1–15) appears broad, because the brain and basicranium develop earlier and faster than the facial composite, as already explained.

FIGURE 1–14. (Courtesy of William L. Brudon. From Enlow, D. H.: *The Human Face.* New York, Harper & Row, 1968.)

FIGURE 1–15

As development continues, **vertical** facial growth then comes to bypass expansion in width to a marked extent, so that a much more narrow facial proportion characterizes the adult. (Even so, some adult faces can still appear rather wide and round. This is because of a wider and rounder type of brain configuration, a common variation. Such brachycephalic faces thus have a more juvenile-like appearance.)

The nasal bridge is quite low in the child. It rises (to a greater or lesser extent in different facial types) to become much more prominent in many adults.

The eyes of the infant seem quite wide-set, with a broad-appearing nasal bridge between them. This is because the nasal bridge is so low, and also because much of the width of the bridge has already been attained in the infant. With continued growth, the eyes spread further laterally, but only to a relatively small extent. Actually, the eyes of the adult face are not much farther apart than those in the child. Because of the higher nasal bridge, the increase in the vertical facial dimension, and the widening of the cheekbones, the eyes of the adult thus **appear** much closer together.

The infant and young child have much more of a pug nose than the adult. It protrudes very little and is vertically quite short (Fig. 1–16). The shape and size of the infantile nose, however, give little indication of what will happen to it during subsequent growth. The lower part of the nose in the adult is proportionately much wider and a great deal more prominent. The extent is related to ethnic origin.

The whole nasal region of the infant is vertically shallow. The level of the nasal floor lies close to the inferior orbital rim. In the adult, the midface has become greatly expanded, and the nasal floor has descended well below the orbital floor. This change is quite marked because of the enormous enlargement

FIGURE 1–16

FIGURE 1–17. (Courtesy of William L. Brudon. From Enlow, D. H.: *The Human Face.* New York, Harper & Row, 1968.)

of the nasal chambers. Note the close proximity of the young child's maxillary arch to the orbit, in contrast to their positions in the adult.

The superior and inferior orbital rims of the young child are in an approximately vertical line (see Fig. 1–16). Because of the unique human forehead, frontal sinus development, and supraorbital protrusion, however, the upper orbital rim of the adult noticeably overhangs the lower. The orbit opening becomes inclined obliquely forward. Supraorbital and glabellar protrusion is particularly marked in the adult male because of the larger nose needed to accommodate larger lungs.

Below the orbit, the nasal chambers in the adult face expand laterally nearly halfway across the orbital floor. In the infant, the breadth of the nasal cavity scarcely exceeds the width of the nasal bridge (Fig. 1–17). During subsequent growth, the inferior portion of the nose expands laterally much more than the superior part.

The tip of the infant's nasal bone protrudes very little beyond the inferior orbital rim. The area **between** the nasal tip and the inferior rim of the orbit (that is, the lateral bony wall of the nose) is characteristically narrow. In the adult, this area becomes markedly expanded. The divergent directions of orbital, nasal, cheekbone, and maxillary arch growth "draw out" the contours among them.

The nasal region of a growing child's midface is, almost literally, a keystone of facial architecture, that is, a key part upon which other surrounding parts, and the multiple arches formed by them, are dependent for placement and stability. If this keystone is malformed for any reason, other facial parts are affected during growth, and facial dysplasia or malocclusion takes place. The facial airway, therefore, is an exceedingly significant component involved in normal versus abnormal facial morphogenesis.

The lateral orbital rim and the cheekbone in the child are more forward-appearing because the whole face is still relatively flat and wide (Figs. 1–18 and 1–19). Because of an actual "regressive" mode of growth, however, these facial parts come to lie in a less prominent position in the adult face. In the infant, the protrusive appearance of the cheekbone is augmented by the characteristic

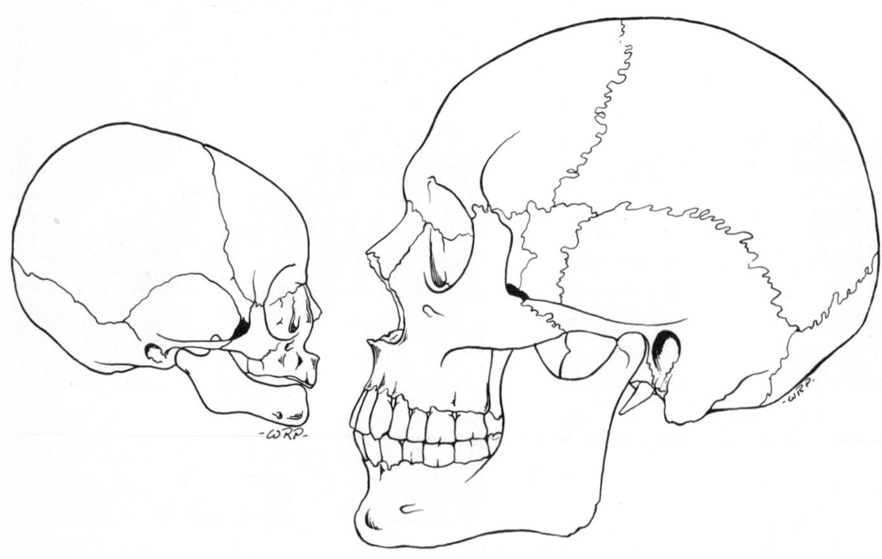

FIGURE 1–18

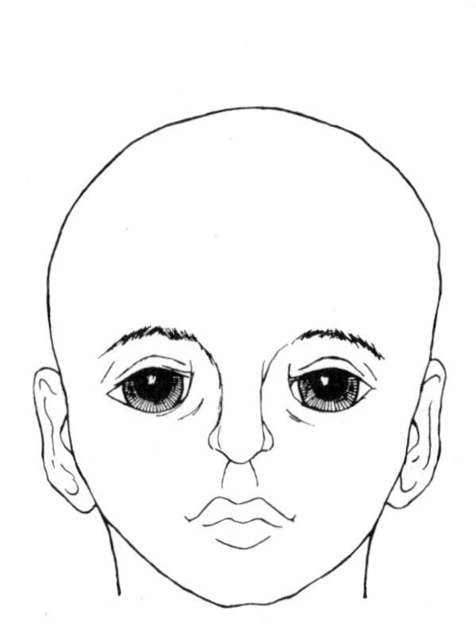

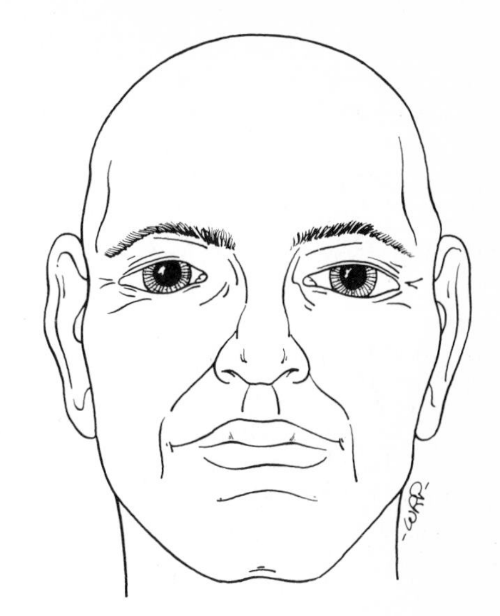

FIGURE 1–19

infantile buccal fat pad in the hypodermis of the cheek. Adults tending to have a relatively wide, short (thus more childlike) type of face typically show an even greater "cherubic" appearance if they are overweight; the buccal region contains adipose tissue resembling the buccal fat pad of infancy.

Although the cheekbone is prominent in early childhood, it is nonetheless quite diminutive and fragile, compared with that of the adult. The malar process and the inferior part of the zygoma enlarge considerably during childhood growth, even though they actually grow in a **backward** direction (as explained in Chapter 3). Because of the differential extents and directions of growth in other parts of the face, these growth increases by the zygoma are often masked. The **protrusive** modes of supraorbital and nasal growth cause the adult forehead and nose to appear progressively more prominent relative to the **retrusively** growing cheekbones and lateral orbital rims, thus drawing out the depth of the face. This feature is more noticeable in the male.

The entire face of the adult is much deeper horizontally because of these divergent directions of growth among all the various regional parts. The whole face is **drawn out** in many directions. The adult face has much bolder topographic features, and it is much less "flat."

As the whole face expands, the frontal, maxillary, and ethmoidal sinuses enlarge to occupy spaces not otherwise functionally utilized. Architecturally, the sinuses are leftover "dead" (unused) spaces. They were not created especially to provide "resonance to the voice," nasal drip, or other special functions, although they have become secondarily involved in such roles.

The mandible of the young child appears quite small and "underdeveloped" relative to the upper jaw and the face in general. It is small not only in actual size but also proportionately, and it is retrusively placed as well. The child's retrusive mandibular position is a normal relationship. The reason is that the anterior cranial fossae directly overlie the nasomaxillary complex suspended from them. Because the anterior cranial fossae are developmentally precocious, the nasomaxillary complex is thereby carried to a more protrusive position than the mandible, which articulates on the ectocranial side of the middle endocranial fossae located more posteriorly. Only later does the mandible catch up. Because of this, it is sometimes difficult to predict during early childhood possible malocclusions that might or might not become fully expressed during later development.

The chin is incompletely formed in the infant; indeed, it hardly exists at all. Because of remodeling changes that gradually take place, however, the chin becomes more prominent year by year. A "cleft" is sometimes formed in the **fleshy** part of the chin (not usually in the bone itself) when the two sides of the lower jaw fuse during early development. The cleft deepens when the soft tissues of the two sides then continue to expand postnatally. For some reason, this facial feature has become adopted in our society as a symbol of masculinity when it is present in the male. Its presence in the female has no social significance one way or the other.

The young child's mandible appears to be pointed. This is because it is wide, short, and more V-shaped. In the adult, the entire lower jaw becomes "squared." With the development of the chin, together with massive growth in the lateral areas of the trihedral eminence, eruption of the permanent dentition, lateral enlargement of each ramus, expansion of the masticatory musculature, and flaring of the gonial regions, the whole lower face takes on a U-shaped configuration, resulting in a **considerably** more full appearance.

In the infant and young child, the gonial region lies well inside (medial to) the cheekbone. In the adult, the posteroinferior corner of the mandible extends laterally out to the cheekbone, or nearly so. This gives the posterior part of the jaw a square appearance.

The ramus of the adult mandible is much longer vertically (see Fig. 1–18). It is also more upright (this refers to the ramus as a whole and not to the misleading "gonial angle" measurement). The elongation of the ramus accommodates the massive vertical expansion of the nasal region and the eruption of the deciduous and then the permanent teeth (see Fig. 1–19).

The premaxillary region normally protrudes beyond the mandible in the infant and young child, and it lies in line with or forward of the bony tip of the nose (see Fig. 1–18). This gives a prominent appearance to the upper jaw and lip. In subsequent facial development, however, the nose becomes much more protrusive, and the tip of the nasal bone comes to lie well ahead of the basal bone of the premaxilla.

The forward surface of the bony maxillary arch in the infant has a vertically convex topography. This is in contrast to the characteristically concave contour of this region in the adult. The alveolar bone in this area of the adult face is noticeably more protrusive and proportionately much more massive (in conjunction with the permanent dentition).

The whole face, vertically, is a great deal longer and more sloping as a result of the many changes outlined above.

The quite small mastoid process of the infant develops into the sizable protuberance of the adult. A bony styloid process is also lacking in the newborn.

Table 1–1. PERCENTAGES OF MALE HEAD GROWTH COMPLETED

Age (years)	Head Length	Head Width	Bizygomatic	Nasal Height
2.5–3.5	85.9	85.3	76.1	65.8
6.5–7.5	90.5	91.7	83.4	79.7
10.5–11.5	93.8	94.4	91.8	88.9
14.5–15.5	99.5	100%	96.2	96.3

Adapted from Goldstein, M. S. Changes in dimension and form of the face and head with age. Am. J. Phys. Anthrop., 22:37, 1936.

Table 1–2. PERCENTAGES OF MALE FACIAL GROWTH COMPLETED

Age (years)	Upper Facial Height (Nasion-Prosthion)	Total Facial Height (Nasion-Menton)	Bizygomatic Width (Skeletal)
Birth	40–61	38–45	58–69
1	63–86	50–66	76–86
3	72–90	66–78	79–88
5	82–100 (mean, 89)	68–82 (mean, 77)	80–90 (mean, 84)
	Nasal Height (N-ANS)	**Ethmoidal (Interorbital Width of Nasal Cavity)**	**Basion-Nasion**
Birth	36	—	54
1	—	75	—
4–7	74	94	80
8–12	81	96	90
	Basion-Sella	**Sella-Foramen Cecum**	**Foramen Cecum–Nasion**
Birth	47	62	25
4–7	74	94	56
8–12	89	98	63

Adapted from Lowe, A. A.: *Growth of Children.* Aberdeen, Scotland, Aberdeen University Press, 1952; and Tanner, J. M.: Aberdeen growth study. Arch. Dis. Child, 31:372, 1956.

The ring-shaped bone around the external acoustic meatus faces downward in the infant but is later rotated during growth into a more vertical position.

At birth, the overall length of the cranium is approximately 60 to 65 per cent complete, and it increases rapidly. By 5 to 7 years, it reaches about 90 per cent of its full size. Also, about 85 per cent of the adult width of the cranium is attained by the second to third year (Tables 1–1 and 1–2).

In the newborn, six fontanelles ("soft spots") are present between the bones of the skull roof. They cover over at different times, but all have been reduced to sutures by the eighteenth month. The sutures of the cranial vault are relatively nonjagged in the baby, and the outer surface of the bone is smooth. A much rougher bone texture characterizes the surface of the adult calvaria, and the suture lines become noticeably much more interlocking. The metopic suture (separating the right and left halves of the frontal bone) usually fuses by the second year and the premaxillary-maxillary suture is mostly fused by the first to second year, with only a trace sometimes remaining. By the third year, the principal cranial and facial suture systems still intact are the coronal, lambdoidal, and circumaxillary. Subsequent closure then begins around the twenty-fifth to thirtieth year, usually in the sequence of sagittal, coronal, and lambdoidal, with those bounding the temporal bone following. The latter can remain partially open even in the aged skull. Traces of the facial sutures often remain through advanced old age.

In the child, the slender neck below a relatively large cranium, particularly in the occipital region, gives a characteristic "boyish" appearance to the whole head. This gradually disappears (to a greater or lesser extent) until about puberty, when the expansion of the neck muscles and other soft tissues causes a proportionate decrease in the prominence of the head relative to neck circumference. This is less noticeable in the female.

The external appearance of the baby's face does not reveal the truly striking enormity of the dental battery developing within it (Fig. 1–20). The teeth are a dominant part of the infant's face as a whole, yet they are not even seen. The parent does not usually realize they are already there at all, much less suspect the massiveness of their extent. In this illustration, one is almost

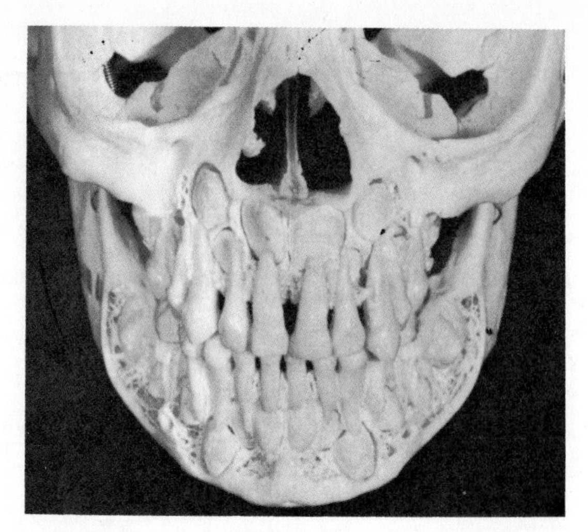

FIGURE 1–20

overwhelmed by the remarkable extent of **teeth** all over the midfacial region. The average person does not appreciate that the mouth of the little child is bounded by a virtual palisade of multitiered primary and permanent teeth in many stages of development. When a crown tip first protrudes through the gingiva as it erupts, the parent naturally believes that the process is just beginning and that the tooth is only a tiny but newsworthy addition to the pink mouth. It is not realized that the whole midface is occupied by a vast magazine of unerupted teeth hidden to the eyes. The thin covering and supporting **bone** of the jaws is a much less commanding feature of the young face.

CHAPTER TWO

Introductory Concepts of the Growth Process

Part 1

Part 1 is an introductory overview. Part 2 provides more in-depth information covering essentially the same range of subjects.

Concept 1

Facial **growth and development** is a morphogenic process working toward a composite state of aggregate structural and functional **balance** among all of the multiple, regional **growing and changing** hard and soft tissue parts. The same underlying process then continues to work in order to sustain ongoing equilibrium throughout adulthood and old age in response to ever changing internal and external conditions and relationships.

An in-depth understanding of facial morphogenesis is essential so that the clinician can properly grasp (1) differences between "normal" and ranges of abnormal; (2) biologic reasons for these differences and the virtually limitless variations involved; (3) reasons for rationales utilized in diagnosis, treatment planning, and selection of appropriate clinical procedures; and (4) the biologic factors underlying the important clinical problems of retention, rebound, and relapse after treatment. For researchers in all aspects of facial growth and development, from the molecular to experimental gross morphology, this subject has exploded into a major field of interest and activity worldwide.

To begin, fundamental concepts with regard to the morphogenesis of the facial bones are presented. The craniofacial skeleton as seen in x-ray headfilms represents a primary tool for evaluating facial morphogenesis. Later, the many other aspects also basically involved follow.

Concept 2

Bones grow by adding new bone tissue on one side of a bony cortex and taking it away from the other side (Fig. 2–1). The surface facing toward the

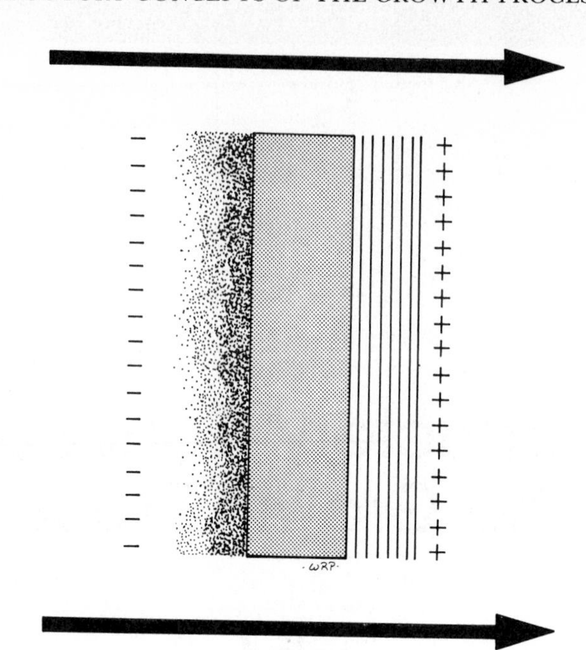

FIGURE 2–1

direction of progressive growth receives new bone **deposition** (+). The surface facing away undergoes **resorption** (−). This composite process is termed "drift." It produces a direct **growth movement** of any given area of a bone.

Concept 3

The outside and inside surfaces of a bone are completely blanketed by a mosaic-like pattern of "growth fields" (Fig. 2–2). Note that the outside surface, however, is **not** all "depository," as one might presume. About half of the periosteal (external) surface of a whole bone has a characteristic arrangement of **resorptive fields** (darkly stippled areas); a characteristic pattern of **depository fields** covers the remainder (lightly stippled areas). If a given periosteal area has a resorptive type of field, the opposite inside (endosteal) surface of that same area has a depository field. Conversely, if the periosteal field is depository, the endosteal field on the opposite side of the cortex is usually resorptive. These combinations produce the characteristic **growth movements** (that is,

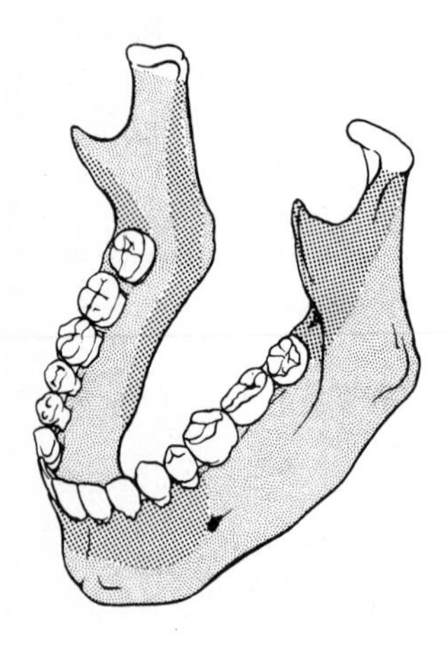

FIGURE 2–2

drift) of all parts of an entire bone. How can a bone enlarge if half of its **outside** surfaces actually undergo resorption? This is explained in the concepts that follow.

Concept 4

Bone produced by the covering membrane ("periosteal bone") constitutes about half of all the cortical bone tissue present; bone laid down by the lining membrane ("endosteal bone") makes up the other half (Fig. 2–3). In this diagram, note how the cortex on the right was formed by the periosteum and the cortex on the left by the endosteum as both sides shifted (drifted) in unison to the right.

Concept 5

The operation of the growth fields covering and lining the surfaces of a bone is actually carried out by the osteogenic **membranes** and other surrounding tissues, rather than by the hard part of the bone (Fig. 2–4). The bone does not "grow itself"; growth is produced by the **soft tissue matrix** that encloses each whole bone. The genetic and functional determinants of bone growth reside in the composite of soft tissues that turn on and turn off and speed up and slow down the histogenic actions of the osteogenic connective tissues (periosteum, endosteum, sutures, periodontal membrane, etc). Growth is not "programmed" within the calcified part of the bone itself. The "blueprint" for the design, construction, and growth of a bone thus lies in the muscles, tongue, lips, cheeks, integument, mucosae, connective tissues, nerves, blood vessels, airway, pharynx, the brain as an organ mass, tonsils, adenoids, and so forth, all of which provide information signals that pace a bone's development.

Concept 6

All the various resorptive and depository growth fields throughout a bone do not have the same **rate** of growth activity. Some depository fields grow much more rapidly or to a much greater extent than others. The same is true for

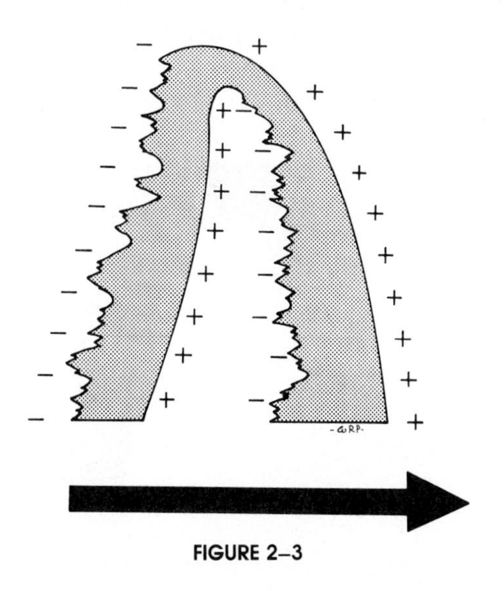

FIGURE 2–3

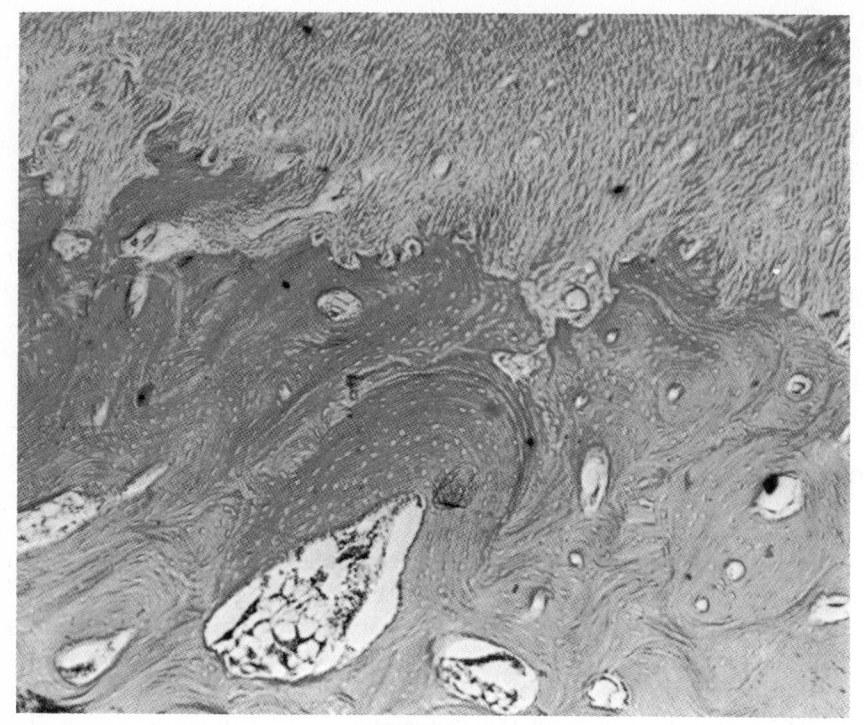

FIGURE 2–4

resorptive fields. Fields that have some special significance or noteworthy role in the growth process are often called growth **sites.** The mandibular condyle, for example, is such a growth site (Fig. 2–5). Remember, however, that growth does not occur just at such special growth sites, as is sometimes presumed. The entire bone participates. **All** surfaces are, in fact, sites of growth, whether specially designated or not. The old term "condylar growth" is still often used. It is misleading, however, because it mistakenly implies that the condyle is **the** growth center largely responsible for overall mandibular growth and development. If only condylar growth were operative, the condyle would sit on an elongated neck as a giraffe's head perches high on its neck. The **entire** ramus, together with its condyle, participates actively and directly.

During remodeling, the extent of bone deposition usually slightly exceeds that of resorption, so that the regional parts of a bone gradually **enlarge** and the cortical plates thicken as they remodel.

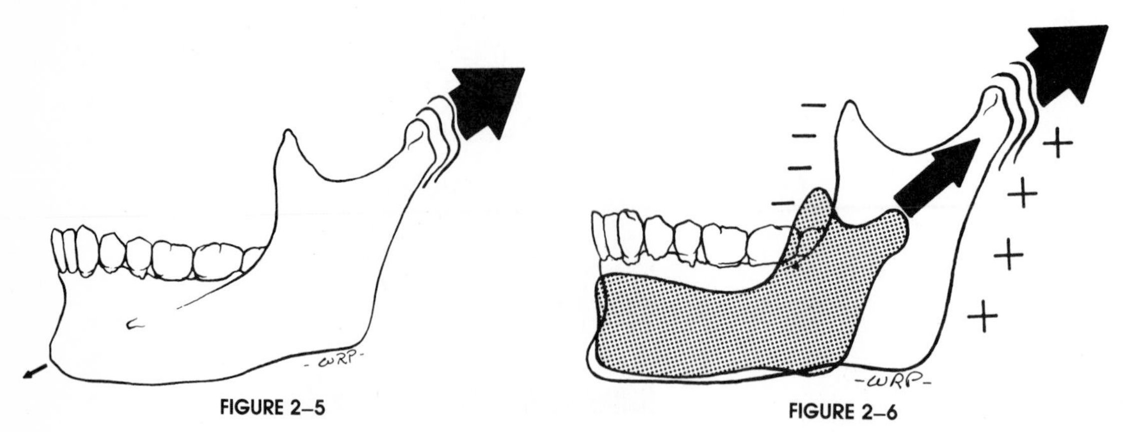

FIGURE 2–5

FIGURE 2–6

Concept 7

Remodeling is a basic part of the growth process. The reason why a bone must remodel during growth is because its regional parts become **moved;** "drift" moves each part from one location to another as the whole bone enlarges (Fig. 2–6). This calls for sequential remodeling changes in the shape and size of each region. The ramus, for example, moves progressively posteriorly by a combination of deposition and resorption. As it does so, the anterior part of the ramus becomes **remodeled** into a new addition for the mandibular corpus. This produces a growth elongation of the corpus. This progressive, sequential movement of component parts as a bone enlarges is termed **relocation.** Relocation is the basis for remodeling. The whole ramus is thus relocated posteriorly, and the posterior part of the lengthening corpus becomes relocated into the area previously occupied by the ramus. Structural remodeling from what **used to be** part of the ramus into what then **becomes** a new part of the corpus takes place. The corpus grows longer as a result.

The same deposition and resorption that produce the overall growth enlargement of a whole bone carry out relocation and remodeling at the same time. Growth and remodeling are, in effect, inseparable parts of the same actual process. It can now be understood why about half of any given bone can and must have a resorptive **external** (periosteal) surface as the bone increases in overall size. The reason is that the bone does **not** simply enlarge symmetrically by new bone deposition uniformly over all external surfaces, as shown in Figure 2–7. Rather, each of the regional parts of the bone becomes relocated into a sequentially new position. Some outside surfaces, therefore, are necessarily resorptive.

In the maxilla, the palate grows downward (that is, becomes **relocated** inferiorly) by periosteal resorption on the nasal side and periosteal deposition on the oral side (Fig. 2–8). This growth and remodeling process serves to enlarge the nasal chambers. What used to be the bony maxillary arch and palate in early childhood are remodeled into what then becomes the nasal chambers of the adult. About half of the palate is thus resorptive, and about half, depository. The nasal mucosa provides the periosteum on one side, and the oral mucosa provides it on the other side.

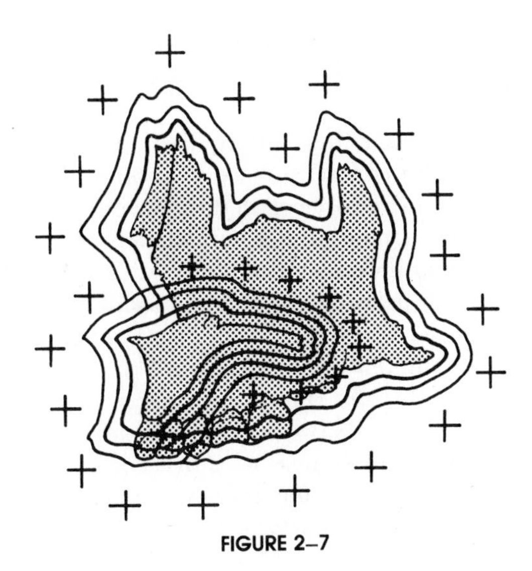

FIGURE 2–7

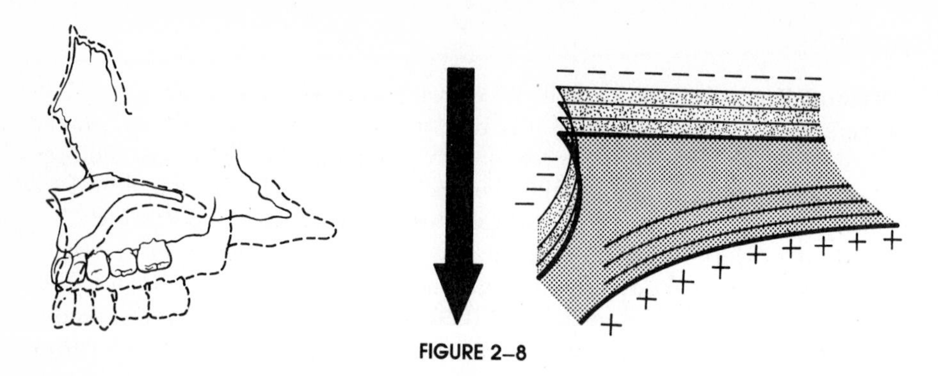

FIGURE 2–8

In summary, the process of growth remodeling is paced by the composite of soft tissues housing the bones, and the functions are to (1) progressively enlarge each whole bone; (2) sequentially relocate each of the component parts of the whole bone to allow for overall enlargement; (3) shape the bone to accommodate its various functions in accordance with the physiologic actions exerted on that bone; (4) provide progressive fine-tune fitting of all the separate bones to each other and to their contiguous, growing, functioning soft tissues; and (5) carry out continuous regional structural adjustments of all parts to adapt to the multiple intrinsic and extrinsic changes in conditions. Although these remodeling functions relate to childhood, most also continue on into adulthood and old age to provide the same ongoing functions. This is what in freshman histology is meant when it is stated that bones "remodel throughout life," without an explanation of the reasons. Another point is that the common clinical problem of rebound and relapse is a normal expression of the same remodeling functions as a biologic means for restoring a pre-existent state of composite morphologic balance that had been altered by clinical intervention.

Concept 8

As a bone enlarges, it is simultaneously carried away from other bones in direct contact with it. This creates the "space" within which bony enlargement takes place. The process is termed **primary displacement** (sometimes also called "translation"). It is a physical movement of a whole bone and occurs while the bone grows and remodels by resorption and deposition. As the bone grows by surface deposition in a given direction, it is simultaneously displaced in the **opposite** direction (Fig. 2–9).

The process of new bone deposition does not cause displacement by **pushing** against the articular contact surface of another bone. Rather, the bone is **carried** away by the expansive force of all the growing soft tissues surrounding it. As this takes place, new bone is added immediately onto the contact surface, and the two separate bones thereby remain in constant articular junction. The nasomaxillary complex, for example, is in contact with the floor of the cranium (Fig. 2–10). The whole maxillary region, *in toto,* is **displaced** downward and forward away from the cranium by the expansive growth of the soft tissues in the midfacial region (Fig. 2–11A). This then triggers new bone growth at the various sutural contact surfaces between the nasomaxillary composite and the cranial floor (Fig. 2–11B). Displacement thus proceeds downward and forward as growth by bone deposition simultaneously takes place in an opposite upward and backward direction (that is, **toward** its contact with the cranial floor).

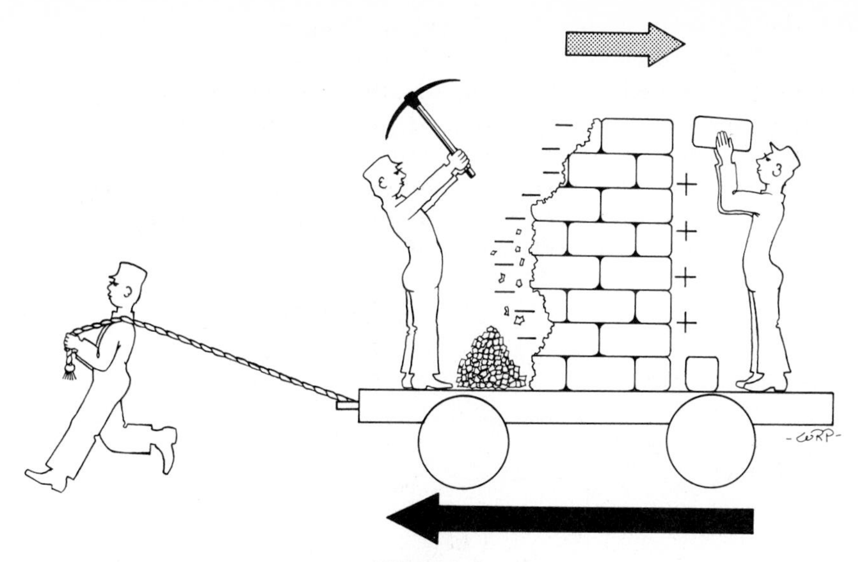

FIGURE 2–9

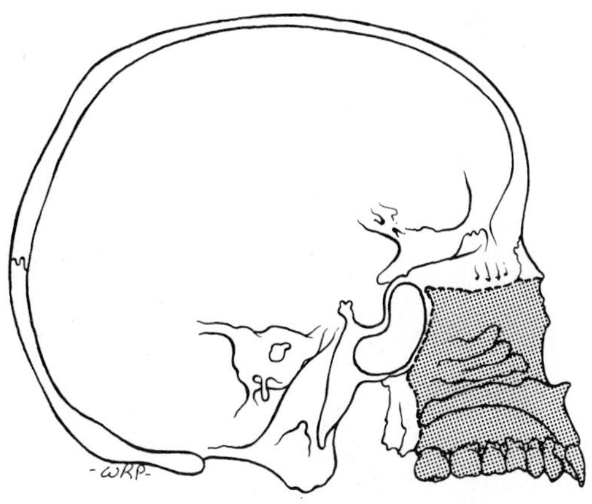

FIGURE 2–10

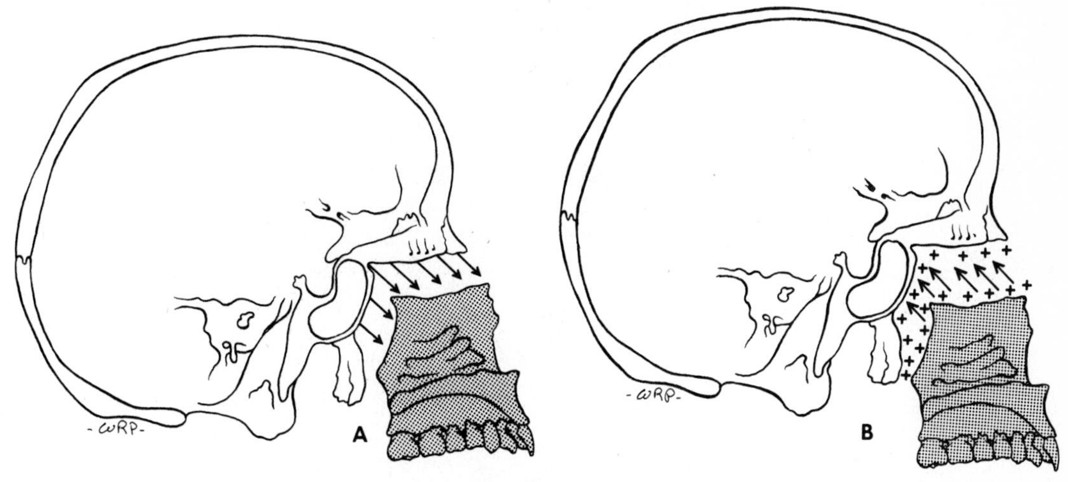

FIGURE 2–11

Similarly, the whole mandible (Fig. 2–12) is **displaced** "away" from its articulation in each glenoid fossa by the growth enlargement of the composite of soft tissues in the growing face. As this occurs, the condyle and ramus grow upward and backward into the "space" created by the displacement process. Note that the ramus "remodels" as it relocates posterosuperiorly. It also becomes longer and wider to accommodate (1) the increasing mass of masticatory muscles inserted onto it, (2) the enlarged breadth of the pharyngeal space, and (3) the vertical lengthening of the nasomaxillary part of the growing face.

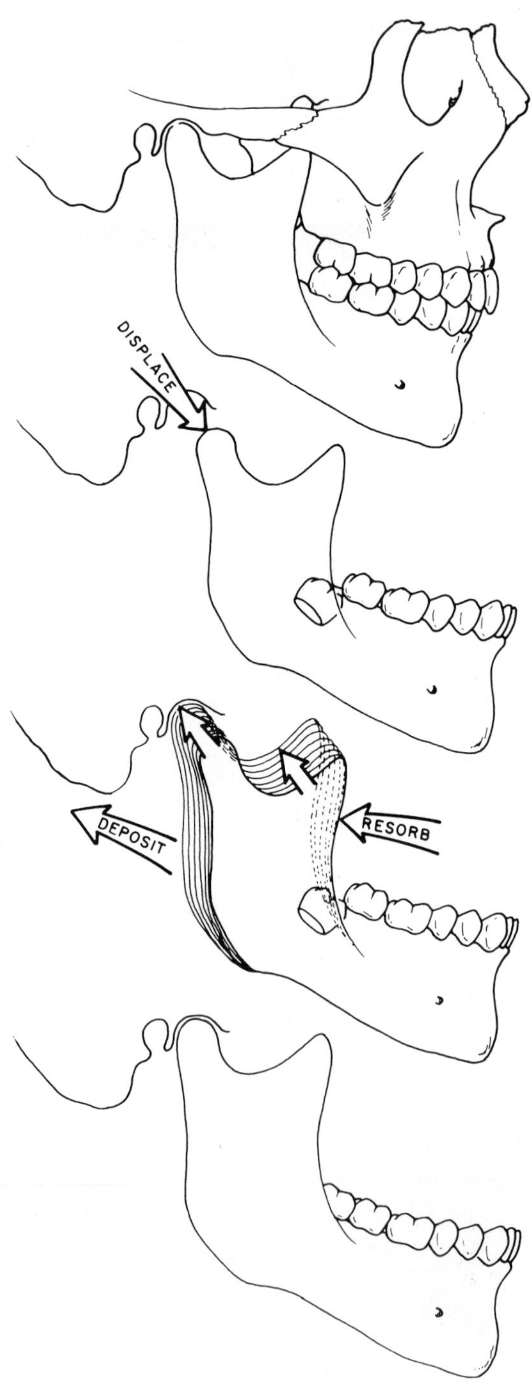

FIGURE 2–12

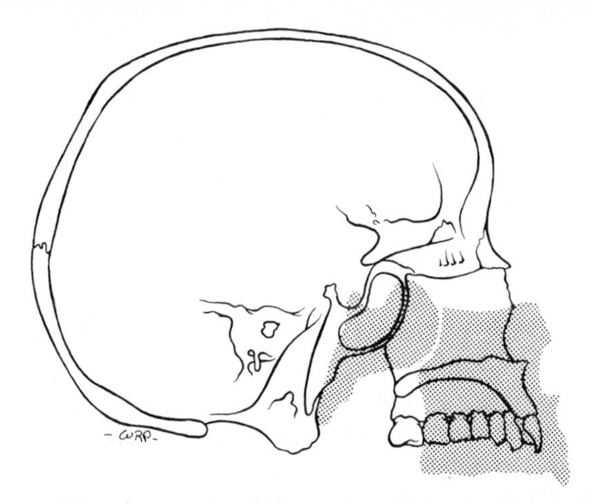

FIGURE 2–13

A beginning student is often confused because he or she repeatedly hears and reads that the face "grows forward and downward." It would seem reasonable that the growth activity of the mandible or the maxilla would be in their anterior, forward-facing parts. However, it is mostly the displacement movement that is forward and downward, thereby complementing the predominantly posterosuperior vectors of remodeling. Another way to view this is that all long bones, such as the humerus and femur, lengthen at their articular ends. The mandible, although curved in a U-shaped configuration, does the same. This is one fundamental reason all joint contacts and bone ends are of basic significance in the growth picture. They are the points away from which displacement proceeds and, at the same time, the sites where remodeling lengthens a given bone.

Concept 9

A process of **secondary displacement** also occurs during growth. **Primary displacement,** just described, is associated with a bone's **own** enlargement. Secondary displacement, however, is the movement of a whole bone caused by the separate enlargement of **other** bones, which may be nearby or quite distant (Fig. 2–13). For example, increases in size of the bones that compose the middle cranial fossa (in conjunction with growth of the brain) result in a marked displacement movement of the whole maxillary complex anteriorly and inferiorly. This is quite independent of the growth and enlargement of the maxilla itself. The displacement effect is thereby of a secondary type. What happens deep in the cranial base thus affects the placement of the bones in the face. The effects of growth activities in relatively distant locations are passed on, and one must take all such changes into account when analyzing the growth processes and the facial characteristics of any individual person.

In summary, the overall skeletal growth process (displacement and remodeling) has two general functions: (1) positioning each bone and (2) designing and constructing each bone and all of its regional parts so that they can carry out that bone's multifunctional role. The functional input to the osteogenic connective tissues (membranes and cartilages) of the bone from the aggregate of soft tissues causes a bone to develop into its definitive morphologic structure and to occupy the location it does.

Concept 10

Facial growth is a process that requires intimate morphogenic **interrelationships** among all of its component growing, changing, and functioning soft and hard tissue parts. No part is developmentally independent and self-contained; this is a fundamental and very important principle of growth. As emphasized earlier, the growth process works toward an ongoing state of composite functional and structural equilibrium. However, the evolutionary design of the human head is such that certain regional "imbalances" are inescapable and normal, for example, relationships established by head form variations, male/female differences, and so forth. The growth process, in response, develops some regional imbalances, the aggregate of which serves to make adjustments to correct the other imbalances. A Class I face is the common result in which the underlying factors that would otherwise lead to a more severe Class II or III malocclusion still exist but have been "compensated for" by the growth process itself, that is, offsetting imbalances, the net effect of which is an overall, composite "balance." Further, it always comes as a conceptual jolt to a beginning student of facial growth to realize, also, that even severe malocclusions and congenital craniofacial dysplasias (birth defects) are actually "in balance," and that clinical treatment can disturb this state of structural and functional equilibrium, resulting in a natural rebound. For example, if a premature fusion of some cranial sutures has resulted in growth-retarded development of the nasomaxillary complex because the anterior endocranial fossae (a template for midfacial development) have been foreshortened, as in the Crouzon or Apert syndromes, the altered nasomaxillary complex itself nonetheless has grown in a balanced state relative to its basicranial template, even though abnormal in comparison with a population norm. Craniofacial surgical correction attempting to more or less create this norm can disturb the former balance, and some degree of natural rebound can be expected as growth attempts to restore the original state of equilibrium among all of the regional parts involved, since some extent of the original underlying conditions can still exist that were not, or could not be, altered clinically.

Part 2

The basic concepts of growth briefly summarized in Part 1 are considered in more detail in Part 2.

Step 1 in understanding how the bones of the face and cranium grow is to realize that they **do not** enlarge in the simplified manner shown in Figure 2–14. A bone does not increase in size merely by direct, symmetrical, outward expansion of all surfaces and contours, as if magnified by a lens.

GROWTH FIELDS

A bone **does not** grow by generalized, uniform deposition of new bone (+) on all outside surfaces, with corresponding resorption (−) from all inside surfaces, as one might erroneously presume (and as has often been incorrectly taught). It is not possible for bones having the complex morphology of, for example, the mandible or the maxilla to increase in size by such a growth process (Fig. 2–15). Because of the topographically complex nature of each bone's shape, the bone must have a **differential** mode of enlargement, in which some of its parts and areas grow much faster and to a much greater extent than others. Many of the **external** surfaces of most bones are actually **resorptive** in nature. How can a bone increase in size, even though many outside (periosteal) surfaces undergo resorptive removal as the bone grows? Keep this question in mind as the processes of facial growth are explained in the pages that follow.

Two basic kinds of **growth movement** occur during the enlargement of each bone in the facial and cranial skeleton: (1) remodeling, which produces the size, shape, and fitting of a bone, and (2) displacement. **Displacement** is a movement of whole bones away from one another, creating the space within which growth enlargement of each of the separate bones takes place. **Cortical drift** is the process that carries out the **remodeling** functions, and it is a direct growth movement produced by deposition of new bone on one side of a cortical plate, with resorption from the opposite side. The process of remodeling is explained first; then that of displacement is explained.

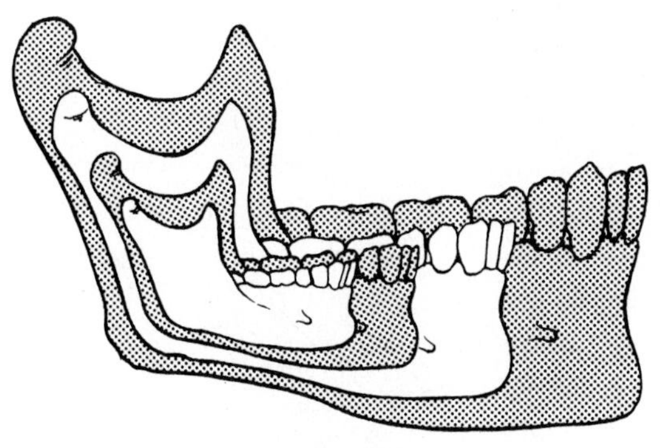

FIGURE 2–14

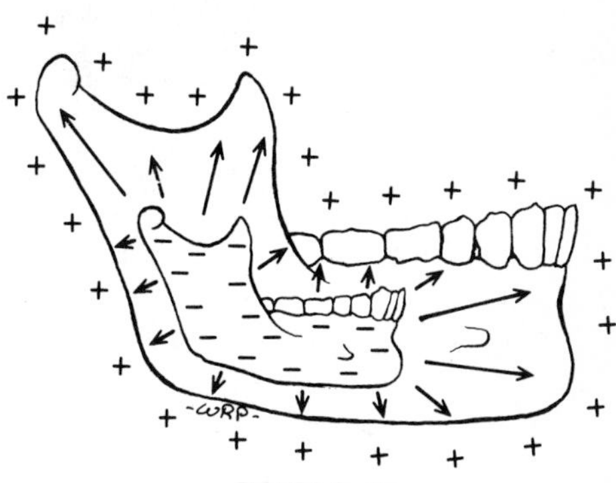

FIGURE 2–15

THE BONE REMODELING PROCESS

Figure 2–16 schematizes a process of remodeling. The bony cortex **moves** from *A* to *B* by cortical drift. The surface that faces **toward** the direction of movement is depository (+). The opposite surface, facing away from the growth direction, is resorptive (−). If the rates of deposition and resorption are equal, the thickness of the cortex remains constant. If deposition exceeds resorption, overall size and cortical thickness gradually increase. It is apparent that the actual bone tissue present in *B* is not the same present in *A*, owing to the continuous process of new additions on one side, combined with removal of the older bone from the other side. Different combinations of resorption and deposition (drift) in a variety of regional directions and amounts throughout the entire bone provide for the remodeling enlargement of the bone as a whole.

If a metallic marker* is implanted on the depository side of a cortex, it becomes progressively more deeply embedded in the cortex as new bone continues to form on the surface and as resorption takes place from the other side. Eventually, the marker becomes translocated from one side of the cortex to the other, not because of its own movement (the marker itself is immobile), but because of the "flow" of the drifting bone around it (Fig. 2–17).

Directions of growth sequentially undergo **reversals** (Fig. 2–18). A **reversal line,** shown by the small arrow pointing to the crossover between the resorptive (−) and depository (+) growth fields, can be seen in microscopic sections wherever this occurs; it is the interface between the layers of bone that were produced first on one side and then on the other as the **direction** of growth turned about.

A given bone has **fields** of resorptive (darkly stippled) and depository activity over all its inside and outside cortical surfaces (Fig. 2–19). This is the basis for the **differential growth** process that produces the irregular shape of

*Metallic implants (tiny pieces of tantalum or some other appropriate metal) are often used as radiographic markers in clinical and experimental work to study bone growth and displacement in headfilms. Using the markers as registration points when superimposing serial headfilm tracings, one can readily determine the amount and direction of remodeling as well as displacement movements.

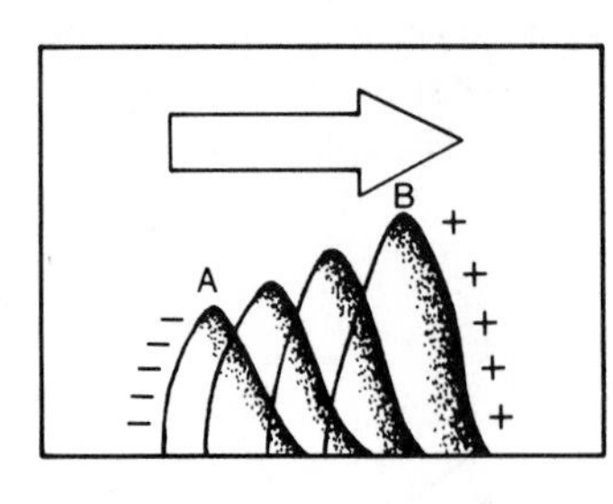

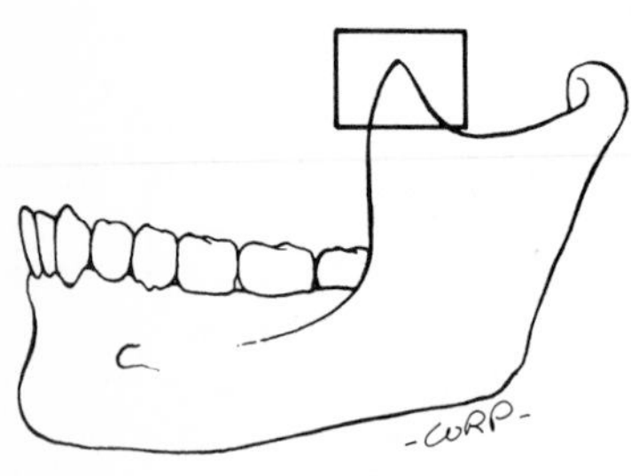

FIGURE 2–16

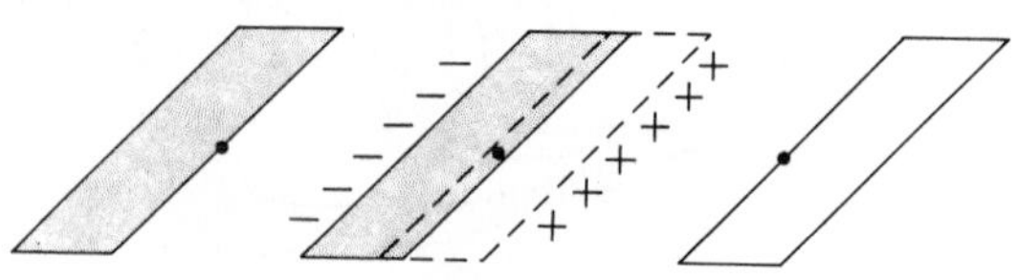

FIGURE 2–17

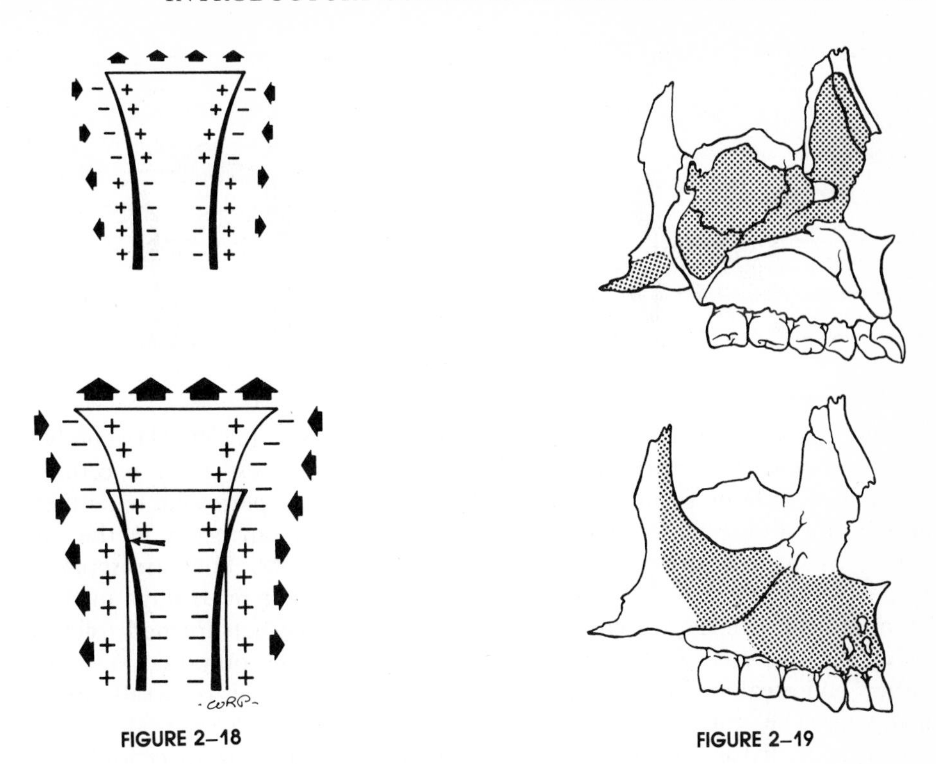

FIGURE 2–18

FIGURE 2–19

a bone. The shape is irregular because of its many varied functions (that is, attachments of many different soft tissue masses and their diverse growth patterns and functional actions, articulations with other bones, support of teeth, and so on).

In Figure 2–20, the pattern of growth fields results in a **rotation** of the skeletal part shown. Such rotations are a significant part of the developmental process of the face and cranium, as will be seen in later chapters.

The activities of the growth fields reside in the soft tissue of the **periosteum** and **endosteum,** not in the hard part of the bone itself (Fig. 2–21). The bone does not control and carry out its own growth. The membranes and other soft tissues that **enclose** the bone (a suture shown here) produce and control the bone's growth in response to a complex of signals from functioning and growing muscles and other organs and tissues that activate the osteoblasts and osteoclasts of the osteogenic connective tissues enclosing all inside and outside bone surfaces. As the bones become separated by the displacement process, propor-

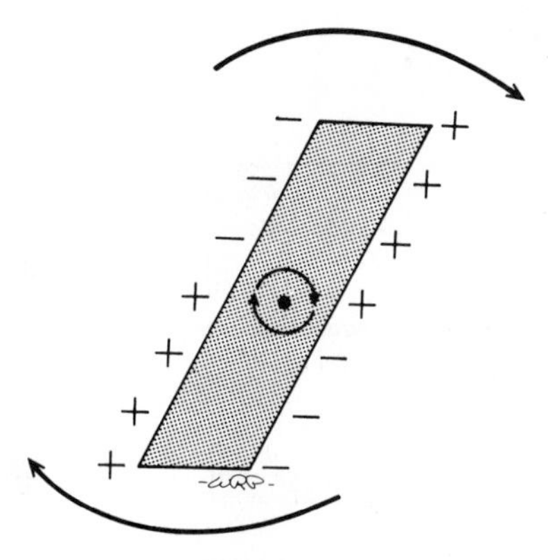

FIGURE 2–20

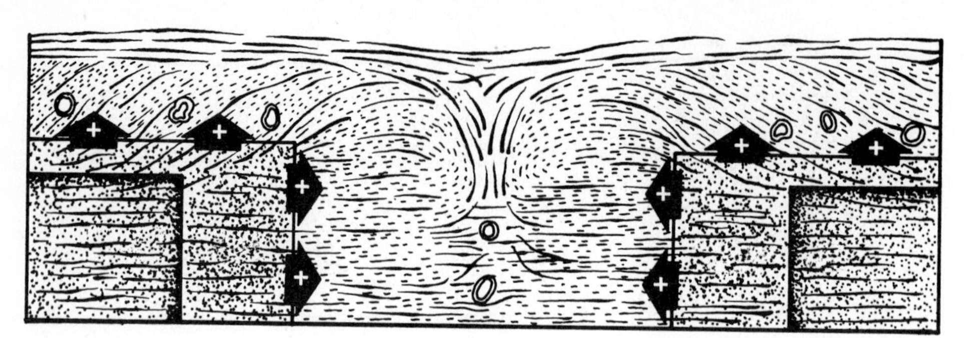

FIGURE 2–21

tionate new bone deposits are laid down ("+" arrows in Fig. 2–21). The bone itself is passive. It is a product of the overall growth process, not a pacemaker.

Layers of bone formed by the covering membrane (periosteal bone tissue) and by the lining membrane (endosteal bone tissue) can occur in the same cortex, as seen by growth stages 1 and 2 in Figure 2–22. They are separated by a reversal line. Layer 1 was produced during a past growth stage involving an endosteal direction of cortical growth. Layer 2 was then formed following reversal in the course of growth. A given cortex can be composed entirely of either endosteal (*a*) or periosteal (*b*) tissue (Fig. 2–23) if reversals are not involved. The entire bony configuration in this example is undergoing a coordinated drifting movement ("relocation") in a progressive direction toward the right.

In most bones of the face and cranium (and most other bones of the body as well), about half of the total amount of cortical bone tissue is of **endosteal** origin and about half of **periosteal** origin. Many of the **periosteal** surfaces are **resorptive,** and others are **depository** in nature (Fig. 2–24). The same is true for endosteal surfaces. This provides two growth functions: the **enlargement** of any given bone and also the **remodeling** of each bone, a process that accompanies its enlargement.

Four different kinds of remodeling occur in bone tissues. One is **biochemical** remodeling, taking place at the molecular level. This involves the constant

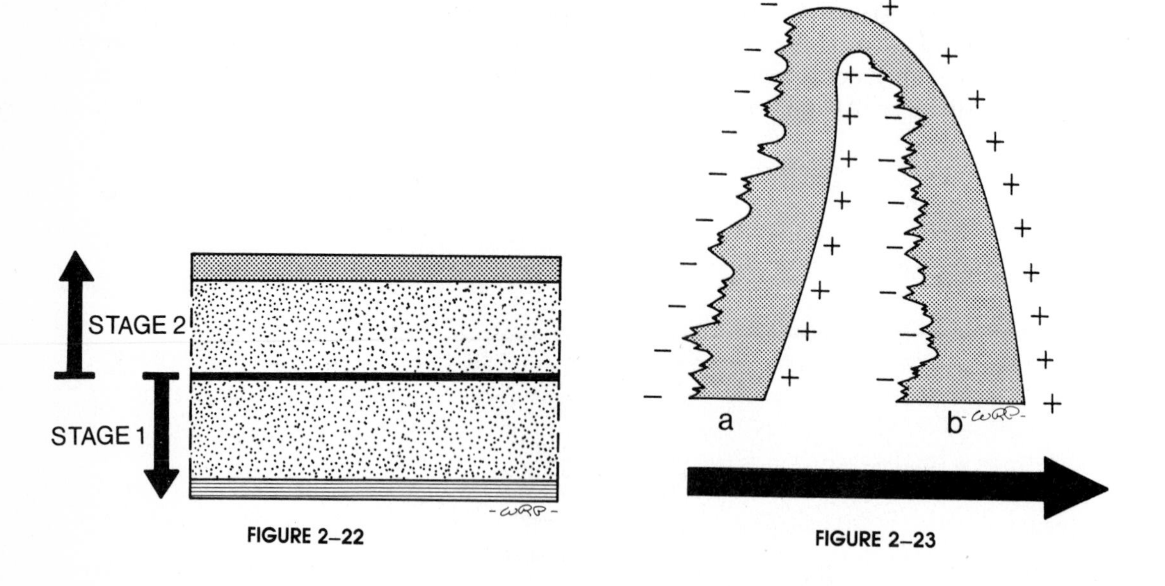

FIGURE 2–22

FIGURE 2–23

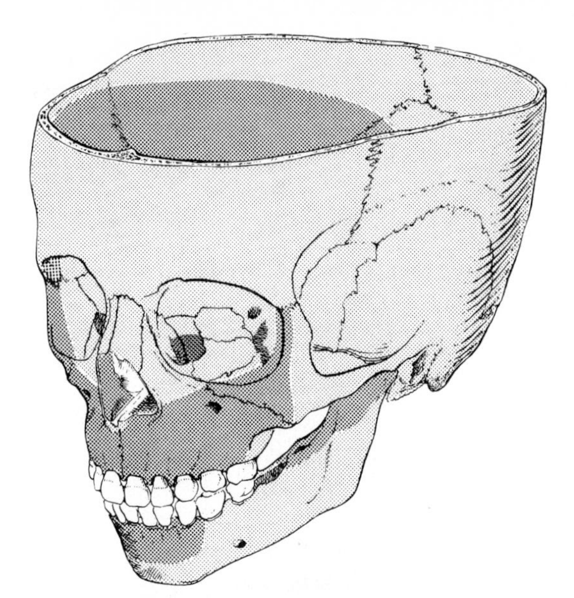

FIGURE 2–24. Resorptive surfaces are indicated by dark stippling; depository surfaces, by light stippling. (From Enlow, D. H., T. Kuroda, and A. B. Lewis: The morphological and morphogenetic basis for craniofacial form and pattern. Angle Orthod., 41:161, 1971.)

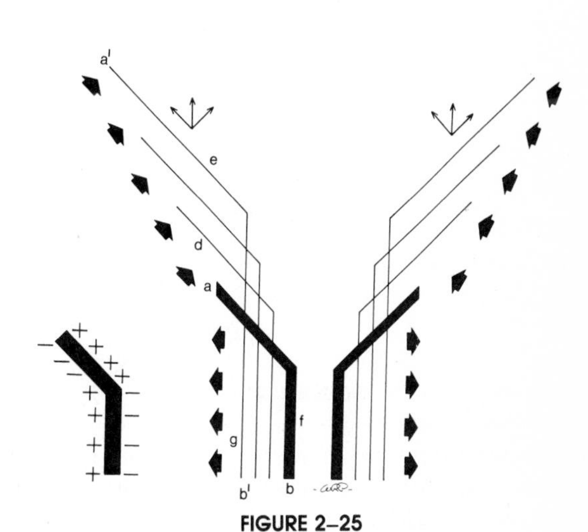

FIGURE 2–25

deposition and removal of ions to maintain blood calcium levels and carry out other mineral homeostasis functions. Another type of remodeling involves the secondary reconstruction of bone by **haversian systems** and also the rebuilding of cancellous trabeculae. A third kind of remodeling relates to the regeneration and reconstruction of bone during or following **disease** and **trauma.** The remodeling process that we are dealing with in facial morphogenesis, however, is **growth remodeling.** In order for a bone to grow and enlarge, it must also undergo a simultaneous process of remodeling.

The V Principle and Remodeling Pattern

The pattern of resorption and deposition seen in Figure 2–25 produces a combination of growth movements in which part *a* moves to *a′* and *b* moves to *b′*. Surfaces *d* and *g* represent the external (periosteal) sides, and *e* and *f* are internal (endosteal) surfaces. Surface *f* is resorptive, and *g* is depository. The **inside** surface at *e* faces the direction of growth; the bone thereby actually enlarges here by an endosteal mode of growth, that is, by the continued adding of new bone on the inside, rather than the outside. Surface *d* is resorptive. The bone as a whole thus increases in size, even though about half of its external surfaces are resorptive.

A most useful and basic concept in facial growth is the V principle (Fig. 2–26). Many facial and cranial bones, or parts of bones, have a V-shaped configuration. Note that bone deposition occurs on the **inner** side of the V; resorption takes place on the outside surface. The V thereby **moves** from position *A* to *B* and, at the same time, **increases** in overall dimensions. The direction of movement is toward the wide end of the V. Thus, a simultaneous growth movement and enlargement occurs by additions of bone on the inside with removal from the outside. The V principle will be referred to many times in later explanations of the facial growth process.

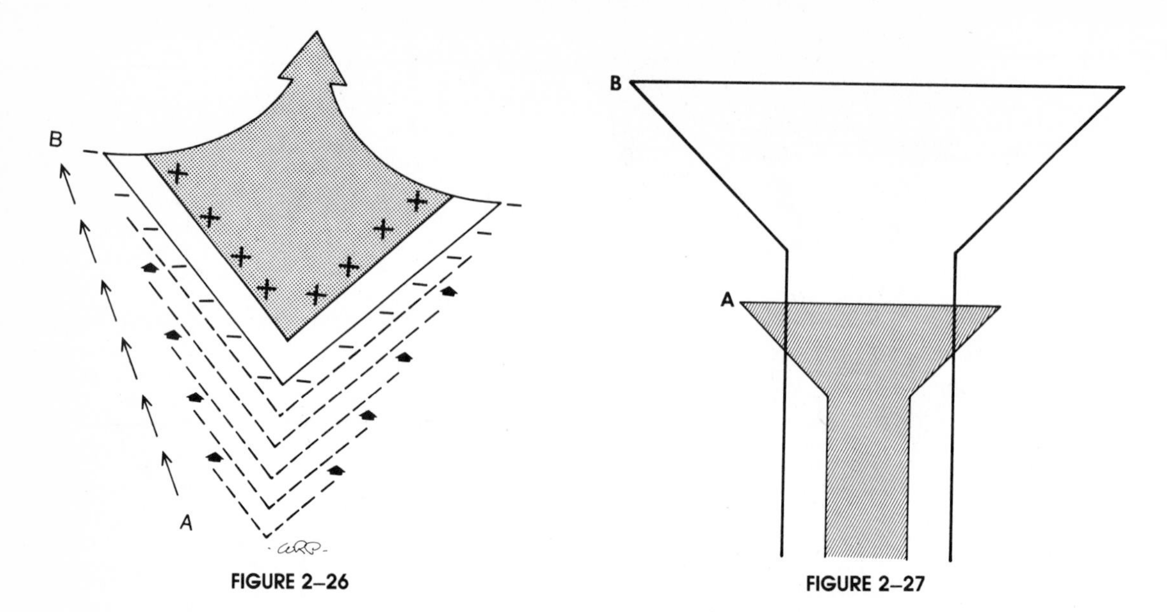

FIGURE 2–26

FIGURE 2–27

Note that the diameter at *A* in Figure 2–27 is **reduced** because the broad part of the bone is relocated to position *B*. This is a remodeling change that converts a **wider** part into a more **narrow** part, as both become sequentially relocated by the V principle. Periosteal resorption and endosteal deposition of bone tissue carry this out.

If a transverse histologic section of the bone is made at *A* in Figure 2–28, it can be seen that the periosteal surface is **resorptive;** bone-removing osteoclasts blanket this surface field during the active period of bone growth. The depository endosteal surface is lined with bone-producing osteoblasts. A transverse section at *B* shows new endosteal bone added onto the inner surface of the cortex. A transverse section made at *C* shows an endosteal layer that was produced during the inward growth phase. This is covered by a periosteal layer of bone following **outward** reversal, as this part of the bone now increases in diameter. A transverse section at *D* shows a cortex composed entirely of periosteal bone. The outer surface is depository, and the endosteal surface is resorptive.

If metallic implant markers are nailed into the bone at *X, Y,* and *Z,* note that marker *X* is subsequently released from the bone because that part becomes removed by periosteal resorption. It will lie free in the surrounding soft tissue. Marker *Z* is also released because of endosteal resorption taking place in this location, and it will lie free in the medullary cavity. Marker *Y* was originally inserted into the cortex on the periosteal side but becomes translocated over to the endosteal side of the cortex. This is because new cortical bone is added to the left, and older bone is removed from the right, thus changing the relative position of the marker from one side of the cortex to the other.

A transverse section through the zygomatic arch in Figure 2–29 demonstrates how a bone grows and remodels laterally as the whole bone simultaneously grows in length. The zygomatic arch is moving and enlarging laterally and also inferiorly as the entire face, brain, and cranium widen and expand into space formerly occupied by the zygomatic arch. It does this by progressive deposition on the lateral-facing and downward-facing periosteal and endosteal surfaces, with resorption from the opposite surfaces. The remnants of the old cortical contours can be recognized in microscopic sections. The right and left

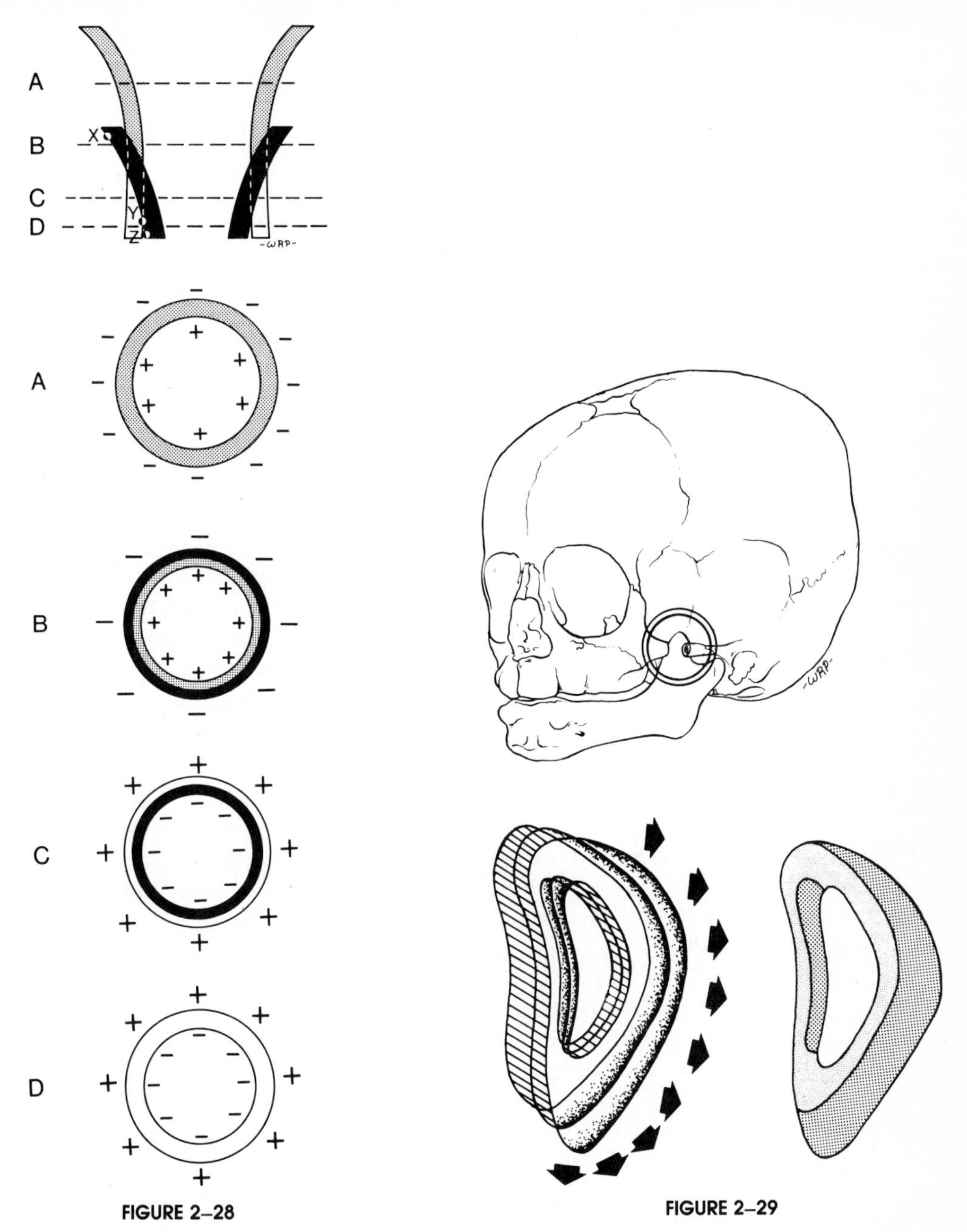

FIGURE 2–28

FIGURE 2–29

zygomatic arches thus grow out and away as the rest of the head enlarges between them. The arches also increase in size to accommodate the muscles attached to them. (Note: Half of the bone tissue is endosteal in origin, and half is periosteal. Half of the inner and outer surfaces are resorptive, and half are depository.)

The Relocation Function of Remodeling

Why do bones remodel as they grow? The basic, key factor is the process of **relocation.** In this stack of chips (Fig. 2–30), the black chip is at the right end in *a*. This black chip has been placed at the level of the condyle in the

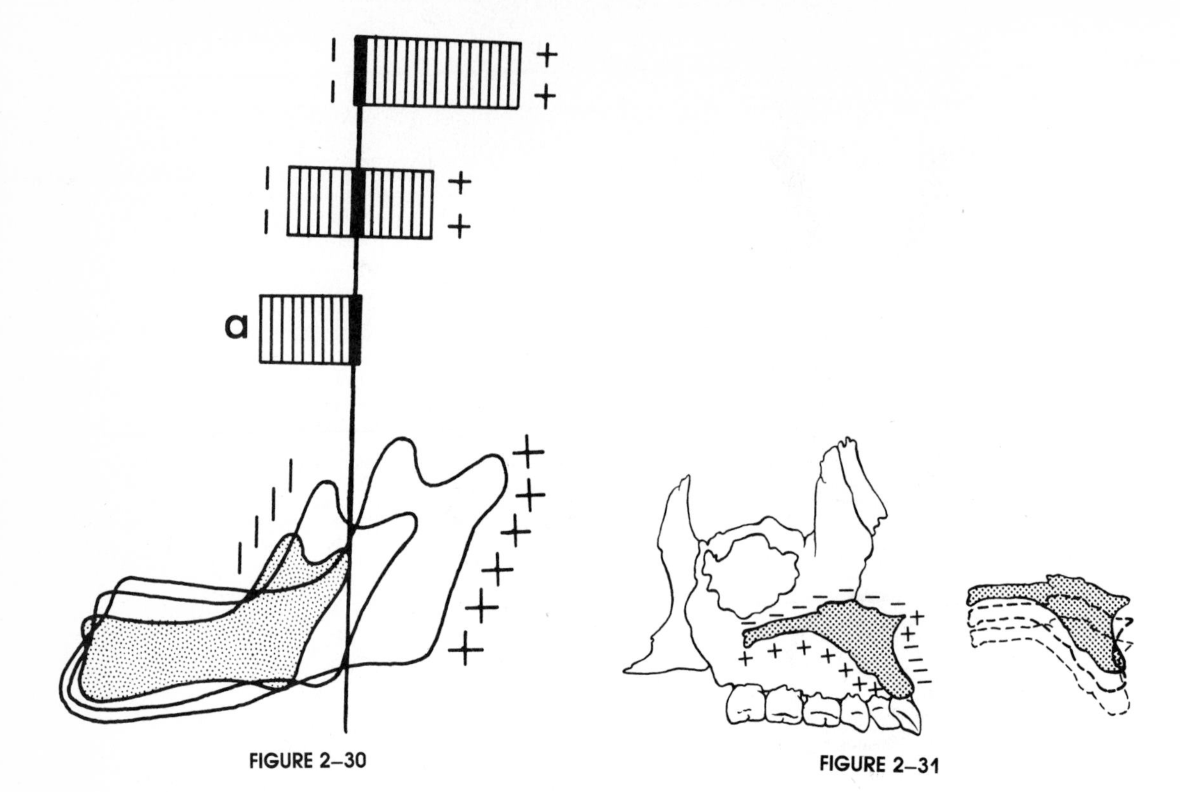

FIGURE 2–30

FIGURE 2–31

smallest mandibular stage to demonstrate how this location becomes translocated "across" the ramus to lie at the level of the anterior margin in the third stage. As "growth" continues to take place, the black chip becomes progressively "relocated"—not by its own movement but because new chips have been added on one side and removed from the other. This changes the **relative position** of the black chip within the stack, even though this chip itself does not move. Let the stack of chips represent a whole growing bone having complex topographic shape, rather than a perfectly cylindrical form. It is then apparent that the changing relative positions of the black chip would require continuous **remodeling** of the shape and sectional dimensions to conform with each successive position the chip comes to occupy. A **sequence** of continuous remodeling changes is required level by level. Remodeling is a process of reshaping and resizing each level (chip) within a growing bone as it is relocated sequentially into a succession of new levels. This is because additions and/or resorption in the various **other** parts cause changes in the relative positions of all levels. Note that the **position** of the condyle in the smallest mandibular stage becomes relocated into the middle of the ramus and then onto the anterior border of the ramus. Continuous remodeling is thus involved as this and all other areas change in relative position.

In the face of a young child, the levels of the maxillary arch and nasal floor lie very close to the inferior orbital rim. The maxillary arch and palate, however, **move** downward. This process involves (in part) an inferior direction of **remodeling** by the hard palate and the bony maxillary arch (Fig. 2–31). Bone deposition occurs on the downward-facing surfaces, together with resorption from the superior-facing surfaces on the palate. The combination results in a downward **relocation** of the whole palate—maxillary arch composite into progressively lower levels, so that the arch finally comes to lie considerably below the inferior orbital rim. The vertical dimension of the nasal chamber is greatly increased as a result.

About half of the external surfaces involved in this growth and remodeling change are resorptive and half depository. About half of the bone tissue of the palate is thus endosteal and half periosteal. (The cortex on the nasal side of the palate is produced by the endosteum of the medullary cavity.)

Because of the **relocation** process, the nasal area of the adult occupies an area where the bony maxillary arch **used to be** located during childhood (Fig. 2–32). What was once the bony maxillary arch and palatal region has been converted into the expanded nasal region. This is "growth remodeling"; the basis for it is **relocation.**

As the palate and arch grow downward by constant deposition of new bone on one side and resorption of previously formed bone from the other, the bone tissue that comes to house the teeth at older age periods is not the same actual bone enclosing them during the succession of former growth levels. This is significant because the growth movement and the exchanges of bone involved are **used** by the orthodontist to "work with growth."

As the mandible grows, the ramus moves in a backward direction by appropriate combinations of resorption and deposition. As the ramus is **relocated** posteriorly, the corpus becomes **lengthened by a remodeling conversion** from what was at one time the ramus during a former growth period (Fig. 2–33). During growth from the fetus to the adult, the "molar" region in the younger mandible, for example, undergoes relocation to occupy the "premolar" region of the larger, older mandible. It is apparent that remodeling is a process of relocation and that the same **deposition and resorption producing growth enlargement also carry out the growth remodeling process.** Growth, indeed, **is** a process of remodeling to provide enlargement. (There is more to growth, however, as will be seen.)

Remodeling maintains the overall morphologic characteristics of a bone during its growth. Any given bone grows differentially; that is, it increases in some directions much more than others and at varying regional rates (Fig. 2–34). If a bone were to grow uniformly in all directions, relocation would not

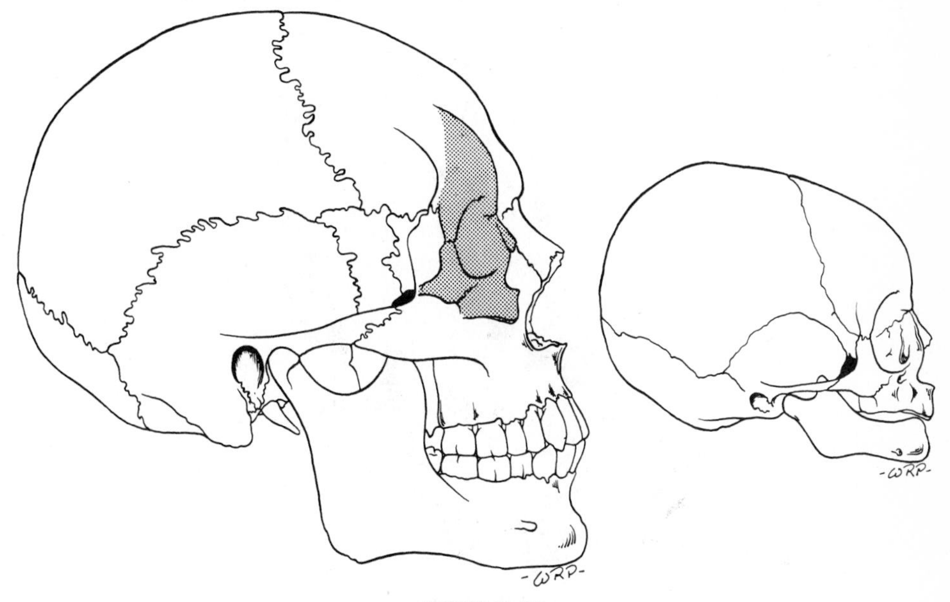

FIGURE 2–32

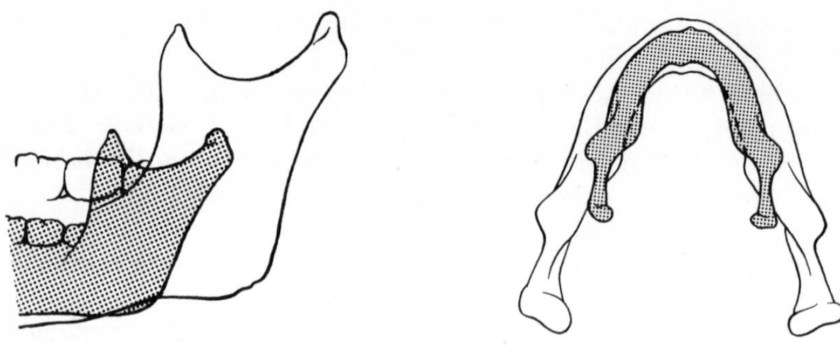

FIGURE 2–33

be involved, and remodeling would not be required as a part of the growth process. Because of the multiple physiologic and mechanical functions of a bone, however, it necessarily has a complex topographic shape. This can only be produced by a differential mode of growth involving the relocation process by remodeling.

The mandible grows differentially in directions that are predominantly posterior and superior. Even though successive remodeling of one part into another constantly takes place as the whole bone enlarges, the form of the bone as a whole is sustained (with some characteristic age changes in shape). It is remarkable that the external morphologic characteristics of any given bone are relatively constant, even though its substance undergoes massive internal changes and all its parts experience widespread alterations in regional shape and size as they are relocated. This is the special function of growth remodeling; it **maintains** the form of a whole bone while providing for its enlargement at the same time. Thus, remodeling is not a process that functions essentially to alter overall shape, although some degree of this is also involved. Although the term "remodeling" implies such change, the actual changes produced by growth remodeling are mostly those that deal with the sequential **relocation** of a bone's component parts.

The entire bone takes part in this process of growth enlargement and remodeling. **All** surfaces, inside and out, participate actively at one time or

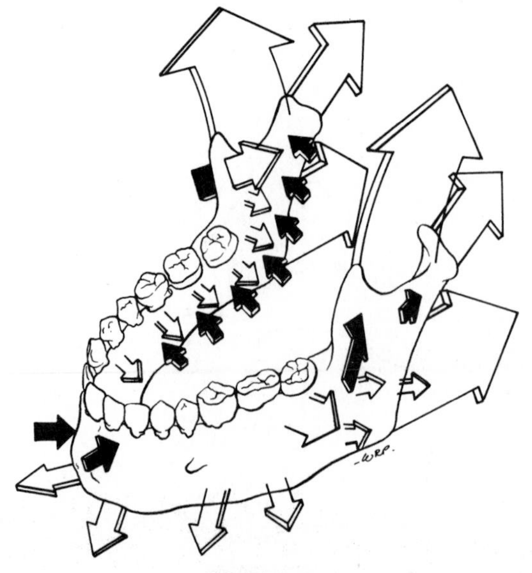

FIGURE 2–34

another. The entire periosteal surface and the entire endosteal surface of the cortex (and all trabecular surfaces of the cancellous bone as well) are involved. The bone does not just grow at special sites or "centers." Many divergent directions of growth take place among the many different regions of a bone. It is a three-dimensional process (see Chapter 3, Figs. 3–128 and 3–142).

Fields of Remodeling

Resorptive and depository **fields** of growth blanket all of the outside and inside surfaces of a bone (Fig. 2–35). This mosaic pattern is more or less constant for each bone throughout the growth period, unless a major change in the shape of a region becomes involved. As the perimeter of these growth fields enlarges, the parts of the bone associated with them correspondingly increase in size. As emphasized earlier, of course, the actual operation of these fields of growth is performed by the enclosing osteogenic connective tissues. The bone itself is the product of this field activity. Thus, during the operation of the relocation process, it is the **growth fields in the osteogenic connective tissues** that first move and control the relocation movements of the underlying bony parts associated with each field. The growth movement of the bone **follows** the pace-setting movement of the overlying growth field. There is virtually no lag time, however, between the two.

Note that resorptive field *a* moves to *a′* in Figure 2–36, and that the corresponding zone of the bone beneath it remodels and moves under its control. The boundaries of this resorptive field are thus changed so that they come to occupy the region that was located just posteriorly. A **reversal line** (*x*) separates field *a* from the area of the ramus behind it, and moves to *x′*. The resorptive field in the larger mandible (*a′*) occupies the same **relative** placement as it did when the bone was smaller during the former growth stage (*a*). The field, however, (*1*) is now larger and (*2*) has moved to a new location, bringing its part of the bone with it by continuous bone deposition and resorption (that is, remodeling growth). The actual bone tissue present in the ramus of the smaller stage has been replaced by a whole new generation of bone in the location occupied by the ramus of the larger stage following relocation. The

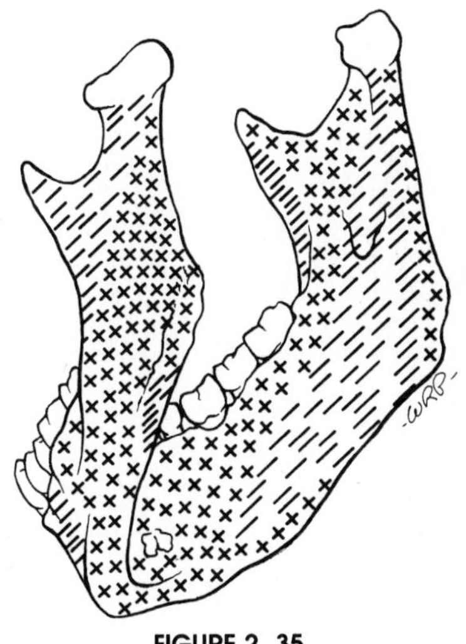

FIGURE 2–35

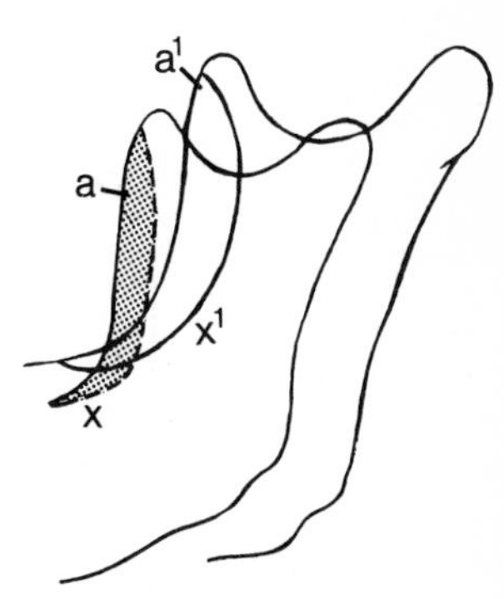

FIGURE 2–36

patterns of distribution of all the various resorptive and depository fields, however, have not changed; the fields have only moved from one position to another as the whole bone enlarged. This requires **reversals** as any one field expands into areas previously occupied by other fields, which in turn have moved on to hold successively new locations.

Although these growth movements are carried out by the osteogenic membranes and cartilages, the bone contributes feedback information to them, so that when the size, shape, biomechanical properties, and so forth, of the bone come into equilibrium with functional requirements, the histogenetic activity then becomes turned off.

Variations in the shape and size of the face are always the rule (Fig. 2–37). No two faces are quite alike. Morphologic variations, normal and abnormal, are produced by corresponding developmental variations that take place during the growth process. Some can be genetically established by characteristic soft tissue relationships that are hereditary determinants of bone growth (and also cartilage relationships that are genetic determinants, according to some investigators). Other variations are largely determined by functional changes in soft tissue relationships within a given individual during his own development. The results, however, are all based on the following factors that establish the nature of anatomic variations in an individual person:

1. Fundamental differences in the **pattern** of the fields of resorption and deposition, that is, the distribution and the configuration of the growth fields in an individual person
2. The specific placement of the **boundaries** between growth fields, that is, the size of any given growth field
3. The differential **rates** and **amounts** of deposition and resorption in each field
4. The **timing** of the growth activities among the different fields

Understanding the regional growth field concept and the operation of the remodeling process, how the complementary process of displacement operates

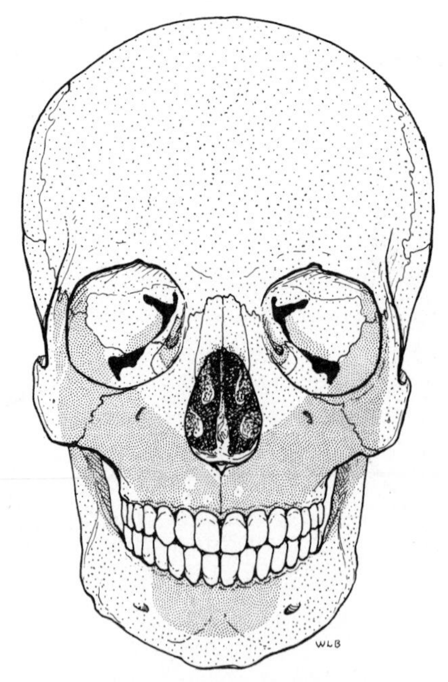

FIGURE 2–37. (After Enlow, D. H.: *The Human Face.* New York, Harper & Row, 1968.)

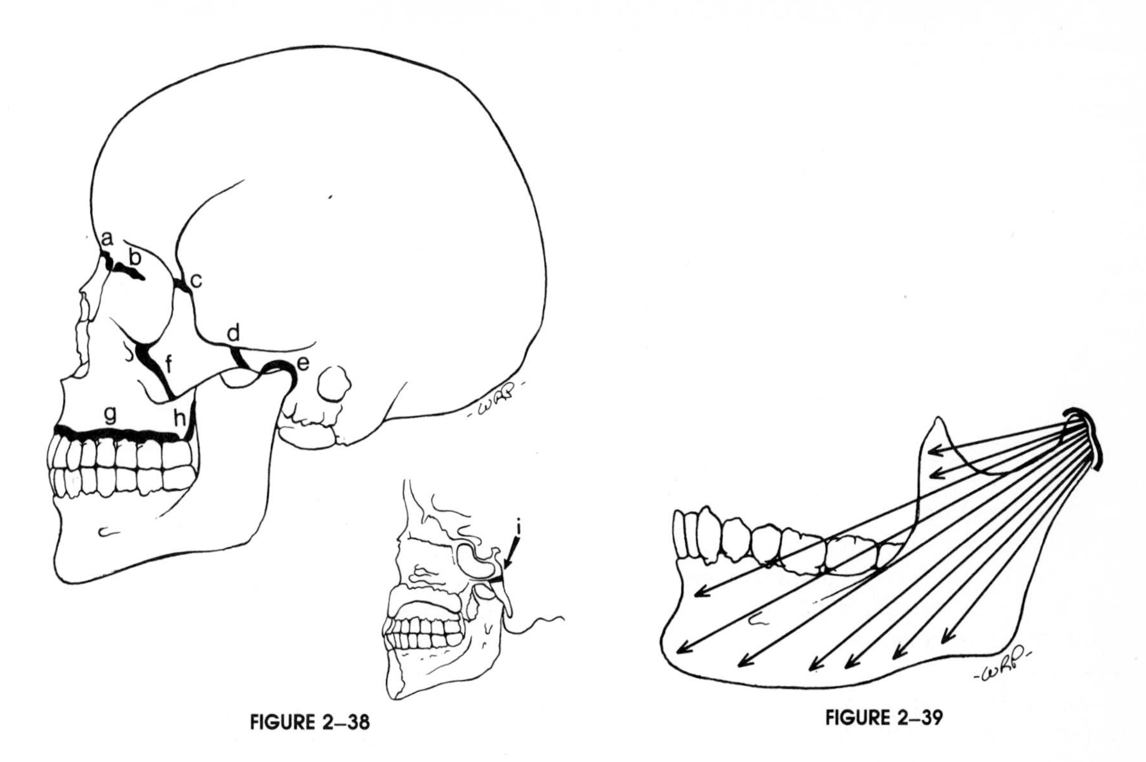

FIGURE 2–38

FIGURE 2–39

(described later), how an individual's growth variations can occur, is basic for the clinician. The question can be one of working "with" or "against" the intrinsic processes in the use of different orthodontic procedures. Does a given procedure, for example, harness the same intrinsic regional remodeling or displacement direction but with alteration of magnitude? Or is a direction actually reversed, perhaps with severe violations of remodeling field boundaries, leading to rebound? Does the clinician presume a given area to be depository when in fact it is intrinsically resorptive and undergoing a relocation directional movement not understood? Or, for the surgeon, is a bone or periosteal-endosteal transplant, or the field to which it is moved, unknowingly "programmed" to be resorptive when assumed to be depository?

Some growth fields have traditionally been singled out for special attention because of their particular role in the growth of a given facial or cranial bone (Fig. 2–38). These special **growth sites** include the sutures of the face and cranium (*a, b, c, d, f*), the mandibular condyle (*e*), the maxillary tuberosity (*h*), the synchondroses of the cranial base (*i*), and the alveolar bone housing teeth (*g*). All these growth sites are described in later chapters, but it is to be understood that such special sites **do not** carry out the **entire** growth process for the particular bones associated with them. **All other** inside and outside surfaces of a given bone also actively participate in the overall growth process. The contributions made by these other growth fields are just as basic and essential as the specially designated sites. This important point has not always been understood.

As mentioned, the mandibular condyle is a specially recognized growth site. The condyle and some other special sites are sometimes also termed growth centers. This label has come into disfavor, however, because it is now believed that such centers do not actually control the growth processes of the bone as a whole (Fig. 2–39). They are not master centers that directly regulate the overall growth process of the entire bone and all its regional parts. Although these centers are important growth sites, they represent only regional fields of growth adapted to the localized morphogenic circumstances in their own particular areas. Such centers, as pointed out previously, do not provide for the entire growth of the whole bone, as has sometimes been mistakenly presumed.

Routine headfilms, of course, are **two-dimensional,** and this is a limitation that presents many troublesome problems. Only the anterior and posterior **edges** of the ramus, for example, can be visualized (as at *A* and *B*) in lateral cephalograms (Fig. 2–40). Important changes on surfaces in the span **between** these edges (*C*) cannot be visualized. This is all the more reason for the clinician and researcher to thoroughly understand what happens when such areas grow in a **three-dimensional** manner not representable by the headfilm itself.

THE DISPLACEMENT PROCESS

In **remodeling,** the bone grows (cortical drift) from *A* to *B* by deposition on the side facing the direction of growth movement, as seen in a suture (Fig. 2–41). In the process of **displacement,** however, the whole bone is **carried** from *B* to *A* by mechanical force as it simultaneously enlarges from *A* to *B*. Drift and displacement are separate processes, but they occur in conjunction with one another.

The growth expansion of a single bone is a process by which the size and shape of the bone develop in response to the composite of all the functional soft tissue relationships associated with that individual bone. The bone does not grow and enlarge in an isolated way, however; its increases in size include articular contacts with **other** bones that are also enlarging at the same time. For this reason, as emphasized before, all articular contacts are important—condyles, sutures, and synchondroses—because they are the sites where displacement is involved. Articulations are the interface surfaces "away" from which the displacement movements occur as all of the whole bones enlarge. The amount of enlargement equals the extent of displacement. That is, a bone grows into

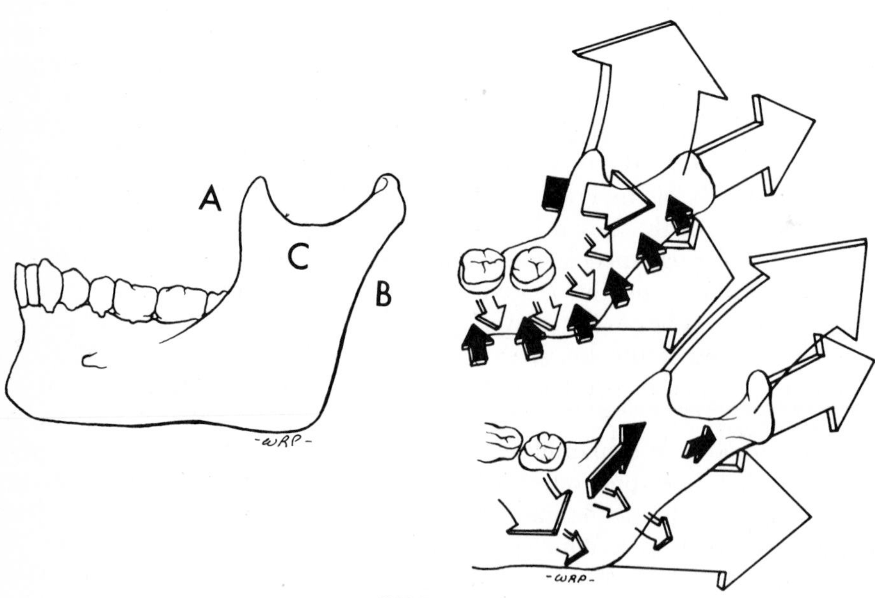

FIGURE 2–40

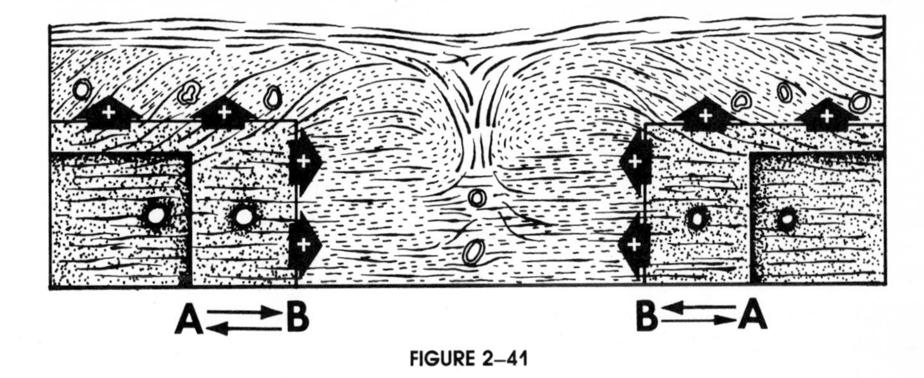

FIGURE 2–41

the space being created as the whole bone is displaced by amounts determined by the extent of surrounding soft tissue enlargement. The growth of each bone thereby keeps pace with that of the soft tissues it serves.

In the analogy shown in Figure 2–42, the expansion of a single balloon does not "compete" for space. However, if **two** enlarging balloons are in contact with each other, a displacing **movement** takes place until their positions become adjusted as either one or both expand. This movement proceeds away from the interface between the two balloons. What happens when the mandible, for example, grows in a direction **toward** its articular contact with the cranium? A "displacement" takes place in which the whole mandible moves **away** as it enlarges toward the temporal bone (Fig. 2–43).

Do the balloons **shove** each other apart **because** of the pushing force produced by the expansion? Or are the balloons **carried** apart by other (outside) mechanical forces, with growth expansion **responding** to the separation, thereby maintaining contact between them (Fig. 2–44)? In the first possibility, the extent of push equals but follows the combined amount of expansion. In the second possibility, the extent of combined enlargement equals but follows (virtually simultaneously) the amount of separation, with the balloons "growing" into the potential space being created. In other words, which is the primary (pacemaker) movement, displacement or remodeling enlargement? The question is more than academic; clinical treatment procedures utilize both kinds of growth

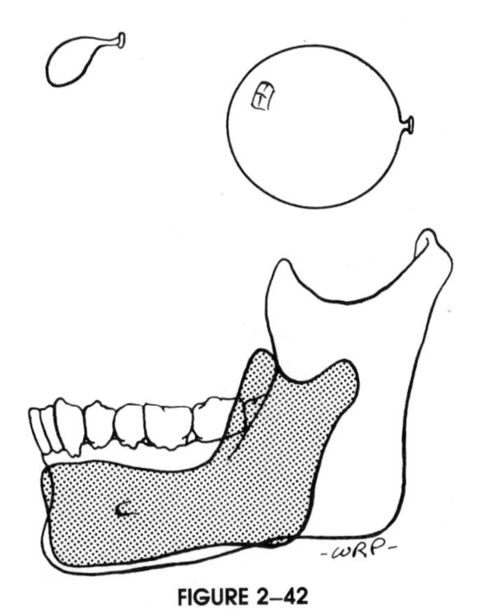

FIGURE 2–42

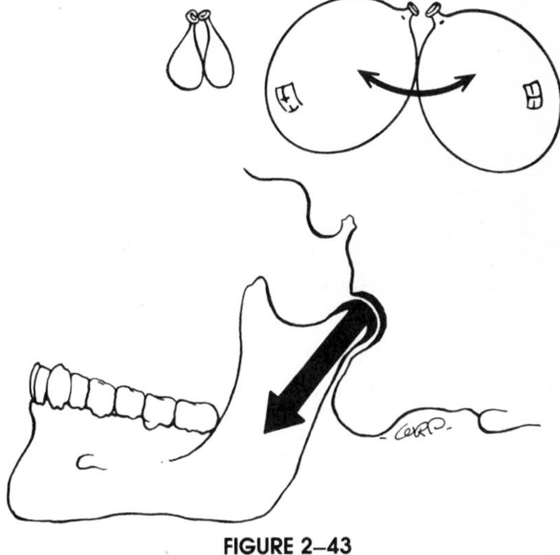

FIGURE 2–43

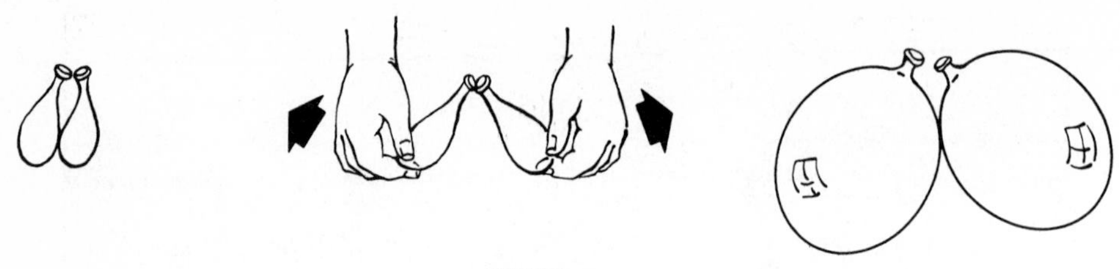

FIGURE 2–44

movements, and the clinician should understand which kind he or she is attempting to control.

This has been, and still is, one of the great historical controversies in craniofacial biology. The mandible grows by deposition and resorption in the manner shown in Figure 2–45 (top). The predominant vectors (direction and magnitude) of growth are posterior and superior. Thus, the condyle grows directly **toward** its articular contact in the glenoid fossa of the cranial floor.

As this takes place, the whole mandible is moved forward and downward by the same amount that it grows upward and backward (Fig. 2–45, bottom). The direction of growth by new bone additions at the condyle and the direction of displacement are opposite to each other.

Is the forward and downward displacement movement of the mandible accomplished by a **shove** against the articular surface caused by the growth of the condyle, or conversely, by a **carry** of the entire mandible away from the cranial base by other mechanical forces (such as the expansive growth of the contiguous muscle and connective tissue mass) (Fig. 2–46)? If the latter is true, bone growth follows secondarily (but almost simultaneously) at the condyle to maintain constant contact with the temporal bone. As "force *a*" carries the mandible anteriorly and inferiorly, the condyle is **triggered to respond** by an equal amount of growth at *b*.

Is condylar growth thus the active cause of displacement or the passive (secondary) response to it? Current theory favors the passive **carry** concept, rather than an active thrust or push (which was popular for many years); the problem, however, has not yet been resolved to everyone's satisfaction. (See the "functional matrix" concept described later.)

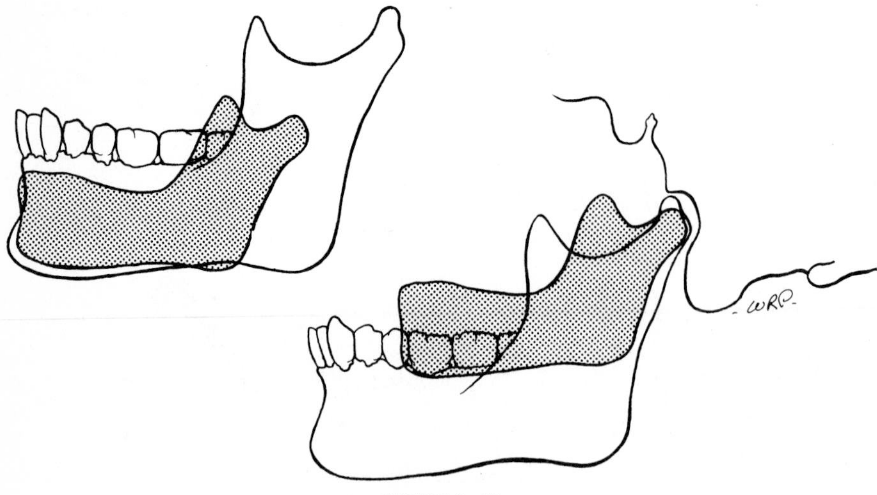

FIGURE 2–45

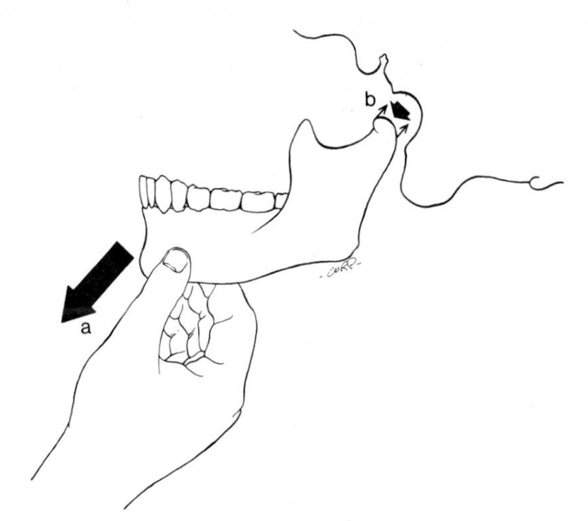

FIGURE 2–46

In summary, two basic modes of skeletal movement take part in the growth of the face and cranium. **Remodeling** involves deposition of bone on the side pointed toward the direction of growth of a given area; resorption usually occurs on the opposite side of that particular bony cortex (or cancellous trabecula). **Displacement** is a separate movement of the **whole bone** by some physical force that carries it, *in toto,* away from its contacts with other bones, which are also growing and increasing in overall size at the same time. This two-phase remodeling-displacement process takes place virtually simultaneously. The displacement movement, however, is presently believed by many researchers to be the pacemaking change, with the rate and direction of bone growth representing a transformative response. Two kinds of displacement occur, **primary** and **secondary.**

In **primary displacement,** the process of physical carry takes place in conjunction with a bone's **own** enlargement (Fig. 2–47A). Two principal growth vectors in the maxilla, for example, are posterior and superior. As this occurs, the whole bone is displaced in opposite anterior and inferior directions. Primary displacement produces the "space" within which the bone continues to grow. The amount of this primary displacement exactly equals the amount of new bone deposition that takes place. The respective directions are always opposite

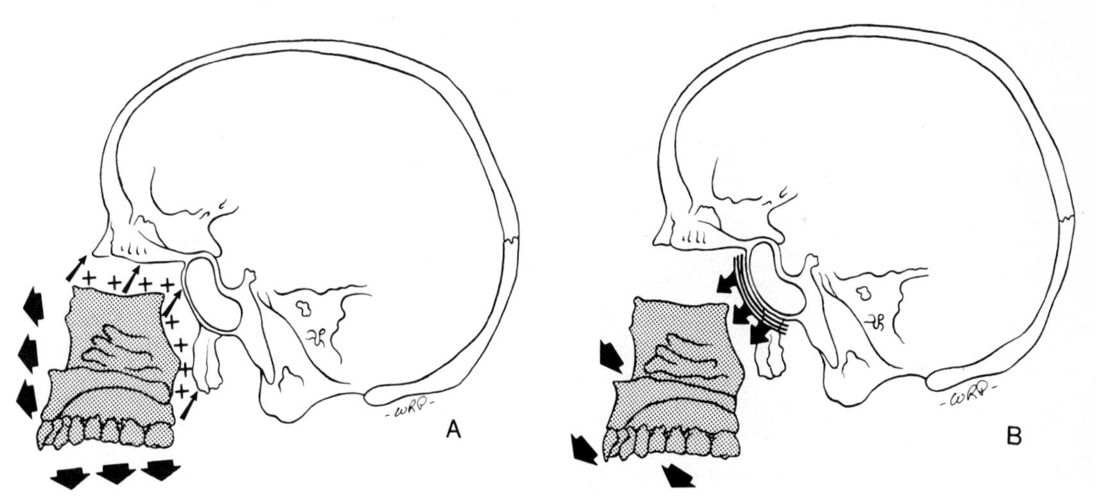

FIGURE 2–47

in the primary type of displacement. Because primary displacement takes place at interfaces with other, contiguous skeletal elements, **joint contacts** are important sites involved in this kind of remodeling change.

In **secondary displacement,** the movement of the bone is not directly related to its own enlargement. For example, the anterior direction of growth by the middle cranial fossa and the temporal lobe of the cerebrum displaces the maxilla anteriorly and inferiorly (Fig. 2–47B). Maxillary growth and enlargement itself, however, are not involved in this particular kind of displacement movement. Thus, as any bone grows, remodels, and becomes displaced in conjunction with its own growth process, it is also displaced, in addition, by the growth of **other** bones and their soft tissues. This can have a "domino effect." That is, growth changes can be passed on from region to region to produce a secondary effect in areas quite distant. Such changes are cumulative.

Note that much of the anterior part of the midfacial region is **resorptive** in nature (Fig. 2–48). Yet the face grows **forward.** How can this be? The face does not simply grow directly anteriorly. The forward movement is a **composite** result of growth changes (a) by resorption and deposition that cause the maxilla to **enlarge backward** and (b) by primary and secondary displacement movements that cause it to be **carried forward.** The resorptive nature of the anteriorly facing surface of the premaxilla is concerned with its downward, not forward, growth (as explained in Chapter 3).

To illustrate the composite nature of these different growth processes, the growth of the arm is used as an analogy (Fig. 2–49). The tip of the finger moves away from the shoulder as the whole arm increases in length. Most of this growth movement of the finger, of course, is not a consequence of growth at the fingertip itself. The aggregate summation of linear growth increases by all the bones in the arm at each particular interface between the phalanges, carpals, metacarpals, radius, ulna, and humerus is involved. The contribution by the tip of the terminal phalanx is only a relatively small part of the total. It is the secondary displacement effect produced by all the **other** bones in the arm that causes most of the growth movement of the fingertip.

Similarly, the greater part of the growth movement of the tip of the premaxilla is produced by the growth expansion of all the bones behind and

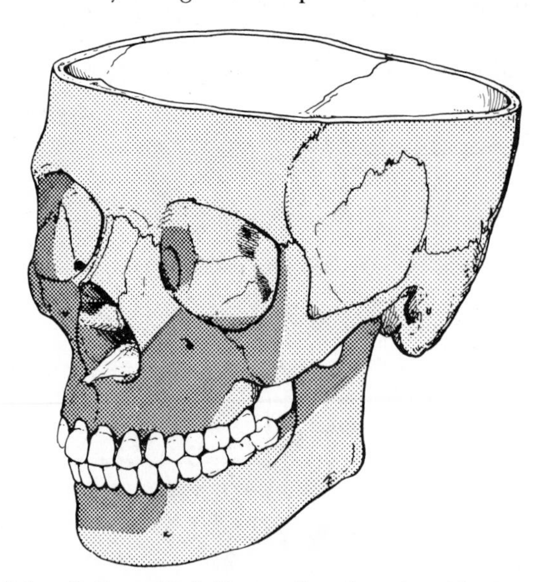

FIGURE 2–48. (From Enlow, D. H., and R. E. Moyers: Growth and architecture of the face. J.A.D.A., 82:763, 1971.)

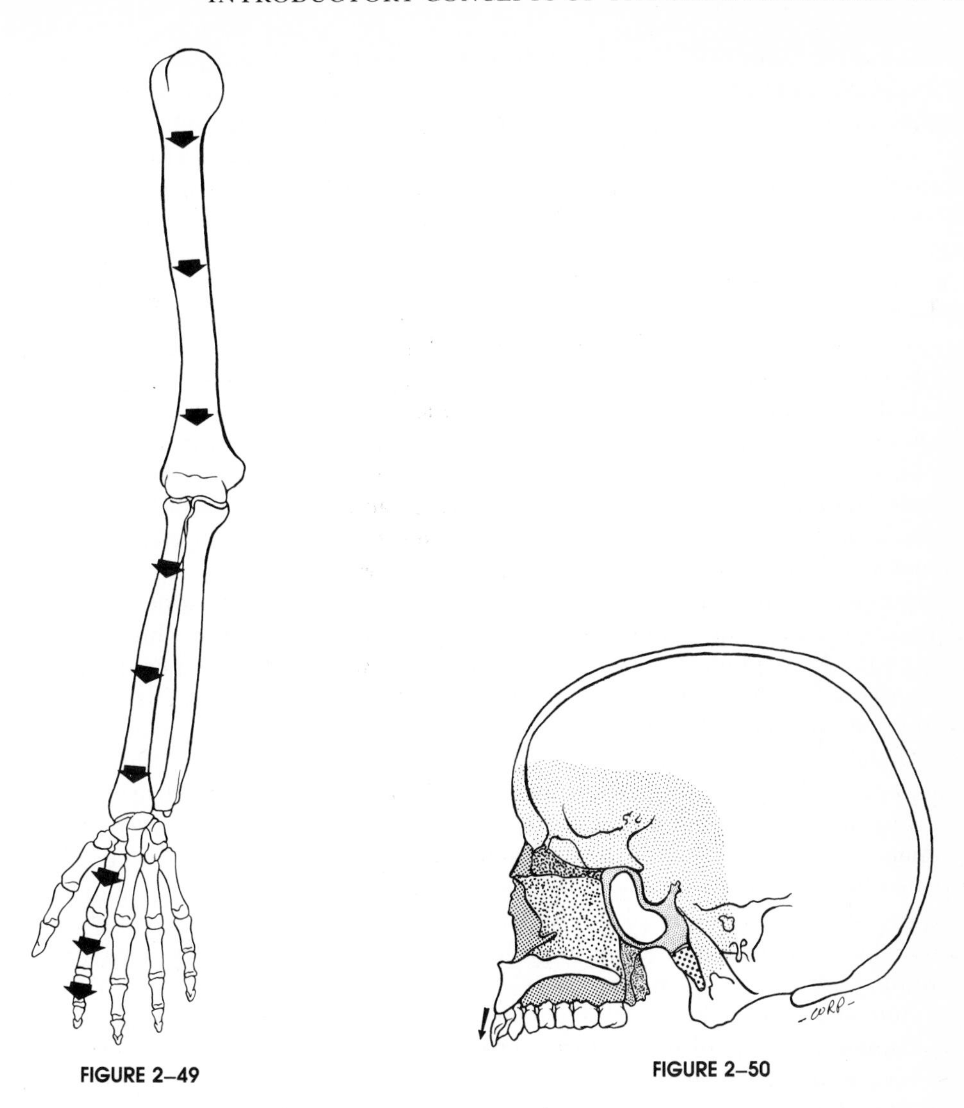

FIGURE 2–49

FIGURE 2–50

above it and by growth in other parts of the maxilla (Fig. 2–50). The premaxillary tip itself contributes only a very small part of its own forward growth movement. The enlargement of the maxilla proper and the frontal, ethmoid, occipital, sphenoid, and temporal bones provides an aggregate expansion, the sum of which is the basis for most of the total forward movement of the premaxilla. (It contributes somewhat more to its own downward movement, however, as illustrated in Chapter 3.) Keep in mind, also, that the biomechanical basis for these displacement growth movements is actually the "carry effect" produced by the expansion of the soft tissues associated with the bones, not a "pushing effect" of bones against bones.

The points made in the previous paragraph are basic and should be understood from Day One in a specialist training program. If one is to "work with growth," yet presuming that "growth" is primarily within, for example, the premaxillary region itself rather than actually elsewhere, a false start has been made that will seriously handicap the student.

The factor of secondary displacement is a fundamental part of the overall process of craniofacial enlargement. Growth effects of skeletal parts far removed are passed on, bone by bone, to become expressed on the resultant topography of the face. Cranial floor–facial growth imbalances often contribute materially to misalignments and improper positionings of the facial bones. Secondary displacement is one of several basic factors involved in the developmental basis for malocclusions and other types of facial dysplasias. In Figure 2–51, note, for example, how a remodeling rotation alignment of the middle cranial fossae, in the manner shown, has the secondary displacement effect of maxillary retrusion and mandibular protrusion.

The subject of **rotations** is a major consideration. Although confused by different terminologies in the literature, there are, simply, two basic categories of rotations: (1) remodeling rotations (see Figs. 3–76 and 3–104); and (2) displacement rotations, either primary or secondary (see Fig. 5–35).

Both primary and secondary displacement as well as remodeling are involved in the growth movements of all bones. A great many different combinations of all three processes are found throughout the craniofacial complex. For example, in Figure 2–52, bones *X* and *Y* are in contact (as by a suture, condyle, or synchondrosis). A growth increment by bone deposition at *a* produces a similar end-effect as deposition at *b,* with accompanying primary displacement of the whole bone to the right. Or increment *c* is added at the contact interface, with accompanying primary displacement of the whole bone to the right. Resorption at *d* occurs, however, producing an end-result equivalent to the two examples above. Or secondary displacement of segment *Y* is caused by separate segment *X,* owing to growth addition *e.* Primary displacement accompanies growth at *f.* With resorption at *g,* it is thus seen that this combination also produces end-results similar to all the examples above.

The analysis of composite growth changes is always difficult in headfilm evaluation because, as just seen, the **same** growth results can be theoretically attained by many different combinations of remodeling and displacement. The purpose of Chapter 3 is to analyze just which of the many hypothetical combinations actually take place in each of the many regions of the face and cranium.

Note this important point. The word "growth" is a general term that is, unfortunately, all too often used in a loose and imprecise sense. For example, it is frequently heard that some given clinical procedure "stimulates growth." One should always attempt to specify, whenever appropriate, just what kind of "growth" is indeed involved. Remodeling? Primary or secondary displacement?

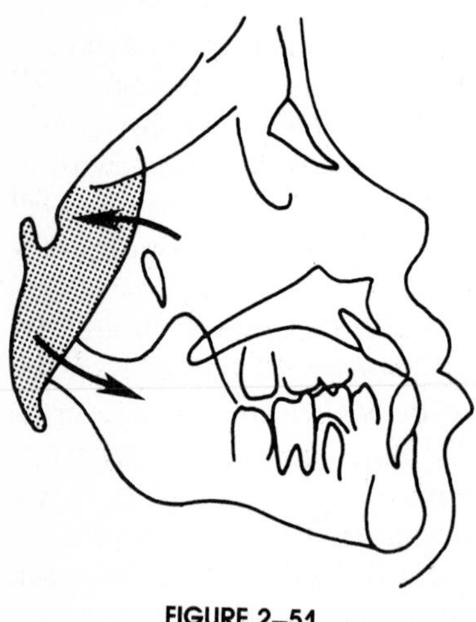

FIGURE 2–51

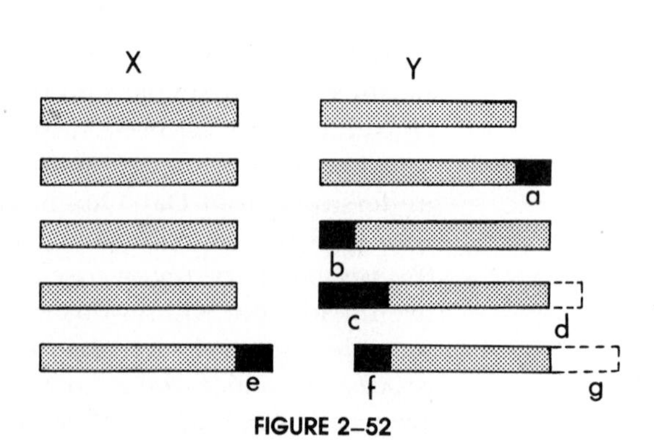

FIGURE 2–52

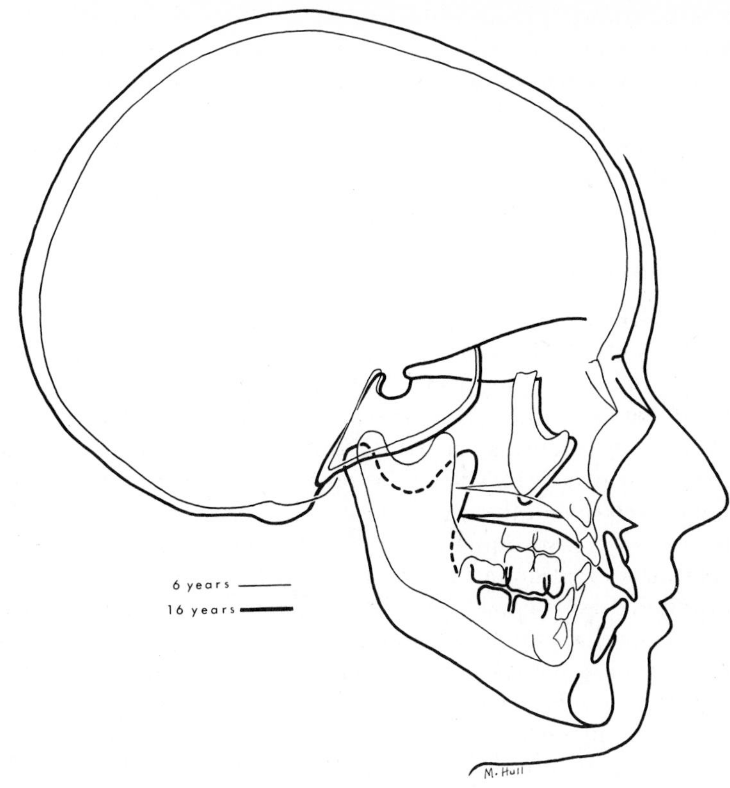

FIGURE 2–53. (From Enlow, D. H.: *The Human Face.* New York, Harper & Row, 1968.)

A particular combination? The biologic reason is apparent. If one is to control the growth process, just what is to be controlled must be understood, and the specific targets involved must be identified.

SUPERIMPOSING HEADFILM TRACINGS

The conventional method used to show facial "growth" is to superimpose serial headfilm tracings (i.e., tracings of the same individual at different ages) on the cranial base as shown in Figure 2–53. Sella is usually used as a registration point for the superimposition. Tracings are ordinarily used instead of the headfilms themselves because superimposed x-ray films pass insufficient light.

Superimposing on the cranial base demonstrates the "downward and forward" (one of the most common clichés in facial biology) expansion of the whole face relative to the cranial base. Great caution must be exercised, however; one must be certain one understands possible misrepresentations of just what this really shows, because multiple, complex combinations of regional remodeling and primary and secondary displacement are involved. This is the subject of Chapter 3.

First, this method of superimposition is appropriate and valid because we all naturally tend to visualize facial enlargement in **relation** to the cranium (and brain) behind and above it. That is, the characteristically small face and the earlier developing, larger brain in early childhood **change** progressively in respective proportions. The face grows and develops rapidly throughout the childhood period; the size and shape of the brain and the cranial vault also change, but much less noticeably. The parent **sees** the structural transformations of the child's face, month by month, as it "catches up" with the earlier maturing brain and braincase (Fig. 2–54). Superimposing on the cranial base (sella, etc.) thereby represents what one actually visualizes by direct observation as the face progressively enlarges.

Superimposing headfilm tracings "on the cranial base" is **not** valid, however, **if** the following **incorrect assumptions** are made:

1. The incorrect assumption that the cranial base is truly stable and unchanging. It is not. This notion has often mistakenly been made. The floor of the cranium continues to grow and undergo remodeling changes throughout the childhood period (although this is much more marked in some regions than in others at different age levels). Properly taken into account, however, this is not necessarily a factor, since the purpose, really, is only to show facial growth changes **relative** to the cranial base, whether or not it is actually stable.

2. The incorrect assumption that "fixed points" actually exist, that is, anatomic landmarks that do not move or remodel. All surfaces, inside and out, undergo continued, sequential growth movements and remodeling changes during morphogenesis (with the exception of no size changes by the ear ossicles). Although the **relative** position of some landmarks can remain relatively constant, the structures themselves actually experience significant growth movements and remodeling changes along with everything else (Fig. 2–55). Sella (*a*) has often been presumed to be a true "fixed" point or one that represents the "zero growth point" in the head. Of course, it is not. Sella changes during continued growth. This, however, does not invalidate the use of sella to represent a registration point on the cranial base **if** these various considerations are properly taken into account. **Nasion** (*b*) is another such landmark. So many marked growth and remodeling variations are associated with this point relating to age, sex, ethnic, and individual differences, however, that the use of nasion as a cephalometric landmark requires great caution. (**Note:** There are other basic reasons why points such as nasion and sella are misleading if improperly used, as explained in Chapters 5 and 6.)

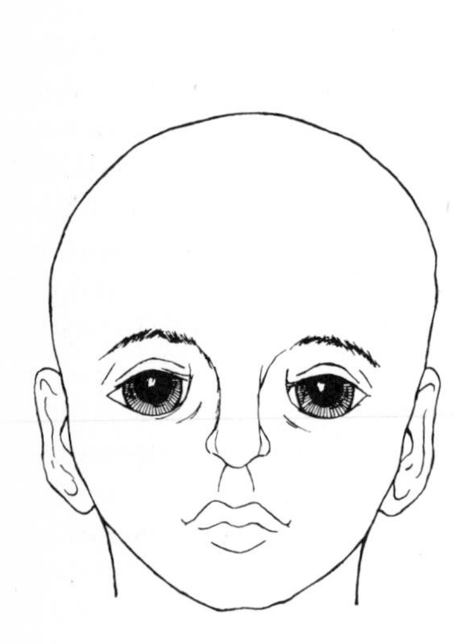

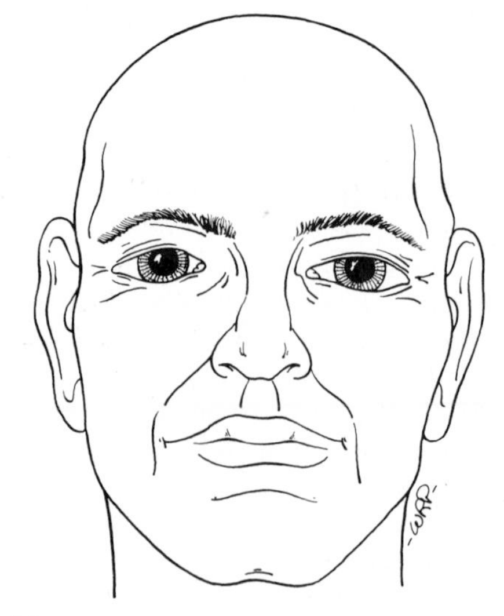

FIGURE 2–54

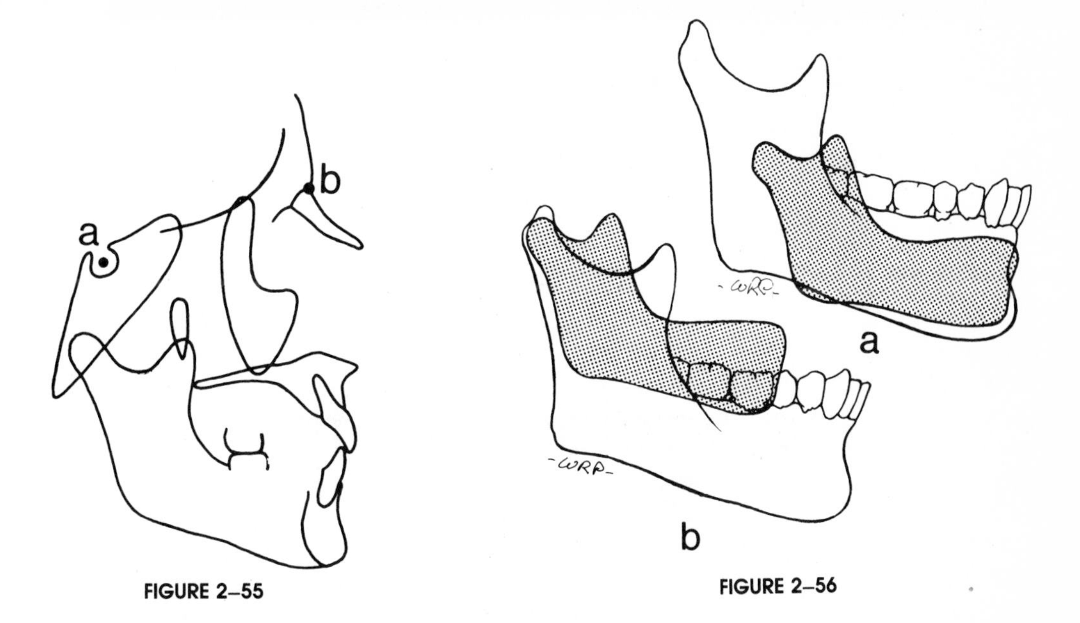

FIGURE 2–55

FIGURE 2–56

3. The incorrect assumption that the traditional "forward and downward" picture of facial enlargement, seen when serial tracings are superimposed on the cranial base, represents the **actual mode of facial growth.** Many workers believe, quite incorrectly, that this is how the face really grows, that is, that the facial profile of the younger stage expands straight to the profile of the older stage by direct growth from one to the other. This has been one of our most common misconceptions and one of the most difficult to overcome. The face does not merely "grow" in the manner represented by such an overlay. Growth is a multifactorial, cumulative composite of changes in the many regions of the head, the summation of which produces the "forward and downward" expansion seen in the overlay.

As mentioned above, superimposing headfilm tracings on the cranial base shows the **combined** results of (*a*) deposition and resorption (remodeling) and (*b*) primary and secondary displacement **relative** to a common reference plane (such as sella-nasion of the cranial base). The superimposing procedure, however, does not provide an accurate representation for either remodeling or displacement in most facial regions. Note that the two placements of the mandible in the preceding Figure 2–53, for example, do not properly represent either its growth by deposition and resorption (*a*) or its primary displacement (*b*), as shown in Figure 2–56. The overlay positions for the mandible in Figure 2–53 (and other facial bones as well) simply indicate their successive **locations** at the two age levels relative to the cranial base, not their actual modes of growth.

One basic problem always encountered with routine methods of superimposing headfilm tracings on the cranial base is that the **separate** effects of growth by deposition and resorption and displacement are **not distinguishable.** This is an important consideration. The purpose of the next chapter is to demonstrate these separate effects and to explain how the process of craniofacial growth is really carried out.

CHAPTER THREE

The Facial Growth Process

Part 1

In the pages that follow, the overall sequence of facial and cranial growth is outlined. Part 1 is a digest version to provide a general, less detailed understanding of the growth process as a whole. In Part 2, the sequence of growth changes is repeated but with more in-depth information about the underlying theory of the growth mechanisms involved. Before beginning with the regional descriptions of the growth process, the rationale behind these processes is briefly explained below; read carefully.

1. The multiple growth processes in all the various parts of the face and cranium are described separately as individual "regions" or "stages." The sequence will begin arbitrarily with the maxillary arch. Changes are then shown for the mandible, followed by growth changes in parts of the cranium and then those of the other regions, one by one. Keep in mind that these regional growth processes all take place simultaneously, even though they are presented here as a sequence of separate stages.

2. Growth increases are shown in such a way that the same craniofacial form and pattern are maintained throughout; that is, the proportions, shape, relative sizes, and angles are not altered as each separate region enlarges. Thus, the geometric form of the whole face for the first and last stages is exactly the same; only the overall **size** has been changed. Each sequential stage incorporates all the stages that precede it. The final stage is a cumulative composite.

3. Facial and cranial enlargement, in which form and proportions remain constant, constitutes "balanced" growth. However, a perfectly balanced mode of growth in **all** the parts of the face and cranium **never** occurs in real life. Because imbalances always occur during the actual developmental processes, **changes** in facial shape and form always take place as the face grows into adulthood. That is, imbalances in the growth process lead to corresponding imbalances in structure. Most of these "imbalances" are perfectly normal and are a regular part of the developmental and maturation process. This is why the face of a child undergoes sequential alterations in profile and in facial

proportions as growth progresses. The mandible of the very young child, for example, is characteristically small in relation to his maxilla but later "catches up" to provide anatomic balance. The forehead is bulbous in the young child but becomes sloping as the frontal sinuses develop. The nasal region is shallow early in postnatal life but later becomes markedly expanded relative to other cranial and facial regions. Many more such progressively imbalanced changes occur. Thus, **imbalanced** growth is **always** involved in the development of any individual's face. This is also why no two faces are exactly alike; the extent, locations, and patterns of growth changes are highly variable and individualized.

4. The reason the facial growth descriptions that follow are presented first as a "balanced" series is twofold. First, they will show just what constitutes the concept of "growth balance" itself and how to understand what this actually means. Second, in order to be able to recognize and explain facial **imbalances,** one must know what constitutes deviations from the balanced mode of development, that is, exactly **where** disproportions develop to cause a given facial pattern, and **how much** is involved in terms of dimensional and angular departures from balanced growth. Only by understanding the balanced process can one accurately identify, measure, and, importantly, account for the imbalanced changes.

5. No face has yet been encountered, as mentioned previously, that has a perfect anatomic and geometric "balance" among all of its regions and parts, although a functional equilibrium usually exists. **Variation in form and balance** is the name of our professional game. By analyzing the growth process and the results of facial growth in any given person, we can identify **where** "imbalances" occur, and we can determine just how these developmental variations have caused a given facial pattern. Any face is a composite of a great many regional imbalances, some slight but others sometimes marked. The face of each of us is the aggregate sum of all the many balanced and imbalanced craniofacial parts combined into a composite whole. Regional imbalances often tend to compensate for one another to provide functional equilibrium. The process of **compensation** is a feature of the developmental process; it provides for a certain latitude of imbalance in some areas in order to offset the effects of disproportions in other regions.

6. Because variations in regional cranial and facial balance exist as a **normal** developmental process, many different kinds and categories of facial form and pattern occur. This underlies the characteristic differences associated with age, sex, ethnic group, and individualized features of the face. Some variations, however, exceed the limits of what can be regarded as "normal." Because we can account for the growth of a balanced face, we can also explain many (but not all) developmental and structural factors that relate to the abnormal face. This is the special subject of a later chapter.

7. The regional descriptions of the growth process outlined below are not randomly presented. Rather, a system is used that, in fact, is the same developmental plan utilized in the growth process itself. This is the **counterpart principle** of craniofacial growth. It states, simply, that the growth of any given facial or cranial part relates specifically to **other** structural and geometric "counterparts" in the face and cranium. For example, the maxillary arch is a counterpart of the mandibular arch. These are **regional** relationships throughout the whole face and cranium. If each regional part and its particular counterpart enlarge to the same extent, balanced growth between them is the result. This is the key to what determines the presence or lack of balance in any region. Imbalances are produced by differences in respective amounts or

directions of growth between parts and counterparts. Many part-counterpart combinations exist throughout the skull, and these provide a meaningful and easy way to evaluate the growth of the face and the morphologic relationships among all its structural components.

The "test" for a part-counterpart relationship in the face and cranium is not difficult. The question is simply asked: "If a given increment is added to a specific bone, **where** must an equivalent increment be added to **other** bones if the same form and balance are to be retained?" The answer to this question then identifies which other specific bones or parts of bones are involved as counterparts. This counterpart concept will be used repeatedly in this chapter as well as in following chapters dealing with facial variations and abnormalities.

8. The growth process for each region is presented as two separate parts. First, the changes produced by **deposition and resorption** (remodeling) are described and are shown by **fine arrows** in the illustrations. Second, the changes produced by **displacement** are described and are represented by **heavy arrows.** These two processes, it is understood, take place at the same time, but they must be described separately because their effects are quite different.* Then the question asked is, "Where do counterpart changes also occur if the same pattern is to be maintained?" This identifies the **next** anatomic region, which is described in turn.

To illustrate the counterpart principle, an expandable photographic tripod is used here as an analogy (Fig. 3–1). The tripod has a series of telescoping segments in each leg; the length of each segment matches the length of its "counterpart" segments in the other two legs. If all the segments are extended to exactly the same length, the tripod retains geometric balance and overall symmetry. If, however, any one segment is not extended equal to the others, that leg as a whole is either shorter or longer, although the remainder of all the segments in that leg match their respective counterparts. One can thus identify **which** particular segment is different and determine the extent of imbalance. Segment *x,* for example, is short relative to *y,* thus causing a retrusion of *z.* The relative (not actual) length of a whole leg can also be altered by changing its alignment.

Many other hypothetical combinations exist. For example, segments *a, b,* and *c* in Figure 3–1 are short with respect to their segment counterparts in the

*The remodeling changes for each region are shown first, followed by displacement, because the sequence is much easier to illustrate this way. As already explained, however, displacement movements are believed to be the actual growth pacemakers.

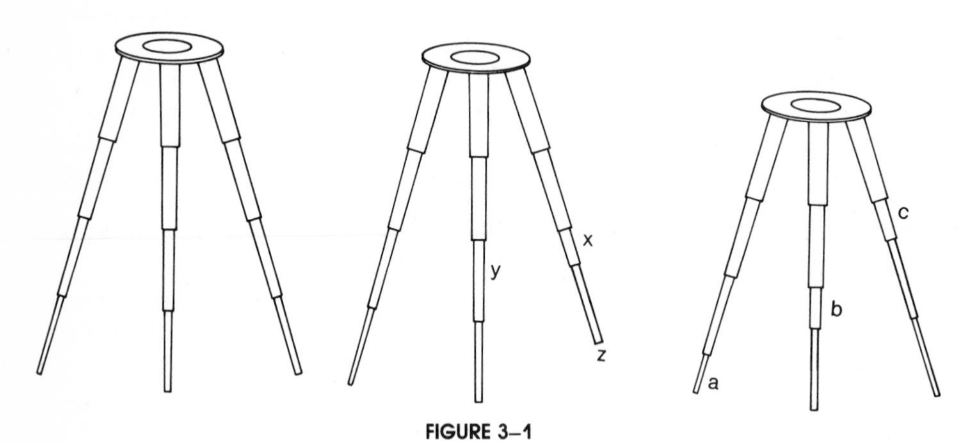

FIGURE 3–1

other legs. Overall symmetry is balanced, nonetheless, because all of these regional inbalances offset one another, and the total length of each leg is therefore the same.

Regional Change 1

Note that two reference lines are used, a horizontal and a vertical,* so that directions and amounts of growth changes can be visualized (Fig. 3–2). The bony maxillary arch lengthens horizontally in a **posterior** direction (this always comes as a surprise to new people in the business, and to some older "pros" as well). This is schematized by showing a posterior movement of the **pterygomaxillary fissure (PTM).** Note its new location behind the vertical reference line (Fig. 3–3).

PTM is the routine radiographic landmark (Fig. 3–4) used to identify the maxillary tuberosity, and it appears on headfilms as an "inverted teardrop" produced by the gap between the pterygoid plates and the maxilla (heavy arrow). The "point" used for headfilm analysis is shown by the fine arrow.

The overall length of the maxillary arch has increased by the same amount that **PTM** moves posteriorly. Bone has been deposited on the posterior-facing cortical surface of the maxillary tuberosity. Resorption occurs on the opposite side of the same cortical plate, which is the inside surface of the maxilla within the maxillary sinus.

*This vertical line is not arbitrary; it is the **PM** boundary, which is one of the most basic and important natural anatomic planes in the head (see Chapter 5). The horizontal line is the functional occlusal plane.

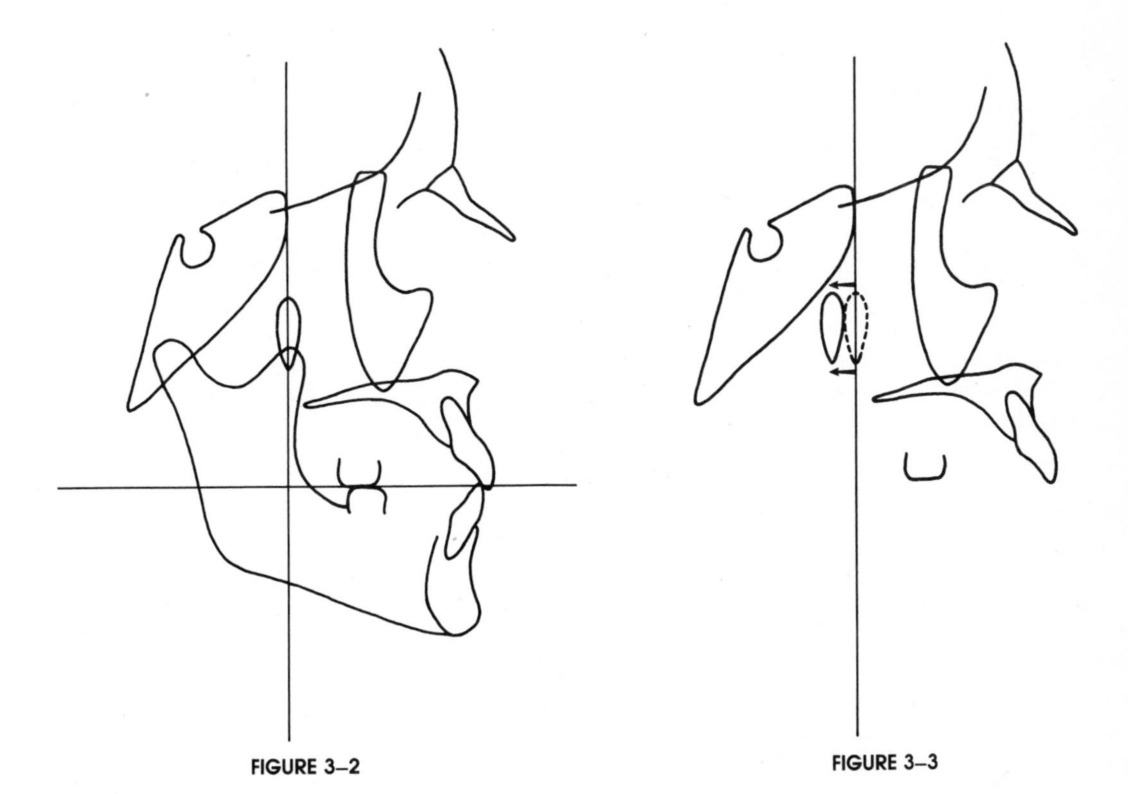

FIGURE 3–2

FIGURE 3–3

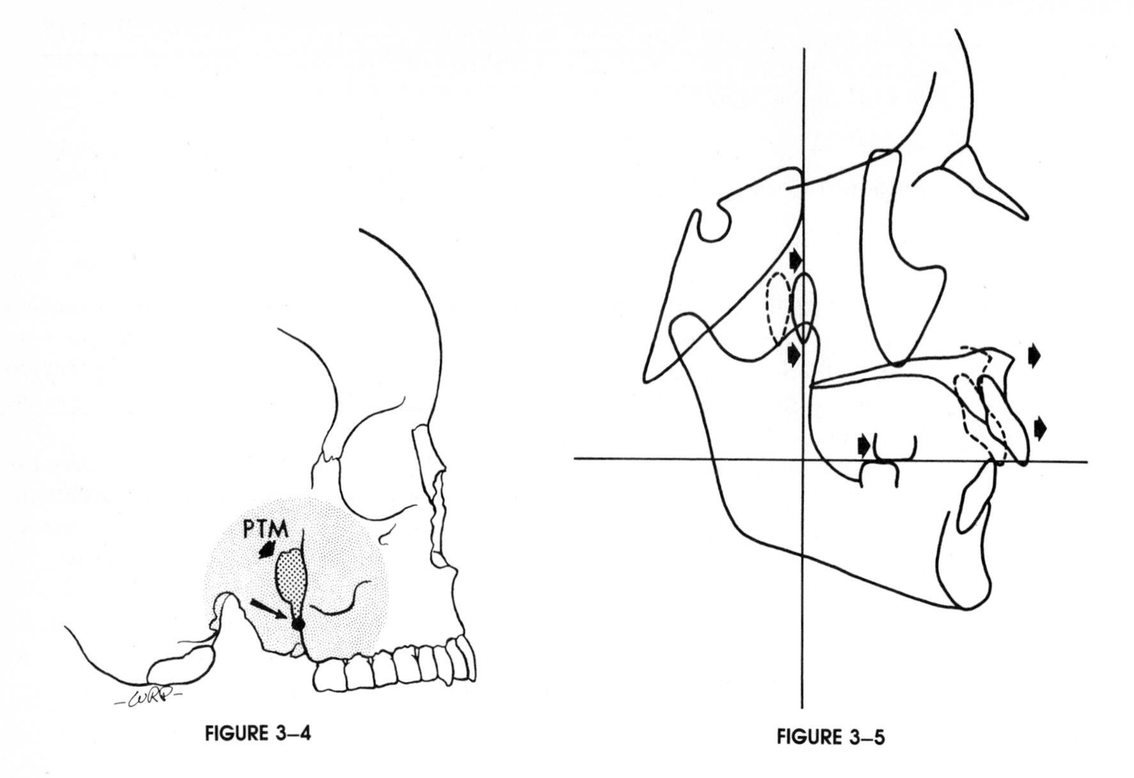

FIGURE 3–4

FIGURE 3–5

Regional Change 2

The preceding stage is the first of the two-part growth process described for each region, that is, growth by deposition and resorption. The second part involves **displacement,** described in the present stage (Fig. 3–5). As the maxillary tuberosity grows and lengthens posteriorly, the whole maxilla is simultaneously **carried** anteriorly. The amount of this forward displacement exactly equals the amount of posterior lengthening. Note that **PTM** is "re-turned" to the vertical reference line. Of course, it never actually departed from this line because backward growth (Stage 1) and forward displacement (Stage 2) occur at the same time. This is a primary type of displacement because it occurs in conjunction with the bone's own enlargement; that is, as the bone is displaced, it undergoes remodeling growth in order to keep pace with the amount of displacement. A protrusion of the forward part of the arch now occurs, not because of direct growth in the forward part itself, but rather because of growth in the **posterior** region of the maxilla as the whole bone is simultaneously displaced anteriorly.

Regional Change 3

The question is now asked: "When the elongation of the maxilla in Stage 1 is made, **where** must equivalent changes **also** be made if structural balance is maintained?" In other words, what are the **counterparts** to the bony maxillary arch? Several are involved, including the upper part of the nasomaxillary complex, the anterior cranial fossa, the palate, and the corpus of the mandible. The mandible is described in this stage. The mandible is not to be regarded as a single functional element; it has two major parts, the **corpus** (body) and the

ramus (Fig. 3–6). These two parts must be considered separately because each has its own separate counterpart relationships with other, different regions in the craniofacial complex.

The bony mandibular arch relates specifically to the bony maxillary arch; that is, the body of the mandible is the structural counterpart to the body of the maxilla. The mandibular corpus now lengthens to match the growth of the maxilla, and it does this by a remodeling conversion from the ramus (Fig. 3–7). The anterior part of the ramus grows posteriorly, a relocation process that produces a corresponding elongation of the corpus. What was ramus has now been remodeled into a new addition onto the corpus. The mandibular arch lengthens by an amount that equals the growth of the maxillary arch (Stage 1), and both elongate in a posterior direction. However, note that the two arches are still offset; the maxilla is in a protrusive position even though upper and lower arch lengths are the same, as seen in Figure 3–8. A Class II type of relationship exists between the maxillary and mandibular molars. The proper Class I position is seen in Stage 1; the mandibular posterior tooth shown in the diagram should normally be about one-half cusp width ahead of its maxillary antagonist, as seen in Figure 3–2.

Regional Change 4

The second of the two growth processes (that is, first, growth by deposition and resorption, and second, displacement) will now be described. Remember that these two changes actually occur **at the same time.** The whole mandible is **displaced** anteriorly, just as the maxilla also becomes carried anteriorly while it simultaneously grows posteriorly. To do this, the condyle and the posterior part of the ramus grow posteriorly (Fig. 3–8). This returns the horizontal

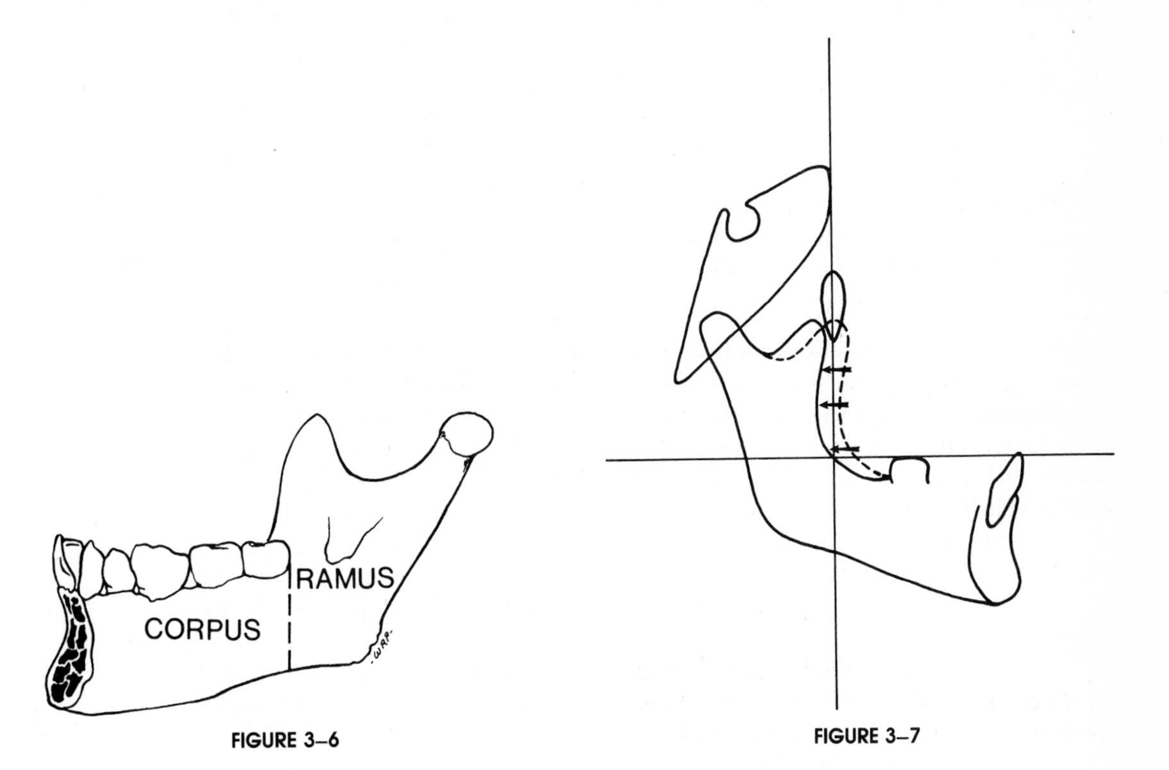

FIGURE 3–6

FIGURE 3–7

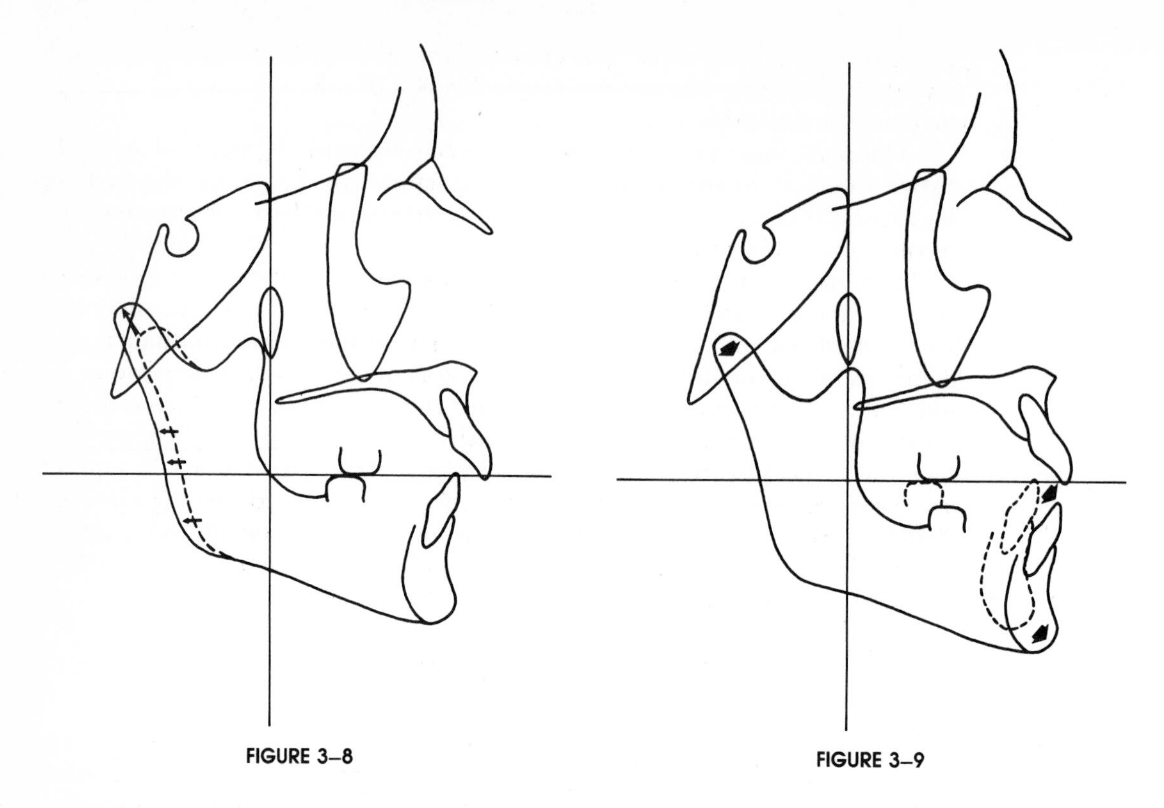

FIGURE 3–8

FIGURE 3–9

dimension of the ramus to the **same breadth** present in Stages 1 and 2 above; the amount of anterior ramus resorption is equaled by the amount of posterior ramus addition. This purpose is not to increase the width of the ramus itself but to relocate it posteriorly for lengthening the **corpus.**

Regional Change 5

The whole mandible, now, is displaced anteriorly by the same amount that the ramus has relocated posteriorly (Fig. 3–9). This is the primary type of displacement because it occurs in conjunction with the bone's own enlargement. As the bone becomes displaced, it simultaneously grows (the stage just described) to keep pace with the amount of displacement. Note the following:

1. The corpus of the mandible elongates primarily in a **posterior** direction, just as the maxilla also lengthens posteriorly (Stage 1). It does this by remodeling from what **was** ramus into what then becomes a posterior addition to the mandibular arch. In this respect, mandibular arch elongation differs from maxillary arch elongation because the maxillary tuberosity is a free surface, unlike the posterior end of the mandibular corpus.
2. The whole ramus has moved posteriorly. However, the only actual change in horizontal dimension involves the mandibular corpus, which becomes longer. The horizontal dimension of the ramus remains constant during **this** particular remodeling stage (the widening of the ramus itself is part of another stage).
3. The anterior **displacement** of the whole mandible equals the amount of anterior maxillary displacement. This places the mandibular arch in proper position relative to the maxillary arch just above it. The arch lengths as well as

the positions of the maxilla and mandible are now in balance, and a Class I positioning of the teeth has been "returned."

4. Note, however, that the obliquely upward and backward direction of ramus growth must also lengthen its **vertical** dimension in order to provide for horizontal enlargement. This separates the occlusion (contact between the upper and lower teeth) because the mandibular arch is displaced inferiorly as well as anteriorly.

5. In both the maxilla and mandible, the type of displacement is **primary** because it takes place in conjunction with each bone's own enlargement.

6. In summary (Fig. 3–10), thus far, the increment of backward growth at the maxillary tuberosity (Stage 1), the amount of forward displacement by the whole maxilla (Stage 2), the extent of remodeling on the anterior part of the ramus and the amount of corpus lengthening (Stage 3), the increment of backward growth by the posterior part of the ramus (Stage 4), and the amount of forward displacement of the whole mandible (Stage 5) are all **precisely equal** in this "balanced" sequence of growth. What happens when they are not all exactly equal (as usually happens) or when differentials in timing occur is described later.

Regional Change 6

While all of the growth and remodeling changes described in the preceding stages have been taking place, the dimensions of the middle cranial fossa have also been increasing at the same time (Fig. 3–11). This is done by resorption of the endocranial side and deposition of bone on the ectocranial side of the cranial floor. The spheno-occipital synchondrosis (a major cartilaginous growth site in the cranium) provides endochondral bone growth in the midline part of the cranial floor.* The total growth expansion of the middle fossa now projects it anteriorly beyond the vertical reference line.

Regional Change 7

All cranial and facial parts lying anterior to the middle cranial fossa (in front of the vertical reference line) become **displaced** in a forward direction as a result (Fig. 3–12). The whole vertical reference line moves anteriorly to the same extent that the middle cranial fossa expands in a forward direction. This is because the line represents the anterior boundary between the enlarging middle cranial fossa and the cranial and facial parts in front of it. The maxillary tuberosity remains in a constant position on the vertical reference line as this line moves forward. The forehead, anterior cranial fossa, cheekbone, palate, and maxillary arch all undergo protrusive displacement in an anterior direction. This is a **secondary** type of displacement because the actual enlargement of these various parts is not directly involved. They are simply moved anteriorly because the middle cranial fossa behind them expands in this direction. The floor of the fossa, however, does not **push** the anterior cranial fossa and the nasomaxillary complex forward. Rather, they are **carried** forward as the interface between the frontal and temporal lobes of the cerebrum becomes "separated" by their respective growth increases.

---

*Note the change in the position of the sella turcica. This is a highly variable structure, however, and other patterns of remodeling movements are also common. See Part 2 of this chapter.

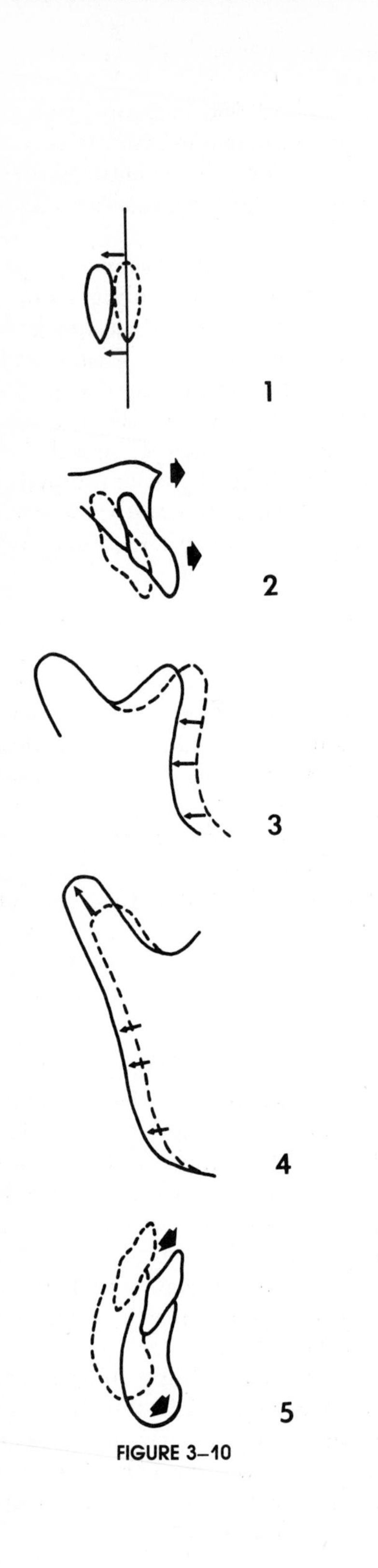

FIGURE 3–10

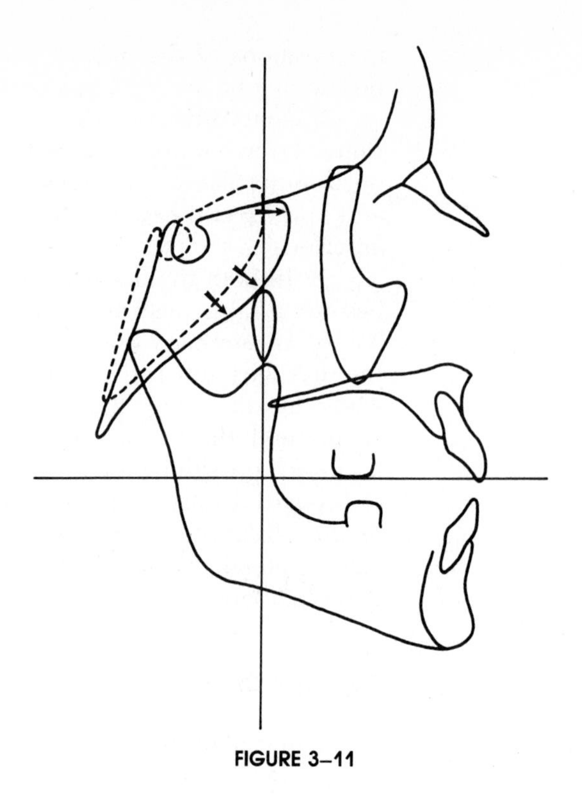
FIGURE 3–11

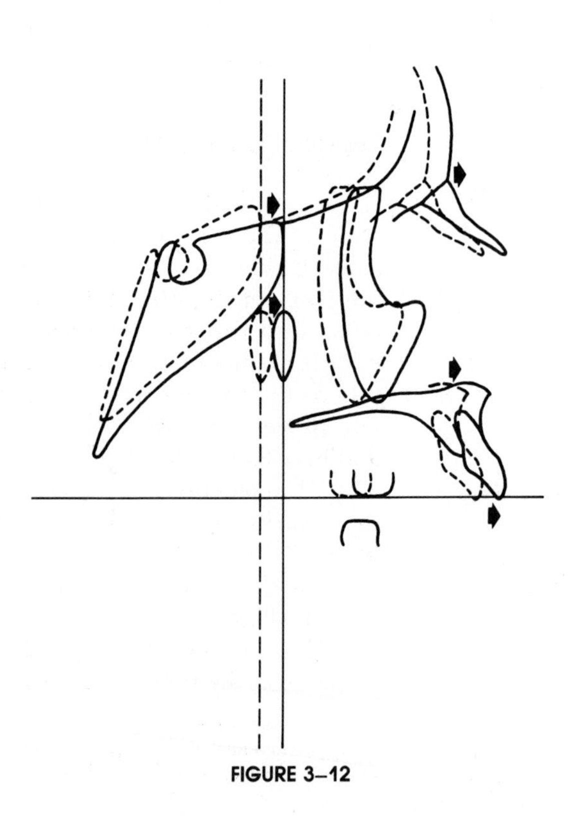
FIGURE 3–12

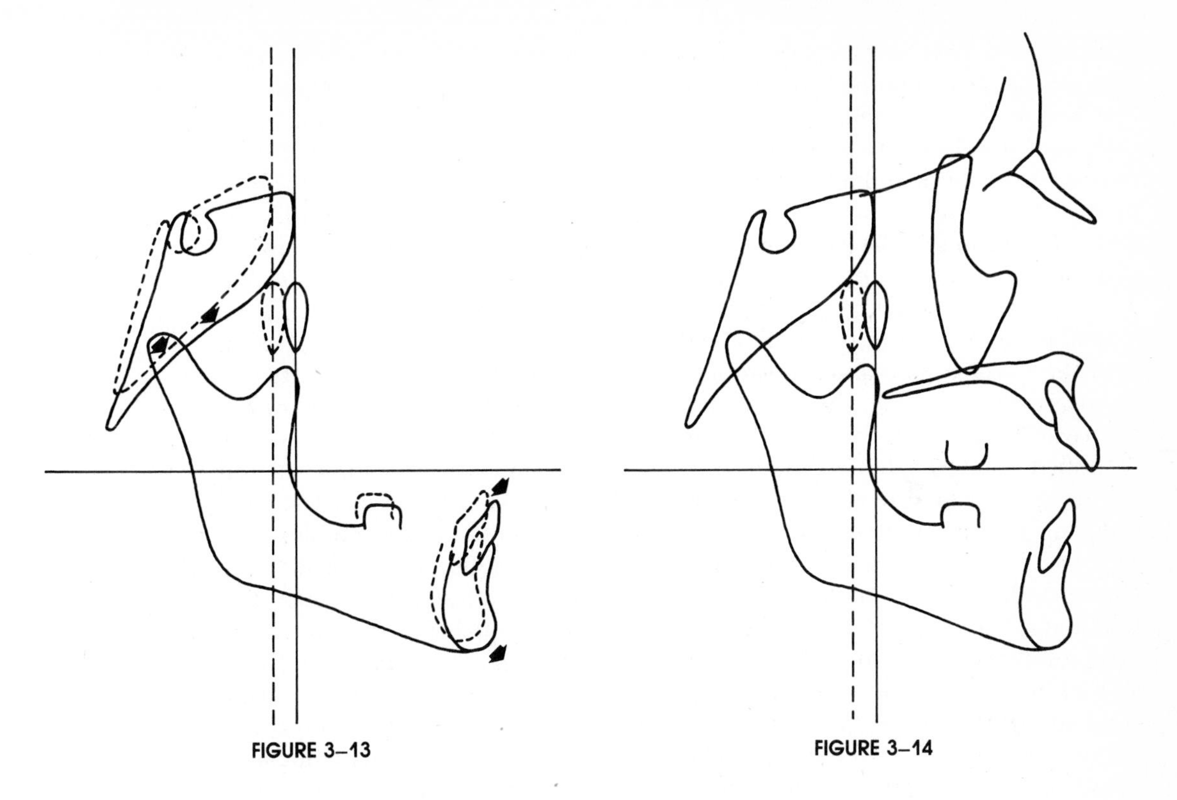

FIGURE 3–13

FIGURE 3–14

Regional Change 8

The expansion of the middle cranial fossa, just described, also has a displacement effect on the mandible (Figs. 3–13 and 3–14). It, too, is a secondary type of displacement. The extent of the displacement effect, however, is much less than that on the maxilla. This is because the greater part of middle cranial fossa growth occurs in front of the condyle and **between** the condyle and the maxillary tuberosity. The spheno-occipital synchondrosis also lies between the condyle and the anterior boundary of the middle cranial fossa. Thus, the amount of maxillary protrusive displacement far exceeds the amount of mandible protrusive displacement caused by middle fossa enlargement. The result is an offset horizontal placement between the upper and lower arches. The upper incisors show an "overjet," and the molars are in a Class II position, even though the mandibular and maxillary arch lengths themselves are matched in respective dimensions. **Sella-nasion** (a much used cephalometric plane) should not be used to represent the "upper face" or "anterior cranial base" dimension in comparisons with the entire mandibular dimension, ramus and corpus, as is often done. The comparison is invalid because dissimilar effective spans are being compared and because sella-nasion itself does not represent any anatomically meaningful dimension, either for the cranial base or for the upper face.

Regional Change 9

The question is now asked: "When this change in the middle cranial fossa takes place, **where** must an equivalent change **also** occur if balance is to be maintained?" This identifies the "counterpart" of the middle fossa and shows where facial growth must take place to match it.

Just as the lengthening of the middle cranial fossa places the maxillary arch in a progressively more anterior position, the horizontal growth of the **ramus** places the mandibular arch in a like position. What the middle cranial fossa does for the maxillary body, in effect, the ramus does for the mandibular body. **The ramus is the specific structural counterpart of the middle cranial fossa.** Both are also counterparts of the **pharyngeal space.** The skeletal function of the ramus is to bridge the pharyngeal space and the span of the middle cranial fossa in order to place the mandibular arch in proper anatomic position with the maxilla. The anteroposterior breadth of the ramus is critical. If it is too narrow or too wide, the ramus places the lower arch too retrusively or too protrusively, respectively. This dimension must be just right. Also, as will be described later, the horizontal dimension of the ramus can become altered during growth to provide intrinsic adjustments and compensations for morphogenic imbalances that can occur elsewhere in the craniofacial complex.

The horizontal extent of middle cranial fossa elongation is **matched** by the corresponding extent of horizontal increase by the ramus (Fig. 3–15). The horizontal (not oblique) dimension of the ramus now equals the horizontal (not oblique) dimension of the middle cranial fossa. The effective span of the latter, as it relates to the ramus, is the straight line distance from the cranial floor-condyle articulation to the vertical reference line. Recall that the ramus was previously involved in remodeling changes associated with corpus elongation (Stage 4), but the actual breadth of the ramus was not increased during that particular stage.

Regional Change 10

The entire mandible is displaced anteriorly at the same time that it grows posteriorly (Fig. 3–16). The amount of this anterior displacement equals (1)

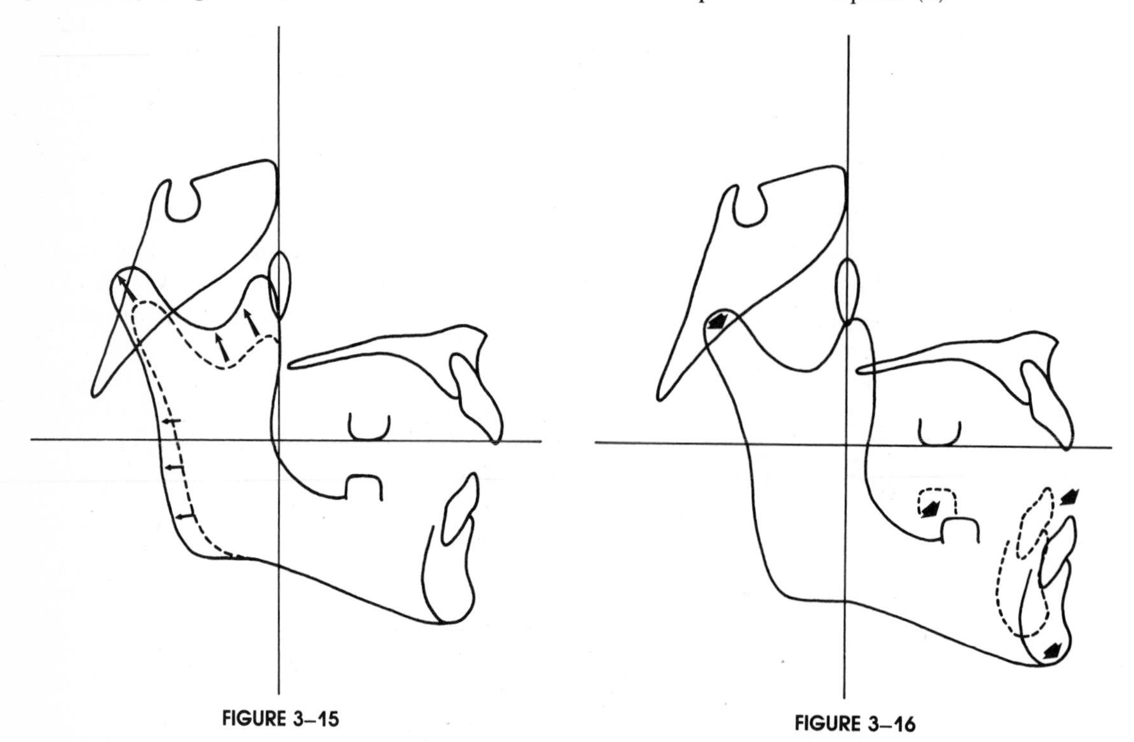

FIGURE 3–15

FIGURE 3–16

the extent of posterior ramus and condylar growth (Stage 9); (2) the amount of middle cranial fossa enlargement anterior to the mandibular condyle (Stage 6); (3) the extent of anterior movement of the vertical reference line; and (4) the extent of resultant anterior maxillary displacement (Stage 7).

The oblique manner of condylar growth necessarily produces an upward and backward projection of the condyle with a corresponding downward as well as forward direction of mandibular displacement. The ramus thus becomes vertically as well as horizontally enlarged. This results in a **further** descent of the mandibular arch and separation of the occlusion (it was also previously lowered during Stages 5 and 8). The total extent of this vertical growth (Figs. 3–8, 3–9, 3–13, 3–15, and 3–16) must **match** the total vertical lengthening of the nasomaxillary complex (Fig. 3–22) and the upward eruption and drift of the mandibular dentoalveolar arch (Fig. 3–23) if the same facial balance is to be achieved.

Note that the protrusion of the maxilla during Stage 7 has now been matched by an equivalent amount of mandibular protrusion. The molars have once again been "returned" to Class I positions, and the upper incisor has no overjet. Note also that the anterior border of the ramus lies ahead of the vertical reference line. The "real" junction between the ramus and corpus, however, is the lingual tuberosity housing the last molar, not the "anterior border." The lingual tuberosity lies on the vertical reference line behind the anterior border, which overlaps this tuberosity (not shown in the figure; observe this on a dry mandible).

Regional Change 11

The **floor** of the anterior cranial fossa and the forehead grow by deposition on the ectocranial side with resorption from the endocranial side (Fig. 3–17).

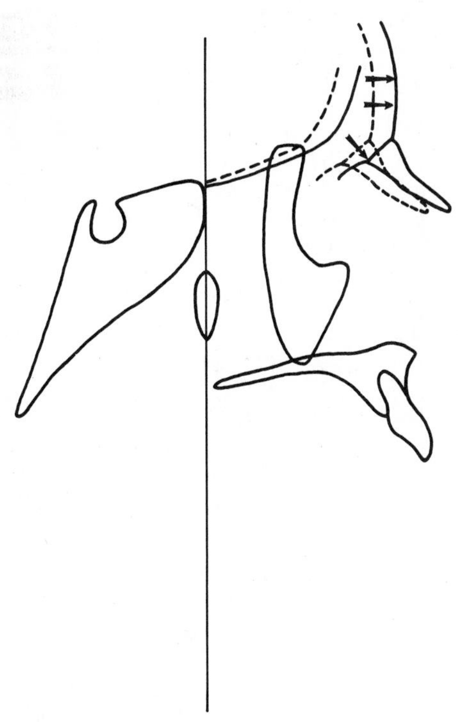

FIGURE 3–17

The nasal bones are displaced anteriorly. The posterior-anterior length of the anterior cranial fossa is now in balance with the extent of horizontal lengthening by its structural counterpart, the maxillary arch (Stage 1). Because these two regions have undergone equivalent growth increments, the profile retains its originally balanced form. (Actually, age differentials, both in timing and amount, always occur, but our present purpose is to describe perfectly "balanced" growth.)

The enlarging brain displaces the bones of the **calvaria** (domed skull roof) outward. Each bone enlarges by sutural growth. As the brain expands, the sutures respond by depositing new bone at the contact edges of such bones as the frontal, parietal, and temporal. This expands the perimeter of each. At the same time, bone is laid down on **both** the ectocranial and endocranial sides to increase the thickness.

The upper part of the face, which is the ethmomaxillary (nasal) region, also undergoes equivalent growth increments. This facial area increases horizontally to an extent that matches (if the same balance is retained) the expansion of the anterior cranial fossa above and the maxillary arch and palate below it. These areas are all counterparts to one another. The growth process involves direct bone deposition on the forward-facing cortical surfaces of the ethmoid, the frontal process of the maxillary, and the nasal bones. Most of the internal surfaces of the nasal chambers are resorptive. In addition, anterior displacement takes place in conjunction with growth at the various maxillary and ethmoidal sutures. The composite of these changes produces an enlargement of the nasal chambers in an anterior (and also lateral) direction.

Regional Change 12

The **vertical** lengthening of the nasomaxillary complex, like its horizontal elongation, is brought about by a composite of (*a*) growth by deposition and

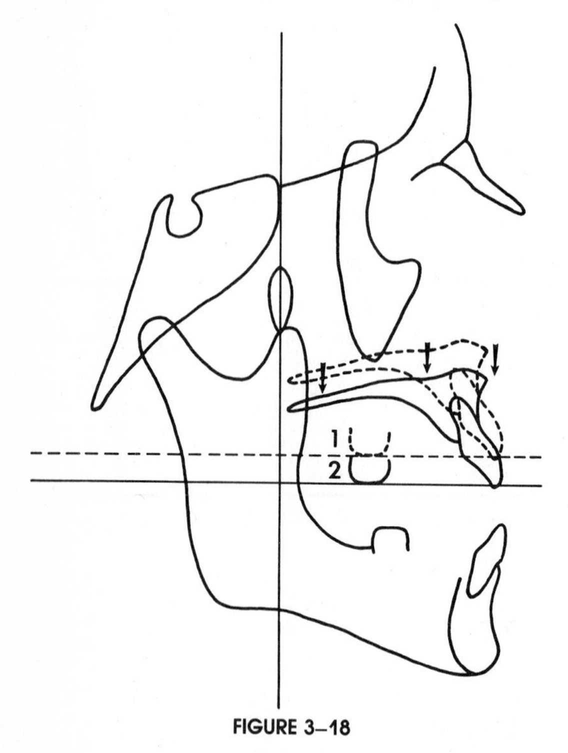

FIGURE 3–18

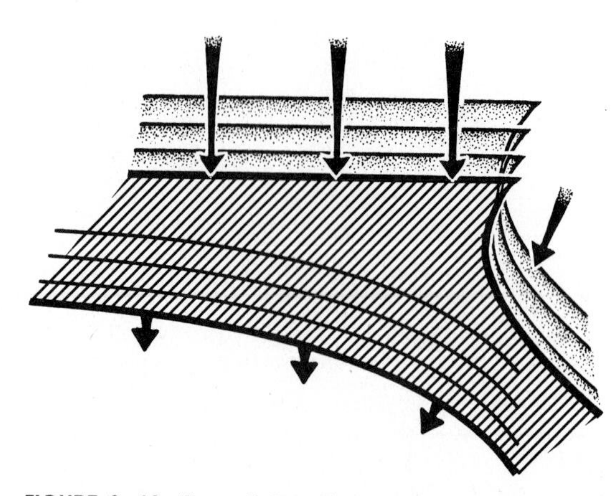

FIGURE 3–19. (From Enlow, D. H., and S. Bang: Growth and remodeling of the human maxilla. Am. J. Orthod., 51:446, 1965.)

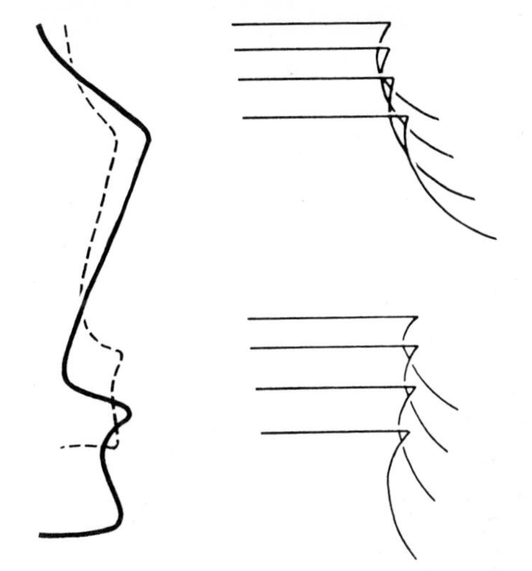

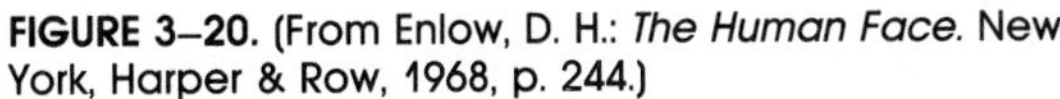

FIGURE 3–20. (From Enlow, D. H.: *The Human Face.* New York, Harper & Row, 1968, p. 244.)

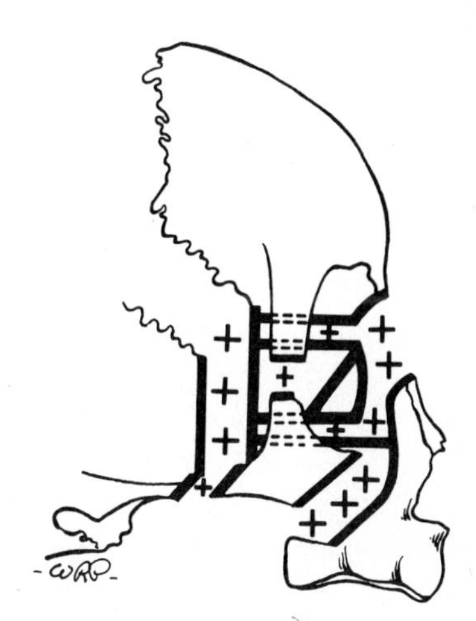

FIGURE 3–21

resorption and (*b*) a primary displacement movement associated directly with its own enlargement. The latter is considered in a later stage. The combination of resorption on the superior (nasal) side of the palate and deposition on the inferior (oral) side produces a downward growth movement of the whole palate from *1* to *2* in Figure 3–18. This **relocates** it inferiorly, a process that provides for the vertical enlargement of the overlying nasal region. The extent of nasal expansion is considerable during the childhood period to keep pace with the enlargement of the lungs. (Note: The extent of downward palatal and maxillary arch remodeling often varies between the mesial and distal parts of the area, as described later. This provides for a range of positional adjustments of the arch to compensate for growth variations and displacement rotations.)

The anterior part of the bony maxillary arch has a periosteal surface that is **resorptive** (the human, with reduced jaws, is the only species that has this), because this area grows **straight downward** (Fig. 3–19). In other species (including primates), the premaxillary region grows forward as well as downward to produce an elongate muzzle.

As seen in Figure 3–20, the labial (external) side of the premaxillary region faces **away** from the downward direction of growth, and it is thus resorptive. The lingual side faces toward the downward growth directions and is depository. This growth pattern also provides for the remodeling of the alveolar bone as it **adapts to the variable positions of the incisors.**

Regional Change 13

Vertical growth by **displacement** is associated with growth at the various **sutures** of the maxilla where it contacts the other separate bones above and behind it. Bone is added at these sutures as the whole maxilla is displaced inferiorly (Fig. 3–21). The addition of new sutural bone does not "push" the maxilla downward. Rather, the maxilla is **carried** inferiorly by other physical growth forces. This triggers bone deposition in the sutures, and new bone is simultaneously laid down on the sutural edges, keeping the bone-to-bone junction intact.

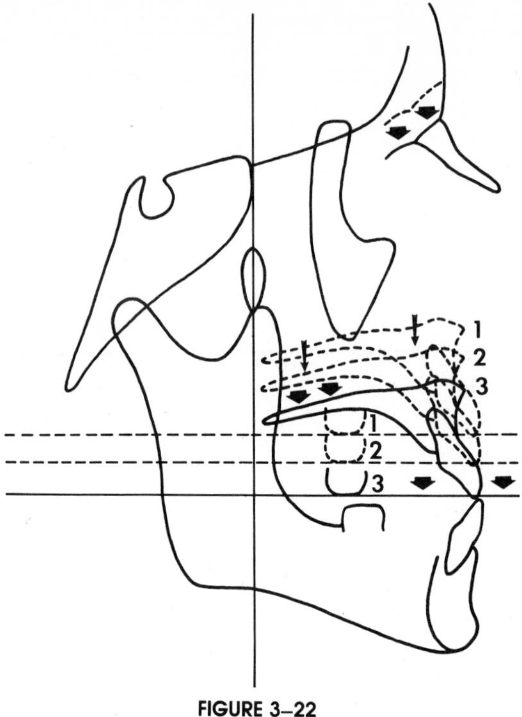

FIGURE 3–22

The increment of bone growth in the suture exactly equals the amount of inferior displacement of the whole maxilla. This is **primary** displacement because it takes place in conjunction with the bone's own enlargement.

Of the total extent of downward movement by the palate and maxillary arch, that part from *2* to *3* is produced in association with sutural growth and primary **displacement** (Fig. 3–22). The part from *1* to *2*, which is about half of the total, is direct cortical growth and relocation by resorptive and depository remodeling. Similarly, the movement of the teeth from *2* to *3* is by the downward displacement of the whole maxilla, carrying the dentition passively with it. The movement from *1* to *2* is produced by each tooth's **own** movement as bone is added and resorbed on appropriate lining surfaces in each socket. This is the **vertical drift** of the tooth, a process that is carried out by the **same** deposition and resorption of alveolar bone that produces the familiar "mesial drift" of the dentition (see Part 2). Vertical drift takes place **in addition to** eruption, which is a separate growth movement. The vertical drift process is important to the clinician because it provides a great deal of growth movement to "work with" during treatment. Tooth movement from *2* to *3* can also be clinically influenced. This involves the use of special appliances intended to either augment or retard the displacement movements of the entire nasomaxillary complex or to alter their directions. This in turn causes remodeling changes in the size or shape of the whole maxilla or of other separate bones (in contrast to remodeling of the alveolar bone supporting the teeth).

Figures 3–18 and 3–22 illustrate the palate and maxillary arch moving inferiorly in an idealized manner, with the anterior and posterior regions growing downward to the same extent. Mesiodistal variations, however, are common. In the displacement movement, a clockwise or counterclockwise **rotation** of the whole palate and arch often occurs. In conjunction with this, the remodeling movement (*1* to *2*) can compensate by producing an opposite direction of rotation, thereby leveling and fine-tuning the palate into definitive adult position. This, indeed, may represent a primary function of the remodeling phase of the composite downward growth process. That is, selective remodeling in the anterior versus posterior parts can serve to adjust and counteract rotations produced during primary nasomaxillary displacement as well as secondary displacement rotations caused by growth of the middle and anterior cranial fossae.

Awareness of the distinction between movements *1* to *2* and *2* to *3* should be a day-one concept in any orthodontic or surgical training program. The clinician addresses one or the other in every patient, or some combination of both. Either the magnitude or the direction can be influenced by substituting "clinical control" for nature's own intrinsic control. The underlying biologic process actually producing the movements is the same, however, whether intrinsic or clinical. If growth movement does not already exist, as in an adult, it must be clinically induced in addition to providing direction, thus representing different biologic as well as "stability" situations, since no subsequent childhood facial growth is to be involved. (See also Fig. 4–48.)

Keep in mind, also, this key point: The teeth themselves have very little capacity for remodeling. They can, essentially, only be moved by the displacement process, either through remodeling of an individual alveolar socket or by displacement of the entire arch as a unit. It is the **bone** that must undergo any remodeling required. (See "Anterior Crowding" in Chapter 6 and Concepts 6 and 7 in Chapter 18.)

Regional Change 14

In three previous stages (5, 8, and 10), it was seen that the mandibular corpus becomes lowered because of the vertical enlargement of both the ramus and the middle cranial fossa. Their combined vertical dimensions represent the growth counterpart of the vertical dimension of the nasomaxillary complex and the dentition. In other words, the amount of vertical separation between the upper and lower arches caused by the vertical growth of both the middle cranial fossa and the ramus must be balanced by an equivalent amount of vertical growth in the nasomaxillary complex and the alveolar-dental region of the mandible. The maxillary arch has grown downward to Level 3 in Stage 13. Now the mandibular teeth and alveolar bone grow **upward** to attain full occlusion (Fig. 3–23). This is produced by a superior **drift** of each mandibular tooth, together with a corresponding increase in the height of the alveolar bone. The extent of this upward growth movement plus that of the downward growth movement by the maxillary arch equals the combined extent of vertical growth by the ramus and middle cranial fossa **if** the pattern of the face is not changed. Note this factor: The extent of downward drift of the maxillary teeth greatly exceeds the extent of upward drift by the mandibular teeth. Much less growth is thus available to work with in major orthodontic movements of the mandibular, as compared with the maxillary, teeth.

Regional Change 15

While the upward growth movements of the mandibular teeth and alveolar sockets are taking place, remodeling changes also occur in the incisor alveolar region, the chin, and the corpus of the mandible (Fig. 3–24). The lower incisors

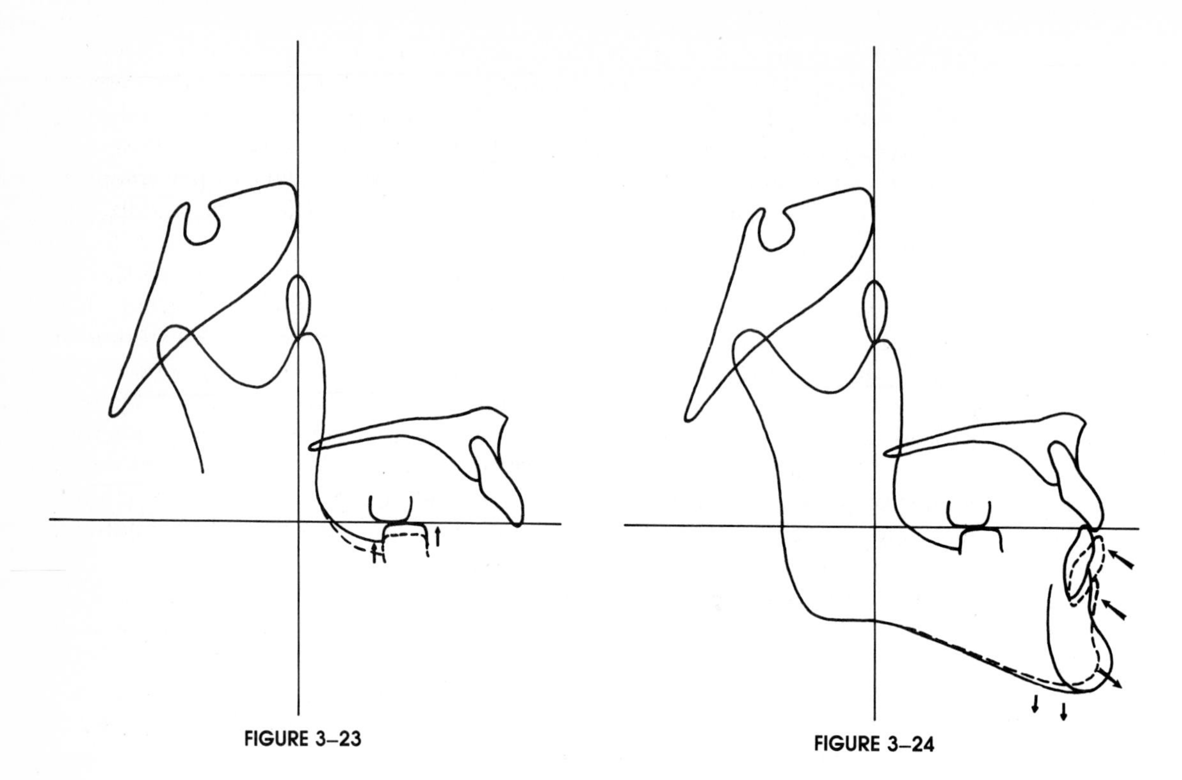

FIGURE 3–23

FIGURE 3–24

undergo a lingual tipping (a "retroclination"), so that the uppers overlap the lowers for the proper **overbite.** This involves a posterior rotation movement of the mandibular incisors as they simultaneously drift superiorly. The movement of the teeth is accompanied by **resorption** on the outside surface of the alveolar region just above the chin (and deposition on the lingual side). The alveolar bone thus moves backward as the incisors undergo lingual drift. This does not occur to the same extent in individuals having an "end-to-end" incisor relationship.

Bone is progressively added to the external surface of the **chin** itself, as well as along the underside and other external surfaces of the corpus. This is slow growth that proceeds gradually throughout childhood. At birth, the mental protuberance is small and inconspicuous. Many anxious parents naturally worry about the chinless appearance of their little child. However, the whole mandible usually tends to lag in differential growth timing and will later catch up to the maxilla in the normal face. The chin takes on more noticeable form year by year. The combination of new bone growth on the chin itself and the posterior direction of bone growth in the alveolar region just above it gradually causes the chin to become more prominent. The whole mandible, meanwhile, is also becoming displaced anteriorly in conjunction with continued growth at the condyle and overall mandibular lengthening.

In Chapter 5, the evolutionary factors of upright (bipedal) body posture and the greatly enlarged human brain related to a marked downward and backward rotational placement of the nasomaxillary complex are described. This midfacial rotation has caught the human mandible as in a closing vice, with the maxilla on one side and the airway and other cervical and pharyngeal parts on the other. The "fitting" of the lower jaw under these conditions has led in part to the overbite and overjet feature of the human face with its unique protruding chin. An anterior crossbite is also a consequence in some individuals.

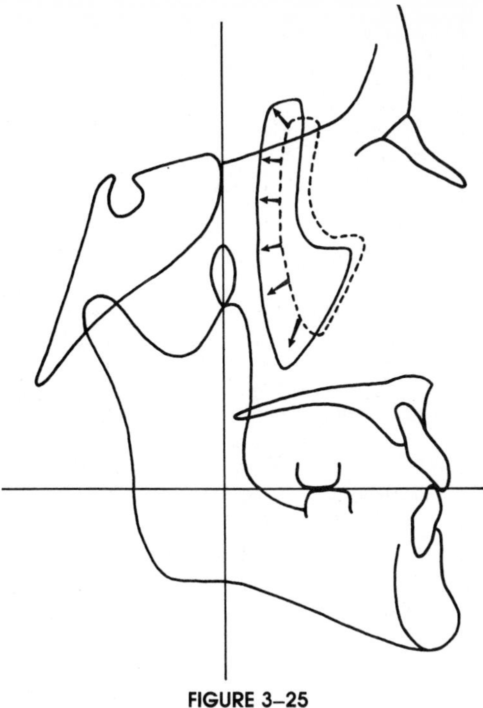

FIGURE 3–25

Regional Change 16

The forward part of the zygoma and the malar region of the maxilla grow in conjunction with the contiguous maxillary complex, and their respective modes of growth are similar. Just as the maxilla lengthens horizontally by posterior growth, the malar area also grows posteriorly by continued deposition of new bone on its posterior side and resorption from its anterior side (Fig. 3–25). The front surface of the whole cheekbone area is thus actually resorptive. This remodeling process keeps its position in proper relationship to the lengthening maxillary arch as a whole. They **both** grow backward, thereby maintaining the proper anatomic positions between them. The amount of deposition on the posterior side, however, exceeds resorption on the anterior surface, so that the whole malar protuberance becomes larger. Another way of understanding the rationale for the growth of the zygomatic process of the maxilla is to compare it with the coronoid process of the mandible. Just as the coronoid process grows backward by anterior resorption and posterior deposition to keep pace with the overall posterior elongation of the whole bone, the zygomatic process similarly grows posteriorly by anterior resorption and posterior deposition.

Note that the vertical length of the lateral orbital rim increases by sutural deposits at the frontozygomatic suture. The zygomatic arch also enlarges considerably by bone deposition along its inferior edge. The arch grows laterally (not seen, of course, in lateral headfilms) by bone deposition on the lateral surface, together with resorption from the medial side within the temporal fossa.

Regional Change 17

Just as the whole maxillary complex is displaced anteriorly and inferiorly as it simultaneously enlarges in overall size, the zygoma is moved anteriorly

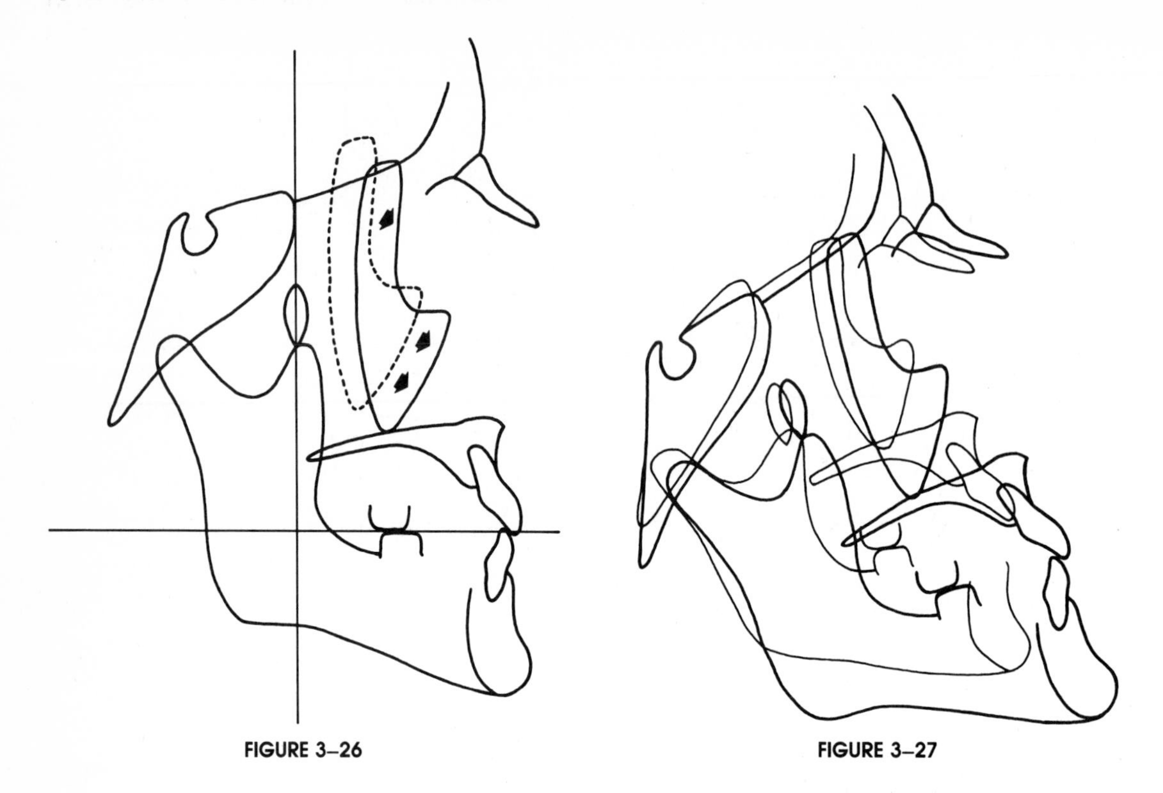

FIGURE 3–26

FIGURE 3–27

and inferiorly by primary **displacement** as it enlarges (Fig. 3–26). The zygoma thereby proportionately matches the maxilla in (1) the directions and amount of horizontal and vertical growth and (2) the directions and amount of primary displacement.

This completes the introductory survey of the regional growth changes taking place in the cranial base and face. The final result is a craniofacial composite that has the same form and pattern present when the first stage was begun. Only the overall size has been altered. All the growth changes among the specific parts and counterparts have been purposefully balanced to give an understanding of the meaning of "balanced growth" and to provide a basis for analyzing **imbalanced** growth changes in a later chapter.

In Figure 3–27, the first and last stages are superimposed with sella as a registration point. When the sequence of changes described in Stages 1 through 17 are considered, it is apparent that the face does not simply grow directly from one profile to the other. Rather, all the regional changes just outlined are involved. The overlay seen here is the traditional way of representing the results of the overall process of facial enlargement. This overlay does **not,** however, represent the actual growth processes themselves—that is, the changes produced (1) by resorption and deposition and (2) by primary and secondary displacement. (Many have not appreciated this basic and important fact.) The overlay shows the cumulative summation of **both.**

Part 2

The regional changes of growth described in Part 1 are now repeated, stage by stage, for explanation of the actual growth processes involved. Three-dimensional representations of whole bones are used to complement the more simplified two-dimensional descriptions previously given.

Stage 1

It will be recalled that the horizontal lengthening of the bony maxillary arch is produced by growth at the maxillary tuberosity. This is represented by a backward movement of **PTM** from the vertical reference line (Fig. 3–28).

The area shown in Figure 3–29 is the specific growth field that carries out the change just described. It is a **depository** field in which the backward-facing periosteal surface of the tuberosity receives continued deposits of new bone as long as growth in this part of the face continues. Because the posterior surface of the maxillary tuberosity points toward the direction of arch elongation, this is the particular surface that is depository. The arch also widens, and the lateral surface is, similarly, depository. The endosteal side of the cortex within the interior of the tuberosity is resorptive. The cortex thus drifts progressively posteriorly and also, to a lesser extent, in a lateral direction. Deep to the tuberosity is the maxillary sinus. It increases in size as a result of the same process. In the newborn, this sinus is quite small, but it becomes greatly expanded as growth continues and eventually occupies the greater part of the large suborbital compartment.

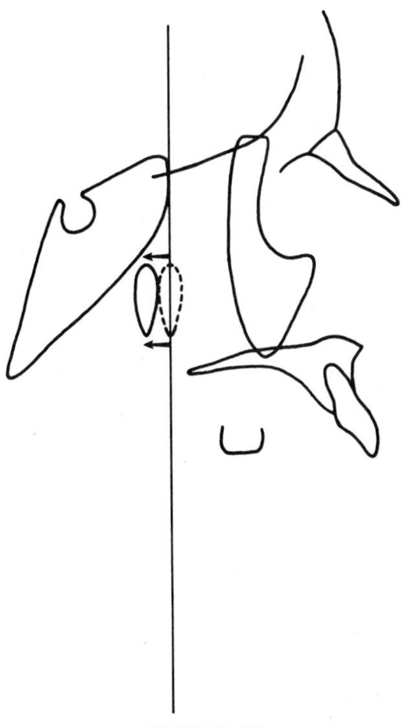

FIGURE 3–28

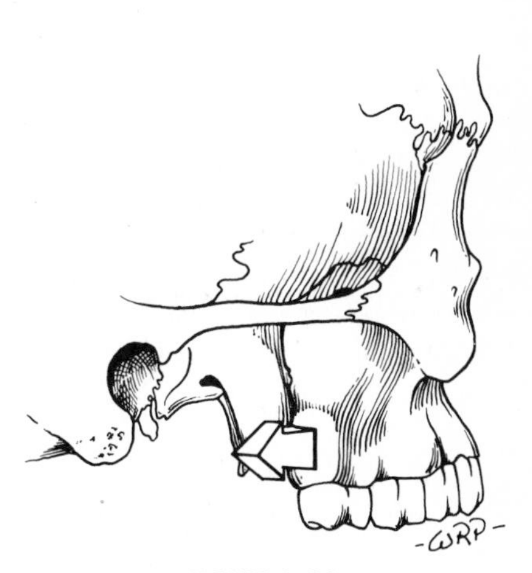

FIGURE 3–29

In Chapter 18, the various kinds of bone tissues are described as they relate to the many different circumstances of regional growth. Included are those types that characteristically grow rapidly. The bone tissue of the maxillary tuberosity is one of these, because this area provides much of the postnatal lengthening of the arch and nearly all of it after about 2 or 3 years of age. This requires a fairly rapid and sizable amount of continued new bone formation. The maxillary tuberosity is a major "site" of maxillary growth. It does not, however, provide for the growth of the whole maxilla but relates only to that part associated with the posterior part of the lengthening arch. Many other basic and important sites of growth also exist throughout the various parts of the maxilla. See Figures 3–128, 3–130, 3–131, and 3–137 to 3–142.

Stage 2

The whole maxilla undergoes a simultaneous process of **primary displacement** in an anterior direction as it grows and lengthens posteriorly (Fig. 3–30). The nature of the force that produces this anterior movement has, historically, been a subject of great controversy. One early theory (long since dropped) suggested that additions of new bone on the posterior surface of the elongating maxillary tuberosity "push" the maxilla against the adjacent muscle-supported pterygoid plates. This presumably would cause a resultant shove of the entire maxilla anteriorly because of its own posterior bone growth activity. The idea was abandoned, however, when it was realized that the bone's membrane is pressure-sensitive and that the bone growth process does not have the physiologic capacity to push the whole bone away from the other bones by itself.

Another theory held that bone growth within the various maxillary **sutures** produced a pushing apart of the bones, with a resultant thrust of the whole maxilla anteriorly (and inferiorly as well). Although this explanation is still

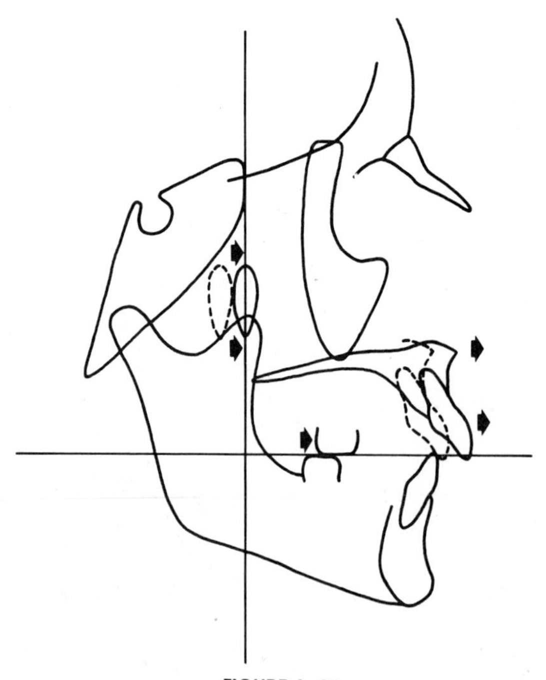

FIGURE 3–30

heard, it has been largely rejected for the reason just mentioned: bone tissue is not capable of growth in a field that requires the levels of compression needed to produce a "pushing" type of displacement. The sutural connective tissue is not adapted to a pressure-related growth process (in contrast to cartilaginous types of bone-to-bone contacts, which are much more compression-tolerant). The suture is essentially a **tension-adapted** tissue. Its collagenous fiber construction is a design for traction resistance across the connective tissue bridge between separate bones. The presence of any unusual pressure on a suture triggers bone resorption, not deposition. The sutural membrane cannot withstand any undue amount of compression because pressure affects its vascular and cellular components. It is believed that the stimulus for sutural bone growth is the tension produced by the **displacement** of that bone. The deposition of new bone is the response to displacement, rather than the force that causes it. (See later in this chapter and also Chapter 18 for further discussion of the sutural growth process.)

Thus, as the entire maxilla is carried forward (and downward) by displacement, tension is produced within the sutural membranes. This, in turn, presumably triggers the sutural membranes to form the new bone tissue that enlarges the overall size of the whole bone and sustains constant bone-to-bone sutural contact. Although the "sutural push theory" is not tenable, some students of the problem are now looking anew at growth mechanisms within sutures, but not in the old conceptual way. The same problem exists for the periodontal membrane and the eruption and drifting process of teeth. We still have much to learn about the growth behavior of sutures and other soft tissues associated with bone, and we may yet find that sutures, the periosteum, and the periodontal membrane participate in the displacement process in a way that is presently unknown or only speculative (such as the presence of actively contractile myofibroblasts that would cause teeth to erupt and drift and sutures to slide over one another).

Another explanation for maxillary displacement is the now famous "nasal septum" theory (Fig. 3–31). This was developed largely by Scott, and the premise for the idea is quite reasonable. It developed from the criticisms of the "sutural theory" described above. The nasal septum hypothesis was soon adopted by many investigators around the world and became more or less the standard explanation, replacing the sutural theory. **Cartilage** is specifically adapted to certain pressure-related growth sites, because it is a special tissue uniquely structured to provide the capacity for growth in a field of compression (see Chapter 18). Cartilage is present in the epiphyseal plates of long bones, in the synchondroses of the cranial base, and in the mandibular condyle, where it

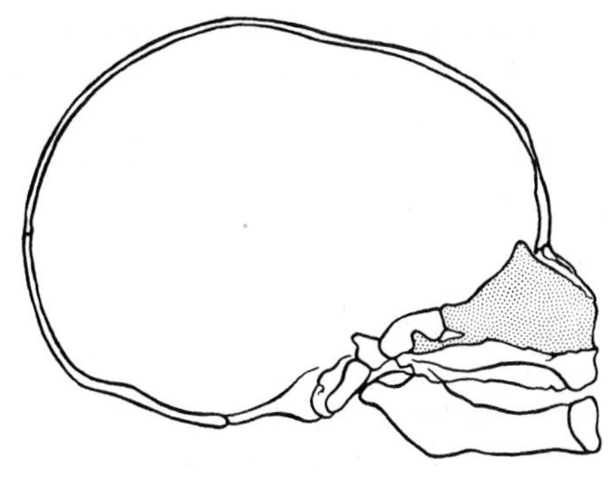

FIGURE 3–31

provides linear growth by endochondral proliferation. Whereas the cartilaginous nasal septum itself contributes only a small amount of actual endochondral growth, the basis for the theory is that the pressure-accommodating **expansion** of the cartilage in the nasal septum provides a source for the physical force that displaces the maxilla anteriorly and inferiorly. This sets up fields of tension in all the maxillary sutures. The bones then secondarily, but almost simultaneously, enlarge at their sutures in response to the tension created by the displacement process.

As with any important explanatory theory, the nasal septum concept has subsequently received a great deal of laboratory study to test its validity. Ingenious experiments, such as those carried out by Sarnat, suggest that the nasal septum does indeed appear to be a factor contributing to the primary displacement of the nasomaxillary complex. Some other experimental studies, however, have not conclusively shown any special role played by the nasal septum in the displacement process. It is held by Latham that the nasal septum, together with the septopremaxillary ligament, between the septum and the premaxilla, is an operative force **early** in postnatal as well as prenatal growth, but that it is replaced by other displacement forces as the face outgrows the capacity to be moved by the septum. Others suggest that the septum functions essentially to support the roof of the nasal chamber and does not actively participate in the displacement movements of the nasal floor.

There are two basic reasons this problem is still unresolved and such divergent opinions are held. The first is that the cause of maxillary displacement is probably multifactorial in nature. If the nasal septum is indeed involved, other factors probably contribute as well, and it is difficult to separate their respective effects in controlled laboratory experiments. The second reason is that experimental studies usually involve surgical removal of parts (such as the septum) to test the nature of their functional roles in growth. It is always difficult in such studies to account for the variables introduced by the experimental procedure itself, such as the destruction of tissues, blood vessels, and nerves that can also play a role in the growth process. Critics of these studies point out that the experimental removal of a given part does not necessarily demonstrate what the actual role of that part is when present *in situ*. It merely shows how the growth process functions in the absence of that part, rather than in its presence. If one experimentally changes some structure, as by surgical deletion, and this in turn affects the growth process, one cannot necessarily conclude that this structure therefore "controls the growth process."

Another important biologic consideration is the concept of "multiple assurance" (Latham and Scott, 1970). The processes and mechanisms that function to carry out growth are virtually always multifactorial. Should any one determinant of the growth process become inoperative (as by experimental deletion of an anatomic part), other morphologic components in some instances have the capacity to "compensate." That is, they can provide an alternative means to achieve more or less the same developmental and functional result, although perhaps with some degree of anatomic distortion. This concept has far-reaching implications, necessitating caution in the interpretation and evaluation of facial growth experiments utilizing laboratory animals.

At present, the nasal septum theory is still accepted as a reasonable explanation by a number of investigators, although it is universally realized that much more needs to be understood. Many teachers use the septum as a "symbol" for the force that causes displacement but make it clear that other factors are involved and that there is much to be explained.

A notable advance was made with the development of the **functional matrix** concept, largely by Moss. It deals with the determinants of bone and cartilage growth in general. The functional matrix concept states, in brief, that any given bone grows in response to functional relationships established by the sum of all the soft tissues operating in association with that bone. This means that the bone itself does not regulate the rate and directions of its own growth; the functional soft tissue matrix is the actual governing determinant of the skeletal growth process. The course and extent of bone growth are secondarily dependent upon the growth and the functioning of pacemaking soft tissues. Of course, the bone and any cartilage present are also involved in the operation of the functional matrix, because they participate in giving essential feedback information to the soft tissues. This causes the soft tissues to inhibit or accelerate the rate and amount of subsequent bone growth activity, depending on the status of functional and mechanical equilibrium between the bone and its soft tissue matrix. The genetic determinants of the growth process reside wholly in the soft tissues and not in the hard part of the bone itself.

The functional matrix concept is basic to an understanding of the fundamental nature of a bone's role in the overall process of growth control. This concept has had great impact in the field of facial biology.

The functional matrix concept also comes into play as a source for the mechanical force that carries out the process of displacement. According to this now popular explanation, the facial bones grow in a subordinate growth control relationship with all the surrounding soft tissues. As the tissues continue to grow, the bones become passively (that is, not of their own doing) **carried** along (displaced) with the soft tissues attached to the bones by Sharpey's fibers. Thus, for the nasomaxillary complex, the growth expansion of the facial muscles, the subcutaneous and submucosal connective tissues, the oral and nasal epithelia lining the spaces, the vessels and nerves, and so on, all combine to move the facial bones passively along with them as they grow. This continuously places each bone and all its parts in correct anatomic positions to carry out its functions, because the functional factors are the very agents that **cause** the bone to develop into its definitive shape and size and to occupy the location it does.

How does the functional matrix explanation relate to the nasal septum theory? This is still highly controversial. The problem, at this writing, is that nobody really understands just how much genetic influence resides within cartilage, and especially the different structural and functional varieties of cartilage. The "forward and downward" displacement of the maxilla could be accounted for solely by the functional matrix mechanism, without direct and purposeful participation by the expanding nasal septum. However, there is still disagreement as to the genetic role of the cartilage in the septum. Does it have an intrinsic genetic capacity to determine the course of its own growth expansion and, with it, resultant displacement of the midface? Or is the growth of the cartilage itself, like that of the bone, controlled by the capsular soft tissue matrix? Further discussion is deferred until the mandibular condyle and the synchondroses are considered in later sections of this chapter and also in Chapters 8 and 18, dealing with growth control.

The functional matrix concept, in general, is established and valid, and it is basic in helping us understand the complex interrelationships that operate during facial growth. It is to be realized, however, that this principle is not intended to explain **how** the growth control mechanism actually functions. This concept describes essentially **what** happens during growth; it does not account for the regulatory processes at the cellular and molecular levels that carry it out.

In summary, Stage 2 of the growth sequence involves a forward **displacement** of the entire maxilla. The amount of this movement equals the increment of new bone growth on the posteriorly facing surface of the maxillary tuberosity. The **timing** of these two growth changes is such that the addition of the new bone coincides with the process of displacement or lags imperceptibly behind it, since the movement of displacement must occur in order to provide the available space for growth expansion. The force that causes the displacement movement, according to current theory, is the "functional matrix" (although some have not entirely abandoned the nasal septum theory).

Stages 3 and 4

Stages 3 and 4 involve the lengthening of the mandibular corpus to an extent that matches its counterpart, the bony maxillary arch (Fig. 3–32). In this growth process, one notable structural difference between the mandible and maxilla exists: the mandible has a **ramus.** The maxillary tuberosity is a **free** skeletal surface. Posterior to it is the oropharyngeal space and the separate (in childhood) pterygoid plates. This maxillary surface grows directly posteriorly. The posterior growth of the mandibular bony arch, however, must proceed into a region already occupied by the ramus. This requires a **remodeling conversion** from ramus to mandibular corpus. That is, the whole ramus becomes relocated posteriorly, and the former anterior part of the ramus is structurally altered into an addition to the corpus, which thereby becomes lengthened by this remodeling process. See Figures 3–129 and 3–132 to 3–136.

The growth movement of the ramus in a backward direction has usually been pictured as essentially a two-dimensional process (Fig. 3–33). This is not merely an incomplete explanation; it is inaccurate as well. The problem is that some of the key anatomic parts that participate in the relocation and remodeling processes of the ramus and corpus cannot be seen or represented in conventional two-dimensional headfilms and tracings. Among these is the **lingual tuberosity.** This is an important structure because it is the direct anatomic equivalent of the maxillary tuberosity (Fig. 3–34). Just as the maxillary tuber-

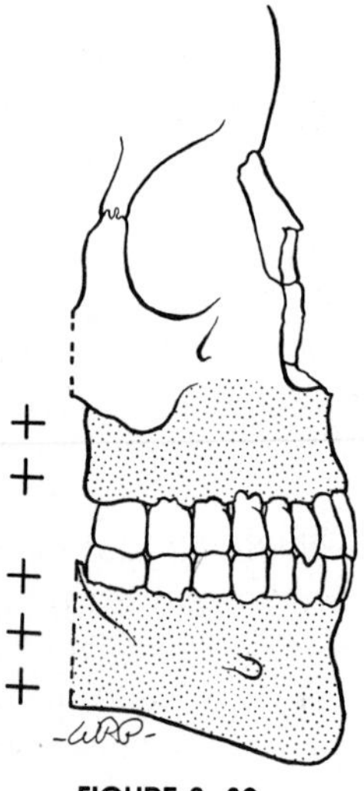

FIGURE 3–32

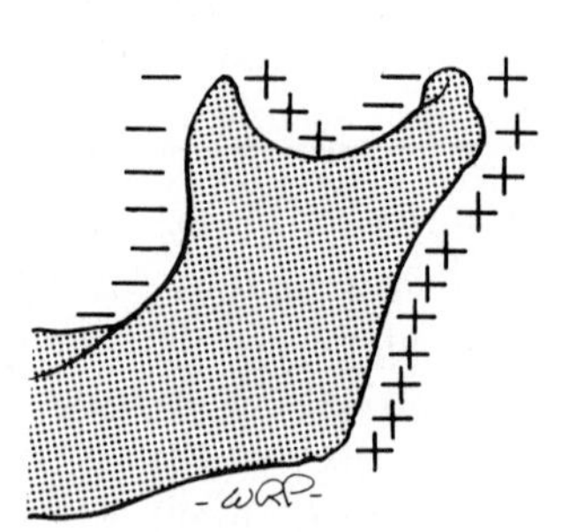

FIGURE 3–33

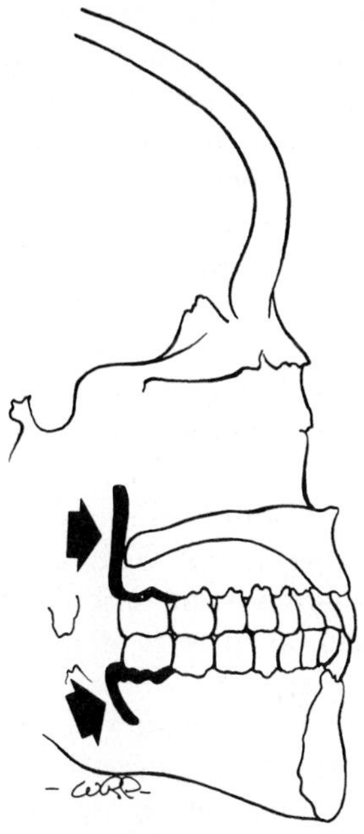

FIGURE 3–34

osity is a major site of growth for the upper bony arch, so is the lingual tuberosity a major site of growth for the mandible. Yet this structure is not even included in the basic vocabulary of cephalometrics. The reason, simply, is that it is not recognizable in the headfilm. This presents a severe handicap, because the lingual tuberosity is not only a major growth and remodeling site but also the effective boundary between the two basic parts of the mandible: the ramus and the corpus. The inaccessibility of the lingual tuberosity for routine cephalometric study is a great loss. Nonetheless, the changes this important structure undergoes during growth **must** be understood, all the more so since it cannot be visualized, at least directly, in headfilms.

The lingual tuberosity grows posteriorly by deposits on its posterior-facing surface, just as the maxillary tuberosity undergoes comparable growth additions (Fig. 3–35). Ideally, the maxillary tuberosity closely overlies the lingual tuber-

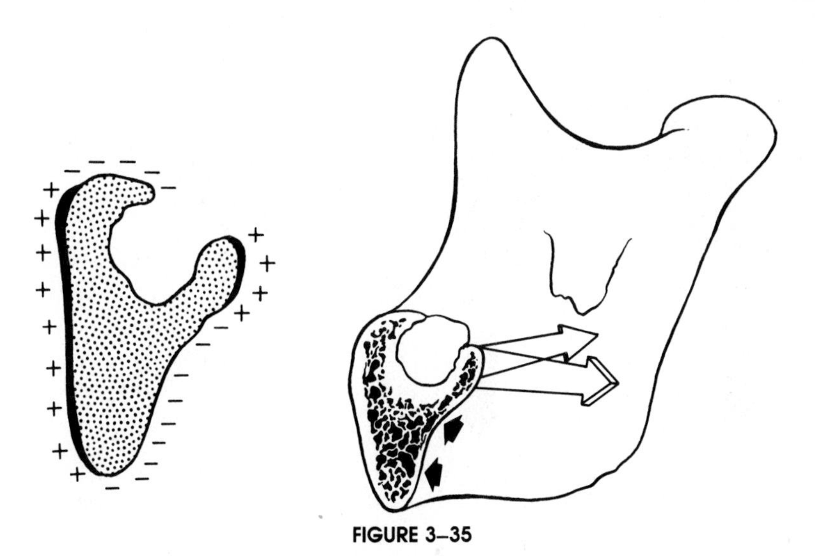

FIGURE 3–35

osity (that is, both are aligned on the vertical reference line). Moreover, the lingual and maxillary tuberosities ideally have equal rates and amounts of respective growth. Although these growth changes often do not match one another, however, such a "balanced" growth process is described here (the consequences of variations are explained later).

Note that the lingual tuberosity protrudes noticeably in a lingual (medial) direction and that it lies well toward the midline from the ramus. The prominence of the tuberosity is increased by the presence of a large resorptive field just below it. This resorptive field produces a sizable depression, the **lingual fossa.** The combination of resorption in the fossa and deposition on the medial-facing surface of the tuberosity itself greatly accentuates the contours of both regions.

The tuberosity grows in an almost directly posterior direction, with only a relatively slight lateral shift (Fig. 3–36). The latter is because bicondylar width does not increase nearly as much as mandibular length during the childhood period, since most of the lateral growth of the cranial base has occurred by about the second and third years.

The posterior growth of the tuberosity is accomplished by continued new deposits of bone on its posterior-facing exposure. As this takes place, that part of the **ramus** just behind the tuberosity grows **medially** (Fig. 3–37). This area of the ramus is coming into line with the axis of the arch in order to join it and thus become a part of the corpus, thereby lengthening it. As pointed out above, the whole ramus lies well lateral to the dental arch.

Deposition on the lingual surface of the ramus just behind the tuberosity produces the medial direction of drift that shifts this part of the ramus into alignment with the axis of the corpus. Keep in mind that the whole ramus is also becoming relocated in a posterior direction at the same time. What has happened, in summary, is that bony **arch length** has been increased and the corpus has been lengthened by (1) deposits on the posterior surface of the lingual tuberosity and the contiguous lingual side of the ramus and (2) a resultant lingual shift of this part of the ramus to become added to the corpus.

The presence of resorption on the anterior border of the ramus (Fig. 3–38) is often described as "making room for the last molar." It is doing much more than this! The resorptive nature of this region is directly involved in the whole process of progressive relocation of the entire ramus in a posterior direction; this movement continues from the tiny mandible of the fetus to the

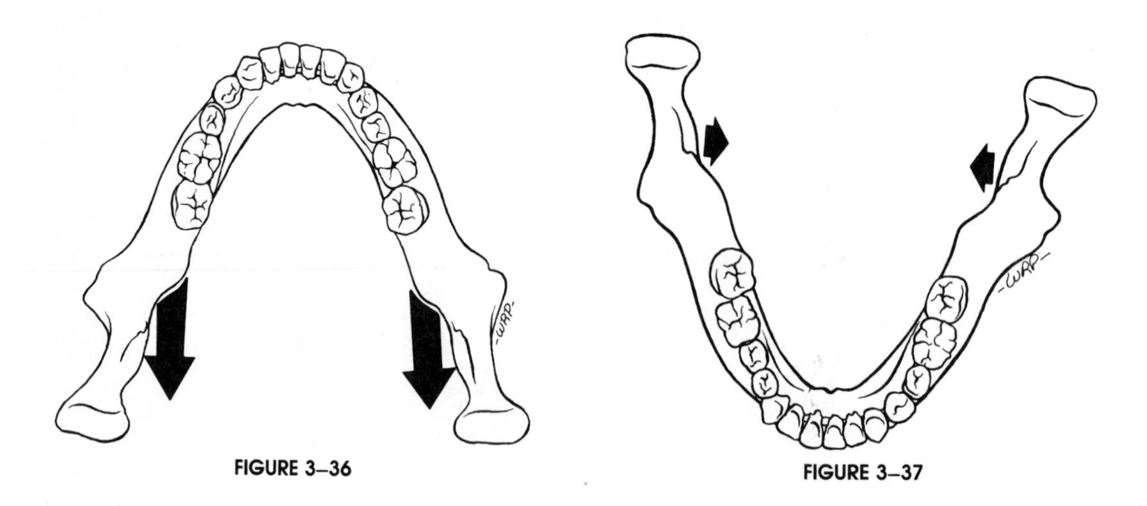

FIGURE 3–36

FIGURE 3–37

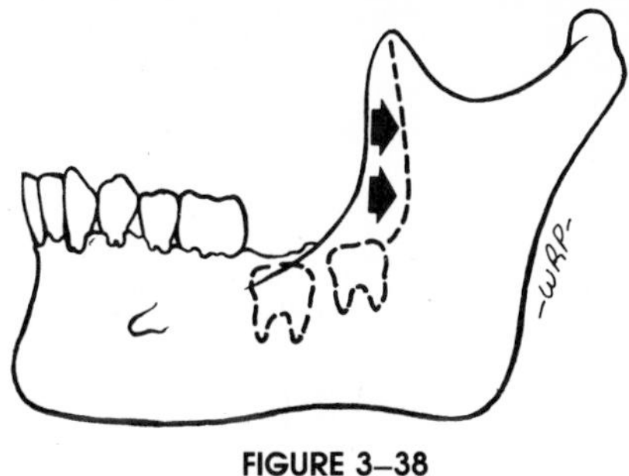

FIGURE 3–38

attainment of full adult mandibular size. The overall extent of ramus movement amounts to several inches, not merely the width of a molar.

Another key point is that the traditional description of posterior ramus movement implies a **straight line** backward growth process in a two-dimensional plane, as represented by *a* and *b* in Figure 3–39. This is not the case at all. Such a picture of ramus growth shows, simply, resorption on the anterior edge and deposition on the posterior edge; this is incomplete. Growth actually takes place as indicated by *c.* In *d,* the growth direction thus follows the *x* arrows, rather than the straight line axis shown by the *y* arrows. Compare with Figure 3–36.

Growth and remodeling activity does not occur **only** on the anterior and posterior margins. Surfaces **between** the anterior and posterior borders, shown by *a, b,* and *c* in Figure 3–40, also directly participate, because the alignment of the ramus is such that these surfaces become directly involved. If growth were to occur simply as schematized in *1,* resorption on one end and deposition on the other (as by activity on only the anterior and posterior margins of the ramus) would move it along the axis of the arrow with no need for activity in the area between the ends. However, the various parts of the ramus are, in fact, oriented so that the span **between** also necessarily comes into play, as shown by *2.* Several different regional directions of alignments are involved in the various parts of the ramus, as described next. The **coronoid process** has a propeller-like twist, so that its lingual side faces three general directions all at once: posteriorly, superiorly, and medially, as shown in Figure 3–40.

When bone is added onto the lingual side of the coronoid process, its growth thereby proceeds **superiorly,** and this part of the ramus thereby becomes increased in vertical dimension (Fig. 3–41). Notice that each coronoid process lengthens vertically, even though additions are made on the medial (lingual) surfaces of the right and left coronoid processes. This is an example of the enlarging V principle, with the V oriented vertically.

These **same** deposits of bone on the lingual side also bring about a **posterior** direction of growth movement, because this surface also faces posteriorly (Fig. 3–42). This produces a backward movement of the two coronoid processes, even though deposits are added on the inside (lingual) surface. This is also an example of the expanding V principle, with the V oriented horizontally. Notice, further, that this enables the whole posterior part of the mandible to **widen** (although not very much except during the period of fetal and early childhood cranial base growth in width), even though deposition occurs on the inside of the V.

These **same** deposits of bone on the lingual side also function to carry the base of the coronoid process and the anterior part of the ramus in a **medial**

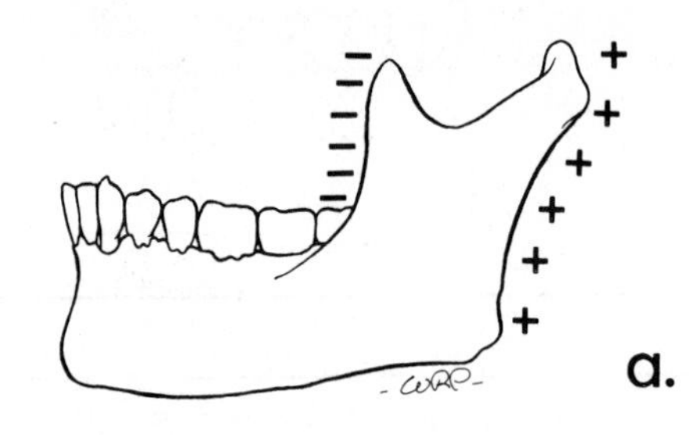

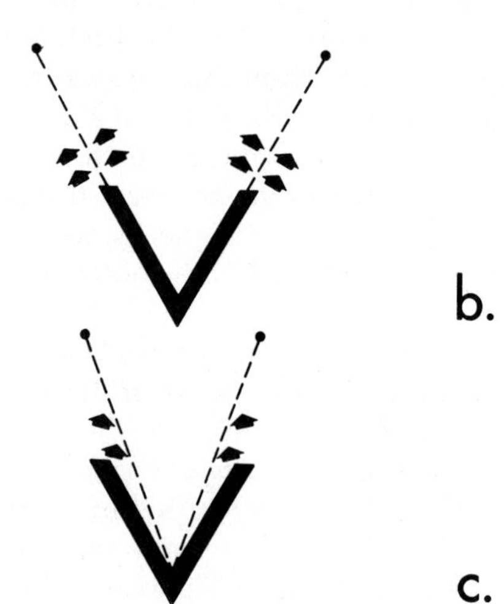

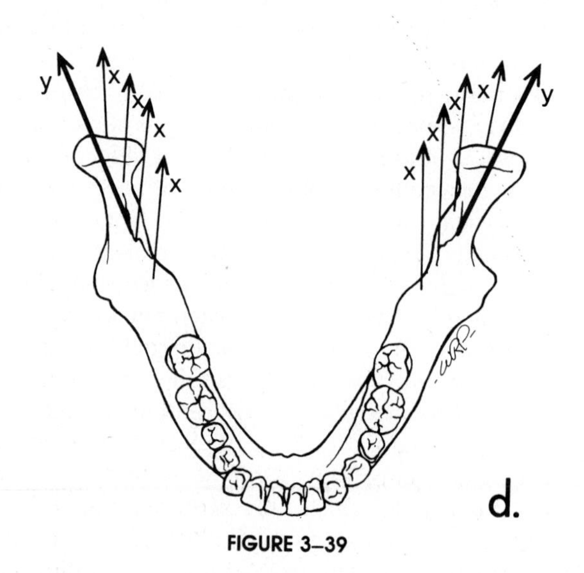

FIGURE 3–39

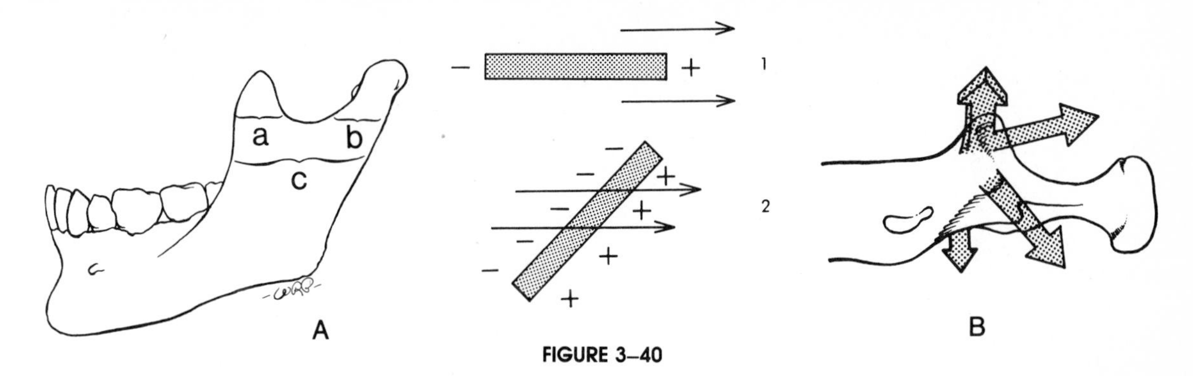

FIGURE 3–40

direction in order to add this part to the lengthening corpus, which lies well medial to the coronoid process. This, again, is an example of the V principle, because a wider part undergoes relocation into a more narrow part as the whole V moves toward its wide end. Thus, the area occupied by the anterior part of the ramus (Fig. 3–43) in *1* becomes **relocated** and remodeled into the posterior part of the corpus in *2*.

In all the above relationships, the **buccal** side of the coronoid process has a **resorptive** type of periosteal surface (Fig. 3–44). This surface faces away from the combined superior, posterior, and medial directions of growth. The remainder of most of the superior part of the ramus, including the whole area just below the mandibular (sigmoid) notch and the **superior** (not lateral or medial) portion of the condylar neck, grows superiorly by deposition on the lingual side and resorption from the buccal side. The lower part of the ramus below the coronoid process also has a twisted contour. Its buccal side faces posteriorly toward the direction of backward growth and thus, characteristically, has a depository type of surface (Fig. 3–45). The opposite lingual side, facing away from the direction of growth, is resorptive.

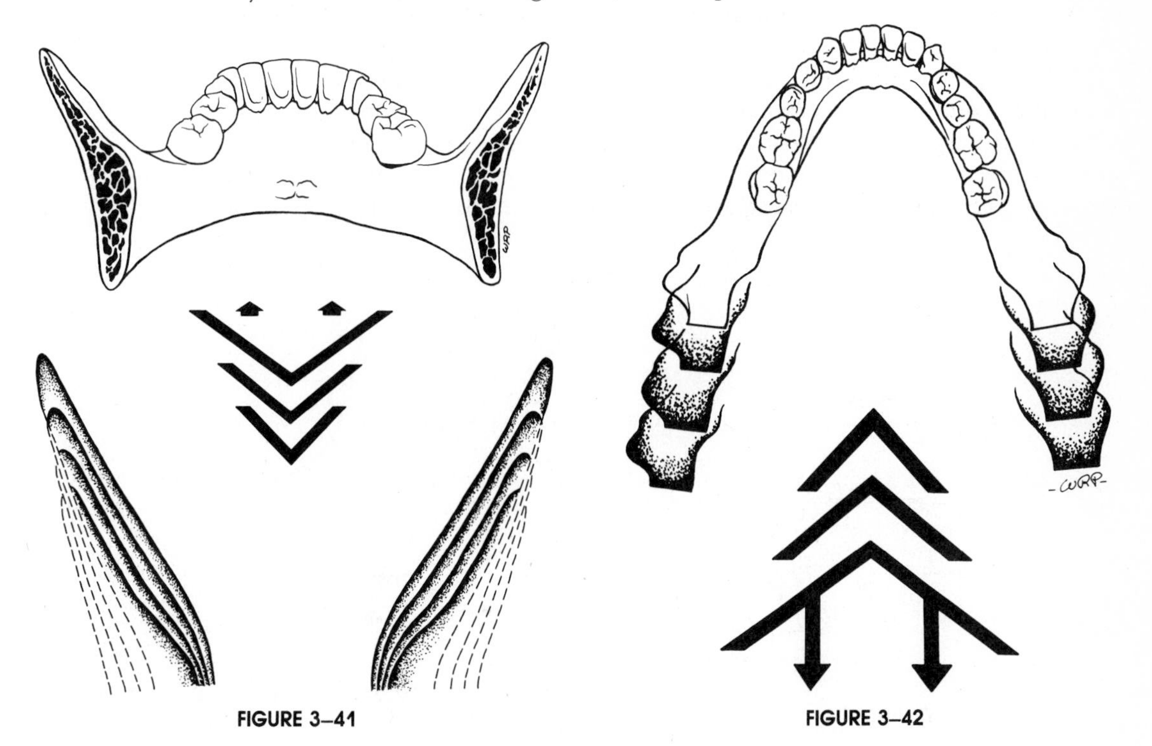

FIGURE 3–41

FIGURE 3–42

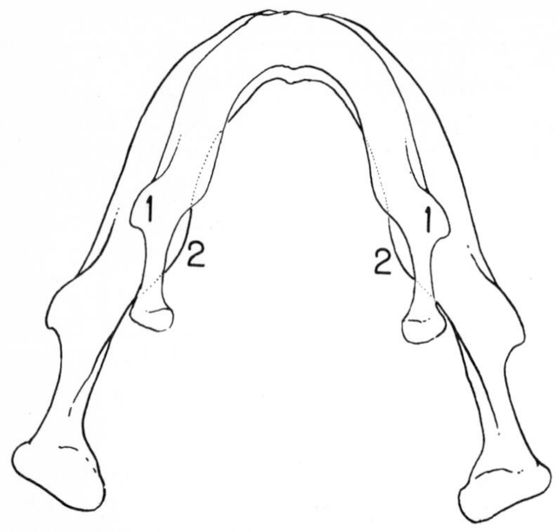

FIGURE 3–43. (Adapted from Enlow, D. H., and D. B. Harris: A study of the postnatal growth of the human mandible. Am. J. Orthod., 50:25, 1964.)

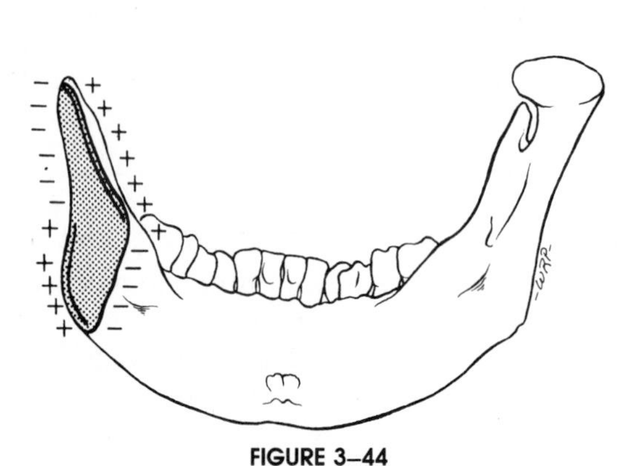

FIGURE 3–44

A single field of surface resorption is present on the inferior edge of the mandible at the ramus-corpus junction. This forms the **antegonial notch** by remodeling from the ramus just behind it as the ramus relocates posteriorly (Fig. 3–46). The size of the notch can be increased whenever a downward rotation of the corpus relative to the ramus takes place (as described later). Other kinds of important mandibular rotations can also involve a sizable resorptive field on the ventral edge of the ramus, as illustrated by Figure 3–76.

The posterior edge of the ramus is a major growth site. The condyle has an obliquely upward and backward growth direction; the angle of growth involved (that is, how much upward and how much backward) is variable and depends on whether an individual is a "horizontal or vertical grower" with respect to the mandible. Such variable directions are produced by selective cell divisions in some parts around the periphery of the condyle with retardation in others. However, the growth of the rest of the ramus necessarily keeps pace with any given amount of condylar growth (Fig. 3–47). Although correlated, these two regional growth sites are essentially separate and develop under

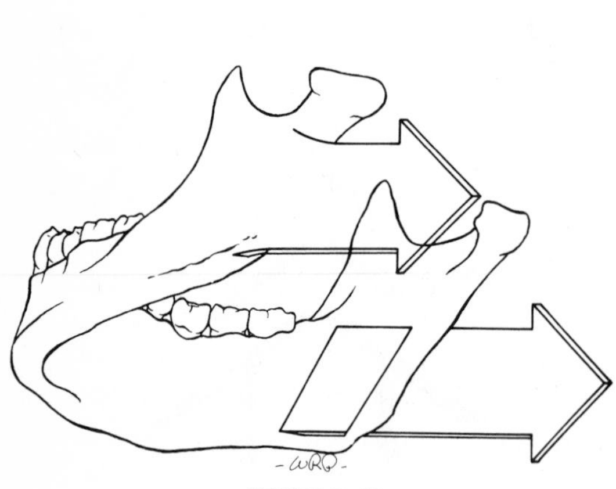

FIGURE 3–45

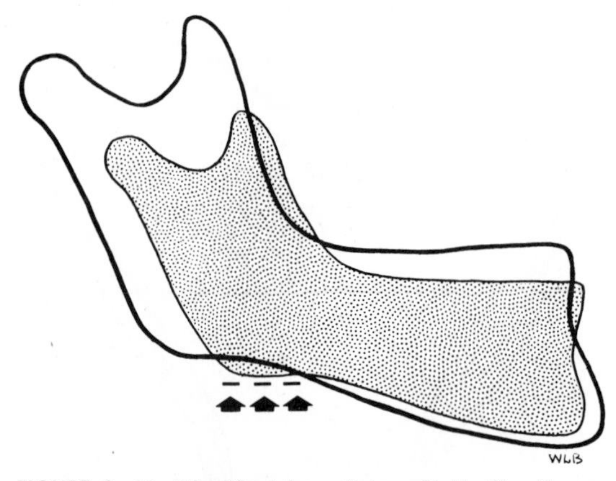

FIGURE 3–46. (Modified from Enlow, D. H.: *The Human Face.* New York, Harper & Row, 1968, p. 232.)

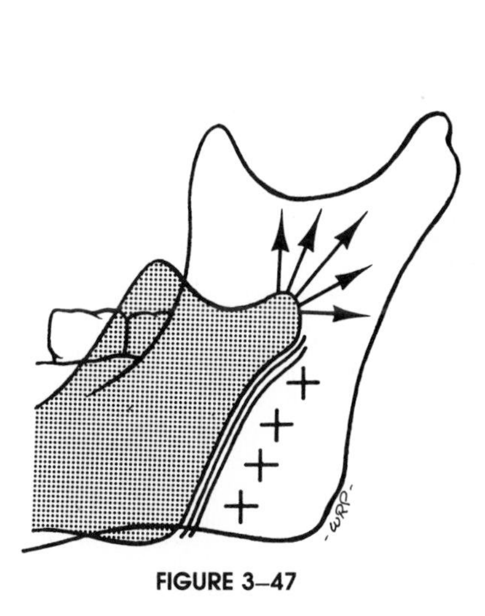

FIGURE 3–47

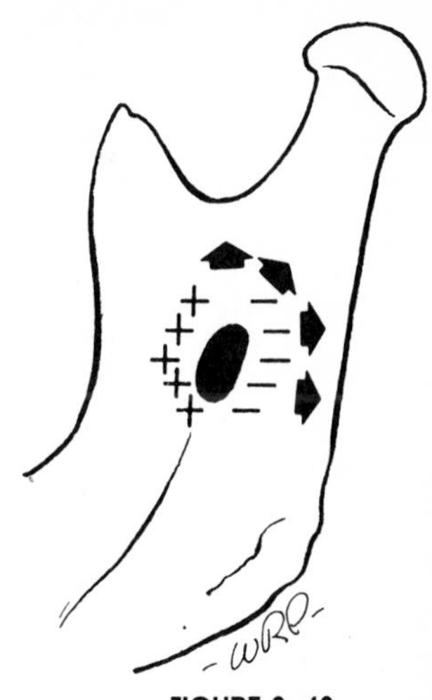

FIGURE 3–48

different regional conditions and control. Together they represent the most active areas during mandibular growth in distance moved and amount of bone deposited. Because of the relatively rapid rate of ramus growth, the bone tissue in the posterior part of the ramus is characteristically one of the more fast-growing types (Chapter 18). Ramus growth often involves a remodeling rotation of the whole ramus, and a resorptive field occurs on the posterior margin below the condyle, as illustrated by Figures 3–74 and 3–76.

The gonial region is anatomically variable, and therefore much variation is involved in its pattern of growth. Depending on the presence of inwardly or outwardly directed gonial flares, the buccal side can be either depository or resorptive, with the lingual side having the converse type of growth. However, many different histogenetic combinations can be encountered because of the topographic complexity and variability of this region.

While the whole ramus grows posteriorly and superiorly, the mandibular foramen likewise drifts backward and upward by deposition on the anterior and resorption from the posterior part of its rim (Fig. 3–48). The foramen maintains a constant position about midway between the anterior and posterior borders of the ramus. Even when the ramus undergoes marked alterations associated with edentulism (during which it may become quite narrow), this foramen usually sustains a midway location.

The **mandibular condyle** is an anatomic part of special interest because it is a major site of growth, having considerable clinical significance. Historically, the condyle has been regarded as a kind of cornucopia (Fig. 3–49A) from which the mandible pours forth. The condyle was believed (and is still believed by some) to be the ultimate determinant that, essentially, establishes the mandibular rate of growth, the amount of growth, growth direction, overall mandibular size, and overall mandibular shape. However, many present-day theoreticians do not regard the condyle as a unique kind of structure actually functioning to "regulate" morphogenesis of the whole mandible, including its many regional parts. It is no longer believed to represent a pacesetting "master center" with all other regional growth fields subordinate to and dependent on it for direct control. The condyle is a major field of growth, nonetheless, and it is an important one. What, however, could be even more important than serving as a master center? That question is answered in the pages that follow.

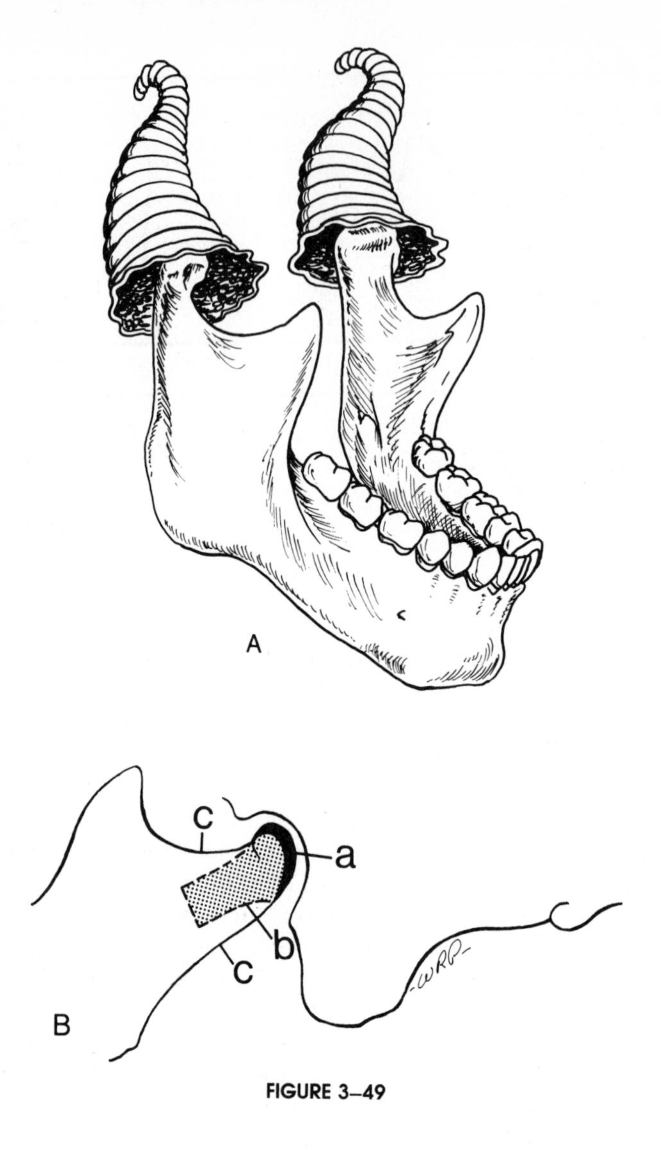

FIGURE 3–49

But during mandibular development, the condyle functions as a **regional** field of growth that provides an adaptation for its own localized growth circumstances, just as all the other regional fields accommodate their own particular (but different) localized growth circumstances. The growth of the mandible is a product of all the different **regional** forces and regional functional agents of growth control acting on it to produce the topographically complex shape of the mandible as a whole. Growth is the aggregate expression of the composite of all these localized factors.

The condylar growth mechanism itself is a clear-cut process. **Cartilage** is present because variable levels of **compression** occur at its articular contact with the temporal bone of the cranium. An endochondral growth mechanism is required, because the condyle grows in a direction toward its articulation in the face of direct pressure. An intramembranous type of growth could not operate, because the periosteal mode of osteogenesis is not pressure-adapted. Endochondral growth occurs only at the articular contact part of the condyle, because this is where pressure exists at levels beyond the tolerance of the bone's soft tissue membrane. As seen in Figure 3–49B, the endochondral bone tissue (*b*) formed in association with the condylar cartilage (*a*) is laid down only in the medullary portion of the condyle. The enclosing bony cortices (*c*) are produced by periosteal-endosteal osteogenic activity; these membranes are not subject to the compressive forces of articulation, but, rather, are essentially tension-related, because of muscle and connective tissue attachments. The real functional significance of the condylar cartilage thus involves an adaptation for regional compression; and this regional, endochondral, bone-forming mechanism develops as a specific response to this particular **local** circumstance. The cartilage itself does not contain genetic programming that directly determines and governs the course of growth in the other areas of the mandible. The pressure-tolerant condylar cartilage, however, also provides another exceedingly basic and significant growth function, as described later in conjunction with Figure 3–56.

The condylar cartilage is a **secondary** type of cartilage, which means that it does not develop by differentiation from the established **primary** cartilages of the skull (that is, the cartilages of the pharyngeal arches, such as Meckel's cartilage, and the definitive cartilages of the cranial base). Phylogenetically, the original cartilage and bone that provided for mandibular articulation (the separate articular bone attached by a suture to the dentary bone) became converted to an ear ossicle (the malleus) in mammals. Thus, a "secondary" cartilage was developed on the dentary bone to provide for lower jaw articulation with the cranium. It is believed that the unique connective tissue covering (capsule) of the condylar cartilage is actually an original periosteum. Its undifferentiated connective tissue stem cells, however, develop into chondroblasts, rather than osteoblasts, because of the compressive forces acting on this membrane. An adventitious type of "secondary" cartilage develops, rather than bone, because of the functional and developmental conditions imposed upon this part of the mandible. It is thus not an "endochondral" bone in the sense that phylogenetically, the bones of the cranial base are endochondral in type. The mandible is essentially a membrane bone in which one part (that is, what has become the condyle in mammals) develops in response to a phylogenetically altered developmental situation. This involves the ectopic presence of pressure that, in turn, causes localized ischemia and anoxia, factors that are known to induce chondrogenesis from the pool of undifferentiated connective tissue cells, rather than osteogenesis.

The condylar cartilage differs in histologic organization from most other growth cartilages involved in endochondral bone formation. It is not directly comparable to an epiphyseal plate. Although different authors have somewhat different working meanings for the terms "primary" and "secondary" cartilages,* it is now generally recognized that the secondary cartilage of the condyle is not the pacemaker for the growth of the mandible. Its contribution is to provide **regional adaptive growth** (that is, "secondary" growth, another definition for the term). It maintains the condylar region in proper anatomic relationship with the temporal bone as the whole mandible is simultaneously being carried downward and forward (see Stage 5). Thus, the condyle is not a "primary" center of growth. It is now believed that the condyle does **not**

*One common definition is that secondary growth cartilages are those, simply, that have a type of structure that puts them in a separate category from the typical epiphyseal growth plates of long bones. Articular cartilages are another example of the secondary type. As shown by Moss, the condylar cartilage is comparable, both in structure and growth behavior, to an articular, rather than an epiphyseal plate, cartilage. It is not actually "articular," however, because of its special fibrous covering.

establish the rate or the amount of overall mandibular growth. The condyle does, however, have a special multidirectional capacity for growth and remodeling in selective response to varied mandibular displacement movements and **rotations** (described below). The special structure of the condyle provides for this, unlike the committed unidirectional linear growth of epiphyseal plates produced by the characteristic linearly oriented direction of chondrocyte proliferation.

A unique **capsular layer** of poorly vascularized connective tissue covers the articular surface of the condyle (*a* in Fig. 3–50). This membrane is highly cellular early in development but becomes densely fibrous with age and function. Just deep to it is a special layer of prechondroblast cells (*b*). This is the predominant site for cellular proliferation, and it is responsible for the feeding process providing cartilage for endochrondral replacement by bone as the deeper layers advance.

The proliferative process produces the "upward and backward" growth movement of the condyle (Fig. 3–51). The condylar cartilage **moves** by prechondroblast cell divisions on the articular side with an equal amount of cartilage removed from the opposite (internal) side. The removal phase involves replacement with endochondral bone. A trail of continually forming new endochondral bone thus follows the moving cartilage, as schematized by layer *d* in Figure 3–50.

The prechondroblast cells are closely packed, and very little intercellular matrix is present. This is due to their rapid proliferative activity. A relatively thin transitional zone of immature hyaline cartilage then occurs deep to the proliferative layer, with a somewhat increased amount of matrix. This zone

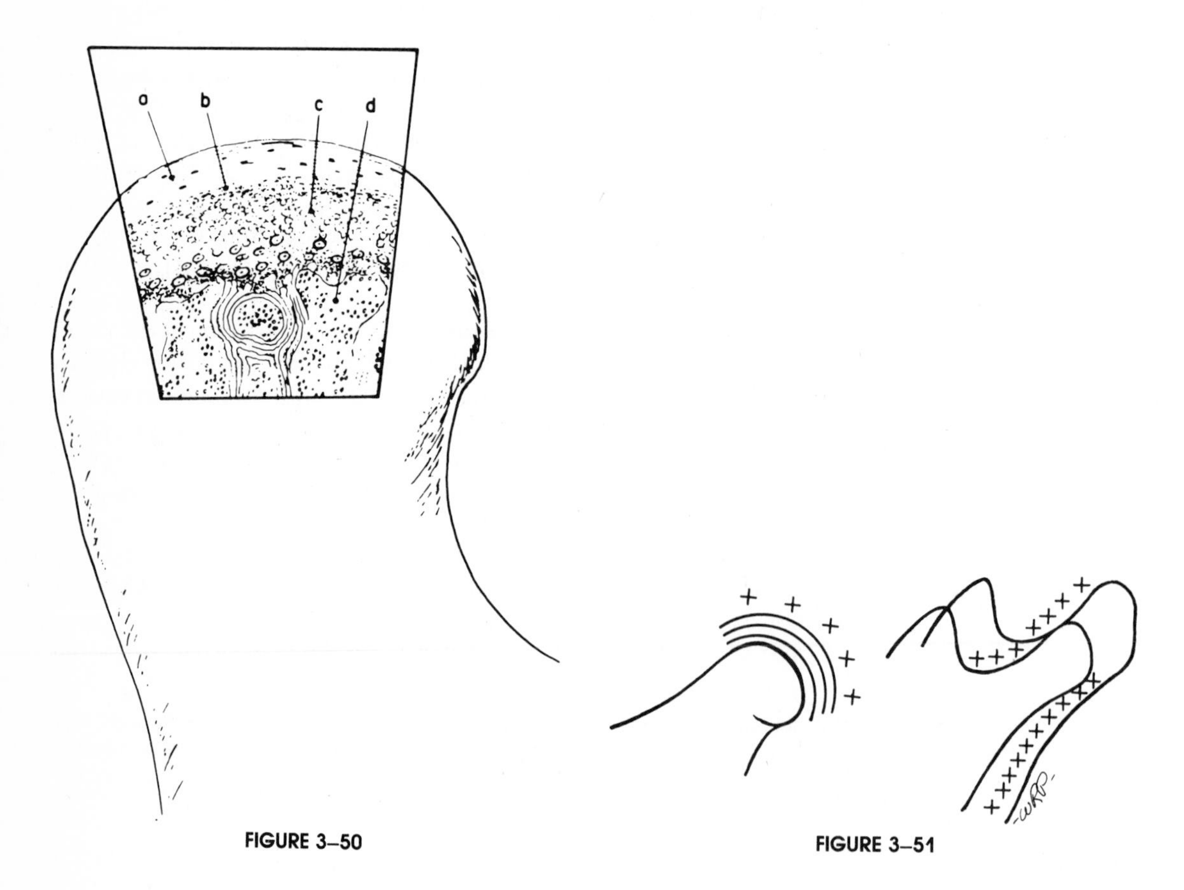

FIGURE 3–50

FIGURE 3–51

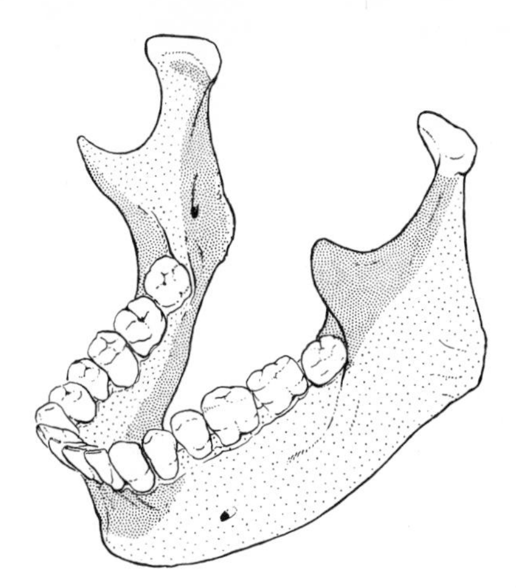

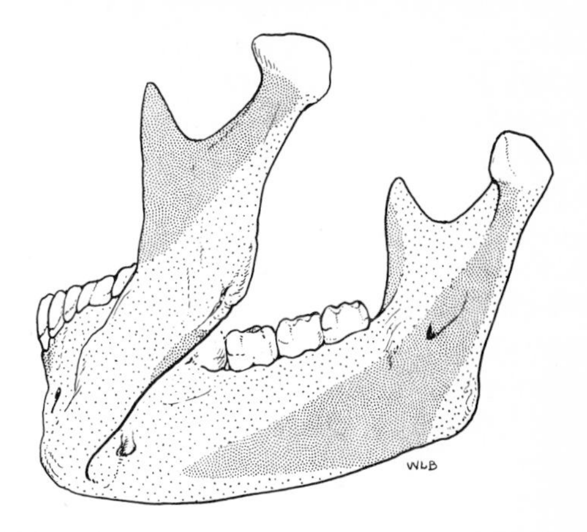

FIGURE 3–52

does not appear to contribute materially to the cell division process. The deeper cells become transformed into the next layer as it "moves up." This layer is composed of densely packed chondroblasts that are undergoing hypertrophy (*c*). The matrix is also noticeably scant.

The small amount of matrix in the deepest part of the hypertrophied zone becomes calcified, and a zone of resorption and bone deposition follows (*d*). Unlike the arrangement in typical **primary** growth cartilage, these various zones **do not have linear columns of daughter cells.** This is a notable histologic difference between primary and secondary types of growth cartilage. As pointed out by Koski, the arrangement of the daughter cells in the condylar cartilage thus does not reflect the direction in which the condyle is growing.

While all this is going on, the periosteum and endosteum are active in producing the **cortical** bone that encloses the **medullary core** of endochondral bone tissue (which takes the brunt of whatever compressive forces are acting on the condyle) (see Fig. 3–51). The cortical layer of intramembranous bone continues down onto the condylar neck. The anterior margin of the condylar neck is depository. This surface is part of the sigmoid notch, the entire edge of which grows superiorly and thus receives new bone.

The posterior edge of the condylar neck, which grades onto the posterior border of the ramus, is also depository and grows posteriorly, as illustrated in Figure 3–51. However, if a mandibular **rotation** is involved, the posterior margin of the condylar neck can be resorptive, as schematized in Figure 3–76.

The lingual and buccal sides of the neck characteristically have **resorptive** surfaces (Fig. 3–52, darkly stippled). This is because the condyle is quite broad and the neck is narrow. The neck is progressively relocated into areas previously held by the much wider condyle, and it is sequentially derived from the condyle as the condyle **moves** in a superoposterior course. What used to be condyle in turn becomes the neck as one is remodeled into the other. This is done by periosteal resorption combined with endosteal deposition. Explained another way, the **endosteal** surface of the neck actually faces the growth direction; the periosteal side points away from the course of growth (Fig. 3–53). This is another example of the V principle, with the V-shaped cone of the condylar neck growing toward its wide end (Fig. 3–54).

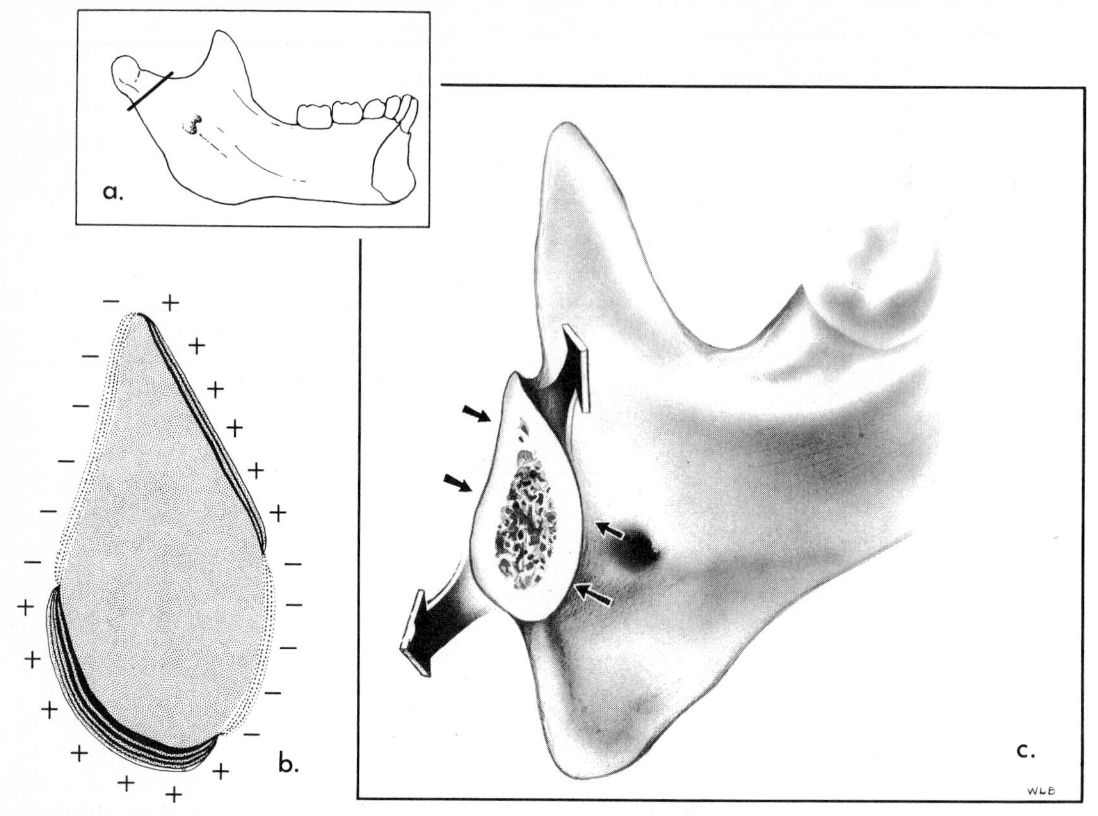

FIGURE 3–53. (Adapted from Enlow, D. H., and D. B. Harris: A study of the postnatal growth of the human mandible. Am. J. Orthod., 50:25, 1964.)

Stage 5

What is the physical force that produces the forward and downward primary **displacement** of the mandible? For many years it was presumed that growth of the **condylar cartilage,** because it is known that cartilage is a special pressure-adapted type of tissue, creates a "thrust" of the mandible against its articular-bearing surface in the glenoid fossa. The proliferation of the cartilage **toward** its contact thereby presumably **pushes** the whole mandible away from it (Fig. 3–55).

We are presently in a period of conceptual transition. Some students of facial biology still accept this explanation. However, a growing number of investigators feel that this is either an incomplete or an incorrect answer. The reasons follow.

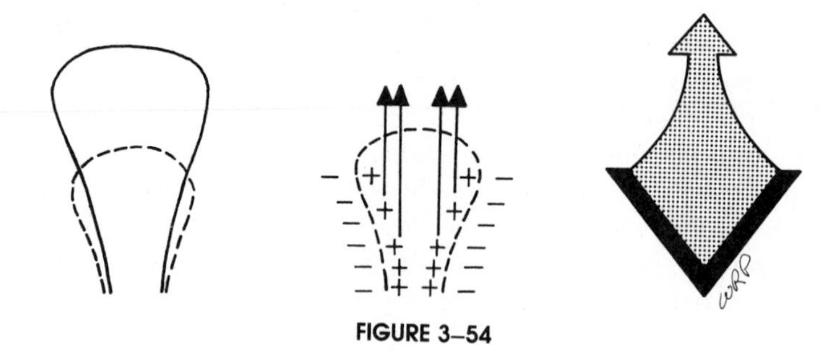

FIGURE 3–54

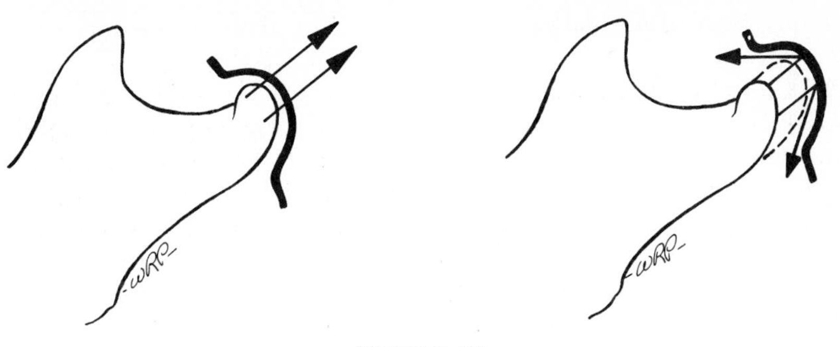

FIGURE 3–55

A great question was created when it was pointed out that mandibles totally lacking condyles exist in nature. Yet their morphology is more or less normal in all other respects; only the condyle and a part of the condylar neck are congenitally missing. Moreover, these **bilaterally** condyle-lacking mandibles occupy a normal **position;** the bony arch is properly placed for occlusion, and the mandible functions (albeit with distress) in movements even though it lacks an articulation. These revealing observations suggested two conclusions. First, the condyles may not play the role of a "master center" regulating the growth processes in the other parts of the mandible. Second, the whole mandible can become **displaced** anteriorly and inferiorly into its functional position without a "push" against the cranial base. Many experimental studies have subsequently been carried out with similar results, although investigators are still arguing about the proper way to interpret their meaning.

These observations led to a consideration of the "functional matrix" by students of facial biology. The idea is essentially that the mandible is **carried** forward and downward, just as the maxilla is presumably carried in conjunction with the growth expansion of the soft tissue matrix associated with it. It is a passive type of carrying in which mandibular growth itself does not directly participate; mandibular enlargement is an effect, rather than a cause, of the displacement movement. Thus, as the mandible is displaced away from its cranial base articular contact, the condyle **secondarily** (but almost simultaneously) grows toward it (Fig. 3–56), thereby closing the potential space without an actual gap being created (unless the condyle does not develop at all, as mentioned previously). There is still, however, actual pressure being exerted on the articular surface; it is presumably a relief of the **amount** of pressure

FIGURE 3–56

that results from the enlarging soft tissue mass that favors and stimulates condylar growth.

The clinical implications are apparent. Just how involved is the condyle as an underlying factor in facial abnormalities? What happens to the mandible if the condyle is injured during the childhood period? To the orthodontist, a key question is whether the condyle itself is the **direct** and **primary** target of any given clinical procedure, or whether it **follows** in response to clinical signals acting on soft tissues (e.g., masticatory musculature), which in turn activate the composite of osteogenic connective tissue of the ramus as a **whole.** How can overall mandibular length be increased or decreased for Class II and III individuals by physiologic or mechanical intervention in this composite growth mechanism?

The current thinking is that the condylar cartilage **does** have a measure of intrinsic genetic programming. This, however, appears to be restricted to a capacity for continued cellular proliferation. That is, the cartilage cells are coded and geared to divide and continue to divide, but extracondylar factors are needed to sustain this activity. The **rate** and **directions** of condylar growth are presumably subject to the influence of extracondylar agents, including intrinsic and extrinsic biomechanical forces and physiologic inductors. It is believed that increased amounts of pressure on the cartilage serve to inhibit the rate of cell division and growth. Decreased amounts of pressure appear to stimulate and accelerate growth. Presumably, forces applied to the mandible in such a way that they increase the level of pressure on the condyle would result in a shorter mandible **if** this were done during the period of active condylar growth. Similarly, a **release** of some of the compressive force acting on the condylar cartilage would produce a larger mandible if done during the active growth period. These conclusions are based largely on animal experiments and are not, at least at present, usable for everyday clinical practice. Moreover, recent research studies (as by McNamara) show that the nature of the condylar stimulus is more complex than simple forces acting directly on the condyle; rather, nerve-muscle-connective tissue pathways are involved, and changes utilize a **composite** of such tissue responses and chain feedbacks with the condyle as well as the other parts of the mandible that also participate. Sensory nerve input from the periodontal membranes and from the soft tissue matrix throughout the facc pick up stimuli that are passed on via motor nerves to muscles that, in turn, alter the displacement and the positioning of the mandible, which then affects the course of growth and remodeling by the condyle and all other areas of the growing mandible. The key point is that **regional** areas within the condyle can be stimulated or inhibited by resultant, localized forces to grow regionally either more or less. This alters the **amount** of ramus growth in different **directions,** thereby continually adjusting both the alignment and the shape of the ramus to accommodate its multiple anatomic and functional relationships.

The random arrangement of the condylar prechondroblasts, described earlier, is in contrast to the linear columns of daughter cells associated with the essentially unidirectional growth of long bones. This is a histogenetic adaptation of the condylar cartilage that provides opportunity for selected, multidirectional growth potential (see Figs. 3–47 and 3–74). Consider the virtually limitless range of anatomic variations that occur in the structural patterns of the nasomaxillary complex and basicranium. There are dolichocephalic and brachycephalic types of head forms, vertically long and short nasomaxillary regions, wide and narrow palates and upper arches, widely separated versus closely placed glenoid fossae, steep versus shallow cranial floor flexures, broad versus narrow pharyngeal regions, large versus small tooth sizes, and so on. If the growth, shape, and dimensions of the mandible were actually "preprogrammed" within the genes of condylar chondroblasts and if the condyle were indeed to function as an ultimate "growth control center," without taking into account structural and developmental vagaries in the rest of the craniofacial complex, there is no way that a fitting of the mandible to the basicranium on one end and to the maxilla on the other could be achieved. If the condyle functioned as a self-contained, independent structure with its growth coded in an isolated cartilage unresponsive to variations and continual changes in the growth and morphology of contiguous regions, the development of functional relationships could not occur. However, it is the adaptive, responsive nature of the condylar growth process that allows for a latitude of morphologic and morphogenic adjustments and a working (if not perfect) functional relationship with all of them. What could be more exalted in status than serving as a "master center of growth"? The answer is the condyle's adaptive capacity.

One specific point, however, needs to be clearly understood. Historically, it has been the condyle that has been given all the glory, whether as the primary determinant of mandibular growth or, as we now see it, as the respondent structure that makes possible adaptive, truly interrelated growth. The problem, however, is that we still use and endure the anachronistic, held-over term "condylar growth." This term, unfortunately, implies an incomplete and inaccurate understanding of the whole picture. True enough, the condyle plays a significant role. It is directly involved as a regional growth site; it provides an indispensable latitude for adaptive growth; it provides movable articulation; it is pressure-tolerant and provides a means for bone growth (endochondral) in a situation in which ordinary periosteal (intramembranous) growth would not be possible; and it can also, all too frequently, become involved in TMJ (temporomandibular joint) pathology and distress. With regard to the growth and adaptive requirement for the mandible, it is not just the condyle, however, that participates as the key component. The **whole ramus** is directly involved. The ramus bridges the pharyngeal compartment and places the mandibular arch in occlusal position with the maxillary arch. The horizontal breadth of the ramus determines the anteroposterior position of the lower arch, and the height of the ramus accommodates the vertical dimension and growth of the nasal and masticatory components of the midface. The dimensions and morphology of the ramus are directly involved in the attachments of the masticatory muscles, and the ramus must accommodate their growth and size. It is the growth and development of the **whole** ramus, not merely the condyle, that accomplish these ends. As already seen, the growth and remodeling of the ramus are complex and involve many regional growth sites, only one of which is the condyle. The term "condylar growth" is misleading and conveys a biologic misconception. More properly, the term needs to be "ramus and condylar growth." In a real sense, the condyle **follows** the growth of the whole ramus and does not lead it. This is important, because studies have shown, and continue to show, that the entire ramus and the muscles attached to it, not just the condyle, are a principal clinical target for many orthodontic procedures. Compare Figure 2–5 (an incomplete picture conveyed by the old term "condylar growth") with Figure 2–40 (whole ramus growth). Significantly, the "adaptive capacity" of the condyle referred to in the preceding paragraph **also** involves the entire ramus. The ramus is an important anatomic part directly involved in growth **compensations,** as described later.

The mandible has often been regarded as less responsive to orthopedic forces than the maxilla because the condyle itself is capped by the more pressure-tolerant cartilage (see Chapter 18) and also because growth preprogramming has been presumed that can resist extrinsic (clinical) forces. However, because the whole ramus is directly involved as well, consider that such clinical forces must overwhelm the massive masticatory musculature, a significant restraining factor less involved in any maxillary response.

Stage 6

It is often assumed that the face is more or less independent of the cranial base, that facial growth processes and the topographic features of the face are unrelated to the size, shape, and growth of the floor of the cranium. This is not the case at all. What happens in the cranial base very much affects the structure, dimensions, angles, and placement of the various facial parts. The reason is that the cranium is the **template** upon which the face develops. The growth of the middle cranial fossa is described below; that of the anterior cranial fossa, later. How differences in the structure of the basicranium as a whole affect facial pattern is explained in other chapters.

As each temporal lobe of the cerebrum grows, the middle cranial fossa (which houses the bottom part of the temporal lobe) correspondingly expands by a like amount. The bony surface of the whole cranial **floor** is predominantly resorptive (darkly stippled) in nature (Fig. 3–57). This is in contrast to the endocranial surface of the **calvaria** (skull roof), which is predominantly depository (lightly stippled; note the circumcranial reversal line indicated by the arrow). The reason for this major difference is that the inside (meningeal surface) of the skull roof is not compartmentalized into a series of confined pockets. The cranial floor, in contrast, has the **endocranial fossae** and other depressions, such as the sella turcica and the olfactory fossae. Why this calls for a difference in the mode of growth is explained below.

As the brain expands (*a* in Fig. 3–58), the separate bones of the calvaria are correspondingly displaced in outward directions (*b*). This is a passive

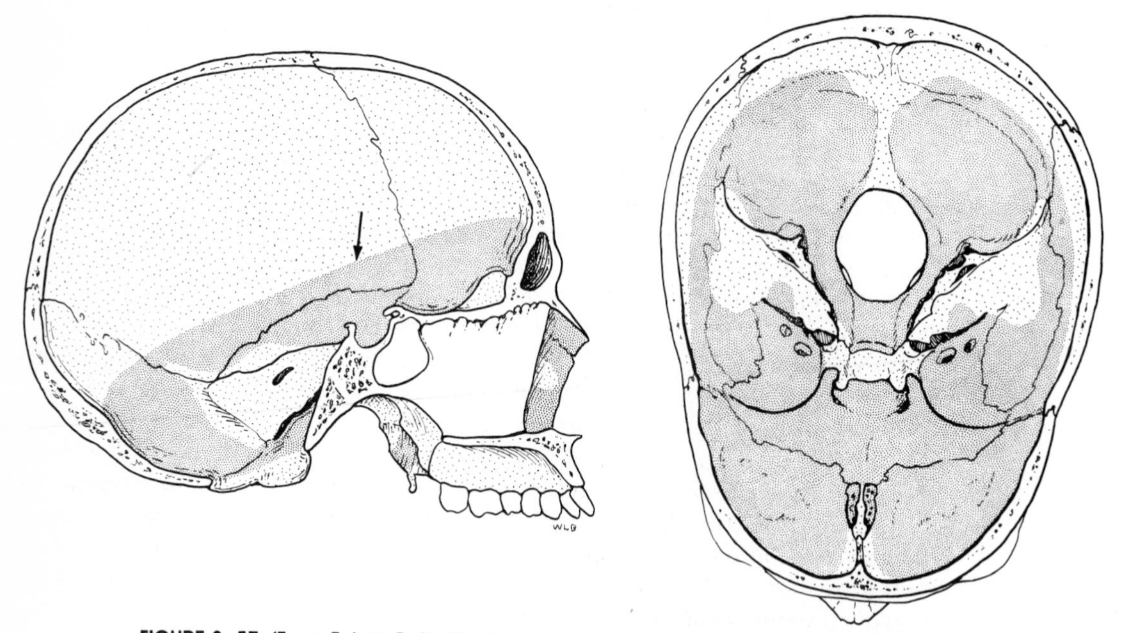

FIGURE 3–57. (From Enlow, D. H.: *The Human Face.* New York, Harper & Row, 1968, p. 197.)

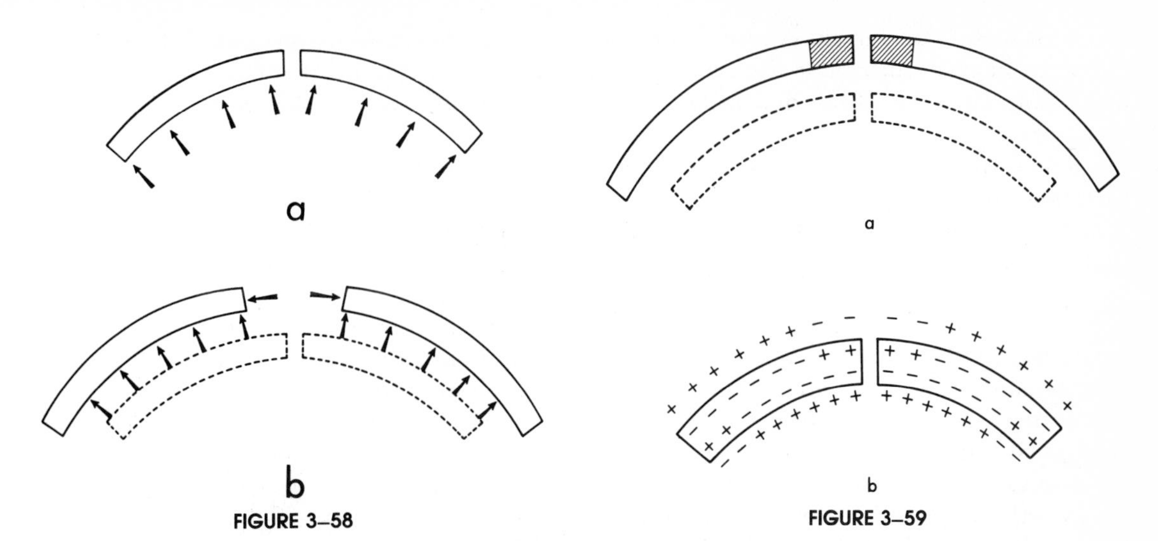

FIGURE 3–58

FIGURE 3–59

movement on the part of the bones themselves in conjunction with brain growth. Brain enlargement does not directly "push" the bones outward; rather, each separate bone is enmeshed within a connective tissue stroma attached to it. This stroma, in turn, is continuous with the meninges inside and the integument outside. As these enclosing connective tissue membranes enlarge with the growing brain, the bones are carried outward (displaced) by them, thereby "separating" all of the bones at their sutural articulations. In Figure 3–59, the primary displacement causes tension in the sutural membranes, which, according to present theory, respond immediately by depositing new bone on the sutural edges (*a*). Each separate bone (the frontal, parietal, and so forth) thereby enlarges in circumference. At the same time, the whole bone receives a small amount of new deposition on the flat surfaces of **both** the ectocranial and endocranial sides (*b*). The endosteal surfaces of the inner and outer cortical tables are resorptive. This increases the thickness of the bone and expands the medullary space between the inner and outer tables. The deposition of bone on the ectocranial surface, however, is **not** the growth change that causes the entire bone to move outward. The arc of curvature of the whole bone decreases, and the bone becomes flatter (Fig. 3–60). Although **remodeling** is not extensive

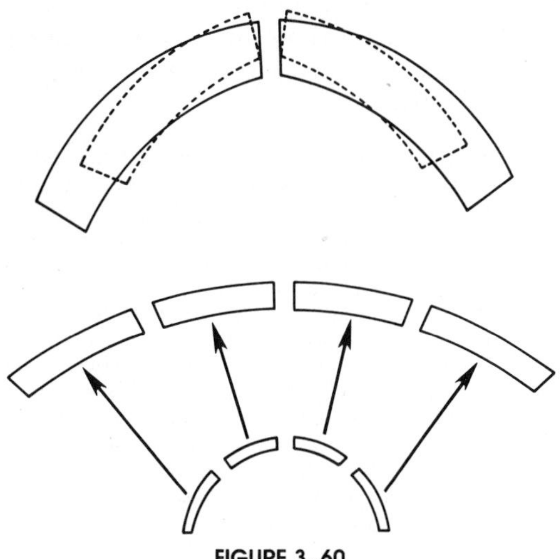

FIGURE 3–60

in any of these "flat" bones because of their relatively simple contours, reversals can occur in areas adjacent to the sutures. Here, either outside or inside surface resorption can take place, depending on the local nature of the changing curvature.

The cranial floor requires an entirely different mode of growth because of its topographic complexity and the tight curvatures of its fossae. The endocranial side (in contact with the dura that functions as a periosteum) is characteristically **resorptive** in most areas (see Fig. 3–57). The reason for this is that the sutures cannot provide for the entire process of growth enlargement. The system of sutures inherited from our mammalian ancestors cannot fully accommodate the markedly deepened endocranial fossae of the human basicranium, and additional widespread remodeling of the cranial floor is required. For example, in Figure 3–61a schematically representing an enlarged human basicranial fossa, sutures are located at *1* and *2*, and these produce growth in the direction of the arrows. However, the two sutures present cannot produce the growth for the **other** directions also needed to accommodate brain expansion, as shown in Figure 3–61b. Growth is accomplished by direct cortical drift, involving deposition on the outside with resorption from the inside; it is the key remodeling process that provides for the direct enlargement of the various endocranial fossae in **conjunction** with sutural (and also synchondrosis) growth.

The various endocranial compartments are separated from one another by elevated bony partitions. The middle and posterior fossae are divided by the petrous elevation; the olfactory fossae are separated by the crista galli; the right and left middle fossae are separated by the longitudinal midline sphenoidal elevation just below the sella turcica; and the right and left anterior and posterior cranial fossae are divided by a longitudinal midline bony ridge. All these elevated partitions, unlike most of the remainder of the cranial floor, are **depository** in nature (Fig. 3–57 and in Fig. 3–62a). The developmental basis for the depository nature of these partitions is schematized in Figure 3–62b. The reason, simply, is that as the fossae expand outward by resorption, the partitions between them must enlarge inward, in proportion, by deposition.*

The **midventral segment** of the cranial floor grows much more slowly than the floor of the laterally located fossae. This accommodates the slower growth of the medulla, pons, hypothalamus, optic chiasma, and so forth, in contrast to the massive, rapid expansion of the hemispheres. Because the floor of the cranium enlarges by remodeling growth in addition to sutural growth, these differential extents and rates of expansion can be carried out. A markedly decreasing gradient of sutural growth occurs as the midline is approached, but direct remodeling growth occurs to provide for the varying extents of expansion

---

*The activity of the bone lining the sella, however, is quite variable and can be either depository or resorptive in different areas. Several reasons apparently contribute to this, including the varying degrees of cranial base flexure and the variable amounts of downward and forward displacement of the midventral segments of the whole cranial base by the different shapes and proportionate sizes of the cerebral lobes. The sella turcica, however, must remain in contact with the hypophysis and also adjust to the variable size of the gland itself. If the pituitary fossa is carried downward by whole cranial base displacement farther than necessary, the floor of the sella will correspondingly rise by surface deposition to maintain contact with the pituitary, or the floor may be partly or entirely resorptive in other individuals to adjust to the balance between cranial base displacement and hypophyseal contact. A common combination is a resorptive posterior lining wall of the hypophyseal fossa and a depository surface on the sphenoidal part of the clivus (see Fig. 3–69). This causes a backward flare of the dorsum sellae to accommodate a pituitary gland that is being displaced to a lesser extent than the sphenoidal body below it. The jugum sphenoidale, like the floor of the sella turcica, shows variations for the same reasons cited earlier. Its dorsal surface may be resorptive in some individuals but depository in others.

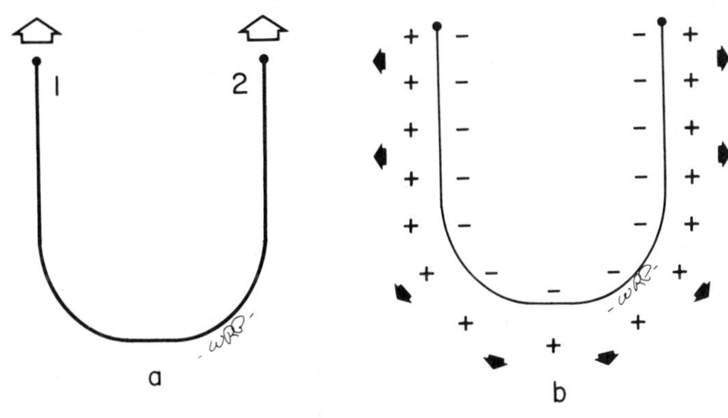

FIGURE 3–61

required among the different midline parts themselves and between the midline parts and the much faster growing lateral regions (see Fig. 3–70b).

Unlike the skull roof, the floor of the cranium provides for the passage of cranial nerves and the major blood vessels. Because the expansion of the hemispheres would cause marked displacement movements of the bones in the cranial floor if only a sutural growth mechanism were operative (as in the skull roof), the process of **remodeling** growth in the cranial base provides for the stability of these nerve and vascular passageways. That is, they do not become disproportionately separated because of the massive expansion of the hemispheres of the brain, as would happen if the basicranium enlarged primarily at the sutures. The foramen enclosing each nerve or blood vessel also undergoes its own drift process (+ and −) to constantly maintain proper position. The foramen at *a* moves to *b* in Figure 3–63 by deposition and resorption, keeping pace with the corresponding movement of the nerve or vessel it houses. This drift movement is much less than the remodeling movement of the lateral walls of the fossa (*c*).

The differential remodeling process maintains the proportionate placement of the spinal cord, even though the floor of the posterior cranial fossa, which rims the cord, expands to a considerably greater extent than the foramen magnum (Fig. 3–64). Note the much larger growth increments of the hemispheres and the squama of the occipital bone, in contrast to the smaller growth increments of the spinal cord and the foramen magnum. Differential remodeling, not merely sutural growth, provides for this.

The midline part of the cranial base is characterized by the presence of **synchondroses.** They are "left over" from the primary cartilages of the early cartilaginous basicranium after the endochondral ossification centers appear

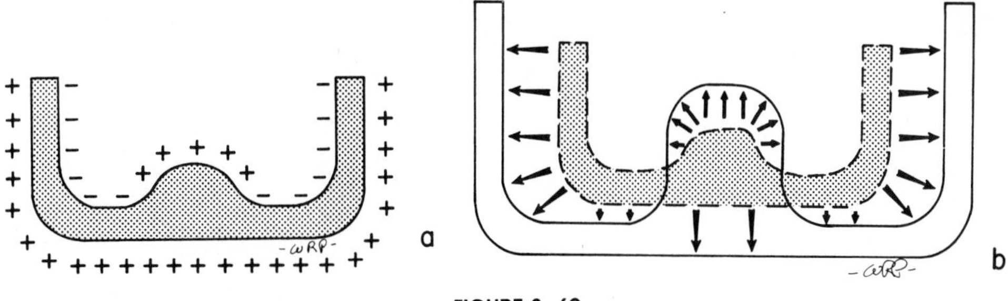

FIGURE 3–62

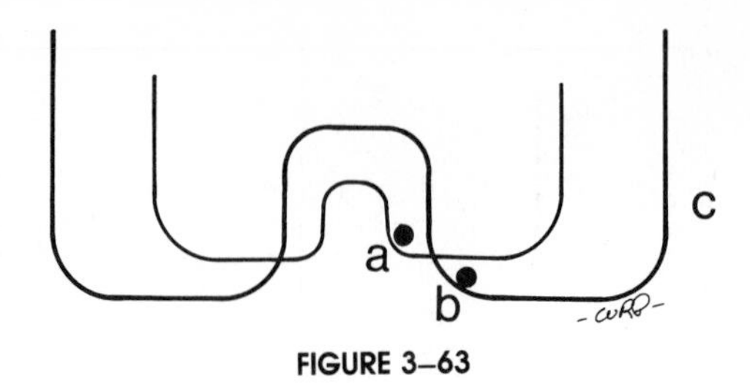

FIGURE 3–63

during fetal development. A number of synchondroses are operative during the fetal and early postnatal periods (see Chapter 12). During the childhood period of development, however, it is the spheno-occipital synchondrosis (arrow in Fig. 3–65) that is the principal "growth cartilage" of the cranial base. Like all "growth cartilages" associated directly with bone development, the spheno-occipital synchondrosis provides a pressure-adapted bone growth mechanism. This is in contrast to the tension-adapted sutural growth process. Compression is involved in the cranial base, unlike the calvaria, presumably because it supports the weight of the brain and the face, which bear down on the fulcrum-like synchondrosis in the midline part of the cranial floor, and also because it is more subject to craniofacial muscle forces. The spheno-occipital synchondrosis is retained throughout the childhood growth period as long as the brain and cranial base continue to grow and expand. It ceases to be active at about 12 to 15 years of age, and the sphenoid and occipital segments then become fused in this midline area before about 20 years of age.

The presence of the spheno-occipital synchondrosis provides for the elongation of the **midline** portion of the cranial base by its pressure-adapted

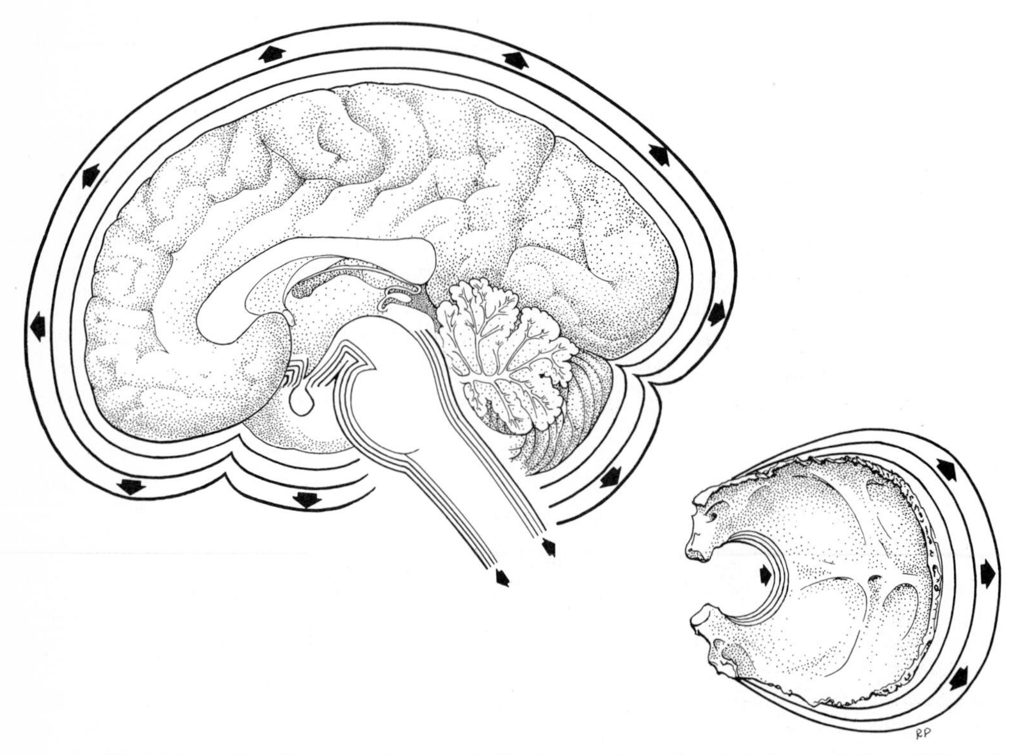

FIGURE 3–64. (Modified from Enlow, D. H.: *The Human Face.* New York, Harper & Row, 1968, p. 202.)

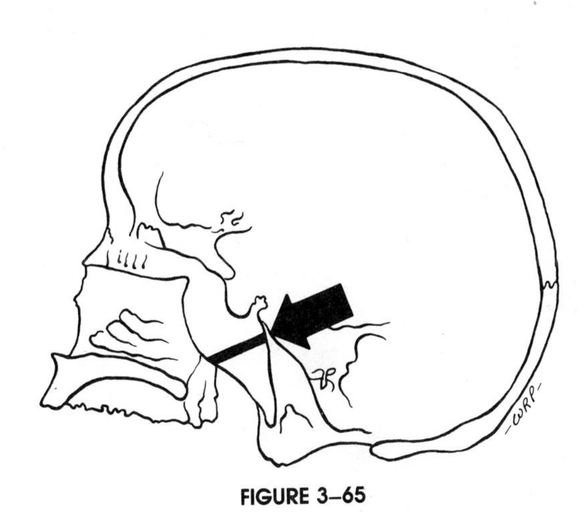

FIGURE 3–65

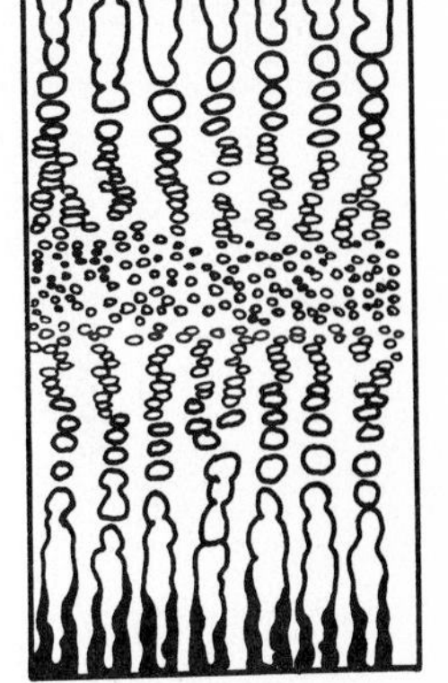

FIGURE 3–66

mechanism of endochondral ossification (see Chapter 18). The floor of the cranium also has sutures in the lateral areas, but (1) the force of the compression is accommodated by the synchondrosis, not the sutures, and (2) the expansion of the laterally located **hemispheres** produces tension in these lateral sutural areas, unlike the more slowly growing midline part of the cranial base not related directly to the hemispheres. Sutures are connective tissue membranes that provide tension-adapted sites of intramembranous bone growth.

Historically, the synchondrosis has been regarded as **the** growth "center" and pacemaker of the cranial base. This overly simplistic notion, however, as with the mandibular condyle, is a conceptual anachronism. The development of the basicranium is **quite** multifactorial and not merely the product of localized, midline cartilages that do not relate to the many regional growth circumstances throughout the basicranium as a whole. Only a very small percentage of the actual bone of the cranial floor is formed in conjunction with the synchondrosis.

The structure of the synchondrosis is similar to the basic plan for all primary types of growth cartilages, in contrast to the secondary variety, which is basically different (see page 91 for a description of the secondary condylar cartilage). As in the epiphyseal plate of long bones, the synchondrosis has a series of "zones," including the familiar reserve, cell division, hypertrophy, and calcified zones (Fig. 3–66). As in the epiphyseal plate but unlike the condylar cartilage, the chondroblasts in the cell division zone are aligned in distinctive columns that point toward the line of growth. Unlike the epiphyseal plate, the synchondrosis has **two** major directions of linear growth. Structurally, the synchondrosis is essentially two epiphyseal plates positioned back-to-back and separated by a common zone of reserve cartilage.

Endochondral bone growth by the spheno-occipital synchondrosis relates to primary **displacement** of the bones involved. Thus, the sphenoid and the occipital bones **move apart** by the primary displacement process (Fig. 3–67). At the same time, new endochondral bone is laid down in the medullary regions of each bone, and cortical bone tissue is formed by the periosteum and/or the endosteum around this core of endochondral bone tissue. Each whole bone (the sphenoid and the occipital) thereby becomes lengthened. Both bones also increase in girth by periosteal and endosteal activity. The interior of the sphenoid bone becomes hollowed to form the sizable **sphenoidal sinus.** This

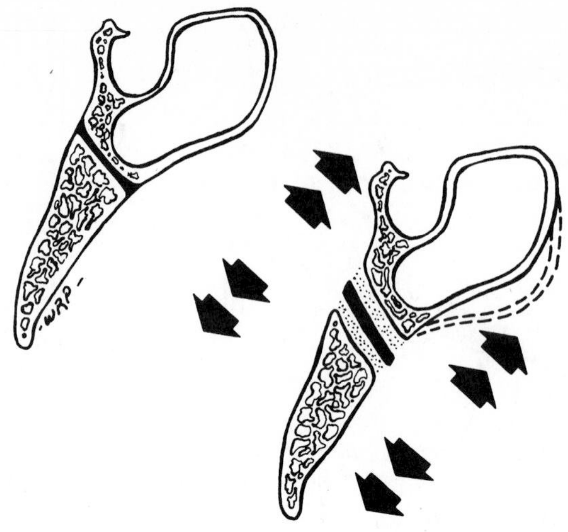

FIGURE 3–67

sinus is just behind and in direct line with the bony nasal septum of the nasomaxillary complex. As the midface is displaced forward and downward, the sphenoid retains contact with it, and the sphenoidal sinus is "drawn out" by the enlargement of this part of the sphenoidal body. Sphenoidal sinus expansion does not "push" the maxilla, however. The sinus is secondarily formed as the body of the sphenoid bone expands, keeping constant relationship with the midface.

Two key questions exist with regard to the lengthening of the cranial base at a synchondrosis and the process of displacement that accompanies the elongation of each whole bone. First, do the synchondroses **cause** displacement by the process of growth expansion, or is their endochondral growth a **response** to displacement caused by other forces (such as, perhaps, brain expansion)? Second, does the cartilage have an intrinsic genetic program that actually regulates the rate, amount, and direction of growth by the cranial base? Or is the cartilage more or less dependent on some **other** pacemaker for growth control and secondarily responsive to it?

Traditionally, the cranial base cartilages (and the whole cranial base in general) have been regarded as essentially autonomous growth units that develop in conjunction with the brain but somehow independent of it. This is difficult to understand; however, cranial base growth has been presumed to be controlled by a genetic code residing within the cartilage cells of the synchondrosis. In other words, the shape, size, and characteristics of the cranial base have evolved in direct phylogenetic association with the brain it supports (that is, a "phylogenetic type" of functional matrix), but the cranial base itself has presumably developed a genetic capacity for its own growth that is at least partially separate and independent of the brain and that may function without it during ontogenetic growth (as can happen in agenesis of the brain). Better and more complete biologic insight is presently needed on the morphogenic relationships involved in basicranial development.

Experimental studies, such as those by Koski, show that the independent proliferative capacity of a synchondrosis does not approach that of epiphyseal plates in long bones. This suggests that whatever capacity the cranial base (not calvaria) does have for continued growth, extrinsic control factors are also required. Just what these factors actually are and to what extent they are involved is not presently understood, since the cranial base **can** develop to a greater or lesser extent, even though the brain is malformed or absent. This is a major problem of our time and one that has more than academic significance to students of facial biology. In contrast, the calvaria appears to be largely dependent on its surrounding matrix for growth control.

As previously pointed out, the contribution of the synchondrosis relates to the midventral axis of the cranium and not the entire cranial floor. Note that the overall enlargement of the midline part of the cranial base (*a* in Fig. 3–68) is much less than the expansion of the more laterally located middle (and posterior) cranial fossae (*b*). This is because the lateral fossae house the various lobes of the hemispheres, which enlarge considerably more than the medulla, pituitary gland, hypothalamus, and so forth. Endocranial resorption occurs on both the endocranial surface of the clivus and, laterally, the floor of the middle cranial fossa. For the clivus, this produces an anteroinferior drift movement **in addition to** the linear growth by the synchondrosis.* For the middle (and also posterior) cranial fossa, it produces a massive enlargement in conjunction with the sutural growth also taking place. The clivus also lengthens by bone deposition on the ectocranial side of the occipital bone at the lip of the foramen magnum. For more detailed descriptions of the regional growth process in each separate bone in the cranium, see Enlow, *The Human Face,* 1968.

Stages 7 and 8

The expansion of the middle cranial fossa has a major **secondary** displacement effect on the anterior cranial floor, the nasomaxillary complex, and the mandible (Fig. 3–69). Because the posterior boundary of the facial complex exactly coincides with the boundary between the anterior and middle cranial fossae, the horizontal enlargement of the middle fossa produces a like amount

*The dorsum sellae, however, shows much variation in shape and size. In some individuals it flares markedly in an upward and backward direction, and the sphenoidal part of the clivus may be correspondingly depository, rather than resorptive, as shown in Figure 3–69.

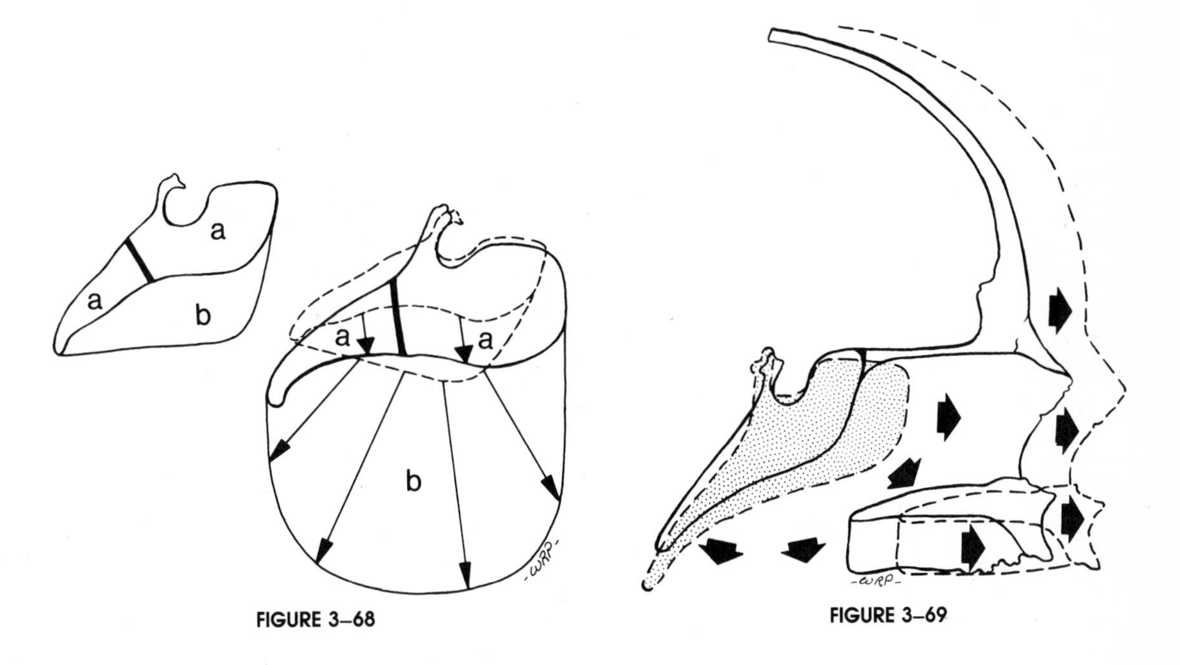

FIGURE 3–68

FIGURE 3–69

of forward displacement for both the anterior cranial fossa and the nasomaxillary complex. The amount of horizontal displacement for the mandible, however, is much less, because most of the enlargement of the middle cranial fossa takes place **anterior** to the mandibular condyle.

The enlarging middle cranial fossa does not in itself **push** the mandible, anterior fossa, and maxillary complex forward. Visualize the enlarging temporal and frontal lobes of the cerebrum as two expanding rubber balloons in contact. They are each displaced **away** from the other, although the net effect is a forward direction. The temporal and frontal "balloons" have fibrous attachments to the middle and anterior cranial fossae, respectively. As **both** balloons expand, these two fossae are thus **pulled away** from each other. This sets up tension fields in the various frontal, temporal, sphenoidal, and ethmoidal sutures, which presumably trigger sutural bone growth (in addition to direct cortical growth by resorption and deposition). Both fossae are thus enlarged, and the nasomaxillary complex is carried along anteriorly with the anterior cranial fossa to which it is attached. At about 5 or 6 years of age, frontal lobe growth and anterior cranial fossa expansion are largely complete. Thus, any further developmental protrusion of the forehead is a result of thickening of the frontal bone with enlargement of the frontal sinus within it. The temporal lobe and middle fossa, however, continue to enlarge for several years. Expansion of the temporal lobe displaces the frontal lobe forward, and this in turn causes tension in the suture systems between these two areas. The anterior fossa and the maxillary complex are carried anteriorly by the frontal lobe, which moves forward because of temporal lobe enlargement behind it. (**Note:** This "tension" trigger in response to **brain** enlargement is the present-day theory. Time may or may not prove this to be correct, but the rationale is at least reasonable as we now see it.)

As schematized in Figure 3–70a, resorption occurs from the lining side of the forward wall of the middle cranial fossa (*1*), deposition takes place on the

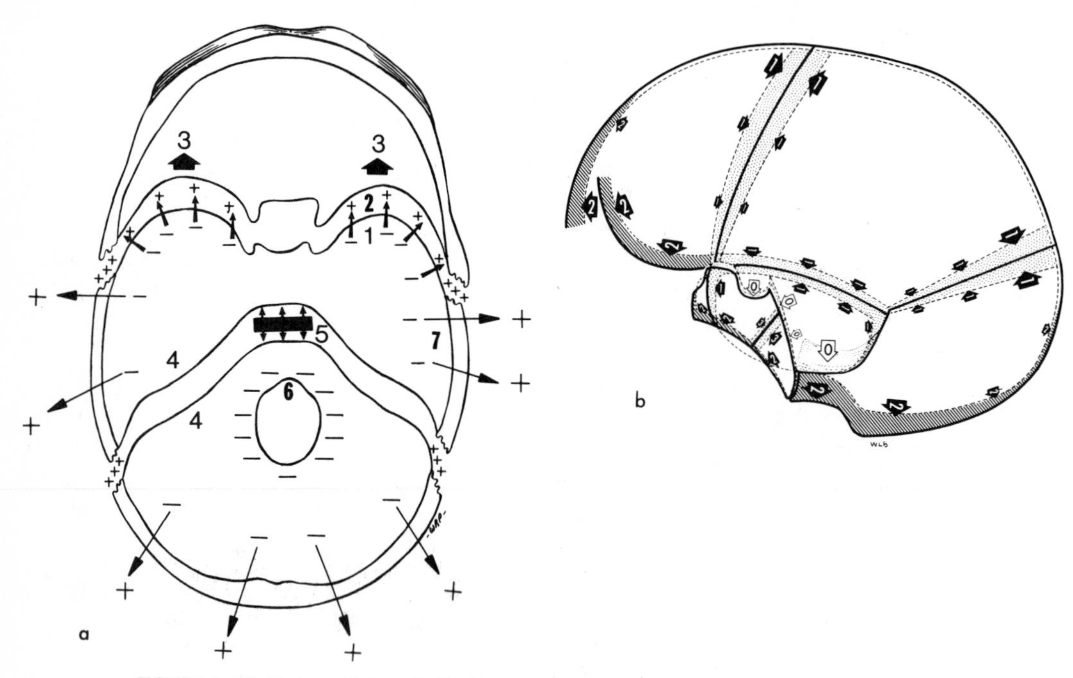

FIGURE 3–70. (*b*, from Enlow, D. H.: *The Human Face.* New York, Harper & Row, 1968.)

orbital face of the sphenoid and in the sphenofrontal suture (*2*), and forward displacement of the anterior cranial fossa occurs as the frontal lobes are displaced anteriorly (*3*). The petrous elevation (*4*) increases by deposition on the endocranial surface, and lengthening of the clivus occurs by growth at the spheno-occipital synchondrosis (*5*). The foramen magnum is progressively lowered by resorption on the endocranial surface and deposition on the ectocranial side. This also contributes to the lengthening of the clivus (*6*). Inferior to the circumcranial reversal line (see Fig. 3–57), the endocranial fossae enlarge by a combination of endocranial resorption and ectocranial deposition (*7*) that occurs in addition to growth at the basicranial sutures.

In Figure 3–70b, the decreasing gradient of sutural growth approaching the midventral part of the basicranium is schematized (lightly shaded areas, *1*). The endocranial fossae enlarge by direct cortical remodeling (outward drift), as shown by the darkly shaded areas (*2*). The clivus lengthens by endochondral bone growth at the spheno-occipital synchondrosis (*3*) and also by direct downward remodeling of the basicranial floor around the rim of the foramen magnum. The sphenoid and occipital complex remodels and rotates anteriorly and inferiorly by endocranial resorption (*0*) and ectocranial deposition.

The vertical enlargement of the middle cranial fossa has a major effect on the vertical placement of both the mandibular and maxillary arches. The effect is a progressive separation of the arches.

Stages 9 and 10

As the horizontal enlargement of the middle cranial fossa and brain growth advance the nasomaxillary complex by forward displacement, the horizontal span of the pharynx correspondingly increases (Fig. 3–71). The skeletal dimension of the pharynx is established by the size of the middle cranial fossa. The

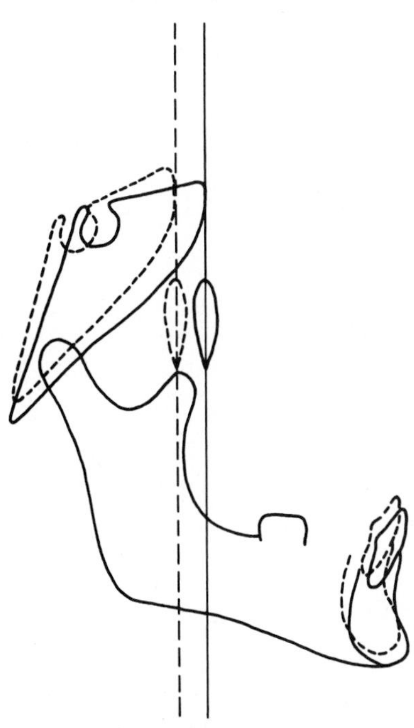

FIGURE 3–71

ramus of the mandible bridges the pharynx; and as this space enlarges, the ramus increases to an equivalent extent to maintain the same facial form. The effective **horizontal** dimensions of the ramus and the middle cranial fossa (not their respective oblique dimensions) are direct counterparts to each other. One structural function of the ramus, in spanning the middle cranial fossa, is to provide a growth capacity for whatever adaptation is required to place the corpus in a continuously functional position relative to the maxillary arch. If this is successful, a normal or Class I occlusion is achieved in a given individual. If less than adequate, the greater or lesser degree of failure of its adaptive or compensatory function contributes, in part, to the basis for a malocclusion.

In Stages 3 and 4, the ramus becomes relocated posteriorly by its own growth movement. The purpose is to horizontally lengthen the **corpus** and to displace it anteriorly. While this is occurring, the middle cranial fossa is also enlarging, however, so that now the ramus correspondingly increases in breadth to equal it (Fig. 3–72). This is done by the **same** process that carried out Stages 3 and 4, except that the amount of posterior ramus growth now **exceeds** the amount of anterior resorption. The horizontal (posteroanterior) breadth of the ramus is thereby increased. As this proceeds, the whole mandible simultaneously undergoes a corresponding extent of anteroinferior **displacement** (Fig. 3–73).

The ramus normally becomes progressively more upright during mandibular development. As long as the ramus is actively growing in a posterior direction, this is accomplished simply by greater amounts of bone addition on the inferior part of the posterior border than on the superior part (Fig. 3–74). A correspondingly greater amount of resorption of the anterior border takes place in the inferior part than in the superior part. A "rotation" of ramus alignment thus occurs. Condylar growth also becomes directed in a more vertical course.

The reason the ramus becomes more upright is that it must lengthen vertically to a much greater extent than it broadens horizontally (Fig. 3–75). In this schematic diagram, the pharynx (and middle cranial fossa) enlarges hori-

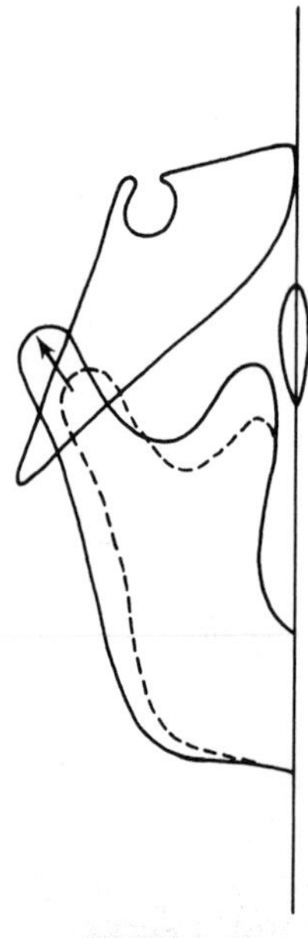
FIGURE 3–72

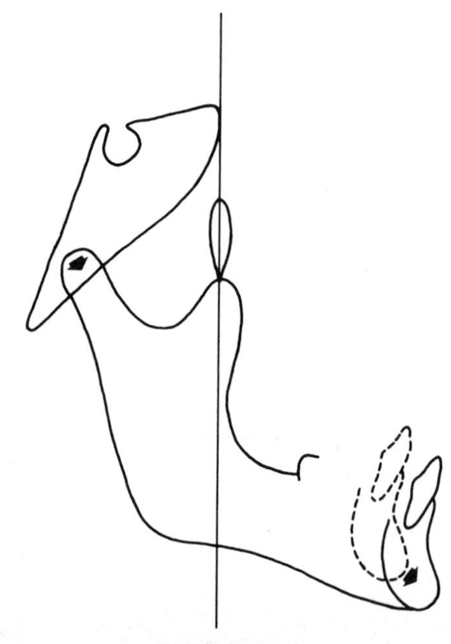
FIGURE 3–73

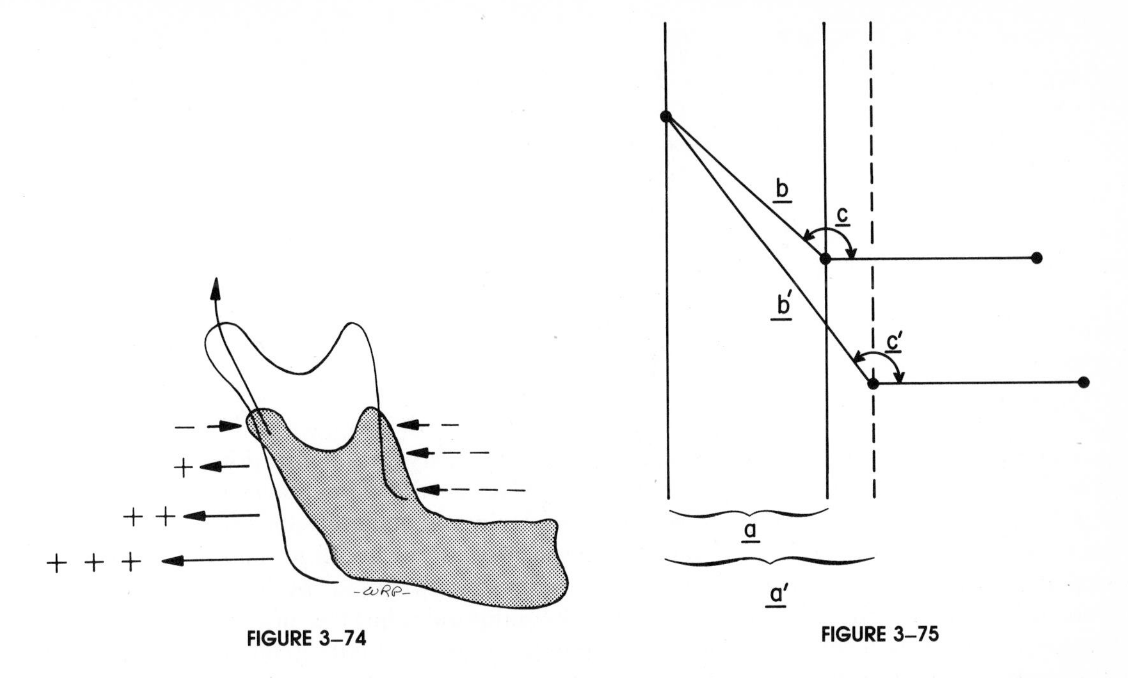

FIGURE 3–74

FIGURE 3–75

zontally from *a* to *a′*. The ramus enlarges, correspondingly, from *b* to *b′* to match it. It also **lengthens vertically,** however. Angle *c* is thereby reduced to *c′* in order to accommodate the vertical increase, which allows for the considerable extent of vertical **nasomaxillary growth** also taking place at the same time. The "gonial angle" thus must undergo change (close) in order to **prevent** change in the occlusal relationship between the maxillary and mandibular arches.

However, vertical lengthening of the ramus continues to take place **after** horizontal ramus growth slows or ceases (when the horizontal growth of the middle cranial fossa begins to stop). This is to match the continued vertical growth of the midface. To achieve this, condylar growth may become more vertically directed, and a different pattern of ramus remodeling can also become operative (Fig. 3–76). The direction of deposition and resorption reverses. A

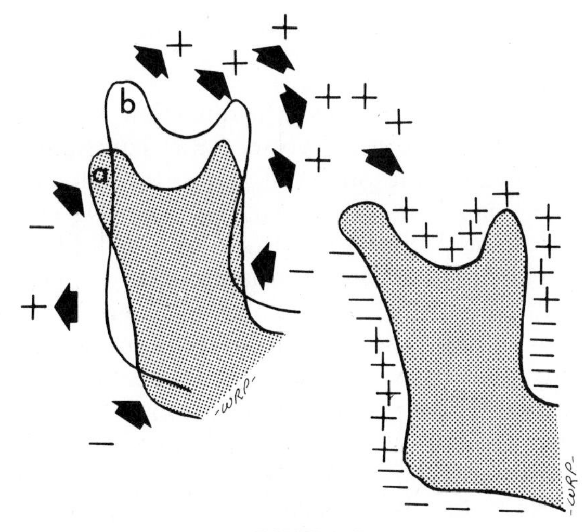

FIGURE 3–76

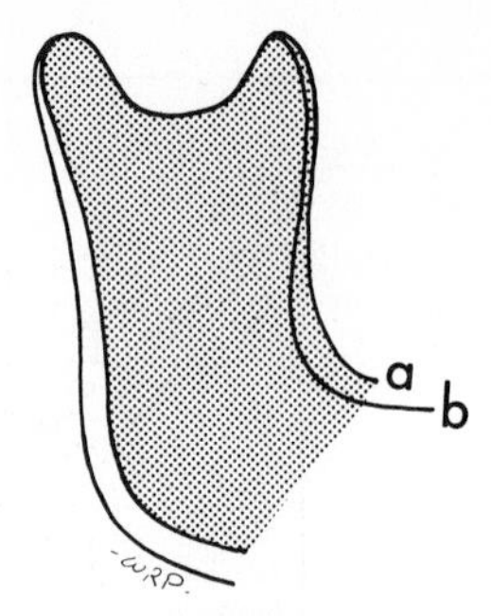

FIGURE 3–77

forward growth direction can occur on the **anterior** border in the upper part of the coronoid process. Resorption takes place on the upper part of the posterior border. A posterior direction of remodeling takes place in the lower part of the posterior border. The result is a more upright alignment and a longer vertical dimension of the ramus without a material increase in breadth. This remodeling change, when it occurs, appears to be more marked in the later periods of childhood, after the enlargement of the temporal lobes has slowed or largely ceased (with a corresponding decrease in the horizontal enlargement of the ramus) and when the backward relocation of the ramus to provide for corpus lengthening has decreased. There are probably other relationships involved, as well, in the range between extremes, including different facial and head form types, although the biological basis is presently not understood.

The ramus thus undergoes a remodeling alteration in which its angle becomes changed in order to retain constant positional relationships between the upper and lower arches. If mandible *a* in Figure 3–76 is superimposed over *b* in the anatomically functional position, it can be seen that all the complex remodeling changes in Figure 3–76 serve simply to change the ramus angle without increasing its breadth, as shown in Figure 3–77. (See also Stage 15 and pages 205 to 214 for further information on mandibular rotations.) This also accommodates the growing muscle sling and muscular adaptations associated with mandibular rotations. In addition, increased space for third molar eruption is provided.

Stage 11

The anterior cranial fossa enlarges in conjunction with the expansion of the frontal lobes. Wherever sutures are present, they contribute to the increase in the circumference of the bones involved. Thus, the sphenofrontal, frontotemporal, sphenoethmoidal, frontoethmoidal, and frontozygomatic sutures participate in a traction-adapted bone growth response to brain and other soft tissue enlargement. The bones all become **displaced** as a consequence. This is a primary type of displacement, because the enlargement of each bone is involved. Together with this, the bones also grow outward by ectocranial deposition and endocranial resorption, as described below. The composite of all these processes produces the growth changes seen in Figure 3–78.

The cranial bones increase in size by sutural bone growth as the forehead becomes displaced anteriorly. The nasomaxillary complex is carried anteriorly as well (Fig. 3–79). Note that the maxillary tuberosity now lies ahead of the vertical reference line. The tuberosity, however, simultaneously grows poste-

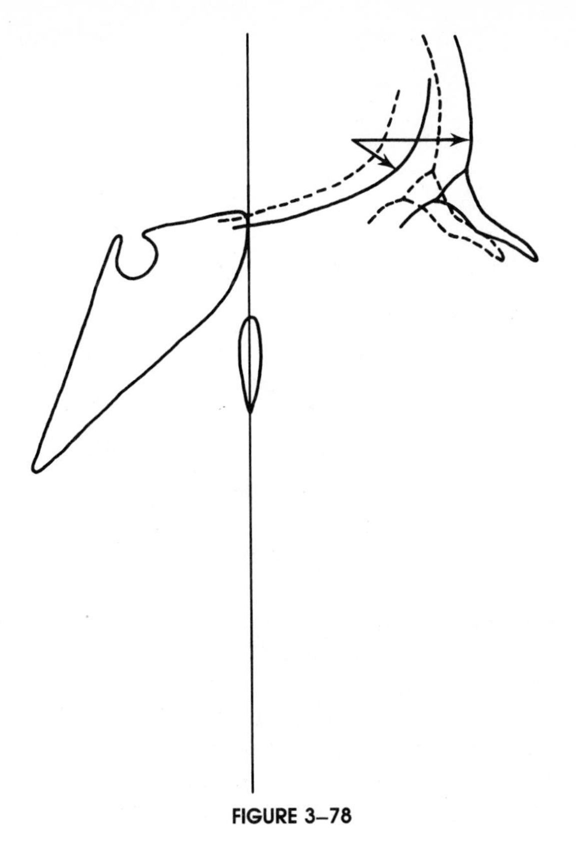

FIGURE 3–78

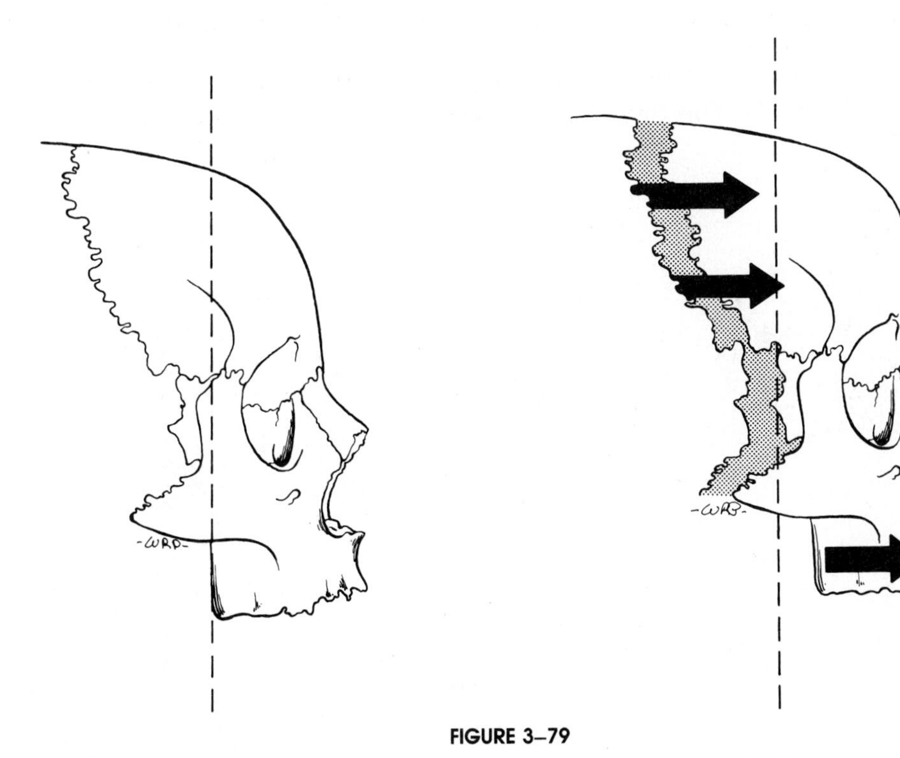

FIGURE 3–79

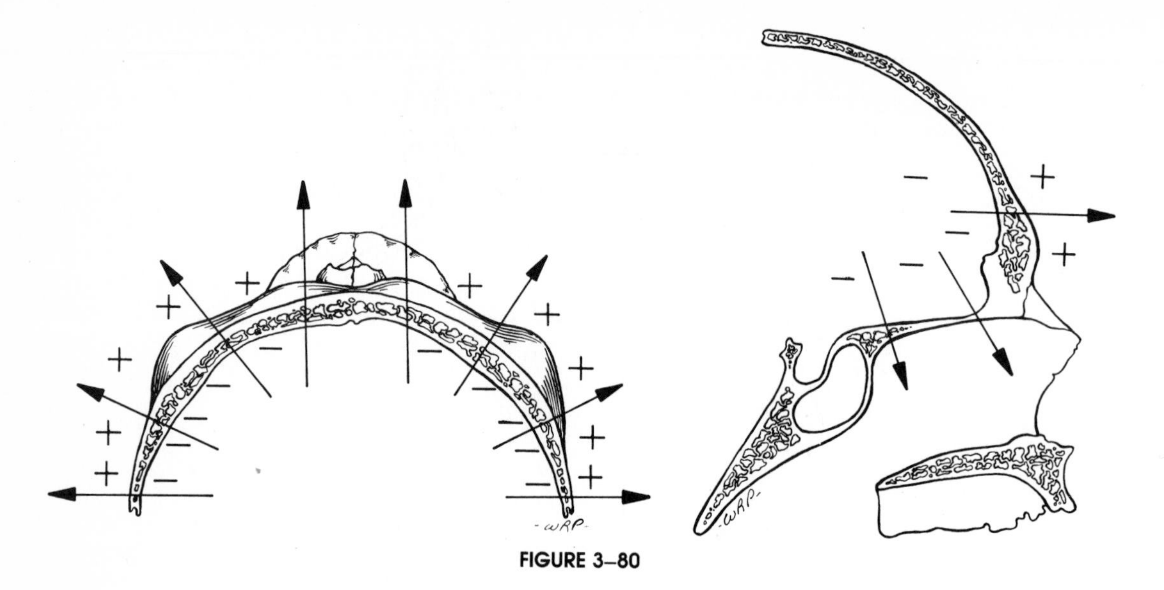

FIGURE 3–80

riorly by an equivalent amount (described previously in Stage 1); the floor of the anterior cranial fossa and the bony maxillary arch are counterparts.

As previously pointed out, sutural growth alone cannot accomplish the extent of cranial fossa expansion required. In addition to bone additions at the various sutures, direct cortical growth also takes place (Fig. 3–80). About midway up the forehead, however, the endocranial side becomes depository rather than resorptive. This reversal line encircles the inner side of the skull and separates the resorptive growth fields of the basicranium from the separate field of the roof (see Figure 3–57).

As long as the frontal lobe of the cerebrum grows, the **inner** table of the forehead correspondingly drifts anteriorly. When frontal lobe enlargement slows and largely ceases sometime before about the sixth or seventh year, the growth of the inner table stops with it. The outer table, however, continues to drift anteriorly (Fig. 3–81). This progressively separates the two tables, and an

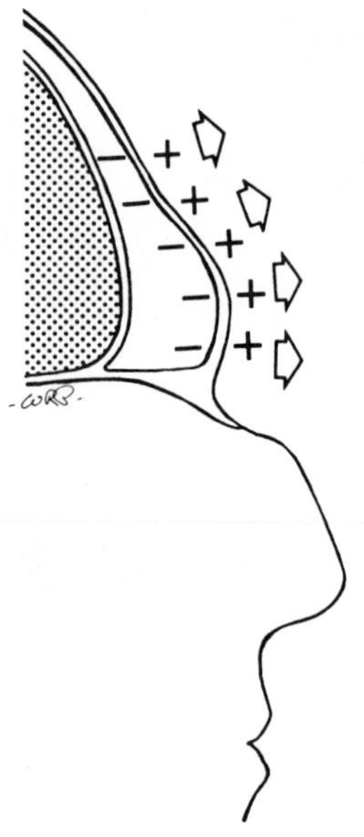

FIGURE 3–81

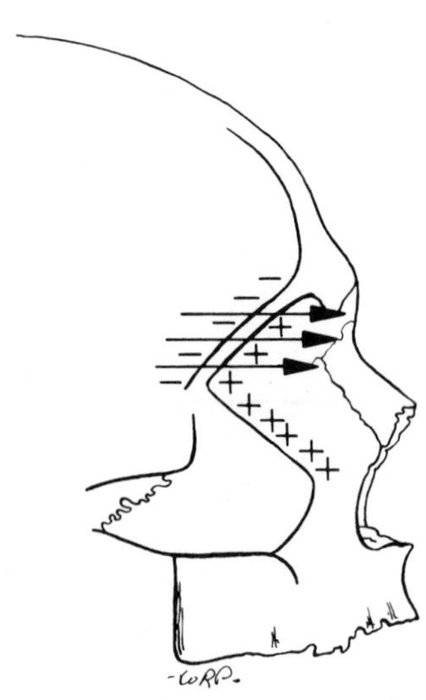

FIGURE 3–82

enlarging frontal sinus results. The size of the sinus, however, and the amount of forehead slope vary considerably according to age, sex, and ethnic characteristics (see Chapter 7). The reason the frontal sinus develops is that the upper part of the nasomaxillary complex continues to grow anteriorly, and the outer table of the forehead remodels with it.

Note that the floor of the anterior cranial fossa is also the roof of the underlying orbital cavity (Fig. 3–82). The endocranial side is resorptive, and the orbital side of the very thin bony plate is depository; it relocates by remodeling progressively downward and outward. While this serves to enlarge the bottom part of the cranial fossa, does it also reduce the size of the orbital cavity? The answer is no, for two reasons. First, the orbits relocate anteriorly by the V principle, which serves to enlarge, not reduce, orbital size. Second, the whole orbit is also becoming **displaced** at the same time in association with growth at the various orbital sutures. This is described later.

Stage 12

The vertical lengthening of the nasomaxillary complex involves (a) remodeling growth (deposition and resorption on the various bony cortices) and (b) displacement. The process of remodeling growth is described in this stage, followed by displacement in Stage 13.

With the exception of the uppermost part of the roof in each nasal chamber, the lining surfaces of the bony walls and floor are predominantly resorptive (Fig. 3–83). The mucosal side of each nasal bone is also resorptive. These regional patterns cause a lateral and anterior expansion of the nasal chambers and a downward movement of the palate; the oral side of the bony palate is depository. The nasal side of the cribriform plate (the roof of the nasal chamber) is depository, and the cranial side is resorptive. This enlarges the small, paired olfactory fossae and lowers them in conjunction with the downward cortical drift of the entire anterior cranial floor.

The ethmoidal conchae (not diagrammed) generally have depository surfaces on their lateral and inferior sides and resorptive surfaces on the superior and medial-facing sides of the thin bony plates. This moves them downward and laterally as the whole nasal region expands in like directions. (The developmentally separate inferior concha, however, can show remodeling variations because it is carried inferiorly, to a greater extent than the others, by maxillary displacement.) The lining cortical surfaces of the maxillary sinuses are all resorptive, except the medial nasal wall, which is depository because it moves laterally during nasal expansion.

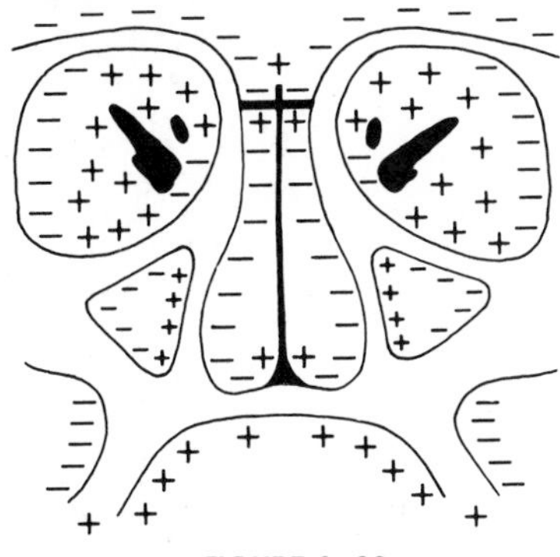

FIGURE 3–83

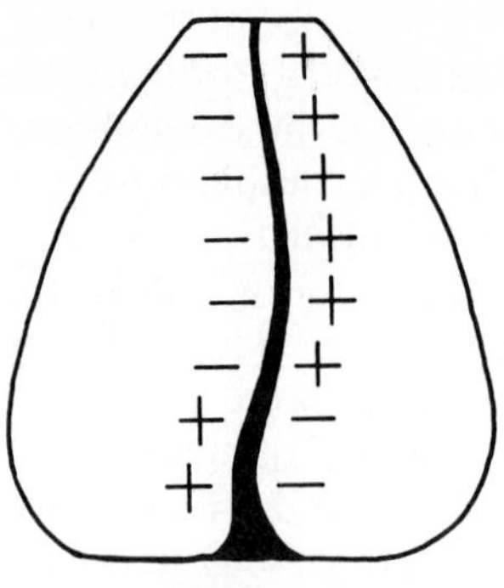

FIGURE 3–84

A basic and important concept of the facial growth process is underscored in Figure 3–83: it is the **entire** facial complex that participates in the growth process. **All** parts and bony surfaces are directly involved, not merely certain special sites and "centers."

The bony portion of the internasal septum (the vomer and the perpendicular plate of the ethmoid) lengthens vertically at the various sutural junctions (and to a much lesser extent by endochondral growth where the cartilaginous part contacts the perpendicular plate of the ethmoid). The bony septum also drifts laterally in relation to variable amounts and directions of **septal deviation** (Fig. 3–84). The remodeling patterns involved are individually variable, and the thin plate of bone typically shows alternate fields of deposition and resorption on the right and left sides, which produce a drift and buckling to one side or the other.

Note that the breadth of the nasal bridge in the region just below the frontonasal sutures does not markedly increase from early childhood to adulthood (Fig. 3–85). More inferiorly in the interorbital area, however, the medial

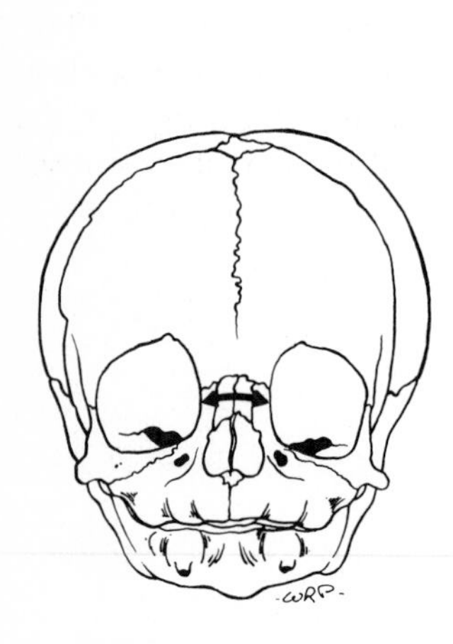

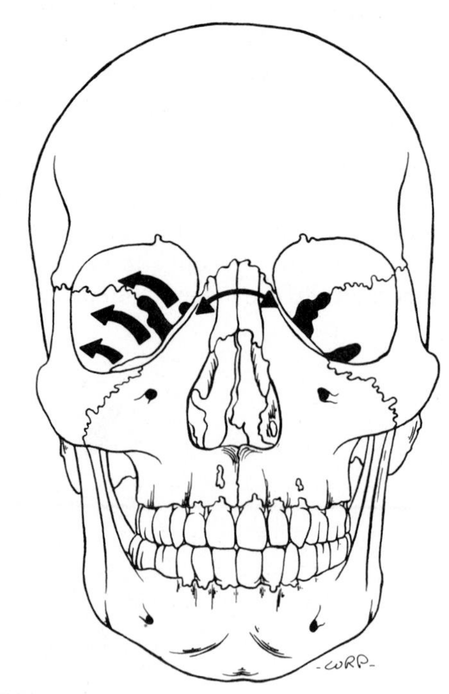

FIGURE 3–85

wall of each orbit (lateral walls of the nasal chambers between the orbits) expands and balloons out considerably in a lateral direction in conjunction with the considerable extent of lateral enlargement of the nasal chambers. The ethmoidal sinuses are thereby enlarged greatly.

The **lacrimal** bone is an important participant during orbital growth and remodeling. This diminutive, thin plate of bone plays a key role in providing adjustments for the major differential movements of all the bones surrounding it. The lacrimal bone is an island of bone with its entire perimeter bounded by sutural contacts separating it from the ethmoid, maxillary, and frontal bones. As these separate bones enlarge or become displaced in many directions and at different rates, the sutural system of the lacrimal bone provides for the "slippage" of these other bones along sutural interfaces as the different bones enlarge differentially. It does this by collagenous linkage adjustments within the sutural membrane (see page 477). The lacrimal bone and its sutures thus make it possible for the maxilla to "slide" downward along the contact with the medial orbital wall as the whole maxilla becomes displaced inferiorly.

The lacrimal bone undergoes a remodeling rotation (Fig. 3–86), because the more medial **superior** part remains with the lesser-expanding nasal bridge, while the more lateral **inferior** part moves markedly outward to keep pace with the great expansion of the ethmoidal sinuses. This remodeling change is illustrated by *a*; the primary displacement that accompanies it is shown by *b*.

In the growth of the bony maxillary arch, area *A* is moving in three directions by bone deposition on the external surface: it lengthens **posteriorly** by deposition on the posterior-facing maxillary tuberosity; it grows **laterally** by deposits on the buccal surface (this widens the posterior part of the arch); and it grows **downward** by deposition of bone along the alveolar ridges and also on the lateral side, because this outer surface slopes (in the child) so that it faces slightly downward (Fig. 3–87). The endosteal surface is resorptive, and this contributes to maxillary sinus enlargement.

In Figure 3–88, note that a major change in surface contour occurs along the vertical crest just below the malar protuberance (small arrow). This crest is called the "key ridge." A **reversal** occurs about here. Although a range of variation occurs in the exact placement of the reversal line, anterior to it most of the external surface of the maxillary arch (the protruding "muzzle" in front

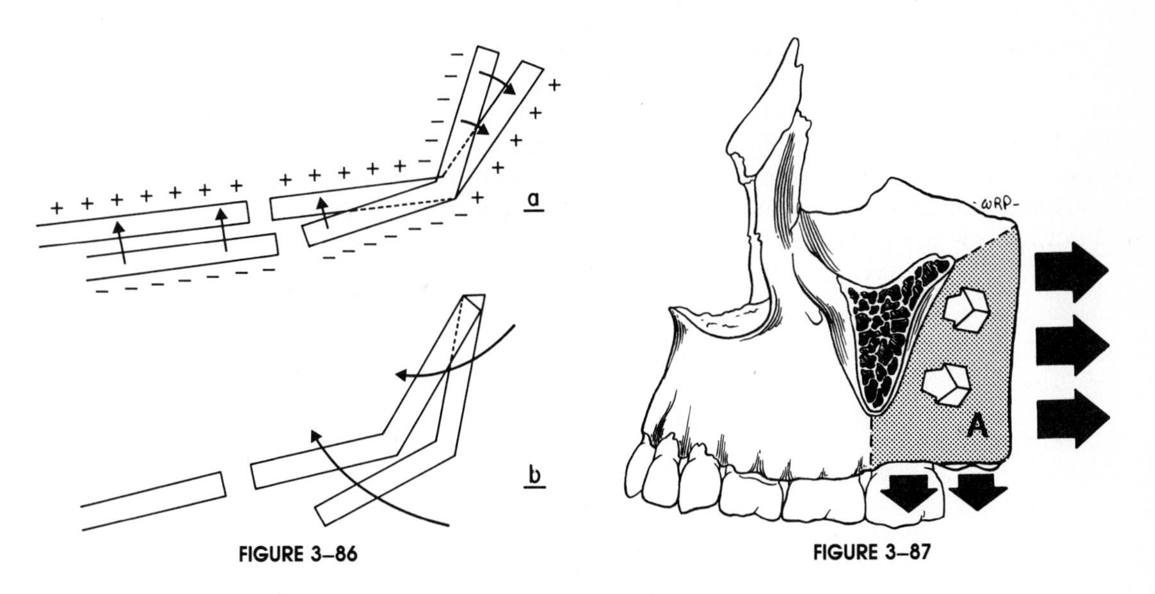

FIGURE 3–86

FIGURE 3–87

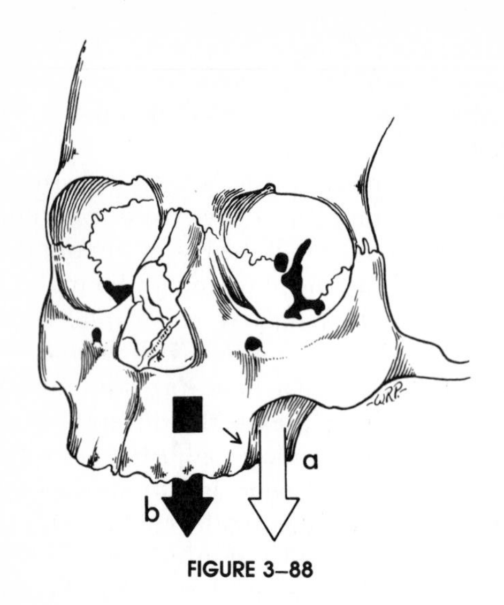

FIGURE 3–88

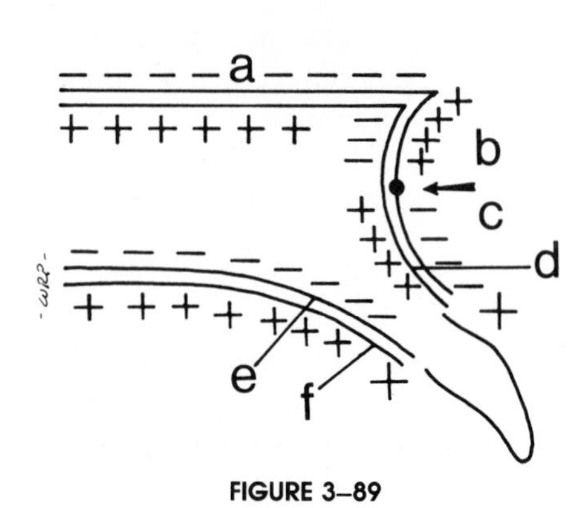

FIGURE 3–89

of the cheekbone) is **resorptive.** This is because that part of the bony arch in area *b* is **concave,** and the labial (outside) surface faces upward, rather than downward. The resorptive nature of this surface provides an inferior direction of arch growth in conjunction with the downward growth of the palate. This is in contrast to area *a,* which grows downward by periosteal **deposition.**

In Figure 3–89, surface *a* is resorptive; *b* is depository. A reversal occurs at about "*A* point" (indicated by arrow; this is a much-used cephalometric landmark). Periosteal surface *c* is resorptive, *d* is depository, *e* is resorptive, and *f* is depository.

Every second-year student of dentistry is familiar with the well-known process of "mesial drift." However, the accompanying and equally important (perhaps more so) process of **vertical drift** has not become a part of our working vocabulary until recently. Yet the vertical movements of teeth are significant in extent and play a key role in maxillary and mandibular development (Figs. 3–90, 3–91, and 3–92). As a tooth drifts mesially (or distally, depending on the species and which tooth), note that the same process of alveolar remodeling (resorption and deposition) relates to a vertical movement of the tooth as well. Any tipping, rotations, or buccolingual tooth movements are also simultaneously carried out by the same remodeling process. As a tooth bud develops and its root elongates, the growing tooth undergoes **eruption,** bringing its crown into definitive position above the bone and gingiva. The vertical drift of a tooth is **in addition to** eruption, and use of the term eruption for this vertical drifting is inappropriate (although too often still used). As the maxilla and mandible enlarge and develop, the dentition drifts both vertically and horizontally to keep pace. The process of drift moves the whole tooth **and**

FIGURE 3–90

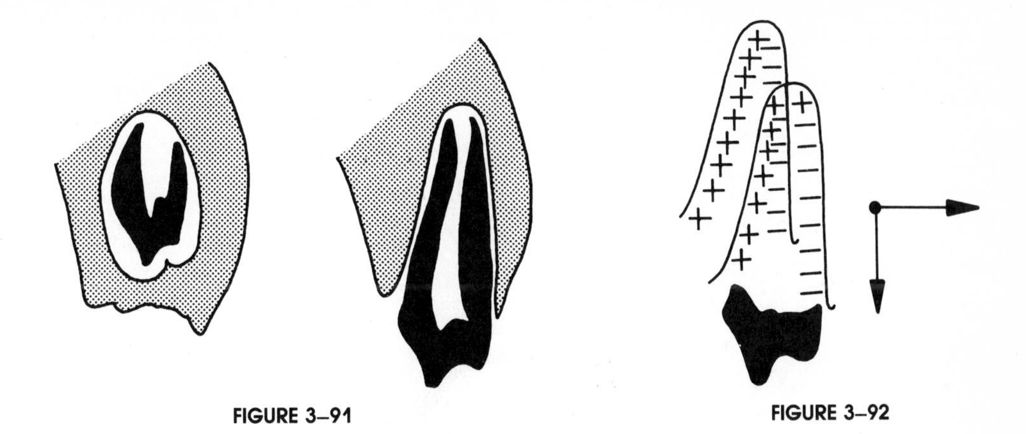

FIGURE 3–91

FIGURE 3–92

its socket; that is, the tooth does not drift vertically out of its alveolar housing as in eruption. Rather, the socket and its resident tooth drift together as a unit. The periodontal membrane also moves, but it does not merely "shift" along with its tooth. Rather, it **grows** from one location to the next, and it undergoes its own process of remodeling (see page 130). It is this important connective tissue membrane that (1) provides the intramembranous bone remodeling that changes the location of the alveolar socket and (2) moves the tooth itself. The horizontal and, especially, the vertical distances moved by the socket, its tooth, and the periodontal membrane can be substantial. By harnessing the vertical drift movement, the orthodontist can more readily guide teeth into new positions, thereby taking advantage of the growth process ("working with growth"). When the teeth are "full banded" by the orthodontist, the objective, even though the whole arch can be wired, is control of each individual tooth's movement. The specific target is each individual periodontal membrane, to bring about the remodeling ("relocation," Chapter 2) of each individual alveolar socket. This target is to be distinguished from others utilized to control the "displacement" movement of the entire maxilla, as described later.

Even though the external (labial) side of the whole anterior part of the maxillary arch (the protruding "muzzle") is resorptive, with bone being added onto the **inside** of the arch, the arch nonetheless increases in width, and the palate becomes wider (Fig. 3–93). This is another example of the V principle. In addition, growth along the midpalatal suture is known to participate to a greater or lesser extent in the progressive widening of the palate and alveolar arch (not shown in this schematic diagram). The extent can vary between the anterior and posterior regions.

As the palate grows inferiorly by the remodeling process, a nearly complete exchange of old for new tissue occurs. At each succeeding level, the palate becomes, literally, a different palate. It occupies a different location and is composed of different bone, connective tissue, epithelia, blood vessels, and so on. When one visualizes the palate of a young child, it should be realized that the palate in that same person at an older age is not the same palate at all.

The rotations, tipping, and inferior drift of the teeth, in combination with the characteristic external resorptive surface of the forward part of the maxilla, sometimes result in a localized protrusion of a tooth root tip through the bony cortex. Such penetration results in a defect (that is, a tiny hole in the bone) called a fenestra.

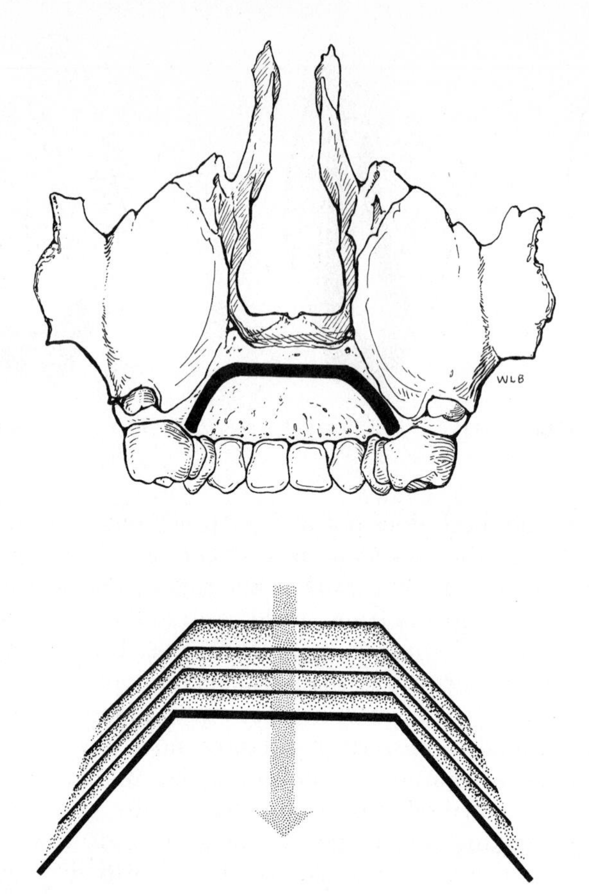

FIGURE 3–93. (From Enlow, D. H., and S. Bang: Growth and remodeling of the human maxilla. Am. J. Orthod., 51:446, 1965.)

Stage 13

This stage involves the **primary displacement** of the whole ethmomaxillary complex in an inferior direction (Fig. 3–94). Its displacement movement accompanies simultaneous enlargement (by the process of resorption and deposition) in all areas throughout the entire nasomaxillary region.

New bone is added at the frontomaxillary, zygotemporal, zygosphenoidal, zygomaxillary, ethmomaxillary, ethmofrontal, nasofrontal, frontolacrimal, pal-

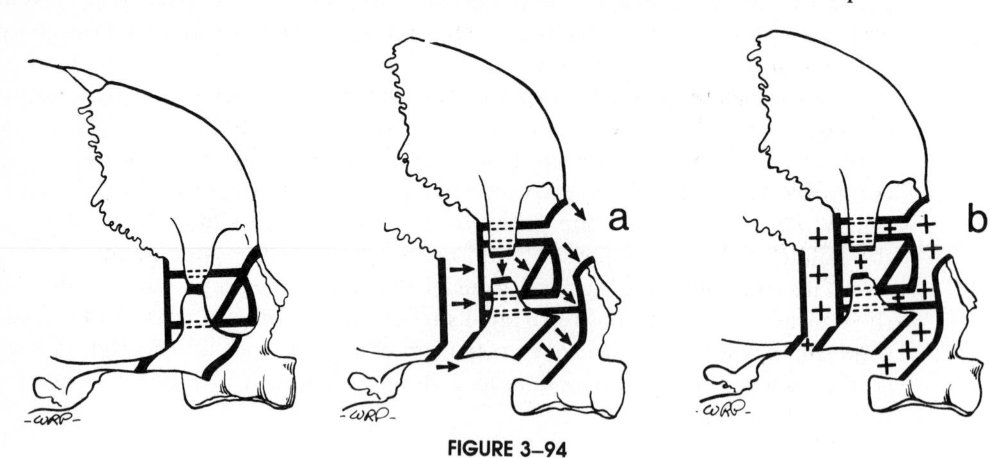

FIGURE 3–94

atine, and vomerine sutures. These sutural deposits, as explained in Stage 2, are believed to be the **response** to displacement and not the cause of it. That is, the whole region is carried inferiorly (and anteriorly), and this is presumed to result in a tension stimulus that triggers sutural osteogenesis. The bones thereby remain in constant sutural contact. The process of displacement produces the "space" within which the bone enlarges. Sutural bone growth does not **push** the nasomaxillary complex down and away from the cranial floor (although future research will probably show that the suture is more active in this process than presently suspected, but not by the traditional "push" idea). The displacement of the bones is a nonparticipating "carry" produced by the expanding soft tissue functional matrix (or, according to older theory, the nasal septum). As the bones of the ethmomaxillary region (Fig. 3–94) are displaced downward (*a*), sutural bone growth (*b*) takes place at the same time in response to it, thus enlarging the bones as the soft tissues grow. This places all the bones in new positions in conjunction with the generalized expansion of the soft tissue matrix and maintains continuous sutural contact as the bones become "separated."

It is believed that more vertical displacement (and thus more vertical sutural growth) takes place in the posterior part of the face than in the anterior portion. This produces a **displacement rotation** of the maxilla. More direct cortical remodeling growth (deposition and resorption), however, occurs in the anterior part. The latter feature also relates to the downward occlusal plane inclination that is often present (page 205). The balance between the greater or lesser amounts of displacement and remodeling growth in the posterior and anterior parts of the maxilla in general is apparently a response, at least in part, to the clockwise or counterclockwise rotatory displacement caused by the downward and forward growth of the middle cranial fossa. The nasomaxillary complex must correspondingly undergo a compensatory **remodeling rotation** in order to retain its proper vertical position relative to the vertical reference (PM) line and to the neutral orbital axis (see also descriptions for Figs. 3–22 and 5–41).

Most sutures in the facial complex do not simply grow in directions perpendicular to the plane of the sutural line. This was pointed out in a previous stage with respect to the lacrimal sutures. Because of the multidirectional mode of primary displacement and the differential extents of growth among the various bones, a slide or slippage of bones **along** the plane of the interface can be involved, as shown by the studies of Latham. As the maxillary complex is displaced downward and forward, or as it grows by deposition and resorption, it undergoes a slide at sutural junctions with the lacrimal, zygomatic, nasal, and ethmoidal bones. This is schematized by a slip of *b* over the sutural front of *a* as shown in Figure 3–95. The process requires relinkages of the collagenous fiber connections across the suture (see Chapter 18).

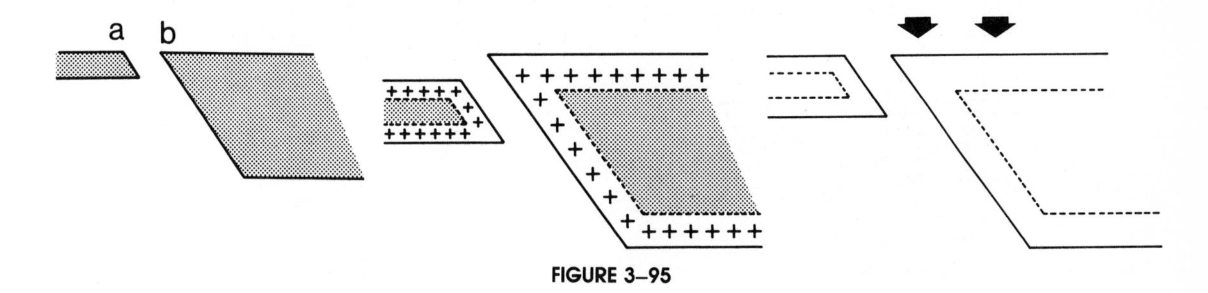

FIGURE 3–95

The present stage deals with the downward, primary displacement of the nasomaxillary complex. However, it is apparent that the downward **and** forward directions of movement occur at the same time and that they are produced by the same actual displacement process. Moreover, the sutures do not represent special "centers" of growth. A suture is just another regional site of growth adapted to its own localized, specialized circumstances, just as all the other parts of the bone have their own regional growth processes. This point is often misunderstood. It is not possible for a bone to grow **just** at its sutures; nor is it possible for a bone to have "generalized surface growth" without sutural involvement (in areas where sutures are present, of course; nonsuture regions may enlarge by direct remodeling). The old idea that "the suture growth system closes down at a given age, but the bone continues to enlarge simply by generalized **surface** deposition" is not valid. For example, bone additions on surface *x* in Figure 3–96 enlarge the surface area of the bone, but additions must **also** be made by deposits at sutural surface *y* in order to maintain morphologic form. It would not be possible for the bone to enlarge in surface area without corresponding additions at the sutural contacts.

As pointed out in Stage 12, the downward movement of the teeth from *1* to *2* (Fig. 3–97) is accomplished by a **vertical drift** of each tooth in its own alveolar socket as the socket itself drifts (remodels) inferiorly by deposition and resorption. The movement of the dentition from *2* to *3*, however, is a passive **carrying** of the maxillary dental arch as a whole, the palate and bony arch, and all of the alveolar sockets as the **entire** maxilla is **displaced** downward as a unit. The *1* to *2* and *2* to *3* movements are shown separately but, of course, actually proceed simultaneously. Recognition and understanding of the biologic difference between them are of basic importance to the clinician and, as always, to researchers and teachers.

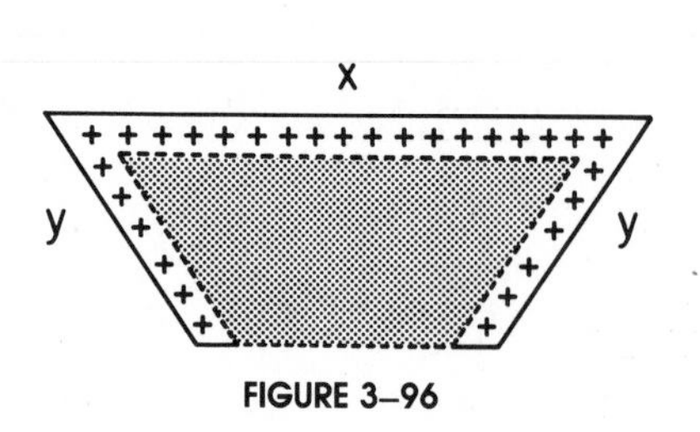

FIGURE 3–96

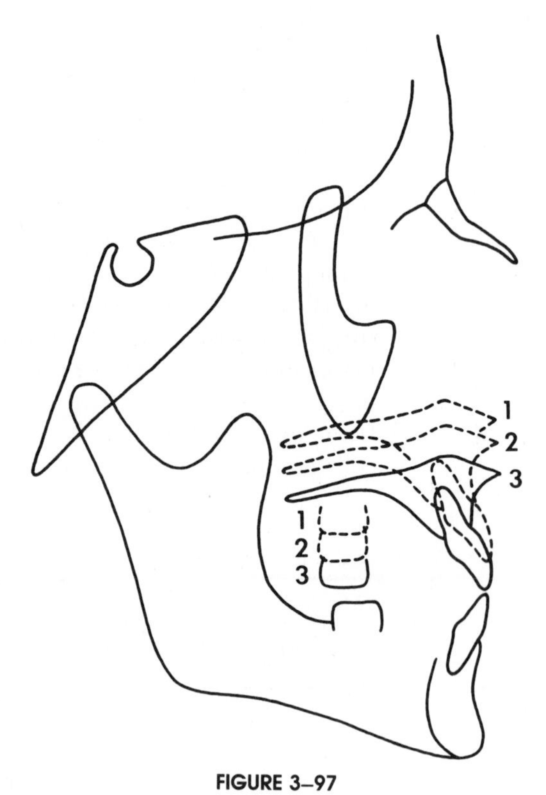

FIGURE 3–97

Some orthodontic procedures are designed to alter the vector of this displacement movement, e.g., to accelerate or restrain it or to change direction. The specific target is thus the growth activity of the various maxillary sutures and other regional growth sites associated with the displacement process. This is in contrast to orthodontic procedures, previously described, in which the periodontal membrane and drift movements of individual teeth (*1* to *2*) are the clinical target. In the mandible, similarly, the displacement movement is one target for treatment (as by a restraining chin cup), and horizontal and vertical drift movements of teeth are another. The former utilizes (it is hoped) regulation of condylar and ramus growth, and the latter involves control of growth movements related to the periodontal membrane. In both the maxilla and the mandible, both types of movements occur most actively during childhood growth, of course. Thus, utilization for clinical purposes can be less effective in adult patients if considerable movement is needed. Also, importantly, any significant extent of clinically induced movement after growth is complete could invite "rebound," because the state of functional equilibrium has been disturbed. Some clinicians hold that so long as growth continues, opportunity for achieving better stability is possible because the growth process itself can help lead to a state of physiologic balance. Others hold that the converse is true. This, however, is a complex, multifactorial biologic problem (see page 192).

As an exercise, see if you can evaluate a variety of different orthodontic and surgical procedures as to whether (1) the remodeling-drift and/or (2) the displacement process is directly involved.

Stage 14

The growth and remodeling changes of both the ramus and the middle cranial fossa (Stages 5, 8, and 10) produce a lowering of the mandibular arch. This accommodates the vertical expansion of the nasomaxillary complex. To bring the upper and lower teeth into full occlusion, the mandibular teeth must drift (not simply erupt) vertically (Fig. 3–98). The amount can vary considerably among different individuals having different facial types, and it can also vary between the anterior and posterior parts of the arch. The latter is involved in occlusal plane rotations (see page 217). Significantly, the amount of upward mandibular tooth drift is much less than the downward drift and displacement of the maxillary teeth. This is one of several reasons that orthodontic procedures, at least at present, often attack the maxillary dentition, even though a

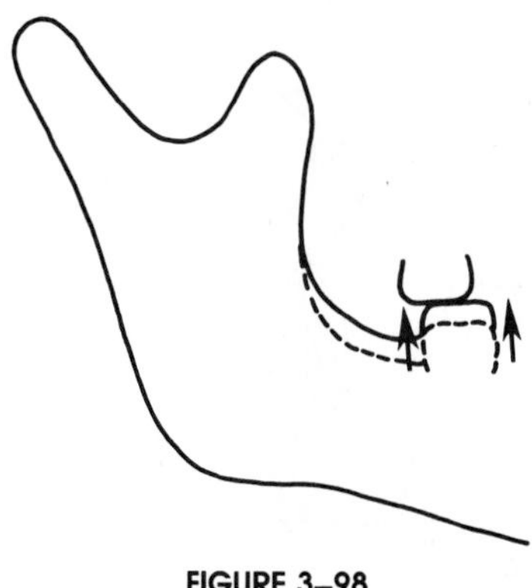

FIGURE 3–98

given malocclusion can be based on an improper positioning of the mandible. That is, an "imbalance" is clinically produced in the maxilla to offset the effect of an existing skeletal imbalance in the mandible (or cranial base), because it is the maxilla that can most readily be altered and controlled. (This paradoxic situation may change with future procedural developments when intrinsic control mechanisms of facial growth are better understood.) Because imbalances may still exist, however, long-term stability can be involved, and retention is a problem.

The composite of the vertical growth changes of the mandibular dentoalveolar arch, the ramus, and the middle cranial fossae must **match** the composite of vertical nasomaxillary growth changes to achieve continuing facial balance. Any differential will lead to a displacement type of mandibular rotation, either downward and backward or forward and upward.

Stage 15

During the descent of the maxillary arch (Stage 13) and the vertical drift of the mandibular teeth (Stage 14), the anterior mandibular teeth simultaneously drift **lingually** and superiorly (Fig. 3–99). This produces a greater or lesser amount of anterior **overbite.** The remodeling process that brings this about (Fig. 3–100) involves periosteal resorption on the labial side of the labial bony cortex (*a*), deposition on the alveolar surface of the labial cortex (*b*), resorption on the alveolar surface of the lingual cortex (*c*), and deposition on the lingual side of the lingual cortex (*d*).

At the same time, bone is progressively added onto the external surface of the basal bone area, including the mental protuberance (Fig. 3–101). The reversal between these two growth fields usually occurs at the point where the concave surface contour becomes convex. The result of this two-way growth process is a progressively enlarging mental protuberance. Man is one of only two species having a **"chin"** (the elephant is the other). Whatever its mechanical adaptations, the chin is a phylogenetic result of downward-backward facial rotation into a vertical position, decreased prognathism (as described in Chapter 5), the marked extent of vertical facial growth, and the development of an overbite (in comparison with an end-to-end type of occlusion).

The human mandible has been "caught" between the inferiorly and posteriorly rotated nasomaxillary complex and the upright body posture positioning of the basicranium and cervical region. Overjet and overbite or anterior

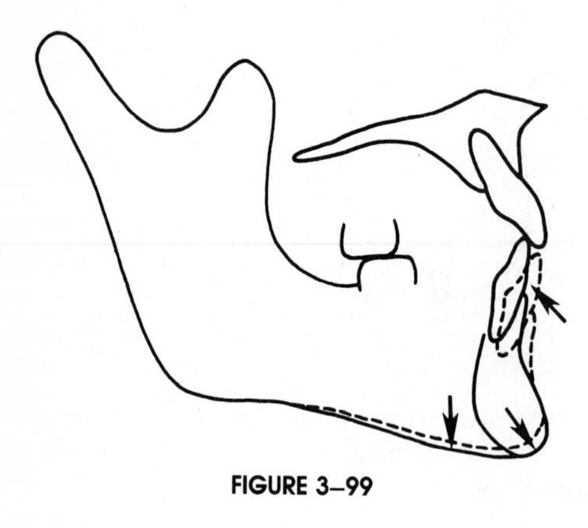

FIGURE 3–99

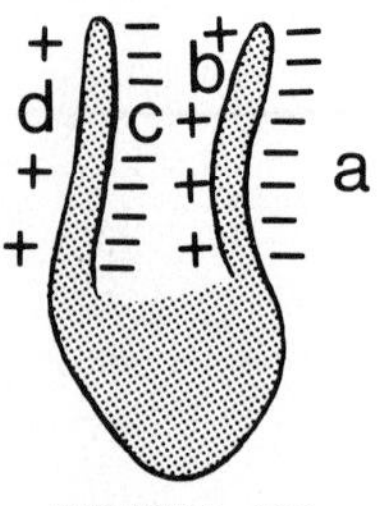

FIGURE 3–100

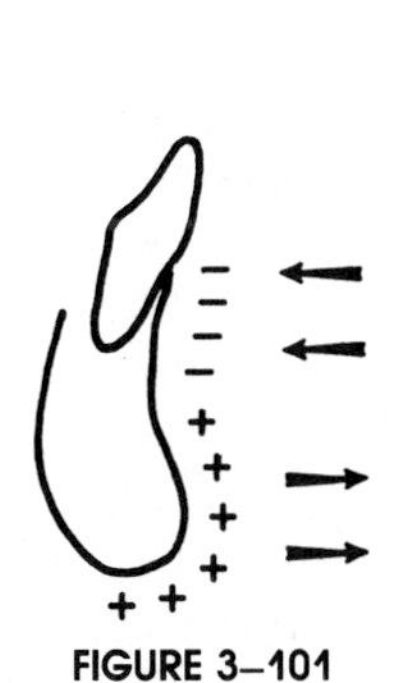

FIGURE 3–101

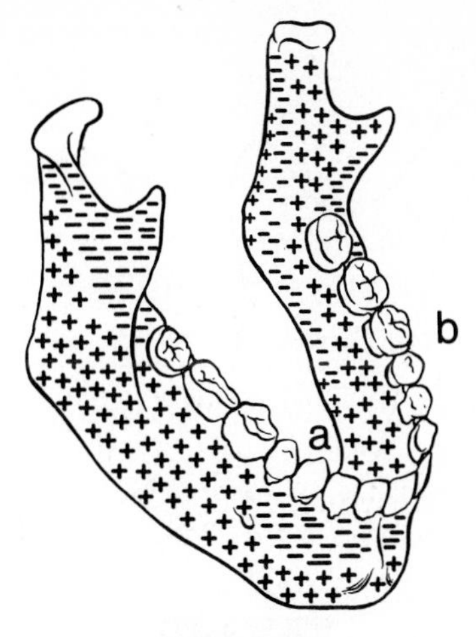

FIGURE 3–102

crowding are phylogenetic means to relieve the mandible and its temporomandibular joint placed in this vulnerable position.

There is considerable variation in the placement of the reversal line between the resorptive alveolar and the depository chin areas; it may be fairly high or low. Variations also occur in the relative amounts of resorption and deposition. There are, correspondingly, marked variations in the shape and the size of the chin among different individuals. It is one of the most variable areas in the entire mandible.

Except for a resorptive zone on the lingual side, the remainder of the perimeter of the mandibular corpus receives progressive deposits of bone (Fig. 3–102). This enlarges the breadth of each side of the corpus; side *b* grows to a slightly greater extent than side *a* because bony arch width increases slightly during postnatal mandibular growth, but not as much as the bony maxillary arch increases in width. The ventral border of the corpus is also depository; this is a slow growth process, however. The amount of upward alveolar growth exceeds the extent of downward enlargement by the "basal bone." (Note: Basal bone is a term sometimes used to denote that part of the corpus not involved in "alveolar" movements of the teeth. This area has a higher threshold of resistance to extrinsic forces than alveolar bone, which is extremely labile. There is no distinct structural line, however, separating basal from alveolar bone tissue. This is more of a physiologic than an anatomic difference.)

Whenever a change in the angle between the ramus and corpus develops, multiple sites of remodeling can be involved. The trajectory of condylar growth is usually a factor (Fig. 3–103), as shown by *a, b,* and *c.* Variable growth directions are produced by selective proliferation of prechondroblasts in some parts around the periphery of the condyle, with retardation of cell divisions in other parts.

If **backward** (but not upward) condylar growth has slowed or largely ceased (Fig. 3–104), combinations such as resorption at *d* and *e* with deposition at *f* and *g* can produce angular changes of the ramus relative to the corpus by direct remodeling. Such remodeling processes can either close or open the "gonial angle." (See Stage 9 and pages 209 and 215 to 217.)

Note: It is the **entire ramus** that is involved, not just "condylar growth." Also, any change in the ramus-corpus ("gonial") angle is largely produced by **ramus,** not corpus, remodeling, as seen next.

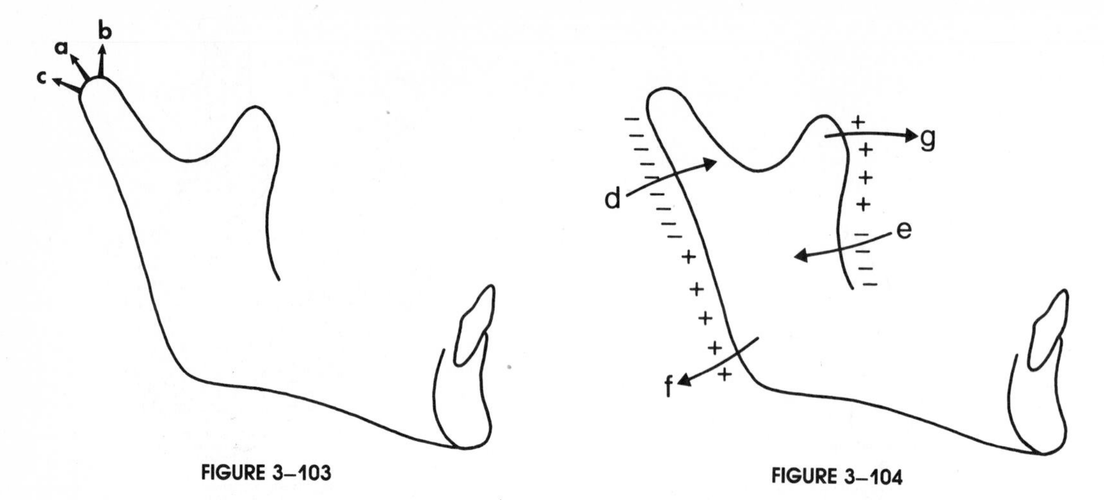

FIGURE 3–103

FIGURE 3–104

The ramus-corpus ("gonial") angle is determined by the growth direction of the ramus and condyle. Although a relatively small extent of downward corpus realignment can be produced by new bone deposition on its anteroinferior surface (Fig. 3–105), it is mostly remodeling combinations such as those shown in Figure 3–104 that are responsible for ramus and corpus alignment relative to each other. Direct upward growth of the corpus, involving resorption on its inferior surface, does not ordinarily occur. A marked superior extent of alveolar bone growth and the drifting of anterior mandibular teeth, however, are possible (see curve of Spee). The size of the antegonial notch is determined largely by the nature of the ramus-corpus angle and by the extent of bone deposition on the underside (inferior margin) of the corpus just posterior or anterior to the notch. The notch itself is also increased in size owing to its resorptive periosteal surface. A mandible characteristically has a less prominent antegonial notch (Fig. 3–106) (*b*) if the angle between the ramus and corpus

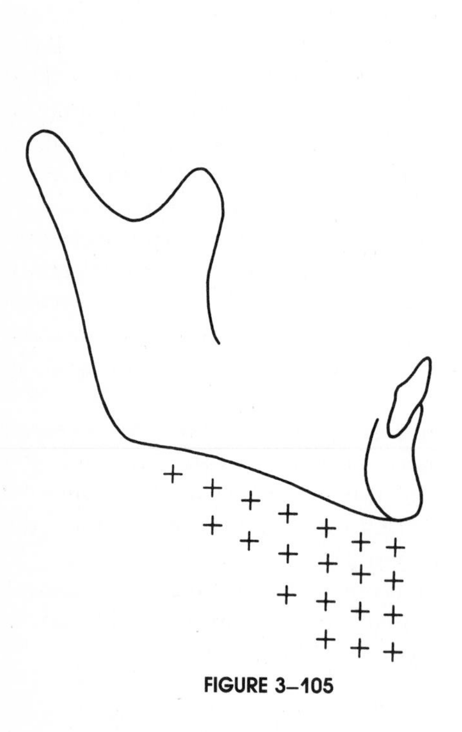

FIGURE 3–105

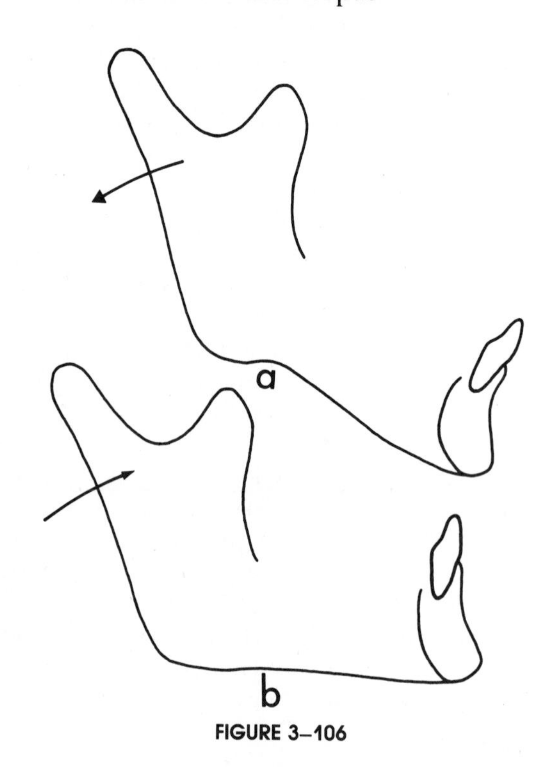

FIGURE 3–106

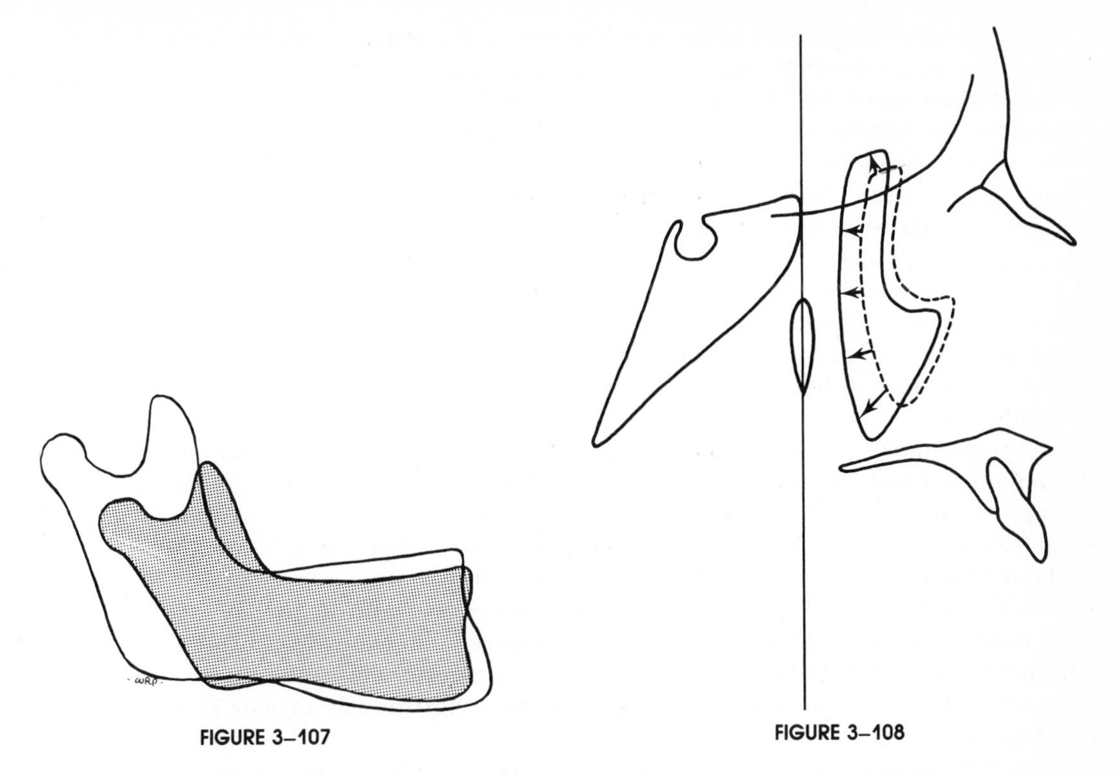

FIGURE 3–107

FIGURE 3–108

becomes closed, and a much more prominent antegonial notch (*a*) if it becomes opened. The antegonial notch itself is resorptive because it is relocated posteriorly, as the corpus lengthens, into the former gonial region of the ramus (Fig. 3–107).

Stage 16

The growth changes of the malar complex are similar to those of the maxilla itself. This is true for the remodeling process as well as the displacement process (Figs. 3–108 to 3–110).

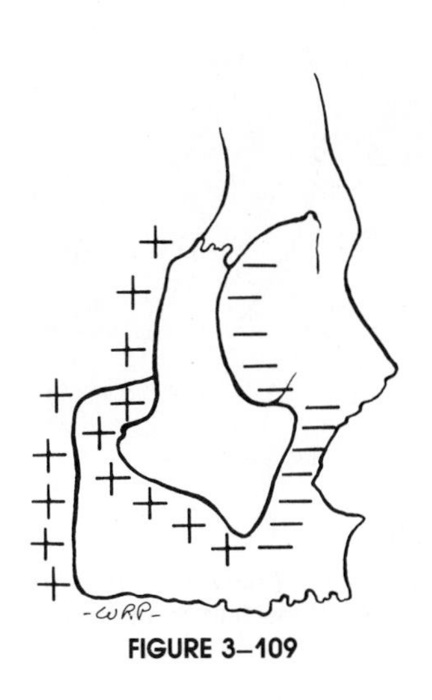

FIGURE 3–109

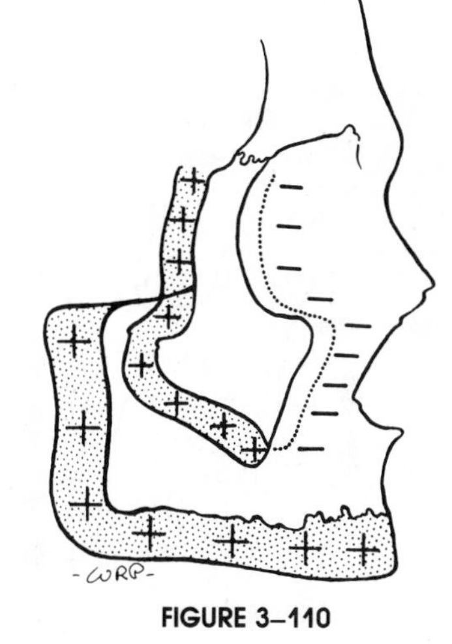

FIGURE 3–110

The posterior side of the malar protuberance is depository. Together with a resorptive anterior surface, the cheekbone relocates **posteriorly** as it enlarges. It would seem untenable that the whole front surface of the cheek area can actually be **resorptive.** However, as the maxillary arch grows posteriorly, the malar region must also move backward at the same time to keep a constant relationship with it. The extent of malar relocation is somewhat less in order to maintain **relative** position along the increasing length of the maxillary arch. The zygomatic process of the maxilla thus behaves in a manner similar to that of the coronoid process of the ramus. Both move posteriorly as the maxillary and mandibular arches also grow posteriorly.

The inferior edge of the zygoma is heavily depository. The anterior part of the zygomatic arch becomes greatly enlarged vertically as the face develops in depth.

The zygomatic arch moves **laterally** by resorption on the medial side within the temporal fossa and by deposition on the lateral side (Fig. 3–111). This enlarges the temporal fossa and keeps the cheekbone proportionately broad in relation to face and jaw size and masticatory musculature. The anterior trough of the temporal fossa moves posteriorly by the V principle.

As the **malar** region grows and becomes relocated posteriorly, the contiguous **nasal** region is enlarging in an opposite, anterior direction (Fig. 3–112). This draws out and greatly expands the contour between them, resulting in a progressively more protrusive-appearing nose and a horizontally deeper face (see Fig. 3–118).

The remodeling changes of the **orbit** are complex. This is because many separate bones are involved in the orbit, including the maxilla, ethmoid, lacrimal, frontal, and zygomatic and the greater and lesser wings of the sphenoid, and because many different rates and amounts of remodeling growth and displacement occur among these different bones and their parts.

The remodeling activities in the medial wall of the orbit, including the

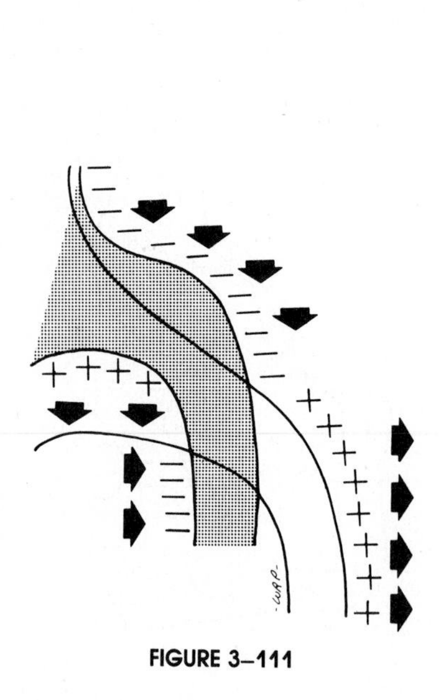

FIGURE 3–111

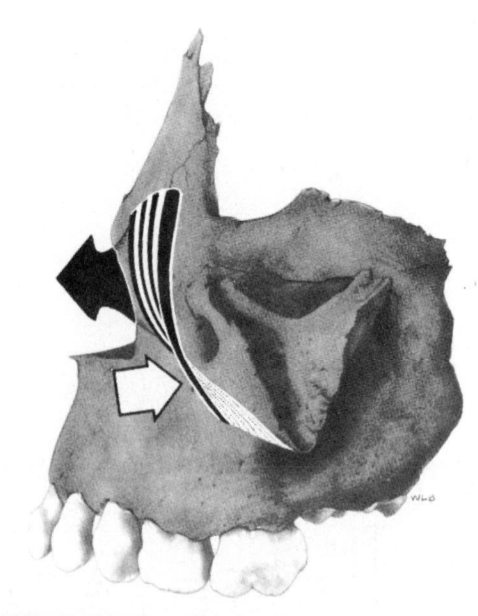

FIGURE 3–112. (From Enlow, D. H., and S. Bang: Growth and remodeling of the human maxilla. Am. J. Orthod., 51:446, 1965.)

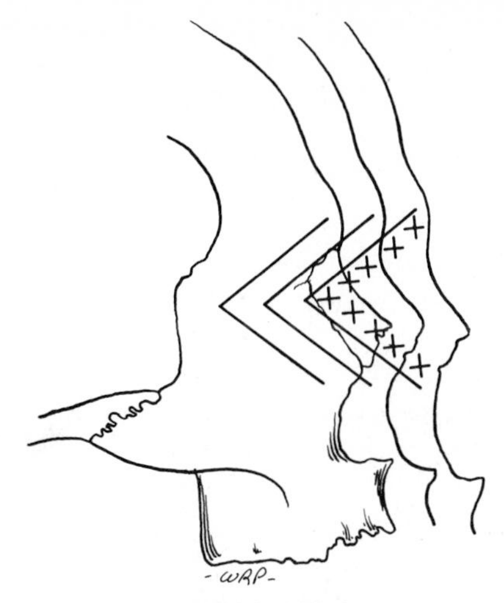

FIGURE 3–113

lacrimal and ethmoid bones, were described in Stage 12. In the remainder of the orbit, most of the roof and the floor are depository. The orbital roof is also part of the floor of the anterior cranial fossa. As the frontal lobe of the cerebrum expands forward and downward, the orbital roof also grows anteriorly and inferiorly by resorption on the cranial side and deposition on the orbital side. It would seem that a depository type of orbital roof and orbital floor would decrease the size of the cavity. However, two changes come into play that actually increase it, although the amount is relatively small in the older child. First, the orbit grows by the V principle (Fig. 3–113). The cone-shaped orbital cavity grows and **moves** in a direction toward its wide opening; deposits on the inside thus enlarge, rather than reduce, the volume. Second, the factor of **displacement** is directly involved. In association with sutural bone growth at the many sutures within and outside the orbit, the orbital floor is displaced in a progressive downward and forward direction along with the rest of the nasomaxillary complex.

Note that the floor of the nasal cavity in the adult is positioned **much lower** than the floor of the orbital cavity (Fig. 3–114). Compare this with the situation in the child. As described earlier, about half the process of palatal descent is produced by downward **displacement** (Fig. 3–115) of the whole maxilla (associated with maxillary sutural growth). The greater part of the orbital floor is a part of the maxillary bone. Thus, both the orbital and nasal floors are regional portions of the **same** bone, and the same displacement process that carries the palate downward **also** carries the floor of the orbit down at the same time (Fig. 3–116). The extent of this downward displacement, however, greatly exceeds the smaller amount required for orbital enlargement; that is, a lesser increase is needed for the eyeball and other orbital soft tissues than for the marked expansion carried out by the nasal chamber. The floor of the orbit offsets this by growing upward. Deposition takes place on the intraorbital side of the orbital floor and resorption on the maxillary sinus side. This maintains the orbital floor in proper position with respect to the eyeball above it. The **downward displacement** movement of the maxilla is thereby compensated for by **upward growth** to an amount that accommodates the relatively small enlargement of the orbital soft tissues. The nasal floor, in contrast, doubles the amount of displacement movement by **additional** downward cortical drift. Thus, the orbital and nasal floors are displaced in the same direction because they are parts of the same bone, but they undergo remodeling growth in opposite directions.

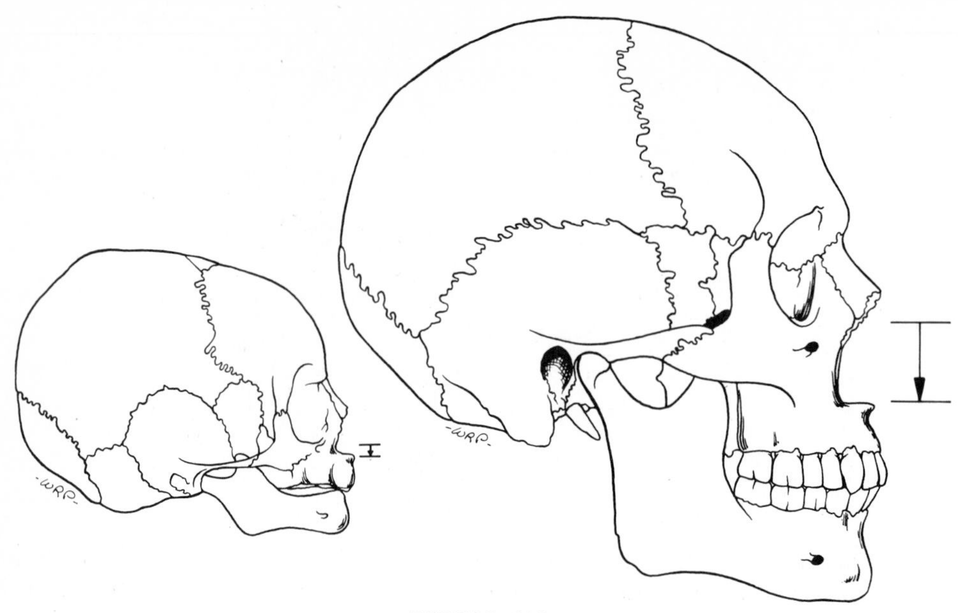

FIGURE 3–114

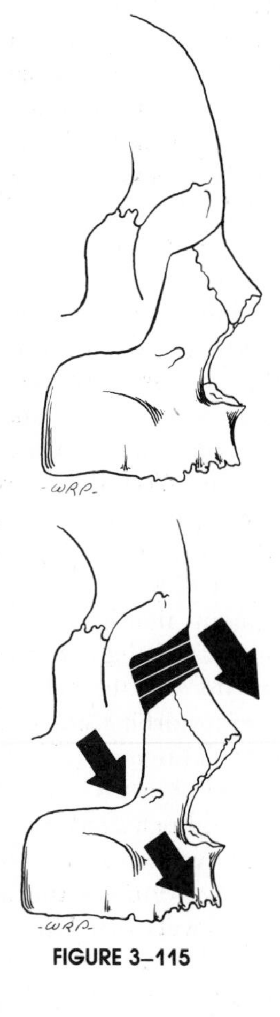

FIGURE 3–115

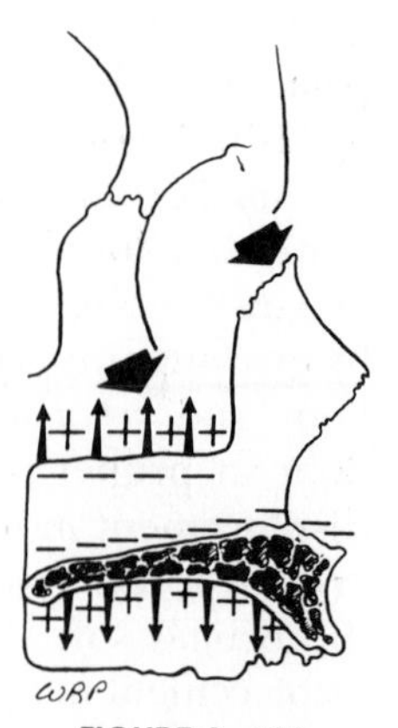

FIGURE 3–116

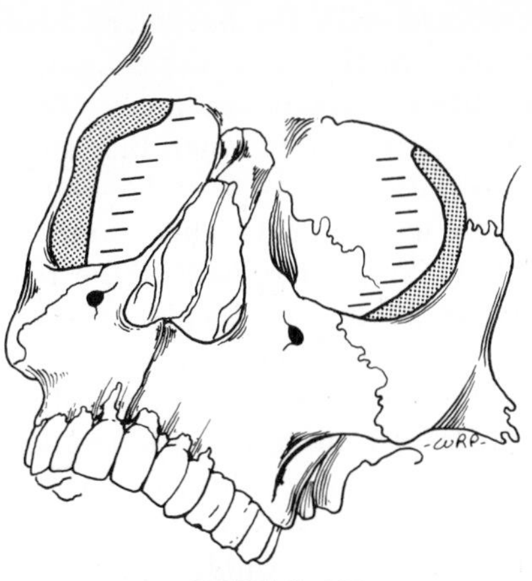

FIGURE 3–117

The floor of the orbit also grows laterally. It slopes in a lateral manner, and deposits on the floor thus move it in this same direction (as shown in Fig. 3–85). The lateral wall of the orbital **rim** grows by resorption on the medial side and by deposition on the lateral side. This intraorbital field of resorption continues directly onto the anterolateral surface of the orbital roof beneath the overhanging supraorbital ridge (Fig. 3–117). This is the only part of the orbital roof and lateral wall that is resorptive, and it provides for the lateral expansion of the domed roof. The cutaneous side of the supraorbital ridge is depository, and this combination causes the superior orbital rim to become protrusive. An upper orbital rim that extends forward beyond the lower rim is a characteristic of the adult face, particularly in the male because of the larger nose associated with larger lungs. **The two-way combination of (1) forward remodeling of the nasal region and superior orbital rim together with (2) backward remodeling growth of the inferior orbital rim and the malar area produces a growth rotation in the topographic alignment of the whole of these middle and upper facial regions** (Fig. 3–118; see Fig. 3–112).

The lateral orbital rim undergoes remodeling growth in a posterior and lateral direction at the same time. The lateral growth change increases the side-to-side dimension of each orbit and also contributes to the lateral movement of the whole orbit involved in the small amount of increase in the interorbital dimension. The backward growth change of the lateral orbital rim keeps it in proper location with respect to the backward direction of growth by the zygoma. **The forward growth of the superior orbital ridge and the whole anterior part**

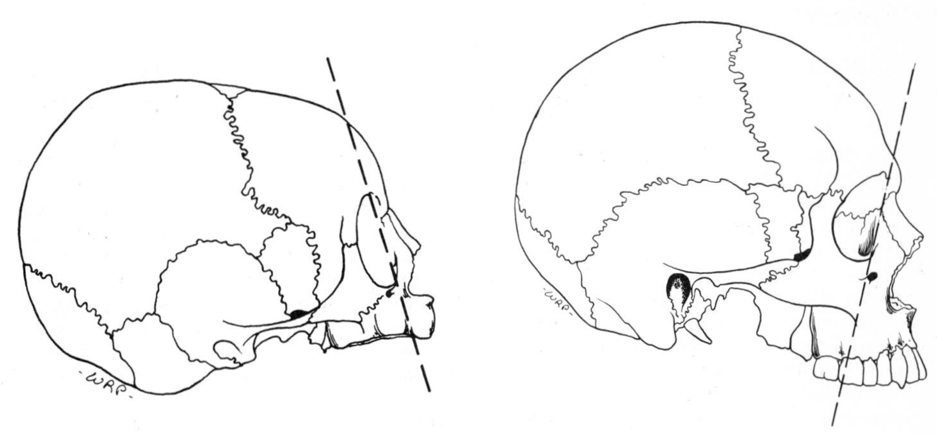

FIGURE 3–118

of the nasal region, combined with the backward growth of the lateral rim and cheekbone, causes the orbital rim in the human face to slant obliquely forward, in contrast to other mammalian faces. This reflects the forward rotation of the entire upper part of the human face and the backward rotation of the lower part (compare with Figs. 4–22 and 8–5).

Note that the resorptive nature of the cheekbone area, combined with the depository nature of the whole nasal region of the maxilla, **greatly expands the surface contour between them and deepens the topography of the face.** The medial rim of the orbit is only slightly in front of the lateral rim in the young child. In the adult, the medial rim has grown forward with the anterior-growing nasal wall, and the lateral rim has grown backward with the cheekbone. The medial and lateral rims are thus drawn apart in divergent posterior-anterior directions as the face deepens. Note the greatly increased depth of the lateral orbital rim and the midface as a whole resulting from these topographic changes.

Stage 17

The zygoma becomes **displaced** anteriorly and inferiorly in the same directions and amount as the primary displacement of the maxilla. The malar protuberance is a part of the maxillary bone and is carried with it. The separate zygomatic bone is displaced inferiorly in association with bone growth at the frontozygomatic suture and anteriorly in relation to growth at the zygotemporal suture. The force that causes it is the same as for the maxilla: the functional matrix or, according to older theory, the nasal septum. The growth changes of the malar process are similar to those of the mandibular coronoid process. Both grow backward, along with the backward elongation of each whole bone, by anterior resorption and posterior deposition. Both become displaced anteriorly along with each whole bone.

Note this important feature of facial growth. In many of the growth and remodeling processes described throughout this chapter, one major difference exists between the female and the male. In the female, skeletal growth changes in the face slow markedly shortly after puberty. In the male, however, topographic and dimensional changes continue through the late adolescent period. The facial similarities that exist between the sexes during childhood, therefore, are substantially altered in the teenage years.

PHYSIOLOGIC TOOTH MOVEMENTS AND ALVEOLAR REMODELING

The **periodontal membrane*** is comparable to both the sutural and periosteal membranes. It is a membrane that, phylogenetically, is the adaptive answer

---

*Also commonly called the **periodontal ligament.** It is indeed a mature ligament in terms of its histologic structure in the more stable, adult form. However, the term "membrane" is much more appropriate for the childhood growth period. The periodontium has a connective tissue membrane that is quite active and dynamic, not one that merely physically supports a tooth (that is, a ligament). It (1) contributes to the growth and development of the tooth; (2) is involved directly in the eruption of the tooth; (3) is involved directly in the drifting, tipping, and rotation movements of the tooth; (4) provides for the formation of the bone tissue lining the alveolar socket; (5) is an active and essential sensory receptor and vascular pathway; and (6) is involved directly in the extensive remodeling of the bone associated with the movements of the teeth. For these reasons, the term periodontal "membrane" is more closely associated with the truly dynamic functions of this connective tissue layer. "Ligament," on the other hand, connotes a more stable, inactive, nonchanging type of tissue that has a single function—fibrous attachment. Alveolar bone, of course, is of intra**membranous** origin, being produced by the periodontal **membrane.**

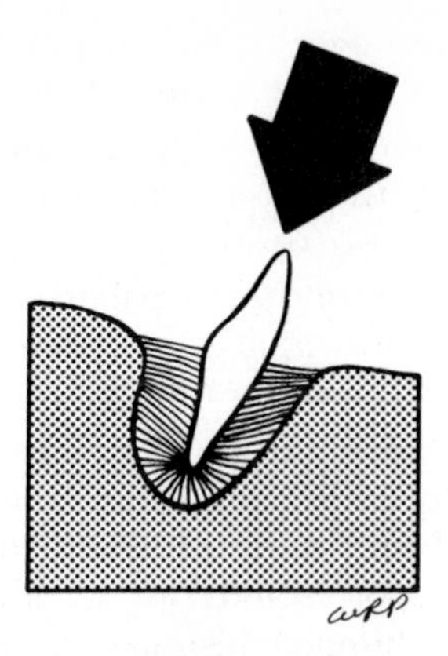

FIGURE 3–119

to a basic functional problem. If the teeth were attached to the jaw in a manner similar to that of a denture riding on the surface of the bone, **pressure** would result on the periosteum during chewing. The bone's membrane is known to be quite pressure-sensitive, however, and resorption would result because of masticatory compression by the teeth. Would cartilage, a pressure-resistant tissue, function satisfactorily as a buffer tissue between the tooth root and the bony surface? No, because cartilage is severely limited in its capacity for remodeling and could not accommodate the dynamic changes required for tooth development, eruption, and drift.

The phylogenetic problem of compression on the bone surface beneath a tooth has been solved in a simple but effective way (Fig. 3–119). Pressure is converted directly into **tension** (which the membranes can handle) merely by the suspension of each tooth in a connective tissue **sling** within a socket.* By this means, the inward direction of compression by a tooth being pushed into its socket is translated, not as pressure, but as direct tension on the alveolar bone. Thus, the periodontal membrane is not exposed to the killing effects of compression as the tooth is depressed into the socket or as it is tipped or rotated in one direction or another by masticatory forces. This relatively simple plan accomplishes several needed functions. It provides effective support for the tooth, gives resilient yet nonbrittle stability, provides a system for eruption, enables each individual tooth to acquire a functional occlusal position, provides for the growth and maintenance of the alveolar bone, and provides for the vertical and horizontal drifting of the tooth and the corresponding remodeling of the alveolar bone.

The teeth drift for two basic, functional reasons. One, as described in all basic oral histology texts, is to close up the dental arch during growth and keep it closed as the contact edges along the interproximal surfaces of the teeth progressively wear. This braces the arch to better withstand masticatory forces. **The second reason, much less known but of great importance, is to anatomically locate the teeth as the whole mandible and maxilla grow and remodel.** Each tooth (and the unerupted tooth buds as well) must drift vertically, laterally, and either mesially or distally in order to retain proper anatomic position. The "molar" region at an early age level, for example, becomes the "pre-molar"

*A violation of this anatomic relationship is the basis for many of the problems encountered by the prosthodontist. Dentures are **pressure**-causing appliances fitted onto bone without a tension-converting sling of periodontal fibers. Uncontrolled resorption is a common consequence.

region of the jaw at a later age as the corpus lengthens posteriorly. The maxillary teeth, for a second example, must drift inferiorly for a considerable distance as the whole bony maxillary arch relocates downward to provide for the vertical enlargement of the nasal chambers. Anterior versus posterior differences in the extents of this vertical dental drift function in conjunction with palatal remodeling to achieve proper palate and occlusal alignment, which varies considerably among individuals (see pages 215 and 219). Thus, the function of **drift** is far more significant than merely "closing up" the dentition. It is one of the basic processes involved in facial growth.

The customary diagram used to illustrate "mesial drift" shows deposition and resorption on the "tension" and "pressure" sides of the alveolar socket, respectively (Fig. 3–120). A posteroanterior section through the jaw showing several tooth roots gives the familiar histologic picture shown here. However, only the **mesial** direction of drift movement is pointed out in the standard textbooks; the important vertical drift movements and the rotations and tipping that **also** take place are not always explained. These other movements are carried out by the **same** alveolar bone deposits and resorption usually associated only with mesial drift. Drift is a three-dimensional growth process, and the oversimplified, two-dimensional, diagrammatic picture illustrated here does not adequately represent it (see page 116).

The "pressure and tension" concept of the mesial and distal sides of the alveolar socket is also an oversimplification. The collagenous fibers of the periodontium on the pressure side are often actually under tension (see Fig. 3–126*A*). Although heavy tooth-to-periodontal membrane-to-alveolar bone surface compression can indeed be involved in extreme tooth movements (as by an orthodontist using heavy forces), such severe levels of force are not usually involved in ordinary physiologic circumstances. On the "tension" side, furthermore, the cells of the periodontal membrane are actually under compression between the taut collagenous fibers.

It has been a major point of controversy for many years whether the pressure that is presumed to trigger alveolar bone resorption acts first on the membrane or directly on the bone, which in turn causes the membrane to respond (see below). One long-standing concept is that very minute **distortions** of alveolar bone by shifting of the tooth's root are required for triggering alveolar remodeling. The piezo effect is now held by some investigators to be the response to this stress trigger, and it is believed that this bioelectric stimulus serves as a "first messenger" that fires the receptor sites on osteoblastic and/or osteoclastic cell membranes within the periodontium. The source of the pressure that causes the active distortion (bending) of the alveolar plate is still not clear.

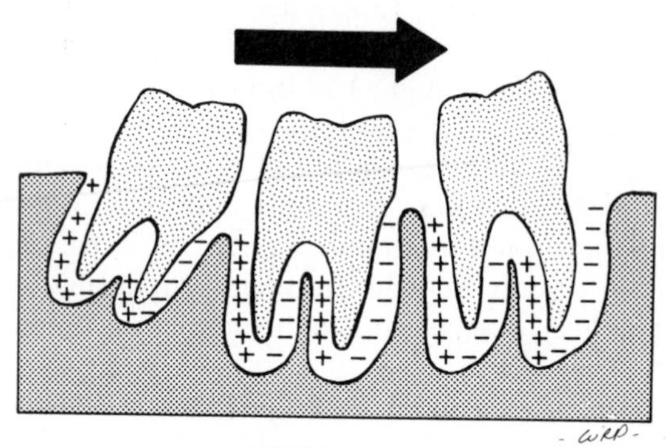

FIGURE 3–120

Some investigators suggest that the viscous fluid matrix functions as the biomechanical intermediary. It presumably acts as a "hydraulic system" that can transmit variable amounts of pressure to the alveolar bone surface by vessel or matrix distention or compression.

If the extent of pressure exerted by a **tooth** on the periodontal membrane, caused, for instance, by heavy orthodontic forces acting on the tooth, causes a **severe** compression of the membrane, a closing off of the blood vessels and cellular necrosis result, and the growth capacity of the membrane is destroyed. Remodeling changes on the periodontal alveolar surface are precluded. This is presumed to trigger **undermining resorption.** In this process, the resorptive changes then proceed from the endosteal cancellous spaces **deep** to the alveolar surface, since the alveolar surface itself is closed off owing to vascular occlusion.

The periodontal membrane is the equivalent of both the periosteum and sutures, its general structure is similar, and its mode of growth is comparable. The notable difference, of course, is that one side attaches to a tooth, rather than to a muscle or another bone. The periodontal membrane is a reflection of the periosteum into the alveolar socket, and these two membranes are directly continuous.

In its "stable," nonremodeling form, the periodontal membrane is essentially a mature ligament composed of dense bundles of thick collagenous fibers with correspondingly few fibroblasts and little ground substance. During the active period of facial growth, dental development, and the establishment of occlusion, however, this membrane has a much more dynamic function, and its histologic structure is adapted to the complex role it plays. During the growth period, the periodontal membrane is much more highly cellular, and more than just ligament fibers are present. Like the actively growing periosteum and sutures, the periodontal membrane has three basic layers. The middle layer, called the "intermediate plexus," is composed of the same slender, precollagenous **linkage fibrils** that are present in the intermediate layers of the periosteum and sutures. Linkage fibrils provide connections and sequential reconnections between the innermost and outermost dense fibrous layers. Their key function is for **adjustments** involved in tooth drift, eruption, rotations, and alveolar bone remodeling. This layer may be poorly differentiated or absent in nonremodeling periods and in regional locations (and species) where tooth movements are relatively slow. Or the distribution of the linkage fibrils may be more diffuse, rather than forming a single, recognizable zone. During active tooth movements, nonetheless, they are necessarily **always** present. During tooth movement and alveolar remodeling relocations, the PDM is not simply "moved," *in toto,* to progressive new positions. Rather, it undergoes its own **remodeling,** just as the bone does, to provide the movement.

It has recently been proposed that the actual source of the propulsive mechanical force that brings about eruption, vertical and horizontal drift, and other tooth movements is provided specifically by an abundant population of actively contractile fibroblasts ("myofibroblasts") on the **resorptive** sides of the socket. The contraction of these special cells (*m* in Figure 3–121) presumably **pulls** the collagenous framework within the periodontal membrane, and thereby the tooth, in the direction of the resorptive bone front. (Or, at least, the contractile cells can pull fibers into new linkage positions as some other unknown force actually moves the tooth.) Simultaneously, special collagen-degrading and collagen-producing cells (*x* and *y*) within the linkage zone provide the fiber remodeling and relinkages described below. This occurs in conjunction with ground substance degradation and synthesis, and the tooth is thus propelled

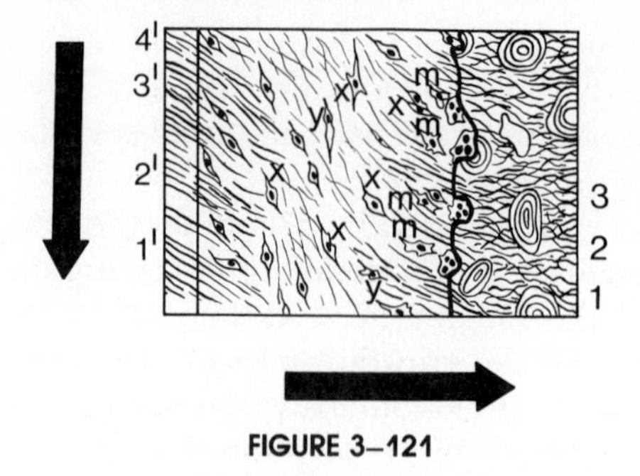

FIGURE 3–121

in horizontal and vertical drift movements (arrows). The same process is also believed to provide for eruption. The fibers at level *1*, formerly linked with *1'*, thus become relinked with fiber level *2'*, and so on. It is suggested, importantly, that these various cells are the specific targets of the clinical forces utilized by the orthodontist to move teeth (see Azuma et al., 1975).

As seen in Figures 3–122 to 3–124, on either side of the zone of linkage fibrils (*b*) are a layer of coarse collagenous fibers that attach to the alveolar bone (*a*) and a layer of coarse fibers attaching to the cementum of the tooth (*c*).

The activity on the tension side is schematized here ("tension" because the pull of the tooth to the right presumably sets up tension on the bone surface by the periodontal fibers). A new layer of bone is deposited on the alveolar surface. This embeds the periodontal fibers of layer *a*. Note that the attachment fibers are not driven into the bone as with a nail; they are progressively enclosed as new bone deposits form around them. It is apparent that the fibers of zone *a* would soon be used up and become completely enclosed. However, the **linkage** fibrils of the intermediate zone *b* become converted into *a*, thereby lengthening *a* in advance of the drifting alveolar wall. The fibers of layer *a* are thus enclosed by new bone on one side while being lengthened by an equal amount on the other. The conversion from *b* to *a* is accomplished by a bundling together of the thin, precollagenous linkage fibrils into the thick, "mature" fibers of layer *a*. Ground substance is believed to be the binding agent, and the process is carried out by the abundant resident population of periodontal fibroblasts. Layer *b* retains its breadth by elongation of the precollagenous linkage fibrils. It is not presently known whether this lengthening process occurs within zone *b* or at the interface between *b* and *c*. New unit fibrils are also constantly added as the tooth grows and as the membrane drifts in conjunction with tooth drift. The fibers of layer *c* are carried in the direction of the tooth's movement. Throughout this membrane remodeling process, continuous attach-

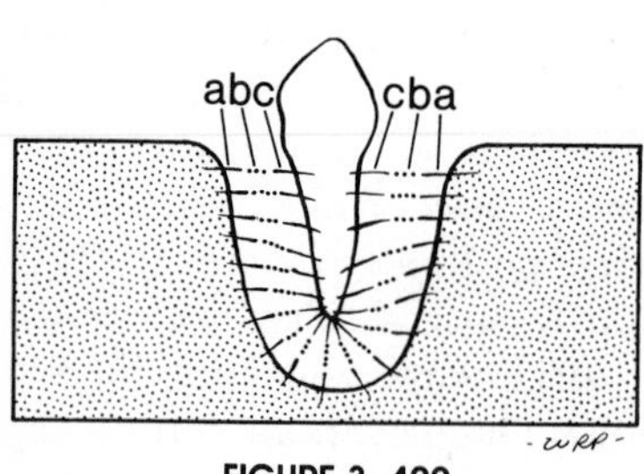

FIGURE 3–122

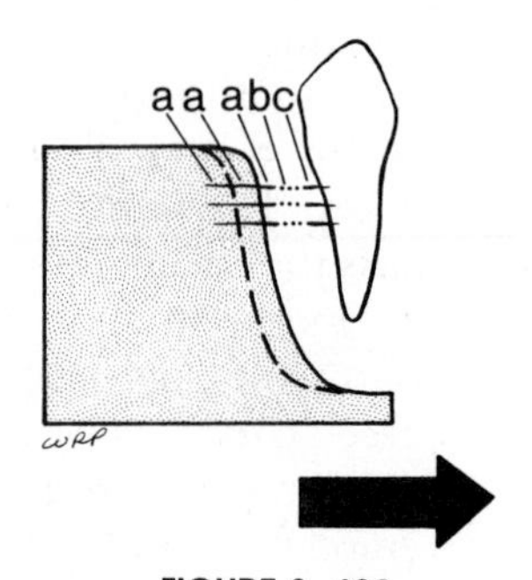

FIGURE 3–123

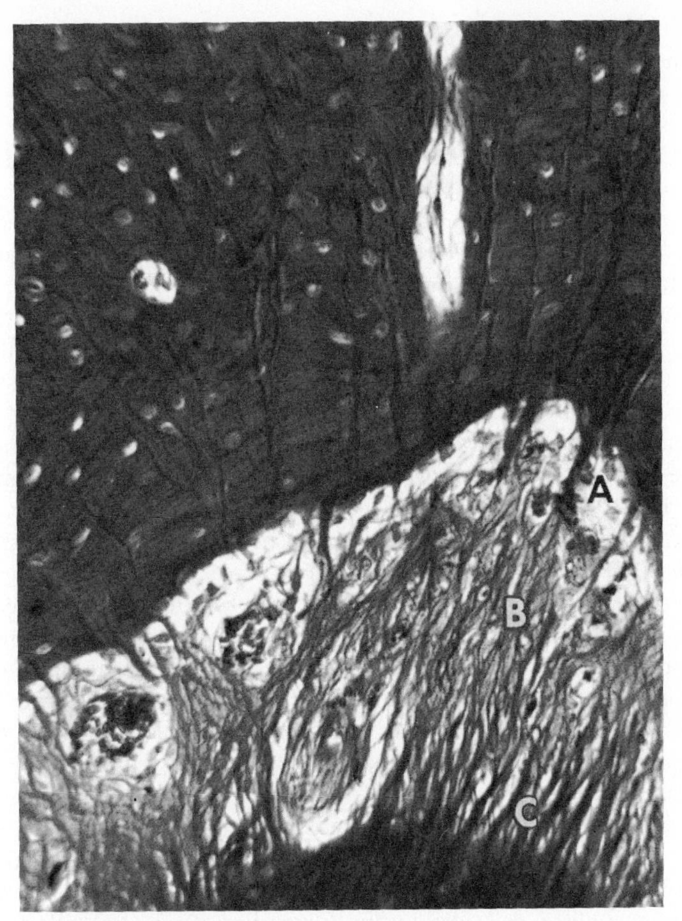

FIGURE 3–124. (From Kraw, A. G., and D. H. Enlow: Continuous attachment of the periodontal membrane. Am. J. Anat., 120:133, 1967.)

ment between tooth and alveolar bone is thereby maintained. Note that the periodontal membrane as a whole is not simply pushed or pulled along as the tooth moves. It **grows** from one location to the next.

As shown in Figure 3–125, the activity on the pressure side of the tooth root ("pressure" because the tooth root, according to long-standing but inadequate theory, exerts direct compression on the periodontal membrane and bony alveolar wall) is the reverse of the remodeling sequence on the opposite ("tension") side. A layer of bone is **resorbed** (*x*′) from the alveolar surface by an abundant sheet of osteoclasts. The resorptive side of the alveolar socket can

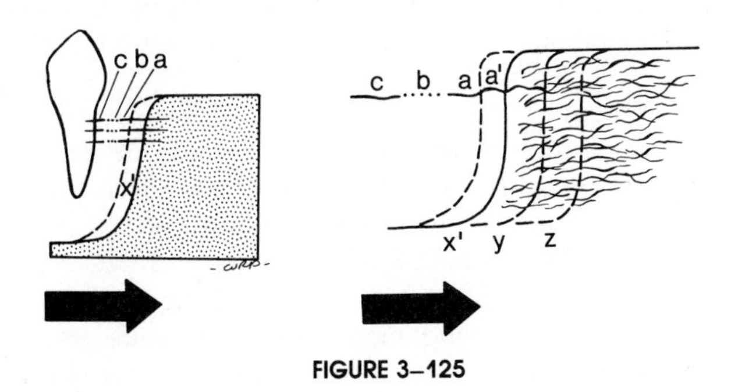

FIGURE 3–125

easily be distinguished microscopically from the depository side by the characteristic chiseled, pitted, eroded appearance of the bone's surface. If the resorptive process was active at the time, numerous osteoclasts are seen in the erosion pits.

Two means of periodontal attachment onto resorptive alveolar bone surfaces are operative. One involves scattered areas in which some bone matrix fibers are converted into periodontal fibers as resorption of bone proceeds. The other, which is the more widespread, involves an adhesive mode of new fiber attachment onto the resorptive bone surface.

For the former mode, as the alveolar resorptive front (Fig. 3–125) proceeds from *x′* to *y* and on to *z*, the linkage fibrils (*b*), the attachment fibers on the bone side (*a* and *a′*), and those on the tooth side (*c*) **remodel** to sustain their proportionate lengths, with new fibers and relinkages sustaining continuous attachments as the tooth moves.The whole membrane thus grows in a direction toward the bone surface and away from the tooth as the tooth simultaneously moves in a like direction. The process is continuously repetitive. The tooth movement itself is presumably provided by myofibroblasts, as described above and schematized in Figure 3–121.

Even though the bone on the resorptive side of the alveolar socket undergoes progressive removal, periodontal connection between bone and tooth is still sustained by the release and conversion of a few bone matrix fibers into a scattering of fibers that become utilized as part of the drifting periodontal membrane. However, this process is supplemented by another, more widespread mechanism that involves waves or periods of transitory reattachment. In many areas on the resorptive alveolar surface, the resorption process is complete, and all fiber anchorage is totally severed in such regions. Periodontal membrane reattachment can be rapidly achieved, however, by deposition of a layer of adhesive ground substance (a component of proteoglycans) on the resorbed bone surface, followed by the formation of new precollagenous fibrils. This can be done almost immediately after the resorptive action of the osteoclasts. Indeed, the fibroblast-like cells involved trail just behind the moving osteoclasts and can re-establish attachment as the osteoclast moves from its Howship lacuna. The new fibrils, embedded in the adhesive proteoglycans secretion on the bone surface, become "stuck" onto the bone. They link with older collagenous fibers deeper within the periodontal membrane, and the transitory attachment between bone and membrane is thereby produced. As the resorptive front continues, such adhesive attachments undergo removal in turn, to be replaced by new ones. Should a reversal occur in which the bone surface becomes depository, rather than resorptive, calcification of the interface proteoglycans layer becomes a "reversal line." Such reversals may be more or less permanent, with substantial new deposits formed. Or they may occur, as frequently seen, as temporary, thin scales of bone ("spot" deposits) that serve to reinforce transient attachments (Fig. 3–126B). In either case, the reversal shows clearly as a refractile line at the interface.

Wherever attachments are made on resorptive alveolar bone surfaces, the fibers of the periodontal membrane are taut between bone and tooth (see Fig. 3–126A). Thus, even though this side of the socket is often referred to as the "pressure side," the fibers are actually under tension.

The periosteal and periodontal membranes are constructed to function in a field of tension (as by the pull of a muscle or biting force on a tooth), not marked surface pressure. Covering membranes are quite sensitive to direct compression because any undue amount causes vascular occlusion and inter-

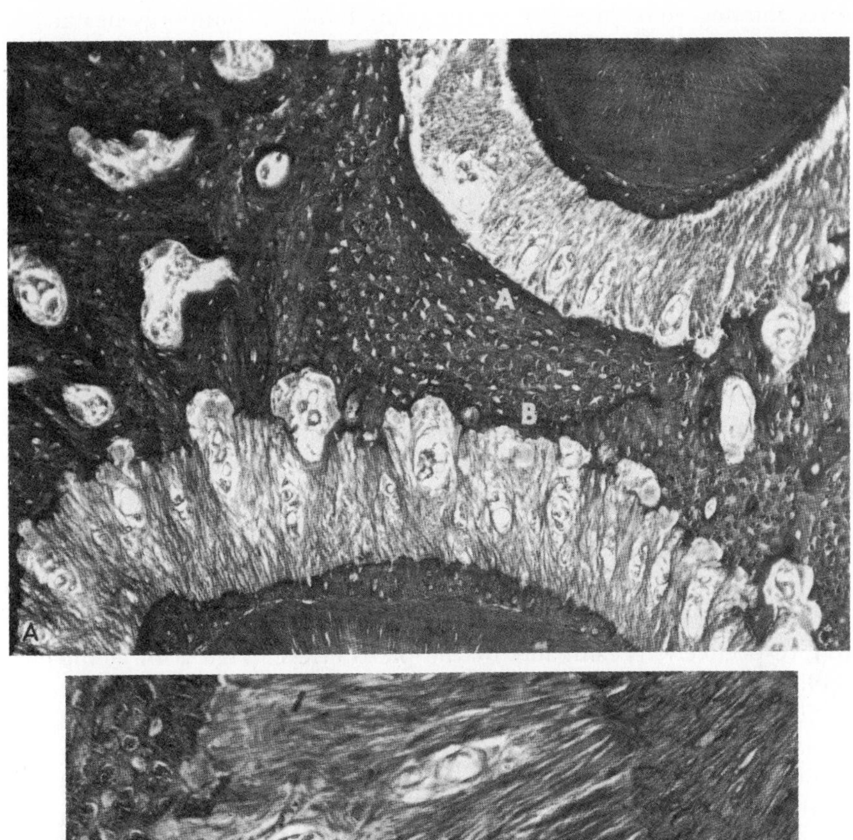

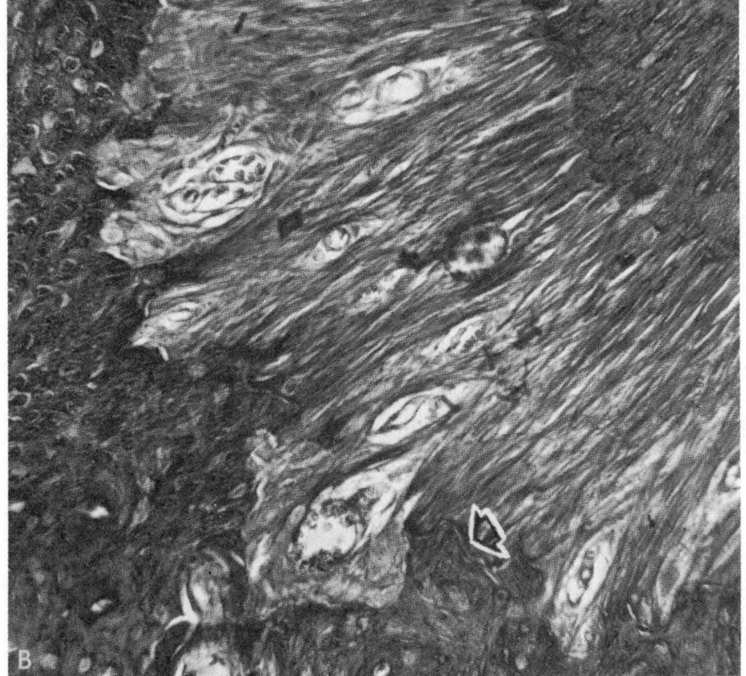

FIGURE 3–126. In the upper photomicrograph, alveolar surface *A* is depository and alveolar surface *B* is resorptive. The entire bony plate was produced by the periodontal membrane of the upper socket. Note that the fibers on the resorptive side (*B*) are taut and under tension, even though this has frequently been referred to as the "pressure" side. In the lower photomicrograph, a "spot deposit" of bone (arrow) has maintained fibrous attachment with the tooth. Note the *reversal line* separating this thin, transient scale of new bone from the resorptive surface just deep to it. (From Kraw, A. G., and D. H. Enlow: Continuous attachment of the periodontal membrane. Am. J. Anat., 120:133, 1967.)

ference with osteoblastic formation of new bone. Osteoclasts, conversely, function to "relieve" the degree of pressure by removing bone. A commonly heard cliché is that "bone" is "pressure-sensitive," and that high-level pressure induces resorption. Actually, it is the covering membrane and not the hard part of the bone itself that responds in such a manner. However, there are two general targets for biomechanical forces acting on bone: the bone's membrane and the bone's calcified matrix. The nature of response is different for each. If surface pressure is exerted on the membrane, the resultant compressive effect is osteoclastic and the response is resorption in the specific, localized area so involved. Tension acting on the membrane is generally osteoblastic, and the response is new bone deposition. These responsive actions presumably continue until physiologic and biomechanical equilibrium is attained, whereupon the blastic and clastic activities are closed down.

In contrast to biomechanical forces acting directly on covering membranes, stresses on a bone's intercellular matrix are believed to have a different mode of physiologic action, as shown schematically in Figure 3–127. The action of a muscle or tooth or the bearing of weight causes minute distortions in a bone (arrows). This leads to regional changes in configuration involving localized surface convexities and concavities. A concavity results in compression and a negative surface charge (*B*), and a convexity causes tension in the bone matrix and a positive surface charge. This is believed to trigger bone deposition and resorption (*C*), respectively, presumably by the piezo effect (page 236) acting on surface cell receptors of osteoblasts and osteoclasts. The bone thereby remodels until biomechanical and bioelectric neutrality is attained (*D*). You will note that the nature of this response is opposite to that seen for forces acting on a bone's covering soft tissue. Pressure on the periosteum or periodontal membrane leads to resorption, and tension can trigger deposition. Pressure in

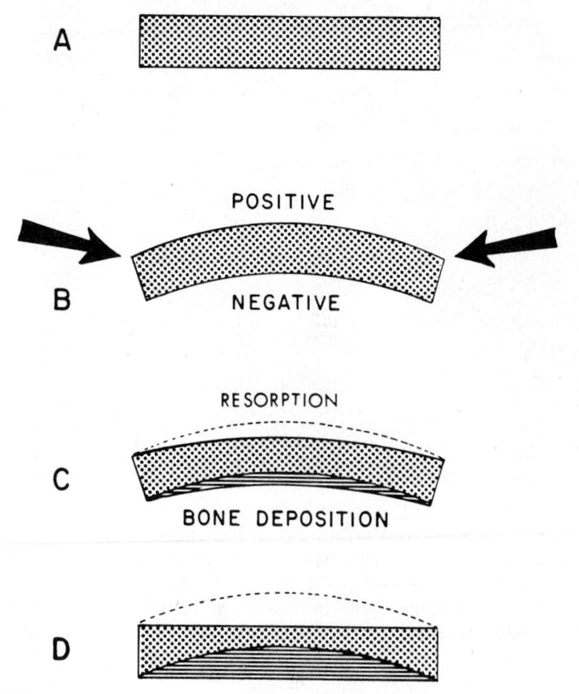

FIGURE 3–127. Piezo response to forces acting on the bone matrix. See text for description.

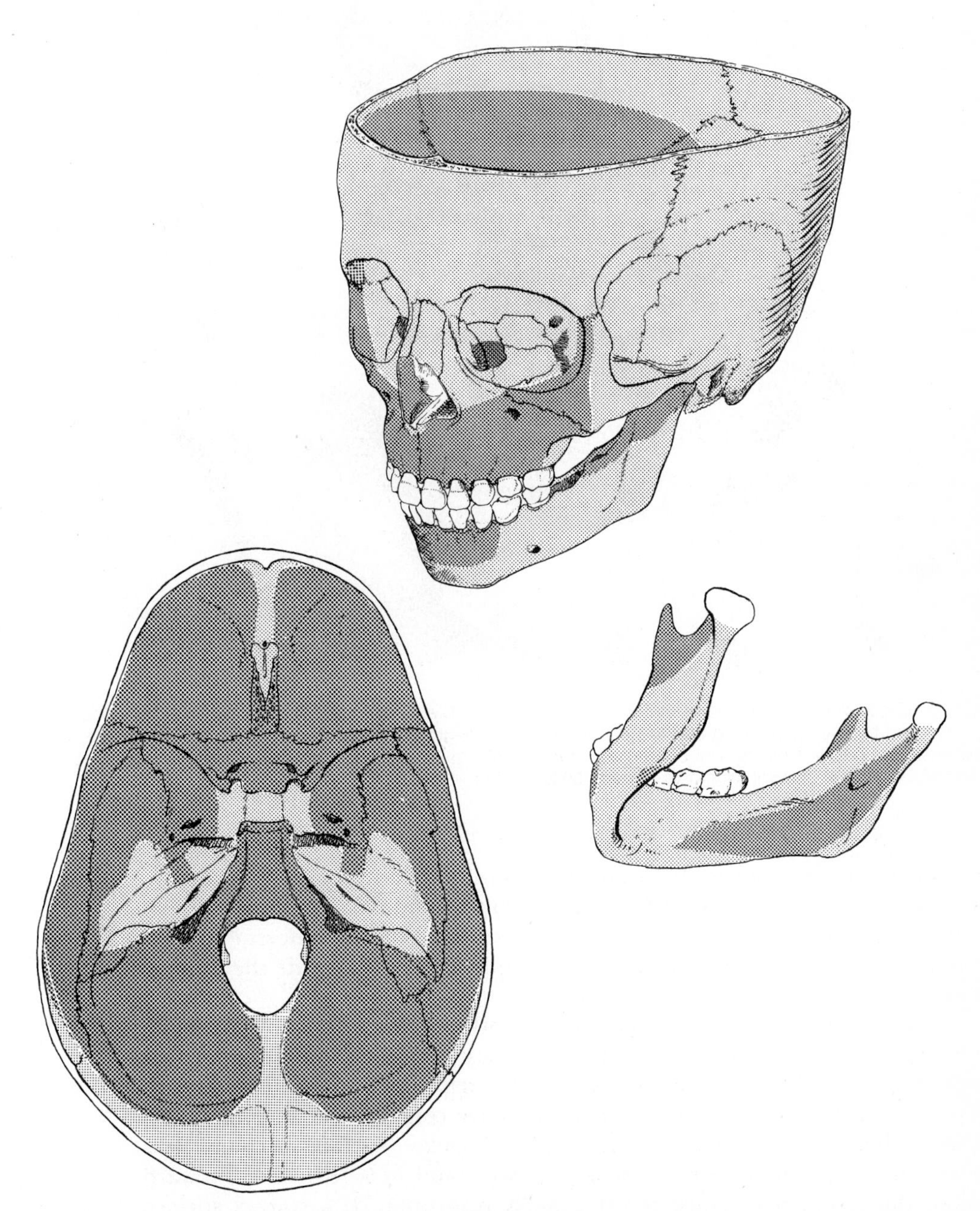

FIGURE 3–128. Summary diagram of the resorptive (darkly stippled) and depository (lightly stippled) fields of growth and remodeling. (From Enlow, D. H., T. Kuroda, and A. B. Lewis: The morphological and morphogenetic basis for craniofacial form and pattern. Angle Orthod., 41:161, 1971.)

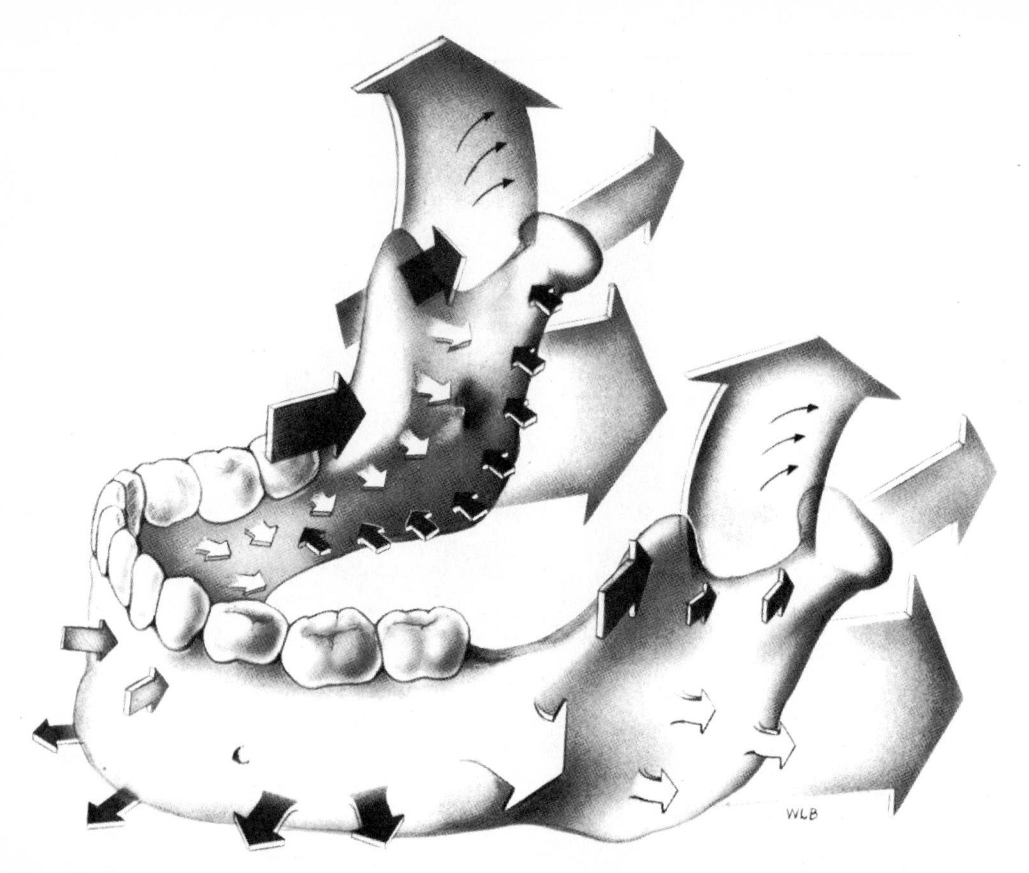

FIGURE 3–129. Summary diagram of the growth of the mandible. Growth directions involving periosteal resorption are indicated by arrows pointing into the bone surface, and growth directions involving periosteal deposition are represented by arrows pointing out of the bone surface. (From Enlow, D. H., and D. B. Harris: A study of the postnatal growth of the human mandible. Am. J. Orthod., 50:25, 1964.)

the bone's matrix, conversely, leads to deposition, and tension relates to resorption. The nature of the operational interplay and balance between these seemingly opposite remodeling effects is not presently understood, and whether, or how, either one can yield to, complement, or override the other is not known.

As an exercise, see if you can account for orthodontic tooth movements on the basis of the membrane/bone matrix information provided in the preceding paragraph. One added theoretical idea will be helpful. If an existing concave surface becomes **more** concave, the effect is active compression and the action response thereby depository. If an existing concave surface becomes **less** concave, however, the action is less compressive and in a "direction" toward tension; the resultant response is presumably resorption. If a convex surface becomes either more or less convex, similarly, the results are believed to be resorption and deposition, respectively.

A final and inspiring thought with regard to the biology of tooth movement: Consider the **remarkable degree of precision** operative among the separate movements of a tooth, its alveolar bone, the periodontal connective tissue, and the remodeling of all the other surrounding hard and soft tissues affected by these movements. It is a lock-step composite of movements that demonstrates

Text continued on page 145

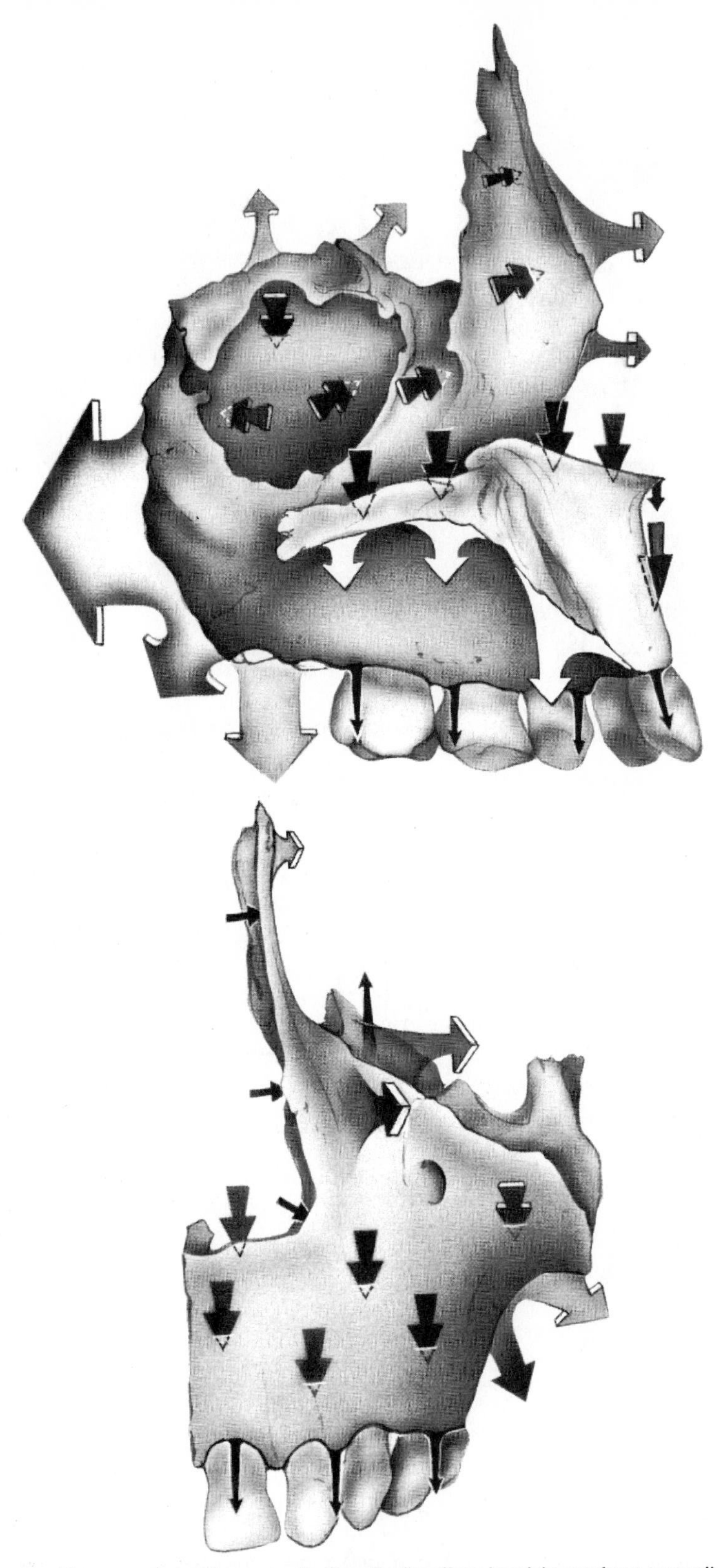

FIGURE 3–130. Summary diagram of maxillary growth. Growth directions involving surface resorption are represented by arrows entering the bone surface. Directions of growth involving surface deposition are shown by arrows emerging from the bone surface. (From Enlow, D. H.: *The Human Face.* New York, Harper & Row, 1968, p. 164.)

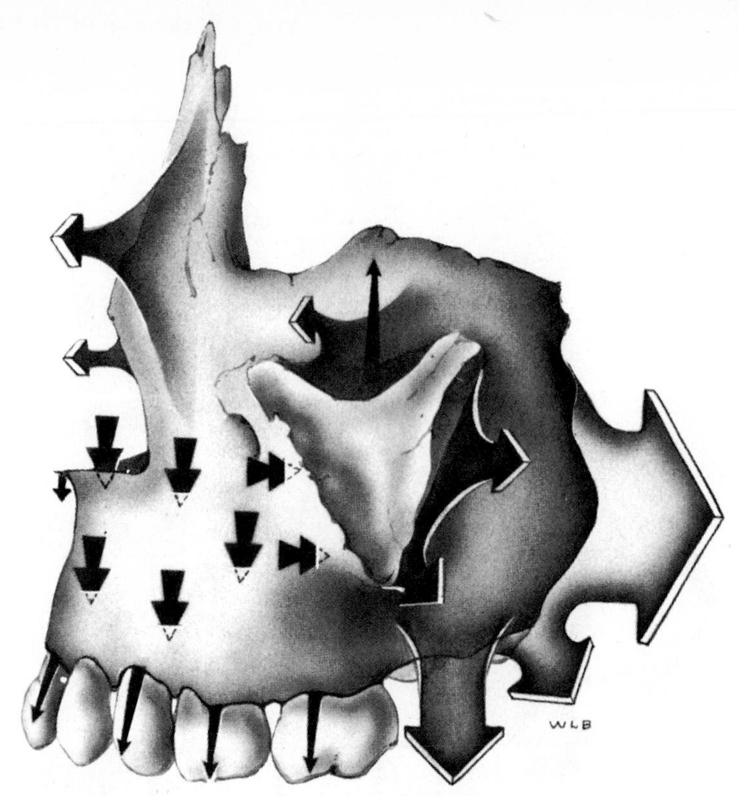

FIGURE 3–131. (From Enlow, D. H.: *The Human Face.* New York, Harper & Row, 1968, p. 164.)

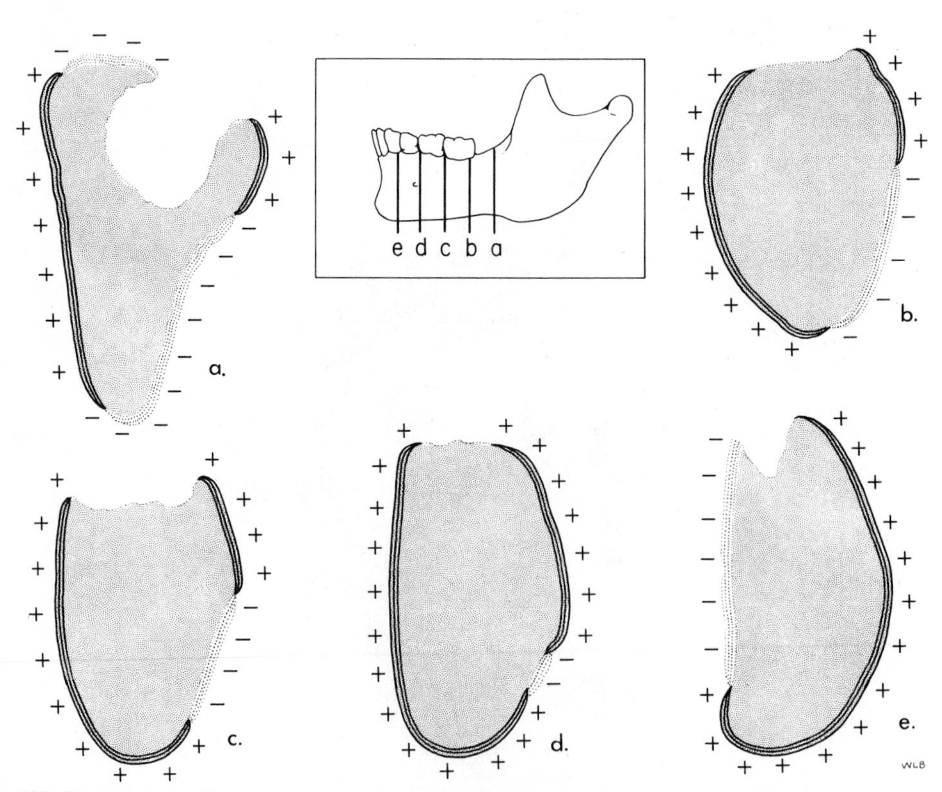

FIGURE 3–132. Transverse sections made at *a, b, c, d,* and *e* are characterized by the growth and remodeling patterns shown. (Adapted from Enlow, D. H., and D. B. Harris: A study of the postnatal growth of the human mandible. Am. J. Orthod., 50:25, 1964.)

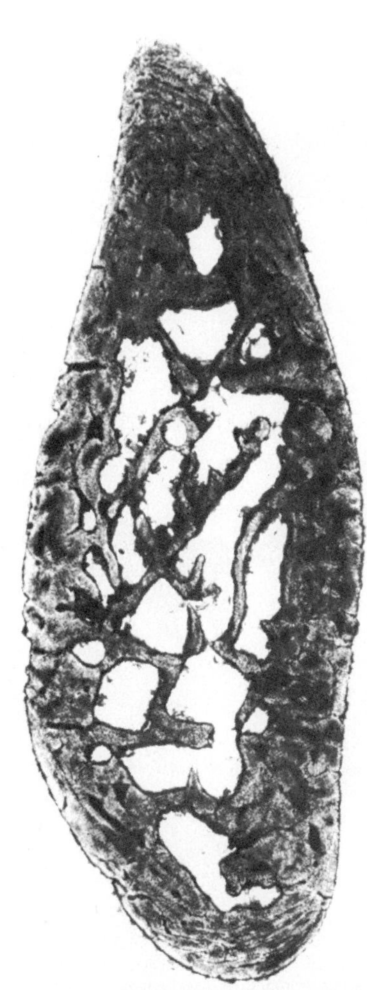

FIGURE 3–133. Transverse section through the neck of the mandibular condyle. The superior margin, which is part of the mandibular (sigmoid) notch, is depository. The lower margin, which is the posterior border of the ramus, is also depository. Both the buccal and the lingual surfaces are resorptive. (From Enlow, D. H., and D. B. Harris: A study of the postnatal growth of the human mandible. Am. J. Orthod., 50:25, 1964.)

FIGURE 3–134. The buccal side of the coronoid process (left) is resorptive, and the lingual side (right) is depository. (From Enlow, D. H., and D. B. Harris: A study of the postnatal growth of the human mandible. Am. J. Orthod., 50:25, 1964.)

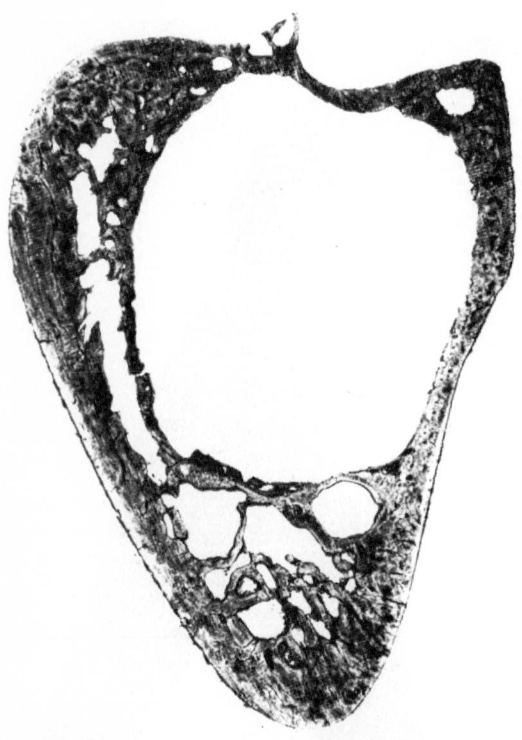

FIGURE 3–135. Transverse section through the posterior part of the mandibular corpus. The lower two thirds of the lingual surface (on the right) are resorptive and form the lingual fossa. The upper third, which is the lingual tuberosity, is depository. Except for a tiny patch of surface resorption near the top, the entire buccal side (left) is depository. Note the reversal between the depository buccal and the resorptive lingual sides on the inferior margin of the corpus. (From Enlow, D. H., and D. B. Harris: A study of the postnatal growth of the human mandible. Am. J. Orthod., 50:25, 1964.)

FIGURE 3–136. Transverse section of the mandible taken from the level of the second molar. On the lingual side (left), the entire surface is resorptive except for a depository zone near the superior border. The entire buccal side (right) is depository. (From Enlow, D. H., and D. B. Harris: A study of the postnatal growth of the human mandible. Am. J. Orthod., 50:25, 1964.)

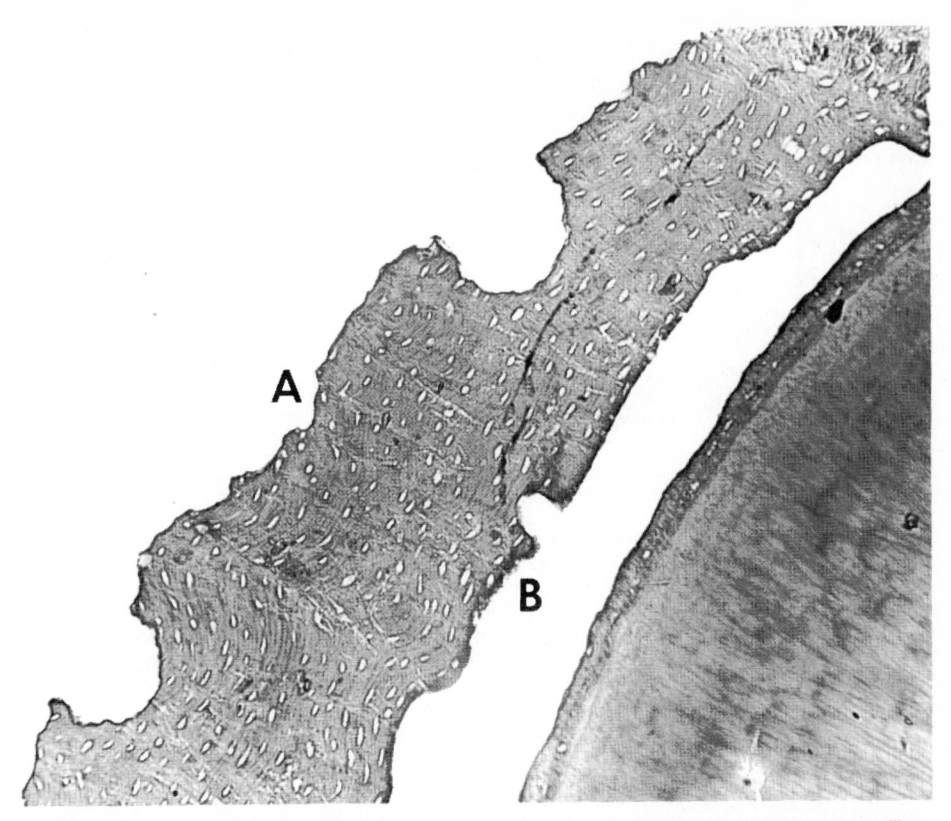

FIGURE 3–137. The resorptive nature of the external (periosteal) surface of the human premaxillary cortex is shown (*A*). The opposite alveolar lining surface is depository (*B*). The bone of the cortex was laid down by the periodontal membrane on the one side and resorbed by the periosteum from the other. This combination **moves** this part of the bony arch, and also the tooth, in a downward direction (see **vertical tooth drift** on page 116). The bone tissue on the outer (labial) surface, therefore, was actually formed by the periodontal membrane on the inside and subsequently translocated to the periosteal side of the moving cortical plate. (From Enlow, D. H.: *The Human Face.* New York, Harper & Row, 1968, p. 152.)

a level of coordination worthy of highest eulogy! First, a tooth cannot move itself; it must be physically moved by the soft tissue surrounding its root. The directions, the amounts, and the timing of the tooth's movement must be **precisely** matched by the peridontal membrane (PDM) and alveolar bone remodeling movements, with attachments sustained all the while. If the tooth moves farther than, faster than, or at a time different from that of the PDM remodeling movement on, for instance, the resorptive bone side, the periodontal space would become lost, and tooth-to-bone ankylosis would result. Similarly, the same symphony of matching growth actions on the depository alveolar bone surface must proceed in precise coordination with both tooth and PDM for the same reason. Too much or too little, in any varying directions, or with off-timing, and the periodontal space would either be lost or enlarged beyond functional tolerance. The whole process works because of the finely tuned, exact operation of the communality of stimulatory signals that activate the closely interrelated developmental responses involving all of these separate parts. This means concordance within a communal organization in which everything happens with precision and harmony. Function continues uninterrupted throughout the exercise.

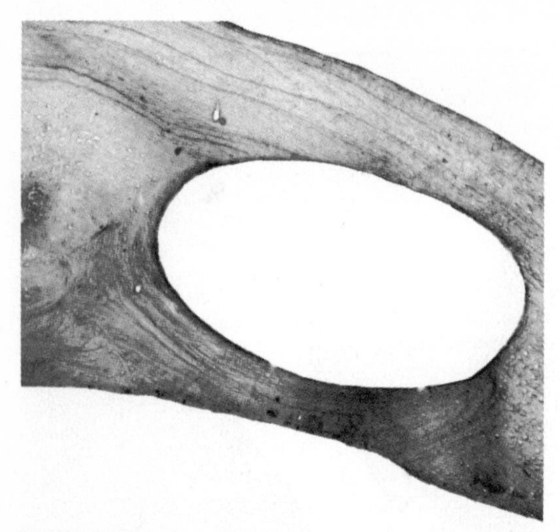

FIGURE 3–138. The frontal process of the maxilla has a depository outer surface (top) and a resorptive mucosal surface within the nasal chamber (bottom). Note the corresponding drift of the cancellous space in a like direction by resorption on the upper side and deposition on the lower side. (From Enlow, D. H.: *The Human Face.* New York, Harper & Row, 1968, p. 156.)

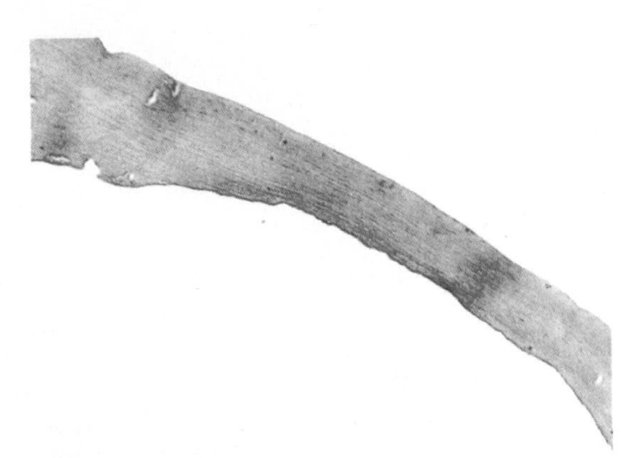

FIGURE 3–139. The maxillary portion of the orbital floor is composed of a single, thin plate of periosteal lamellar bone. The upper (orbital) surface is depository, and the opposite side (bottom), which lines the maxillary sinus, is resorptive. (From Enlow, D. H., and S. Bang: Growth and remodeling of the human maxilla. Am. J. Orthod., 51:446, 1965.)

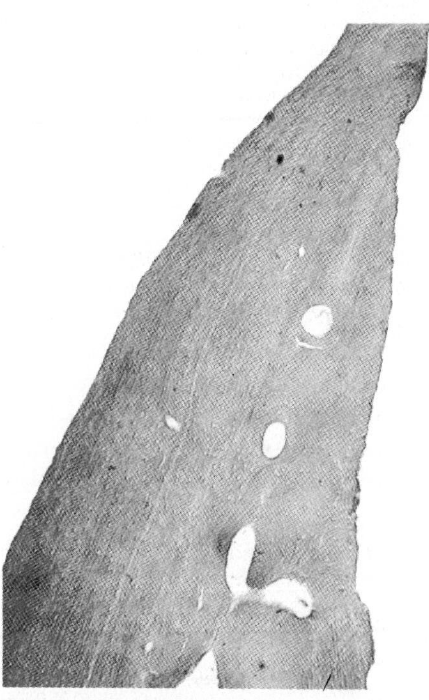

FIGURE 3–140. Transverse section through the lateral nasal wall (part of the maxillary bone). The external side (left) is depository, and the opposite side within the nasal chamber (right) is resorptive. (From Enlow, D. H.: *The Human Face.* New York, Harper & Row, 1968, p. 156.)

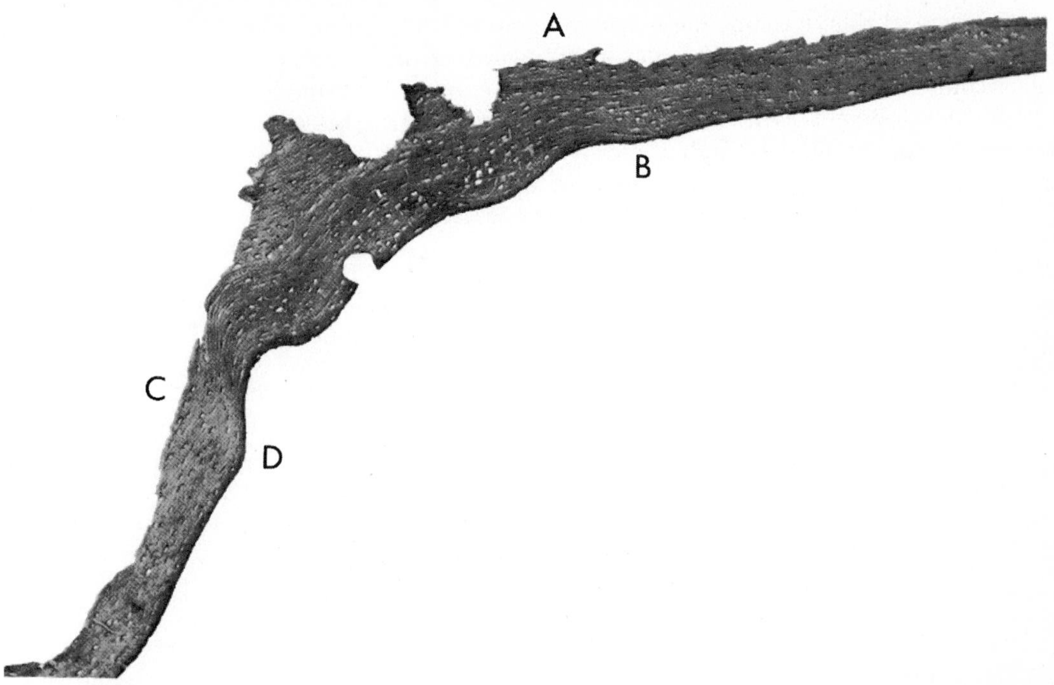

FIGURE 3–141. Lacrimal bone. Surface *A* is the external side of the cortex that forms a portion of the medial lining wall of the orbital cavity. In the cephalic part of the lacrimal bone, the surface is characteristically resorptive, as seen in this section. The contralateral side within the nasal chamber is depository (*B*). A similar combination is seen within the lacrimal groove (*C* and *D*). (From Enlow, D. H.: *The Human Face.* New York, Harper & Row, 1968, p. 162.)

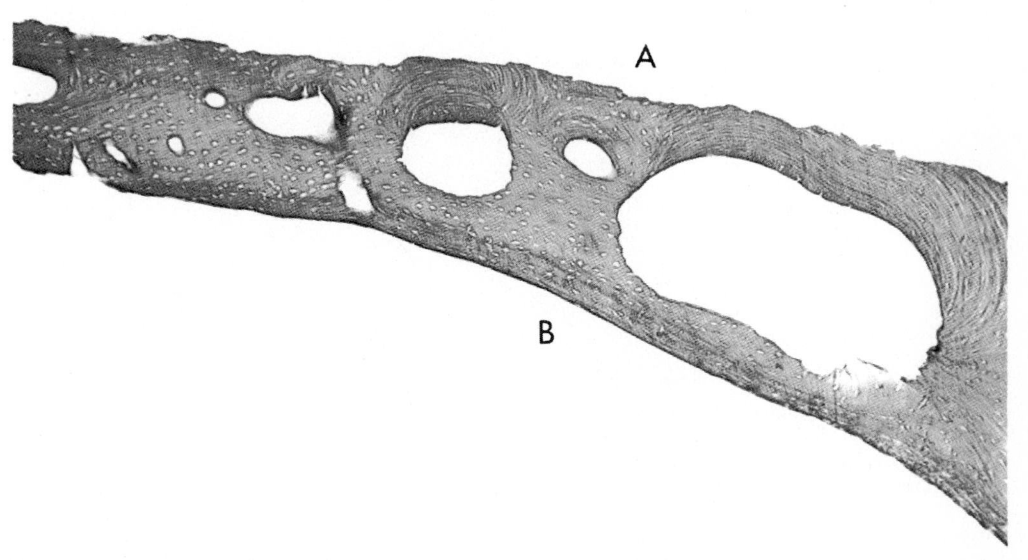

FIGURE 3–142. This section of the anterior part of the palate overlies a tooth socket. The nasal floor (*A*) is resorptive, and the roof of the socket is depository (*B*). Note the downward movement of the medullary spaces by bone additions on superior lining surfaces and removal from the inferior lining surfaces. (From Enlow, D. H.: *The Human Face.* New York, Harper & Row, 1968, p. 162.)

This remarkable developmental system operates under an intrinsic system of control and implementation responsive to growth and functional conditions and circumstances that spread throughout the craniofacial complex during ordinary development. **Orthodontic** tooth movement harnesses and manipulates this control system through clinically induced signals that override and replace the intrinsic signals. The system of operation itself, however, is the same and utilizes the same intrinsic histogenetic mechanism.

CHAPTER FOUR

Introduction to the Temporomandibular Joint

J. M. H. Dibbets, D.D.S., Ph.D.

The temporomandibular joint (TMJ or the TM joint) is a bilateral synovial diarthrosis. This learned expression means that we have one freely movable joint on each side, left and right, surrounded by a capsule whose internal lining produces a viscid synovial fluid. The joint permits the mouth to be opened and closed, and the jaw to be protruded, retruded and shifted laterally. During these movements the joint capsule, together with the lateral and the sphenomandibular ligaments, provides structural stability. The stylomandibular ligament is considered of minor importance in this respect and therefore accessory.

Because the dynamic performance of all joints normally covers the functional demands placed upon them, it is not logical to assign to the TM joint additional structural/functional complexities, as some authors do. Any joint of the body may be regarded as an intricately interrelated functional structure as well as a natural adaptation to environmental needs and constraints.

The TM joint has received much attention in the literature during recent years, and this special interest is still increasing. What makes this joint so interesting? Large-scale research has revealed that this articulation often generates signs and symptoms that may indicate a dysfunction of the system. Clicking, snapping, crepitation, locking, pain, and instability are reported with high frequency. Numerous, too, are the etiologic factors assigned to dysfunction. Whether or not the resulting therapeutic regimens are oriented to the cause of the symptom is a question still open to debate. However, from information available at this moment, obtained from conscientiously conducted studies, it appears that simple mechanical explanations for dysfunction, such as occlusal irregularities, do not hold.

In this chapter developmental and structural aspects of the TM joint will be briefly recapitulated. Emphasis will be placed on those aspects that are unique for the components of the temporomandibular articulation. Among these are the "secondary" character of the joint and the specific analysis of the processes and mechanisms involved in the growth of the components.

STRUCTURAL ASPECTS

In the sixth week of intrauterine life, a condensation of mesenchyme develops lateral to Meckel's cartilage. The development of this condensation into a lower jaw proceeds rapidly. Within 1 week a complete membranous bony plate, albeit fragile, is formed, paralleling and locally enveloping the bilateral Meckel's cartilaginous rods. At 10 weeks the bony mandible has recognizable form, and Meckel's cartilage starts to be resorbed. This branchiomeric cartilage does not contribute to the newly formed mandible by means of endochondral bone formation. During the same period condylar fields develop at the cranial ends of the mandible. Within 2 weeks the condylar processes are clearly recognizable, and (secondary) cartilage production will have begun. By another 2 weeks, during the fourteenth week, endochondral ossification of this new cartilage will start centrally in the ramus, proceeding upward. From the twentieth week on there occurs an equilibrium between the production of cartilage and subsequent replacement by bone, creating the typical picture of a growing mandibular condyle.

At 10 weeks the lateral pterygoid muscle is already formative, and its two heads may be distinguished. One head attaches to the condyle and one to the formative disk. The disk emerges from a mesenchymal field that develops between the developing condyle, the temporal squama, and Meckel's cartilage. Some authors assume continuity as a single system, extending from the lateral pterygoid muscle to the malleus, which is also arising from the cartilage.

The cartilage of both the condyle and the tubercle consists of cartilage cells and of matrix, which is composed of a network of collagen fibers and hydrophilic proteoglycans containing bound water. The collagen fibers mechanically prevent a continuous swelling, resulting from osmotic absorption of water by the proteoglycans, and thus allow pressure to be generated within the network. Functional loading of the joint is counteracted by this pressure. When this loading exceeds the pressure, liquid is expelled into the interstitium. This fluid provides lubrication—"weeping lubrication"—and metabolic support; unloading makes the fluid return into the cartilage matrix.

It has been suggested that increased numbers of elastic fibers, replacing collagen, mark a beginning of joint pathology. One of the first signs of degeneration of the cartilage, therefore, consists of increased water absorption by proteoglycans, probably because of failing restriction by the remaining collagen fibers.

The temporal component of the joint does not acquire its characteristic sigmoid shape until after birth. As may be seen in Figure 4–1, an 8-month human fetus, a straight zygomatic arch occurs before birth. The mandible may slide forward and backward horizontally without being displaced vertically. This situation will change quickly after tooth eruption has started.

At 4 years of age, as seen in Figure 4–2, the temporomandibular articulation has achieved many of its adult characteristics. A tubercle has formed, and the condylar process and mandibular shape have clearly progressed beyond the neonatal state. The external meatus still occupies a low position relative to the condylar head, but with further maturation this position will change in a vertical direction.

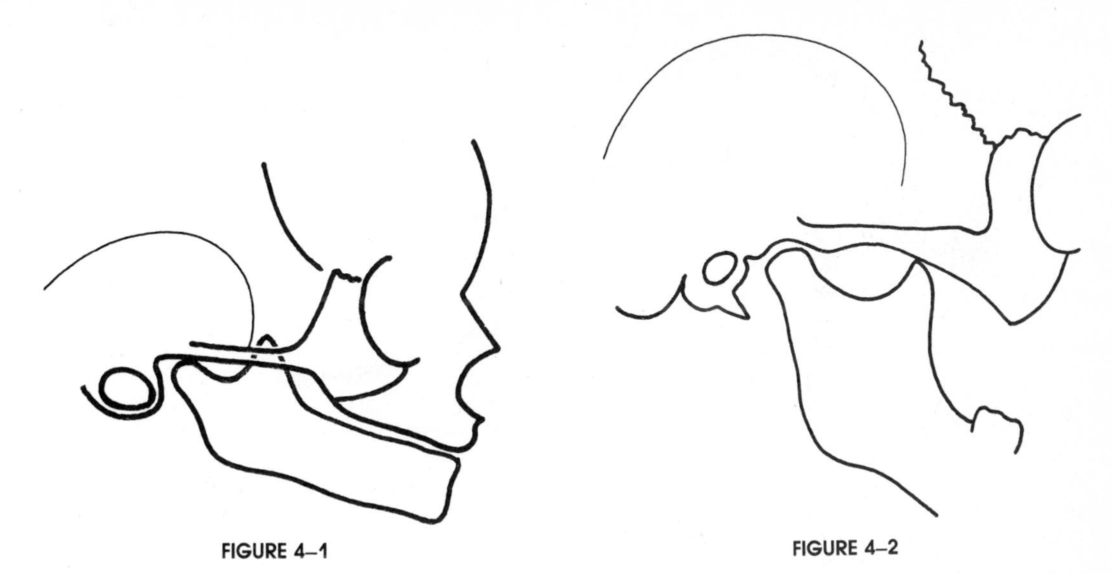

FIGURE 4–1

FIGURE 4–2

In the adult edentulous situation, shown in Figure 4–3, the slope of the temporal tubercle is more vertical. The external meatus occupies a vertical position at the same level as the condyle. This change is not due to remodeling of the structures of the middle ear and their bony housing, but can be accounted for by downward remodeling of the temporal part of the joint.

Figure 4–4 is an abstraction of the adult temporomandibular joint in lateral view. The condyle is indicated, together with its covering of cartilage. The temporal counterstructure, namely, the tubercle and its cartilaginous covering, form the cranial component of the joint. The bony plate that separates the median upper part of the joint from the inferior surface of the temporal lobe of the brain is paper thin.

Between the tubercle and the condyle in Figure 4–4 a compact connective tissue disk is situated that divides the joint cavity into upper and lower chambers.

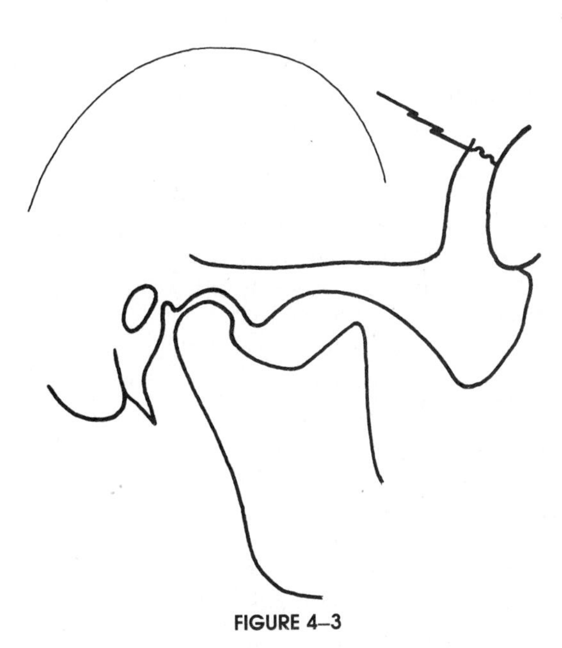

FIGURE 4–3

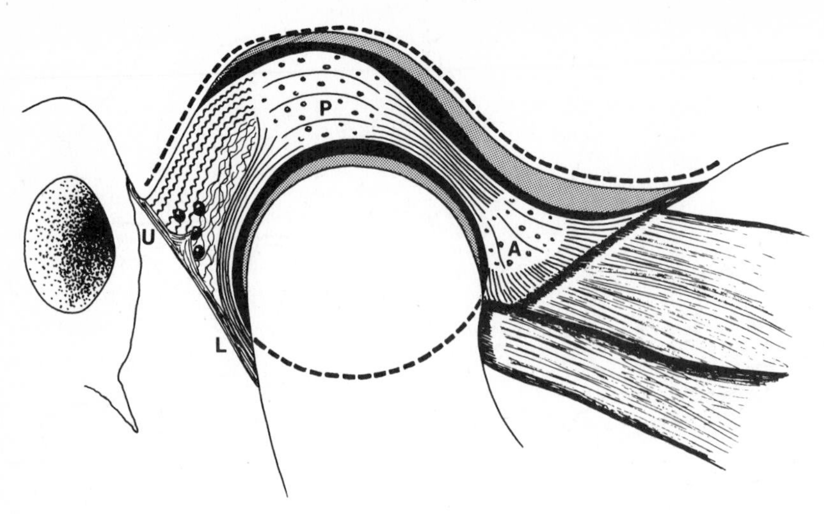

FIGURE 4–4

These cavities are filled with synovial fluid and are indicated in black in the illustration. Centrally this disk is composed of dense avascular tissue, with fibers oriented in a sagittal direction. Above and in front of the condyle the dotted areas in the figure indicate cross sections through the so-called posterior and anterior bands. These bands are part of a continuous system and may be compared to a stretched ring. This ellipsoid ring, viewed from above, runs superiorly along the long axis of the ovoid condyle. In the figure its cross section *(P)* forms the larger or posterior band. Sliding downward and forward at the lateral and medial poles to the anterior face of the condyle, it forms the smaller or anterior band *(A)*. The ellipsoid ring, incorporating both the anterior and posterior bands, is completely integrated within the disk. Dorsally, the posterior band continues into a bilaminar zone. The upper zone *(U)* consists of highly elastic tissue, permitting displacement of the disk during opening and closing. The lower zone *(L)* is much less elastic and assures positional stability between disk and condyle.

Between the two zones, and in particular near the posterior capsule, there is loose connective tissue with a rich blood supply. During forward shift of the condyle these vessels significantly enlarge their cross-sectional area. The articular capsule and ligaments run from the temporal mandibular fossa and tubercle to the neck of the condyle. The insertions, in the lateral view shown, are indicated by broken lines. The capsule and disk are continuous at the dorsal, medial, and lateral aspects. As a result, the disk in reality is a three-dimensional structure, similar to a hat capping the condylar head. The anterior band is attached to the superior head of the lateral pterygoid muscle; the inferior head connects with the condyle itself.

In Figure 4–5, an individual with the mouth in closed and widely opened positions, several basic observations are made. With the mouth open, the condyle has slid forward and downward along the tubercle. Two separate actions, in fact, have occurred simultaneously: the condyle rotated on the disk while the disk slid along the slope of the tubercle. At the same time the entire head has been tilted upward. In this way a considerable mouth opening can be obtained, and without danger of compressing the pharyngeal space dorsal to the mandible. In more than 40 per cent of individuals the condyle will occupy a position anterior to the tubercle at maximum mouth opening. This position consequently is regarded as normal.

Another (more controversial) aspect depicted in Figure 4–5 is the mandibular center of rotation during opening and closing. From the observation of a forward and downward displacement of the condyle, it must be deduced that this center has to be located somewhere posterior to the mandible. Goniometric constructions for successive opening positions reveal the path of the center. For the first and a very small opening movement, the mandible will hinge around a center somewhere within the condyle. Then, as the condyle shifts forward and downward, the location of this center will drop rapidly vertically, as indicated by the bold line. The bold line represents the successive shift of the center during opening. It is indicated as C_d the dynamic center of the opening movement. The center of rotation constructed from the static situations of both extremes, open and closed, is indicated by C_s. Note the considerable discrepancy between a static center and the dynamic center trajectory. The difference is accounted for by the sigmoid shape of the tubercle.

SECONDARY JOINT

The vertebrate temporomandibular joint is known as a secondary joint. The adjective "secondary" refers to several separate properties of the joint that are not original; that is, there were "primary" characteristics that have been replaced phylogenetically by "secondary" ones. Let us find out which was first and what came thereafter.

In Chapter 3 it is explained that during phylogeny a new joint evolved, replacing the original, or primary, articulation. This original joint developed within the branchial arch system at the junction of the floor of the housing of a nerve cell concentration and the first gill arch in primitive fishes. In our

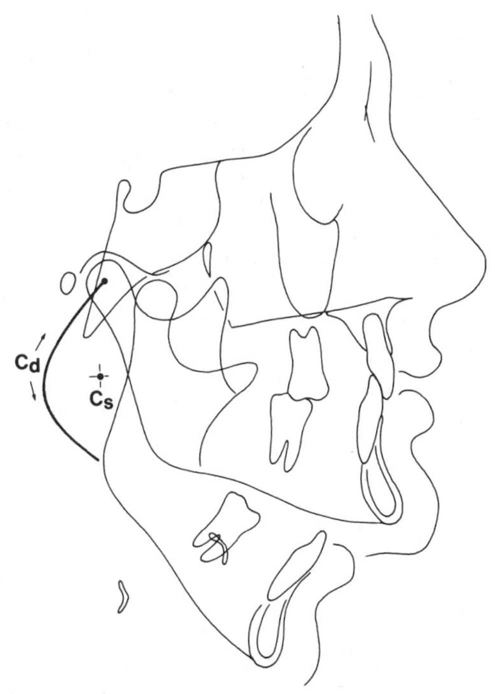

FIGURE 4–5

phylogenetic history this was the primary mandibular joint. During evolution neurons became concentrated at one end, and structural stability was provided by the bilateral arch of the first branchial cartilages. Through many evolutionary stages and continuing time this articulation, even at the same time, combined the functions of the jaw joint and the middle ear in some Amphibia. After that stage a new or secondary association between the skull and the existing teeth-bearing structure, the dentary bone, came into being in front of the original joint. "Secondary," therefore, may apply to the joint in that it is a later development in our phylogenetic history.

During ontogeny some major events in our phylogenetic history are reproduced. This is also true for the original articulation, between Meckel's cartilage (the primary mandible) and the cranium, which is replaced by a new joint. By this time in prenatal development most other synovial joints of the body have already been formed. The new articulation between the temporal bone and the mandible is therefore indicated as secondary, in contrast to primary joints formed earlier. "Secondary," therefore, may apply to the joint because of its being late in our ontogenetic development.

There is a third and again important reason to attach to the TM joint the adjective "secondary." It refers to the "second" appearance of the cartilaginous components of the joint. Subsequent to all other true primary cartilages that form, then, within the mesenchymal blastema of what will be the future mandible, new cartilage formation begins as a secondary event in four regions: the condylar process, coronoid process, the symphysis, and the gonial area. The latter three will have disappeared around birth. Condylar secondary cartilage, however, will remain for the rest of our life. Probably due to articular functioning, cartilage is induced and maintained within the membranous components of both the squamous portion of the temporal bone and the condylar process of the mandible. "Secondary," therefore, applies to the new cartilage tissue that only comes to exist in addition to the primary cartilages that have fully appeared.

A fourth reason to justify the label "secondary" to the TM joint may be found in the origin of the cartilaginous tissue. Being late in ontogeny, the new cartilage develops within a mesenchymal blastema, as explained in the previous paragraph. The secondarily induced differentiation of membranous bone will testify to its inheritance for the rest of its life cycle by its connective tissue covering. Primary cartilages are covered by a thin perichondrium. Secondary cartilage, in contrast, is covered by a fully developed, however thin, mesenchymal tissue layer. The source from which condylar cartilage is derived is found within this covering. First there are mesenchymal cells, and these cells differentiate into cartilage only as a secondary event. "Secondary," therefore, may apply to the late differentiation of original mesenchymal tissue from which the cartilage originates. We will expand on that shortly.

This mesenchymal-like covering of the condyle underlies a fundamental characteristic of the condylar cartilage. Primary epiphyseal cartilage reacts during development primarily to overall systemic growth stimuli such as hormones. In contrast to this, condylar cartilage only secondarily follows these overall stimuli after additional modulation by local factors. This is substantiated by numerous laboratory experiments. Condylar cartilage cannot be cultured *in vitro* easily, as can epiphyseal or other primary cartilages. We will also come to that point later. "Secondary," therefore, may apply to the characteristic response of the condyle during growth.

TISSUE PROLIFERATION

The notion of a mesenchymal-like covering of condylar cartilage is fundamental for understanding the growth system for the condyle. This growth process is different from epiphyseal growth and is a unique feature of secondary cartilage.

Primary cartilage growth is presented in a highly schematized fashion in Figure 4–6. From left to right, the symbols indicate the following. The left and large circle represents a cartilage cell within the central germinative layer of an epiphyseal plate. The arrow indicates the transition to the subsequent developmental condition, a normal mitosis, which is shown by the large slash. As a result of the mitosis, two daughter cells will originate, together containing the total amount of organic substance from the original mother cell. Each will inherit half of the duplicated maternal chromosomes, and each cell will be smaller than the original. The next phase during epiphyseal growth is enlargement of the two daughters, each to the full size of their mother. At this stage both now-mature cells produce and secrete extracellular matrix, which will make these cells drift apart. One may remain within the germinative layer and probably be a new mother; the other may drift away and subsequently be eroded and replaced by bone. This very schematic series highlights one of the essential elements of primary cartilage growth: cleavage of previously differentiated mature cartilage cells. The translation of the diagram to reality is facilitated by the autoradiogram in Figure 4–7. Autoradiography is a technique of exposing a photographic emulsion to a radioactive source that is incorporated within a tissue. Here the radioactive-labeled nucleotide thymidine was injected into a 49-day-old rat 2 hours before sacrifice. The thymidine was incorporated rapidly into cells preparing for division. Its radioactivity causes black spots in the emulsion. The circles point to their positions. Apparently cell divisions are taking place in the middle part of an epiphyseal plate of a tibia. The vertical bar equals 50 μ. The type of growth in which new material is formed within existing tissue is **interstitial growth**. This interstitial mode contrasts sharply with secondary cartilage growth.

The next diagram, Figure 4–8, represents a highly schematized cycle of secondary condylar cartilage growth. The double contour indicates the mesenchymal-like tissue covering of the prenatal or postnatal condyle. It consists, among other components, of a thin layer of undifferentiated cells directly overlying the cartilage of the condyle. This differentiated condylar cartilage is represented by the large circle and is situated in the proximity of the covering membrane. The arrow indicates the transition to the subsequent developmental condition, the birth of a new cartilage cell. The place of labor is the undifferentiated soft tissue layer, in which one small cell splits itself into two even smaller new cells. This very special event may be seen in Figure 4–9, showing a 49-day-old rat condyle. The technique of autoradiography was described above, and the vertical bar still equals 50 μ. Here, within the soft tissue covering and surrounded by undifferentiated cells, mitosis is about to occur: the cell

FIGURE 4–6

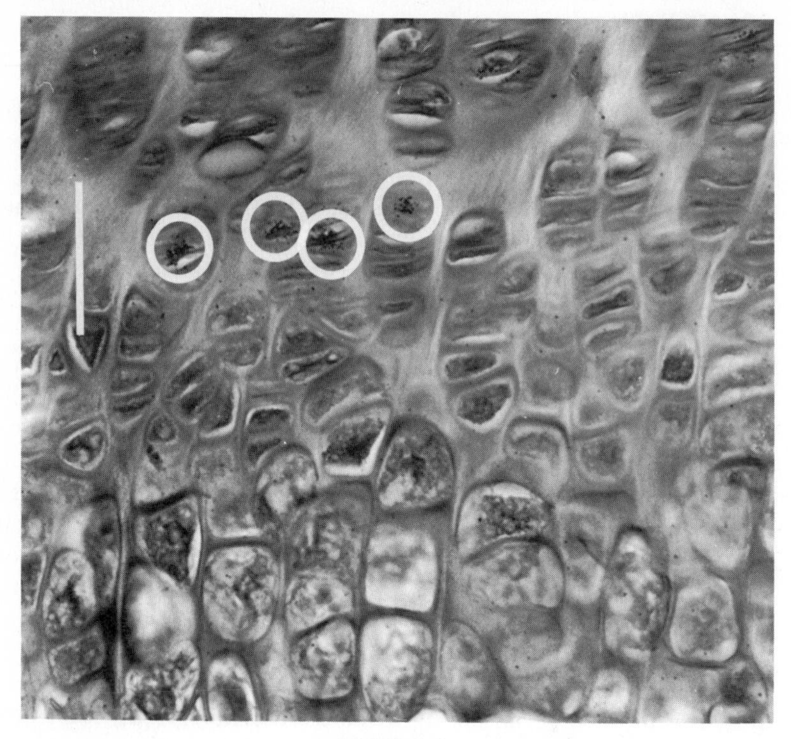

FIGURE 4–7

preparing for division is identified by the black spots above the nucleus. The circle locates this happy moment. Let us return to schematic Figure 4–8. Both cells shortly thereafter will come to full size, resulting in the migration of one of them out of the covering membrane in the direction of the interior of the condyle. This is schematized by the black dot directly underneath the soft tissue covering, but still within the condylar membrane. Here, and at this specific moment, a differentiation takes place in which the mesenchymal-like cell becomes an immature cartilage cell. This is evidenced by the small circle positioned between the membrane and the larger, more mature cartilage cell in the third step. A new member of the cartilage family has been added **without** the **mitosis** of an **existing cartilage mother cell,** but through mitosis of an undifferentiated mesenchymal cell. The far-reaching consequences of this will be discussed later on. To complete the present sequence we have to note that the new cell after differentiation will expand to mature size and subsequently will start to produce extracellular matrix, although the amount of matrix may be small, compared with primary growth cartilage. Cells will drift apart, and the process of endochondral bone formation will finish the life cycle. The mode of growth in which new cells are added from the exterior is **appositional growth.**

One of the properties of secondary cartilage growth depicted in the diagram is the quantitatively equivalent result, compared with primary cartilage. As may be deduced from the figure, there is no reason to assume a difference in

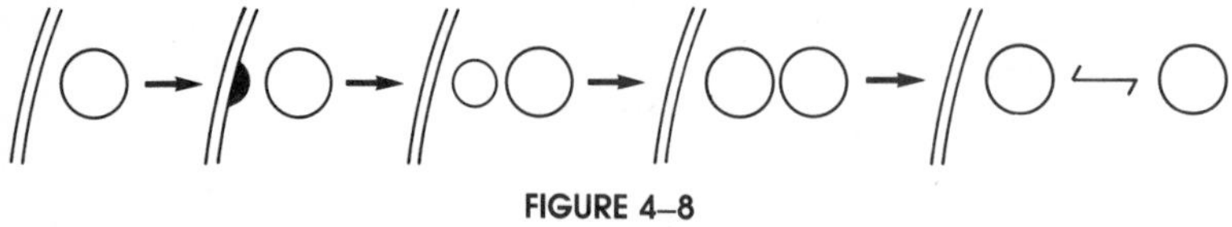

FIGURE 4–8

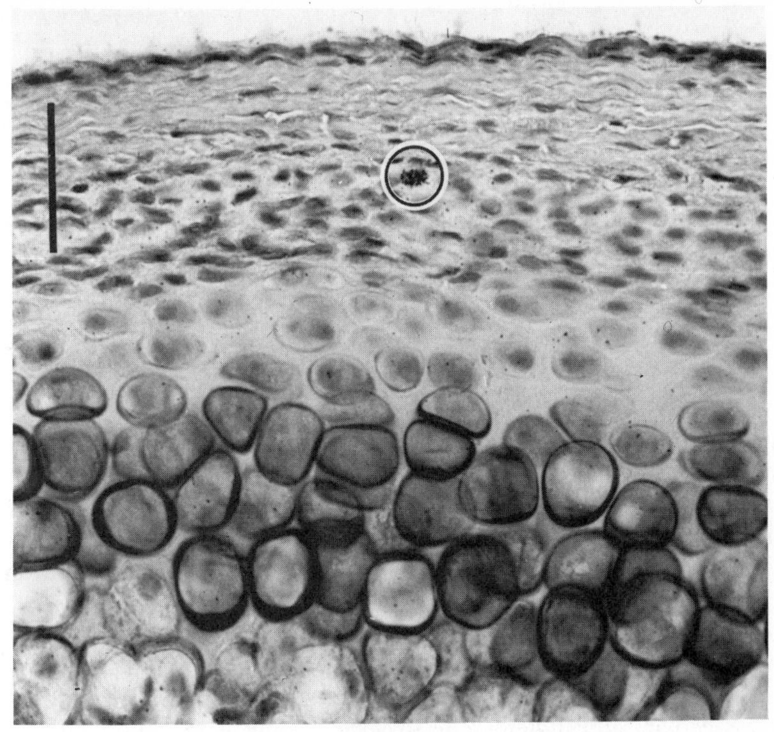

FIGURE 4–9

capacity for producing new tissue. Both schematic series may occur within equal time spans, and both sequences may respond at comparable quantitative mitotic rates during growth.

Another and most vital property depicted in the diagram is related to the source from which new cells are derived. There can exist in the human and some other species an inherited dysplasia that prevents mitoses of differentiated cartilage cells (achondroplasia). This necessarily results in dwarfism, because interstitial proliferation of cartilage cells within epiphyseal growth plates is inhibited. However, because for the mandibular condyle the cartilage originates appositionally from within the cells of the soft tissue covering, chondrogenesis is not affected. In cases of achondroplasia, the mandible has a normal growth tendency. This is evidenced dramatically by some breeds of dogs, as in bull dogs and the King Charles spaniel (Fig. 4–10). Here the cartilaginous development of the cranium is disturbed exclusively at the synchondroses (a primary, not secondary, cartilage), which leaves the mandible unaffected. The resulting profile is characteristic for the differences between the modes of interstitial cartilage growth and appositional growth. The cranium has become dome-shaped because of a failing response of the synchondroses within the cranial floor to environmental enlargement. To accommodate for the expanding brain, the desmocranial covering has reacted like an inflated balloon. The mandible, on the other hand, has developed normally. This once again demonstrates the unique character of secondary, compared with primary, cartilage.

GROWTH OF THE TUBERCLE

At birth the temporal component of the human TM joint is essentially flat or shallow. This early developmental stage of its anatomy thus facilitates

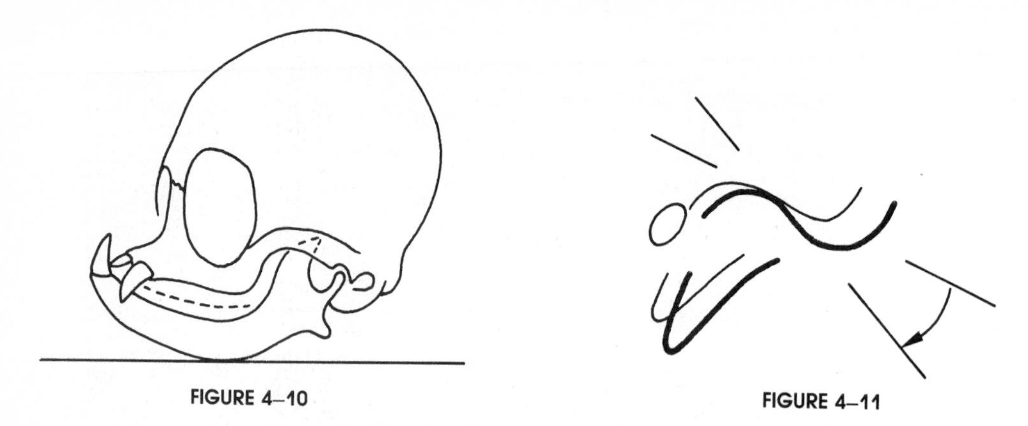

FIGURE 4–10

FIGURE 4–11

horizontal mandibular excursions during breast-feeding. The capacity for horizontal excursions then remains to some extent for the rest of the life. Unlike most other diarthrotic joints, the TM joint has a considerable degree of translational freedom. On opening, the condylar component not only rotates relative to its temporal counterstructure but also translates forward and downward.

During the first years of life the most notable morphologic change in the joint, except for dramatic size changes, is the development of the temporal tubercle. Depicted in Figure 4–11 are angular developmental changes of the tubercle as seen in lateral view. The superimposing of the older and younger tracings is on the external acoustic meatus and aligned along the Frankfort horizontal. This horizontal plane is established by the meatus and the inferior border of the orbit. The meatus was chosen because it is unlikely for the entire middle ear to remodel extensively. Therefore, changes between the petrous and squamous portions of the temporal exterior are assigned to the squamous region.

The temporal bone anterior to the condyle is progressively lowered relative to the posterior part. The development of the tubercle is characterized by its increasing slope, as indicated in the illustration by an arrow. At birth the surface is practically horizontal, and a slope hardly exists. When the primary teeth are present, permitting the first forceful chewing actions, this slope has become steeper and has already attained more than 40 per cent of its adult alignment. At the time of the first transitional period, when the first molars and front teeth have emerged, the inclination has reached 70 per cent of its adult value. When the transition of the premolars starts, 90 per cent of adult angulation is attained. The total change postnatally amounts to about 40 degrees.

The tubercle is covered by a thin layer of secondary cartilage. This cartilage is derived from cell divisions in the mesenchymal tissue covering, with subsequent differentiation, and thus is analogous to the condyle. In this way the articulating areas of the temporal bone, and consequently of the tubercle, are products of endochrondral bone formation. In contrast to this, the region posterior and anterior to it are subject to processes of intramembranous bone formation and remodeling. In Figure 4–12, such membranous bone deposition is shown by shading. As the tubercle grows by endochondral bone formation, its anterior aspect thus enlarges. Meanwhile, the glenoid fossa deepens, and the tangent to the posterior slope changes orientation into a more upright position. Because of the downward relocation of the superior aspect, the relative vertical position of the meatus also changes. In adulthood the condyle and meatus come to occupy about the same horizontal level.

Also indicated in Figures 4–11 and 4–12 is the position of the posterior part of the middle basicranium (the occipital part of the clivus and the anterior margin of the foramen magnum with the basion). Here a median structure is superimposed onto a lateral one, and this median region deepens during growth relative to the lateral middle cranial fossa.

GROWTH OF THE CONDYLE

One of the very important functions of the TM joint postnatally is to provide the amount, direction, and timing of its own regional growth responses in relationship to the ongoing and widespread changes in the surrounding craniofacial regions. It is demonstrated throughout this book that remodeling is capable of maintaining form and proportions while it simultaneously provides changing size. The growing mandible as a whole is dependent for the bulk of its substance on the process of intramembranous bone formation and remodeling. The endochondral contribution of the condyle in the actual amount of new bone tissue produced is, by far, of lesser magnitude.

Growth everywhere in the face and cranium is partly governed at regional levels. When components throughout the craniofacial complex continue to enlarge and remodel, counterparts proceed proportionally. If the upper arch displaces and remodels inferiorly, for example, the ramus lengthens vertically to move the corpus likewise. When molars erupt, the corpus elongates as the ramus relocates posteriorly. If the pharynx deepens, the ramus widens. All these and many more examples represent local regional changes and concomitant adaptations. For the mandible the situation is such that any change of each such component **always** involves condylar growth.

One aspect of TM joint growth consists of **interrelated** enlargements of its various components, in addition to the developmental interrelationships of the facial and cranial parts. The condyle enlarges in harmony with the disk and the glenoid fossa as a tubercle undergoes development at the temporal part. These changes involve both intramembranous and endochondral bone formation and continuous reattachments of the connective tissues of the associated ligaments and the capsule. The fossa simultaneously enlarges by means of anterior remodeling relocation and a vertical development of the tubercle. The condyle simultaneously expands by appositional (and some interstitial) growth. The capsular ligaments and disk also enlarge and grow over the bony surfaces with new attachment locations. With these changes, growth proceeds in a more

FIGURE 4–12

or less comparable fashion for other joints in the body. They all grow larger, whether knee, finger, or chewing joints; and they all continue to function as they grow and develop postnatally.

In Figure 4–13 the condyle is seen to remodel and relocate only in a posterior direction without any contribution to a vertical lengthening. It does this by selective direction of appositional divisions of the prechondroblasts. The whole ramus simultaneously remodels and relocates in a like manner, with posterior deposition (and medial—remember the V principle; see Chapter 3) and anterior resorption. If the ramus were to grow only vertically, on the other hand, without any contribution to widening of the ramus, as illustrated in Figure 4–14, the endochondral mode of bone growth would create a track of new bone one condyle wide. The bulk of the vertically lengthening ramus and the cortical bone of the condylar head and neck are formed by intramembranous bone production. These extremes of horizontal and vertical growth can have infinite intermediate combinations, as seen in Figure 4–15, for example. Whatever the combination, the condyle becomes progressively relocated by appositional cell divisions, differentiation into cartilage, and expansive endochondral growth. That the condyle moves by this process from a small mandible all the way to adult size is a notion of utmost importance, as will be explained.

Because of this sizable trajectory of growth of the condyle all surrounding structures attached to the condylar neck and elsewhere have to relocate in proportionate amounts. Capsular ligaments, rigid and strong in their function of stabilization against disarticulation, relocate by detachments and continuous reattachments. This capsular relocation can be done with substantial speed since condylar growth undergoes spurts every now and then. The process certainly requires a very sophisticated biologic mechanism in order to provide firm attachments **and** a changing interface simultaneously, often on a **resorptive** surface of cortical bone (see Chapter 18).

By reason of two conditions—instantaneous, versatile condylar directional response and intracapsular appositional endochondral growth—TM joint de-

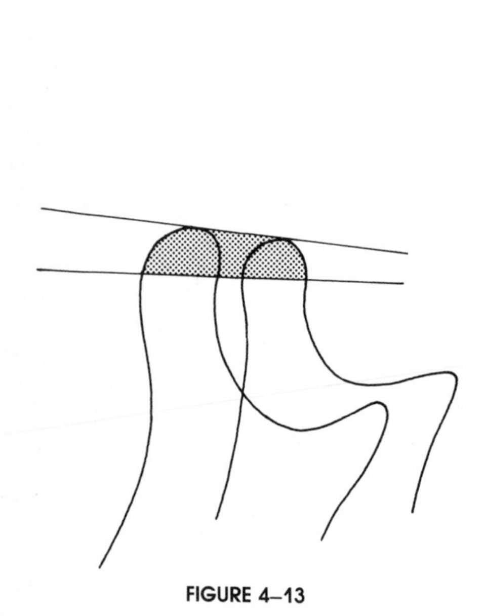

FIGURE 4–13

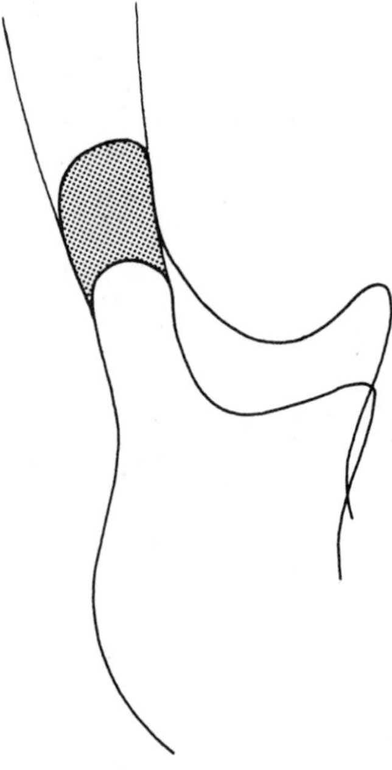

FIGURE 4–14

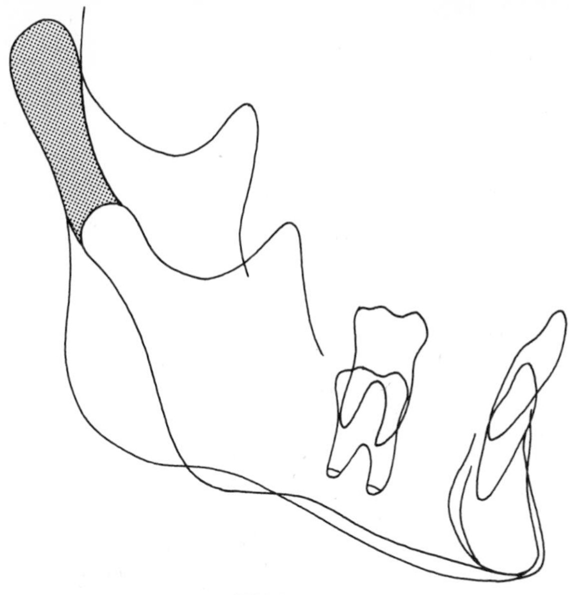

FIGURE 4–15

velopment occupies a special position among joints. Not so much because of biomechanical properties related to mandibular movements as well as its translation during growth, as often cited, is the condyle so special, not even because of its bilateral construction, which certainly is not really more noteworthy than the construction of, for example, the atlanto-occipital joint. Unique, indeed, however, is the impressive trajectory of growth that the condylar surface has to achieve, and the ongoing and imperative reorientation of the stabilizing elements of this joint. This contrasts with other joints in which integrity and stability are less affected during growth because their site of cartilage proliferation occurs within the **separate** epiphyseal growth plates. These plates are removed from the articulating area and are outside the proximal attachments of the capsules and ligaments. This is exemplified in the illustration of a growing tibia as compared with the condyle in Figure 4–16. The articular periphery is indicated above the broken line, and the growth zone by a bold line. It is apparent that in the tibia there exists a considerable epiphyseal area for attachment of ligaments between the broken and the bold lines (hatched). This area, except for changes related to enlargement of the area itself, provides firm hold during growth. For the TM joint the situation is quite different. All growth has to take place between the articulating area and the proximal attachments of capsule and ligaments.

Overall mandibular dimensional enlargement, being incorporated within a common system of overall body growth, conforms roughly to the well-known general somatic growth pattern. The same holds true for condylar growth. Fundamental disagreements, however, exist with regard to the condylar growth spurt. A spurt in this respect is a temporary acceleration of the growth velocity.

One pubertal condylar growth spurt is accepted by most authors for the male mandible. There is less consensus on a female mandibular condylar growth spurt, and severe disagreement is roused by the notion that multiple growth spurts might exist. If we reduce the complex biologic three-dimensional mandibular structure to a triangle (Fig. 4–17), with condylion at the top and gonion together with pogonion at the base, simple mathematical calculations can be

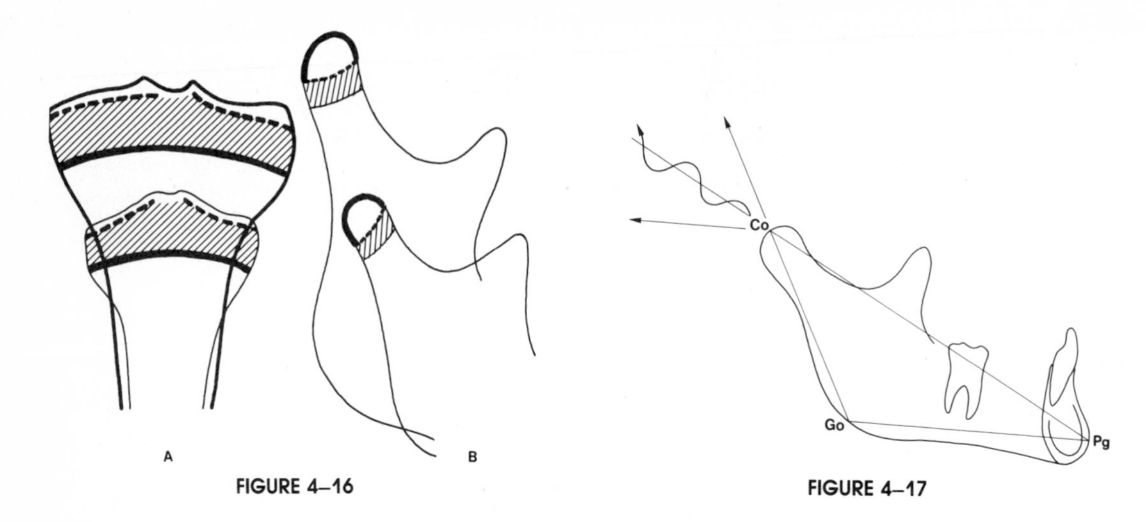

FIGURE 4–16

FIGURE 4–17

made, and these show that it is impossible for a single pattern of enlargement to change form by uniform somatic growth along the three sides of this triangle simultaneously. Biologic analyses of longitudinal growth studies, furthermore, show that only the development of mandibular depth, from condylion to pogonion, is closely associated with general somatic timing. This is, of course, to be expected, because this dimension is associated with the position of the chin and thus determines pharyngeal depth. Development of the ramus and corpus separately, consequently, is, to a certain extent, somewhat independent from overall timing. This is also to be expected, in this instance by application of the counterpart principle (see Chapter 3). Ramus and corpus both relate to different counterparts. And even for one component such as the ramus, there are separate anatomic counterparts for vertical, horizontal, and transverse dimensions. It is most unlikely that all these parts, in combination, can have one uniform overall growth pattern. The condyle as a servant of all masters will have to find a mode to satisfy them all. Timing for each may be different; but in addition, the amount and direction of growth may differ for the ramus and corpus separately.

One of the possibilities to accommodate the contradictory constraints placed upon condylar growth is a capacity for changing the direction of growth. This is depicted in Figure 4–17. Rather than proceeding along a smooth curve, condylar growth may frequently alternate directions slightly in order to efficiently fulfill divergent developmental demands placed upon the various regional mandibular components. On a theoretical basis this postulate may seem logical, but in actual practice it may prove difficult to document.

Thus far condylar growth has been portrayed as a dimensional contribution to the enlarging mandible. It has been postulated, however, that the condyle can be one of the initial structures to react to changing, concomitant mandibular functional and developmental conditions. Although complex in its association with surrounding structures, and apparently unique in its mechanism for tissue production, the condylar growth process can be logically viewed in terms of timing, amount, and direction. At this point in our understanding and with the information presently available, overall mandibular remodeling and condylar growth are perceived as geared together to accommodate housing for the developing dentition, to adapt to the mandibular displacement movements, and to adapt as well to the complex growth changes occurring throughout the whole head. This perception is essentially a sophisticated elaboration of John Hunter's pioneering concept of mandibular growth and has not been modified in theory by subsequent laboratory and animal experiments.

To accommodate the complex conditions that exist with respect to changing vertical versus horizontal middle cranial fossa enlargement, progressive horizontal and vertical adaptations of mandibular form and position are necessary in order to place the lower arch in correct juxtaposition with the upper arch. These complex conditions, among others, include pharyngeal and nasal expansion, displacements of the palate and maxillary arch, remodeling adjustments of the palate and upper alveolar structure, drifting of primary and permanent dentitions (more specifically the vertical drift component), basicranial angular changes, concomitant secondary ethmomaxillary displacement, and nasomaxillary primary rotations. In addition, there are to be noted significant facial and cranial variations related to head form and morphologic and morphogenetic differences during ascending ages. Obviously the condyles (and whole rami) must have an exceedingly versatile capacity for composite adjustments and adaptations for all of these multiple conditions. It is essential that throughout all of this, the mandibular arch is continuously positioned in functional occlusion with the maxillary arch and that a functional articulation in the TM joint is sustained, all simultaneously and without developmental interruption.

These essential relationships are sustained by appropriate responses of the condyles to biologic signals in the amount, direction, and timing of its growth in conjunction with corresponding remodeling of the rami by **their** osteogenic connective tissues. It is primarily the mandibular rami, not the corpus, that provide for these mandibular adjustments (Fig. 4–15). The mandibular dentition drifts vertically to provide for these adjustments. Primary and secondary rotational and positional adjustments accompany these remodeling adaptations as the whole mandible simultaneously enlarges, as described in Chapter 3.

Acknowledgment

The author is indebted to Dr. H.W.B. Jansen, University of Groningen, for providing Figures 4–4, 4–7, and 4–9 and valuable suggestions on content and presentation for this chapter.

CHAPTER FIVE

The Plan of the Human Face

Part 1

The human face is certainly different from that of other mammals. The long, narrow, functional muzzle that slopes gracefully onto the streamlined cranium of a typical mammal is in marked contrast to the muzzleless, broad, vertical, flattened human face, enveloped by an enormous balloon-shaped cranium with a bulbous forehead overhanging tiny, retrusive jaws, a small mouth, a chin, and the curious vestige of a narrow fleshy snout—an owl-eyed face showing changing expressions. Although somehow beautiful to our eyes, this has to be an "odd" design by ordinary mammalian standards. What are the factors that underlie the functional, developmental, and phylogenetic reasons for this specialized facial configuration? There have been many theories over the years; but regrettably, we may never know for sure what the **primary** factors were that initiated the long evolutionary chain of interrelated adaptations throughout the whole body that relate to the many design features of our facial heritage. We can, however, partially explain the anatomic, developmental, and functional meaning of each factor in this series of mutually dependent changes. And we can propose some pretty reasonable phylogenetic explanations as well. More than merely being interesting, this helps all of us to better understand the basic plan of facial construction. We can thereby evaluate more meaningfully the various facial dimensions, planes, angles, and so on, that are so important to the clinician and to basic science researchers.

Concept 1

Man is one of the few truly bipedal mammals (Fig. 5–1). Our upright posture involves a great many anatomic and functional adaptations throughout every part of the body, and no one of these would work without all the others. We have "feet," and the human foot stands by itself, as it were, as a unique anatomic feature of man. The designs of the toes, foot bones, arch of the foot, ankle, leg bones, pelvis, and vertebral column all interrelate in the anatomic

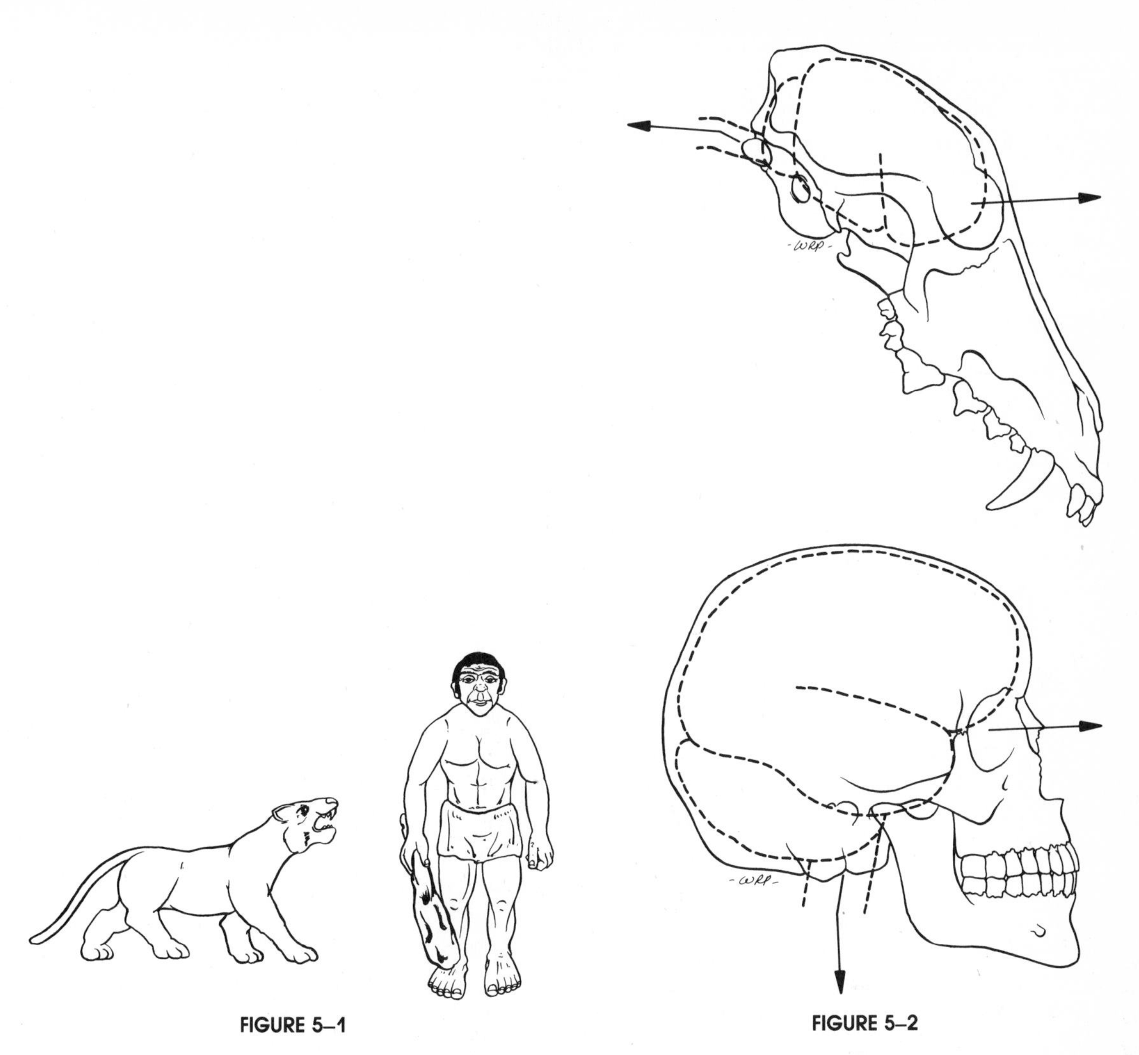

FIGURE 5–1

FIGURE 5–2

composite that provides upright body stance. The head is in a balanced position on an upright spine. The arms and hands have become freed. The manipulation of food and other objects and defense, offense, and so forth, utilize primarily the hands, rather than the shortened jaws.

Concept 2

The enormous enlargement and the resultant configuration of the brain have caused a "flexure" (bending) of the human cranial base (Fig. 5–2). This relates to two key features. First, the spinal cord is aligned vertically, a change that permits upright, bipedal body stance with free arms and hands. Second, the orbits have undergone a rotation in conjunction with frontal lobe expansion. This aligns them so that they point in the forward direction of upright body movement. The body has become vertical, but the neutral visual axis is thereby still horizontal, as in other mammals. (Note: The muzzle of a typical animal points obliquely downward in the "neutral" position, not straight forward. This positions the orbital axis approximately parallel with the ground and toward the direction of body movement.) The cranial base of the typical mammal is flat, in contrast to the human cranium, and the spinal cord passes into a horizontally directed vertebral column.

Which particular anatomic or functional change "came first" in this evolutionary chain has long been argued. Upright stance? Brachiation? Enlarged brain? Downward-rotated and decreasingly prognathic dental arches and jaws? Basicranial flexure? Development of hands and binocular vision? An important concept, however, is that these changes are functionally interrelated. They developed as a phylogenetic "package," regardless of which one (or combination) of them led off as a primary evolutionary step.

Concept 3

The large size of the human brain also relates to a rotation of the orbits toward the midline (Fig. 5–3). This results in a binocular arrangement of the orbits, a feature that complements finger-controlled manipulation of food, tools, weapons, and so forth. The absence of a long, protrusive muzzle does not block the close-up vision of hand-held objects. The human **mind** directs the **free hands** that can work with **three-dimensional** perspective in an **upright** stance on feet. The enormous size of the human brain and the human cranial base flexure are key factors, but **all** these changes are required, and they are all mutually interdependent.

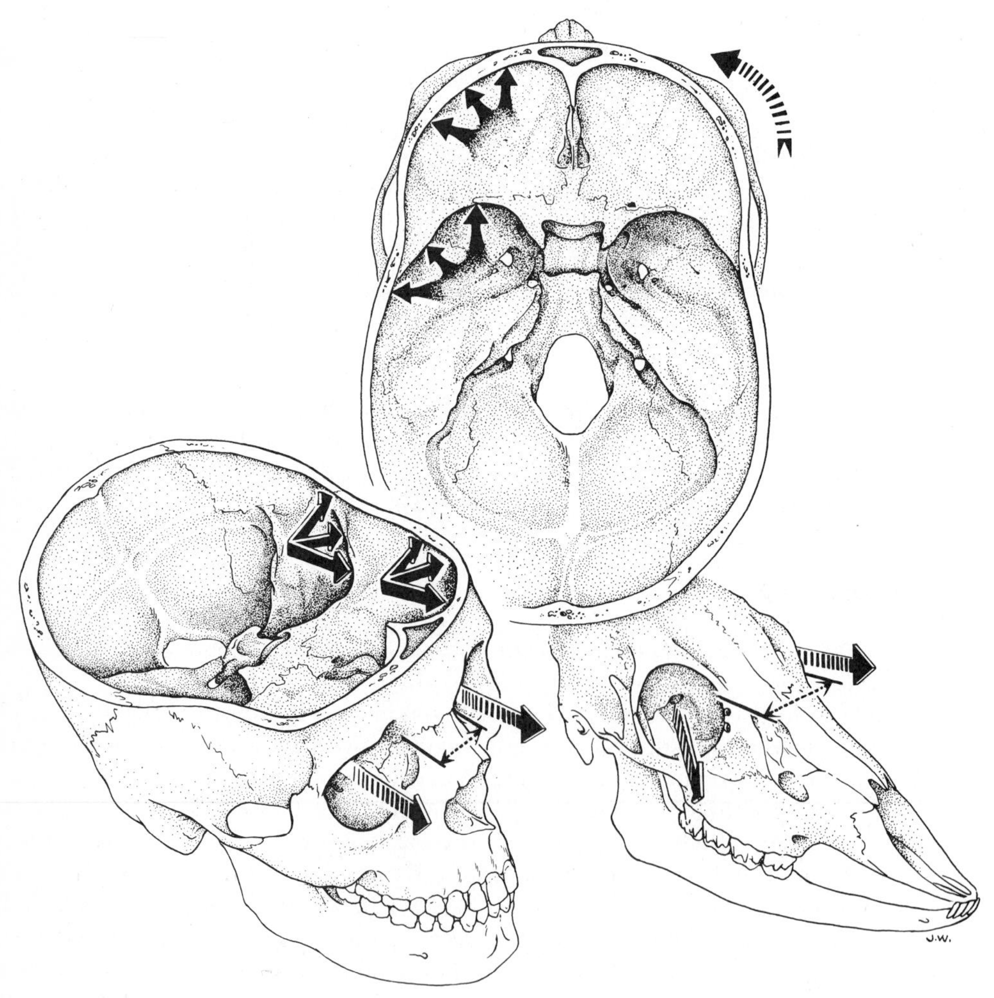

FIGURE 5–3. (From Enlow, D. H.: *The Human Face*. New York, Harper & Row, 1968, p. 190.)

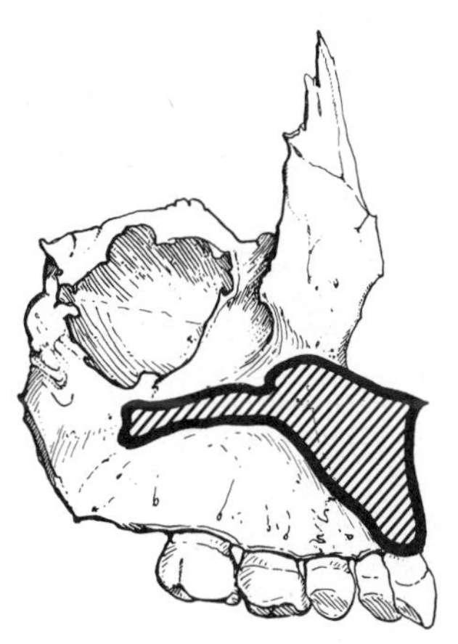

FIGURE 5–4. (From Enlow, D. H., and S. Bang: Growth and remodeling of the human maxilla. Am. J. Orthod., 51:446, 1965.)

Complete orbital rotation into a forward-pointing direction, however, has also caused a marked reduction in the interorbital part of the face. This is significant, because the area involved is the root of the nasal region, and the result of man's close-set eyes is a **narrow nose**. Because the nose is so thin, it is also necessarily quite short. The much broader nasal base of most other mammals supports a correspondingly much longer snout.

The olfactory sense in *Homo* has become a much less dominant factor in environmental awareness and is far exceeded by many other mammalian groups. In addition to proportionate down-sizing of the human nasal and olfactory mucosae, olfactory receptors in the mucosae of the frontal sinuses are also lacking, which is in contrast to other forms more dependent on aromatic sensations for food-getting or protection.

Concept 4

The nasal region above and the oral region below are two sides of the same coin, that is, the palate (Fig. 5–4). Reduction in nasal protrusion is accompanied by a more or less equivalent reduction of the jaw (nasal reduction apparently paced this evolutionary process). The whole face has necessarily become reduced in horizontal length as a result. However, the face has also been **rotated** into a nearly **vertical** alignment in relationship to the massive enlargement of the brain and the flexure of the cranial base. The downward rotation of the **olfactory bulbs** and the whole **anterior cranial floor** by the enlarged frontal lobes of the cerebrum has caused a corresponding downward rotation of the nasomaxillary complex.

Facial rotation has led to the development of the human **maxillary sinus** beneath the orbital floor and above the shortened maxillary arch (Fig. 5–5). Because of its adaptation to facial rotation, the human maxilla is uniquely rectangular, rather than triangular like that of most other mammals. It is a distinctively shaped upper jaw. An orbital **floor** has also been added to the human maxilla because the middle and lower parts of the face have been rotated to a position **beneath the eyes**.

The nasal mucosa is ordinarily an active site involved in temperature regulation in most mammals. Vasoconstriction and vasodilation of the vessels

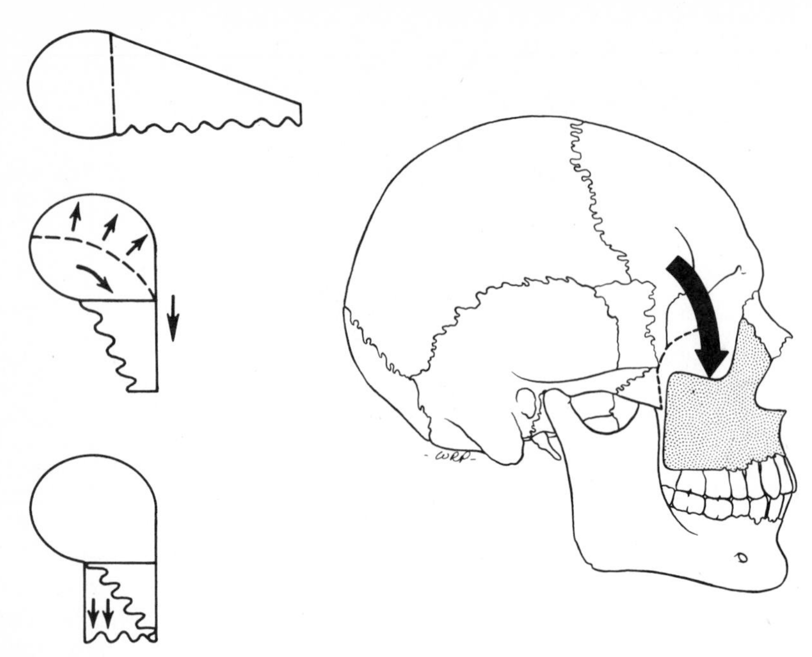

FIGURE 5–5. (From Enlow, D. H.: Postnatal growth and development of the face and cranium. In: *Scientific Foundations of Dentistry.* Ed. by B. Cohen and I. R. H. Kramer. London, Heinemann, 1975.)

in the massive spread of mucosa covering the turbinates control the amount of heat retention or loss. Because of marked nasal reduction in man, however, this function has been largely taken over by the relatively hairless and sweat gland–loaded integument. Control of blood flow in the dermis, combined with sweat gland activity, provides the equivalent for nasal thermoregulation. This is possible in man (and in a very few other species, such as the pig) because of a near-naked skin. In thick-furred animals, thermoregulation is carried out by regulating heat transfers in the nasal mucosa, panting to release excess heat, limited perspiration in hairless areas (such as the pads of the paws), and a fluffing of the fur to increase dead air insulation. The latter also makes the animal look larger to prospective enemies, and it increases the nonvital part of his anatomy for them to bite. We have only the holdover: goose bumps.

All mammalian forms have reinforcement "pillars" built into the architectonic design of the craniofacial complex. These pillars are parts of bones that provide a buttress for structural support and biomechanical stress resistance that balances the physical properties of the skull against the composite of forces acting within it, including growth itself, just as the framework of a building provides generalized support and stability. Although customarily described with reference to tooth positions, the nature of support goes well beyond just accommodation to masticatory forces. In the human face, one of these is the "key ridge," which is a vertical column of thickened maxillary bone approximately centered above the functionally important area around the upper first molar. Mechanical support then continues from this ridge into and through the lateral orbital rim on to the supraorbital-reinforced frontal bone. The second maxillary molar is reinforced by a vertical sheet of bone, the posterolateral orbital wall, which extends directly above this tooth. Except for the very thin bone enclosing the posterior part of the large maxillary sinus, the third molar is situated behind the orbit, and it has no further bony support above this. It has thus become effectively disfranchised mechanically and phylogenetically. The incisors are supported by an arch, the bony rim of the overlying nasal opening, with which it shares a common embryonic development, and also by the vertical nasal septum. Each canine is reinforced by a marked thickening of the lateral nasal wall, toward which the canine root points, and thence on to and through the thickened frontal process of the maxilla into the glabellar enlargement of the forehead.

Concept 5

The human face is exceptionally wide because the brain and cranial floor are wide. However, the face has been almost engulfed by the massive brain behind and above it (Fig. 5–6). Note the wondrously, incredibly colossal size of the human cranium, in comparison with that of the typical mammal. The expanded frontal lobes of the human brain lie **above** the eyes and almost the whole remainder of the face, rather than behind, and a forehead has thus been added. This also relates to the rotation of the orbits into vertical, forward-facing positions as well as to the rotation of the face as a whole into a downward-backward position.

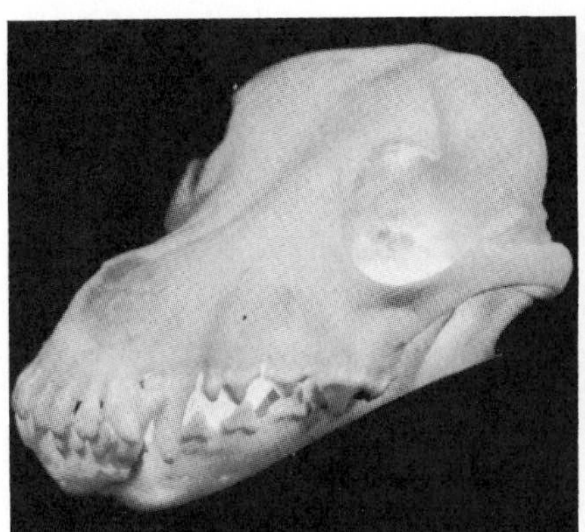

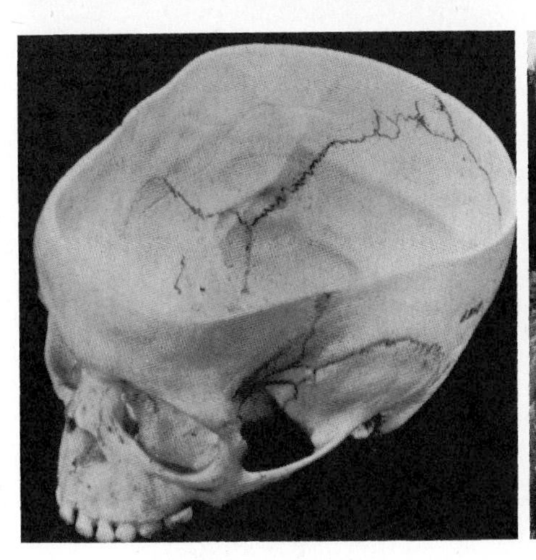

FIGURE 5–6

Concept 6

The expansion of the various parts of the cerebral hemispheres has created sizable pockets in the cranial floor (Fig. 5–7). Each of these **endocranial fossae** relates to specific lobes of the brain on the inside of the cranial floor and to specific parts of the face, pharynx, and so forth, on the outside. We can utilize our knowledge of these brain-cranial floor-facial relationships to advantage in analyzing the structure of the face and the basis for its many variations in form and pattern.

Concept 7

The nasomaxillary complex relates **specifically** to the anterior cranial fossa (Fig. 5–8). The anterior boundary of this fossa establishes where the anterior boundary of the nasomaxillary complex will be. The posterior boundary of the anterior cranial fossa determines the corresponding posterior boundary of the nasomaxillary complex. This is a key anatomic and functional rule.

The general configuration and proportions of the hard palate are a projection of the anterior cranial fossa. In turn, the apical base of the maxillary dental arch is established by the perimeter of the palate. Dental arch configuration thus relates indirectly to basicranial form, brain shape, and a person's head form type.

The pharynx relates specifically to the middle cranial fossa. Because of the human cranial floor flexure, the size of the middle cranial fossa in man determines the horizontal dimension of the pharyngeal space (Fig. 5–9). The dimension of the middle cranial fossa **should** be equaled by the breadth of the mandibular ramus. The function of the ramus is to span the pharynx and middle cranial fossa in order to place the lower arch in occlusion with the

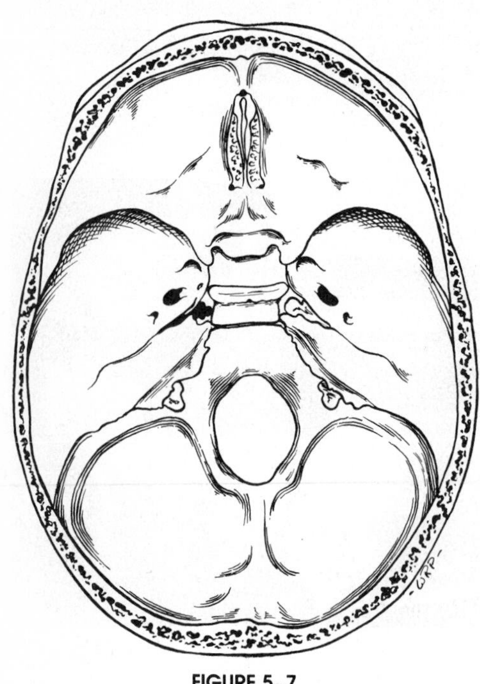

FIGURE 5–7

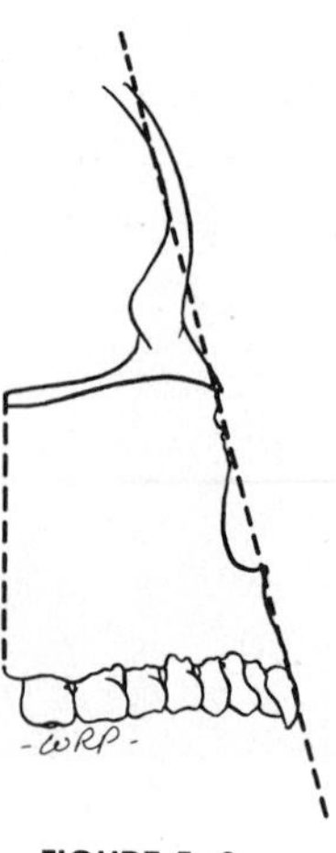

FIGURE 5–8

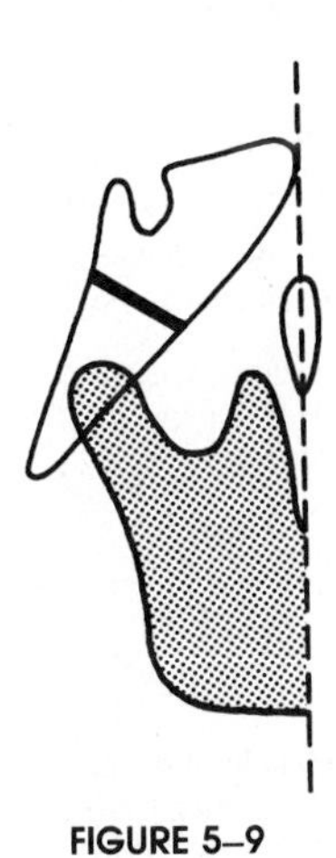

FIGURE 5–9

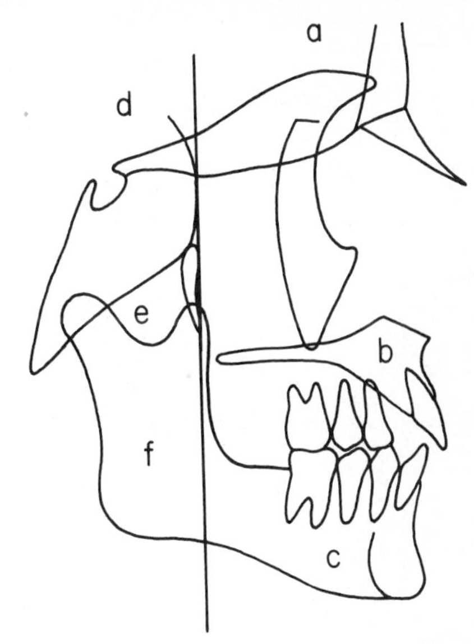

FIGURE 5–10. (From Enlow, D. H.: Postnatal growth and development of the face and cranium. In: *Scientific Foundations of Dentistry.* Ed. by B. Cohen and I. R. H. Kramer. London, Heinemann, 1975.)

upper. The length of the corpus of the mandible should also match the size of the bony maxillary arch. The mandible is a separate bone, however, joined to the skull by a movable articulation, and the size and the placement of its parts are independently variable. This is a major factor in the variations of facial form and profile. The maxilla, in contrast, is joined directly to the anterior cranial floor by sutures, and the growth of the cranium directly influences the corresponding growth of the midface, because common boundaries are shared by their respective growth fields.

Concept 8

One of the most basic and important planes in the whole head is the posterior maxillary **(PM)** plane. This is a natural anatomic boundary that represents the contact interface among certain key facial and cranial sites of growth, remodeling, and displacement. It will be recalled that a vertical reference line was used in Chapter 3 to visualize various major growth changes. This is the PM line, and it is a fundamental axis of growth activity involved in many relationships that exist during the overall growth process.

The PM is a natural boundary line (Fig. 5–10) that separates *a, b,* and *c* in the diagram from *d, e,* and *f.* The PM line delineates and establishes the boundary for these "counterparts" of the face and cranium. Thus, parts *a, b,* and *c* are all structural and growth counterparts to one another. The same is true for parts *d, e,* and *f.* Note that the boundary between the anterior and middle cranial fossae is the **exact** posterior-superior corner of the nasomaxillary compartment. The growth and size of the ethmomaxillary complex relate specifically to the frontal lobe of the cerebrum. They are counterparts. The floor of the anterior cranial fossa is the skeletal platform between them, and the anterior floor is thus a counterpart to the frontal lobe on one side and the upper part of the nasomaxillary complex on the other side.

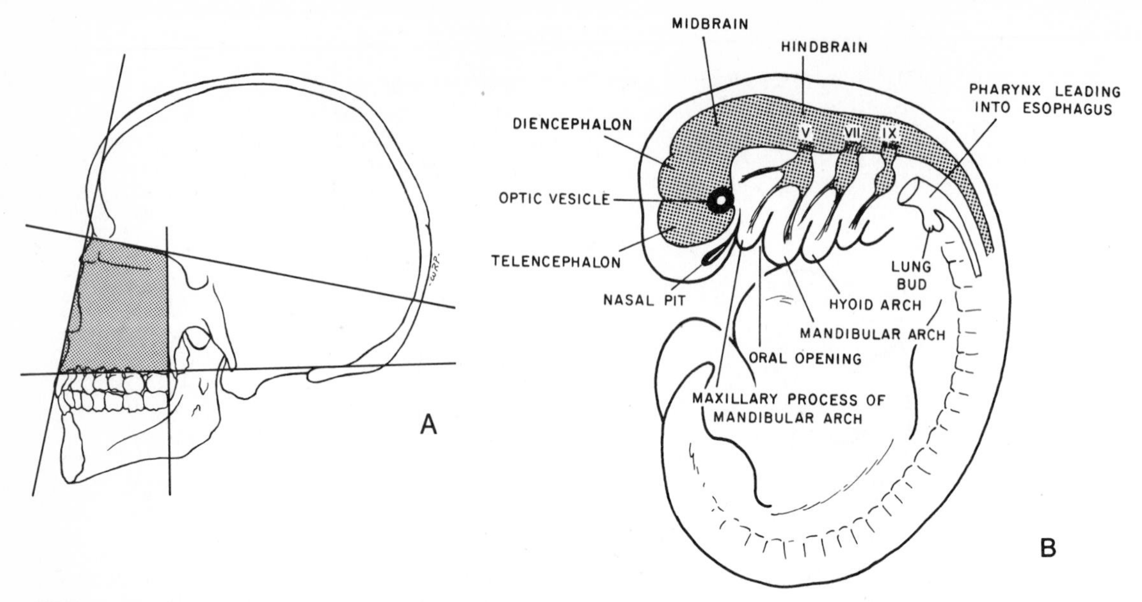

FIGURE 5–11. (*A* from Enlow, D. H., and M. Azuma: Functional growth boundaries in the human and mammalian face. In: *Morphogenesis and Malformations of the Face and Brain.* Ed. by D. Bergsma. Birth Defects Orig. Art. Ser., Vol. XI, No. 7. New York, Alan R. Liss, Inc., for The National Foundation—March of Dimes, White Plains, New York.)

Concept 9

As shown in Figure 5–11A, other major boundaries of the brain are shared by corresponding facial boundaries. These brain boundaries are established by growth fields, within which facial growth also takes place. The face has a prescribed perimeter of maximum growth, and this is the same as the growth perimeter for the brain. Some of the major **directions** of facial growth are established by the **special senses** located within the face itself (olfaction and vision). The two basic growth factors, amount and direction, constitute the growth "vector."

The anterior border of the nasomaxillary region corresponds to the anterior edge of the frontal lobe. The posterior margin of the nasomaxillary region corresponds to the posterior border of the frontal lobe (and in each case the anterior cranial fossa). The front and back vertical planes of the midface are **perpendicular** to the olfactory bulb and the orbit, respectively. The lower border of the midfacial compartment corresponds to the inferior level of the brain. All these various borders and planes represent the maximum normal growth boundaries for the nasomaxillary complex, and the growth directions relate to two major special senses that are directly involved in basic functions of the face.

Concept 10

The positional relationships between the frontal lobes of the cerebrum (anterior cranial fossae) and facial components, and also between part of the middle cranial fossae and the pharynx, are established early in embryonic development. In Figure 5–11B, note that the cephalic flexure places the maxillary and mandibular arches in direct juxtaposition to what will become the frontal lobes and the anterior cranial fossae.

Part 2

Although the human face is topographically "different" from the faces of other mammals, no special violations of the general mammalian plan for facial construction seem to have occurred. The face of man conforms to the same basic morphologic and morphogenetic rules complied with by most mammals in general. Differences have to do mostly with proportionate sizes of component parts and their rotational placements as related to body stance, head posture, and brain size and configuration, but not with any basic departures from the standard guidelines.

BRAIN ENLARGEMENT, BASICRANIAL FLEXURE, AND FACIAL ROTATIONS

If a short piece of adhesive tape is affixed to a rubber balloon and the balloon is then inflated, it will expand in a curved manner (Fig. 5–12). The balloon bends because it enlarges around the nonexpanding segment. The enormous human cerebrum similarly expands around a much smaller enlarging midventral segment (the medulla, pons, hypothalamus, optic chiasma). This causes a bending of the whole underside of the brain. The **flexure** of the cranial base results (Fig. 5–13). The foramen magnum in the typical mammalian skull is located at the posterior aspect of the cranium (Fig. 5–14). In man, it is in the midventral part of the expanded cranial floor at an approximate balance point for upright head support on a vertical spine (Fig. 5–15).

The expansion of the frontal lobes displaces the frontal bone upward and outward (Figs. 5–16 and 5–17). This results in the distinctive, bulbous, upright "forehead" of the human face, although it is really part of the neurocranium and not the face proper. The frontal lobes also relate to a rotation of the human orbits into new positions. As the forehead is rotated into a vertical plane by the brain behind it, the superior orbital rim is carried with it. The eyes now point at a right angle to the spinal cord. The spine is vertical, and the orbital axis is horizontal. Vision is directed toward forward body movement.*

*In some anthropoids, such as the gorilla, the massive supraorbital ridges may also rotate vertically independent of the frontal lobe. In the human face, however, the orbits **must** rotate into a vertical alignment because of the expanded size of the frontal lobes.

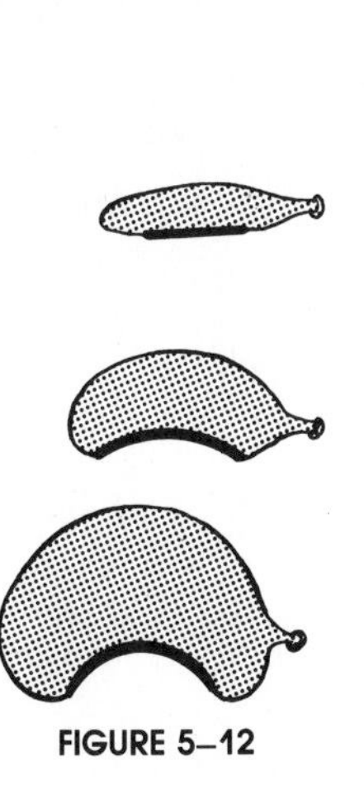

FIGURE 5–12

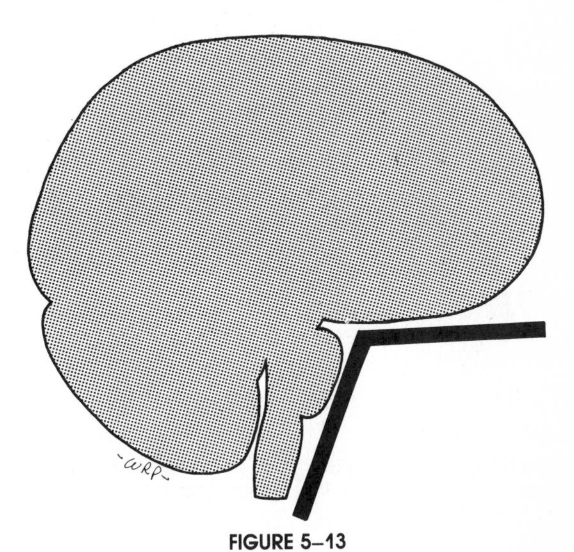

FIGURE 5–13

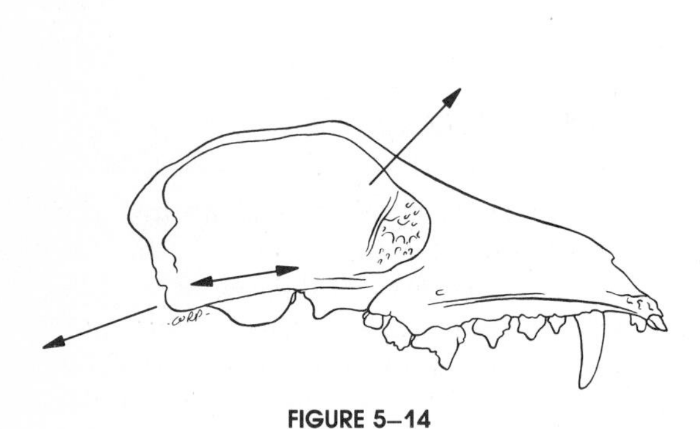

FIGURE 5–14

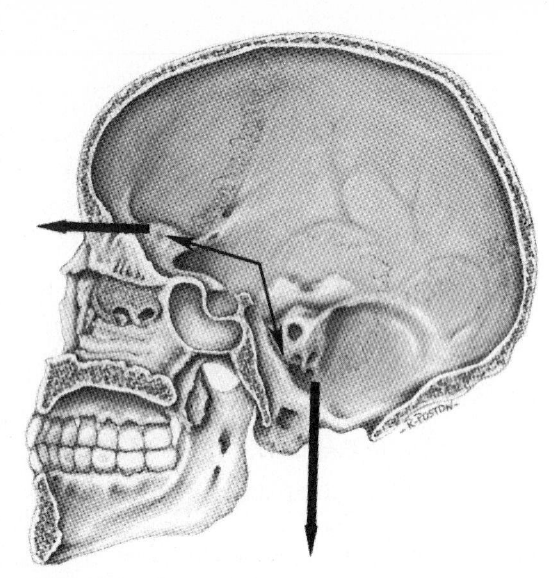

FIGURE 5–15. (From Enlow, D. H.: Postnatal growth and development of the face and cranium. In: *Scientific Foundations of Dentistry.* Ed. by B. Cohen and I. R. H. Kramer. London, Heinemann, 1975.)

The expansion of the frontal and, particularly, the temporal lobes of the cerebrum relates to a rotation of the orbits toward the midline (Figs. 5–18 and 5–19). The eyes come closer together. Two separate axes of orbital rotation are thus associated with the massive expansion of the cerebrum. One displaces the orbits vertically, and the other carries them horizontally in medial directions into a binocular position. Different extents of these two separate rotational movements are seen among different primate species. In the monkey, for example, the extent of upright orbital alignment and frontal bossing is much less than in man, as determined by the relative sizes of the frontal lobes. The simian orbits are quite close-set, however, and this relates to the proportionate sizes of the temporal lobes.

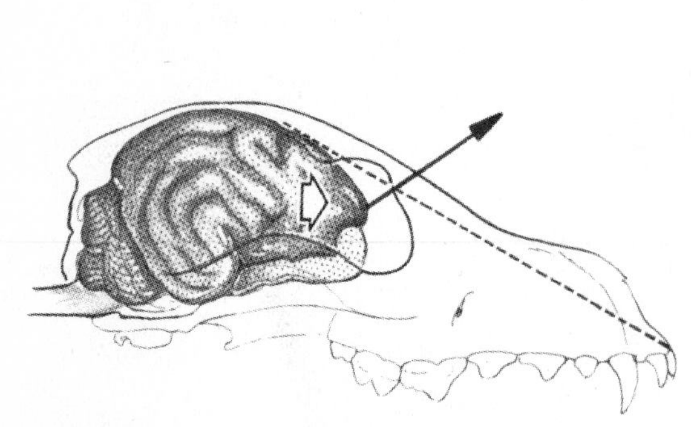

FIGURE 5–16. (From Enlow, D. H., and J. McNamara: The neurocranial basis for facial form and pattern. Angle Orthod., 43:256, 1973.

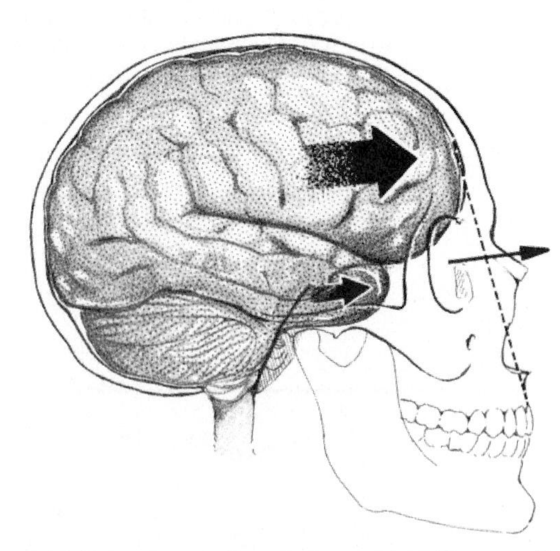

FIGURE 5–17. (From Enlow, D. H., and J. McNamara: The neurocranial basis for facial form and pattern. Angle Orthod., 43:256, 1973.)

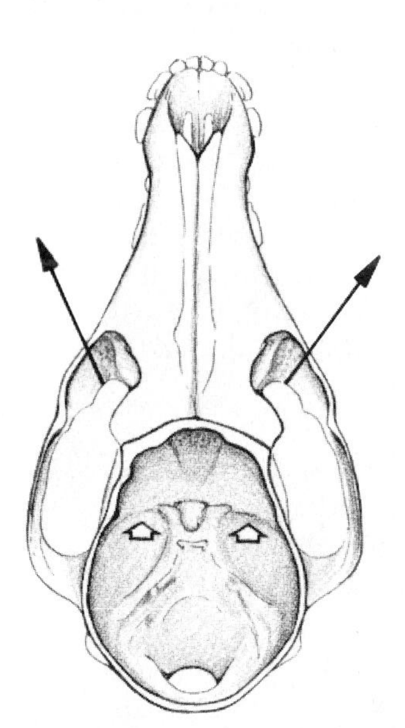

FIGURE 5–18. (From Enlow, D. H., and J. McNamara: The neurocranial basis for facial form and pattern. Angle Orthod., 43:256, 1973.)

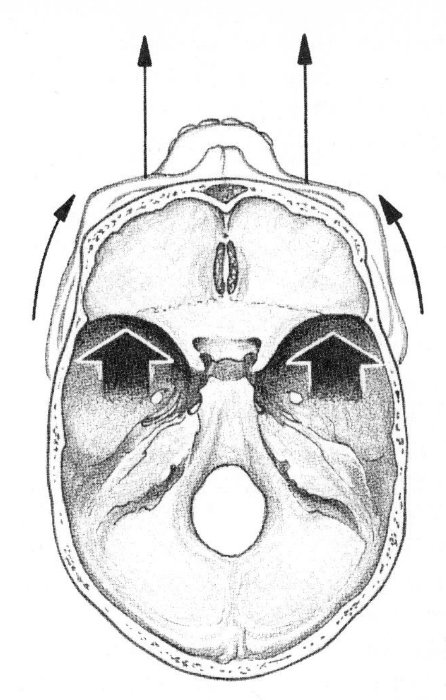

FIGURE 5–19. (From Enlow, D. H., and J. McNamara: The neurocranial basis for facial form and pattern. Angle Orthod., 43:256, 1973.)

Orbital rotation toward the midline, importantly, significantly reduces the dimension of the interorbital space (Fig. 5–20). This is one of two basic factors that underlie reduction in the extent of snout protrusion in man and some other (but not all) primates. Because the interorbital segment is the root of the nasal region, a decrease in this dimension reduces the structural (and also the physiologic) base of the bony nose. A wide nasal base can support a proportionately longer snout. A narrow nasal base, however, reduces the architectural

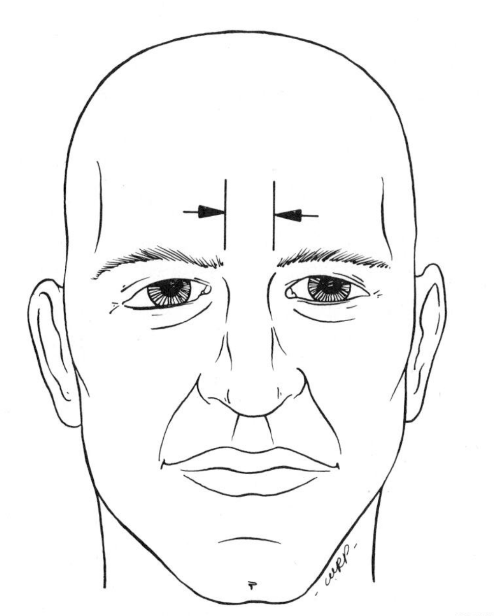

FIGURE 5–20

limit to which the bony part of the nose can protrude, and the snout is thereby shorter. The second basic factor involved in the extent of reduction of nasal protrusion deals with the rotation of the olfactory bulbs (see below).

Herbivores have orbits widely set to each facial side, thus providing a much greater peripheral range of vision to detect some approaching carnivore. The carnivore, on the other hand, can rotate its eyeballs into more forward-looking positions, even though the orbits are aligned obliquely laterally, thus favoring a stereoscopic chase. Carnivores generally tend to have a shorter muzzle and snout because of the greater degree of medial rotation of the orbits, producing a more narrow interorbital nasal base. Herbivores generally tend to have greater snout and muzzle protrusion, because their orbits are much more wide-set, with a broader interorbital nasal root.

Note that the enlarged human cerebrum has caused a downward rotational displacement of the olfactory bulbs (Fig. 5–21). In all other mammals, they are nearly upright or obliquely aligned, depending on the size and configuration of the frontal lobes. In man, the bulbs have been rotated into **horizontal** positions by the cerebrum. This is a significant factor in the basic design of the human face.

The olfactory bulbs relate directly to the alignment and the direction of growth of the adjacent nasal region (Fig. 5–22). The long axis of the snout in most mammals is constructed so that it necessarily points in the general direction of the sensory olfactory nerves within it. The plane of the nasomaxillary region is thereby approximately **perpendicular to the plane of the olfactory bulbs**. This is a major anatomic and functional relationship involved in the basic plan of the face in any mammal. As the bulbs become rotated progressively from a vertical position to a horizontal one because of increases in brain size or because of its shape *(1, 2, 3)*, the whole face is similarly rotated from a horizontal to a vertical plane *(1a, 2a, 3a)*. Or, stated another way, the face is rotated down by the expanded anterior cranial floor as it rotates downward as a result of the enlargement of the frontal lobes.

NASOMAXILLARY CONFIGURATION

The maxilla of most mammals has a triangular configuration. In man, it is uniquely rectangular (Figs. 5–23 and 5–24). This is caused by a rotation of the

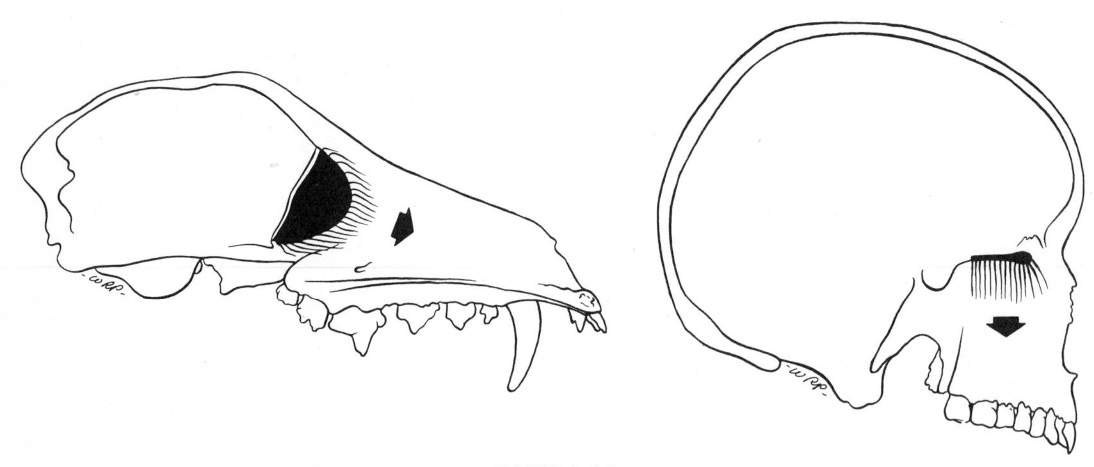

FIGURE 5–21

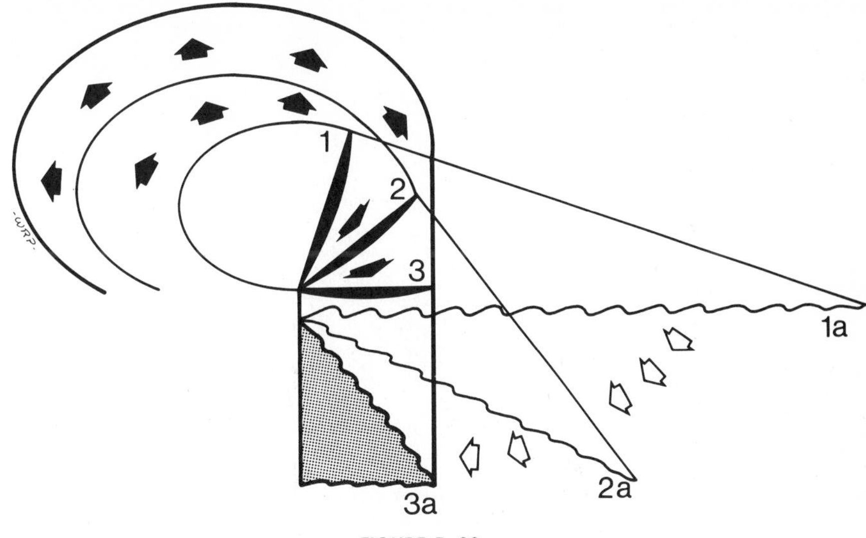

FIGURE 5–22

occlusion into a horizontal plane to adapt to the vertical rotation of the whole midface. The occlusal plane in most mammals, including man, is approximately parallel to the Frankfort plane (a plane from the top of the auditory meatus to the inferior rim of the orbit). This aligns the jaws in a functional position relative to the visual, olfactory, and hearing senses. In the human maxilla, the design change that allows for this resulted in the creation of a new arch-positioning facial region, the **suborbital** compartment. Most of this phylogenetically expanded area is occupied by the otherwise nonfunctional maxillary sinus (uses such as air warming, nasal drip, and voice resonance are secondary). An **orbital floor** was also newly created in conjunction with this added facial region. Compare also with Figure 5–22.

The nasal region is thus **vertically** disposed in the human face (Fig. 5–25). The neutral axis of the spread of the sensory olfactory nerves is vertical, and the vertical vector of nasomaxillary growth has become a major feature of

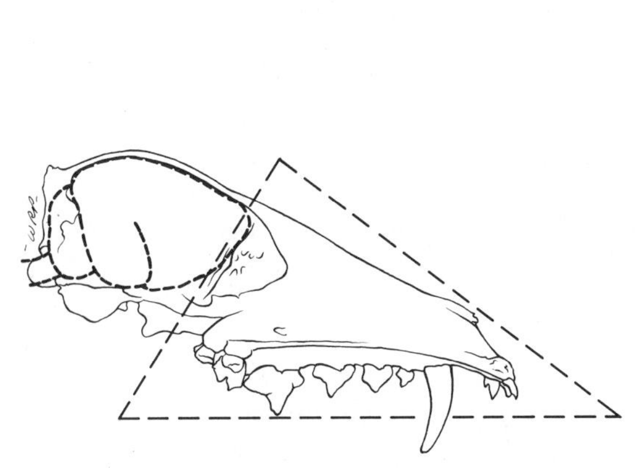

FIGURE 5–23. (From Enlow, D. H.: Postnatal growth and development of the face and cranium. In: *Scientific Foundations of Dentistry.* Ed. by B. Cohen and I. R. H. Kramer. London, Heinemann, 1975.)

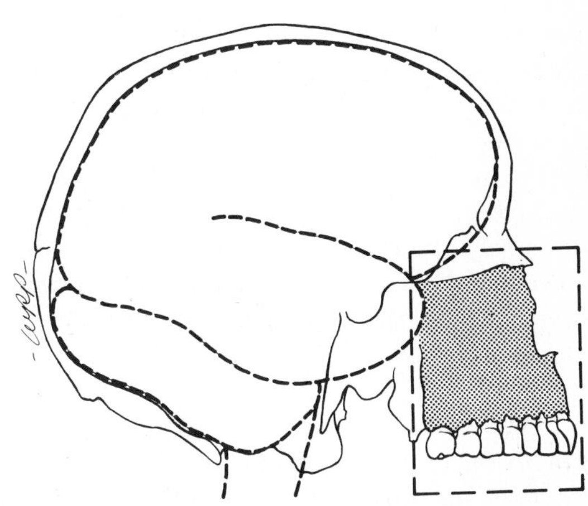

FIGURE 5–24. (From Enlow, D. H.: Postnatal growth and development of the face and cranium. In: *Scientific Foundations of Dentistry.* Ed. by B. Cohen and I. R. H. Kramer. London, Heinemann, 1975.)

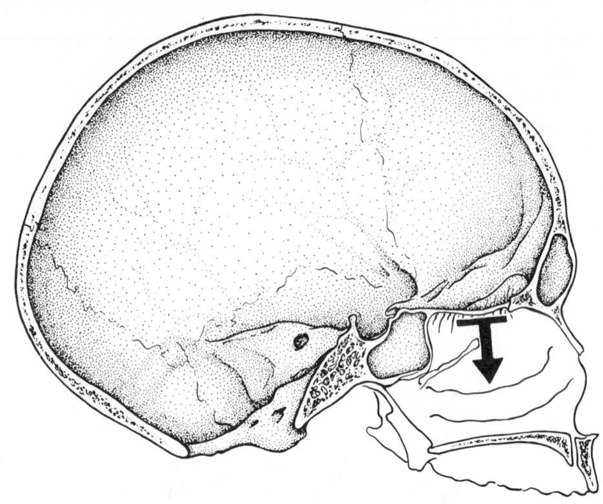

FIGURE 5–25. (Modified from Enlow, D. H.: *The Human Face.* New York, Harper & Row, 1968, p. 187.)

human facial development. The characteristic vertical human facial profile is a composite result of (1) a bulbous forehead, (2) rotation of the nasal region into a vertical plane, (3) reduction of snout protrusion in conjunction with medial orbital convergence, (4) rotation of the orbits into upright positions, (5) rotation of the maxillary arch downward and backward, and (6) bimaxillary reduction in the extent of prognathism matching nasal reduction. The face also becomes markedly widened because of the increased breadth of the brain and cranial floor and because the orbits and cheekbones are rotated into forward-facing positions. The face of man lies **beneath** the frontal lobe of the brain; in contrast, in other mammals the face is largely in front of the cerebrum. The nasal chambers are housed largely **within** the face, between and below the orbits, rather than projecting forward with a protrusive muzzle. The human snout itself houses very little of the mucosal part of the nasal chambers. The whole face has been "reduced" to a quite flat topographic configuration as a combined result of these various alterations.

Reduction of the nasal region associated with orbital convergence and olfactory-anterior cranial fossa rotation must necessarily also be accompanied by a more or less equal reduction in maxillary arch length, because the floor of the nasal chamber is also the roof of the mouth (Fig. 5–26). Only a relatively slight degree of horizontal divergence between the two can exist. The palate is shared by both regions. Whether the phylogenic chain of events that led to nasal reduction "caused" a shortening of the maxillary arch or whether, conversely, the chain went in the other direction we shall probably never know. However, if either one becomes reduced in length, so must the other. This refers only to the bony part of the nasal region; some species have a fleshy proboscis protruding well beyond the jaws and palate (such as man and the elephant).

Why does the human face have an overhanging, fleshy "nose"? The protrusion of the cartilaginous and soft tissue portion of the nasal complex provides for **downward**-directed external nares (Fig. 5–27). This aims the inflow of air obliquely upward into the vertically disposed nasal chambers toward the vertically aligned sensory nerves of the olfactory bulbs located on the **ceiling** of the chambers. This is in contrast to the anteriorly directed

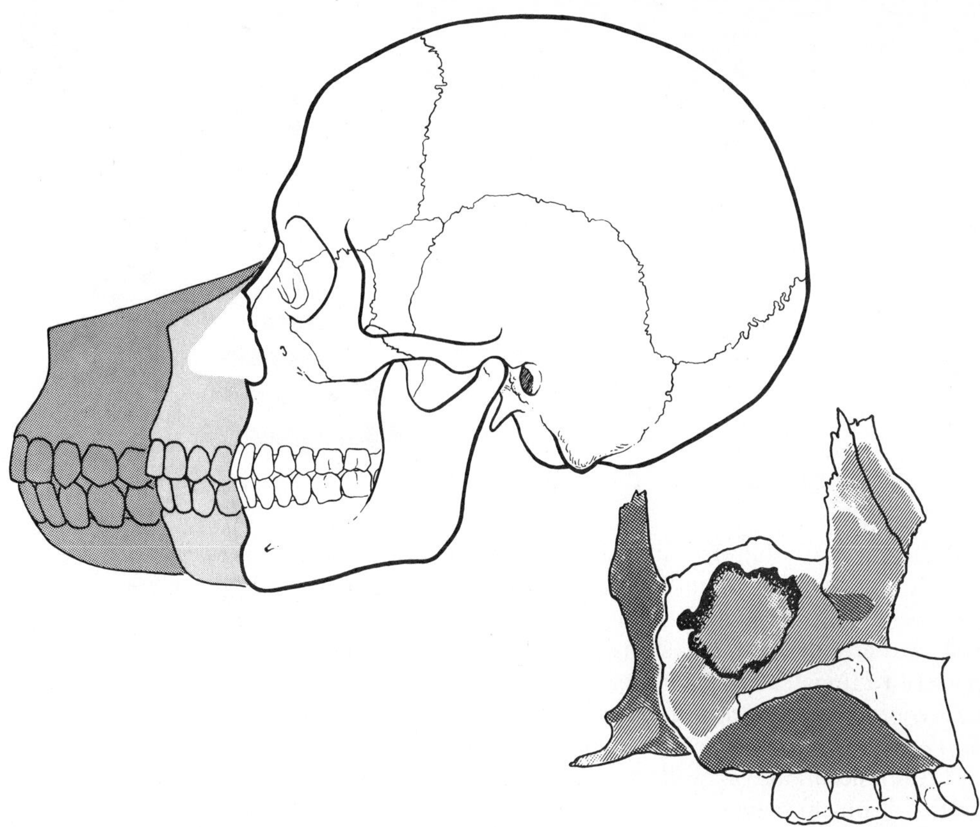

FIGURE 5–26. (Lower figure from Enlow, D. H., and S. Bang: Growth and remodeling of the human maxilla. Am. J. Orthod., 51:446, 1965.)

external apertures of other mammals taking air into more horizontal nasal chambers having the cribriform plates located as part of the posterior wall. Thus, the human face has a fleshy, protuberant "nose."

The rotation of the whole face downward and **backward** has resulted in a facial placement within the recess (the "facial pocket") created by the cranial base flexure (Fig. 5–28). What will happen if the brain **continues** to enlarge

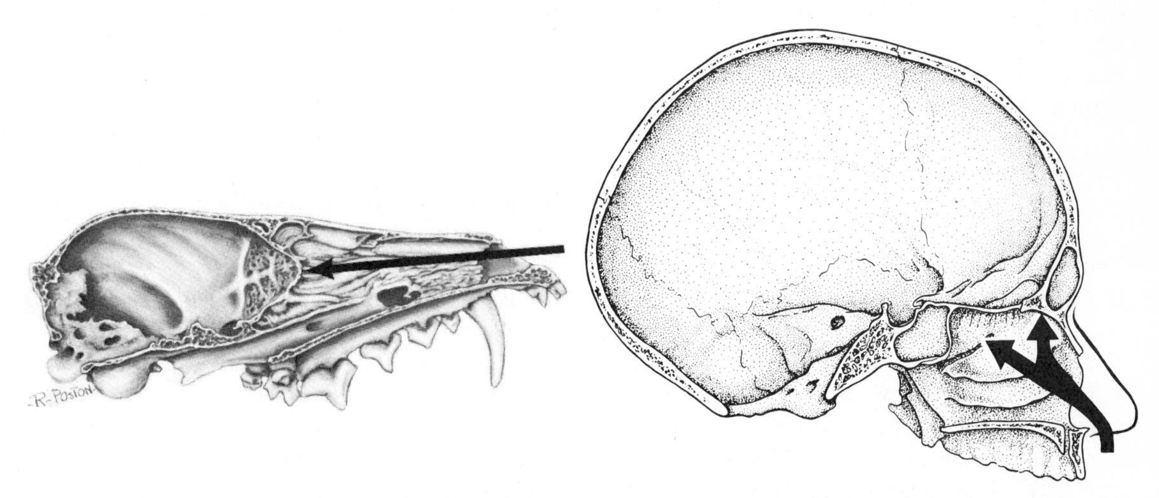

FIGURE 5–27. (Figure at right from Enlow, D. H.: *The Human Face*. New York, Harper & Row, 1968, p. 188.)

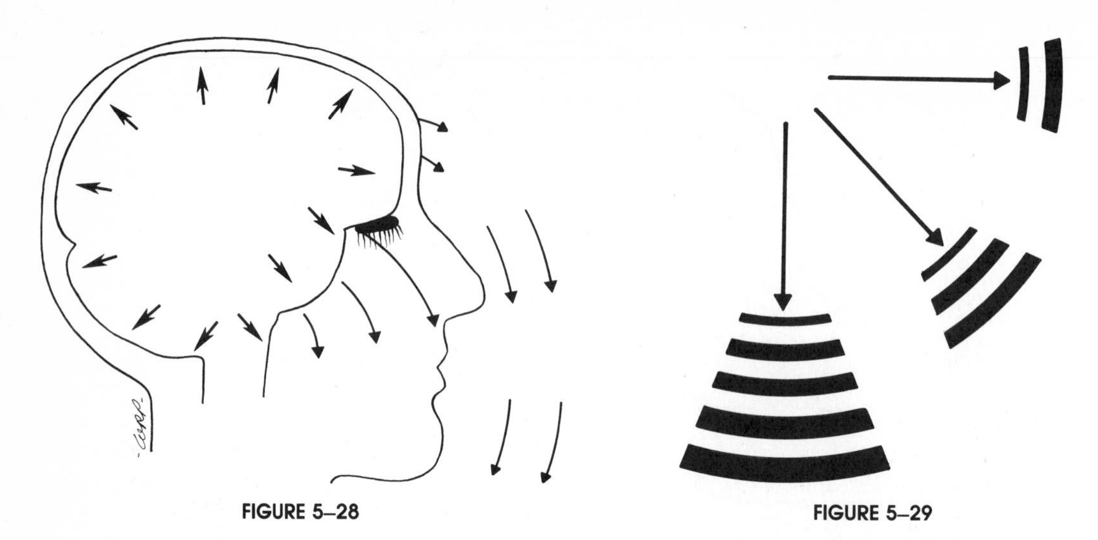

FIGURE 5–28

FIGURE 5–29

phylogenetically and thereby produce an even further extent of backward rotation? It has virtually no more room left to "rotate into," at least given the present design and arrangement of the various soft tissue and skeletal parts involved. Already the face has almost reached the airway because of rotation. The posterior cranial floor, vertebral column, and face, as in a closing vise, are coming together; there are important parts in between. What further phylogenetic facial adjustments or compensations can you think of? With all the science-fiction situations, as well as legitimate anthropologic prognostics, why has not **this** problem been considered? (A study of the porpoise skull can show how evolutionary craniofacial adjustments have occurred in a way that is unconventional among mammals.)

GROWTH FIELD BOUNDARIES

The growth of each part of the face involves two basic considerations. The first is the **amount** of growth, and the second is the **direction** of growth (Fig. 5–29). These two factors constitute the growth "vector."

The amount of growth involves fields of growth and the boundaries of these fields (Fig. 5–30). There is a prescribed perimeter of maximal growth capacity for each major part of the face, and growth does not ordinarily exceed this perimeter. Significantly, the forward, downward, backward, and lateral growth boundaries that exist for the face are **shared** by the brain. The perimeter for the field of brain growth and the perimeter for the field of facial growth have become established in common.

The reason for this basicranial and facial relationship is that the brain has evolved in conjunction with the cranial floor. The form, size, topographic features, and angular characteristics of one conform to the other. The floor of the cranium, in turn, is the **template** upon which the face is built. The junctional part of the face cannot be significantly **wider**, for example, than the maximal width of the cranium. There would be nothing to which it could be attached. Similarly, the **length** and **height** of specific parts of the cranial floor are expressed as equivalent dimensions for the face.

The face is **not** independent of the basicranium structurally or developmentally. The whole face has often been described as a genetically and developmentally separate region, the only tie between them being that they happen to be placed in juxtaposition, as a picture hangs in juxtaposition to a wall. No cause-and-effect relationships were believed to exist between the cranium and the size or shape of the face. This is certainly not the case. **Many** structural features and dimensions of the face are based on brain-cranial base-facial relationships, as will be seen. This is an important concept, because a number of normal and abnormal variations in facial form relate, at least in part, to underlying circumstances present in the cranial floor (see Chapter 6).

The floor of the cranium has developed in **phylogenetic** association with the brain. Whatever "independent" genetic control actually exists in the cranial base (this is controversial), the shape and size of the floor of the cranium have become established because its genes were especially adopted by the natural selection process to accommodate an **interdependent** association with the developing brain. Which came first will probably always be argued, although the brain gets the nod by most present-day theoreticians. Thus established, however, the basicranium presumably has a measure of genetic independence. The face similarly develops **in conjunction with** the cranial floor (and the brain), and the genetic control of facial development (by its soft tissues) has become established to accommodate its own functional association with the cranium. In all areas, however, a developmental **latitude** exists that provides **adjustments** during growth to accommodate variations in one part or another. This is involved in the operation of the "functional matrix," and it is the factor that allows for a unified coexistence of the many separate, developing parts all growing in relation to one another. There are regional differences, however, in the **capacity** for such developmental adjustment. Some areas, such as the bony alveolar sockets, are extremely labile and responsive to variable circumstances. Other areas, such as the basicranium, are much less sensitive and adaptable. The "intrinsic programming" for the latter is presumed to be greater than for the former because of their different levels of developmental independence and whatever the different, poorly understood factors are that determine and control this.

The endocranial side of the cranial floor is adapted to the configuration and topographic contours of the ventral surface of the brain. The topography of the ectocranial side of the basicranium, however, is structurally adapted to

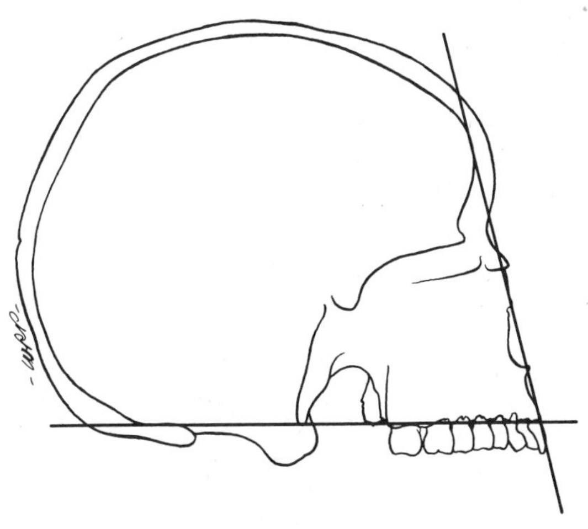

FIGURE 5–30

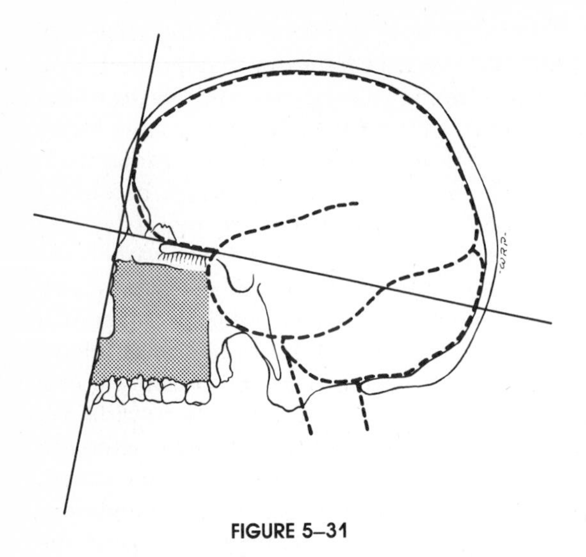

FIGURE 5–31

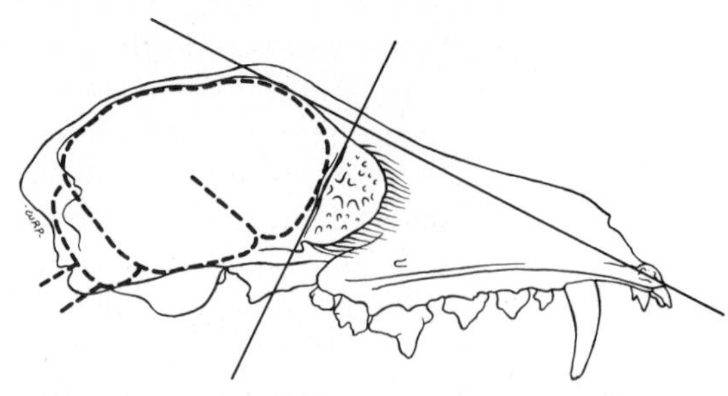

FIGURE 5–32. (From Enlow, D. H., and M. Azuma: Functional growth boundaries in the human and mammalian face. In: *Morphogenesis and Malformations of the Face and Brain.* Ed. by D. Bergsma. Birth Defects Orig. Art. Ser., Vol. XI; No. 7. New York, Alan R. Liss, Inc. for The National Foundation—March of Dimes, White Plains, New York.)

the composite of facial, pharyngeal, and cervical components related to the external part of the skull. A measure of morphogenic divergence thus occurs between these two sides of the cranial floor.

The forward **boundary** of the brain is shared by the forward border of the nasomaxillary complex. The **direction** of growth by the nasal part of the face is established by the olfactory bulbs and the sensory olfactory nerves. These two factors underlie the "vector" of midfacial growth, that is, the amount and the direction. To show this, a line is drawn from the forward edge of the brain down to the anterior-most, inferior-most point of the nasomaxillary complex (superior prosthion; Figs. 5–31 and 5–32). This represents the **midfacial** plane.* Note that the midfacial plane is perpendicular to the **olfactory bulb** (or the cribriform plate, as seen in lateral headfilms). In the human face, this plane also frequently touches the anterior nasal spine. The long axis of the nasal region points in the same general direction as the neutral axis of its sensory nerve spread. The amount of growth is established by the prescribed perimeter of its growth field. **The nasomaxillary complex grows as far forward as the edge of the brain in a direction approximately perpendicular to the olfactory bulbs.**

The olfactory bulb and nasomaxillary alignment relationship exists among mammals in general. In species or groups having a smaller brain and, as a result, a more upright olfactory bulb, the snout and muzzle tend to be correspondingly more horizontal and much more protrusive. As the olfactory bulbs become rotated downward because of increasing brain size (or shape, as in more round-headed species), the muzzle correspondingly rotates down with them and becomes less protrusive. In man, the olfactory bulbs have become virtually horizontal because of the massive growth of the frontal lobes. The

*"Nasion" is often used in cephalometric studies as a point for drawing the facial plane, but this is a poor selection because nasion is so variable in relation to distinct male/female and head form differences, which are seldom taken into account. Moreover, the purpose of the **midfacial** plane described above is to show the relationship between the **brain** and the **nasomaxillary complex**. Thus, the edge of the brain is used, rather than nasion. Also, the above descriptions presume that the alignment of the nerves is the lead factor that determines the direction of midfacial growth. Of course, it may be the converse. Whichever, the important point is that they are established **together** in a constant relationship to the olfactory bulbs, which, in turn, are placed according to the size and shape of the brain.

nasal part of the face is thus vertically aligned in conjunction with the neutral vertical axis of olfactory nerve distribution (Figs. 5–33 and 5–34).

The nasomaxillary complex, as mentioned earlier, is specifically associated with the anterior cranial fossa. The posterior boundary of this fossa establishes the corresponding posterior boundary for the midface. This is essentially a nonvariable anatomic relationship. The **direction** of growth in this region is established by the particular special sense located in this part of the face, which is the visual sense. The posterior maxillary tuberosity is located beneath the floor of the orbit, and the orbital floor is the roof of the maxillary tuberosity. The tuberosity is aligned approximately perpendicularly to the neutral geometric axis of the orbit (Fig. 5–35). **The posterior plane of the midface extends from the junction between the anterior and middle cranial fossae (and the inferior junction between the frontal and temporal lobes) downward in a direction perpendicular to the neutral line of the orbit.** This plane passes almost exactly along the posterior surface of the maxillary tuberosity.

The boundary just described represents one of the key anatomic planes in the face. This is the PM plane (Fig. 5–36). There are many "cephalometric planes" in the face and cranium. Most of these, however, do not represent (and are not so intended) (1) key sites of growth and remodeling or (2) functional relationships among the various parts of the skull, including soft tissue associations. Most conventional cephalometric planes, such as sella-nasion, bypass the

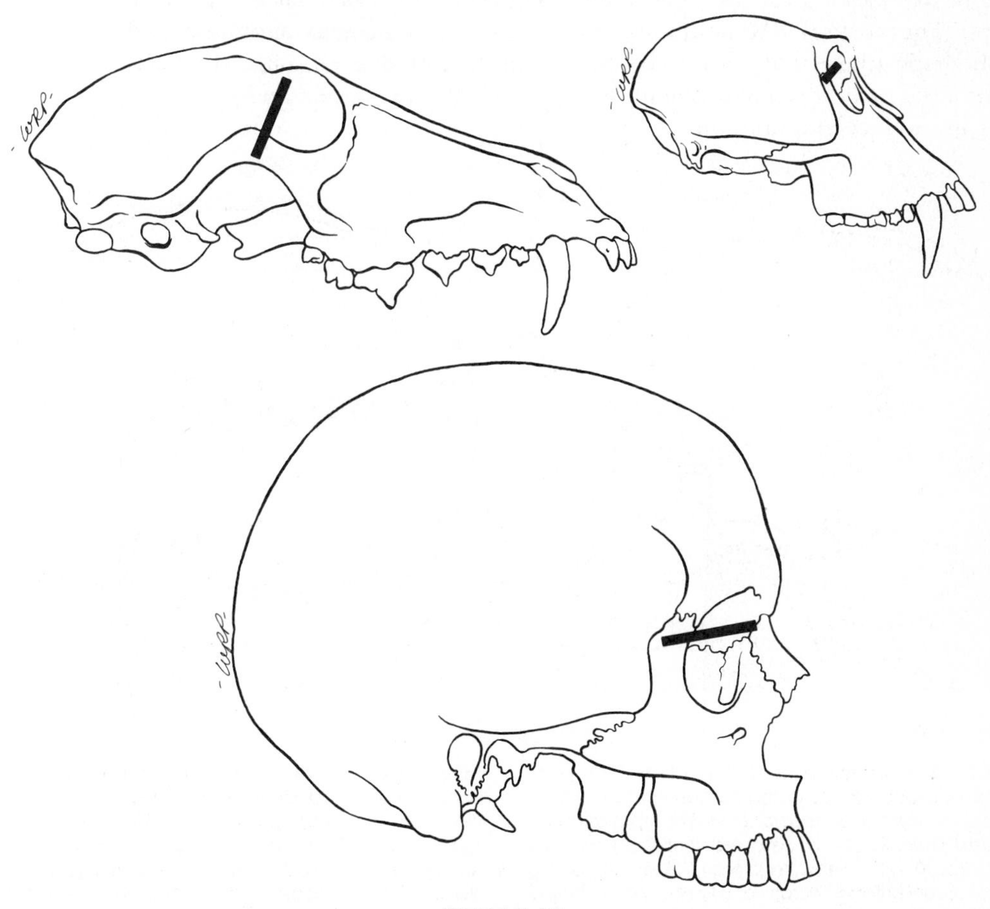

FIGURE 5–33

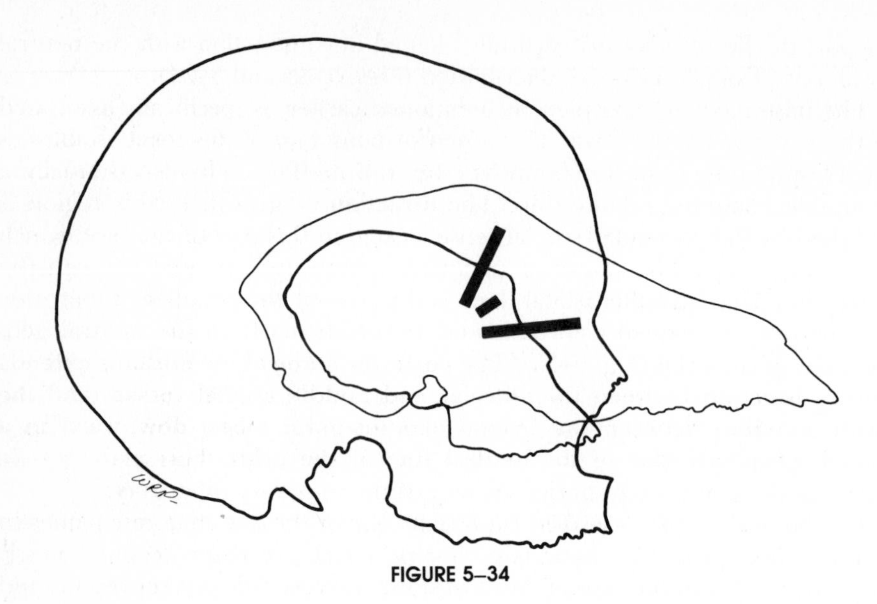

FIGURE 5–34

really important key sites of growth without recognizing them. Sella itself, for example, is a "landmark of convenience" because it can be readily and reliably located. But trying to use it to determine real morphogenetic relationships would be like looking for lost keys under a lamp post because that is where the light is. The vertical PM boundary, in contrast, is a **natural anatomic and morphogenic** plane that relates directly to the factors that establish the basic design of the face. It is one of the most important developmental and structural planes in the face and cranium.

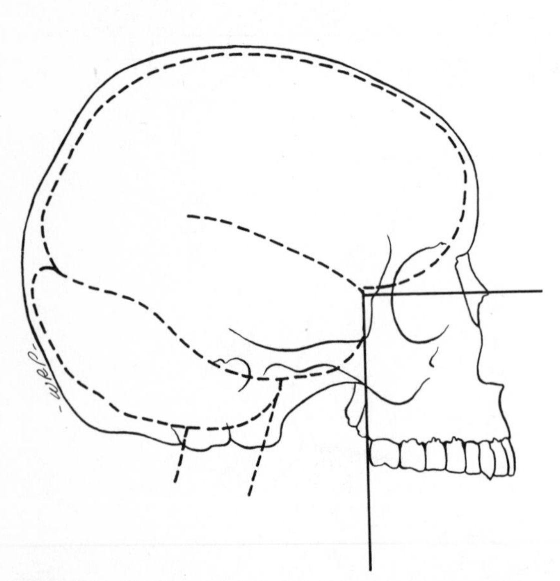

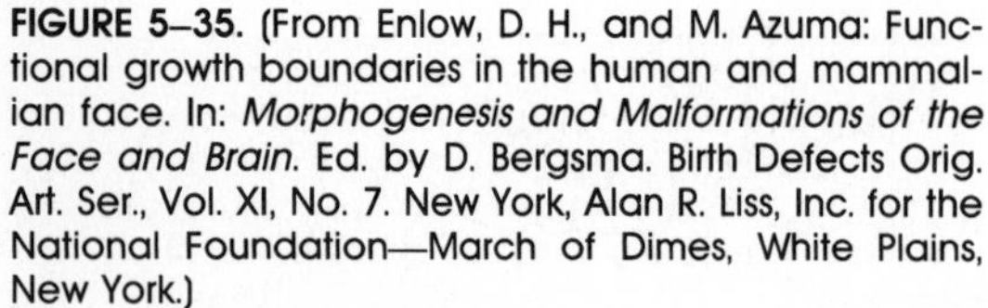

FIGURE 5–35. (From Enlow, D. H., and M. Azuma: Functional growth boundaries in the human and mammalian face. In: *Morphogenesis and Malformations of the Face and Brain.* Ed. by D. Bergsma. Birth Defects Orig. Art. Ser., Vol. XI, No. 7. New York, Alan R. Liss, Inc. for the National Foundation—March of Dimes, White Plains, New York.)

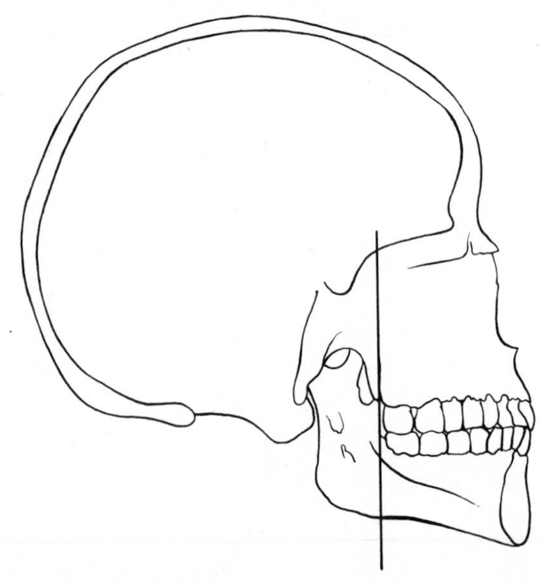

FIGURE 5–36. (From Enlow, D. H., and M. Azuma: Functional growth boundaries in the human and mammalian face. In: *Morphogenesis and Malformations of the Face and Brain.* Ed. by D. Bergsma. Birth Defects Orig. Art. Ser., Vol. XI, No. 7. New York, Alan R. Liss, Inc. for The National Foundation—March of Dimes, White Plains, New York.)

The **PM** plane delineates naturally the various anatomic **counterparts** of the craniofacial complex. The frontal lobe, the anterior cranial fossa, the upper part of the ethmomaxillary complex, the palate, and the maxillary arch are all mutual counterparts lying anterior to the PM line (Fig. 5–37). All these parts have posterior boundaries that are placed along this vertical plane. Similarly, the temporal lobe, the middle cranial fossa, and the posterior oropharyngeal space are mutual counterparts located behind the PM plane. The anterior boundaries of these parts are precisely positioned along this vertical line. The PM plane is a **developmental interface** between the series of counterparts in front of and behind it. This key plane retains these basic relationships throughout the growth process.

The **corpus** of the mandible is a counterpart to those parts lying in front of the PM plane. The **ramus** is a counterpart of the parts behind the PM plane. The placement of the mandible and the size of its parts, however, are somewhat more independently variable than those of the ethmomaxillary complex. The posterior boundary of the corpus **should** lie on the PM line. This is the "lingual tuberosity," which is the direct mandibular equivalent of the maxillary tuberosity. The forward boundary of the ramus, where it joins the lingual tuberosity, should also lie on the PM line. (Note: The anterior edge of the obliquely aligned ramus overlaps the lingual tuberosity, but this edge does not represent the actual forward point of the effective ramus dimension. The lingual tuberosity itself is the functional junction between the corpus and the ramus.) Because the mandible is a separate bone not attached directly to the cranium by sutures, its latitude for structural variation is not subject to the same degree of developmental and structural communality that occurs between the growth fields shared by the cranial floor and the maxilla. Independent variations can

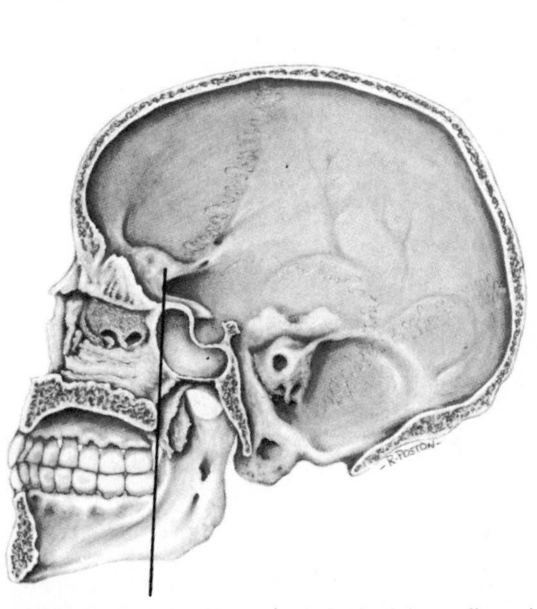

FIGURE 5–37. (From Enlow, D. H.: Postnatal growth and development of the face and cranium. In: *Scientific Foundations of Dentistry.* Ed. by B. Cohen and I. R. H. Kramer. London, Heinemann, 1975.)

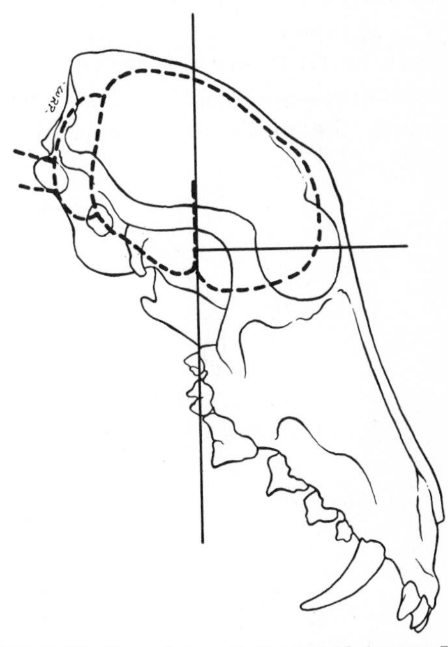

FIGURE 5–38. (From Enlow, D. H., and M. Azuma: Functional growth boundaries in the human and mammalian face. In: *Morphogenesis and Malformations of the Face and Brain.* Ed. by D. Bergsma. Birth Defects Orig. Art. Ser., Vol. XI, No. 7. New York, Alan R. Liss, Inc., for The National Foundation—March of Dimes, White Plains, New York.)

thus exist in the dimensions and the placement of both the ramus and the corpus. The ramus, for example, may fall short of the PM plane, or it may protrude well forward of it. This variability feature is often **compensatory**. That is, a narrow or broad ramus can offset a tendency for an imbalance or a malocclusion caused by factors in **other** parts of the face and cranium. In any event, such mandibular variations can be recognized, and in analyzing the basis for any given individual's facial pattern, normal or abnormal, the part played by the mandible can be determined by noting where the lingual tuberosity is located with respect to the maxillary tuberosity (or with the PM plane), as described in Chapter 6.

The above relationships hold true for mammals in general. Thus, the posterior boundary of the midfacial compartment in most species coincides with the posterior boundary of the anterior cranial fossa (the frontal lobe–temporal lobe junction on the floor of the cranium). The posterior midfacial plane extends downward from this point in a direction approximately perpendicular to the neutral axis of the orbit (Fig. 5–38). This line passes along the posterior surface of the maxilla. Note that the head is aligned in its neutral position with the orbital axis pointed straight forward in the direction of body movement.

Just as the other boundaries of the midface coincide with respective brain boundaries, the inferior boundary of the nasomaxillary complex is established by the bottom-most surface of the brain and cranial floor (Figs. 5–39 and 5–40). Note that a line from the floor of the posterior cranial fossa passes through, or very nearly so, both the inferior corner of the posterior maxillary tuberosity and the inferior corner of the front part of the bony maxillary arch (prosthion). This relationship is achieved only after facial development is complete, because neurocranial growth precedes facial growth.

The alignment of the hard palate conforms closely to the alignment of the maxillary nerve before this nerve enters the orbital floor (Fig. 5–41). This relationship exists in most mammals, including man. Passing from the foramen rotundum, the maxillary nerve crosses the pterygopalatine fossa and then

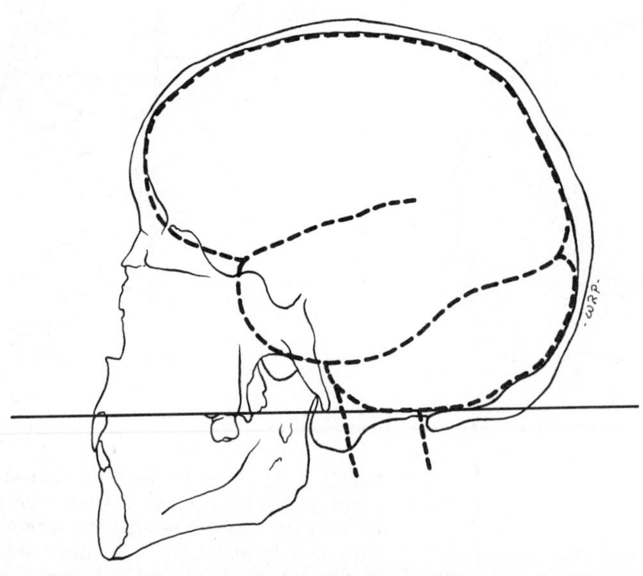

FIGURE 5–39. (From Enlow, D. H., and M. Azuma: Functional growth boundaries in the human and mammalian face. In: *Morphogenesis and Malformations of the Face and Brain.* Ed. by D. Bergsma. Birth Defects Orig. Art. Ser., Vol. XI, No. 7. New York, Alan R. Liss, Inc., for The National Foundation—March of Dimes, White Plains, New York.)

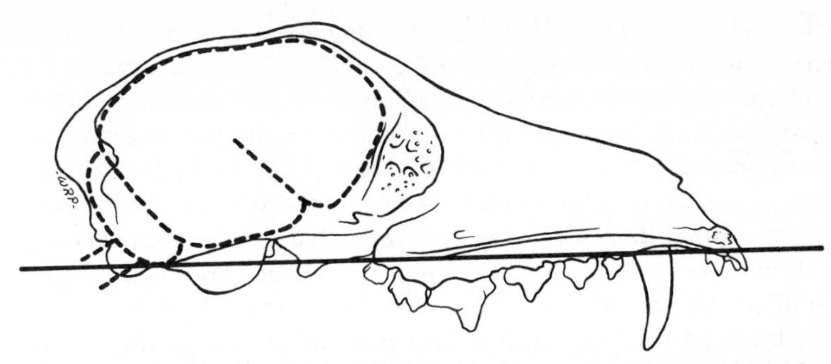

FIGURE 5–40. (From Enlow, D. H., and M. Azuma: Functional growth boundaries in the human and mammalian face. In: *Morphogenesis and Malformations of the Face and Brain.* Ed. by D. Bergsma. Birth Defects Orig. Art. Ser., Vol. XI, No. 7. New York, Alan R. Liss, Inc., for The National Foundation—March of Dimes, White Plains, New York.)

enters the inferior orbital fissure. It is this segment of the nerve *(F)*, prior to its downward turn through the infraorbital canal and out through the foramen *(G)*, that closely parallels the plane of the palate. An embryonic relationship exists. Any variation in the alignment of the nerve is usually accompanied by a corresponding upward or downward rotational alignment of the palate. In its neutral position, the plane of the palate *(C)* projects approximately to the inferior-most point *(E)* in the occipital fossa (± 3 mm), as does the alveolar plane *(D)* of the maxilla, as also seen in Figures 5–39 and 5–40. If the palate has undergone a severe clockwise or counterclockwise rotation during development, the palatal plane will project well above or below, respectively, the occipital point. Such rotations are commonly observed as one of several skeletal features characterizing deep bite and anterior open bite* (see page 220). Note: In lateral headfilms, the course of the nerve can be approximated as a plane from the posterosuperior corner of the PTM to a point 4 mm above orbitale.

*Some simians and anthropoids have an **established vertical hypoplasia** in the anterior part of the maxillary arch. In the rhesus monkey, for example, the premaxillary region is "high," or, at least, the posterior part of the nasomaxillary complex is vertically "long." A differentially greater extent of downward displacement takes place in the posterior part of the arch as compared with the anterior part. This in effect causes an "upward" rotation of the anterior region, and direct downward bone growth by this area does not move it fully to the inferior level attained by the posterior part of the arch. Anterior open bites are quite frequent, much more so than in the human being. A similar rotation does, in fact, occur in the human maxilla, but the anterior part of the arch grows downward to an extent that fully offsets it. (See also Chapter 6.)

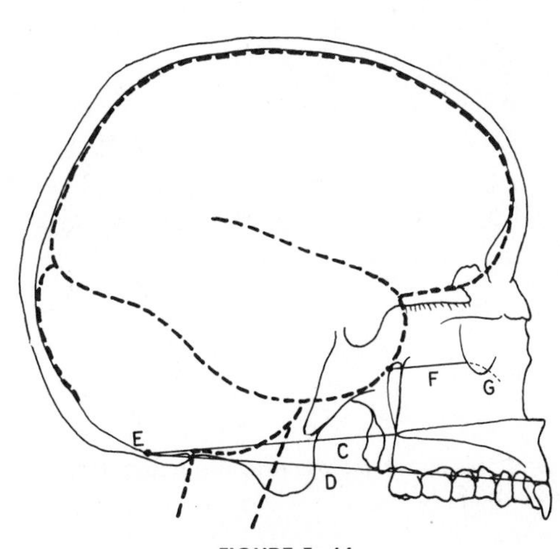

FIGURE 5–41

If Class II and Class III headfilm tracings are superimposed on the cribriform plates (representing the olfactory bulbs), it is apparent that the anterior plane of the nasomaxillary region in both conforms closely to the normal, perpendicular olfactory relationship. Note the similarity of the midfacial plane alignments (Fig. 5–42). In this particular Class II individual (and most others as well), it is not the basal bone of the maxilla itself that "protrudes"; rather, it is the **mandible** that is actually **retrusive**. In the Class III individual, it is not always the maxilla that is retrusive; the **mandible** can be **protrusive**. In both individuals, the nasomaxillary complex is located where it is supposed to be, and its horizontal dimensions are not out of line as they relate to the brain.

In summary, the growth in each region of the face involves two basic factors: (1) the amount of growth by any given part and (2) the direction of growth by that part. The brain establishes (or at least shares) the various **boundaries** that determine the amount of facial growth. This is because the floor of the cranium is the template upon which the face is constructed. The **directions** of regional growth among the different parts of the face are inseparably associated with the special sense organs housed within the face. These two factors establish a prescribed growth perimeter that defines the borders of the growth compartment occupied by the nasomaxillary complex (Figs. 5–43 and 5–44). All the many components that constitute the midface,

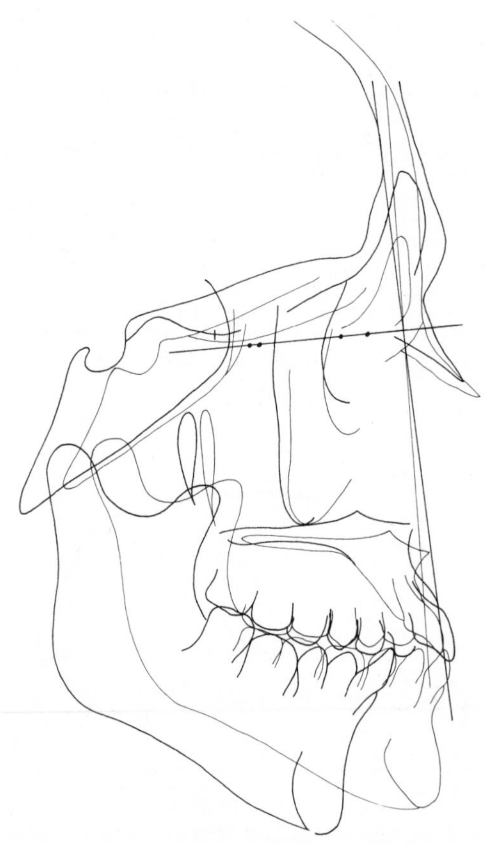

FIGURE 5–42. (From Enlow, D. H., and J. McNamara: The neurocranial basis for facial form and pattern. Angle Orthod., 43:256, 1973.)

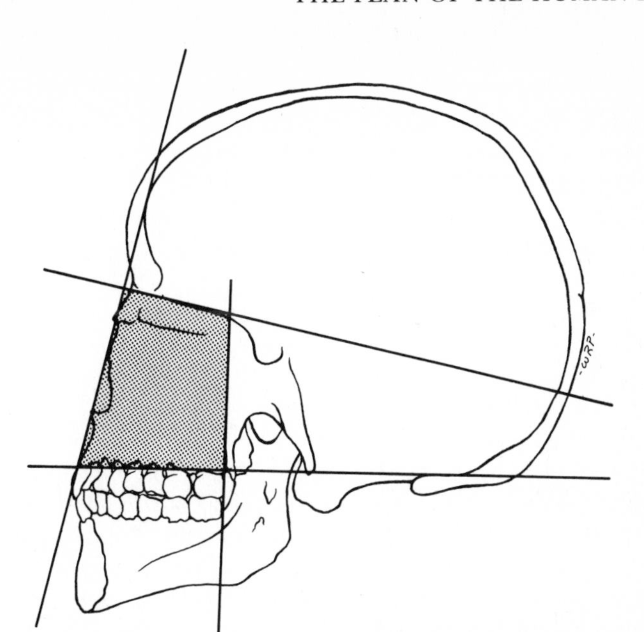

FIGURE 5–43. (From Enlow, D. H., and M. Azuma: Functional growth boundaries in the human and mammalian face. In: *Morphogenesis and Malformations of the Face and Brain.* Ed. by D. Bergsma. Birth Defects Orig. Art. Ser., Vol. XI, No. 7. New York, Alan R. Liss, Inc., for The National Foundation—March of Dimes, White Plains, New York.)

including the bones, muscles, mucosae, connective tissues, cartilage, nerves, vessels, tongue, teeth, and so on, contribute to a composite expression of growth, the sum of which can produce enlargement up to the maximum, as determined by the midfacial growth boundaries. The growth of the **midface** is not limitless, and it is not independently and randomly determined entirely within itself.

Superior prosthion thus comes to lie in a predetermined position that has been programmed by the brain-cranial base-sense organ-soft tissue composite of developmental factors. Superior prosthion is composed of alveolar bone, which is a highly labile and responsive type of bone tissue. Traditionally, this area of bone is regarded as quite unstable and subject to a wide range of variations according to the many forces that act on it. This is quite true, as will

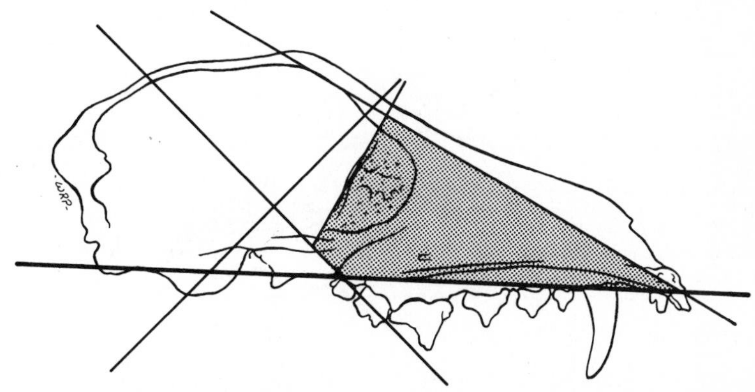

FIGURE 5–44. (From Enlow, D. H., and M. Azuma: Functional growth boundaries in the human and mammalian face. In: *Morphogenesis and Malformations of the Face and Brain.* Ed. by D. Bergsma. Birth Defects Orig. Art. Ser., Vol. XI, No. 7. New York, Alan R. Liss, Inc., for The National Foundation—March of Dimes, White Plains, New York.)

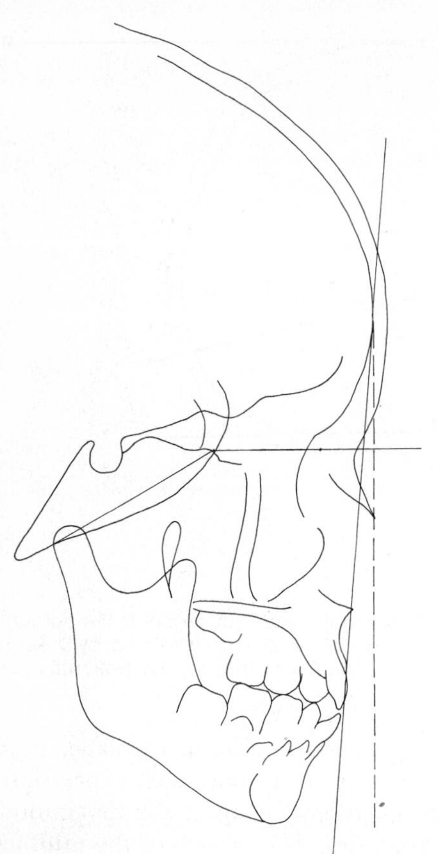

FIGURE 5–45. (From Enlow, D. H., and J. McNamara: The neurocranial basis for facial form and pattern. Angle Orthod., 43:256, 1973.)

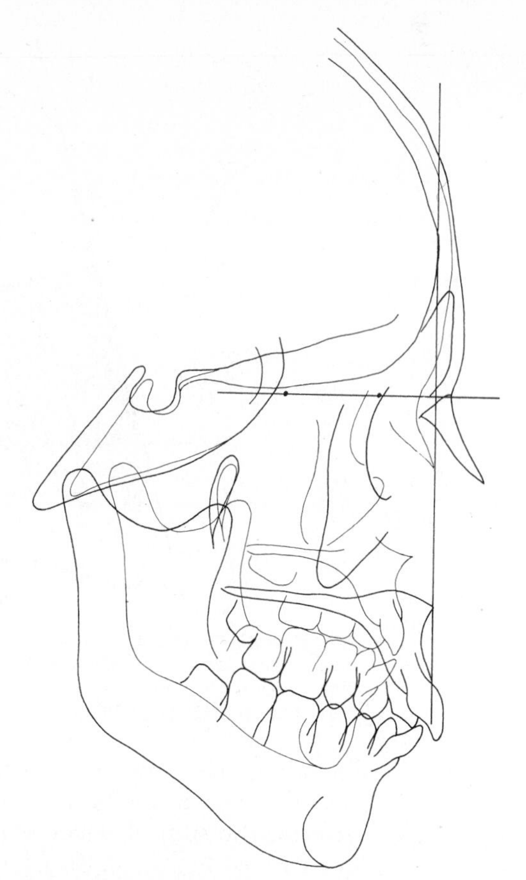

FIGURE 5–46. (From Enlow, D. H., and J. McNamara: The neurocranial basis for facial form and pattern. Angle Orthod., 43:256, 1973.)

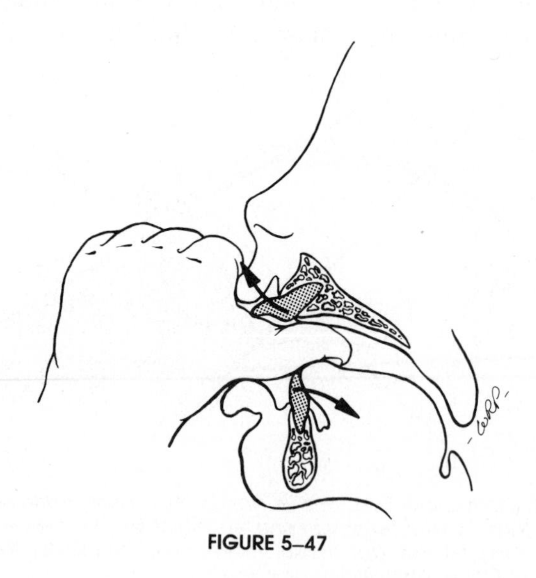

FIGURE 5–47

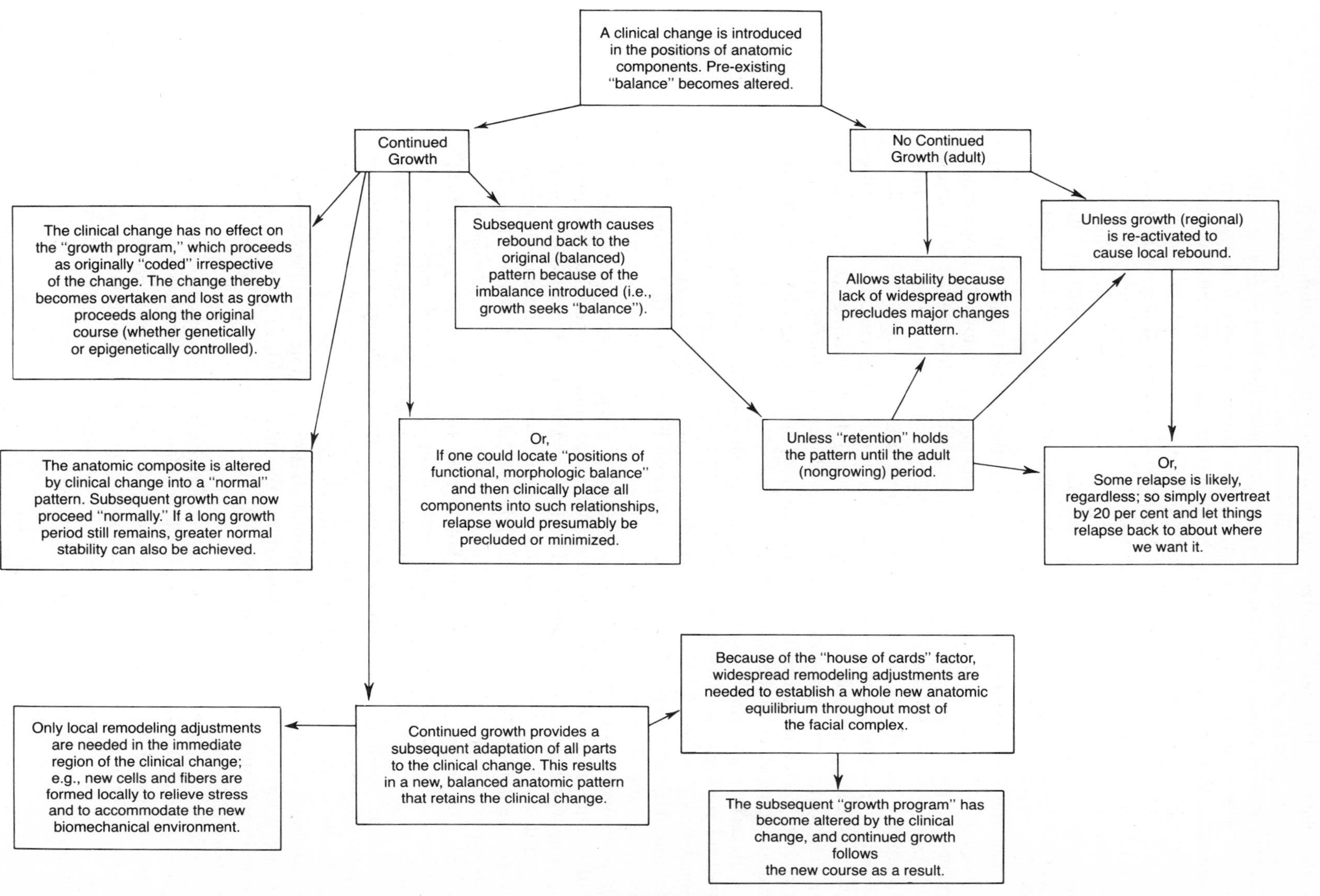

A clinical change is introduced in the positions of anatomic components. Pre-existing "balance" becomes altered.
Continued Growth
No Continued Growth (adult)
The clinical change has no effect on the "growth program," which proceeds as originally "coded" irrespective of the change. The change thereby becomes overtaken and lost as growth proceeds along the original course (whether genetically or epigenetically controlled).
Subsequent growth causes rebound back to the original (balanced) pattern because of the imbalance introduced (i.e., growth seeks "balance").
Allows stability because lack of widespread growth precludes major changes in pattern.
Unless growth (regional) is re-activated to cause local rebound.
The anatomic composite is altered by clinical change into a "normal" pattern. Subsequent growth can now proceed "normally." If a long growth period still remains, greater normal stability can also be achieved.
Or, If one could locate "positions of functional, morphologic balance" and then clinically place all components into such relationships, relapse would presumably be precluded or minimized.
Unless "retention" holds the pattern until the adult (nongrowing) period.
Or, Some relapse is likely, regardless; so simply overtreat by 20 per cent and let things relapse back to about where we want it.
Because of the "house of cards" factor, widespread remodeling adjustments are needed to establish a whole new anatomic equilibrium throughout most of the facial complex.
Only local remodeling adjustments are needed in the immediate region of the clinical change; e.g., new cells and fibers are formed locally to relieve stress and to accommodate the new biomechanical environment.
Continued growth provides a subsequent adaptation of all parts to the clinical change. This results in a new, balanced anatomic pattern that retains the clinical change.
The subsequent "growth program" has become altered by the clinical change, and continued growth follows the new course as a result.

FIGURE 5–48

be seen below. However, prosthion has a specific target location that it will occupy if the growth process is not disturbed by intrinsic or extrinsic imbalances. The target point is not programmed within prosthion itself, or even just within the maxilla. It is determined, rather, by the composite of all the growth-establishing factors mentioned above. In most cases, prosthion will have settled in, when growth is complete, right on or very close to its target point.

In the headfilm tracing shown in Figure 5–45 it is seen that prosthion falls short of the predetermined midfacial plane; growth is incomplete, however. In the same individual, when facial growth is largely completed, prosthion will have arrived at its place on the perpendicular adult (dashed) midfacial line. In Figure 5–46, the two headfilms are superimposed on the cribriform plane to show the "before" and "after" growth stages.

Can the brain-sense organ relationship with the face be violated? Of course; it frequently happens. For example, thumb sucking and various developmental defects can move the teeth and alveolar bone to places that are out of bounds with respect to the normal growth process (Fig. 5–47). The forces and factors of ordinary growth become overridden by extrinsic forces, and the prescribed boundary and the usual limit of growth are thereby overrun. However, this produces a structural and functional imbalance. If the overriding ectopic factors are removed, the normal balance of functional intrinsic forces work toward a greater or lesser return to the normal position, conforming with the natural anatomic boundary of the growth field.

Because many anatomic boundaries, large and small, exist throughout the face and cranium, the factor of boundary "security" is a major and important consideration to the clinician. If one given facial growth field is made to overrun the boundary of another field, either by clinical intervention or because of a developmental abnormality, one or the other will necessarily have to become compromised. A competition for the same space by the two overlapping growth fields occurs, and one field will necessarily become subordinate. This has great meaning with regard to the **stability** of a region and the functional "equilibrium" among different structural parts. If, for example, a given treatment procedure causes a violation of some growth boundary, will hard-earned treatment results subsequently be lost because functional stability and balance have been disturbed? Or, perhaps, will results be lost because the activity of a growth field that has been imposed upon subsequently causes a return ("rebound") toward the original structural pattern when treatment is stopped? Another similar question is whether a treatment procedure can actually **change** the long-term growth program. If it does not, subsequent growth, after treatment is ceased, can erase the treatment results, because growth then proceeds along its original, unaffected course. These are fundamental clinical questions, and they have to do with priorities of growth control among the many fields of growth and the natural integrity of their boundaries. (Read the sections on "retention" in orthodontic texts.)

There are many theoretically possible alternatives with regard to stability versus rebound/relapse, as summarized in Figure 5–48. Some seem more feasible than others, and still more possibilities probably exist that are not included. Some may hold true for one given clinical or growth circumstance, and others for different circumstances. Try comparing and evaluating them, and also see if you can invent one or more new ones.

CHAPTER SIX

Normal Variations in Facial Form and the Anatomic Basis for Malocclusions

Part 1

Variation is a basic law of biology. The pool of structural, functional, and genetic-based variations that are always present within a population of any species provides the capacity for adaptation to a changing environment. This increases the probability of survival for those individuals having features most suitable to the needs of the time. The human face, like most of our other "specialized" anatomic parts, certainly has its share of variations. Indeed, there are probably more basic, divergent kinds of facial patterns among humans than among the faces of most other species. This is because unusual facial and cranial rotations have occurred in relation to human brain expansion. A greater latitude for facial differences exists because the brain, proportionately, is so large and so variable in configuration. There is also a much greater likelihood for different kinds of malocclusions in the human face than in the faces of most other species for the same reasons. In fact, actual tendencies toward malocclusions are **built into** the basic design of our faces because of the unusual relationships that are inherent in its design.

Concept 1

How does one "size up" a person's face to determine just what kinds and combinations of facial variations are present for that **individual** person? There are many sophisticated cephalometric "analyses" that can be used to appraise the structural details of any given face and cranium; these are briefly explained later. However, one quick and effective way to evaluate the general features of a face is simply to visualize that face as it would be represented by a skillful caricature artist. What are the topographic idiosyncrasies that the artist would seize upon to portray that person by **greatly exaggerating** these characteristics in cartoon style? The caricaturist always has a difficult time of it with a really well-balanced, attractive face, since there are no particularly special features to emphasize to make a caricature clearly recognizable. For most of us, on the other hand: Pug nose? Retrusive mandible? Sloping forehead? Heavy eyebrow ridges? Hollow cheekbone area? Moon-faced? Horse-faced (long)? Dish-faced (concave profile)? Mouse-faced (convex profile)? Flat-faced? Pointed mandible? Protruding teeth? Heavy chin? Aquiline nose? Narrow-set eyes? Criminal tendencies? High nasal bridge? Fat cheeks? And so on. Find a mirror.

Concept 2

There are two basic extremes in the shape of the head: (1) **dolichocephalic** and (2) **brachycephalic** (a third, mesocephalic, lies between). The oval-shaped dolichocephalic head form is horizontally long and relatively narrow, in contrast to the more rounded brachycephalic head form, which is horizontally shorter and broader (Figs. 6–1 and 6–2). The **cephalic index** is the ratio between overall head length and breadth: dolichocephalic, up to 75.9; mesocephalic, 76 to 80.9; and brachycephalic, over 81. Specific facial and occlusal types relate to these head form shapes, as explained below.

Concept 3

Three general types of facial profile exist: **orthognathic, retrognathic,** and **prognathic** (Fig. 6–3). The orthognathic ("straight-jawed") form is the everyday standard for a good profile, and it is the type common to most Hollywood and television big names. It is easy to "eyeball" a person's face, without actual need for headfilms or precision anthropometric instruments, to see what his or her profile type is. Simply visualize a line extending straight out from the center of the orbit looking straight forward with the head and this line angling neither upward nor downward *(a)*. The head and body can be in any position, lying down, standing up, leaning over, and so forth, when you are visualizing this neutral orbital axis line. Now visualize a **vertical** line **perpendicular** to the

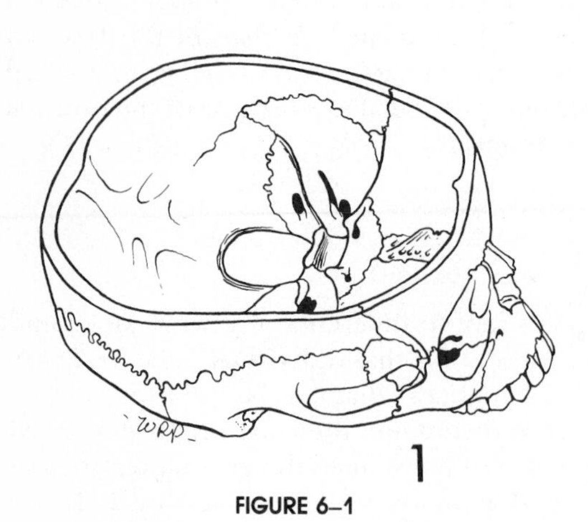

FIGURE 6–1

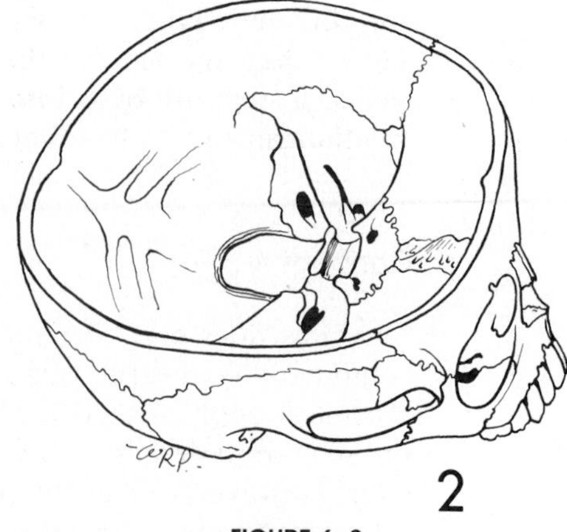

FIGURE 6–2

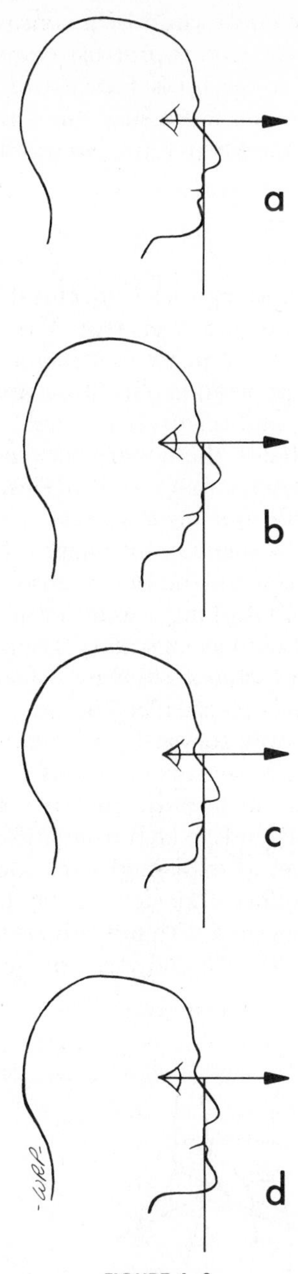

FIGURE 6–3

orbital line extending down along the surface of the upper lip. This line will just touch the lower lip and the tip of the chin in a person with an orthognathic profile. Time otherwise thrown away waiting around air terminals, sitting out classes, or standing in lines can be put to interesting use quietly studying people's profiles.

The retrognathic face has a characteristic convex-appearing profile. The tip of the chin lies somewhere behind the vertical line, and the lower lip is retrusive. The chin may be 2 or 3 centimeters behind the line in a severely retrognathic face *(b)*. Among many Caucasians, however, it is common to have about a half-centimeter or so of chin retrusion *(c)*. The reason is explained later.

The prognathic face is characterized by a concave-appearing ("dished-in" midface) profile. The tip of the chin is protrusive and lies somewhere in front of the vertical line *(d)*. The lower lip is forward of the upper. This kind of profile is far less frequent among Caucasians (in comparison with some other ethnic groups) than is the mandibular retrusive profile type.

Concept 4

There are several general categories of **occlusal** patterns: "normal," Class I, Class II, Class III, open bite, and deep bite. The morphogenic factors that underlie them are explained in the pages that follow.

In individuals (or whole populations) having a **dolichocephalic** head form, the brain is horizontally long and relatively narrow. This sets up a cranial base that is somewhat more flat; that is, the flexure between the middle cranial floor and the anterior cranial floor is more open (Figs. 6–4 to 6–6). It is also horizontally longer. These factors have several basic consequences for the pattern of the face. First, the whole nasomaxillary complex is placed in a more **protrusive** position relative to the mandible because of the forward basicranial rotation and, also, the horizontally longer anterior and middle segments of the cranial floor. Second, the whole nasomaxillary complex is lowered relative to the mandibular condyle. This causes a downward and **backward** rotation of the entire mandible. Third, the occlusal plane becomes rotated into a downward-inclined alignment. The **two-way** forward placement of the maxilla and backward placement of the mandibular corpus results in a tendency toward mandibular retrusion, and the placement of the molars results in a tendency toward a Class II position. The profile tends to be retrognathic. However, compensatory changes are usually operative, as explained later. Because of the more open cranial base angle and the resultant trajectory of the spinal cord into the cervical region, this type of face is associated with individuals having a greater tendency toward a somewhat stooped posture and anterior inclination of the head and neck.

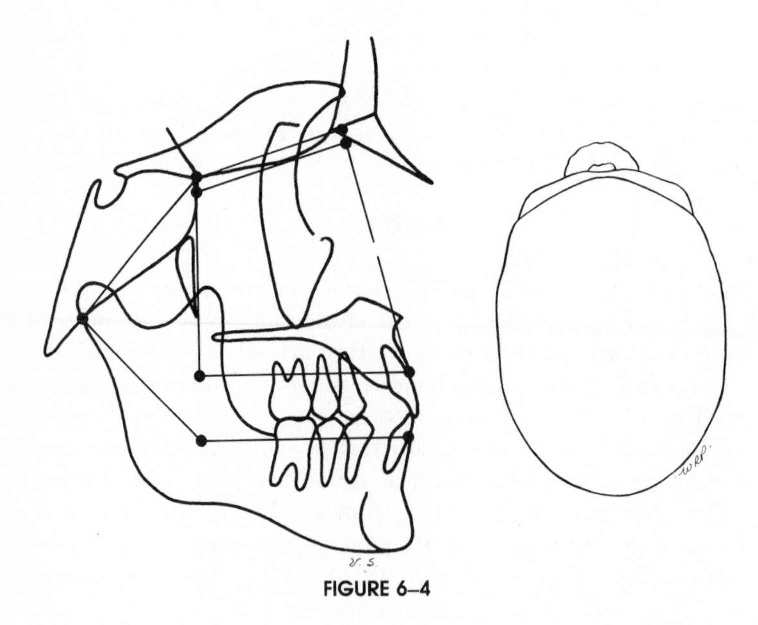

FIGURE 6–4

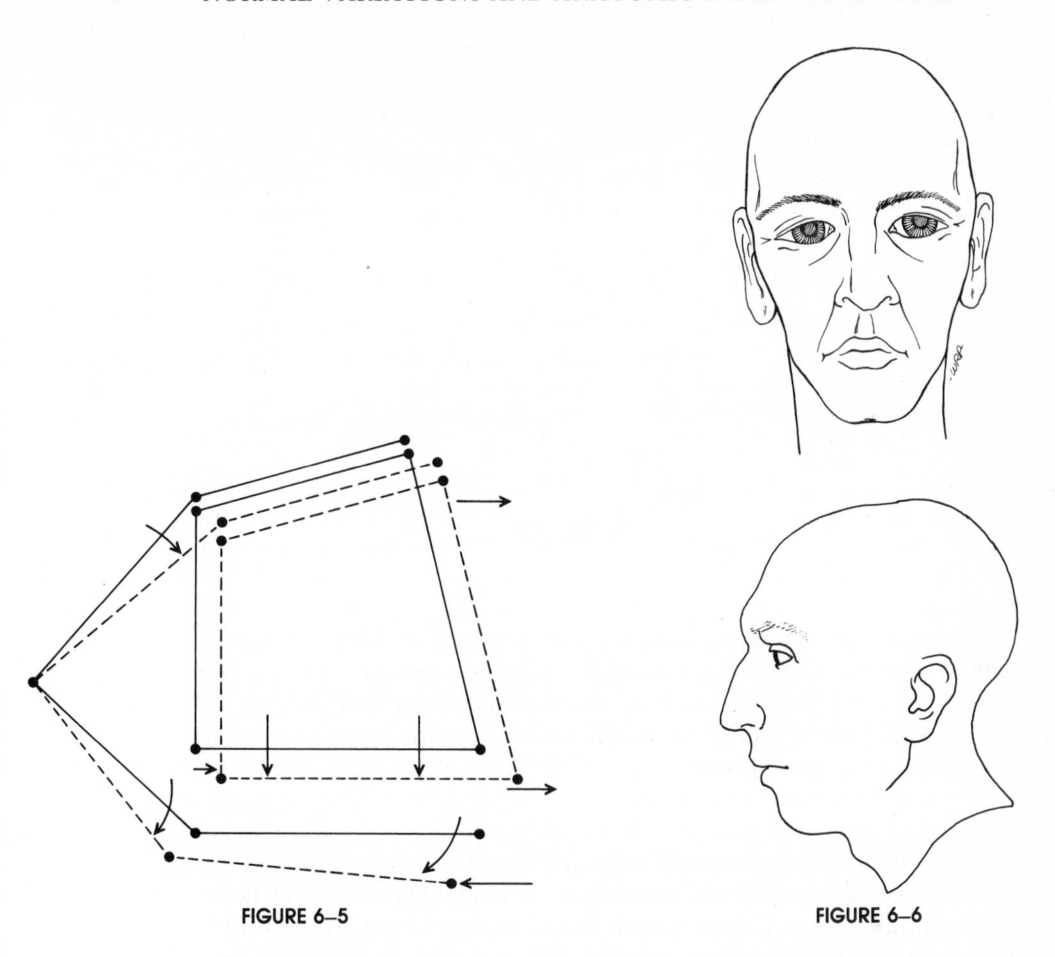

FIGURE 6–5

FIGURE 6–6

Individuals or ethnic groups with a **brachycephalic** head form have a rounder, shorter (horizontally), and wider brain. This sets up a cranial base that is more upright and has a more closed flexure, which decreases the effective horizontal dimension of the middle cranial fossa (Figs. 6–7 and 6–8). The facial result is a posterior placement of the maxilla. Furthermore, the horizontal length of the nasomaxillary complex is also relatively short. Because the brachycephalized basicranium is wider but less elongate in the anteroposterior dimension, the middle and anterior cranial fossae are correspondingly foreshortened (not shown in the schematic diagram). The anterior cranial fossa

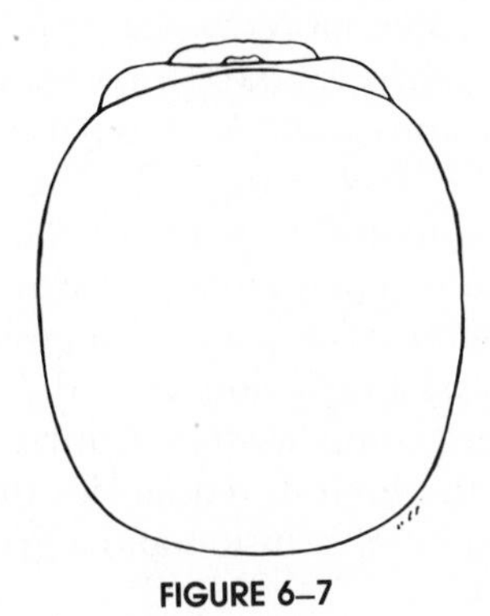

FIGURE 6–7

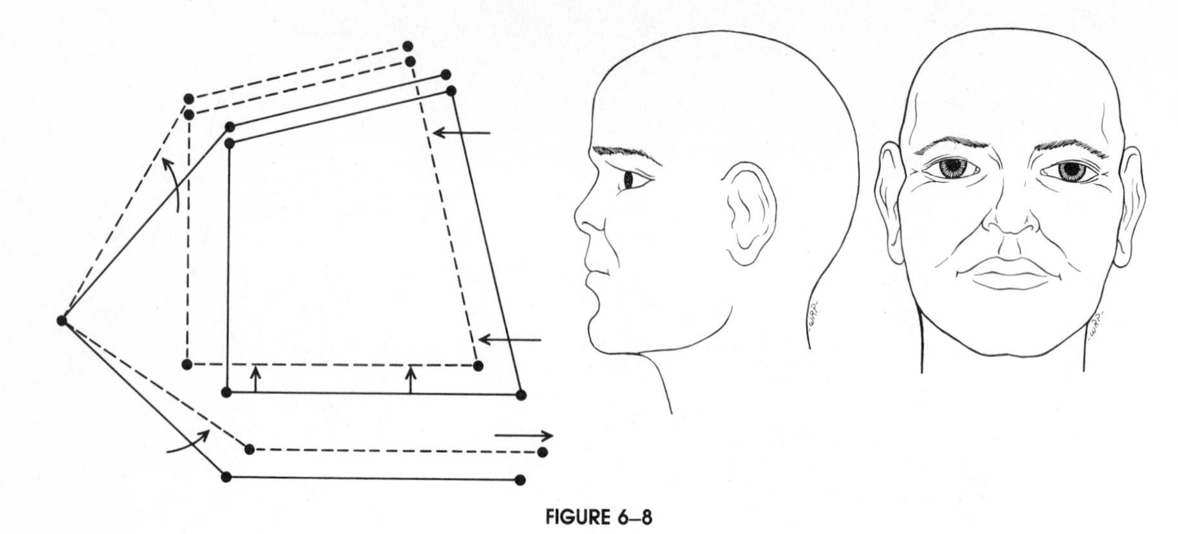

FIGURE 6–8

provides the template that establishes the horizontal length and bilateral width of the nasomaxillary complex, which is thereby also shorter but wider. The composite result is a relative retrusion of the nasomaxillary complex and a more forward relative placement of the entire mandible. This causes a greater tendency toward a prognathic profile and a Class III molar relationship. The occlusal plane as well as the ramus of the mandible may be aligned upward, but various compensatory processes usually result in either a perpendicular or a downward-inclined occlusal plane and slight backward rotation of the ramus. Other compensatory changes are also operative, as explained next, and these tend to counteract the built-in Class III tendencies. Because of the more upright middle cranial fossa and the more vertical trajectory of the spinal cord, individuals with all these various facial features also have a tendency for a more erect posture with the head in a more "military" (at braced attention) position.

Concept 5

The basic nature of interrelationships among (1) brain form, (2) facial profile, and (3) occlusal type, as just seen, causes a predisposition toward characteristic facial types and malocclusions among different types of populations. Some Englishmen, for example, and some other Caucasian groups with a tendency for a dolichocephalic head form, have a corresponding **tendency** toward Class II malocclusions and a retrognathic profile. The Japanese, having mostly a brachycephalic head form, have a correspondingly greater tendency toward Class III malocclusions and a prognathic profile. These respective tendencies are built into the basic plan of facial construction. However, most of us also have intrinsic structural features that have compensated for these tendencies. **If** we have such compensatory features, the built-in tendencies are offset, to a greater or lesser extent, and we thereby have at least reasonable facial proportions with a Class I occlusion, even though the underlying tendencies are still present. **If** these compensatory features do not develop, however, or if they are insufficient, the built-in tendencies then become expressed, and we have a more or less severe malocclusion and a greater extent of retrognathia or prognathia.

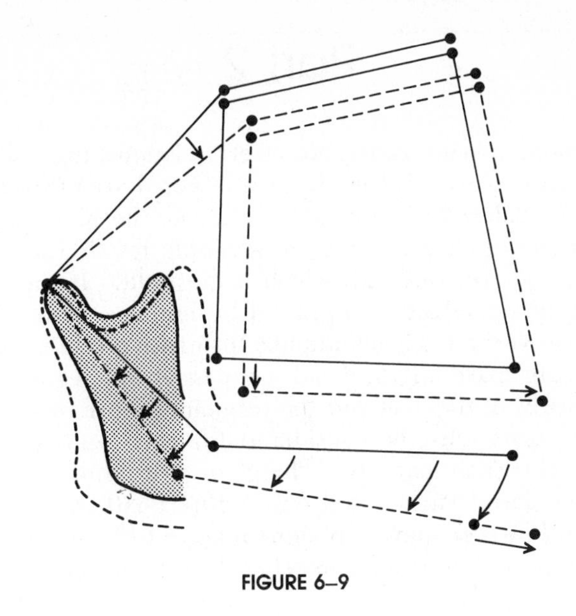

FIGURE 6–9

Concept 6

How does a face undergo intrinsic compensations during its development? One example that is very common is shown here (see Part 2 for others). In Concept 4 above, the mandible was placed in a retrusive (retrognathic) position owing to its downward and backward rotation resulting from the more open type of cranial base flexure (and/or a vertically long nasomaxillary complex). The mandibular **ramus**, however, can compensate by an increase in its horizontal dimension (Fig. 6–9). This places the whole mandibular arch anteriorly into a proper position beneath the maxilla, and it positions the teeth in a "normal" or a Class I type of molar relationship. The mandibular retrusion that would otherwise be present thus becomes partially or completely eliminated, and a profile in which the chin lies on or within a half-centimeter or so of the orthognathic profile line results. The downward mandibular rotation is compensated for by an upward drift of the anterior mandibular teeth and a downward drift of the anterior maxillary teeth. This causes a curved occlusal plane, the curve of Spee (see Part 2 for details).

The face on each of us, virtually without exception, is the composite of a great many regional "imbalances." Some of these offset and partially or completely counteract the effects of the others. The wide ramus cited above, for example, is actually an imbalance, but it serves to reduce, as a normal compensatory process, the effects of some other angular or dimensional imbalances caused by the built-in tendencies toward malocclusions. The particular feature of a wide ramus is very common among the dolichocephalic Caucasians. When this and other compensatory factors are present, the underlying stacked deck toward retrognathia and a Class II malocclusion is removed or made less severe. Thus, many of us have a slightly retrognathic profile and a little anterior tooth crowding.

Part 2

In this section, specific cause-and-effect relationships underlying differences in facial pattern are explained. Each regional area throughout the face and cranium is considered separately. To evaluate the structural and developmental situation for each given region, a simple test is used: that region is compared with other regions with which it must "fit." If they do not have a good fit, the resultant effect is appraised by noting whether it causes (1) a mandibular retrusive or (2) a mandibular protrusive effect. As will be seen, imbalances in many parts of the head are passed on, region by region, and affect the placement of the jaws and the resultant nature of the occlusion.

Two basic factors must be considered for each region. The first is the **dimension** of a particular part. Is it "long" or is it "short" with regard to its fitting with other parts? Parts *a* and *b* in Figure 6–10 should "fit," as should parts *c* and *d*. If, however, part *b'* is short relative to *a'*, it causes part *d'* to be "retrusive," even though it actually matches the dimension of *c'*.

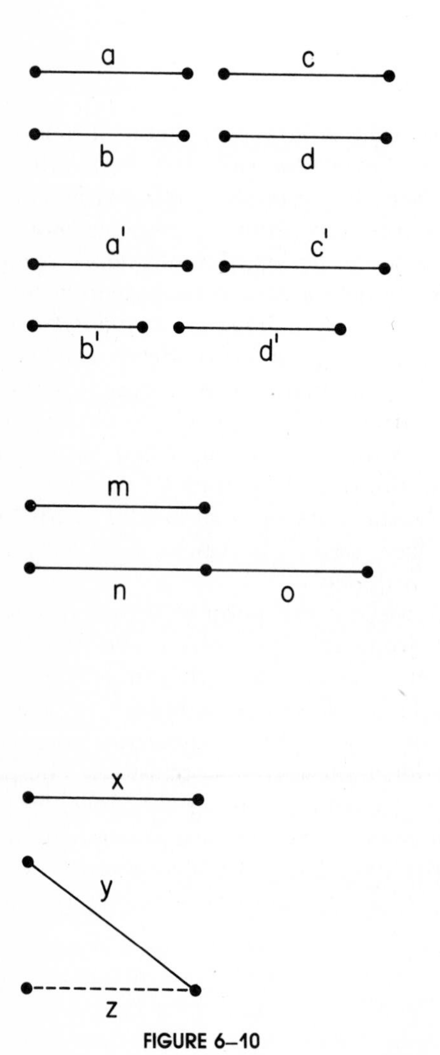

FIGURE 6–10

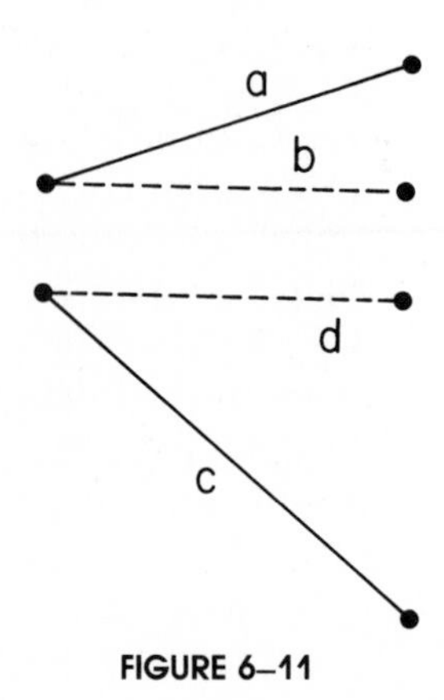

FIGURE 6–11

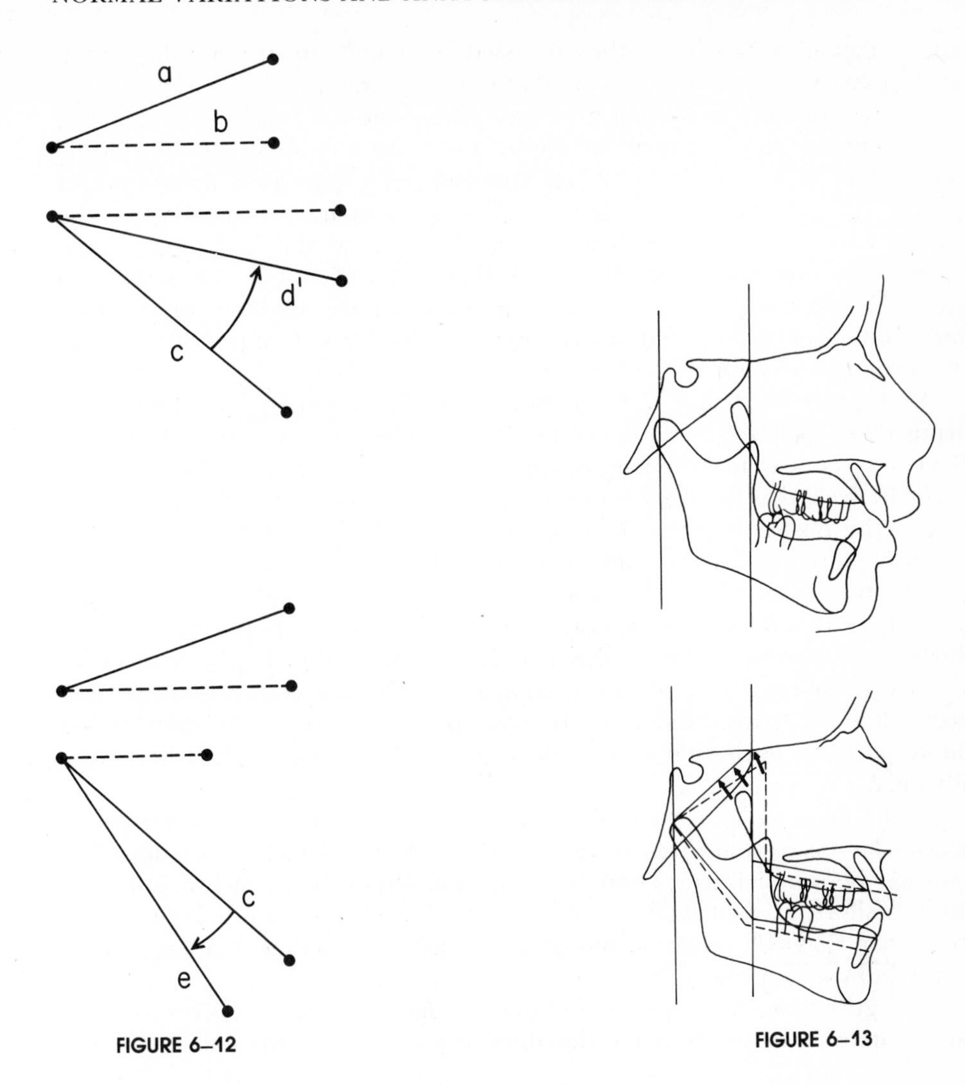

FIGURE 6–12

FIGURE 6–13

Great care must be used to evaluate only that particular span or dimension of a bone specifically involved in the actual, direct fitting of the bones one to another. This is the **effective dimension**. For example, bone *m* in Figure 6–10 relates only to dimension *n*; dimension *o*, even though part of the same bone, is not involved in the counterpart relationship with *m*. Or, the "effective" straight-line dimension of oblique part *y* is actually the horizontal dimension *z*, which relates to the composite fit with separate bone *x*.

The second fundamental consideration is the **alignment** of any given part. This must also be included (although many cephalometric studies do not), because any rotational change either increases or decreases the **expression** of a dimension. For example, separate parts *a* and *c* in Figure 6–11 match, although their real oblique lengths are different; the straight line parallel dimensions *b* and *d* are equivalent.

Note what happens, however, if the **alignment** of bone *c* in Figure 6–12 is changed to *d'*. Its actual dimension is exactly the same, but the **expression** of this dimension has been altered; the expressed horizontal dimension has been increased, and the expressed vertical has necessarily been decreased at the same time. Even though the actual dimensions have not been changed, parts *a* and *c* no longer fit because their effective dimensions do not now match. If part *c* is aligned (rotated) to *e*, the expressed horizontal dimension is decreased, but the expressed vertical is increased at the same time.

To illustrate the important effects of **alignment** as a basic factor involved in determining facial pattern, in Figure 6–13 the alignment of the middle cranial fossa in a Class II child was changed (on paper) to a more upright position. All the other facial regions, including the mandible, maxilla, and the anterior cranial fossa, were then reassembled around the realigned middle cranial fossa. **No** changes in the actual dimensions of any of the parts were made. The horizontal and vertical **expression** of the middle cranial fossa dimension, however, resulted in a change from the Class II pattern into a Class I pattern, even though all the individual bones were exactly the same size.

In Figures 6–14 and 6–15, if horizontal dimension of the mandibular corpus *(b)* is short relative to its counterpart, the bony maxillary arch *(a)*, the effect is, of course, mandibular retrusion (probably with anterior crowding of the teeth). Note that this does not necessarily cause a Class II molar relationship, because the **posterior** parts of the upper and lower bony arches can still be properly positioned. It is emphasized that these are **relative** comparisons between two parts within the **same** individual. The mandible is not being compared with a norm or an average value derived from a population sample. Whatever the actual value of this mandibular dimension happens to be in millimeters, or regardless of how it compares with some statistical mean, it is short when compared with the dimensional value that really matters—its counterpart, the horizontal dimension of the maxillary body in that particular individual.

If the mandibular corpus is dimensionally long, the effect, of course, is mandibular protrusion (Figs. 6–16 and 6–17). A horizontally short maxillary arch has the same effect. (There are anatomic ways to tell which is long and which is short, as explained in Chapter 14.) Whether or not a long corpus produces a Class III molar relationship depends on whether it is long mesial or distal to the first molars.

In Figure 6–18, the upper part of the nasomaxillary complex is horizontally long **relative** to its counterparts, the anterior cranial fossa, the palate, and the

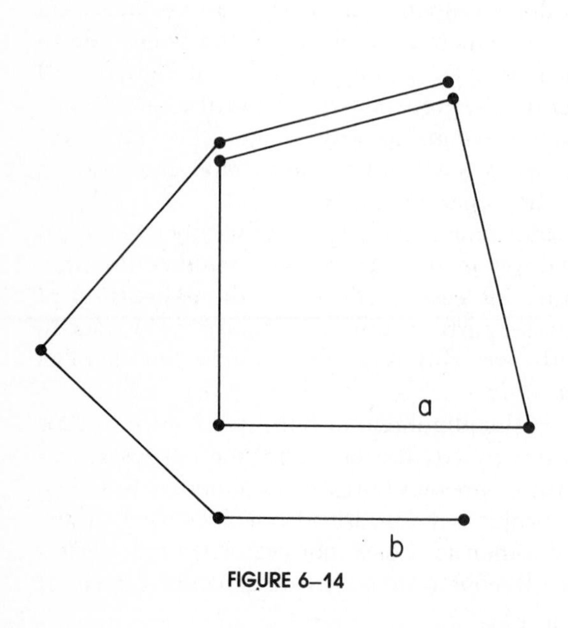

FIGURE 6–14

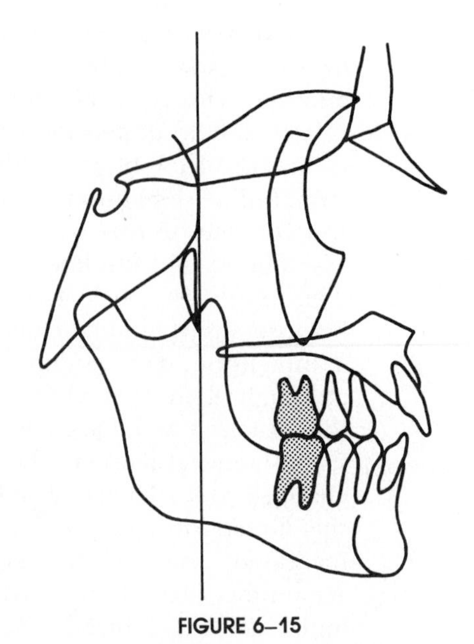

FIGURE 6–15

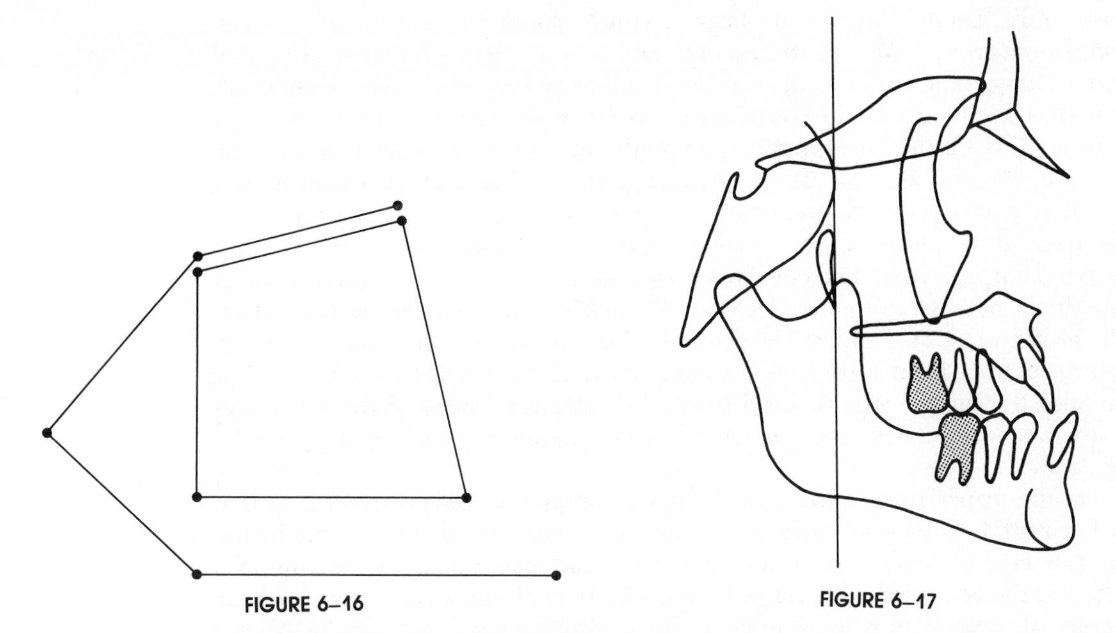

FIGURE 6–16

FIGURE 6–17

maxillary and mandibular arches. Note that this has no effect on the occlusion. The individual can **appear** retrognathic, but this is a result of the protrusive nature of the upper part of the face and not the jaws themselves. Because the superior part of the ethmomaxillary region is protrusive, the outer table of the frontal bone remodels with it. The result is a sizable frontal sinus, heavy eyebrow ridges and glabella, sloping forehead, high nasal bridge, and long nose. The cheekbone area appears retrusive because of the prominent nasal region and forehead.

If this upper part of the nasomaxillary complex is quite protrusive, the upper edge of the nose will often be curved or bent into a classic aquiline (eagle beak), Roman nose, or Dick Tracy configuration **if** the nose is **also** vertically long (Fig. 6–19). The longer the vertical dimension of the nose, the more its

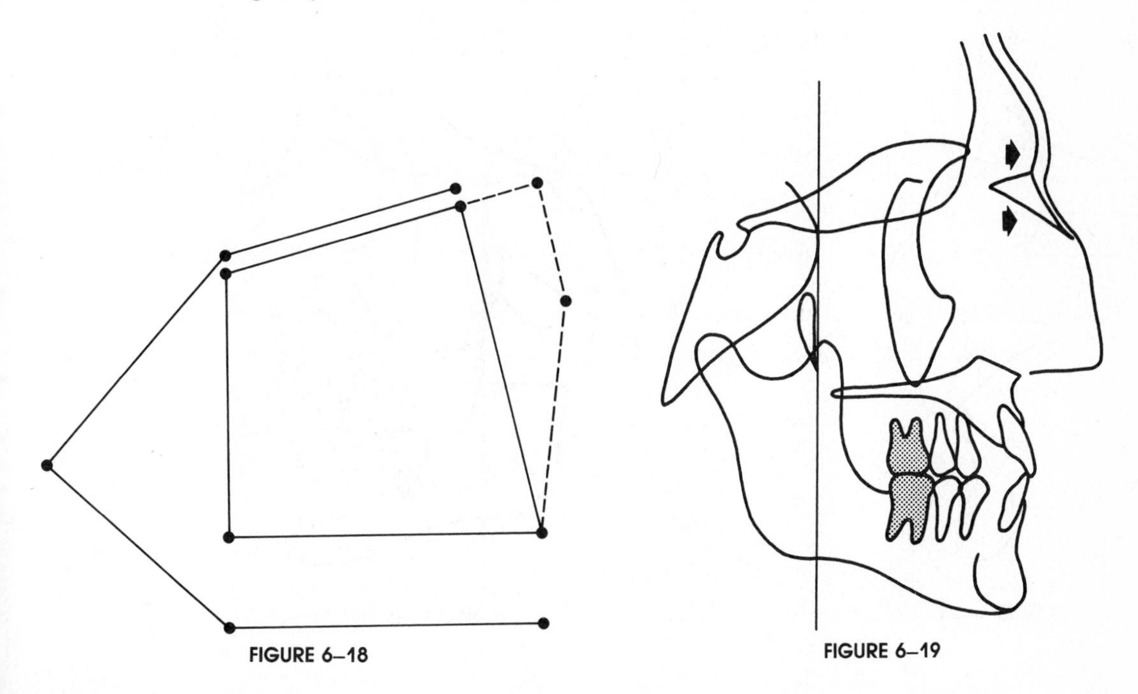

FIGURE 6–18

FIGURE 6–19

slope must bend. This nasal shape is quite common in some European population groups, such as the French, and typically has a rather narrow and sharp configuration. The ventral edge of this nose type may be horizontal but often has a tendency to tip downward, in contrast to the vertically shorter type of nose in which the lower margin can angle upward. In another type of nasal bending, the **middle** part of the nasal region may be quite protrusive; this produces a characteristic and gracefully recurved (sigmoid) configuration of the nasal slope as the lower portion grades and curves back onto the less protrusive upper part. The cheekbone area in this type of face is often notably prominent because this entire level of the midface also tends to be protrusive. All the above facial features, in general, characterize the long, narrow-faced, dolichocephalic head form found among many (but not all) Caucasian groups and also the dinaric type of head form. The extent of some of these features is sex- and age-related, such as frontal sinus expansion and the slope of the forehead.

If the upper part of the nasomaxillary complex is **not** protrusive, so that its horizontal dimension more nearly matches counterpart dimensions in the anterior cranial fossa, palate, and maxillary and mandibular arches, quite a different facial profile results (Fig. 6–20). The frontal sinuses are comparatively smaller, the forehead is more upright, the eyebrow ridges and glabella are not as prominent, the nose is not nearly as protrusive, and the nasal bridge is much lower. The jaws appear more prominent because the upper nasal region is less protrusive. The cheekbones also appear more prominent for the same reason. The whole face is much flatter. This composite of facial features is typically found in the broad-faced, brachycephalic type of head form that characterizes many Oriental individuals. Some Caucasian populations are also broad-faced, with a shorter nose, more prominent mandible, lower nasal bridge, and so forth, including, for example, many individuals having a facial heritage from middle regions of Europe, parts of southern Ireland, and a scattering of geographic locations elsewhere in the world. It has become a common type of Caucasian face in North America. A shorter but wider nose and nasal chambers

FIGURE 6–20

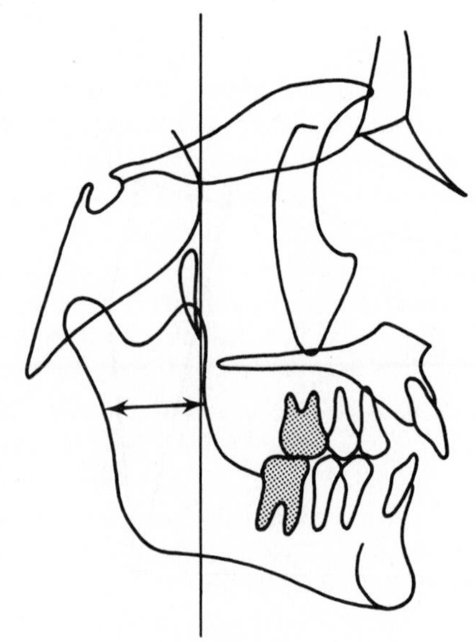

FIGURE 6–21

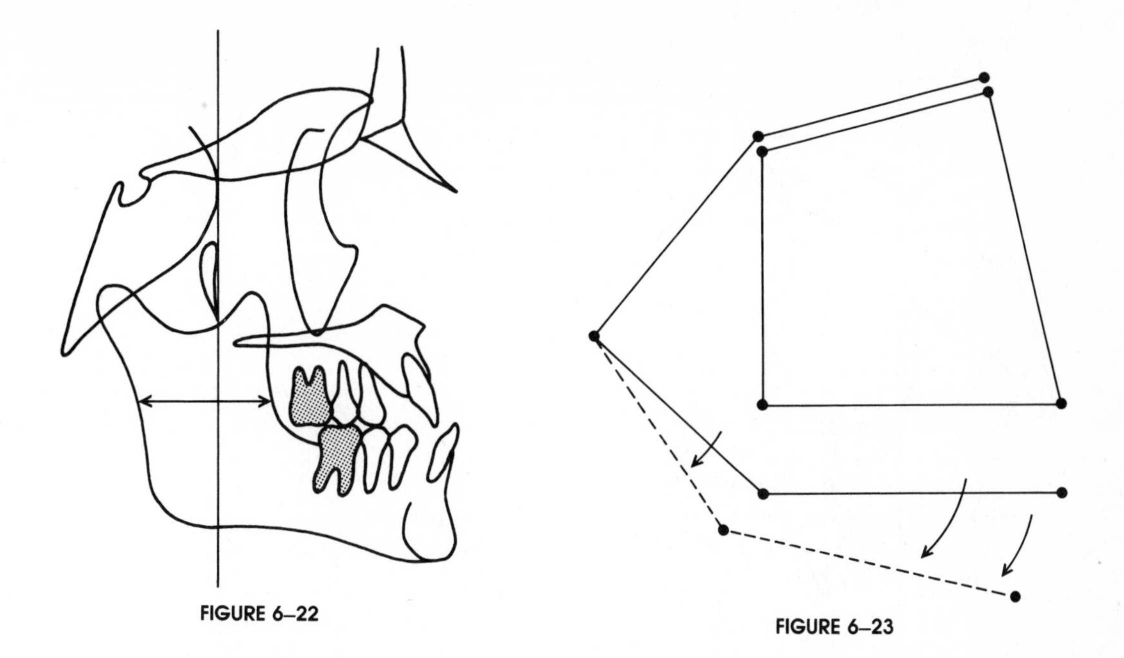

FIGURE 6–22

FIGURE 6–23

provide approximately equivalent airway capacity in comparison with the narrower but longer and more protrusive nose of the dolichocephalic type of head form.

If the effective horizontal (not oblique) dimension of the ramus is narrow relative to its counterpart, which is the effective horizontal (not oblique) dimension of the middle cranial fossa, a mandibular retrusive effect is produced (Fig. 6–21). Note that the mandibular arch lies in a resultant **offset position** relative to its counterpart, the maxillary arch. Even though the upper and lower arches themselves are actually matched in dimensions, the profile is retrognathic. The arches are in offset positions because the parts **behind** them are "imbalanced." Note that the posterior part of the maxillary arch lies well anterior (mesial) to the posterior part of the mandibular arch. This is one (of several) of the basic skeletal causes that underlie a **Class II molar relationship**. Remember, the "real" anatomic junction between the ramus and corpus is the lingual tuberosity, rather than the oblique "anterior border" where it overlaps the corpus. Because the lingual tuberosity cannot be directly visualized in headfilms, it is not represented here. However, it is located distal to the vertical reference line because of the narrow ramus in this individual.

In Figure 6–22, the effective horizontal (not oblique) dimension of the ramus is broad relative to the middle cranial fossa. Or, the cranial fossa is horizontally narrow relative to the ramus (either way because this is a relative comparison). The effect is mandibular protrusion due to the resultant offset positions between the upper and lower arches, even though the horizontal dimensions of the arches themselves can match. This is one (of several) of the basic skeletal causes for a **Class III molar relationship**. The lingual tuberosity (not shown) is mesial to the vertical reference line.

If the ramus has a more upright alignment (as a result, for example, of a vertically long nasomaxillary region), the effect is mandibular retrusion (Figs. 6–23 and 6–24). While this increases the expression of the vertical ramus dimension, the horizontal is necessarily decreased at the same time. The whole mandible is rotated downward and **backward**. As a result, the mandibular arch

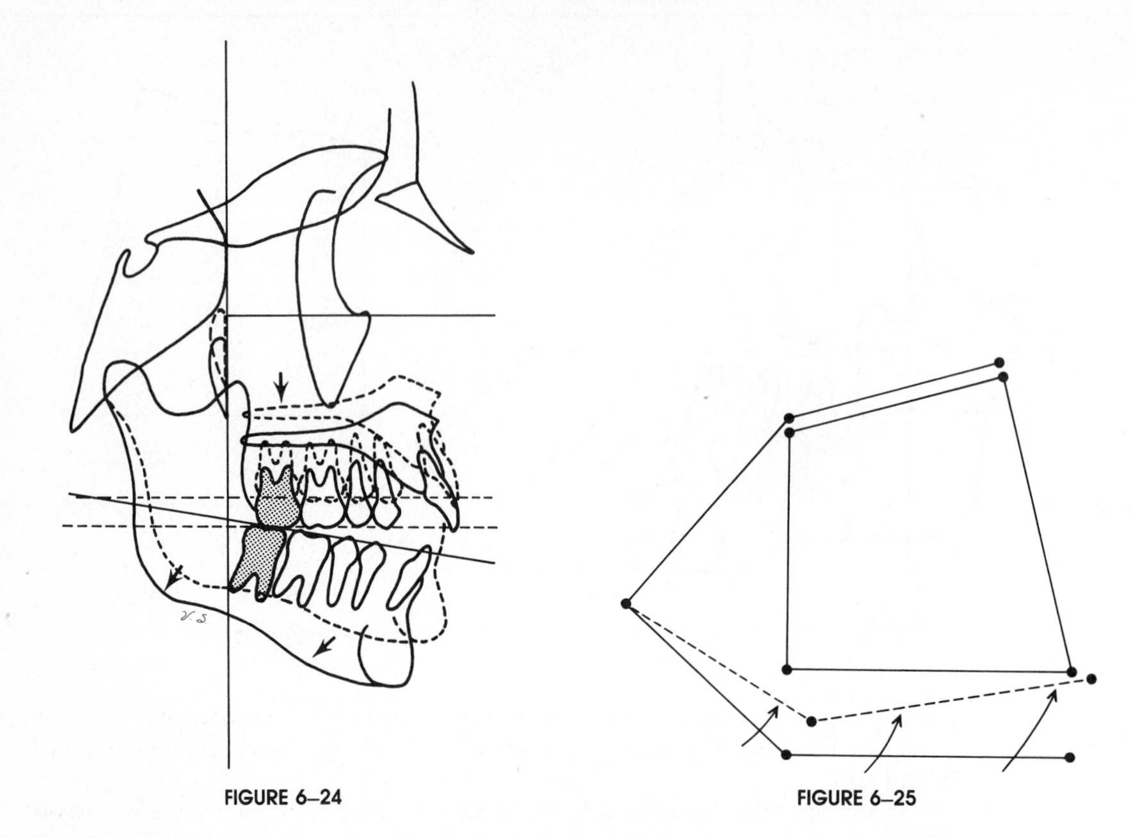

FIGURE 6–24

FIGURE 6–25

becomes offset relative to the upper arch. The profile is retrognathic, and the offset placement of the arches causes a Class II molar relationship. Note that the mandibular corpus is rotated downward, causing a downward-inclined mandibular occlusal plane (see page 217 for an explanation of dental compensations).

If the ramus has a more forward-inclined alignment (as a result of a vertically short midface), the effect is mandibular protrusion, because the expression of the horizontal dimension is increased (Fig. 6–25). The vertical is decreased. Or, stated simply, the ramus rotates forward and upward, causing the mandible to protrude. The arches are offset, and the molars have a resultant Class III relationship. The occlusal plane has an upward inclination relative to the neutral orbital axis or to the maxillary tuberosity (that is, the vertical posterior maxillary [PM] line). The posterior teeth can drift inferiorly and/or the gonial angle can open (compensatory adjustments) to provide proper occlusion.

If the **corpus** (not ramus) has an upward alignment (that is, a closed "gonial angle"), a mandibular **retrusive** effect is produced. These various alignment relationships have often been misunderstood, and the whole subject of mandibular "rotations" has been perplexing to many workers. There are two basic and separate kinds of mandibular skeletal rotations (exclusive of dentition rotations, which will be described separately).

1. In the closed position, the **whole mandible** can rotate up or down at the condylar pivot. This was described above as a "ramus" rotation because it involves changes in ramus alignment at the condylar pivot. The corpus is carried with it. The primary reason that this kind of rotation takes place is to adjust the ramus, and thereby the corpus, to whatever vertical position exists

for the midface. The ramus rotates forward and upward to meet a short midface and/or an upright cranial base flexure, and it rotates down and back to accommodate a vertically long midface and/or a more open cranial base flexure.* This is a **displacement** type of rotation (see page 210).

2. The angle between the ramus and the corpus also can become increased or decreased as a separate kind of rotation. This does not refer merely to the conventional "gonial angle" but, rather, to the alignment between the whole of the ramus and the corpus. This is a **remodeling** type of rotation, in contrast to the displacement type (page 109). The axis of the ramus can thus be **more upright**, with the ramus-corpus angular relationship thereby "closed." Or the converse can occur by an opening of the ramus-corpus angle. In either case, the corpus is aligned up or down **relative** to the ramus. Both parts can participate in the bony remodeling changes involved in the opening and closing of the angle between them, although it is necessarily the **ramus**, rather than the corpus, that carries out most of them because this is where most of the active remodeling processes occur. It would not be possible, for example, for the entire corpus (not merely the dentoalveolar portion) to rotate upward by its own remodeling to close the gonial angle.

There are two basic reasons ramus-corpus remodeling rotations occur. The **first** was described on page 109 and deals with the need for a progressively more upright ramus to accommodate a vertically lengthening midface. The remodeling changes that carry this out were also outlined. The result is a ramus-corpus alignment that naturally and normally becomes more closed as the midface grows. The **second** reason is to accommodate the results of whole-mandible (displacement) rotation. When the entire mandible rotates forward and upward (see previous sections for the reasons), the mandibular corpus normally rotates downward to some extent in order to compensate. This helps to keep the mandibular arch in a constant or proper occlusal plane relationship. In addition, the posterior maxillary teeth may drift inferiorly. The occlusal plane can be brought to a perpendicular position relative to the PM plane, or it may still have a slight upward inclination. When the ramus (and whole mandible) rotates backward and downward, the ramus-to-corpus angle can also

---

*Another type of whole-mandible rotation has been reported by some investigators and is believed to involve a pivot located at some point along the occlusal plane. This axis of rotation may occur at the bicuspid level or the first erupted molar. The whole mandible thus "rocks" around this pivot as the maxillary arch grows downward. This will lower or raise the posterior and anterior ends of the mandible, depending on the direction of rotation. Yet another type of whole-mandible growth rotation is also believed to occur in some individuals, and the rotational axis for this variation occurs in the incisor region. Compensatory remodeling changes by the mandible involved in these various rotations are essentially the same combinations of resorption and deposition described in Chapter 3. That is, the ramus and condyle will grow either upward and forward or upward and backward in order to sustain proper articular position of the condyle. The mandibular condyle has a type of cartilage lacking the linearly oriented **columns** of proliferating chondrocytes. This is a special feature that gives the condylar cartilage a **multidirectional** growth capacity, in contrast to the restricted unilinear growth direction of a typical epiphyseal plate. This feature allows the condyle to adapt to the wide variety of rotational situations encountered among different individuals with different facial types, as well as the mandibular rotations that take place as a normal part of the growth process. Remember, the growth behavior of the condyle is **secondary**; it is not a pacemaker (see pages 89 to 91). As the whole mandible becomes displaced by whatever vectors are involved at different ages and by whatever variations occur among different individuals, the condylar cartilage and the contiguous membranes forming the intramembranous bone of the condylar cortex grow in whatever directions and amounts are required to sustain constant functional position and articulation with the cranial floor. The variable capacity of condylar growth thus provides adaptation to different facial types, different basicranial configurations, different occlusal patterns, and the normal structural changes occurring during progressive growth (such as the "rotations" that the ramus undergoes at different age levels).

close to some extent, thereby compensating for it. The respective amounts of these counteracting rotations are not always equal, however. If they are equal, or if no rotations at all occur, the occlusal plane is almost exactly perpendicular to the vertical **PM** plane. Often, however, the occlusal plane has a noticeable downward angulation because the amount of upward corpus realignment falls short of the downward rotation of the whole mandible. When you are observing the faces of your patients, you can easily "eyeball" how much downward occlusal plane rotation exists by visualizing it relative to the neutral horizontal axis of the orbit. If the two are parallel, the occlusal plane is perpendicular to the **PM** plane. In many of your patients the occlusal plane will angle downward, to a greater or lesser extent, and in a few it will angle upward. Persons with a vertically shorter nasal region tend to have a perpendicular or an upward occlusal plane alignment, or at least a much lesser amount of downward rotation. The occlusal plane in long-faced and long-nosed individuals tends to be downward-rotated to a greater extent.

A closed alignment between the ramus and the corpus **shortens** overall mandibular length and thereby has a mandibular retrusive effect. An open alignment increases it and has a protrusive effect. There are two ways to illustrate why this occurs. First, the straight-line dimension (overall mandibular length) from *a* to *b* in Figure 6–26 is decreased; the dimension from *a* to *c* is increased. Second, if the upper and lower arches *M* and *N* in Figure 6–27 are aligned upward, *M* protrudes beyond *N* by the distance *x* relative to the **occlusal plane** (not the vertical facial profile). When aligned downward, *N* protrudes by the distance *y* relative to the downward-inclined occlusal plane.

If the ramus-corpus angle is opened, the prominence of the **antegonial notch** is increased. This is caused by the downward angulation of the mandibular body at its junction with the ramus. If the ramus-corpus angle is **closed**, the size of the antegonial notch can be reduced or obliterated entirely because of the upward alignment of the corpus relative to the ramus. (See Stages 9 and 15 in Part 2 of Chapter 3 for additional factors involved and for an account of the remodeling processes that produce these various rotations.)

Note especially that the effects of whole-mandible rotations and ramus-to-corpus rotations are **opposite**. When the entire mandible is aligned downward, a mandibular retrusive effect is produced; but when just the corpus is aligned downward, a mandibular protrusive effect results (Fig. 6–28). An upward whole-mandible alignment is mandibular protrusive, and an upward alignment of the corpus only is mandibular retrusive.

An individual can have a retrognathic profile and **not** have a Class II malocclusion. Even though the underlying skeletal factors are the same for both, the effects for one may be more severe than for the other. This is because different planes of reference relate separately to the profile and to malocclusions. In Figure 6–29, note that the profile *(a)* is retrognathic, but the lower incisor region, relative to the downward-inclined **occlusal plane**, is not retrusive *(b)*.

A forward-inclined middle cranial fossa has a maxillary protrusive and a mandibular retrusive effect (Figs. 6–30 and 6–31). Because the expression of the effective horizontal (not oblique) dimension of the middle fossa is increased, the maxilla becomes offset anteriorly with respect to the mandibular corpus. The midface is also lowered, and this causes the whole mandible to rock down and back. The maxilla thus is carried forward, and the mandible is rotated backward in this composite, two-way movement. Mandibular retrusion results, even though the arch lengths of the upper and lower jaws can have equivalent

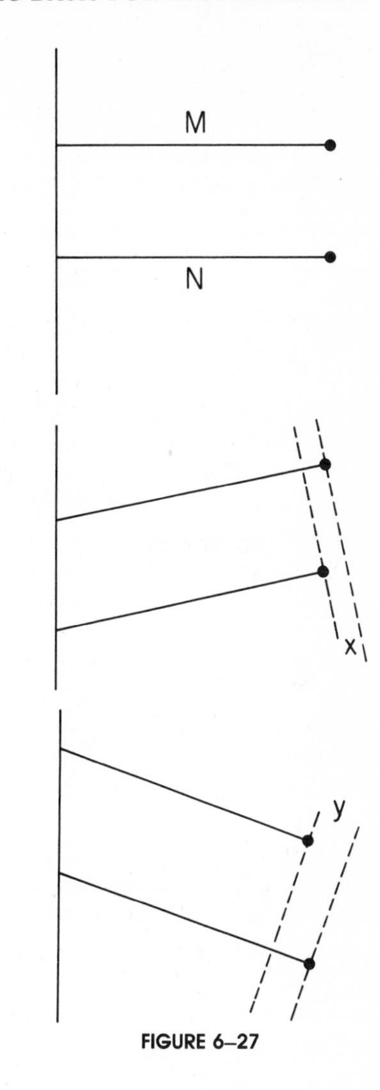

FIGURE 6–27

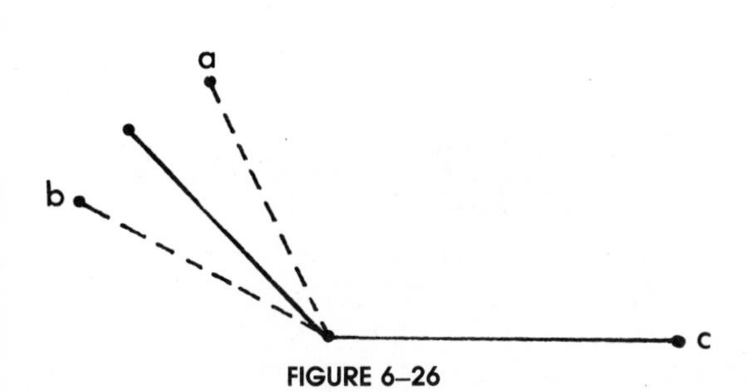

FIGURE 6–26

dimensions, as shown here. These various changes in skeletal pattern cause a Class II molar relationship because the lower bony arch is posteriorly offset.

A backward-inclined middle cranial fossa* has a mandibular protrusive effect (Figs. 6–32 and 6–33). This contributes to a Class III type of molar relationship. The maxilla is placed backward, and the mandible rotates forward

*Note this important point. The conventional way to represent the "cranial base angle" is by a line from basion to sella to nasion. This is not the anatomically meaningful way to do it. The **real** relationship (so far as the face is concerned) involves the contact between the condyle and the cranial floor (thus not basion), and the junction corner between the cranial floor and the nasomaxillary complex (thus not sella). **This** is the relationship that **directly** determines the anatomic effects of the three-point contact among the cranial floor, the mandible, and maxillary tuberosity. Basion-sella-nasion only indirectly reflects this. These three traditional landmarks have nothing to do with the actual anatomic fitting of the key junctions involved. They are removed **midline** structures that do not relate directly to the **lateral** positions of the upper and lower arches, the lateral contacts between the mandibular condyles and the cranial floor, and the lateral effects of the angle between the lateral parts of the floor of the middle and anterior cranial fossae relative to the maxillary tuberosities. Sella, basion, and nasion themselves can be almost **anywhere** along the midline axis, within normal variation limits, and not affect the "angle" that really counts: the angle from the condyle–glenoid fossa articulation to the point of junction between the middle and anterior cranial fossae, that is, the point where the nasomaxillary complex joins the cranial floor.

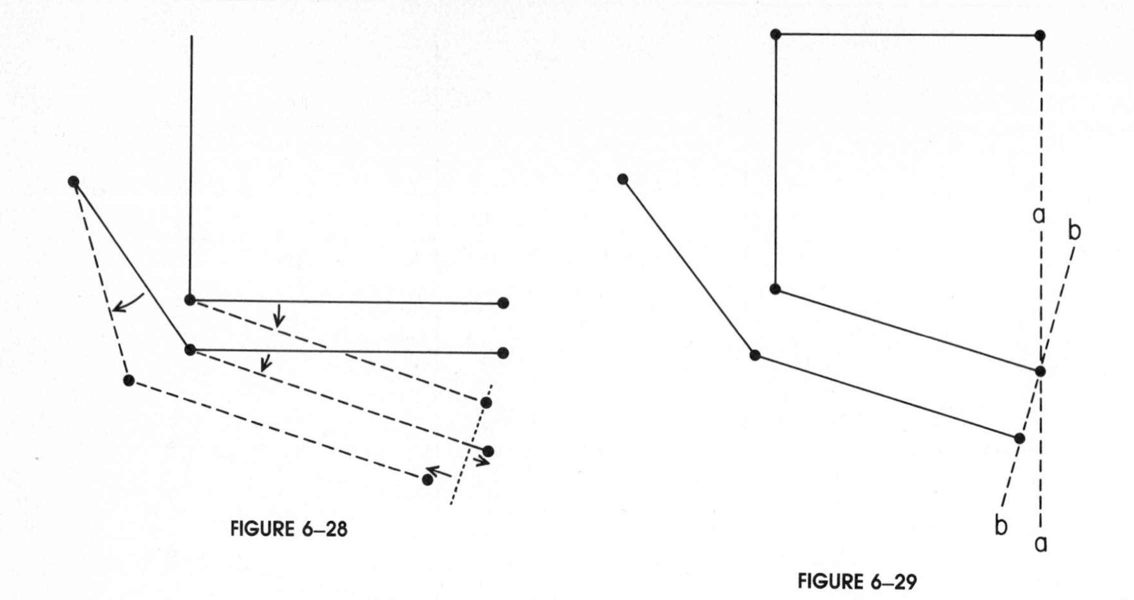

FIGURE 6–28

FIGURE 6–29

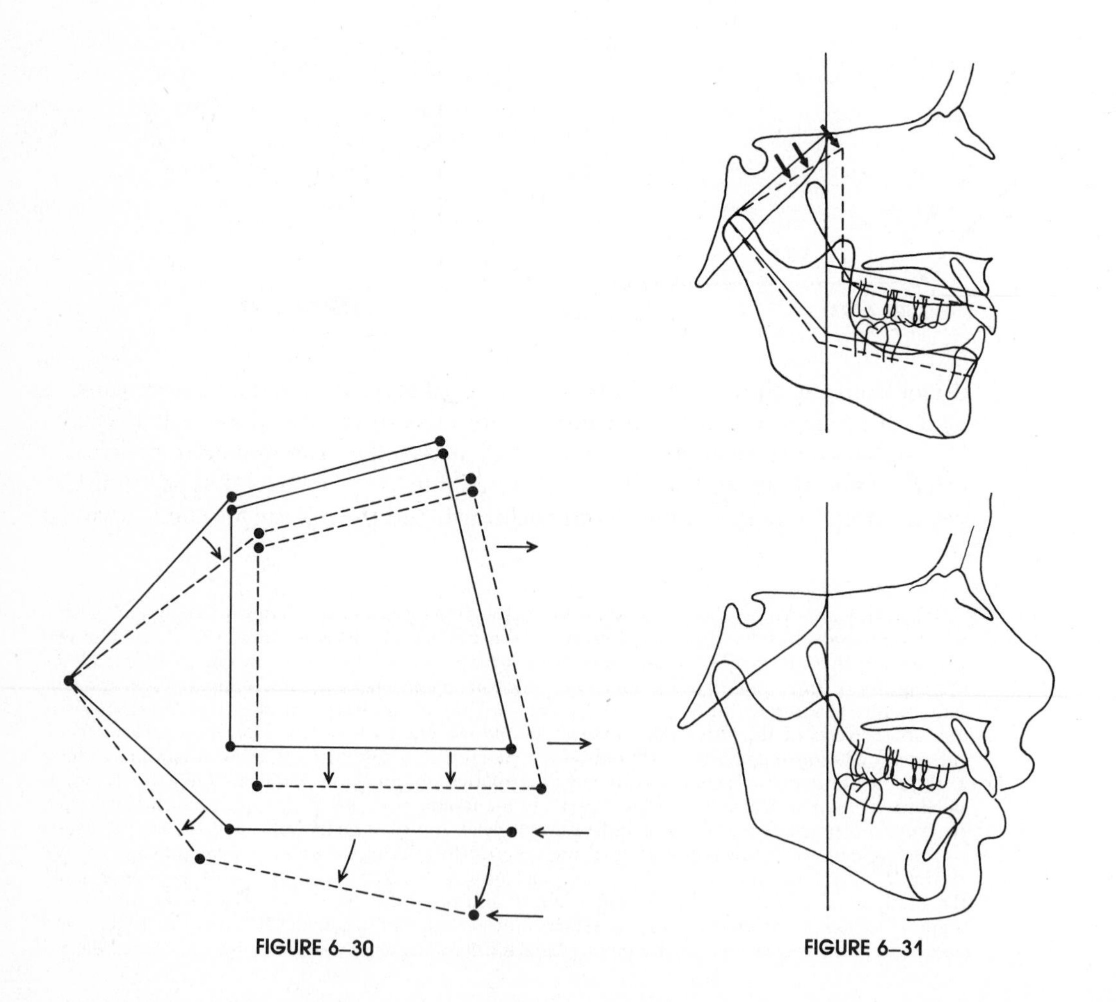

FIGURE 6–30

FIGURE 6–31

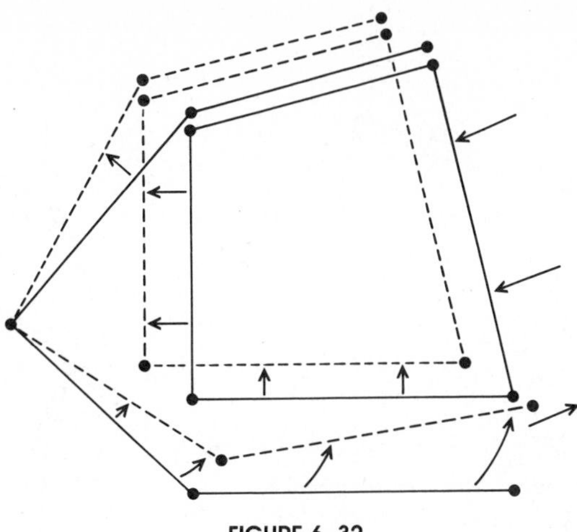

FIGURE 6–32

into a protrusive position. Note that the mandibular occlusal plane is rotated into an upward-inclined position. To compensate, the posterior maxillary teeth descend (drift inferiorly) or the ramus-corpus angle is opened, or both.

The nasomaxillary region in many (but not all) individuals tends to be vertically long relative to the ramus and middle cranial fossa. The result is a downward and backward placement of the whole mandible to varying degrees in different faces (Figs. 6–34 and 6–35). Note the resultant mandibular retrusive effect, the retrognathic profile, and the skeletal basis for a Class II molar relationship. It will be recalled that a forward alignment of the middle cranial fossa also causes a similar kind of mandibular rotation. If **both** occur in the same individual, the total extent of mandibular rotation is the sum of the two. (Dental changes occur to preclude an anterior open bite; see discussion of curve of Spee below.)

If the nasomaxillary region is vertically short, as noted earlier, a mandibular protrusive effect is produced (Fig. 6–36). The mandible rotates forward and upward, and the resultant offset positions between the maxillary and mandibular arches can contribute to a Class III type of molar relationship. Note that a **vertical** imbalance has resulted in a **horizontal** structural effect. It is incorrect

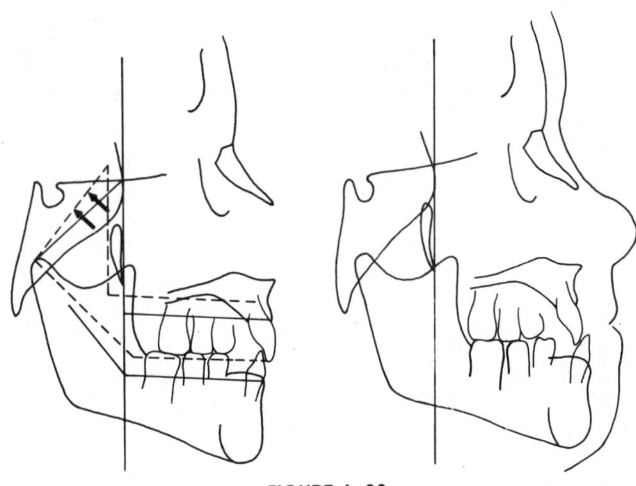

FIGURE 6–33

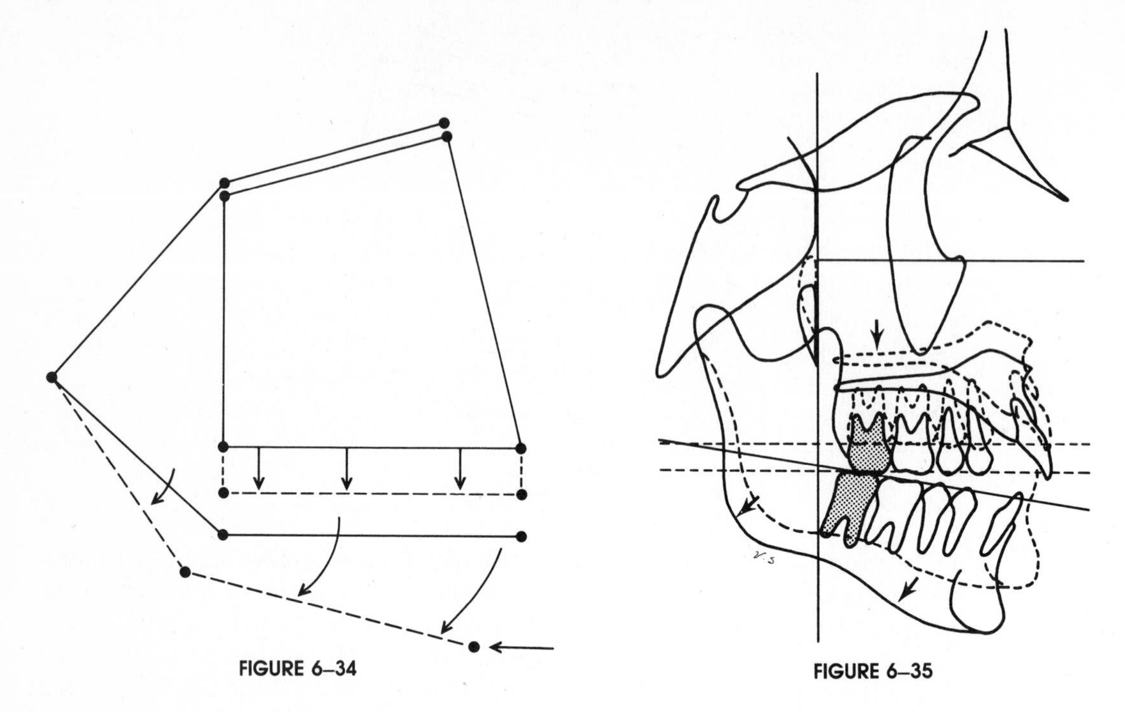

FIGURE 6–34

FIGURE 6–35

to assume, as many do, that malocclusions are based, essentially, only on horizontal dysplasias.

All the above relationships illustrate the various effects of changes in the dimensions or the alignment of any **one** given region, as for the ramus, middle cranial fossa, maxillary arch, and so on. The skull of any given individual, however, is a composite of many combinations of such relationships among **all** the regional parts. Outlined below are examples of several different combinations of various regional dimensional and alignment imbalances and balances.

To review a key point, how many specific anatomic factors have been described thus far that can contribute to the multifactorial, composite basis for a Class II or Class III molar relationship?

In the combination shown in Figure 6–37, the horizontal dimension of the

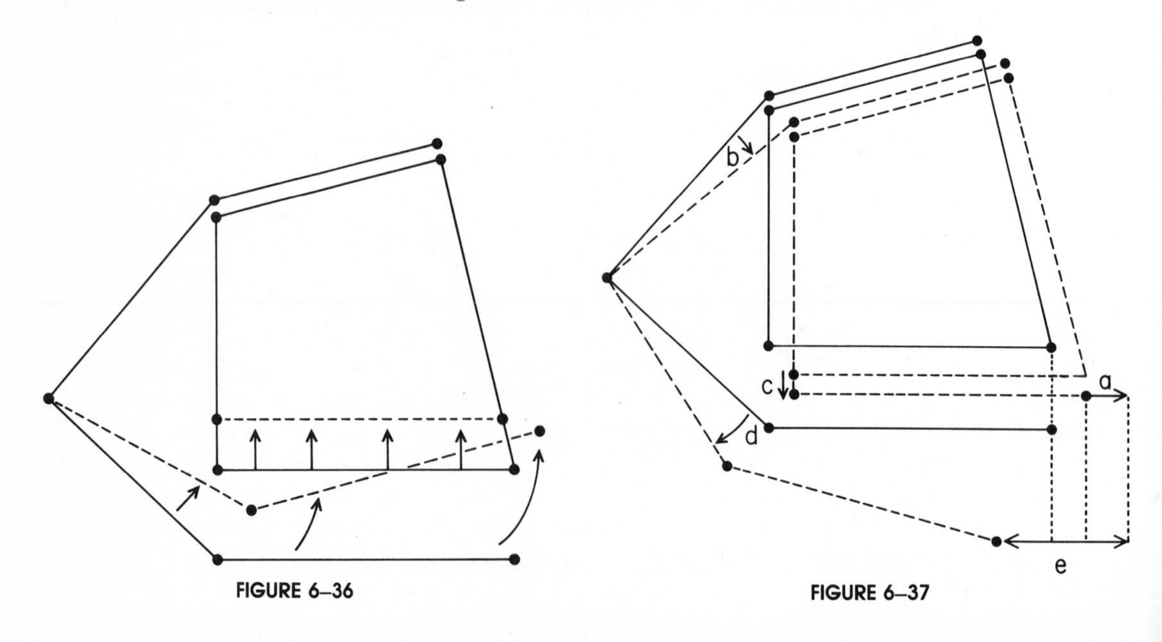

FIGURE 6–36

FIGURE 6–37

maxillary arch exceeds that of the mandibular arch *(a)*. The middle cranial fossa has a forward-inclined alignment *(b)*, and the midface *(c)* is also vertically long. The ramus is rotated backward *(d)*. **All** of these features have mandibular retrusive effects, and their combined sum *(e)* results in a severe Class II malocclusion and extreme retrognathia.

In the combination shown in Figure 6–38, the horizontal dimension of the middle cranial fossa exceeds that of its counterpart, the ramus. The mandibular corpus, however, is long relative to the horizontal dimension of the bony maxillary arch. The composite result actually produces a prognathic profile, but with a Class II molar relationship. Such individuals are encountered now and then.

The combination schematized in Figure 6–39 illustrates a horizontally short mandibular corpus (relative to the individual's maxillary arch) in combination with a backward-rotated middle cranial fossa, a forward-rotated ramus, and a downward-rotated mandibular corpus. The composite result is an individual with a Class II type of lower arch, a Class II molar relationship, and a Class I (orthognathic) type of profile.

The combination in Figure 6–40 involves a horizontally long mandibular corpus, a forward alignment of the middle cranial fossa, and a backward rotation of the ramus. The cumulative result is a Class I orthognathic profile, a Class II molar relationship, and a Class III type of bony mandibular arch. The corpus-ramus angle has closed somewhat, but there is still a downward-inclined alignment of the occlusal plane.

The factor of morphologic **compensation** during facial development is a basic and important biologic concept. Compensatory adjustments involve a latitude of morphogenetic give-and-take among the various regional parts as all grow in close interrelationship. The composite result is a state of functional and structural "balance" (equilibrium, homeostasis). Indeed, growth is a constant compensatory process striving toward ultimate balance as a bone grows in relation to its developing muscles, as connective tissue grows in relation to both bone and muscle, and as blood vessels, nerves, epithelia, and so forth, all develop in relation to everything. When the growth process is complete, a state of compromise equilibrium has been achieved, even though a malocclusion or some other dysplasia may exist. There nearly always exist a number of regional

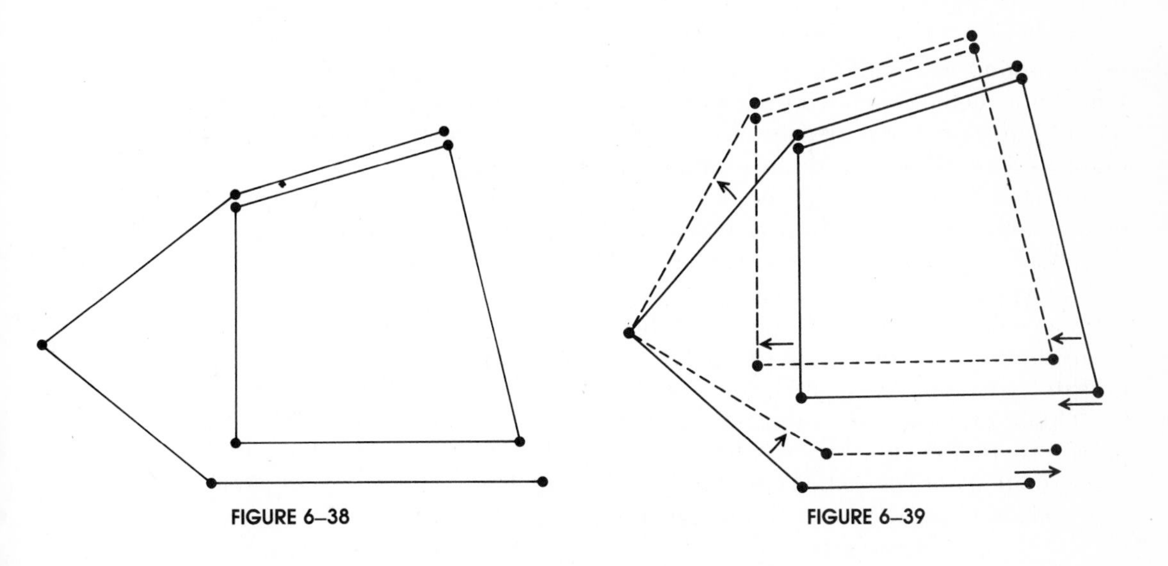

FIGURE 6–38

FIGURE 6–39

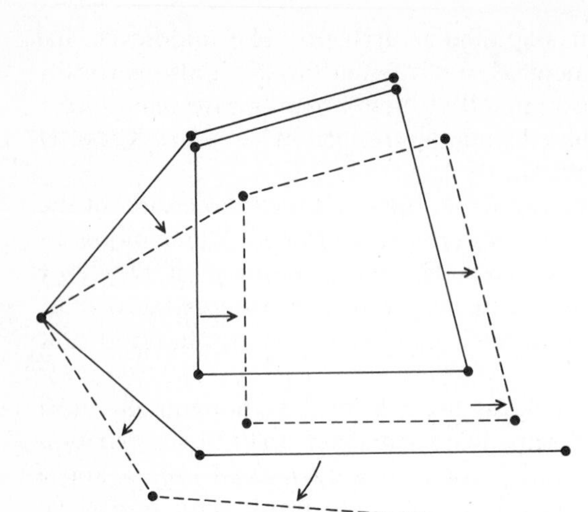

FIGURE 6–40

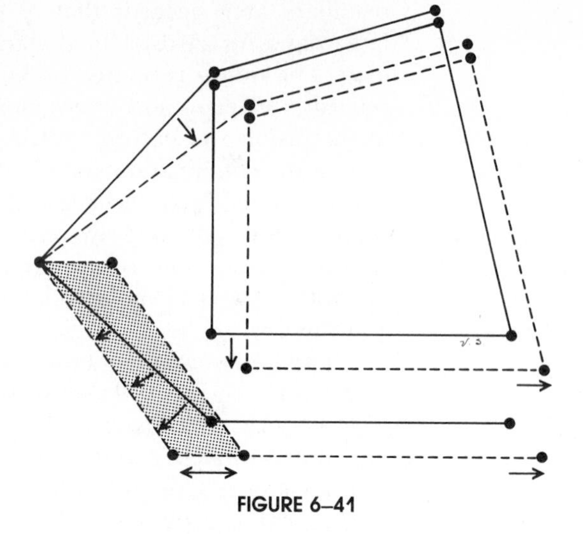

FIGURE 6–41

morphologic imbalances to greater or lesser degrees of severity, but the aggregate construction of the craniofacial composite as a whole is more or less "functional," albeit with some regional variations from the ideal or from the population mean.

A frequently encountered compensatory combination involves the ramus of the mandible (Fig. 6–41). When the nasomaxillary complex is vertically "long" and/or the middle cranial fossa has a forward-downward rotational alignment, the whole mandible consequently becomes rotated into a downward-backward placement. As described previously, these factors underlie the skeletal basis for a retrusion of the mandible and a Class II molar relationship. However, developmental processes can respond by a widening of the horizontal breadth of the ramus. This compensatory adjustment places the mandibular arch more protrusively, thereby partially or totally counteracting the extent of its backward rotation. What would have been a Class II malocclusion and a retrognathic profile has been converted into a Class I occlusion and a more orthognathic profile. Should the extent of compensation fall somewhat short, at least the severity of the potential malocclusion has been reduced. Should compensation fail entirely, the malocclusion becomes fully expressed.

Understand that in carrying out its compensatory role, the ramus does not itself respond as though it has a brain of its own and somehow elects to do something good. As pointed out earlier, growth is a prolonged process striving toward functional and structural equilibrium. The skeletal response by the ramus is a result of growth and continuous remodeling actions paced by the growth and function of the masticatory muscles, the airway, the pharyngeal mucosa and muscles, connective tissue, and so on, all of which develop in a composite, interrelated manner that has a latitude for adjustment to the growth and morphology of other, contiguous regions, e.g., the basicranium and ethmomaxillary complex. If such latitude is not exceeded, at least a partial compensatory relationship can be achieved during the growth period. When growth has become completed, the capacity for compensation diminishes.

Other examples of compensatory developmental adjustments, including palatal rotations, anterior crowding, gonial angle remodeling, and occlusal plane rotations, are described elsewhere in this and other chapters. It is apparent that a Class II malocclusion is **not** merely caused by a "long maxillary arch."

Malocclusions are multifactorial. Examples of Class II and Class III headfilm tracings are shown next for comparisons of their respective structural characteristics according to the different kinds and combinations of relationships outlined above.

SUMMARY OF CLASS II VERSUS CLASS III SKELETAL FEATURES

In the Class II individual (Fig. 6–42), note that the mandibular arch is short relative to the maxillary arch. The mandibular arch in the Class III individual (Fig. 6–43), conversely, is horizontally long relative to the maxillary arch.

The middle cranial fossa in the Class II individual has a forward and downward-inclined alignment. In the Class III individual the middle cranial fossa is aligned backward and upward. This places the nasomaxillary complex more retrusively in the Class III individual and more protrusively in the Class II individual. It also contributes to rotations of the mandible (see below).

The nasomaxillary complex in the Class II individual is vertically long relative to the vertical dimension of the ramus (or the ramus is short relative to the maxilla). This long midface, together with the downward-forward alignment of the middle cranial fossa, causes a downward-backward rotational alignment of the ramus (and whole mandible) in the Class II individual.* The Class III ramus, conversely, is rotated forward in conjunction with an upward-backward middle cranial fossa rotation and a vertically short nasal region. The midface is short relative to the vertical dimension of the ramus (or the ramus is long relative to the maxilla). Although the face of the Class III individual "looks" quite long, it is usually the lower face (mandible), not the nasal region, that causes this (see below).

---

*The dashed lines in the figures represent neutral alignment positions. See Enlow et al, 1971*a*, for the rationale involved.

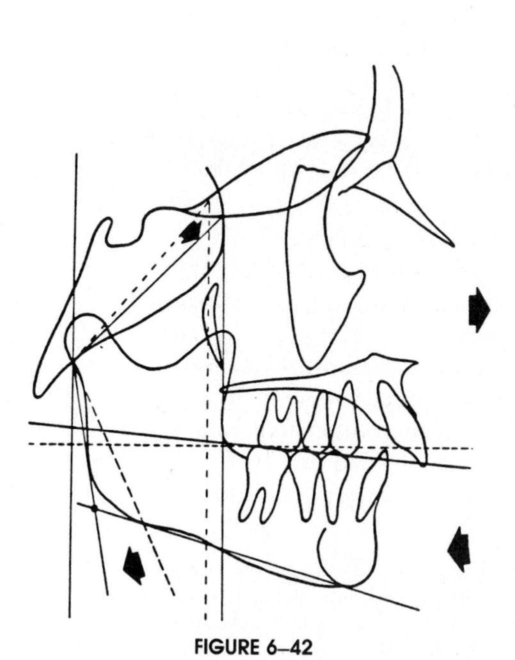

FIGURE 6–42

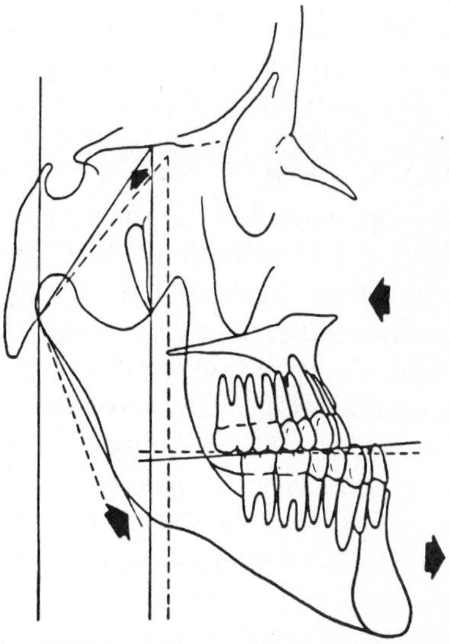

FIGURE 6–43. (From Enlow, D. H., T. Kuroda, and A. B. Lewis: The morphological and morphogenetic basis for craniofacial form and pattern. Angle Orthod., 41:161, 1971.)

In Class II individuals, the head form type is often dolichocephalic. The anterior cranial fossa is thereby long and narrow; and because it is the template for the nasomaxillary complex, the palate and maxillary arch are correspondingly elongate and narrow. In Class III individuals, conversely, the anterior cranial fossa tends to be wider and shorter (brachycephalic), and this thereby establishes a foreshortened but wider palate and maxillary arch.

The ramus-corpus (gonial) angle is more closed in the Class II but open in the Class III face, thereby shortening and lengthening, respectively, overall mandibular length. In the Class III face this produces the characteristic steeply angled alignment of the mandibular corpus.* Note that the anterior mandibular teeth in the Class III individual have drifted upward to a considerable extent (a compensatory adjustment), so that the occlusal plane is not angled as sharply downward. This causes the characteristically elongate, high alveolar region above a prominent-appearing chin observed in many Class III faces. Compensatory downward drifting of the maxillary dentition may also occur in some cases (as seen in Figure 6–43). In the Class II face, conversely, it is the nasal region that appears relatively elongate vertically, with a much shorter-appearing vertical depth in the region of the chin. However, some Class II individuals may show a steeply inclined mandibular plane (not present in Figure 6–42), causing a lengthened, but retrusive, appearance of the lower face. This results when the downward-backward rotation of the whole mandible (produced by the Class II long face/short ramus/open middle cranial fossa relationships described earlier) is not accompanied by closure of the gonial angle. A deeper compensatory curve of Spee often develops (page 220). For the Class III face, a variation can occur in which the forward rotation of the ramus is not accompanied by an opening of the gonial angle. The corpus (mandibular plane) then has a correspondingly more shallow alignment.

To date, the combination of these features contributes to the composite skeletal basis for mandibular retrusion in the Class II individual and mandibular protrusion in the Class III individual. However, note that the Class II ramus is horizontally broad and that the Class III ramus is narrow. These are, as explained earlier, compensatory features that partially counteract the other characteristics that combine to cause mandibular retrusion and protrusion, respectively. The resultant malocclusions are thereby less severe than they would have been had the ramus in each been of "normal" dimension. Had the ramus actually been narrow in the Class II individual and wide in the Class III individual, it would, of course, have added to (rather than subtracting from) the composite basis for the malocclusion.

Most Class II individuals thus have a horizontally short mandibular corpus, a vertically long nasomaxillary complex, a downward- and backward-aligned ramus, a forward middle cranial fossa alignment, a closed ramus-corpus (gonial) angle, and (in severe malocclusions) a narrow ramus and horizontally broad middle cranial fossa. The **converse** of all these regional relationships characterizes the Class III malocclusion. Each such feature occurs in about 70 per cent or more of Class II and III individuals. What about the other 30 or so per cent? This is where "offsetting penalties" come into play. Instead of a forward-inclined, dolichocephalic type of middle cranial fossa causing mandibular

*Note that the Class II face can have a steep mandibular plane angle, thereby appearing similar to the downward-inclined mandibular corpus of some Class III faces. The underlying reasons, however, are different and should not be confused. In the Class II, it is a **whole mandible displacement** rotation. In the Class III, it is a **ramus-to-corpus remodeling** rotation. The former relates to a **long** midface, whereas the latter relates to a **short** midface.

retrusion, for example, a given individual can have a **backward**-aligned, brachycephalic type of fossa. This feature may then combine with one or more other regional mandibular **protrusive** features, such as, perhaps, a broad ramus, a long corpus, or a downward-rotated corpus, to partially counteract the various mandibular retrusive factors that are also present. In any given individual, the sum of the dimensional values for all the mandibular protrusive features weighs against the sum of the values for all the mandibular retrusive features. Either they come into an effective balance or one or the other wins. If the mandibular retrusive features dominate, the **severity** of the resultant Class II malocclusion and retrognathic type of face depends, first, on how much (in millimeters) the total of these retrusive features amounts to and second, how much the counteracting features subtract from this total.

Each of us has a face and cranium that represent such a composite mixture of regional counteracting imbalances. Here and there in the different regions of the face and cranium, something can be balanced, but no one of us has a head that is regionally in balance throughout. Add to this the topographic facial features involving the frontal sinuses, upper face protrusion relative to the anterior cranial fossa, corresponding variations in the nasal bridge, the shape and size of the nose, and the narrowness or widening of the whole face in relation to brain shape and head form; and the almost limitless variations that occur in overall facial form can readily be appreciated.

Each of us has a natural, normal predisposition toward either mandibular retrusion (Class II) or protrusion (Class III). There is no such thing, in a sense, as a "separate" Class I facial category. All Class I individuals have a predominant tendency one way or the other toward a malocclusion. Most narrow- and long-faced Class I Caucasians have the **same** underlying facial and cranial features that are present in the long-faced Class II Caucasians; the same 70 or so per cent of the various mandibular retrusive relationships described above also occur in the Class I individual. This is why a Class II "tendency" usually exists to a greater or lesser extent. The difference between the Class I and II malocclusions, however, is the **extent** of the imbalances and the number and extent of counteracting features. If the compensating characteristics are adequate, a more or less normal face results. If they partially or totally fail, marginal to severe malocclusion and facial disproportion result. A person having an attractive, well-proportioned face, for example, with an orthognathic (or nearly so) profile and only relatively minor occlusal irregularities **also** has, unsuspected by him or her and deep within the face and cranium, the **same** underlying characteristics that caused a cousin to have a noticeably retrognathic profile and a Class II malocclusion. Our hero, however, has a particularly broad ramus and some other happy characteristics that are winners for him as an individual. Most of us have at least a reasonable-appearing face, although somewhat short of perfect, for the same reasons.

A fundamental principle to keep always in mind is that the growth process is continuously creating imbalances as, for example, the muscles and the airway continue to differentiate. At the same time, however, the growth process is **working toward** an aggregate state of composite balance.

DENTOALVEOLAR COMPENSATIONS

During the development and establishment of the occlusion, compensations occur involving dentoalveolar remodeling as well. The placement of the teeth interrelates with the many other skeletal and soft tissue growth processes taking place in the face and cranium. Figures 6–44 to 6–48 explain some common changes involved.

In the first diagram (Fig. 6–44), the vertical and horizontal dimensions among the various skeletal parts and counterparts are in balance. The alignments of all the parts also are in "neutral" positions. That is, the nature of the alignments is such that neither protrusion nor retrusion of the upper or lower jaws is produced; the angular relationships are balanced so as not to increase or decrease the "expression" of any of the various key dimensions. Note that the occlusal plane is perpendicular to the vertical reference line (the PM plane) and parallel to the neutral orbital axis (shown here below the orbit, rather than within its geometric center).

The nasomaxillary complex in the next stage (Fig. 6–45) has become lengthened vertically to a greater extent. This is common, as mentioned earlier. The amount of midfacial growth has **exceeded** the vertical growth of the ramus-middle cranial fossa composite. The result is downward and backward alignment of the whole mandible to accommodate the longer nasomaxillary complex. A vertical "imbalance" has thus been introduced, and the expression of the vertical ramus height has been increased to match it by a downward rotation. (This same effect on the mandible can also be caused by a proclination of the middle cranial fossa, as previously described.) Note especially that the mandibular corpus, and with it the lower teeth, now has a consequent downward inclination relative to the vertical **PM** line. This "opens" the anterior bite; only the first and second molars are in occlusal contact. The amount of occlusal separation increases toward the incisors.

Note also the retrusion of the mandible, overjet, and the Class II molar relationship caused by the ramus rotation. These resultant effects, however, may be partially or completely offset by a widening of the ramus or other compensatory changes.

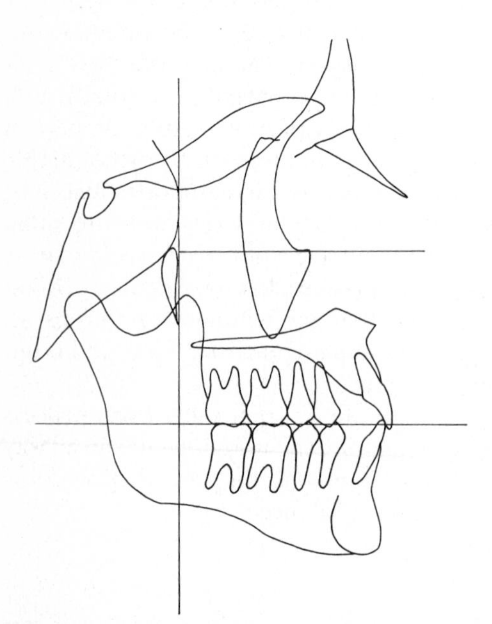

FIGURE 6–44. (From Enlow, D. H., T. Kuroda, and A. B. Lewis: The morphological and morphogenetic basis for craniofacial form and pattern. Angle Orthod., 41:161, 1971.)

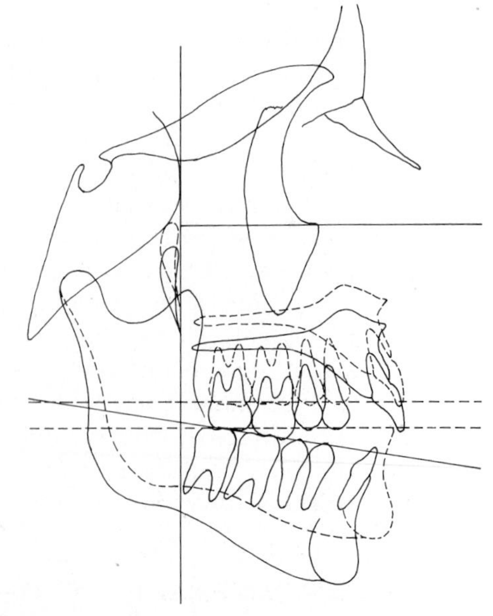

FIGURE 6–45. (From Enlow, D. H., T. Kuroda, and A. B. Lewis: The morphological and morphogenetic basis for craniofacial form and pattern. Angle Orthod., 41:161, 1971.)

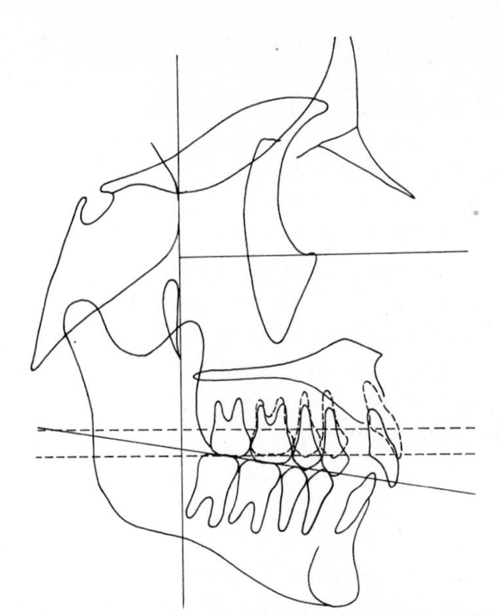

FIGURE 6–46. (From Enlow, D. H., T. Kuroda, and A. B. Lewis: The morphological and morphogenetic basis for craniofacial form and pattern. Angle Orthod., 41:161, 1971.)

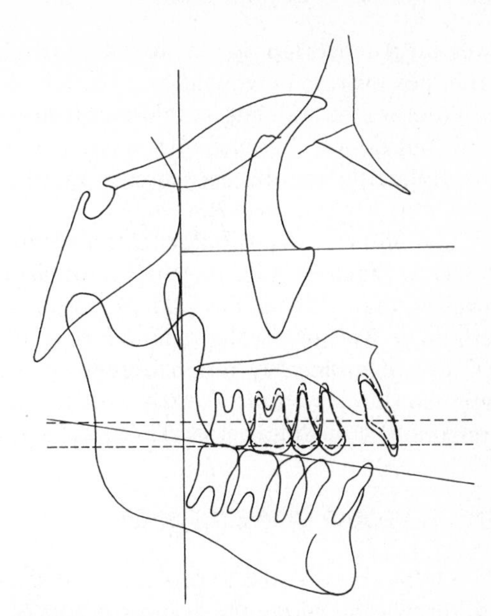

FIGURE 6–47. (From Enlow, D. H., T. Kuroda, and A. B. Lewis: The morphological and morphogenetic basis for craniofacial form and pattern. Angle Orthod., 41:161, 1971.)

The upper teeth "drift" (**not** erupt) inferiorly until each comes into contact with its antagonist (Fig. 6–46). The molars were already in contact; the second premolar must drift downward only a short distance. The first premolar drifts inferiorly even more because of the greater gap potential involved. The central incisors move down the greatest distance. As a final result, full arch-length occlusal contact is attained. The occlusal plane is **straight** (not curved, as in other variations described below). The occlusal plane bisects the upper and

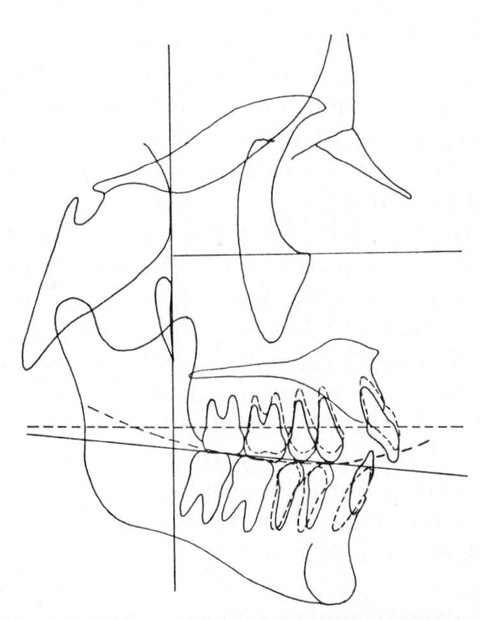

FIGURE 6–48. (From Enlow, D. H., T. Kuroda, and A. B. Lewis: The morphological and morphogenetic basis for craniofacial form and pattern. Angle Orthod., 41:161, 1971.)

lower incisor overlap, just as it did in the first, "balanced" stage. The occlusal plane, however, is now inclined obliquely downward.

Another remodeling combination may occur. The upper teeth drift inferiorly, but the canines and incisors do not move down to the full extent needed to completely close the occlusion, only to about the extent that the premolars drift inferiorly (Fig. 6–47).

The anterior **mandibular** teeth now drift superiorly until full arch occlusal contact is reached (Fig. 6–48). The incisors must move upward much more, however, than the canines and premolars. Note that the roots of the anterior teeth have become realigned and that the cusps of the lower incisors and canines are **noticeably much higher** than the premolars and molars. Palpate your mandibular anterior teeth with the tongue to determine if this common growth and adjustment process has occurred in your own dentition.

DENTOALVEOLAR CURVE (OF SPEE)

There are two ways to represent the **occlusal plane**. The traditional method is to draw a line along the contact points of all the teeth to the midpoint of the overlap between the upper and lower incisors. In the first two examples cited above, this line is straight. In the last, however, note how the line is curved as it exactly bisects the overlap of the upper and lower incisors. This is called the **curve of Spee**, and the reason for its development was outlined in the previous paragraphs. The second way to represent the occlusal plane is to run a line from the posterior-most molar contact point straight to the anterior-most premolar contact point. The incisors are not considered. This is termed the "functional occlusal plane," and it is always a straight line whether or not a curve of Spee exists.

In the first and second examples of occlusal development (Figs. 6–44 and 6–46), a curve of Spee has not developed, and the two methods for representing the occlusal plane result in the same line. In the last example, however, the curved occlusal plane bisecting the incisor overlap and the straight functional occlusal plane are divergent. Note how the mandibular incisors rise considerably above the level of the functional occlusal plane. The maxillary incisors, however, fall well short and do not even touch this straight line functional occlusal plane. In individuals having a marked curve of Spee, the alveolar region of the mandible just above the chin is characteristically more elongate because the incisors have drifted superiorly for several millimeters or more.

The dentoalveolar curve (of Spee) is a common developmental adjustment that can provide intrinsic compensation for an **anterior open bite**. A combination of several factors underlies the skeletal basis for this type of malocclusion. If airway (or other) problems lead to an opening of the ramus-corpus (gonial) angle and a counterclockwise rotational alignment of the palate and maxillary arch, the vertical drifting (not eruption) of the anterior mandibular teeth can close what would otherwise be a skeletal (not merely dental) open bite. Should this intrinsic process fail, the open bite thereby expresses itself fully. Conversely, if a **deep overbite** occurs, it is usually seen in patients who have a horizontally short mandible in conjunction with a closed ramus-corpus angle, a clockwise rotational alignment of the palate and maxillary arch, and a deep curve of Spee.

Another kind of dental compensation also commonly occurs. It was pointed out earlier that the teeth have only a very limited capacity for remodeling (particularly after they have become fully formed). That is, a tooth cannot become markedly reshaped by selective remodeling resorption and deposition of dentin and enamel throughout its various areas to accommodate spatial and functional relationships; only a relatively limited extent of root resorption, deposition of cementum, trajectory of root growth, and crown wear is possible in this regard. This means that most adaptive adjustments for a tooth must be carried out by the "displacement" process. Any marked extent of resorptive and depository remodeling required is thus a function of the housing alveolar bone, not the tooth itself. If, however, the capacity for this bone remodeling is exceeded, as for example by an alveolar arch that is too small for the teeth it must support, the recourse then is displacement of some of the teeth. Thus, **anterior crowding** is, in effect, a compensatory means by which the teeth are housed beyond the limit provided by the available bone and its growth and remodeling potential (see also page 189). The compromising result, while successfully accommodating the teeth, also unfortunately creates a type of malocclusion.

CHAPTER SEVEN

The Structural Basis for Ethnic Variations in Facial Form

Age, sex, and population differences in the pattern of facial structure have been pointed out in the preceding chapters. The purpose of this section is to summarize this information briefly and add to it as a separate topic. Although this is an interesting subject in its own right, it is quite important for the clinician to realize that population norms derived from a given sample are not necessarily valid or accurate for other samples or groups, especially if ethnic variations are involved.

The phylogenetic basis for the form and pattern of the human face was outlined in Chapter 5. It will be recalled that both the shape and the size of the brain are key factors relating to the structure of the face. Because the cranial base is the bridge between them, and because the floor of the cranium is the template upon which the face is constructed, variations in the shape of the brain in **any** species are associated with corresponding variations in the form of the face. For example, the junctional part of the midface can only be as wide as the floor of the cranium. It cannot be wider because there is nothing to attach it to. Thus, narrow-brain species or subgroups are correspondingly narrow-faced. Compare the face of the long, narrow-brained collie dog with that of the short, round-brained boxer or bulldog.* Man has an exceptionally wide face, in comparison with the typical mammal, because of his colossal brain size and the shape of the brain. The various rotations of the olfactory bulbs, orbits, and so forth combine with the boundaries of the brain to establish, in all species, the amount and the principal directions of facial growth. Because of these factors, the shape and the size of the brain are involved in the variations of facial pattern **within** any given species as well as between species. There are, however, other factors that come into play, as will be seen.

Human population groups having a dolichocephalic head form naturally have a proportionately more narrow and longer face than those with a

*It has long been argued as to which is the pacemaker, the brain or the cranial floor (see Chapter 5). Regardless of which one is primary and which secondary, the effects on facial configuration are nonetheless real.

brachycephalic type of head form. The wider brain (with no special difference in overall volume) has the wider face. It has been claimed that there is an evolutionary (secular) trend toward the brachycephalic type among "long-headed" human groups. If this is happening, there will also be related long-term changes in facial structure, the nature of built-in tendencies toward malocclusions, and profile type.

The more open ("flat") cranial base flexure that usually characterizes the dolichocephalic head form in many Caucasian groups sets up a more protrusive upper face and a more retrusive lower face (Figs. 7–1 and 7–2). The whole nasomaxillary complex is placed in a more forward position, and it is lowered relative to the mandibular condyle. Because the condyle is "higher," there is the tendency for a downward and backward rotation of the whole mandible. The posteroanterior dimension of the pharynx is relatively large because of the longer and more horizontally aligned middle cranial fossa. Because the anterior cranial fossa is relatively elongate and narrow, the palate and maxillary arch are correspondingly long and narrow. For these various reasons, there is a greater frequency of the retrognathic type of profile and a Class II tendency among groups with a dolichocephalic type of head. There is also a high incidence of a "broad" ramus to compensate for the built-in tendency toward mandibular retrusion.

Among most dolichocephalic Caucasian groups, the upper part of the ethmomaxillary region characteristically is even **more** protrusive, thus enlarging the entire nasal region, including its fleshy proboscis, to compensate for relative narrowness (Figs. 7–3 and 7–4). This adds to the forward manner of upper face placement caused by the elongate and narrow brain. The anterior cranial floor and the frontal lobe *(a)*, the superior part of the ethmomaxillary complex *(b)*, the palate and maxillary arch *(c)*, and the mandibular corpus *(d)* are all structural "counterparts" to each other. Depending on (1) the actual horizontal dimension each part attains by its own regional growth, (2) the alignment of that part, and (3) the direction and the extent to which each part is **displaced**

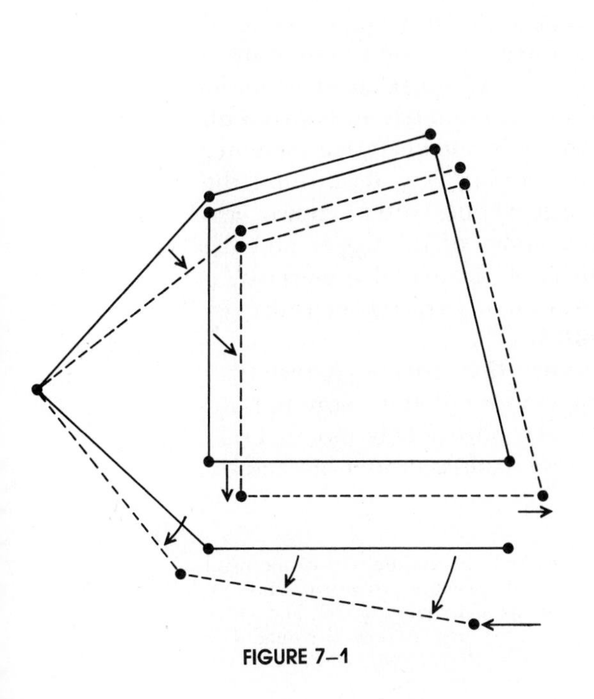

FIGURE 7–1

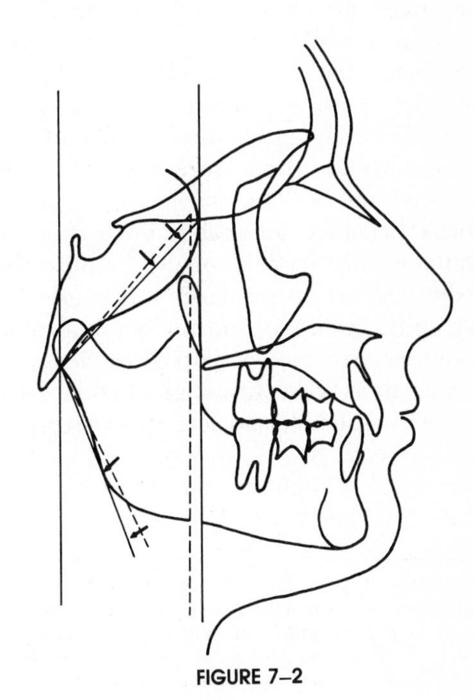

FIGURE 7–2

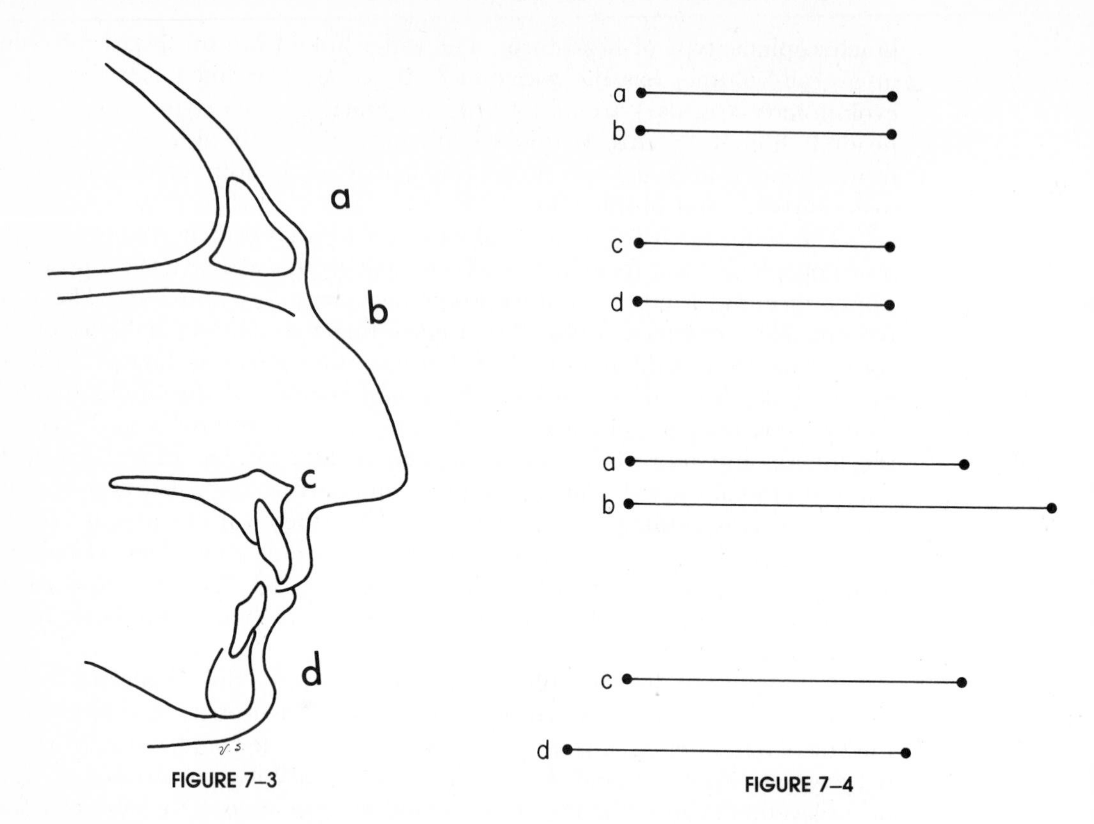

FIGURE 7–3

FIGURE 7–4

by other regions, the form and profile of the face are changed accordingly. Because the anterior cranial fossa *(a)* is horizontally long and narrow, the upper face *(b)* is also correspondingly long and narrow. In many dolichocephalic Caucasians, however, part *b* adds to the extent of upper face protrusion by continued horizontal expansion beyond *a* and *c*. This causes a characteristically high, sharp nasal bridge and a sizable nose. The nasal part of the long, narrow airway, as a whole, thereby has approximate equivalence in volume to that of other facial types having a wider but shorter nasal region. If the extent of nasal protrusion is quite marked, the nose can bend in order to relate structurally with part *c*.* The outer cortical table of the forehead remodels anteriorly with the nasal bridge, and a large frontal sinus is thereby formed between the inner and outer tables. The forehead is much more sloping as a result, and the glabella becomes noticeably protrusive. The cheekbones often appear less prominent and more "hollow" because the remainder of the upper and the middle face are so protrusive. Because the mandible is rotated posteriorly, it tends to be retrusive, and the whole profile takes on a characteristic convexity for all these reasons. A Class II tendency is **built in**.

The more closed, upright cranial base flexure that usually characterizes the brachycephalic head sets up a correspondingly wider, flatter, more upright type of face (Figs. 7–5, 7–6, and 7–7). The rounder, horizontally shorter brain and correspondingly shorter anterior cranial fossa establish a wider but antero-

*In another type of facial configuration, as previously described, the **middle** part of the nasal region anterior to the cheekbones, rather than the upper part, is the more protrusive. This also causes a bent nasal shape, but the slope is characterized by a sigmoid configuration. The whole cheekbone area is also protrusive. In either case, the curved type of nose is **vertically** long. The longer it is, the more sharply it must curve.

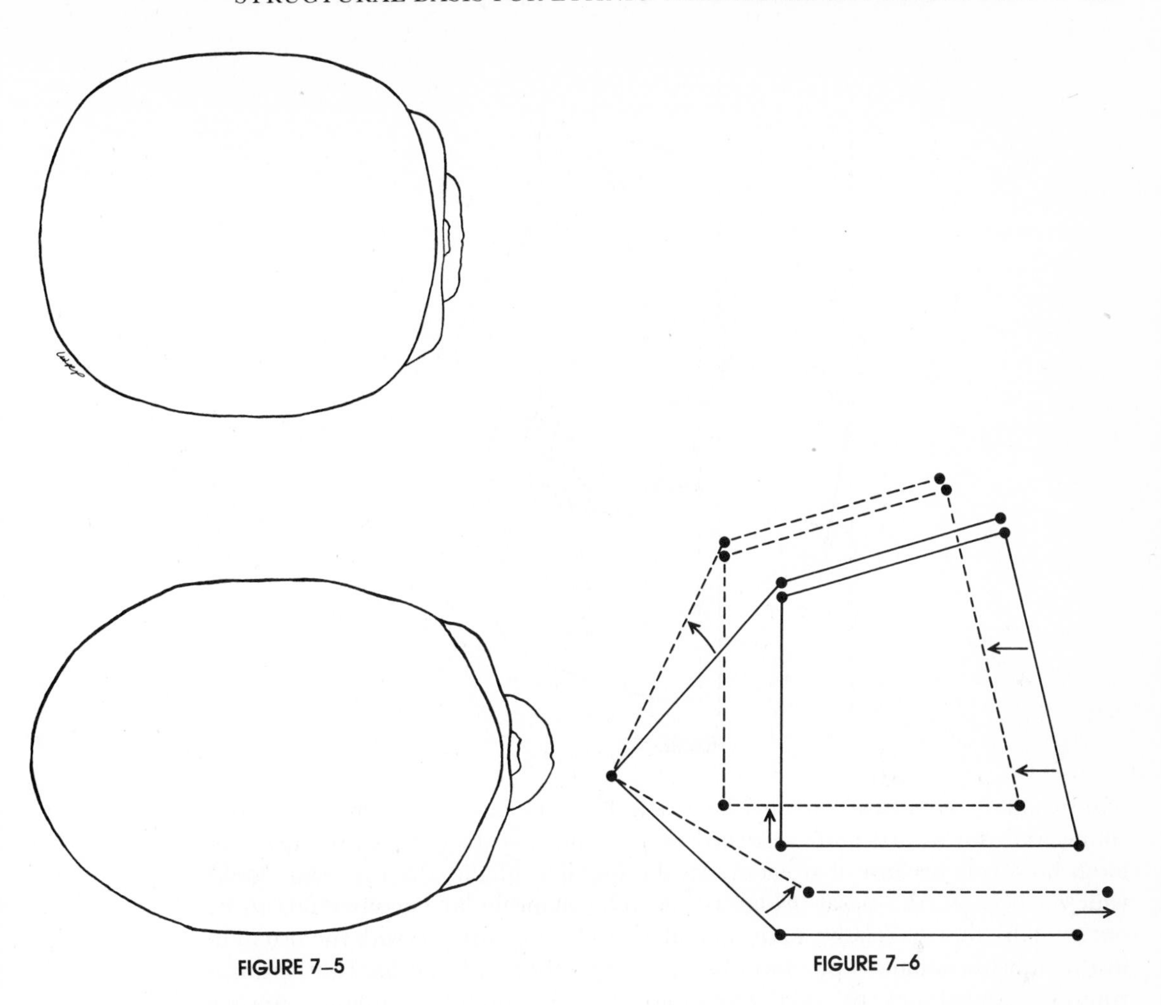

FIGURE 7–5

FIGURE 7–6

posteriorly shorter upper and midfacial region. The palate and dental arches are thereby foreshortened but relatively wide. The whole upper and midfacial region is also placed less protrusively because of the more upright middle cranial fossa. The middle cranial fossa, and therefore the pharyngeal region, is horizontally shorter for the same reasons. This further decreases the relative extent of the upper and midfacial protrusion. In addition, the upper part of the ethmomaxillary complex does not expand anteriorly to nearly the same extent described for the previous facial type. The wider, shorter nasal and pharyngeal airway is approximately equivalent to those of other facial types having a much greater extent of nasal and maxillary protrusion but with a more narrow passageway. The composite result is a more upright and bulbous forehead, a lesser protrusion of the glabella and eyebrow ridges, a thinner frontal sinus, a much lower nasal bridge, a shorter pug-type nose, and a tendency for a forward rotation of the entire mandible (unless offset by a vertical lengthening of the midface, which is a common feature). These features give a quite "vertical" character to the whole face, and the face appears much flatter, broader, and squared. The cheekbones are more prominent-appearing because the remainder of the upper and middle face is not as protrusive. There is a greater likelihood for an orthognathic (straight) profile, and the chin appears prominent and the mandible quite full. A greater tendency for a Class III type of malocclusion and a prognathic mandible exists. However, the

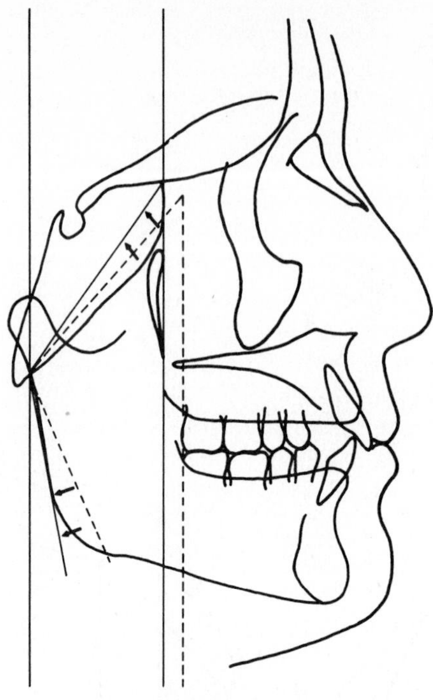

FIGURE 7–7

nasomaxillary complex can also be vertically long, and the ramus can thus rotate well downward and backward as a result. The face does not "look" as long, however, because it is wider. In the brachycephalic, the eyes can "look" wide-set because the nasal bridge is low. The mandibular corpus tends to be horizontally shorter relative to the maxillary arch, as compared with the situation in the dolichocephalic type, and this factor, together with the backward ramus rotation, contributes to a compensation for the built-in tendency toward prognathism and bimaxillary protrusion.*

It is important to realize that any predominantly brachycephalic population embodies a range of variation from the "typical," as outlined above, to a mix of facial features that grade toward the dolichocephalic/leptoprosopic form. The Oriental "race," for example, does not represent a single, homogeneous grouping, but, rather, a composite assemblage of many geographically, environmentally, and morphologically diverse subgroups that have evolved into quite distinctive and dissimilar craniofacial types. In contrast to the very round and flat-faced pattern, a more leptoprosopic, angular, long, and thin-nosed form also exists. The extent to which such facial variation relates to different anatomic types of malocclusions, and responses to different clinical procedures, is presently not at all well known. Thorough anatomic and morphogenetic descriptions of these variations are yet very scant.

The above features characterize the Oriental face, as well as certain Caucasian groups that also have a rounder brachycephalic ("Alpine") type of head form with many of these same facial features. (This does not include the dinaric head form, which is a fundamentally separate brachycephalic category.) The brachycephalic Caucasian type of face, like many Oriental faces, is wider,

*The information in this section is based on previously unpublished work carried out by the author in collaboration with Dr. Takayuki Kuroda of the Tokyo Medical and Dental University.

the nasal bridge lower, the nose flatter and shorter, the midface variably shorter, the forehead more upright, and the mandible more prominent. There are fewer underlying Class II tendencies in this basically different type of Caucasian face. Class I individuals having this composite facial structure tend toward a more orthognathic type of profile. When a Class II malocclusion does develop, however, it is a different kind (see Bhat and Enlow, 1985). Care must be taken by the orthodontist because there are often stronger mandibular protrusive factors within this face, and Class II treatment procedures can sometimes produce unexpected and undesired results. However, this type of Class II malocclusion is characteristically less severe than the others, and proper treatment results are often gratifying.

Black individuals, like some Caucasians, tend to have an elongate, dolichocephalic head form, although there are wider-faced individuals, just as among Caucasians (Fig. 7–8). The middle cranial fossa has an anteriorly inclined (open) alignment, even more so than in Caucasians. This factor, together with a vertically long nasomaxillary complex in some black groups, causes the ramus (and whole mandible) to rotate markedly down and back. The mandibular corpus tends to be horizontally long relative to the bony (not dental) maxillary arch. This is similar to the Caucasian pattern but unlike the Oriental. Unlike the typical "long-headed" Caucasian facial type, however, the upper part of the face in the black expands much less and is therefore not nearly so protrusive. In this respect, the face of the black corresponds to that of the Oriental. The forehead is more upright and bulbous than in most Caucasians, the frontal sinus proportionately less expanded, the nasal bridge lower, the nose flatter, wider, and less protrusive, and the cheekbones more prominent. Although the upper part of the nasal region in narrow-faced black individuals tends to be correspondingly narrow (as it is in narrow-faced types of Caucasians), it is not as protrusive as the dolichocephalic Caucasian nasal region. In many blacks, however, approximately equivalent airway capacity is achieved by a somewhat wider dimension in the more inferior part of the nasal passageway in conjunction with a flaring of the nasal alae. One special feature characterizes the black

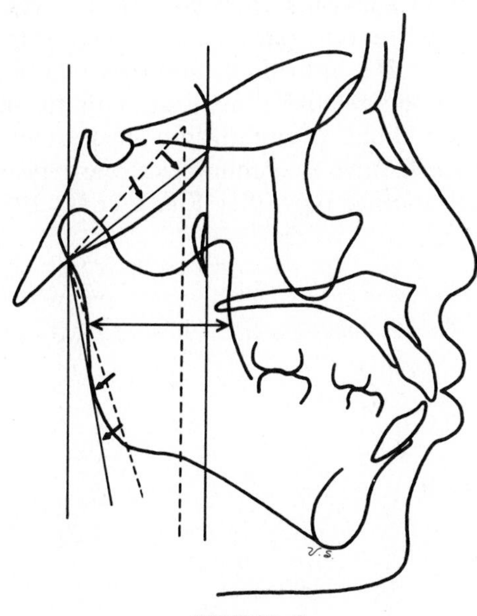

FIGURE 7–8

face; the mandibular **ramus is quite broad** proportionately. In a previous chapter, it was pointed out that the horizontal dimension of the ramus is a site that commonly participates in compensations for structural imbalances in other parts of the face and cranium. The forward inclination of the middle cranial fossa that characterizes many Caucasian groups, for example, is partially or completely counteracted by the development of a wider ramus, thereby offsetting an intrinsic tendency for mandibular retrusion and a Class II malocclusion. The mandible of the black also has this feature, but the amount is characteristically **much** greater. The very broad ramus places the mandibular corpus (which can also be long relative to the bony maxillary arch) in a resultant protrusive position. This, in turn, causes the maxillary incisors to tip labially, and a **bimaxillary** protrusion is thereby produced. This is an advanced feature that for the dolichocephalic black often forestalls severe Class II malocclusions. They are usually of the Class II "B" type. That is, mandibular B point lies well ahead of maxillary A point, in contrast to the more severe Class II "A" type, in which A point is the more protrusive relative to the occlusal plane (see Enlow et al., 1971).

The anatomic basis for the Class III type of malocclusion in blacks, however, has a different structural pattern. The basicranium of the black Class III (or the related bimaxillary protrusive tendency) does not usually have a posterosuperior alignment of the middle cranial fossa, unlike the brachycephalic Oriental Class III (and bimaxillary protrusion). This difference also exists between blacks and the wide-faced type of some (not all) Caucasian Class III individuals. Rather, the black Class III malocclusion tends to have a forward-downward rotated middle cranial fossa, and the ramus is aligned backward, not forward. The nasomaxillary complex is thereby placed more anteriorly, not posteriorly. The basicranium, thus, is not a principal factor among blacks that contributes directly to the protrusive placement of the mandible in Class III malocclusions, as it does in the other groups. As pointed out above, the **wider ramus** of the black is a key anatomic compensatory feature that effectively reduces the frequency and severity of Class II malocclusions. However, the same broad nature of the ramus **also** exists in most black Class III individuals as well. In Orientals and Caucasians, the Class III ramus is often "narrow" and reduces the extent of, and thereby partially compensates for, lower jaw protrusion. In the black Class III individual, conversely, not only is the ramus noncompensatory, its broad relative dimension adds to, rather than subtracts from, the extent of mandibular prognathism. Thus, the mandibular ramus of the black is an effective feature that minimizes one type of malocclusion but that tends to aggravate another type. (See Enlow et al., 1982.)

CHAPTER EIGHT

Control Processes in Facial Growth

Many students of facial growth feel that we are in the midst of a great conceptual revolution dealing with the basic processes that govern morphogenesis. It is believed that we are now at the threshold of major, exciting breakthroughs with regard to one of the most important biologic problems of our time—how the local growth control process actually operates at the tissue and cellular levels.

Until fairly recently, some explanations for the growth control process were regarded as more or less complete, with the theories underlying them sound and secure. This has all changed. We are now beginning to recognize what we do not really understand at all, and we think we can begin to define the problems involved. This in itself is a breakthrough. So many advances in so many diverse, interdisciplinary fields relating to the general subject of growth control have been made that we are dazzled by what is happening. Is there clinical significance? Can you imagine what could be done with treatment procedures if we could effectively control, at the most basic level, the local growth control process itself? This is the great goal.

The explanations of the growth control process that prevailed until just a few years ago were straightforward, easy to understand, and so plausible that they were adopted and used for many years as the basis for a number of clinical concepts. Most seemed to center on control of **bone** growth. The entire process of growth control was no particular puzzle and seemed to be readily explainable as follows. First, the growth of bone tissue by cartilage growth plates was presumed to be regulated entirely and directly by the intrinsic genetic programming within the cartilage cells. Intramembranous bone growth, however, was believed to have a different source of control. This type of osteogenetic process is particularly sensitive to biomechanical stress and strain, and it responds to tension and pressure by either bone deposition or resorption. Tension, as traditionally believed, specifically induces bone formation. Pressure, if it exceeds a relatively sensitive threshold limit, specifically triggers resorption. When tension is exerted on a bone, as at places of muscle attachment, the bone grows locally in response. Thus, sites of muscle insertion are usually marked by tuberosities, tubercles, and crests that form because of direct, localized fields of muscle traction. Because many muscles attach near the ends of a bone, rather than on its shaft, the epiphyses are much larger than the diaphysis, because this is where the muscles apply the most tension and where the bone thereby expands. As long as a muscle continues to grow, the bone is also stimulated to grow. This is because of the continuing biomechanical imbalance between them due to the expansion in muscle mass and resultant increasing force. The growing muscle exceeds the capacity of the bone to support it, and the osteoblasts are thereby triggered to form new bone in response. When muscle and overall body growth is complete, the bones attain biomechanical equilibrium with the muscles (and body weight, posture, and so forth). The forces of the muscles are then in balance with the physical properties of the bone. This turns off osteoblastic activity, and skeletal growth ceases. If any future circumstances cause departure from this sensitive state of bone–soft tissue equilibrium, such as major changes in body weight, loss of teeth, or the fracture of a bone, the process is revived until once again mechanical equilibrium subsequently becomes attained.

It is easy to understand why the preceding explanations were attractive and almost universally adopted by earlier workers. These concepts served to explain almost everything then known about bone and its growth. More recently, the realization that a number of shortcomings exist with regard to these reasonable concepts led to a re-evaluation of the whole process of growth

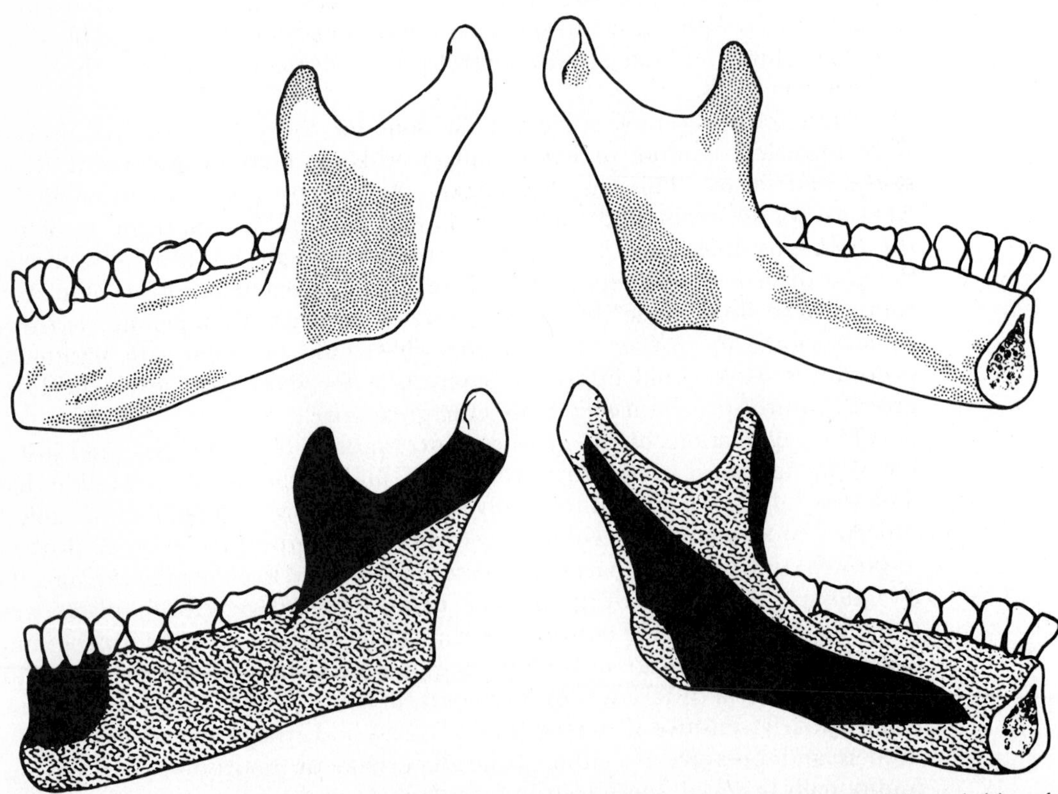

FIGURE 8–1. The top figures show the distribution of muscle attachments on the buccal and lingual sides of the mandible. The bottom figures illustrate the pattern of surface resorptive (dark) and depository (light) growth and remodeling fields. Note that there is no one-to-one correlation between these respective patterns. As described in the text, this does not mean that muscle forces are not involved in growth control; it does show, however, that the old "muscle tension—direct bone deposition" concept is invalid. (From Enlow, D. H.: Wolff's law and the factor of architectonic circumstance. Am. J. Orthod., 54:803, 1968.)

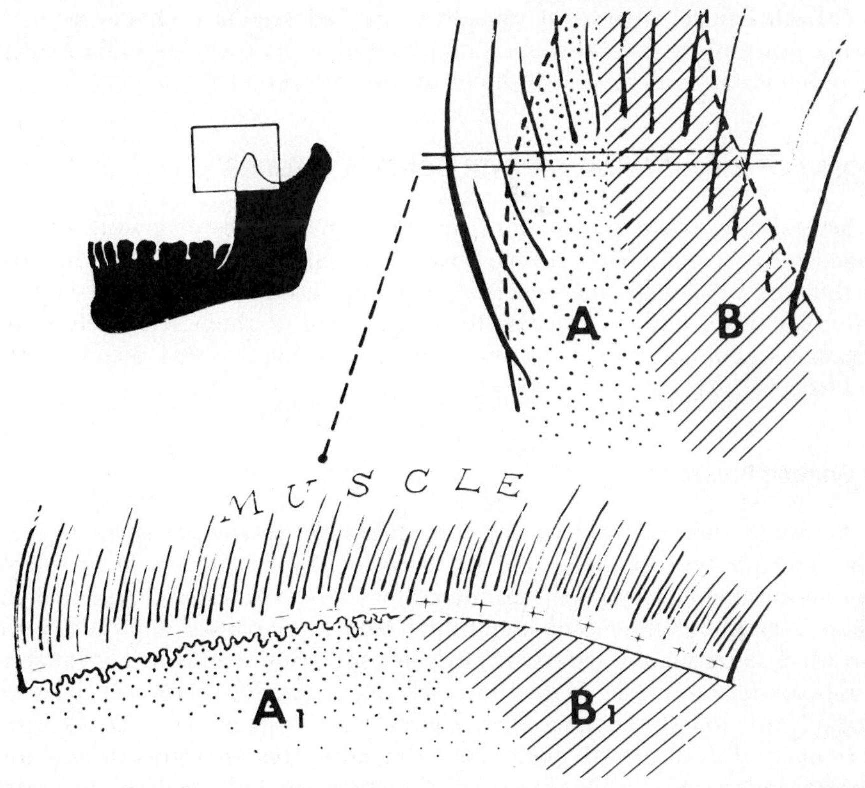

FIGURE 8–2. The temporalis muscle attaches to surface *A* and *B* on the lingual side of the coronoid process. In microscopic sections, it is seen that the attachment on A_1 involves a resorptive bone surface; the same muscle is also inserted on surface B_1, which is depository. (From Enlow, D. H.: Wolff's law and the factor of architectonic circumstance. Am. J. Orthod., 54:803, 1968.)

control. The subject has become a "new" frontier in facial biology. It is perhaps the most important problem that now faces us.

First, there is no one-to-one correlation between places of muscle attachment and the pattern of distribution of resorptive and depository fields (Fig. 8–1); growth control is more complex. Moreover, it is now known that there is no direct, one-to-one correlation between tension-deposition and pressure-resorption (this pressure-tension concept is greatly oversimplified; see pages 137 and 467). This is important. About half of all bone surfaces to which muscles attach are actually **resorptive,** not depository. Many muscles have widespread attachments, and within these surface areas, some growth fields are resorptive and others are depository. Yet these different surfaces are subject to the same pull by the same muscle, supplied by the same blood vessels, and innervated by the same nerves. The temporalis muscle, for example, inserts onto the coronoid process of the mandible (Fig. 8–2). As shown in Chapter 3, parts of this mandibular region have external surfaces that are resorptive. The muscle exerts tension, but the bone to which it directly attaches undergoes resorption. Other surfaces of temporalis muscle attachment are characteristically depository.

Furthermore, some muscles pull in one direction, but the bone surfaces into which they insert grow in other directions. The pterygoid muscle, for example, attaches onto the posterior part of the ramus. The muscle pulls anteriorly, but this part of the bone grows posteriorly.

Growth control involves a cascade of graded feedback chains from the systemic down to the local tissue, cellular, and molecular levels and back again. The problems at hand deal with the **local** control process.

SYNOPSIS OF CRANIOFACIAL GROWTH CONTROL THEORIES

Several alternative explanations for the ultimate basis of growth control, or some of its component aspects, have historically dominated the attention and thinking of biologic theoreticians. Although each such working theory is separate, a trend has always been to merge some of them selectively into a composite scheme in order to account for the baffling array of poorly understood issues.

The Genetic Blueprint

Always at the forefront of any growth control discussion is the old and perplexing question of the extent of "genetic" control. The role of genetic preprogramming has long been presumed by many to have a fundamental and perhaps overriding influence in establishing basic facial pattern and the features upon which internal and external "environment" then begins to play at some yet-to-be-understood level. Many contemporary researchers, however, have not been able to accept the idea that, simply stated, genes are the exclusive determinants for all growth parameters, including regional growth amounts, velocities, and minute details of regional configuration. Fully realized, of course, is the understanding that genes are indeed a basic participant in the operation of any given cell's organelles leading to the expression of that cell's particular function—for example, an osteoclast, a prechondroblast, or a contractile fibroblast. At issue is the mechanism by which intercellular conditions **activate** a given intracellular process, and just how the complex array of many different cell types and tissue combinations interact as a composite whole. Selective activations of specific genes within a cell's full genetic complement, however, are presumed to be one answer. Another factor is the recognition that **epigenetic** regulation can determine, to a substantial extent, the behavioral growth activities of certain tissue types, such as bone and other connective tissues. This means that such tissues or some of their parts do not entirely govern their own differention; rather, their growth is controlled by genetic influences, or the cause-and-effect growth and physiologic actions, of **other** tissue groups.

Biomechanical Forces

A powerful line of reasoning has historically focused on the play of physical forces acting on a bone to regulate its development, morphologic configuration, histologic structure and relationships, and physical properties. Wolff's law of bone transformation, introduced in 1899, quickly became a leading and most useful working concept—still quite valid if not misused. Essentially an application of the old and trusted idea that form interrelates with function, this cornerstone principle states that a bone grows and develops in such a manner that the composite of physiologic forces exerted on it are accommodated by the bone's developmental process, thereby adapting its structure to its complex of functions. This descriptive perspective, however, has often been misinterpreted; it has been presumed that it explains the actual biologic process of control, that is, **how** control of growth is carried out, rather than simply what is happening. One principal omission, and a major flaw, in most attempts to apply Wolff's law has been a lack of distinction between physical forces acting on a bone (i.e., its hard part) and forces acting on the osteogenic connective tissues (periosteum, growth cartilages, sutures, etc.) that actually form and remodel the bone.

Many experiments have been carried out in which muscles were severed, or the soft tissues otherwise altered, and in which artificial mechanical forces were exerted on a living bone. Because such procedures always result in some kind of response with a resultant change in the form of the bone, it has often been concluded that stress is therefore the principal factor controlling bone growth. Such experiments, however, do not "prove" such a role for mechanical forces, since certain critical variables necessarily exist that cannot be controlled in the experimental design. These include vascular interruption, nerve severance, temperature changes, alterations in pH and oxygen tension, and so on, all of which are known to affect bone growth. The fundamental question must be asked: Do extrinsic or unusual factors that can affect the course of bone development also necessarily represent the same factors that actually carry out the direct, primary control of the basic processes of growth and differentiation? Nonetheless, there can be no doubt whatsoever that mechanical forces indeed represent one (of many) of the "messengers" (see below) involved in the activation of osteogenic connective tissues.

Sutures, Condyles, and Synchondroses

In the 1930s, a then new model for growth control began to emerge that flourished through the 1940s and 1950s, with some holdover even today. Many of the ideas within these pioneering explanations have since been set aside and replaced by more biologically complete understandings, but they did indeed introduce issues that led to much basic research and new, more on-target questions to be asked. History was served.

It was presumed, quite reasonably at the time, that the growth, form, and dimensions of a bone are governed by intrinsic programming residing within that bone's own bone-producing cells of the periosteum, sutures, and bone-related cartilages. While influences such as hormones and muscle actions could augment these gene-dominant determinants, bones such as the mandible or maxilla, and all of their morphologic features, were held to be largely self-generated products. The displacements of bones as they enlarge were also attributed to the expansive forces residing within some of these osteogenic membranes (sutures) and cartilages and the new bone tissues they produce. The idea grew to include a concept of "centers" that provide inclusive growth regulation for the whole bones they serve. Today, most researchers discount the notion of such a center, replacing it with a concept of "sites" of growth, each of which is a localized region having its own regional circumstances and conditions and which operates under its own regional process of growth control, with a feedback system that allows reciprocal adaptations with the other sites.

The Nasal Septum

Aware that "centers" such as the facial sutures cannot drive the nasomaxillary complex into its downward and forward displacement because a suture is a traction-adapted (not pressure-adapted) type of tissue, James Scott, a well-known Irish anatomist, later reasoned that the cartilaginous nasal septum has features and occupies a strategic position that might answer the question of what "motor" causes the midface to displace anteriorly and inferiorly as it grows in size. Because cartilage is a more pressure-tolerant tissue than sutures, it presumably has the developmental capacity to expansively **push** the whole nasomaxillary complex downward and forward. Scott's famous nasal septum theory was born.

The laboratory testing of this theory has been a concerted interest of many researchers over many years but has encountered many experimental obstacles because of difficulties in controlling the multiple developmental variables involved. This has led to differing interpretations of the results. One problem is that animal experiments produce conditions in which functions normally carried out by some given structure, when that part is altered or removed, can be performed to some extent by other structures as compensations. Then, too, it cannot be assumed that any given structure functions in the same manner when conditions have been experimentally altered as when they exist in an undisturbed state. Moreover, the actual physical force for the maxillary displacement movement may be, at least in part, a pulling action of the septopremaxillary ligament resulting from septal enlargement, rather than a pushing action (Latham, 1970b). Such an effect can be noted in a bilateral palatal cleft: the embryonic nasomedial process ("premaxilla") is displaced protrusively, but without maxillary-to-premaxillary sutural attachment, the maxillae are not drawn forward and thus left behind retrusively.

Whether or not the nasal septum operates as the essential pacemaker for maxillary displacement, it nonetheless has membership as a component of the "functional matrix" and thereby contributes some share of developmental participation in combination with all of the other components necessarily also involved.

An important and most fundamental conceptual advance is that the old notion in which any single, presiding agent, such as the nasal septum, has sole responsibility for pacesetting the growth process has been pre-empted by concepts involving **multifactorial** interrelationships, as described later.

The Functional Matrix

Basic form/function principles proposed by van der Klaauw were greatly elaborated by Melvin Moss and evolved into a landmark concept having great impact among practicing clinicians and academic theoreticians alike. Although some of the deeper issues involved have been heatedly controversial, one very valued outcome has been intensive and productive debate and a great deal of new research.

Simply stated and omitting some details, the concept deals primarily with the ultimate source of osteogenetic regulation; and although many aspects are clouded by operational uncertainties, the core of the idea itself is straightforward. To begin, any genetic predetermination of a bone's morphologic characteristics by self-contained chromosomal design, at least beyond early histogenetic establishment, is largely bypassed. Acknowledged, however, is the role of genes in cellular organelle functioning (e.g., production of specific tissue protein types, enzymes, etc.) in response to extracellular messengers that activate a given cell's physiologic part in the grand scheme. Stimuli emanating from the growth and the actions of any and every source within the growing head and body (the functional matrix), directly or indirectly, function to turn on or turn off cellular organelle activity in the bone-producing cells. This yields a growing, changing, custom-fitted bone having regional dimensions and configurations that accommodate the changing developmental conditions and biomechanical circumstances in each localized region of each separate bone and the aggregate of all bones in an interrelated system. Each bone is precisely adapted to these multiple developmental conditions because it is the composite of the conditions that regulate a bone's configuration, size, fitting, and the timing involved.

The functional matrix concept is not intended to explain **how** the actual morphogenic process works, but, rather, describes **what** happens to achieve the combination of actions, reactions, and feedback interplay that occurs. The nature of the signals involved and how they operate are separate, but quite significant, issues dealt with later.

Also, the term "functional matrix" can be misleading, because it connotes primarily the function of a soft tissue part (e.g., muscle contraction). **Growth enlargements** are also directly involved in giving the signals that activate osteogenic connective tissues (see page 247).

Composite Explanations

Many experimental studies, together with observations of certain congenital craniofacial dysplasias ("nature's experiments") and much theoretical reasoning, have led to combinations of various growth control theories in attempts to confront and account for the many complexities embodied in regulating growth. These have been summarized by van Limborgh in the construction of a model that distinguishes the composite of factors influencing chondrocranial versus desmocranial (intramembranous) craniofacial growth and development. With the chondrocranium serving as an early but ongoing template, intrinsic genetic cell multiplication capability, general epigenetic influence (e.g., hormones), and general environmental factors (food and oxygen supply, etc.) are all proposed as agents within an interplay scheme for the endochondral part of basicranial developmental control. Desmocranial development, separately, is described as a morphogenic response to some balance among most of these factors, but with local epigenetic and local environmental factors (mechanical forces) playing a greater regulatory role.

The elegant studies of Petrovic and his colleagues have had great impact on the directions of thinking among contemporary craniofacial biologists. Emerging from their long-term experimental work have been elaborate cybernetic models that illustrate, in extensive detail, many of the complex developmental interrelationships that have been found among almost all of the multiple factors briefly outlined throughout this chapter, and more, involved in growth control. Professional students going beyond the present chapter's introduction will need to utilize their work as a springboard into the shadows yet remaining.

Control Messengers

Growth control is essentially a localized process complemented by systemic support. This is because growth is carried out by specific, restricted, regional **fields** of differing growth activity in amounts, directions, velocities, and timing as described in earlier chapters. The diverse cell populations within each of these fields respond to activating intracellular or extracellular signals. "First messengers" are extracellular activators for which specific cell-surface receptors are sensitive. Reception fires a cascade of "second messengers" within a given cell type that results in the function of that cell and its organelles, such as fiber and proteoglycan production, calcification, acid or alkaline phosphatase secretion, and rate and duration of mitotic cell divisions. Mechanical forces, bioelectric potentials, oxygen concentrations, and specific hormones are examples of first messengers. Adenyl cyclase and adenosine 3′:5′-cyclic phospate (cyclic AMP) are second messengers leading to cytoplasmic and nuclear DNA-RNA transfers.

In the **immediate** environment enclosing an osteoblast or osteoclast, a first-messenger hormone or enzyme, a bioelectric potential change, or a pressure/tension factor acting on the cell's outer membrane receptors can activate second-messenger, membrane-bound adenyl cyclase, which in turn accelerates the transformation of adenosine triphosphate (ATP) to cyclic AMP within the cytoplasm, which then activates the synthesis of other specific enzymes relating specifically to bone deposition or resorption. Ionic calcium is mobilized from mitochondrion storage, and inner and outer membrane permeability is altered that selectively controls the flux of other ions in the synthesis and discharge of the products secreted by the cell.

During bone formation, the osteoblast takes in amino acids, glucose, and sulfate for the synthesis of the glycoproteins and collagen in the formative organic part of the bone matrix. The cytoplasmic organelles within the osteoblast participate in the formation, storage, and secretion of tropocollagen, the ground substance, and also ions that form the inorganic (hydroxyapatite) phase of the bone matrix. Alkaline glycerophosphatase is related to bone formation (in contrast to acid phosphatase, which relates to resorption) and is associated with the collagen fibril as it is released from the osteoblast. High levels of alkaline phosphatase are also involved in the formation of the hydroxyapatite. The critric acid cycle and glycolytic enzymes provide generalized energy sources for all these activities.

The osteoclast contains an abundance of mitochondria in addition to lysosomes and an extensive endoplasmic smooth membrane system. The osteoclast produces, stores, and secretes enzymes (such as collagenase) and acids that relate to the breakdown of both the organic and inorganic components of bone. The lysosomes are involved in acid phosphatase storage and transport. First messengers, such as parathyroid hormone or bioelectric changes, stimulate receptor sites on the cell membrane. This activates adenyl cyclase, which in turn causes increases in cytoplasmic cyclic AMP. The latter then increases the permeability of the lysosomal membrane. By an exocytosis of the lysosomal contents, the resorption of both the organic and inorganic parts of the bone is carried out through the activity of the acid hydrolases, lactates, and citrates. The endoplasmic smooth membrane system is also involved in this process of enzyme transport and release.

Bioelectric Signals

The piezo factor has been one of the great bone growth-control hopes since the mid-1960's and has promised to explain just how muscle action can be translated into precisely regulated bone remodeling responses. The idea, in brief, is that distortions of the collagen crystals in bone, caused by minute deformations of the bone due to mechanical strains, generate bioelectric changes in the area of deformation (that is, the piezo effect). These altered electrical potentials appear to relate, either directly or indirectly, to the triggering of osteoblastic and osteoclastic responses (see page 138).

To put this factor in perspective, one key point must be understood. There are two separate **target categories** for the mechanical actions of muscles, and also muscle and soft tissue enlargements, gravity, and all other such physical sources. One target is the osteogenic *connective tissue membrane* that covers a bone. Its cells are loaded with plasmalemma surface receptors that can receive the **direct** effects of first messenger agents and forces. The second basic target category is the calcified part of the bone itself, in contrast to the connective tissue membranes just mentioned. Mechanical forces produced by growth and function acting on it cause small distortions of the collagenous matrix that create **localized** convexities and concavities, which in turn relate to the generation of resultant positive or negative polarities. While response thresholds of regional force levels are still poorly understood, a concavity under **active** distortion is known to emanate a negative bioelectric charge, and a convexity generates a positive charge. Negative charges then transmit to the osteogenic cells within the contiguous connective tissue membrane to fire osteoblasts into depository activity; a positive charge activates an osteoclastic response. The result is **coordinated regional** remodeling throughout multiple localized surfaces, inside and outside surfaces alike, that shapes the bone and enlarges its overall size during growth. When mechanical equilibrium is achieved between the bone and the composite of growth and functional forces playing on it, the polarities are neutralized, and the remodeling activity is thereby turned off.

This scheme appears to explain nicely the basis for coordination between the periosteal side and the contralateral endosteal surface in the remodeling of a given region of a bone; i.e., one side can be convex, and the other concave, with the common forces acting within this region thus resulting in complementary deposition/resorption responses. Also, note that the pressure/tension and deposition/resorption responses charcterizing the osteogenic connective tissue membrane versus the bone matrix itself are opposites. For the membrane, pressure appears to relate to resorption and tension to deposition. For the bone matrix, the relationships are reversed. (See pages 130 to 137 for further discussion.)

There is a key question, however, that contemporary researchers appear not to have asked, at least to date: an explanation of the nature of the **balance** between growth-affecting mechanical forces acting directly on a bone's osteogenic membranes (the first category) versus the bone matrix itself (second category).

Other Factors of Growth Control

The neurotropic factor involves the network of nerves (all kinds, motor as well as sensory) as possible links for feedback interrelationships among all the soft tissues and bone. The nerves provide pathways for stimuli that presumably can trigger some bone and soft tissue responses. It is not believed, however, that this process is carried out by actual nervous impulses. Rather, it appears to function by transport of neurosecretory material along nerve tracts (analogous perhaps to the neurohumoral flow from the hypothalamus to the neurohypophysis along tracts in the infundibulum) or by an exoplasmic streaming within the neuron. In this way, feedback information is passed, for example, from the connective tissue stroma of a muscle to the periosteum of the bone associated with that muscle; and the "functional matrix" thereby operates to govern the bone's development.

A great many laboratory studies are now being conducted in attempts to clarify relationships between, for example, cyclic AMP and biomechanical forces. Other studies are now dealing with the role of important substances such as the cascade of prostaglandins, somatomedins, osteonectines, leukotrienes, and possible neurotropic agents. Still other work is under way for seeking out possible chalone-like agents in bone (that is, localized tissue-level, hormone-like substances that are believed to accelerate or retard cell divisions). Day by day, new information and new links are being added. To help place all these old and new factors in some kind of working perspective, one can use the following relatively simple "test" to classify their respective roles:

1. Is a given factor the sole, primary agent that is directly responsible for the master control of growth? Of course, such a single, ubiquitous agent does not exist. Historically, bone investigators have searched for such a master control factor, perhaps a special "hormone" or inductor, that does it all. On the basis of our present knowledge, however, it is now realized that the control process must necessarily be multifactorial. Control involves a chain of regulatory links. Not all of the individual links are involved in all types of growth changes. Rather, a selected combination exists for different specific control pathways that can follow many routes and involve many different agents.

2. Does a given factor function as a "trigger" that induces or turns on other specific, selected agents that then launch the process of control response? Is it the first link in the chain? Is it the initial agent in the process of "induction"? Is it a "first messenger"? Biomechanical forces presumably represent such a trigger. It is still justifiably believed that pressures and tensions are indeed among the basic agents involved in growth control (although not following the traditional but oversimplified historical explanations). Different kinds of bone, however, appear to have variable thresholds of response to physical forces (e.g., basal bone versus alveolar bone in the mandible and maxilla, which are relatively nonsensitive versus quite labile, respectively, to physical strains). It is also evident that biomechanical forces are not the sole agents participating in control. Even when involved, many other links are also required as second and third level messengers.

3. Is a given factor, in effect, the title for some biologic process without accounting for the actual operational mechanism involved? This is an important category. Such a title describes what happens but not how it happens. It does not explain the implementation of the control process it represents. It is just, in effect, a synonym for "control process" without explaining how it works. The reason an awareness of this category is so important is that we often tend to use such titles as though they do indeed explain the mechanism involved. With continued use, we delude ourselves into believing that we actually understand the real basis for the control process. "Wolff's law," "functional matrix," and "induction" are all such descriptive labels for biologic control systems. They do not, of course, explain how the system works (they were actually never intended for this ultimate purpose, even though many have used them as such). Be ever alert to this deceptive conceptual pitfall.

4. Does a given factor function essentially in a supportive role or as a catalyst? Many nutrients would be examples of this category.

5. Does a given factor accompany the control process but not actually take part? It has been necessary for laboratory researchers studying the piezo effect, for example, to establish whether bioelectric potentials are actually first messengers or whether they simply "occur" in a nonparticipating role.

6. Is a given factor an actual cause, rather than an effect, in the growth control process? The piezo factor also serves as an example of this category. Do bioelectric changes in bone directly trigger remodeling responses, or are the responses merely the incidental result of the changes? Present evidence is believed to support the former.

One fundamental feature of the control process is now clear. No given tissue, such as bone, grows and differentiates in an isolated, independent manner by a wholly intrinsic regulatory process. Control is essentially a system of feedback pathways, informational interchanges, and reciprocal responses. Tissues, organs, and part of organs develop in conjunction with one another. A given bone and all its muscles, nerves, blood vessels, connective tissues, and epithelia represent an interdependent, developmental composite. Bones have specific mechanisms for increasing in length (e.g., an epiphyseal plate, synchondrosis, condyle), and they have another specific mechanism for increasing in width (subperiosteal, intramembranous growth). Correspondingly, muscles also have a specific growth mechanism for increases in length and another for increases in width. Both of these conjoint muscle growth processes proceed in concert with respective growth mechanisms for the bone. There are reciprocal feedback interrelationships between the muscle and bone, as well as the various other tissues, and they all enlarge together, not as "separate" and independent units. For example, input to the sensory nerves in the periodontal membrane can trigger growth responses to occlusal signals from the teeth. These signals can be passed on through an arc to motor nerves supplying the muscles of mastication. In conjunction with muscle adaptations to the individualized nature of the occlusion, then, the bones of the face can remodel in association with the muscle and soft tissue matrix that encloses them and affects their course of growth. The old concept that a "growth cartilage" serves as the primary regulator for the overall development of a musculoskeletal composite is now regarded as an incomplete and unacceptable explanation, because many more input factors are now known to be involved. However, we still have a long way to go in understanding the whole of the growth control system. History will probably judge this one of the great problems of our time.

Final Exam Exercise

The ultimate clinical goal is to be able to "control the local growth control process" at a highly sophisticated biologic level. Such a fanciful breakthrough will probably be made sooner or later and will involve advanced-level tissue growth control techniques that might replace all or some of today's conventional procedures (banding, functional appliances, headgear, surgery, retainers).

As a theoretical exercise, assume that such a conceptual and procedural breakthrough had been made, and that by means of a "black box" in your possession you could actually induce and regulate localized cellular and tissue growth, remodeling, and displacement activities among all key areas throughout the facial complex with an effective and exceedingly fine degree of control. Develop a hypothetical treatment protocol that you might use for a Class II or III patient starting at about 3 or 4 years of age (or even younger), another at about 11 or 12 years, and a third as a young adult. Keep in mind that normal growth is "differential" (different areas or parts develop at different times, in different directions, and at different magnitudes). Remember the need for functional and morphogenetic interdependence among the different parts and developmental coordination among them. Keep in mind, further, that any alterations made clinically by you would likely create "imbalances" in one or more places elsewhere throughout the facial complex (orbit, nasal region, hyoid complex, muscle trajectories) that would require intrinsic or extrinsically induced compensations, which in turn could cause additional imbalances. Assume, also, that in your "black box" you have an effective means for monitoring the ongoing growth/clinical changes with a high degree of accuracy and sensitivity (i.e., replacing conventional two-dimensional headfilms). With your sophisticated new instrumentation, you can also determine the existence of balance versus imbalance among all the regional parts and can measure the changing magnitudes of any imbalances that exist or will develop.

How would you proceed with clinical treatment? What major problems would you expect to encounter? To what extent should an individual's inherent facial features be sustained? Should everyone be made into a beautiful person? Should everyone be made to a look the same with only a few standard variations? All clones? Should only potential or real dysplasias and malocclusions be addressed, or just a wish for improvement of appearance? What are the ethical cautions?

THE BIG PICTURE

A diverse collection of factors that need to be taken into account whenever one is dealing with facial "growth control" has been drawn from previous chapters. These are presented below as a series of separate points and issues with emphasis on the "architectonics" of growth control—simply, the dynamics of the developmental interrelationships among all of the growing parts, soft and hard tissues alike, and how this enters into the Big Picture.

There was a time, not long ago, when attempts to understand how facial growth is regulated were at a much simpler level. Most of the discussions seemed to settle on intrinsic "genetic" control versus everything else, such as biomechanical forces and hormones, and the lability of intramembranous versus the presumed preprogrammed (i.e., genetic) control of endochondral growth. A common approach was, and often still is, the naming of some supreme force or sovereign power to explain something, a kind of theologic disclaimer. From primitive civilizations to today, if some particular phenomenon is not understood, a "deity" can be contrived to account for it, that is, a graven image, a fanciful image from human imagination. Thus, we have "condylar growth," and there has been abiding faith. Genetics itself has been such a deity, often used to cover our shortfall and to delude ourselves into believing we understand what we do not. A common variation is the giving of some descriptive title, quite legitimate as far as it is intended to go, to explain the "how" when only the "what" is partially revealed; e.g., the functional matrix. With regard to genetics, the old and compelling idea that there exist specific genes for virtually every structural detail throughout the craniofacial complex has been yielding to the newer concepts outlined herein. Further, how selective gene activations in **response** (effect, rather than cause) to extracellular signals can be involved is a direction receiving increasing understanding and emphasis.

Following is a brief selection of architectonic factors that play into the Big Picture.

1. Growth is a **differential** process of progressive maturation. Different parts have different schedules in which changing growth velocities occur at different times, by different regional amounts, and in different regional directions. For example, vertical facial development has a different morphogenic timetable than the transverse because basicranial width is precocious, compared with facial airway enlargement and tooth eruption. The airway relates to whole body and lung development, whereas the bicondylar and bizygomatic dimensions relate to the earlier transverse maturation of the cerebral temporal and frontal lobes with their basicranial fossae. This creates many developmental complexities. For example, the architectonics of mandibular growth control must thereby provide for a differential timing of whole ramus (not just condylar) development to adapt to differences in vertical versus anteroposterior maturation of the pharyngeal compartment as well as, at the same time, differential anteroposterior versus vertical growth enlargements, remodeling, and displacement of the nasomaxillary complex. This indeed requires fancy developmental footwork if precision of dimensional changes, fitting of separate parts, and proper timing are to be collectively achieved.

2. **Development is a process working toward an ongoing state of aggregate, composite structural and functional equilibrium.** Growth, of course, involves constant changes in size, shape, and relationships among all of the separate parts and the regional components of each part. Any change in any given part must be proportionately matched by appropriate growth changes and adjustments in many other parts, nearby as well as distant, to sustain and progressively achieve functional and structural balance of the whole. In short, growth anywhere in any region, local area or part, is **not isolated.** This seems quite obvious; yet the complex of **interrelations** that exist is often bypassed in our thinking and in the literature. "Balance" is a developmental aggregate involving close interplay throughout. For example, the shape and size of one's external nose and facial airway are not determined solely by genetic blueprint (or any other kind of control) just within these parts themselves, since other parts elsewhere establish rigid developmental conditions. Interorbital width and nasomaxillary boundaries presented by the basicranium, for example, require reciprocal compliance of any genetic, epigenetic, or soft tissue growth determinants within the separate nasal region.

Certain regional **imbalances** exist in everyone that are natural and architectonically inescapable. Individual or population differences in head form configuration, for example, establish corresponding differences in the basicranial template for many facial dimensions, growth field boundaries, and component alignments. These variations set up many different kinds of facial configurations within which some "imbalances" have necessarily been introduced because of architectural complexity. However, growth and development work toward a state of aggregate architectonic equilibrium, so that emerging from the process of development are certain other regional "imbalances" that are offsetting and compensating. Groups of localized imbalances thus can balance other groups of imbalances as a normal part of the growth control process, the composite of which is more or less in functional equilibrium. Nonetheless, virtually limitless morphologic variations have been introduced, most of which we regard as more or less normal. For some others, the latitude for developmental adjustment is exceeded, and a malocclusion or some other structural dysplasia results, even though "balanced" in itself. In a sense, the process of growth and development is nature's own clinician, and it usually works quite well, even though various malocclusion **tendencies** exist in all of us.

In addition to different phylogenic lines (e.g., head form variations) leading to developmental facial variations, ontogenic factors are involved as well. Mouth breathing is an example. A broken lip seal requires different muscle actions for mandibular posturing, and an open-jaw swallow similarly requires different muscular combinations. These factors produce different signals to the osteogenic, chondrogenic, myogenic, and fibrogenic components that revise the course of development, thus leading to adjustive morphologic variations to create a developmental balance among parts that had become morphogenetically imbalanced.

3. The goodness of fit among contiguous anatomic parts and groups of parts is remarkable. Consider the extreme precision of shape and size fitting between, for example, the temporalis muscle and its bony coronoid process; the fit between a tooth's root and its alveolar socket; or one bone in sutural articulation with another. Developmental interplay establishes this, while all the time sustaining continuous growth and function. This requires a kind of "servo" system to create the architectonic exchange of signals that turn on or off, up or down, the cytogenic responses that drive the remodeling progression. Consider a nerve and its foramen in the basicranium. The size, shape, location, and progressive relocations as everything grows are primary for the nerve but secondarily adjustive for the foramen. This is an example of the many delicate, very significant kinds of developmental relationships that all too often have gone unsung in our appreciation of the Big Picture.

4. If we consider only bony components, we see that each bone and all of its regional parts participate directly and actively. Most of us readily acknowledge this. Why, then, do we persist in highlighting just a mandibular condyle, for example, while largely ignoring the rest of the ramus, which is just as significant and developmentally noteworthy in the Big Picture. It is the **whole** ramus that is a respondent to the growth-influencing muscles of mastication, and it is the development of the whole ramus and all of its parts, in teamwork, that establishes mandibular fitting with the maxilla on one side and the basicranium on the other.

It is important to understand that there is no common, overriding and centralized control force that regulates the developmental and anatomic details for each individual region throughout a whole bone. Rather, the different regional areas have different **local** developmental, functional, and structural conditions and circumstances that require appropriate regional signals to the osteogenic connective tissues (periosteum, endosteum, cartilages, sutures, periodontal membranes) for precision operation of their responses in the growth control system. This provides the **regional** balances and goodness-of-fit among parts, as highlighted herein.

5. The nature and capacity for interrelated growth adjustments among separate parts vary among different tissue types. Bone, for example, has a wide latitude for responsive remodeling adaptations through intrinsic manipulation of the osteogenic connective tissues to produce size and shape conformation in precision fitting. A tooth, in contrast, has a much lower potential for adjustive remodeling, and hence teeth undergo post-eruptive growth movements and location changes mostly by a displacement process. Thus, anterior crowding is actually an adjustive compensation to provide fitting of the dentition within a prescribed growth field as established by other, multiple relationships. One

imbalance has balanced others to achieve a kind of aggregate structural equilibrium. An occlusal curve is another such intrinsic developmental adjustment of the dentition and its alveolar bone to conditions imposed on the dental arches from without.

The capacity for the remodeling of cartilage either interstitially or by its chondrogenic connective tissue is also more constrained than for bone. Cartilage can, however, undergo alterations in growth direction and magnitude by differential turning on and turning off of prechondroblastic proliferation around its periphery, thus producing adaptive growth vectors in response to changing architectonic conditions.

6. **Displacement** and **remodeling** represent one of the most fundamental of bone growth activities. Yet to this day, this important facet, more often than not, is disregarded when we try to account for how a given appliance or other clinical procedure is presumed to work. Both types of movement are usually clumped together simply as "growth" without distinguishing between them. This is a regrettable short-sell. The reason distinction is very important is that each represents a separate and distinct target in the intrinsic control process utilized for different clinical procedures. Headgear, for example, manipulates directions and magnitudes of the displacement type of movement (whole bone movements), with bony and soft tissue remodeling then providing responsive adjustments to altered whole-bone placements. Periodontal connective tissue responses to fixed appliances activate alveolar remodeling adjustments in response to tooth displacements. Functional appliances presumably activate altered combinations of displacement and remodeling up and down the line. Some orthognathic procedures involve the surgical moving of bones or their parts followed by osteogenic transplants, thus paralleling that which natural growth did not fully achieve—displacement (surgical movements) and remodeling (sizing and shaping by transplants). This also highlights the reason bony articulations, including sutures, movable joints, synchondroses, and tooth junctions, need a full share of attention in understanding the growth process and its control. They are the places from which displacement movements emanate as a given bone simultaneously undergoes its remodeling by the enclosing osteogenic connective tissues. Displacement moves whole bones away from each other at their contacts, thus complementing their enlargements. Remodeling, at the same time, produces the enlargement, constructs the configuration and its progressive changes, provides precision fitting with contiguous soft tissues and with other bones, and creates the ongoing "compensations" or adjustments leading to a composite balance among all the separate parts.

Another example of the displacement/remodeling combination is the placement and development of the incisor part of the maxilla. By far, most of the actual downward-forward growth movement of the premaxillary region is by whole-maxilla displacement in conjunction with enlargements of all the other craniofacial parts above and behind, not just intrinsic remodeling growth within that localized premaxillary region itself. Localized remodeling produces "only" the size and shape of that regional area, not most of its considerable extent of growth movement over the years.

7. Variations in head form, as previously mentioned, are an important factor. The reason is that the dolicho, brachy, or dinaric types establish quite different basicranial templates for facial development. Whatever genetic or epigenetic growth control resides within the mandibular and ethmomaxillary components themselves must necessarily yield and conform to a higher level of predetermination in a number of respects. For example, the shape and

proportions of the palate are projections of the anterior cranial fossa, and can include any basicranial asymmetries that exist. The apical base of the maxillary dentition, in turn, is established by the configuration and size of the palatal perimeter. For another example, the breadth of the middle cranial fossa establishes the bicondylar sites for mandibular articular positioning. The dimensions and the configuration of the basicranium and its flexure, further, establish the anteroposterior placements of the maxilla relative to the mandible and the field of growth for the maxilla. Because it is known that (1) variations in head form set up corresponding variations in facial type and pattern, (2) head form pattern variations predispose specific, corresponding malocclusion tendencies, (3) pattern variations involving different anatomic combinations respond differently to different treatment procedures, and (4) different rebound tendencies probably exist in different pattern combinations, **much** closer attention should be given to head form consideration than at present. A Class II dinaric has a different anatomic combination than a Class II brachycephalic, and both are basically different than a Class II dolichocephalic. The intrinsic control of facial growth, therefore, is strongly influenced by factors external to the face itself, and this must be taken into account in our Big Architectonic Picture. Predispositions for mandibular retrusive versus protrusive tendencies are **built into** the phylogenetic heritage of different population groups and underlie predisposed imbalances that must be more or less offset by the wondrous process of facial morphogenesis and its remarkable system of intrinsic control.

Beginning students are inclined to perceive the Class I category as an anatomically separate and distinct type having good balance for each regional part with only minor departures from an ideal. Not so. In the assembly of all the multitude of craniofacial parts, **all** of us have phylogenic and head form–established imbalances as briefly summarized above. Some of these can be mandibular retrusive and others protrusive in their composite effects. If one or the other type dominates, the anatomic aggregate can be either Class III or II, respectively. If the aggregate is offsetting (i.e., a balancing of imbalances), a Class I results, but with a likely "tendency" one way or the other. The Class I is in the middle span between the extremes of this spectrum involving multiple composites of different architectonic combinations but with two sides grading toward the two opposite directions.

8. With regard to our phylogenic facial heritage, the factor of bipedal body posture interrelating with our enormous human brain and marked basicranial flexure have led to an inferoposterior rotational placement of the nasomaxillary complex. The midface has come to lie **below** the anterior cranial fossa, rather than protruding largely forward from it. This has caught the mandible in a closing vise between the midface above and the pharyngeal airway, gullet, and cervical column behind. Overjet, overbite, anterior crossbite, and an unprecedented developmental problem situation for the human temporomandibular joint (TMJ) are some consequences. The adaptive remodeling capacity of the ramus, with its condyle, is especially noteworthy in adjusting to these severe conditions. The wonder is not that so much TMJ distress exists, but, rather, that there is not much more, and the adaptive ramus is our hero.

Further, the basicranial suture system inherited from smaller-brained ancestors cannot provide for the full range of sutural growth adaptations needed to accommodate our grossly enlarged brain, so that unusual basicranial remodeling not found in any other known mammal, including anthropoids, has been added to the human growth package. Still another evolutionary factor is that of orbital convergence toward the midline in conjunction with enlarged temporal lobe expansion. This affects nasal configuration and placement, since the interorbital (nasal) compartment has become significantly reduced. This in turn has required a facial growth control system that has allowed nasomaxillary developmental accommodation for our species as well as ontogenic conformation for individual variations in magnitude.

9. With respect to growth rotations, this timely subject has justifiably become of great interest. There exist basicranial and facial component rotations within rotations within still other rotations. Some are "displacement" types, and some are "remodeling" rotations. The latter represent one of the developmental adjustments emphasized herein. For example, basicranial growth can predispose one to a nasomaxillary rotation that would "unbalance" the alignment of the palate. By differential remodeling of the anterior versus posterior parts of the palate, however, the palate as a whole can be progressively leveled into a functional position as the basicranium grows. Similar mandibular rotations exist in relation to variable basicranial proportions as well as midfacial size, configuration, rotations, and alignment variations. The architectonic factor of developmental rotations thus enters into our Big Picture, within the composite of the multiple factors involved in the operation of selective, regional growth control.

10. Considering architectonic feedback communication and give-and-take regional adaptations in the interrelationships involved in growth control, three examples in which component parts each play a particularly significant role stand out. These are in addition to the many other localized responses and adjustments already mentioned. One is the mandibular ramus; the second is the periodontal connective tissue membrane; and the third is the insignificant-appearing little lacrimal bone.

In a typical gross anatomy course for freshmen, the ramus is usually dismissed just as a base for masticatory muscle attachments, which is important enough in itself. But how much more dynamic and interesting for the student if its **other** very interesting and essential (developmental) functions were also to be dramatized. Consider, first, that the ramus bridges the pharyngeal space to functionally position the lower arch in occlusion with the upper. The enlargement and the size of this pharyngeal space are progressively established by its ceiling, which is formed by the enlarging middle cranial fossae and their growing temporal lobes, a morphogenic process continuing long into childhood. Now, the anteroposterior breadth of the ramus must **match** this basicranial developmental progression, with multiple and diverse basicranial, maxillary, and mandibular rotations also taken into account, by equivalent growth amounts and with corresponding timing—an elaborate architectonic interrelationship of separate parts. Otherwise, either excessive anterior crossbite or mandibular retrusion would ensue. Further, the vertical height of the ramus must match the vertical height and progressive increases of the nasal and dental parts of the ethmomaxillary complex, taking into account the vertical lengthening of the middle cranial fossae as well and, importantly, marked differences in vertical versus horizontal timing within the nasomaxillary complex and basicranium. Too much or too little and too early or too late sets up an anterior deep or open bite, and the latitude for mismatch is very slight. All of this requires a precision and coordination of signals given to the connective tissues that enclose the ramus, which respond by enlarging, relocating, rotating, shaping, and constantly adjusting the whole ramus (and its condyle). This is a most remarkable, complex, and demanding developmental interplay among separate parts and parts within parts.

The periodontal membrane (PDM) is relatively thin connective tissue that shapes, sizes, and constantly remodels and relocates the alveolar bone to match its moving, resident teeth and also carries the teeth in vertical and horizontal drifting movements in addition to their initial eruption. The PDM is a dynamic, complex connective tissue membrane with many and diverse component elements (often demeaned as merely a ligament) responsive to the multiple signals that activate its multiple cell types to carry out these architectonic functions. In addition, the PDM contributes to tooth formation and provides vascular pathways and proprioceptive and other innervations. Consider the high degree of precision required of the intrinsic control process in coordinating tooth movement and alveolar remodeling. The direction, amount, and timing must be absolutely precise, with virtually no divergence. It is the wondrous PDM that carries this out. Clinically, this elaborately coordinated growth process is manipulated by substituting clinical control to override the intrinsic control. But the histogenic process itself is the same. The remarkable PDM and its symphony of movement are a workhorse for the orthodontist.

The tiny, thin flake of a lacrimal bone receives little, if any, attention in a standard gross anatomy course; it and its canal are merely identified. Yet phylogenetically this seemingly insignificant little bone has survived as a discrete element even when many of the other more robust craniofacial bones have lost their individual identity through multiple fusions. The reason for its evolutionary retention is that it has a special and essential architectonic role in facial development. It is an island of bone surrounded by osteogenic (remodeling-capable) sutural connective tissue responsive to growth control signals emanating from all around it. It is strategically situated with the ethmoid, the nasal part of the maxilla, the frontal bone, and the orbital part of the maxilla, all of which are growing in different directions, at different times, by different amounts, and with different functional relationships. It was pointed out that precision of fitting is an essential part of growth control; by virtue of the lacrimal bone's adjustive suture system, all of these separate parts can undergo their differential displacements and their own enlargements, relocations, and remodeling, yet continuously fit with one another as they all grow and function.

There should be marble and bronze monuments glorifying the ramus, the PDM, and the lacrimal bone, all prominently displayed in the atrium of every dental school; and students should be expected to doff their caps in reverence when passing!

11. Clinical intervention into the growth process and its control is by either one of two approaches, both of which are analogous to the intrinsic growth process itself. The first approach is by surgical substitutions for the natural displacement and remodeling processes that were incomplete or derailed. The second approach is by overriding intrinsic control signals with clinically induced (e.g., orthodontic) signals that overwhelm the intrinsic regulation of osteogenic, chondrogenic, myogenic, and fibrogenic systems. Then, the same actual biologic operations of these systems proceed but under the control-revised directions. However, in all cases, if the same conditions that created the original intrinsic signals still persist after treatment, then architectonic rebound growth naturally adjusts back to the former, balanced pattern.

12. There is another fundamental consideration that, more often than not, is conceptually bypassed. When the muscles of facial expression contract (function), the mechanical effect is an **upward** and **backward** retrusive force exerted on the maxilla. Yet everyone knows that the maxilla "grows forward and downward." Does this not contradict the functional matrix principle?

Similarly, when the masticatory muscles function, the net mechanical effect on the mandible is also upward and backward, not downward and forward. Does **this** not thereby also violate belief that "function" of the functional matrix is the basic driver for growth control? However, two basic factors were omitted in presenting these comments, with the intention of misleading one's thinking. First, the important distinction between the **displacement** type of growth movement versus **remodeling** growth movement was not made. Second, importantly, the **growth enlargements** of the respective muscles were not included, only their contractile functions (Fig. 8–3).

With respect to the displacement movements, the connective tissue stroma of each muscle is directly or indirectly continuous with fibers attaching to the bones, and **enlargements in diameter** of mandibular muscles such as the masseter and temporalis have an anteriorly displacing effect on the whole mandible. Their **enlargements in length** have an inferiorly displacing and carrying effect. As the facial expression muscles, oropharyngeal soft tissues, and facial integument undergo outward growth expansion, there is a similar outward and downward carrying movement of the nasomaxillary bony parts.

At the same time, the **functioning** of all the muscles (contractions) and all other soft tissue parts is proceeding, and the osteogenic connective tissues (condylar cartilages and the sutural, periosteal, endosteal, and periodontal membranes) respond to the signals produced by the functioning, growing systems everywhere around the mandible and maxilla to bring about the changing regional sizes, progressive regional configurations, and ongoing adjustments involved throughout all regional parts of each whole bone. Thus, the maxilla and mandible "separate" (displacement) at their sutures and at the

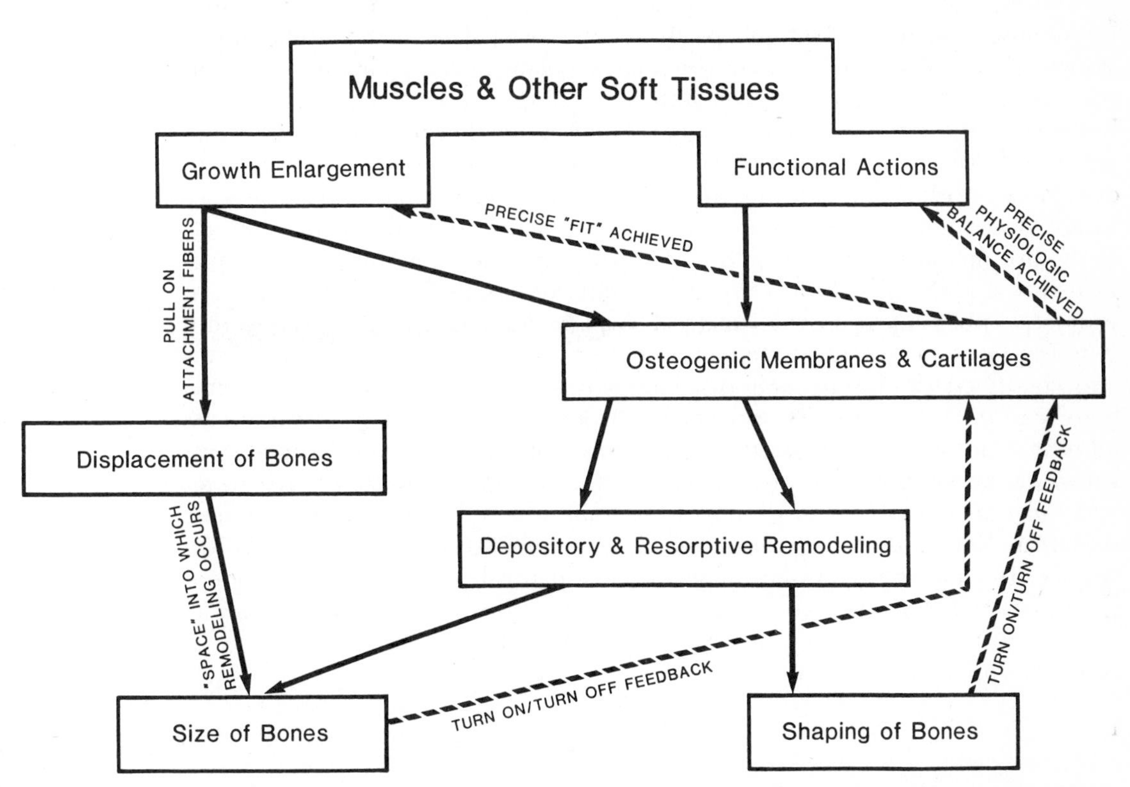

FIGURE 8–3. See text for discussion. (From Enlow, D. H.: Structural and functional "balance" during craniofacial growth. In: *Orthodontics: State of the Art, Essence of the Science.* Ed. by L. W. Graber, St. Louis, C. V. Mosby, 1986.)

TMJ, and this is simultaneously accompanied by overall enlargement of each bone into the "spaces" created. The coronoid process, the gonial region, the lingual tuberosity, and so forth, are all formed and continuously enlarged to **precisely fit** with the muscles and other soft tissues they serve. They fit because of feedback control among them involving the turning on and off of the regional osteogenic connective tissues. Tooth roots fit their sockets for the same reason. Bony edges at the sutures fit precisely. Nerves and vessels to and from a bone fit their foramina in size and location. The condylar cartilage fits its fossa. And so on. To do this, (1) the growth and size changes of the muscles and other soft tissues, (2) the displacements of the bones, (3) the functions of the various soft tissues, and (4) the bony remodeling processes are all inseparable. All are required as an architectonic package. They are artificially isolated here so that we can better group their respective roles, but in real life they cannot be so separated. One of the reasons many animal experiments intended to "prove or disprove" for all time, for example, the functonal matrix or the condyle as a master growth center, have often been less than fully successful is that these four factors were not **each** recognized and included within a complete, interrelated growth combination. Indeed, such experiments play against a stacked deck because of the actual inseparability of these factors as independent and controllable experimental variables.

13. All of the above 12 points converge to underscore the fundamental point that any given craniofacial component is not entirely self-contained, is not self-regulated and growth-controlled, and is not developmentally separate and unrelated to other parts and the composite of all of them. Although most of us will readily agree on this, some might perceive, for example, the developing palate as essentially responsible for its own intrinsic growth and positioning, and that an infant's palate is the same palate in the adult simply grown larger. **You** know, of course, that a palate in later childhood is not composed of the same tissue (but with more simply added), and that it does not occupy the same anatomic position. Many things happen to that growing palate from without, such as rotations, displacements in conjunction with growth at sutures far removed, and multiple remodeling movements that relocate it to progressively new positions and adjust its size, shape, and alignment continuously throughout the growth period. Similarly, for the mandible, the multiple factors of middle cranial fossa expansion, anterior cranial fossa rotations, tooth eruptions, pharyngeal growth, bilateral asymmetries, an enlarging tongue, lip muscle actions, head form variations, an enlarging nasal airway, changing infant and childhood swallowing patterns, adenoids, head position, sleeping habits, body stance, and an infinite spread of morphologic and functional variations all have input in creating constantly changing states of structural balance. But **growth** is an architectonic process leading to an aggregate state of structural and functional equilibrium, with or without an imposed malocclusion or other dysplasia. Very little, if anything, can be exempted from the Big Picture of factors affecting the operation of the growth control process, and no region can be isolated. The details of the architectonic cascade of interrelated cytogenic controls, functions, signals, and responses that mediate all of this are today's research opportunities.

CHAPTER NINE

Heredity in the Craniofacial Complex

W. Stuart Hunter, D.D.S., Ph.D.

WHY DOES HERITABILITY MATTER?

What is the allure of genetics for orthodontists? Or, indeed, for all dentists?

When planning treatment, orthodontists believe that it would be helpful to know one or all of the following:

1. How much will a child's mandible grow? When the father has a large jaw, one might wonder whether his son will also have a large jaw.
2. When will the mandible grow? Is there growth to come, or is mandibular growth completed? When the patient has a severe Class II malocclusion, it is believed that good growth during treatment will lead to a good result.
3. In what direction will the face grow? If both parents have long faces, one might expect the daughter to have a long face.

That progeny resemble one or both parents is often apparent. We know that an individual has half of his or her genes from each parent. Therefore, in theory at least, there ought to be some similarity between progeny and parents. And the basis for that similarity is genetic.

Again, identical twins are often difficult to distinguish from one another, especially if they have been raised together. Identical twins have identical genes—which is why they are called monozygous (MZ) twins. In other words, the primary basis for the similarity of MZ twins is genetic.

Twins are of no use in answering the questions posed above, because one is not growing in advance of the other. By the time mandibular growth is found in one twin, the same growth has occurred in the other. Hence, although twins are useful for estimating the relative heritability of the components of the face, we turn to parents for knowledge of what is to come.

The matter of final or attained size of the mandible is of great interest, because if we could make an accurate estimate for a growing child of the final size of the mandible, we could then derive the amount of mandibular growth yet to come. However, we have the problem of whose genes are controlling the amount of mandibular growth for the patient in the chair. In this context, it is of interest that stature has been forecast by Tanner et al. (1983) from the subject's present height and developmental stage, with no reference to the parents. However, Tanner et al. (1975) have also used mid-parent height (average of the two parents) in an effort to enhance such forecasts and found the improvement to be about 6 per cent (or 0.2 cm when the standard deviation [SD] of the forecast was on the order of ± 6.0 cm).

Next, we see that timing of growth cannot be estimated from parents because their recollection of changes in rate of growth in their faces will probably be nonexistent.

Although direction of mandibular growth might be derived from each parent's present face shape (for that parent), we then confront the dilemma of what proportion of the child's direction of mandibular growth is controlled by genes from the mother and what proportion by genes from the father.

Finally, if direction of facial growth is inherited, is the inheritance mendelian or polygenic? The differences between mendelian inheritance and polygenic (or multifactorial) inheritance are pertinent and are discussed below. Thus, the idea that family similarities can be put to clinical use rests upon genetic reality and is motivated by clinical need, but the complexities of utilizing those realities are quickly encountered.

When we observe a child who resembles one or the other parent, we say, "Aha! Genetics!" and to some extent we are correct. Each of us has made similar observations, and we remember those situations that substantiate that observation. Do we in fact keep track of the other situations, those that do not support the genetic thesis?

In a sample of 38 families, the family shown in Figure 9–1 was noted, in which one daughter's facial dimensions appeared to resemble her mother's and the other daughter's, the father's. For another family, there was a remarkable resemblance between the father and his two sons in head and neck outlines as seen from behind. One tends not to remember the absence of such superficial similarities in the other 36 families.

Another facet of family resemblance has to do with such matters as expression, shadows, and coloring. We have all noted family similarities in matters of facial expression, mode of laughter, frowns, and such. It is tempting again to say, "Aha! Genetics!" when we know that such matters are usually learned as a result of living together. A report by Garn (1979) documents dimensional similarities in fatness in families as a cohabitational effect and suggests that having the same climate and consuming the same food does result in measurable similarities in such dimensions as fatness and stature within families.

In other words, what we sometimes assume to be genetic is learned or acquired; and although it could be superimposed on a genetic foundation common to both parents and progeny, the similarity we see may in fact be a modification of dissimilar genetic foundations in parents and progeny.

INHERITANCE IN THE FACE

General Concepts of Inheritance in the Face

At this point, we might examine what is known about heredity in the craniofacial complex. A cautious sort of generality is the statement that size is partly inherited in the craniofacial complex.

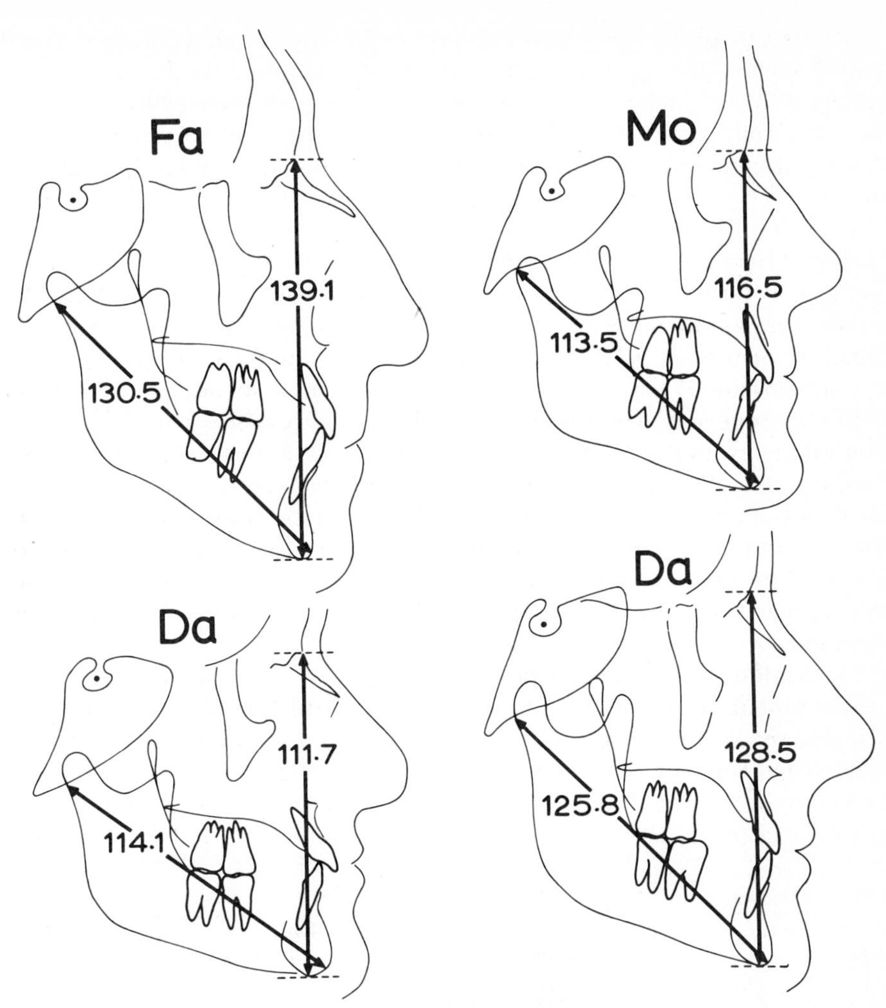

FIGURE 9–1. Family in which the father and daughter *(left)* are long-faced, whereas the mother and daughter *(right)* are, while not short-faced, less long-faced.

1. Size influences shape, if one considers the size of the components of the face individually as they influence the overall shape of the face. Change the length of the mandible only, and the shape of the face is altered. The size of the mandible has a most important effect on the shape of the face, which means that changing only one dimension (that of the mandible) can alter the entire face shape and how the parts fit together. (See Chapter 6.)
2. The key word in the general statement is "partly," and that means that theoretically we can account for 10 to 36 per cent of the size of various dimensions in the face of a subject on the basis of the size of those dimensions in the faces of the parents of that subject.

The genetic theory that underlies that statement will be considered in greater detail subsequently. Initially, however, it might be useful to consider traditional mendelian inheritance. When we think of inheritance, we tend to remember that when both parents have Type O blood, all of their progeny will have Type O blood. There are six genotypes for the three genes, A, B, and O. Because of dominance, only four blood "groups" can be detected; but each person belongs to only one group.

It is as though the gene for Tall produced only people who grew exactly 6 feet 2 inches; the gene for Short produced people 5 feet 2 inches; and Medium, 5 feet 8 inches—with no one in between (or shorter than Short or taller than Tall). If that were so, one could chart the frequency of occurrence of Short, Medium, and Tall for Europeans as opposed to Asian Indians, as is shown for the approximate frequency of occurrence of A, B, and O genes in the same populations in Figure 9–2. That is what is meant by discontinuous variation. There is no one between A and B or between B and O. Of course, blood type is rather unusual, in that environment has no effect at all upon it, whereas most conditions for which the genetic mechanism is known are variable in intensity or time of onset often because of environmental factors. Nevertheless, a discontinuous distribution is evidence for simple mendelian inheritance.

In the same way, continuous distribution of a variable with most of the values falling around the mean is taken as evidence for inheritance on several or many genes. We know of no measures in the face that are discontinuous in their distribution. For example, face height measured from nasion to menton (Figs. 9–3 and 9–4) could involve simple genetic control of symphysis height, tooth size, and maxillary height, which, when measured from nasion to menton, would appear as a continuous distribution and would, of course, be a polygenic dimension.

In addition to multiple genetic control, there is environment and the effects of environment on the results of the genetic controls. At one time, there was much interest in the relative roles of heredity and environment, and it was hotly debated whether environment was more important than heredity. It is no longer a matter of which is more important. Rather, it is agreed that environment changes the dimensions of almost all size components, at least in the face.

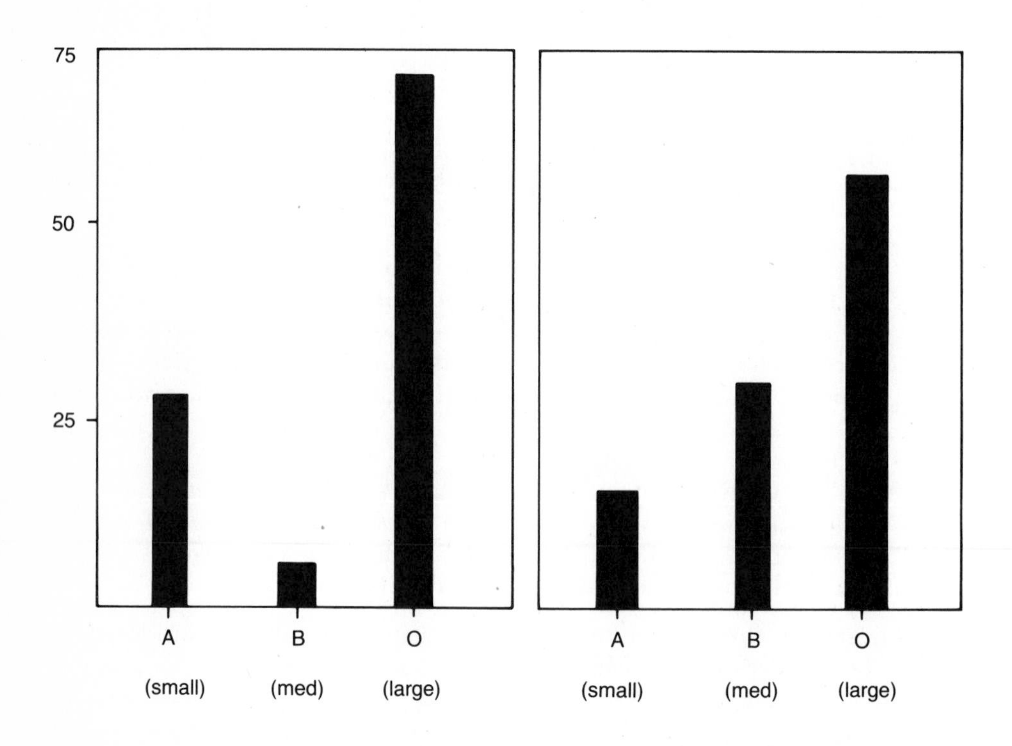

FIGURE 9–2. Approximate frequencies for A, B, and O genes for Europeans and Asian Indians. See text for explanation of small, medium, and large.

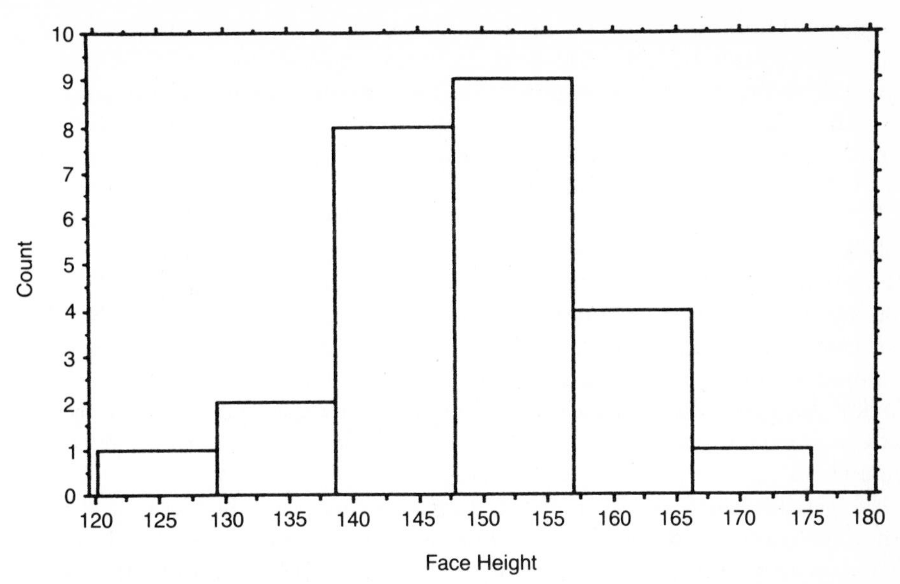

FIGURE 9–3. Distribution of face height from a sample of twins (unpublished data) showing an approximately normal distribution.

Inheritance and Genetic Control Mechanisms

A simplified explanation of genetic control in the face is as follows:

1. There are probably primary controls for initiation and formation. Do this and then that. Tooth buds calcify in certain tissues. Mandibles form in faces, not in elbows. According to Slavkin (1988), it is now postulated that the fundamental structural control is but part of the genetic mechanism contained

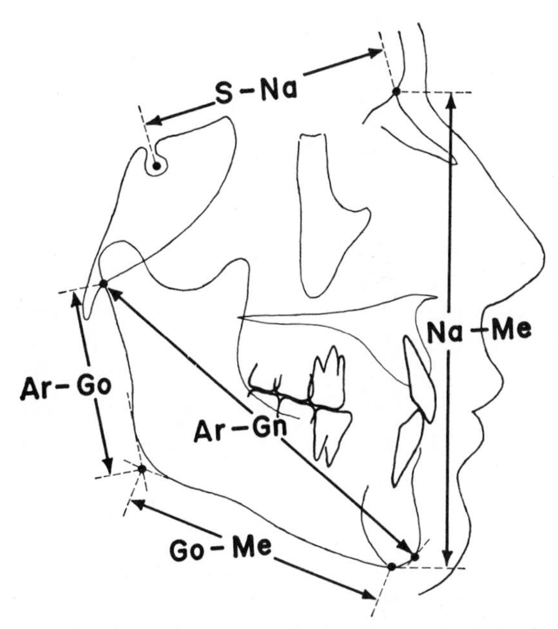

FIGURE 9–4. Measurements made for the study of 38 families. *S,* sella; *Na,* nasion; *Ar,* articulare; *Go,* gonion; *Gn,* gnathion; *Me,* menton. (From Hunter, W. S., D. R. Balbach, and D. E. Lamphiear: The heritability of attained growth in the human face. Am. J. Orthod., 58:128, 1970.)

in genes. There are many kinds of modifier genes, such as regulator genes (which start or stop the activity of the structural genes), oncogenes (which cause cell proliferation), orthogenes, pseudogenes, provirus genes, and so on.

An important point to remember is that a copy of all of an organism's genes exists in every cell of the organism. Only the parts needed by each cell are used by that cell (Beardsley, 1988).

A major difficulty in understanding what happens at the molecular and cellular level is the exotic and often recondite terminology used. What Slavkin (1988) calls the "regulatory genes" are also called "homeotic genes" both by him and by Beardsley. Homeotic genes contain short sequences of DNA called homeoboxes. A homeobox is a protein-encoding sequence in a gene that has been shown to control such things as the body shape of fruit flies or of mice. Slavkin suggests that "homeotic-like gene sequences may serve to regulate pattern formation during mammalian embryonic and fetal craniofacial development."

Beardsley says: "The picture that emerges is of a control hierarchy, with genes at each level passing instructions down the chain. After maternal genes have established an anterior/posterior gradient in the embryo, different types of embryonic genes are activated in sequence, each reacting to the activity of the level above (and probably other levels) and specifying in increasing detail how the cells within its domain should develop."

2. There are certainly local feedback and intercommunication mechanisms between individual cells and tissues that continue throughout the life of an organism. Three subdivisions may be identified: First, there are molecules that control adhesion of specific cells to other cells (cell adhesion molecules, or CAMs) and molecules that control adhesion of specific cells to substrate (substrate adhesion molecules, or SAMs). These adhesion-controlling molecules are thought to permit the regulation of cell division, mobility, and shape, according to Slavkin (1988). He says, "It is now evident that ECM [extracellular matrix], plasma membrane, cytoskeleton, and nuclear matrix are coupled through physical-chemical interactions." Second, there are general "humoral" chemicals synthesized by a cell to regulate its own activity (autocrine) and chemicals secreted by a cell to regulate the activity of adjacent cells (paracrine). The paracrine substances are like endocrine hormones, except that they affect cells only locally and do not have the widespread impact of the endocrine hormones. Third, under local mechanisms, is the gap junction that allows direct intercellular communications via the formation of connecting "pores" between adjacent cells through which small molecules and electrical signals pass from one cell to the other (see Sheridan, 1977). At the local level, then, most communication is accomplished by molecular messengers, the gap junction also permitting changes in electrical potential to transmit information.

At the macroscopic level, analogous activity occurs. The teeth talk to the bone. The muscles talk to the bone. As a result, the genes for muscle have a modifying effect on the bones that is not controlled by the genes for bones. Although not likely, if there were a simple one- or two-gene control for mandibular formation and size, the end-result (after the teeth and the muscles and the stresses of mastication have "massaged" the bone) would appear to be polygenic (or multifactorial). Geneticists tend to disguise that very simple concept by calling it "expressivity" or "penetrance," and it is almost always "variable." In the above example, what is environmental to the bone is genetic to the muscle. Van Limborgh (1970) has enhanced the idea by calling the muscle-molding factors epigenetic.

3. In addition, there is probably another overriding sizing mechanism, because although maxillary and mandibular teeth seldom match perfectly in size, large lower teeth always come with upper teeth in the large range. The correlation between upper and lower tooth size overall is approximately 0.8 (Moorrees and Reed, 1964).

Polygenic Inheritance

Given such multifactorial controls, it can be shown, as the geneticists say, that the highest correlation between parents of one sex and their progeny (of either sex) can only be 0.5 (Fisher, 1918). Or, if the mid-parent value is used, the correlations may increase to 0.70 because of the regression to the mean of parental dimensions experienced by the progeny (see also Penrose, 1949). Consider now that the correlation for blood type is 1.0 between parents and progeny. Comparatively speaking, given polygenic inheritance, there are major difficulties in forecasting progeny size from parent size.

Finger ridge counts (from fingerprints) have been correlated between generations, and the data support the polygenic hypothesis very nicely (Holt 1961). As can be seen in Table 9–1, the parent-child, father-child, mother-child, sibling, and dizygous (DZ, fraternal) twin correlations follow the predictions closely indeed; and all fall between 0.48 and 0.50. The mid-parent/progeny correlation is increased to 0.66 because of the regression to the mean of the parents' values by the progeny. Obviously, fingerprints are not altered by environmental factors.

"Explaining" the Variance with Correlations

At this time, a sleight-of-hand relationship with which most are familiar is useful. That is, by squaring the correlation between two variables, one can arrive at the amount of variation "explained" or predicted for one variable in the correlation by the other. Thus, with a correlation of 0.5 between parent and offspring for mandibular length, one should be able to "explain" 25 per cent of progeny mandible size from parent mandible size for a sample of reasonable magnitude (i.e., larger than 20). Or, to turn the statement around,

Table 9–1. CORRELATIONS BETWEEN VARIOUS RELATIVES FOR FINGER RIDGE COUNTS

Relationship	Correlation	Number of Pairs
Parent-child	0.48	810
Mother-child	0.48	405
Father-child	0.49	405
Parent-parent	0.05	200
Mid-parent-child	0.66	405
Sibling-sibling	0.50	642
MZ twins	0.95	80
DZ twins	0.49	92

Adapted from Holt, S. B.: Quantitation genetics of finger-print patterns. Br. Med. Bull., 17:247, 1961.

75 per cent of the variation (or departure from the mean for progeny mandible size) must be attributed to the vagaries of nongenetic factors (environmental). If the variance (SD) of the mean is ±5.0 mm, then 66 per cent of mandibular lengths will fall within a range of 5.0 mm less than the mean to 5.0 mm more than the mean for progeny mandibular lengths. Knowing the parent value would reduce that 10.0 mm range by 2.5 mm.

Another way of looking at the parent/progeny correlation is to use the correlation of 0.64 plotted in Figure 9–5 to predict progeny mandibular size from parent mandibular size. If one looks only at the father's mandibles, which are 120 mm long, one can see that the mandibles of their sons could be as small as 123 mm or as large as 128 mm if one considers only the cases touching the 120 mm line. If the three next closest cases are added, the range increases to 115 to 128 mm. Because the range for sons whose fathers' mandibles were 125 ± 3.0 mm long is from 120 to 130, the larger range is probably more appropriate. Statistically (and for some relationships) a correlation of 0.64 is usually significant and useful for descriptive purposes. From a clinical point of view, because no one knows which son will have a smaller mandible than his father and which a larger, a correlation of 0.64 is obviously not very helpful.

Therefore, in theory, the examination of parents for the purpose of forecasting progeny size is not likely to be clinically as useful as other less expensive procedures (see Predicting from the Subject Himself). In other words, in order to justify obtaining radiographs of parents and then tracing

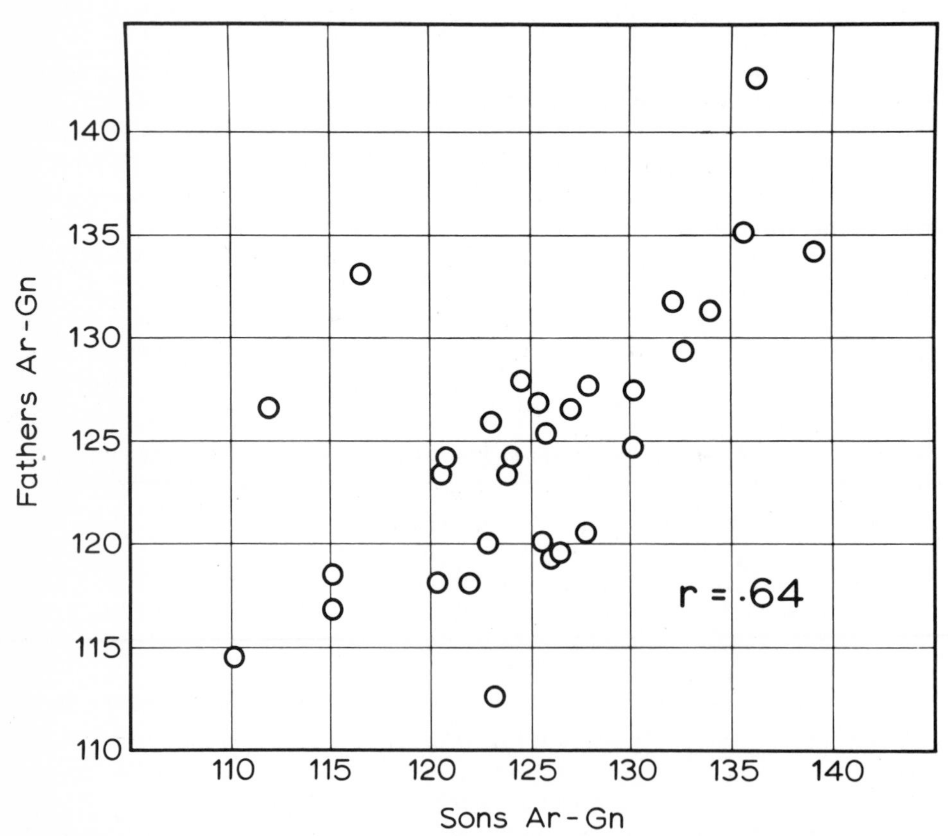

FIGURE 9–5. Scattergram of relationship between 31 fathers and sons for mandibular length from articulare to gnathion (r = 0.64).

Table 9–2. CORRELATIONS FOR TWO OF THE VARIABLES MEASURED IN A STUDY OF 38 FAMILIES

	Fathers	Mothers	n
Articulare-gnathion			
Sons	0.64**	0.16	32
Daughters	0.62**	0.20	28
Nasion-menton			
Sons	0.32	0.35*	32
Daughters	0.46**	0.35	28

From Hunter, W. S., D. R. Ballach, and D. E. Lamphiear: The heritability of attained growth in the human face. Am. J. Orthod., 58:128, 1970.

and measuring them, the improvement in prediction should be more than a few percentage points. However, the possibility exists that heritability in the face might not be entirely polygenic. That is the most rational explanation for the fact that so many investigations of heritability in the face have been undertaken for study of the matter—and continue to be undertaken.

Studies of Polygenic Inheritance in the Face

The most common two-generation studies of the face are of adolescent children and their parents, because both are readily available during the extended course of orthodontic treatment. The progeny are obviously at some point short of complete growth when the records are taken. The amount of growth to maturity is not known, and even shape comparisons can only be of a general nature, because face shape (see Chapter 1, Figure 1–14) is known to change from birth to maturity. Hence, for the investigation used as an example here (Hunter et al., 1970), all but 7 of the 38 families involved progeny 21 years of age or older. No subject was less than 17 years of age.

The measures made are shown in Figure 9–4, and two of the correlations resulting are summarized in Table 9–2. Scattergrams of the highest correlations are shown in Figures 9–5 and 9–6. Of particular interest, as discussed above, is the possibility of forecasting progeny mandibular length (Ar-Gn) from parent mandibular length using data for females (r = 0.62, Figure 9–6) and data for males (r = 0.64, Figure 9–5).

The findings were reported in 1970 by Hunter et al. We noted the particularly high correlation for Ar-Gn for fathers to progeny (both male and female) and concluded that sample bias was a likely source of the high correlation. However, Nakata et al. (1973) reported similar findings for a sample of twins and their parents. An earlier report by Stein et al. (1956) reporting similar findings, using a less rigorous methodology, then gained some credence. However, Saunders et al. (1980), reporting on 147 families, did not concur. They reported no difference between father/progeny and mother/progeny for mandibular length. It should be noted that all their progeny were 18 to 21 years of age.

It is possible then that both the Ann Arbor and the Indianapolis (Nakata et al., 1973) samples were biased in the same direction. Also, it is possible that there is, in fact, something different about the inheritance of mandibular length as compared with other facial dimensions.

In 1980, Houston and Brown summarized the whole area of family studies by first accepting the fact that the subject himself is the best predictor for that subject. They proposed that for parent records to be useful, the progeny must

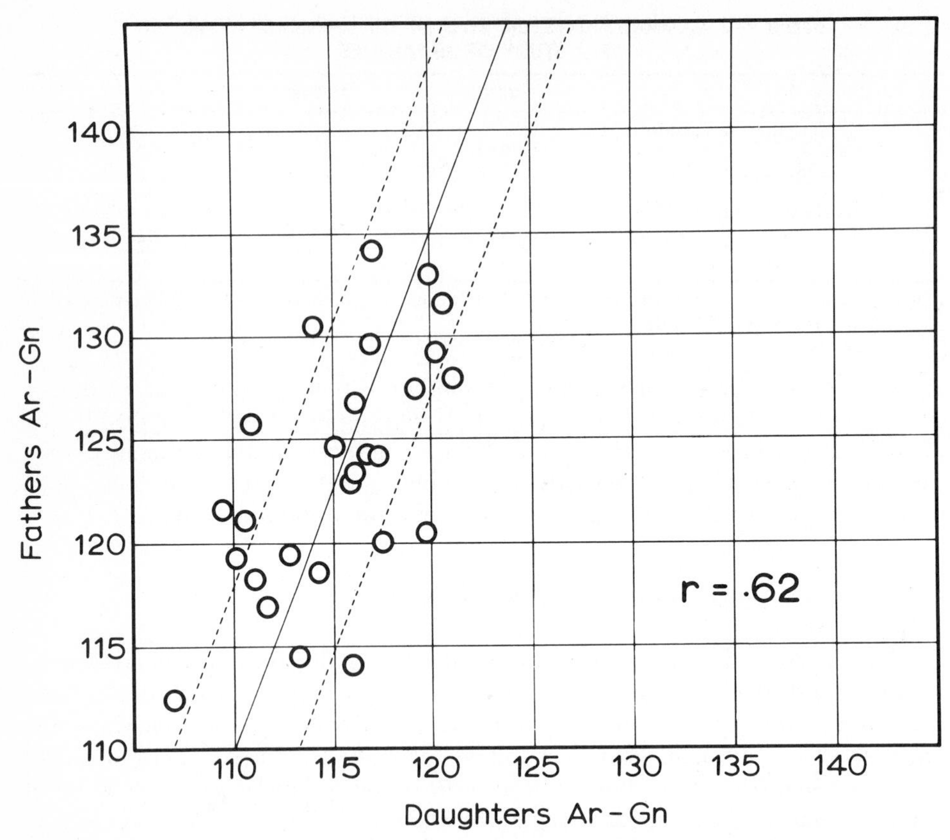

FIGURE 9–6. Scattergram of relationship between 27 fathers and daughters for mandibular length from articulare to gnathion (r + 0.62).

grow to resemble the parents more and more. They tested that theory with 45 families using their 9-year-old versus their 18-year-old progeny and found the older cohort to be no more like the parents than the younger ones in mandibular, maxillary, cranial, and cranial base dimensions.

The problem has also been examined in what is generally considered a less rigorous fashion by several investigators with results that may be of some interest.

Lundstrom (1963) looked at the occurrence of the three classes of occlusion in MZ and DZ twins (Table 9–3). Of particular interest is the Class II category, for which he found that if one DZ (fraternal) twin had a Class II, the other DZ twin had a Class II 24 per cent of the time. However, for the genetically

Table 9–3. CONCORDANCE OF TYPE OF OCCLUSION IN TWINS

	Number of Pairs	Zygosity	Percent Concordance
Class I	55	MZ	87.3
	26	DZ	84.6
Class II	31	MZ	67.7
	42	DZ	23.8
Class III	6	MZ	83.3
	10	DZ	10.0

When one twin has a Class II malocclusion, the chances of the other twin having the same class of malocclusion are shown in the right hand column.

Adapted from Lundstron, A.: *Tooth Size and Occlusion in Twins.* Basel, S. Karger, 1948.

identical twins (MZ), when one twin had a Class II, 68 per cent of the second twins were also Class II. That is nearly three times the percentage for DZ twins (and supports the idea of genetic controls of occlusion) but is far short of the 95 per cent concordance found for finger ridge counts (Holt, 1961). The data can be used to support the importance of both inheritance and environment.

The explanation for the difference between 95 per cent concordance for finger ridge counts and 68 per cent concordance for Class II occlusion is that the dentition is the area in the face most affected by environment. That is based upon a twin study of 37 MZ and 35 DZ twins for whom 26 measures were examined (Hunter, 1965). Those 26 measures could be divided into 14 for depth (anteroposterior dimensions) and 12 for height (vertical). See Figure 9–7. The heavy lines show the statistically significant measures. The depth measures related to the dentition such as maxillary length, mandibular length, overjet, and relationship between point A and point B are not statistically significant. In other words, the anteroposterior dimensions associated with the dentition are more affected by environmental factors than any other area of the face.

Such a concept is supported by the observations of Harris and Smith (1980) on the dental arches and teeth of more than 60 father-offspring, over 100 mother-offspring, and over 200 sibling relationships. They reported that on average, 90 per cent of the variation in overjet, overbite, crowding, rotations, and molar relationship resulted from environmental effects. Note that overjet, molar relationship, and, to a considerable extent, overbite are measures of the anteroposterior relationship of the mandible to the maxilla. If Solow and Tallgren's (1976) hint that there is a relationship between head posture and

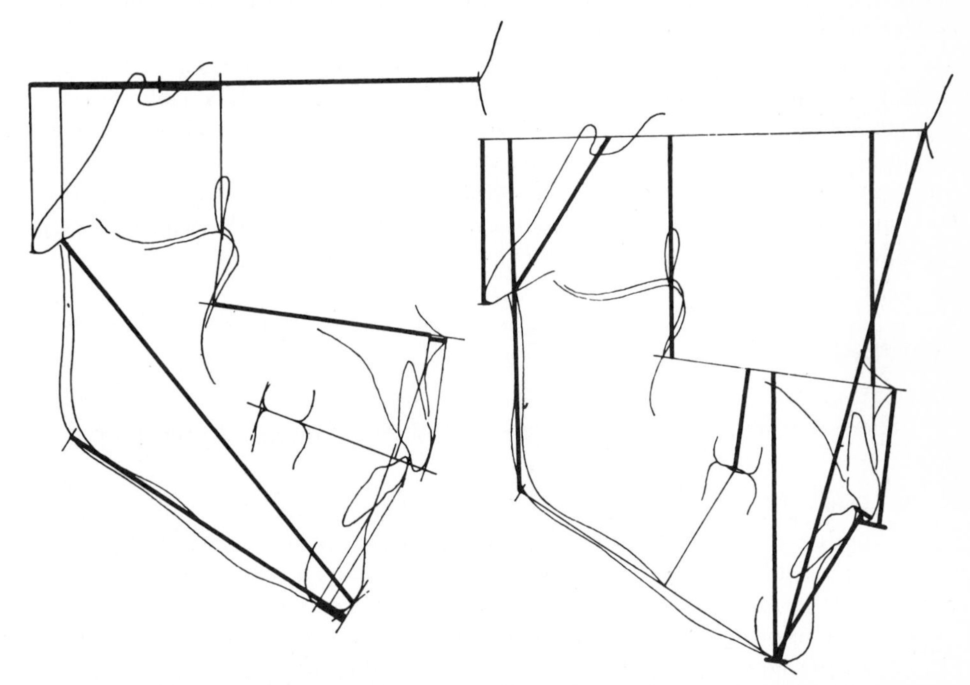

FIGURE 9–7. Statistically significant depth and height measures. On the *left* are the 14 measures of depth with the 8 for which the *F* ratios are significant heavily lined. On the right are the 12 measures of height with the 11 for which the *F* ratios are significant heavily lined. (From Hunter, W. S.: A study of the inheritance of craniofacial characteristics as seen in lateral cephalograms of 72 like-sexed twins. Trans. Europ. Orthop. Soc., 59, 1965.)

growth direction is correct, then it is possible to see how environment could affect those relationships. On the other hand, Harris and Smith (1980) found that only about 40 per cent of the variation for such measures as arch width (at canines and molars) and arch length was due to environment.

Lundstrom and McWilliam (1988), using a much more sophisticated twin analysis (which permits the partitioning of error and cultural factors) on 56 pairs of like-sexed twins, were not able to establish that horizontal factors are more susceptible to environmental influences than vertical factors (using 6 horizontal and 6 vertical measures). Although in their study there was a tendency for the horizontal measures to show more environmental influences than vertical measures, it was not statistically significant.

If vertical facial measures may be considered as longitudinal relative to the rest of the body, then one can relate the findings to those of Susanne (1975), Osborne and DeGeorge (1959) and Howells (1949), who reported longitudinal measures to be more highly heritable than anteroposterior or circumferential measures.

A sibling investigation of Hawaiian school children was reported in 1975 by Chung and Niswander. They asked the question: "If the first sibling has a Class II malocclusion, what are the chances of the second sibling having a Class II malocclusion?" That is a standard genetic question one can ask about any inherited condition. For a sample of 113 first-borns with Class II, they found that the second siblings were almost twice as likely (14 per cent) to have Class II malocclusion as the unrelated population (7 per cent). Of course, the incidence of Class II in first siblings for that population is remarkably low, compared with that for North American populations, for which we expect about 26 per cent Class II, rather than 7 per cent, according to Fisk (1960), or between 30 and 40 per cent for North American white populations according to McLain and Proffit (1985).

Perhaps the most important thing to note is that the chances of having a Class I occlusion are overwhelming no matter what your older sibling has if you are Hawaiian. Nevertheless, when the first child has a Class II, the chances of the next child having a Class II are increased or may even be doubled, but the population incidence of Class II will be quite important. If the population incidence is between 10 and 30 per cent, as it is in North America, then "double" becomes quite a good chance.

Limitations

The conclusions that can be drawn from these investigations seem inescapable:

1. Inheritance of size in the face is polygenic, and therefore only about 25 per cent of the variability of any dimension in progeny can be explained by consideration of the same dimension in parents.
2. Further, because the usual correlation between parents and progeny dimensions is about 0.3, we are more realistically looking at about 10 to 15 per cent of progeny dimensions explicable by parent dimensions.

Because as has been shown above, correlations of 0.6 and 0.7 are of questionable clinical utility, a correlation of 0.3 is not helpful at all. What are the alternatives?

Predicting from the Subject Himself

The prediction problem was examined by Johnston (1968), and he noted that we can explain 30 to 54 per cent of the variability of a patient's Class II malocclusion by using data from that patient. Balbach (1969) obtained a slightly higher 60 per cent. Hence, the alternative is the use of the patient himself (as recently noted by Houston and Brown, 1980), and probably the most efficient predictor of future mandibular length is the existing size of the subject's mandible.

THE INTERACTION OF HEREDITY AND ENVIRONMENT

Generally, it is assumed that environment factors of one sort or another are constantly modifying whatever has been inherited. The genotype is inherited and, having been inherited, remains the same. Blood type does not change and is not susceptible to environmental alterations. Many other observable (phenotypic) factors are very susceptible to environmental effects. Thus, the basic minimum of 28 permanent teeth (excluding third molars from consideration) is often reduced by extractions to facilitate orthodontic alignment of the teeth. Finger sucking is known to alter the alignment of the incisor teeth (Larsson, 1987).

Cleft Palate and Environment

The interaction of heredity and environment is seen dramatically in MZ twins who are discordant for cleft lip and/or palate. The various analyses of a sample of 45 pairs of such twins (20 MZ and 25 DZ) were reviewed by Hunter in 1981. The fact that one of a pair of genetically identical twins can have a cleft lip and/or palate, while the other does not, is illustrative of the role of environment in the etiology of clefting. Because the relatives of subjects with clefts have a higher incidence of clefting than the population as a whole (Ross and Johnston, 1972), there is a heritable component to clefting that is thought to predispose the possessor to clefting when an appropriate environmental stimulus occurs. There is not much agreement as to what such stimuli are for humans, although for mice, oxygen deprivation and various chemicals will cause clefting in susceptible strains.

One rather interesting finding of the cleft twin studies is that the cleft twins do not fall behind population standards for height and weight until puberty (Hunter and Dijkman, 1977). Although the phenomenon was first noted in the rather small cleft twin sample under discussion, review of another much larger sample (Menius et al., 1966) shows the same effect.

In a summary of differences between 19 male subjects with unrepaired unilateral cleft of the lip and palate and 30 male noncleft subjects, Ross (1987) concluded that the cleft subjects had more retruded maxillae and a related more open mandibular posture (larger mandibular plane angle). He noted that the deficiency and the differences are "exacerbated" a relatively small amount by contemporary surgical repair procedures. Indeed, surgical repositioning of the maxilla can almost eliminate the deficiencies.

However, the differences between unrepaired cleft subjects and non-cleft subjects need not be the genetic component predisposing to clefting. The identification of such differences requires a comparison of the close relatives of cleft subjects with non-cleft subjects. Regarding that type of investigation, it has been suggested by Kurisu et al. (1974) that the parents of cleft children have less convex faces than the population as a whole.

Secular Trend and Environment

Another environmental factor which influences the phenotype is nutrition, and it is seen in a phenomenon called secular trend. Secular trend is any change in size, shape, or timing that is consistent in direction over several generations or "successive periods of time" (Garn, 1987). The simplest form of secular trend has to do with height and weight. Harvard men and Wellesley women were weighed and measured for four or five generations (Bakwin and McLaughlin, 1964). There was a steady increase in height and weight until the cohort born in 1915 to 1918, when the college entrants who came from private schools showed no gain over the previous generation. The previous increases have been ascribed to improved nutrition. Because the increases have ceased, it has been suggested that there is a limit beyond which increases will not occur. On the other hand, Garn (1987) suggests that the relatively small samples used by Bakwin & McLaughlin may not truly reflect the population at large and that, generally, the secular increase has not yet reached a plateau.

For the face, which is of more interest to orthodontists than height and weight, we have investigated Secular Trend (Hunter & Garn, 1969) utilizing the same sample of families used by Hunter et al. (1970). Variables around the perimeter of the face were included so that a facial polygon could be constructed for average components of the sample as shown in Figure 9–8. The progeny were, on average, taller than their parents and showed some small but interesting differences in face shape from their parents. Although the sons' faces were slightly more square than the fathers', and although the differences were statistically significant, it can be seen that clinically the trend is not particularly remarkable. The daughters, on the other hand, showed markedly more prognathic mandibles than their mothers.

Behrents (1983) has shown that from 17 years to 50 years, both male and female mandibles grow about 2.0 mm, which makes the secular trend findings quite remarkable. That is, because the parents were on average 25 years older than their progeny, the parent's mandibles were more prognathic than those of their progeny by virtue of adult growth. Hence, the progeny had to exceed that amount and then be more prognathic as well to show more prognathism than their parents.

Evolutionary Trends

Indeed, one may be permitted to wonder about the basis for the popular concept that jaws are getting smaller. The idea appears to be based upon the observation that various species of pre-hominids had larger teeth and jaws than *Homo*. That is true but does not mean that the jaws of a different species, *Homo*, are getting smaller. After all, horses' teeth are also larger than those of *Homo*. Then the "fact" that contemporary populations have more third-molar impactions is always "quoted" with no reference to any research data that will stand close inspection. The idea is also usually offered that since we do not need third molars, we are losing them—which is just simply wrong. That is not the way evolution works—you have to die because of your third molars before

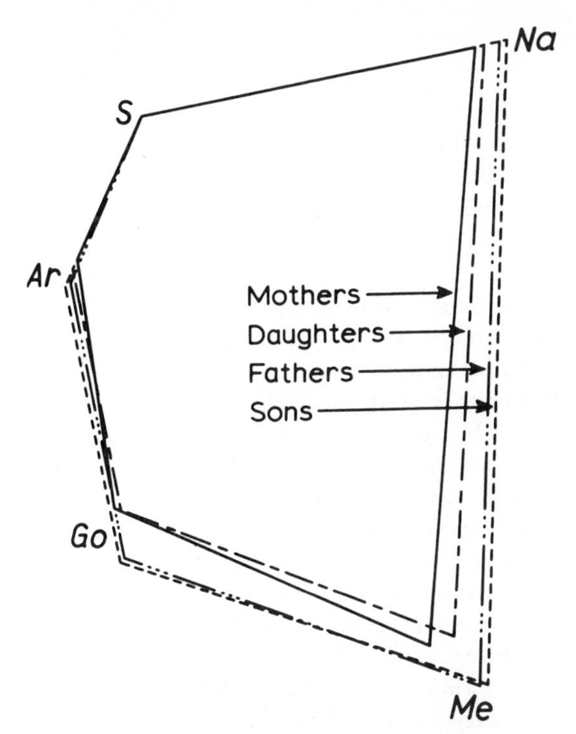

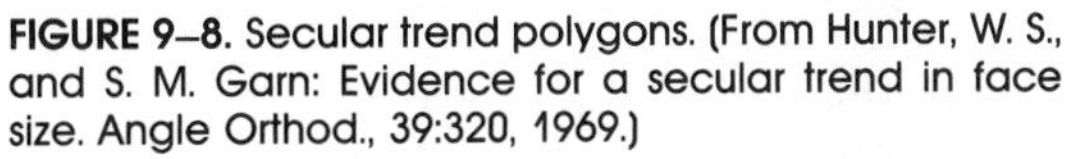
FIGURE 9–8. Secular trend polygons. (From Hunter, W. S., and S. M. Garn: Evidence for a secular trend in face size. Angle Orthod., 39:320, 1969.)

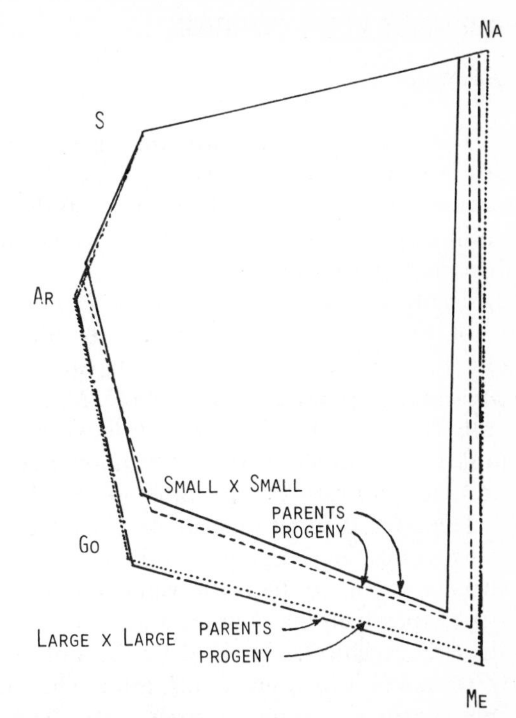

FIGURE 9–9. Differential secular increase polygons. (From Hunter, W. S., and S. M. Garn: Differential secular increase in the progeny of small faced parents. J. Dent. Res., 52:614, 1973.)

reproducing for there to be any evolutionary pressure on third molars. Simply not needing them does not produce evolutionary pressure. In any case, the evidence suggests that, within limits, we have larger jaws than our ancestors 200 generations ago. Teeth are also getting larger, as Garn et al. (1968) have shown.

Differential Secular Increase in Face Size

A subsequent study, also by Hunter and Garn (1973), addressed the question of whether all sizes of faces behave the same way, because we were aware of the fundamental genetic principle that progeny tend to regress toward the mean relative to those parents who are at the extremes.

The parents who were both large-faced were compared with their progeny, and the parents who were both small-faced (Figure 9–9) were compared with their progeny. "Large" and "small" were defined in terms of the sample. The large-faced parents were simply those who had large faces relative to the rest of the sample and whose spouses were large-faced, compared with the rest of the sample. That left the combinations of large/small-faced parents, which have not yet been analyzed. It was found that the progeny of the small-faced parents showed a much greater increase in size and change in shape than the decrease seen in the progeny of the large-faced parents. In other words, the regression to the mean occurred as expected but was modified by the secular trend, which amplified the regression for the progeny of small-faced parents and diminished it for the progeny of large-faced parents.

INHERITANCE IN THE DENTITION

Tooth Size

The heritability of tooth size has been examined with the use of twins by many investigators. The most frequently quoted study seems to be that of Horowitz et al. (1958). Because that study utilized only the 12 maxillary and mandibular anterior teeth, a study by Hunter (1959) that used 24 teeth per subject is reported here.

The most important concept to remember about twin studies is that they do not and cannot tell how a trait is inherited but merely whether the trait seems more or less likely to be inherited insofar as the sample of twins used represents the population as a whole. More specifically, twin studies permit one to rank traits in terms of their relative likelihood of being inherited for the sample being studied. It should be noted that sibling studies are equally effective in ranking and rating traits. Siblings are also in greater supply than are twins.

The management of twin data involves careful determination of zygosity, usually using blood typing techniques and some relatively simple statistics. For a variable being studied, the value for one twin is subtracted from the value for the other twin maintaining the sign. The sum of the differences for the DZ twins (DZ variance) is divided by the sum of the differences for the MZ twins (MZ variance), which gives an *F* ratio. The higher the *F* ratio, the greater is the difference in variability between monozygosity and dizygosity and the more highly inherited the trait.

For the mesiodistal size of the crowns of the teeth, the earlier the crown is completed, the higher the *F* ratio (the correlation is −0.49). That is interpreted to mean that all teeth have about the same amount of heritability as far as crown size is concerned but vary in size according to when the crown is completed. The theory is that the earlier the crown is completed, the less serious is the impact of environment. Older children are more active and have more colds and the usual childhood diseases than newborn children. That concept was examined by Sofaer and MacLean (1972) and found to be a creditable interpretation for their findings as well. Because tooth size is continuously distributed, there is no question that it is inherited in a polygenic fashion.

Dental Caries

From studies of twins and families, Grahnen (1962) stated that it seems likely that genetic factors account for a small but statistically significant part of the variation in susceptibility to dental caries. Muhleman (1972) concluded that hereditary factors are of only minor importance in dental decay. The issue is not particularly significant in today's world of fluoridation.

Oligodontia

The term "oligodontia" is synonymous with "hypodontia." "Anodontia" means "no teeth at all." "Partial anodontia" is a media term, analogous to "partly pregnant."

In 1962 Grahnen noted that for 171 children with one or more missing teeth, the incidence of missing teeth among parents and siblings was 26 per cent (the population incidence was about 5.6 per cent) or 4.5 times the expected

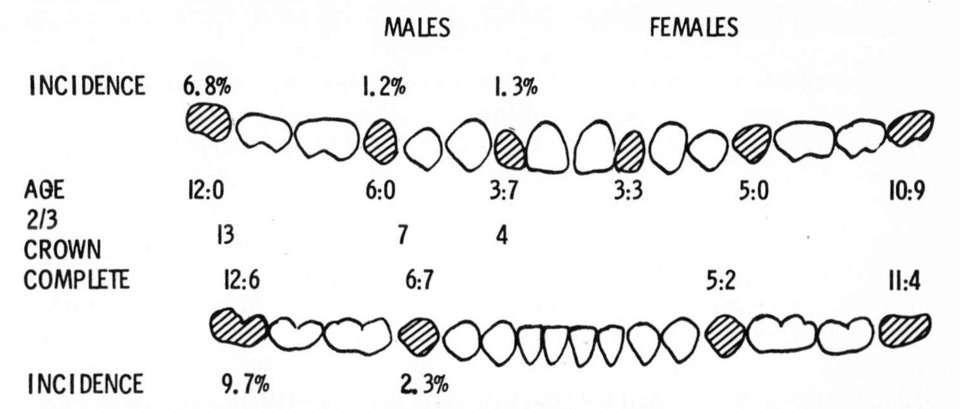

FIGURE 9–10. Incidence of missing permanent teeth related to the times when two thirds of the crowns may be expected to be completed. See text. (After Rose, J. S.: A survey of congenitally missing teeth, including third molars, in 6000 orthodontic patients. Dent. Pract., 17:107, 1966; and Nolla, C. M.: Development of permanent teeth. J. Dent. Child., 27:254, 1960.)

amount, which means that there is a genetic component to oligodontia. He also observed that a low positive correlation exists between missing third molars and missing teeth elsewhere in the mouth. Subsequently, Garn et al. (1963) related missing third molars not only to agenesis of other teeth but also to delayed formation of other posterior teeth, to differences in eruption sequences, and to delayed formation of the teeth of siblings.

The general incidence of oligodontia is around 6 per cent. Authorities differ on whether maxillary lateral or mandibular second bicuspids are the most frequently missing teeth (other than third molars). See Figure 9–10 for a summary of the incidence of missing teeth. The three round numbers in the middle line suggest ages (easily remembered) when teeth may be presumed missing if not seen radiographically.

There is some evidence that the incidence of oligodontia is higher in females than in males (Graber, 1978), but Rose (1966), in a sample of 6000 subjects, found 4.9 per cent of females to have missing teeth, whereas 3.4 per cent of males were so bereft—a difference of 1.5 per cent.

In his excellent summary of congenital absence of teeth, Graber (1978) concluded that oligodontia "appears to be the result of one or more point mutations in a closely linked polygenic system, most often transmitted in an autosomal dominant pattern with incomplete penetrance and variable expressivity."

SUMMARY

Although unlikely, the possibility that a component of the facial skeleton could be inherited in a mendelian fashion has been investigated rather carefully, because if such were the case, examination of parents would probably permit us to predict something useful about the faces of their progeny. The evidence strongly supports multifactorial or polygenic inheritance in the face, and therefore observation of parents can explain about 25 per cent of the size of various dimensions in the faces of progeny at best. Further, if the size of a bone in the face, such as the mandible, is in reality inherited in a mendelian fashion, that inheritance is masked by overriding sizing mechanisms and by environmental influences (some genetic and some general) to such an extent that the underlying genetic mechanisms cannot be detected.

We should stop looking. Although it is possible that the examination of siblings or parents could be used in improving our prediction of future facial growth in patients beyond the 50 to 60 per cent that we obtain by evaluating the patient himself, the gain accomplished would probably not be worth the effort expended, even for the most difficult Class II malocclusion.

As Smith and Bailit (1977) have noted, "Contrary to common opinion, the extent to which genes determine a trait has no relationship whatsoever with the success of environmental intervention. In addition, it will have no bearing on the type of treatment, nor will it be of much use in genetic counselling. For the specific orthodontic problem of predicting growth changes, the value of variables can be determined without any reference to their heritability."

We should stop, as Lewontin (1974) has said, "the endless search for better methods of estimating useless quantities." Nevertheless, it is probably worthwhile to observe that contemporary methods of prediction can tell us a great deal about the things we hoped to discover from genetic studies:

1. The direction of facial growth does not change significantly unless we change it. What exists is what it will be unless we modify it.
2. Although the timing of growth will probably always be an enigma, various pragmatic procedures offer some help.
3. Despite the fact that the amount of mandibular growth yet to come for any specific case is one of the most difficult and least well understood of the many things we want to know, we have a good beginning through the use of average population values.

CHAPTER TEN

Maturation of the Orofacial Neuromusculature

Robert E. Moyers, D.D.S., Ph.D.
David S. Carlson, Ph.D.

Part 1

More attention has been given to the study of the growth of the craniofacial skeleton and the dentition than to the neuromusculature that activates the masticatory region. The methods of study of the neuromusculature are much more difficult; consequently, we know less about the facial and jaw muscles, and we are less certain of what we do know than we are about our knowledge of bones and teeth. Nonetheless, the basic rules of biology apply. There is just as much variability in the morphologic features and action of the muscles as there is in tooth anatomy or craniofacial profiles. Muscles grow, develop, and mature in a planned and scheduled way, even as teeth calcify and erupt and bones form and grow. Many malocclusions may have their origins in abnormal neuromuscular behavior, and many an orthodontically treated malocclusion is not stable because occlusal stability, in the last analysis, could not be maintained by the muscles.

CONCEPT 1: CLASSES OF NEUROMUSCULAR ACTIVITIES

Unconditioned reflexes or responses are present at birth, having appeared as a normal part of the prenatal maturation of the neuromusculature. It is necessary for certain unconditioned congenital reflexes to be operative in the oropharyngeal region before birth in order for the neonate to survive. Conditioned reflexes are of two types: those reflexes that appear with normal growth and development and those, desirable or undesirable, reflexes that have been learned as a singular part of the development of one child. Of course, no conditioned reflex is capable of being learned until all the necessary parts of the central nervous system and musculature have matured sufficiently to make that learning possible. In the orofacial region, the mature swallow and mastication are good examples of reflexes that normally appear with growth and development, while thumb sucking is an example of an undesirable conditioned reflex. Voluntary activities are willful acts under cortical control. Such volitional activities must, of course, be separated from the congenital unconditioned responses and the conditioned reflexes that are learned. Some things we do because we willfully choose to do them; other activities in the orofacial region we carry out because we have learned to do them that way. All mammals display instinctive, primitive, unconditioned neuromuscular activities over which they have little control.

CONCEPT 2: PRENATAL MATURATION

During prenatal life, the musculature associated with the orofacial region matures well ahead of that of the limb regions, since the mouth is the site of a number of vital functions that must be fully operative at the time of birth, such as respiration, nursing, and protection of the airway. Respiratory reflexes, jaw closure reflexes, gag reflex, suckling, and the infantile swallow are all developed in a patterned way between the fourteenth and thirty-second weeks of intrauterine life.

CONCEPT 3: NEONATAL ORAL FUNCTIONS

The mouth at birth is a very active perceptual system. The infant uses the mouth and face for perceptual functions even more than the hands, and this then continues throughout life. The oral area presents in man the highest level of sensory-motor integrative functions.

Infantile Suckling and Swallowing

The infantile swallow is a part of the highly complicated suckling reflex. Both suckling and swallowing must be developed by birth so that the infant can feed. The infantile swallow is characterized by (1) positioning of the tongue between the gum pads, holding the jaws apart as swallowing is completed; (2) stabilization of the mandible by contractions of the facial muscles and the interposed tongue; and (3) swallowing that is initiated and to a great extent guided by the sensory interchange between the lips and the tongue. The infantile swallow is given up normally sometime during the first year of life for the mature swallow (Moyers, 1964).

Maintenance of the Airway

The orofacial and jaw musculature is responsible for the vital positional relationships that maintain the airway. The physiologic maintenance of patency of the airway is of vital importance from the first day of extrauterine life. All learned jaw functions are built around and accommodated to the mandibular and tongue positions that make possible a clear airway.

CONCEPT 4: EARLY POSTNATAL DEVELOPMENT OF ORAL NEUROMUSCULAR FUNCTIONS

Mastication

Mastication is a learned neuromuscular activity, but it cannot be learned until craniofacial growth has enlarged the intraoral volume, the teeth have erupted into occlusion, the musculature and the temporomandibular joint have matured, and the central nervous system's integrative and coordinative functions are possible. Like the early stages of any new motor skill, the first chewing movements are irregular and poorly coordinated. Sensory guidance during this learning period is provided by receptors in the temporomandibular articulation, periodontal membrane, tongue, oral mucosa, and to some extent the muscles. The individual's jaw movements during the chewing cycle are a developed, integrated pattern of many functional elements and are highly adaptive in the young child. Masticatory adaptive changes in later years are much more difficult.

Facial Expression

Although many muscle patterns of facial expression are learned, largely through imitation, some facial responses are unlearned and are very similar to basic primitive reflexes seen in certain lower primates.

Speech

While the reflex infant cry is an unlearned activity, purposeful speech is much more complicated, for it must be performed on a background of stabilized and learned positions of the mandible, pharynx, and tongue. Speech requires complicated, sophisticated, varying sensory conditioning elements during learning. The infant cry is primitive and unlearned.

Swallowing

The mature swallow usually begins to appear in the latter half of the first postnatal year of life. The arrival of the erupted incisors cues the more precise opening and closing movements of the mandible, compels a more retracted tongue posture, and initiates the learning of mastication. The infantile swallow is related to suckling; the mature swallow, to chewing. The transition from the infantile to the mature swallow takes place over several months, depending upon the timing of the maturation of important developmental neuromuscular events, but most children achieve the mature swallow by 1½ to 2½ years. Several features characterize the mature swallow: (1) the teeth are together; (2) the mandible is stabilized by contractions of the mandibular elevators (rather than the facial muscles); (3) the tongue tip is held against the palate, above and behind the incisors; and (4) there are minimal contractions of the lips and facial muscles (Moyers, 1964).

Neural Regulation of Jaw Positions

Jaw position, like a number of other automatic somatic activities, normally is largely reflexly controlled, even though it can be altered voluntarily. The receptors of the temporomandibular capsular area are far more important in the control and guidance of jaw function and positions than it has previously been thought and ordinarily taught. Much of our knowledge of jaw position and its regulation has been derived from studies of adults, and it is hazardous to transfer casually to the growing child concepts that may be correct for older persons. Knowledge about many aspects of developmental jaw neurophysiology is most incomplete at this time.

CONCEPT 5: OCCLUSAL HOMEOSTASIS

The goal of most occlusal treatment by the dentist is to achieve a self-stabilizing occlusal relationship. Modern clinicians have abandoned mechanistic models of occlusion and more practically view the occlusion as the relatively stabilized result of varied and discontinuous dynamic forces operating against the teeth. The sensory receptors of the temporomandibular joints, periodontal membrane, and other parts of the masticatory system provide a constant feedback mechanism that controls forces against the teeth. Factors such as the growth of the facial bones, the force of muscle contractions during mastication, and the natural tendency of the teeth to drift are probably all far more important in maintaining occlusal homeostasis than the oft-mentioned cuspal anatomy.

CONCEPT 6: EFFECT OF NEUROMUSCULAR FUNCTION ON FACIAL GROWTH

The role of neuromuscular function in the growth of the craniofacial skeleton has been brought into better perspective only in recent years. Such factors as the growth of the muscles, their migration and attachment, variations in neuromuscular function, and abnormal function (e.g., mouth breathing) are now known to influence markedly some features of craniofacial growth and form.

Moss's functional matrix concept came to influence thinking in craniofacial growth: " . . . in summary form, the functional matrix hypothesis explicitly claims that the origin, growth and maintenance of all skeletal tissue and organs are always secondary, compensatory, and obligatory responses to temporally and operationally prior events for processes that occur in specifically related non-skeletal tissues, organs for functioning spaces (functional matrices)" (Moss, 1981). Some have misunderstood and misstated or misapplied Moss's ideas. Moss's theory is difficult to prove or disprove, but it has been provocatively useful and probably has done more in modern times than any other single new idea to alter the thinking of those interested in craniofacial growth.

CONCEPT 7: EFFECTS OF ORTHODONTIC TREATMENT ON THE MUSCULATURE

The orthodontist's appliances and therapy are not solely employed for improving positions of the teeth and altering skeletal relationships; orthodontic treatment also influences the neuromusculature. Orthodontic intervention should (1) obliterate all neuromuscular reflexes adversely affecting the dentition and the craniofacial skeleton and (2) create a favorable occlusal relationship that is repeatedly stabilized reflexly by unconscious swallowing. The clinician removes disharmonious occlusal influences and utilizes the primitive reflex positions of the mandible to stabilize the clinical result. Obversely, because it has been found that severe malocclusion provokes changes in the temporomandibular articulation and the neuromusculature, proper orthodontic treatment usually results in an alteration in the range of mandibular positions and an improvement in the precise control of mandibular movements. Other adaptive muscular changes that may follow orthodontic treatment include altered lip posture, tongue posture, mandibular posture, chewing stroke, and method of breathing.

Part 2

PRENATAL MATURATION

During prenatal life the human neuromuscular system matures unevenly. It is not accidental that the orofacial region matures (in the neurophysiologic sense) ahead of limb regions, because the mouth is the primary site of respiration, nursing, and protection of the oropharyngeal airway. In the human fetus, by about the eighth week, generalized uniform reflex movements of the entire body can be elicited by tactile stimulation. A few spontaneous movements, in response to as yet unidentified stimuli, have been observed as early as 9½ weeks. Localized specific and more peripheral responses can be produced before 11 weeks. At this time, stimulation of the nose-mouth region causes lateral body flexion. By 14 weeks, the movements have become much more individualized, and very delicate activities can be executed. When the mouth area is stimulated, general bodily movements no longer are seen; instead, facial and orbicular muscle responses are produced. Stimulating the lower lip, for example, causes the tongue to move. Stimulation of the upper lip causes the mouth to close and, often, deglutition to occur.

Respiratory movements of the chest and abdomen are first seen at about 16 weeks. The gag reflex has been demonstrated in the human fetus at about 18½ weeks (menstrual age). By 25 weeks, respiration is shallow but may support life for a few hours if established.

Stimulation of the mouth at 29 weeks has elicited suckling, although complete suckling and swallowing are not thought be developed until at least 32 weeks.

Davenport Hooker and Tryphena Humphrey have shown us that there is an orderly, sequential staging of events in prenatal orofacial neuromuscular maturation—a staging seen throughout the body, but which is much more advanced in the oropharyngeal region. All this has to be established by the time of birth in order for the child to survive (Humphrey, 1970).

NEONATAL ORAL FUNCTIONS

At birth, tactile acuity is much more highly developed in the lips and mouth than it is in the fingers. The infant carries objects to his mouth to aid in the perception of size and texture; later they go into the mouth as a part of teething. The neonate slobbers, drools, chews toes, sucks thumbs, and discovers that gurgling sounds can be made with the mouth.

Freudians consider all of this oral eroticism, as they do adult smoking; but in the infant surely it is also exploratory and exercises the most sensitive perceptual system in the body at that time. Oral functions in the neonate are guided primarily by local tactile stimuli, particularly those in the lips and the front part of the tongue.

The tongue at this age does not guide itself; rather, it follows superficial sensation. The posture of the neonate's tongue is between the gum pads, and it is often far enough forward to rest between the lips, where it can perform its role of sensory guidance more easily. The young infant, to a great extent, interprets the world with the mouth, and the integration of oral activities is therefore by sensory mechanisms.

If you touch a young child's lips or tongue and have him follow your finger, both the head and body turn. A little bit later the head turns separately from the body, and still later the mandible is moved without moving the head. It is only last of all that the neonate can follow with the tongue, while not moving the mandible. These stages appear in a natural sequence, just as teeth erupt on a kind of schedule.

The infant uses the mouth for many purposes. The perceptual functions of the mouth and face are combined with the sensory functions of taste, smell, and jaw position. The neonate's primary relationship to its environment is by means of the mouth, pharynx, and larynx. Here a high concentration of readily available receptors becomes stimulated and modulates the already matured brain stem coordinations that regulate respiration and nursing and determine head and neck positions during breathing and feeding.

The sensitivity of the tongue and lips is perhaps greater than that of any other body area. The sensory guidance for oral functioning, including jaw movements, is from a remarkably large area. These sensory inputs are compounded by many dual contacting surfaces, such as the tongue and lips, the soft palate and posterior pharyngeal wall, and the compartments of the temporomandibular articulation. A great array of sensory signals is required for the integration, coordination, and interpretation of this complex system.

Infantile Suckling and Swallowing

The effectiveness of these activities is a good indication of the neurologic maturation of premature infants. It has been found that a child will follow the same patterns in certain oral reflex movements years after initial learning. For example, a study was made of children whose records had been kept from infancy. As long as 9 years after weaning, if given a bottle from which to suckle, they produce the same suckling, swallowing, and respiratory rhythms they had when infants. If they swallowed in a suckle-suckle-swallow type of pattern, i.e., two suckles for one swallow, two-for-one, this same rhythm appeared years later. It may be a three-for-one or even a four-for-one ratio, but the pattern is maintained. Such primitive reflexes are difficult for us to change. How foolish it is for us, with our present ignorance about conditioning such basic mechanisms, to try to alter some of these reflexes. We must spend more time with those problems that we have at least a theoretical chance to condition.

Rhythmic elevation and lowering of the jaw provide sequential changes in positions of the tongue in coordination with its suckling contractions. The activities of suckling are closely related temporally to the motor functions of positional maintenance of the airway.

Electromyographic studies in our own laboratory have confirmed visual observations reported in England by a number of people, revealing that while the mandibular movements are carried out by the muscles of mastication, the mandible is primarily stabilized during the actual act of infantile swallowing by concomitant contractions of the tongue and the facial (rather than masticatory) muscles (Moyers, 1964). At the actual time of the infantile swallow, the tongue lies between the gum pads and in close approximation with the lingual surface of the lips. Thus, the infantile swallow is neuromuscularly a different mechanism from the mature swallow.

Characteristic features of the infantile swallow are that (1) the jaws are apart, with the tongue between the gum pads; (2) the mandible is stabilized primarily by contractions of the muscles of the seventh cranial nerve and the interposed tongue; and (3) swallowing is guided, and to a great extent controlled, by sensory interchange between the lips and the tongue.

Maintenance of the Airway

The oral-jaw musculature is responsible for the vital positional relationships that maintain the oral pharyngeal airway. While the infant is resting, a rather uniform diameter for the airway is provided by (1) maintaining the mandible anteroposteriorly and (2) stabilizing the tongue and posterior pharyngeal wall relationships.

The axial musculature around the vertebrae is also concerned. These primitive neonatal protective mechanisms provide the motor background upon which, with growth, all the postural mechanisms of the head and neck region are developed. Physiologic maintenance of the airway is of vital, continuing importance from the first day throughout life.

This little neonate who cannot focus his eyes, who cannot make a purposeful movement of his limbs, who cannot hold his head upright, who has absolutely no control of the lower end of the gastrointestinal tract has absolutely exquisite control of some functions in the orofacial regions. Why? Such control is necessary for survival!

Infant Cry

When the aroused baby is crying, the oral region is unresponsive to local stimulation. The mouth is held wide open, while the tongue is separated from the lower lip and from the palate. The steady stabilization of the size of the pharyngeal airway is given up during crying; and there are irregular, varying constrictions during expiration of the cry and large, reciprocal expansions during the alternating inspirations.

Gagging

Gagging, the reflex refusal to swallow or accept foreign objects in the throat, is an exaggeration of the protective reflexes guarding the airway and alimentary tract. The gag reflex is present at birth, but it changes as the child grows older in order to accommodate visual, acoustic, olfactory, and psychic stimuli that are remembered and thus condition it.

EARLY POSTNATAL DEVELOPMENT OF ORAL NEUROMUSCULAR FUNCTIONS

Mastication

The interaction between the rapidly and differentially growing craniofacial skeleton and the maturing neuromuscular system brings about sequentially progressive modifications of the elementary oral functions seen in the neonate (Moyers, 1964, 1965, 1988). Mandibular growth, downward and forward, is greater during this time than midfacial growth and is associated with a greater separation of the thyroid bone and thyroid cartilage from the cranial base and mandible.

Maturation of the musculature and delineation of the temporomandibular joint help provide a more stable mandible. Although mandibular growth carries the tongue away from the palate and helps provide differential enlargement of the pharynx, patency of the airway is maintained—a most important point.

The soft palate and the tongue are commonly held in apposition, but as the tongue is no longer lowered by mandibular growth, its functional relationship with the lips is altered, an alteration aided by the vertical development of the alveolar process. So the morphologic relationship of the tongue and lips is strained. At rest now, the tongue is no longer in generalized apposition with the lips, buccal wall, and soft palate. The lips elongate and become more selectively mobile; the tongue develops discrete movements that are separate from lip and mandibular movements. The labial valve mechanism is constantly maintained during rest and feeding so that food is not lost.

The development of speech and mastication as well as facial expression requires a furthering of the independent mobility of the separate parts. In the neonate, however, the lips tightly surround a plunger-like tongue, moving in synchrony with gross mandibular movements. Speech, facial expression, and mastication require the development of new motor patterns as well as greater autonomy of the motor elements. Not all the developmental aspects of these functions are known. But mastication certainly does not gradually develop from the infantile nursing. Rather, it seems that the maturation of the central nervous system permits completely new functions to develop. These functions are triggered to an important extent by the eruption of the teeth.

One of the most important factors in the maturation of mastication is the sensory aspect of newly arriving teeth. The muscles controlling mandibular position are cued by the first occlusal contacts of the antagonistic incisors. Serial electromyographic studies at frequent intervals during the arrival of the incisors have demonstrated conclusively that the very instant the maxillary and mandibular incisors accidentally touch one another, the jaw musculature begins to learn to function in accommodation to the arrival of the teeth (Moyers, 1964).

Thus, since the incisors arrive first, the closure pattern becomes more precise anteroposteriorly before it does mediolaterally. All occlusal functions are learned in stages. The central nervous system and the orofacial and jaw musculature mature concomitantly, and usually synchronously, with the development of the jaws and dentition.

The earliest chewing movements are irregular and poorly coordinated, like those during the early stages of the learning of any motor skills. As the primary dentition is completed, the chewing cycle becomes more stabilized, using more efficiently the individual's pattern of occlusal intercuspation. In the very young child, sensory guidance for masticatory movement is provided by the receptors in the temporomandibular articulation, the periodontal membrane, the tongue, and the oral mucosa and muscles; of these, it seems by far that the most important are those of the temporomandibular articulations, and next those of the periodontal membrane. Cuspal height, cuspal angle, and incisal guidance (which is usually minimal in the primary dentition) play a role in the establishment of chewing patterns in the infant. However, condylar guidance is not important at this age, since the eminentia articularis is ill-defined and the temporal fossae are shallow. Rather, it may be supposed that the bone of the eminentia articularis forms where temporomandibular function permits (or causes) it to develop. In a similar fashion, the plane of occlusion is established by the growth of the alveolar process, during eruption of the teeth, to heights permitted by the configuration and functioning of the neuromusculature.

The individual's movements during the chewing cycle are a developed, integrated pattern of many functional elements. In the young child, at the time of completion of the primary dentition, masticatory relationships are nearly ideal, since all three systems (bone, teeth, and muscle) still show the adaptability characteristic of development. Cusp height and overbite in the primary dentition are more shallow, bone growth more rapid and adaptive, and neuromuscular learning more easily cued because pathways and patterns of activity are not yet well established. Adaptations to masticatory change are much more difficult in later years, as every dentist knows.

Facial Expression

In a not dissimilar way, most subtle facial expressions are learned, largely by imitation, so we think, and begin about the time the primitive uses of the seventh nerve musculature for infantile swallowing are abandoned. Those of us who are parents imagine all sorts of facial expressions in the young neonate. Actually, observing the infant objectively, we must admit that the expression is often rather blank. The reason is that the facial muscles are busy being used for the massive efforts of mandibular stabilization necessary during infantile swallowing. Eventually the mandible becomes controlled and stabilized more by the muscles of mastication, particularly during unconscious reflex swallowing, and the delicate muscles of the seventh cranial nerve become truly "muscles of facial expression."

Although many facial expressions are learned through imitation, some facial responses are not learned and can be traced back to reflexes of earlier primates. Similar facial displays have evolved in the four lines of modern primates in which monkey-like forms have developed. Comparative studies have been made revealing similar reflex expressions of basic protective anger, for example, in various primates—the same primitive instinctive expressions you have seen on your best friend.

Speech

Purposeful speech is different from the reflex infant cry. Infant crying is associated with irregular tongue and mandibular positions related to sporadic inspirations and expirations. Speech, on the other hand, is performed on a background of stabilized and learned positions of the mandible, pharynx, and tongue. The infant cry is usually a simple displacement of parts, accompanied by a single explosive emission, whereas speech can only be carried out by polyphasic and sequential motor activities synchronized closely with breathing. Speech is regular; the infant cry is sporadic. Speech requires complicated, sophisticated, varying sensory conditioning elements during learning; the infant cry is primitive and not learned.

Speech consists of four parts: (1) language—the knowledge of words used in communicating ideas; (2) voice—sound produced by air passing between the vibrating vocal cords of the larynx; (3) articulation—the movement of the speech organs used in producing a sound, that is, the lips, tongue, teeth, mandible, palate, and so forth; (4) rhythm—variations of quality, length, timing, and stress of a sound, word, phrase, or sentence. If there is no impairment of hearing, sight, or oral sensation, the child will learn to speak from the speech that is heard. Speech defects are a loss or disturbance of language, voice, articulation, and rhythm or combinations of such losses and disturbances.

Mature Swallow

During the latter half of the first year of life, several maturational events usually occur that alter markedly the orofacial musculature's functioning. The arrival of the incisors cues the more precise opening and closing movements of the mandible, compels a more retracted tongue posture, and initiates the learning of mastication. As soon as bilateral posterior occlusion is established (usually with the eruption of the first primary molars), true chewing motions are seen to start, and the learning of the mature swallow begins. Gradually, the fifth cranial nerve muscles assume the role of muscular stabilization during swallowing, and the muscles of facial expression abandon the crude infantile function of suckling and the infantile swallow and then begin to learn the more delicate and complicated functions of speech and facial expressions. The transition from infantile to mature swallowing takes place over several months, aided by maturation of neuromuscular elements, the appearance of upright head posture, and hence a change in the direction of gravitational forces on the mandible, the instinctive desire to chew, the necessary ability to handle textured food, dentitional development, and so forth. Many children achieve features of the mature swallow at 12 to 15 months, but there is a great variability. Characteristic features of the mature swallow are as follows: (1) the teeth are together (although they may be apart with a liquid bolus); (2) the mandible is stabilized by contractions of the fifth cranial nerve muscles; (3) the tongue tip is held against the palate above and behind the incisors; and (4) minimal contractions of the lips are seen during swallowing.

Neural Regulation of Jaw Positions

Jaw position, like a number of other automatic-somatic activities, normally is largely reflexively controlled, even though it can be altered voluntarily. A surprising number of jaw functions are carried out at the subconscious level, even though conscious control is possible and sometimes necessary. Receptors in the temporomandibular capsule area are far more important than previously thought.

Since more research on the neurophysiologic regulation of jaw position and function has been done on the adult, there has been a tendency to transfer prosthodontically oriented concepts, based on sound adult clinical practice, to children. Our knowledge about the developmental aspects of orofacial and jaw neurophysiology is incomplete at this time, although much research is under way. We must remember that many of our attitudes are victims of our experience with degenerating occlusions in adults, and the critical clinical factors that apply under those circumstances may not be present in the child or may have different relative significance during development.

Unconditioned jaw positions and functions include mandibular posture for the maintenance of the airway and unconscious or reflex swallowing. The neural mechanisms that determine mandibular posture are important to the dentist, because mandibular posture (sometimes in dentistry called the rest position) is a determinant of the vertical dimension of the face. In the opinion of many, the position of the mandible during unconscious swallowing is an important factor in occlusal homeostasis, because every time a person swallows unconsciously, the occlusal relationship is stabilized or, because of tooth interferences, shifts interfering teeth by lower jaw movement until a stable occlusal relationship finally is obtained.

Conditioned jaw positions and functions include all those of mastication, the mature swallow, and speech and most of facial expression.

OCCLUSAL HOMEOSTASIS

Occlusal stability at any moment is the result of the sum of all forces acting against the teeth. Some of these forces have been measured in research, but it is not yet possible to describe precisely in summation all the forces and counterforces that produce occlusal homeostasis. Occlusal homeostasis is dependent upon elaborate and sophisticated sensory feedback mechanisms from the periodontal membrane, temporomandibular joint, and other parts of the masticatory system. Such sensory feedback serves as a regulating mechanism helping to determine the strength and nature of muscle contractions. Each individual tooth is positioned between contracting sets of muscles. It is also in contact with adjacent teeth and in occlusion with the teeth of the opposite arch. A number of physiologic forces determine the tooth's position occlusally, including eruption, the occlusal force during swallowing, the forces of mastication, occlusal wear of the crown of the tooth, and so forth. Occlusal interferences in or near the unconscious swallowing position of the mandible tend to diminish reflexively the force of muscle contractions during swallowing. Because reflex swallowing occurs so frequently, it plays an important role in occlusal homeostasis. Other factors involved in occlusal homeostasis include the natural mesial drifting tendencies of the teeth, the anterior component of force, the growth of bones of the craniofacial complex, and alveolar bone growth and remodeling. It is now believed that the neuromuscular mechanisms and bone growth factors are far more important in the nature of occlusal relations than are the oft-mentioned factors of cuspal inclination, cusp height, condylar guidance, and so on. The occlusal relationships are now generally held to be nowhere near as stable as depicted in some dental textbooks, if for no other reason than that occlusal adaptations must occur constantly to accommodate, in their way, changes in the neuromusculature and the craniofacial skeleton. Occlusal homeostasis is achieved and maintained in a complex system of responses and adaptations in several tissue systems.

EFFECT OF NEUROMUSCULAR FUNCTION ON FACIAL GROWTH

From the earliest periods of embryonic growth, an intimate functional relationship exists between muscles and the bones to which they are attached. Obviously, as the bones grow, the muscles must also change their size. Therefore, a relationship exists between the overall growth of any bone and the muscles attached to that bone; and adjustments between muscle and bone are a normal part of growth and development. During growth, muscles must migrate to occupy relatively different positions with time. As the skeleton grows, there is a constant adjustment of the attachment relationships between muscle and skeleton.

Functional use and disuse determine to some extent the thickness of the cortical plate of limb bones. However, the relationship of muscle function and bone form and growth in the craniofacial skeleton is much more difficult to assess. Certain parts of some of the facial bones are very dependent on function—for example, the alveolar process around the roots of the teeth and the coronoid process to which the temporal muscle is attached. In a more general way, the conformation of the bone and the craniofacial relationships are determined by such factors as mouth breathing, excessive masticatory function, and so forth. In the case of the calvaria, cranial base, and nasomaxillary complex, functional features other than those of muscle apparently play an important role in development and growth—namely, the growth of the brain, the eyeballs, cartilage growth, and so on.

The mandible, with its important condylar cartilage, holds a special interest for dentists, particularly orthodontists. Although there is general agreement that variations in muscle function affect markedly the areas of muscle attachment and that the development and use of the dentition affect the alveolar process, there is some dispute over whether or not muscle function can have a more general effect on the size and form of the mandible. The point is a very important one for orthodontists treating Class II malocclusions in children who are still growing.

Although the evidence is still not complete, most workers now believe that function plays a more dominant role in the determination of mandibular size and conformation than was previously thought. For example, extensive experimental research by Petrovic and associates and by Carlson and associates has shown that the masseter and lateral pterygoid muscles may play a major role in the growth of the mandibular condylar cartilage. It remains unclear, however, whether this effect is a direct one, or whether muscle function influences condylar growth simply by alteration of the biomechanical environment.

EFFECTS OF ORTHODONTIC TREATMENT ON THE MUSCULATURE

It is known that severe malocclusion is often associated with pathologic changes in the temporomandibular articulations, which in turn impair the sensory receptors within the joints, causing such orthodontic patients to have a less precise ability to determine mandibular position than persons who have normal occlusion. After malocclusions have been treated orthodontically, there is a significant change in the range of mandibular movements and an improvement in the precision of the determination of mandibular positions. Occlusal equilibration on treated orthodontic patients has been shown to change significantly teeth-apart swallowing to teeth-together swallowing (Moyers, 1988). Thus, orthodontic treatment, including occlusal equilibration, conditions swallowing reflexes, which in turn help stabilize the orthodontic occlusal result. Occlusal disharmonies, at the end of orthodontic treatment, have been shown to be disruptive to the stability of treated orthodontic occlusions and thus an important cause of relapse in treated malocclusions. Other adaptive muscular changes following orthodontic therapy may include an altered lip posture, tongue posture, mandibular posture, chewing stroke, and method of breathing.

Suggested Readings

Bosma, J. F.: Fourth Symposium on Oral Sensation and Perception. DHEW Publication No. (NIH) 73-546. U.S. Department of Health, Education, and Welfare, National Institutes of Health, Bethesda, Maryland, 1973.

Carlson, D. S.: Craniofacial biology as normal science. In: *New Vistas in Orthodontics.* Ed. by L. E. Johnston. Philadelphia, Lea & Febiger, 1985.

Carlson, D. S., and J. A. McNamara, Jr. (Eds.): *Clinical Alteration of the Growing Face.* Monograph 8, Craniofacial Growth Series. Ann Arbor, University of Michigan, Center for Human Growth and Development, 1978.

Carlson, D. S., J. A. McNamara, Jr., L. W. Graber, et al.: Experimental studies of the growth and adaptation of the temporomandibular joint. In: *Current Concepts in Oral Surgery.* Ed. by W. G. Irby. St. Louis, C. V. Mosby, 1980.

Carlson, D. S., J. A. McNamara, Jr., and K. A. Ribbens (Eds.): *Developmental Aspects of Temporomandibular Joint Disorders.* Monograph 16, Craniofacial Growth Series. Ann Arbor, University of Michigan, Center for Human Growth and Development, 1985.

McNamara, J. A., Jr. (Ed.): *Control Mechanisms in Craniofacial Growth.* Monograph 3, Craniofacial Growth Series. Ann Arbor, University of Michigan, Center for Human Growth and Development, 1975.

Moss, M. L.: Genetics, epigenetics, and causation. Am. J. Orthod., 80:366, 1981.

Moyers, R. E.: The development of occlusion and temporomandibular joint disorders. In: *Developmental Aspects of Temporomandibular Joint Disorders.* Carlson, D. S., J. A. McNamara, Jr., and K. A. Ribbens (Eds.): Monograph 16, Craniofacial Growth Series. Ann Arbor, University of Michigan, Center for Human Growth and Development, 1985.

Moyers, R. E.: Analysis of the orofacial and jaw musculature. In: *Handbook of Orthodontics.* Chicago, Year Book Medical Publishers, 1988.

Moyers, R. E., and D. H. Enlow: Growth of the craniofacial skeleton. In: *Handbook of Orthodontics.* Ed. by R. Moyers. Chicago, Year Book Medical Publisher, 1988.

Petrovic, A. G.: Experimental and cybernetic approaches in the mechanisms of action of functional appliances in mandibular growth. In: *Malocclusion and the Periodontium.* Ed. by J. A. McNamara, Jr., and K. A. Ribbens. Monograph 15, Craniofacial Growth Series. Ann Arbor, University of Michigan, Center for Human Growth and Development, 1984.

Storey, A. T.: Maturation of the orofacial musculature. In: *Handbook of Orthodontics.* Ed. by R. Moyers. Chicago, Year Book Medical Publishers, 1988.

CHAPTER ELEVEN

Masticatory Function and Facial Growth and Development

Ordean J. Oyen, Ph.D.

The purpose of this chapter is to present a way of thinking about and making sense of changes that occur in the human facial skeleton. Although emphasis is placed upon the role played by mastication, the effect of a variety of factors on the facial skeleton are taken into consideration. In this discussion some basic notions about jaw biomechanics are reviewed and evaluated. These notions are then applied in assessments of the structural-functional aspects of the facial skeleton at various stages of growth. Natural as well as induced changes in the face and masticatory apparatus are considered.

The reasoning behind this chapter is as follows. We know a lot about **how** the craniofacial skeleton grows. We also are quite knowledgeable about growth and function of the facial muscles. Much remains to be learned, however, about the **relationships** among the bones and muscles of the face and masticatory function, especially when changes take place. In addition to our need for more factual information, we must also become more thoughtfully critical when making use of whatever information is available. Now seems to be a good time to strongly advocate the idea that a thorough understanding of the face will be forthcoming only if we treat it as a **dynamic** phenomenon whose structural-functional significance at any given time is a transient expression of **past** and **future** as well as present conditions. This view stands in contrast to more conventional notions in which the face is largely accounted for in terms of temporally immediate needs and demands. This view also calls to attention some of the problems that are involved when the face is described and analyzed in terms of "maturation" (see Chapter 10). In the view that is being advocated, there is really no single point or phase to which facial growth is directed, e.g., "maturity," when structural-functional changes diminish to quiescent insignificance. On the contrary, it will be shown that with regard to the masticatory system, flux is virtually constant, and the face is continuously responding to these fluctuations. Building on insights provided by Enlow, Hylander, Moss, Moyers, and others, this chapter captures the dynamic nature of craniofacial morphogenesis and provides a sense of the part played by mastication in this phenomenon.

PRINCIPLES OF MASTICATORY BIOMECHANICS*

The Jaw as a Lever System

There is an overwhelming body of evidence showing how the jaw functions as a lever. In this system, during symmetrical jaw action, such as incisal **biting,** the left and right temporomandibular joints function as a single fulcrum, just as the widely spaced hinges on my office door collectively serve as a single fulcrum because they share a common axis of rotation (Figs. 11–1 and 11–2).

During asymmetrical jaw actions, such as normal **chewing,** more complex movements occur. Nevertheless, the mandible still functions as a lever. In this instance, however, the jaw rotates around the functional axis (fulcrum) formed by the jaw joint on the balancing, i.e., unloaded, side of the mandible, and the food bolus on the working, i.e., loaded, side of the jaw (Fig. 11–3).

---

*The essence of this narrative is based on material by Hylander and especially A. C. Walker. The reader is strongly encouraged to refer directly to this material, which is identified in the References.

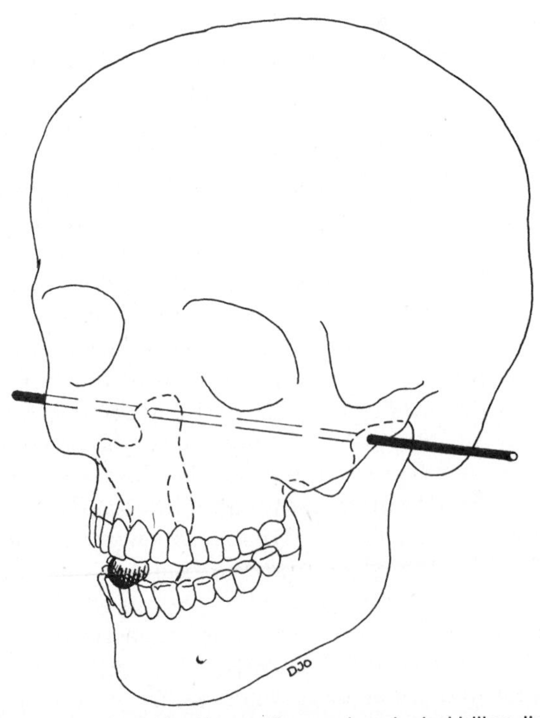

FIGURE 11–1. During normal symmetrical jaw action, such as incisal biting, the separate jaw joints act as a single fulcrum. Forces generated by the jaw elevator muscles are directed mostly into the object being bitten. Smaller "reaction forces" are **symmetrically** directed posteriorly along the left and right sides of the mandible to the mandibular condyle, disk, and articular eminence of each jaw joint. The distribution of forces, especially with regard to the jaw joint, alters significantly during asymmetrical loading.

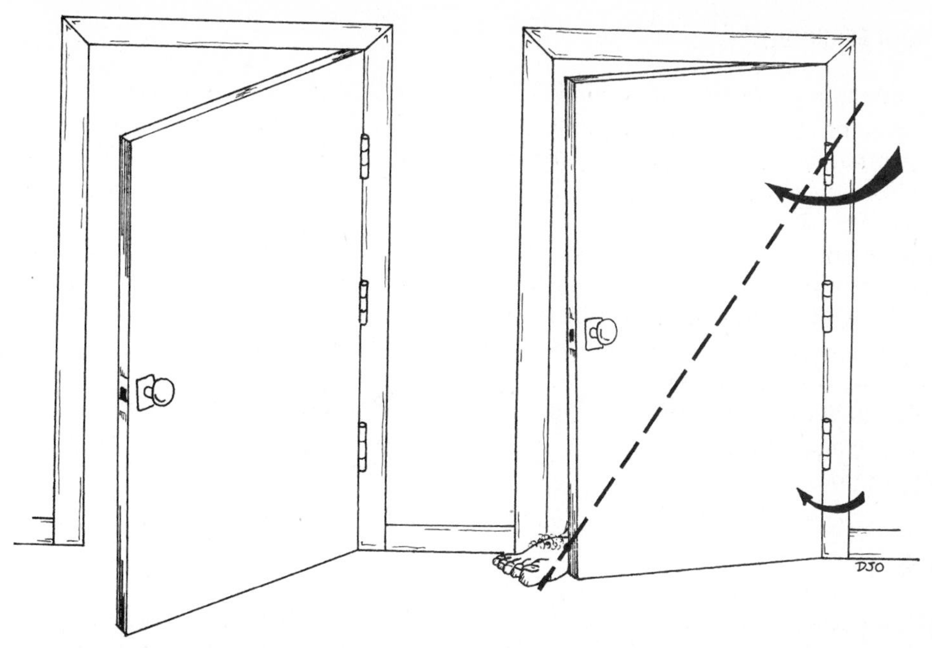

FIGURE 11–2. The multiple hinges of a door function as a single fulcrum because they share a common axis of rotation. When forces are applied to the door in the region of its doorknob under normal circumstances, those forces are dissipated through the door to its hinges more or less symmetrically; and the door moves, rotating on its hinges. If an obstruction is placed in the path of the door near its lower free edge, the position usually taken by the proverbial "foot in the door," and force continues to be applied to the door, those forces will be distributed asymmetrically throughout the door as well as to its hinges. The door, its frame, and its hinges must be so constructed as to allow normal movement and to resist a range of forces that vary in both pattern and magnitude—and so must the jaw and the components of its joints be constructed.

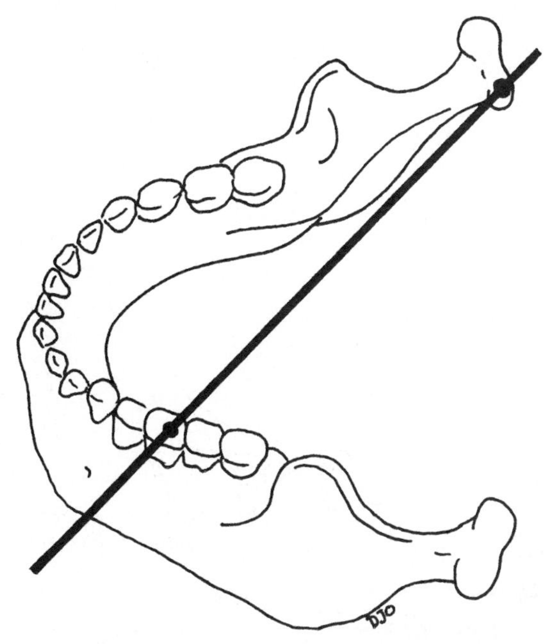

FIGURE 11–3. When the jaw is loaded asymmetrically, as during unilateral chewing, the left and right jaw joints do not function in unison as a single fulcrum. Rather, under these circumstances the functional axis of rotation of the mandible is formed by two points, the food bolus being bitten on the "working side" and the jaw joint on the contralateral "balancing" side. Forces generated by the jaw muscles in this instance are dissipated primarily by the food bolus and secondarily by the balancing side joint; little, if any, force is directed to the jaw joint on the working side.

Mechanical Efficiency

The human jaw functions as a **third order** lever during elevation, regardless of the degree of symmetry of the action involved. This is so because the lines of action of the jaw elevator muscles fall between the fulcrum and the point of force application, i.e., the "bite point." Consequently, in this system the **load arm length,** i.e., the distance from the fulcrum to the bite point, exceeds the **lever arm length,** i.e., the distance from the fulcrum to the line of action of a given jaw elevator muscle, drawn perpendicular to the line of action of that muscle. The **simple mechanical efficiency** of the jaw system can thus be expressed in terms of load arm and lever arm lengths relative to each other: the longer the load arm relative to the lever arm, the less efficient the system; conversely, the longer the lever arm, the more efficient the system; **any** change in load arm length will directly affect the ability to generate bite forces (Figs. 11–4 and 11–5).

Bite Force

The amount of bite force a person can generate is really the product of two basic factors, simple mechanical efficiency of the jaw system and size (cross-sectional area) of the jaw elevator muscles. Thus, in a jaw system with relatively low mechanical efficiency, relatively larger musculature will be required to

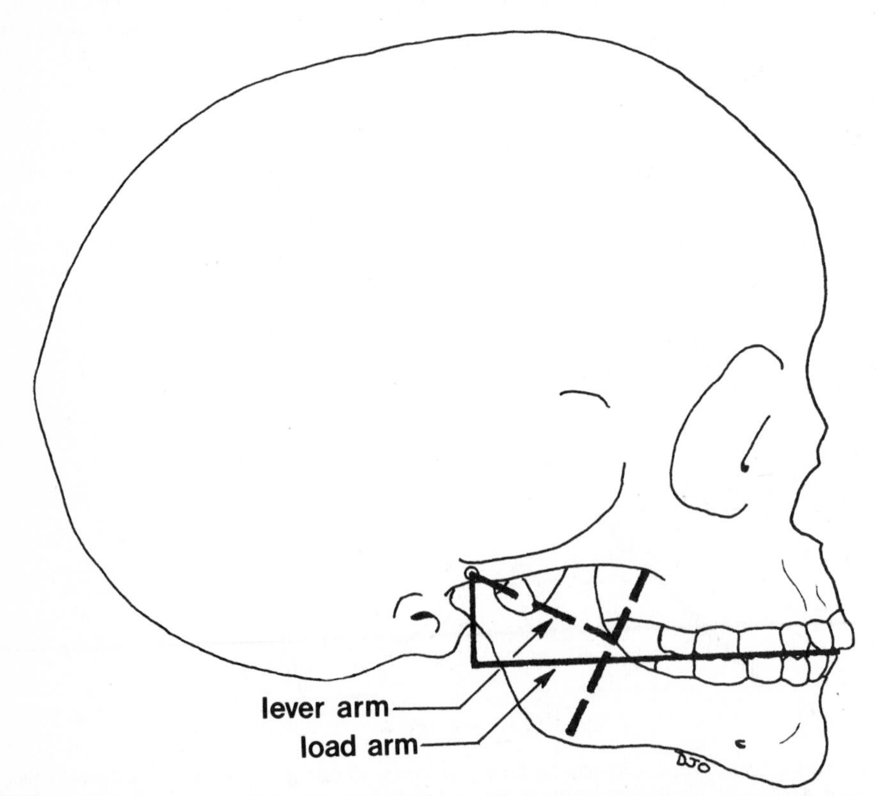

FIGURE 11–4. The simple mechanical efficiency of a masticatory system can be expressed in terms of the relative lengths of the lever and load arms. In a child, for example, the masseteric lever arm length is approximately equal to the load arm length measured to the most posterior maxillary tooth, but this lever is only about half the length of the load arm measured to the incisors. This means that on the basis of these relative lengths alone, the amount of force generated at the incisors would be only about half that generated at the molars.

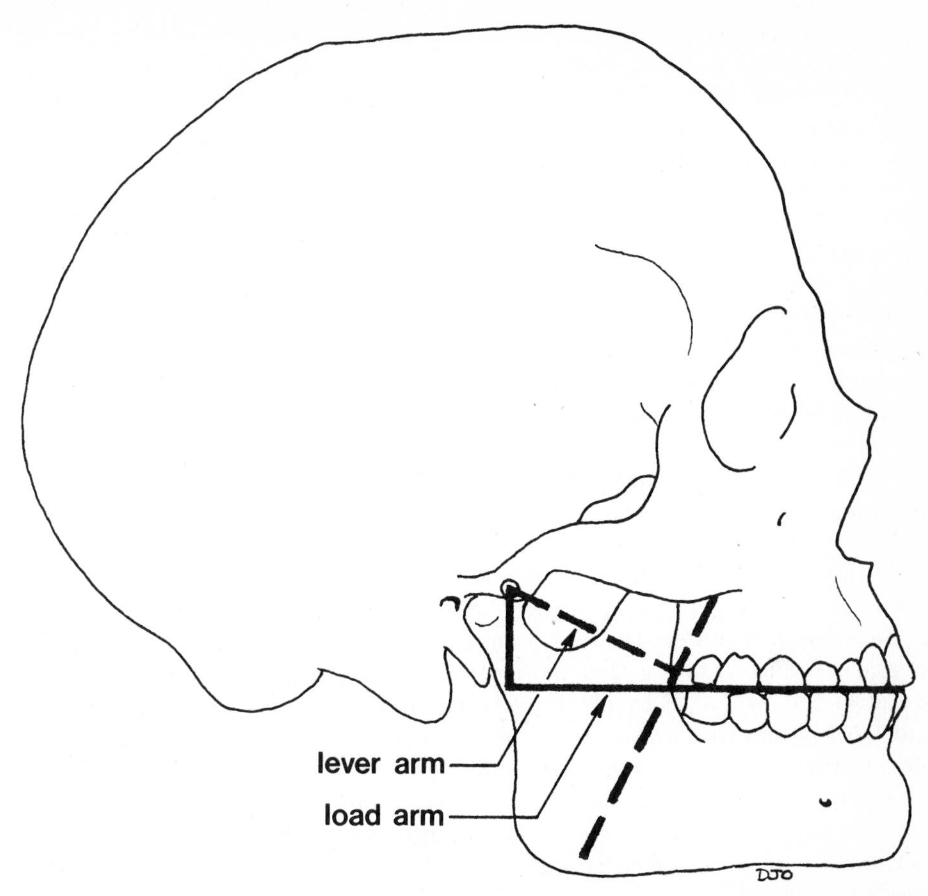

FIGURE 11–5. The efficiency of any masticatory system will change as alterations take place in the system so long as those alterations affect masticatory lever or load arm lengths. Changes in efficiency that accompany growth do not necessarily herald improvements or deterioration in mechanical efficiency relative to some previous state. While human beings usually become more prognathous as the face grows, a reduction in the ability to efficiently generate bite forces does not necessarily ensue. For example, in an adult the masseteric load arm length measured at the second permanent molar can commonly be about half the masseteric load arm length, measured at the incisors, as it is in a child. Moreover, because the second and third molars experience relatively little anterior displacement (mesial drift) after their attainment of occlusion, adults often show a relatively more efficient, shortened load arm length than is seen in the child.

produce the level of force that could be generated by a more efficient system with relatively smaller muscles. If the size of the musculature is held constant and the mechanical efficiency is varied by altering the length of the load arm, the resultant bite force will fluctuate in accordance with the changes in mechanical efficiency. This latter instance accounts for how one can exert more force with the molars than with the incisors.

Changes in mechanical efficiency or in cross-sectional area of the jaw elevator muscles directly affect the bite force. If changes in either of these parameters are of sufficient magnitude, it may be necessary for compensatory alterations to take place elsewhere in the craniofacial musculoskeletal system for masticatory efficiency or structural-functional integrity to be maintained.

Effects of Actions of the Jaw Elevator Muscles

In this probe into the relationships between masticatory function and craniofacial anatomy, it is important to take into consideration in the broadest sense the effects of contraction of the jaw elevator muscles on the mandible into which these muscles insert and on the bones from which these muscles originate. With this information in hand, we can then more intelligently speculate about the effects of jaw use on facial growth.

The basic **actions** of the mandible that are brought about as a consequence of contraction of the jaw (elevator) muscles are shown in Figure 11–6. During the symmetrical loading, the lateral pterygoid muscles propel the mandible and the articular disk forward and work to maintain contact between the condyles and the articular eminence, thereby ensuring a stable axis of rotation (fulcrum) for the jaw joint. In concert, the masseter, medial pterygoid, and temporalis muscles cause the mandible to elevate as it rotates at its fulcrum. The rotation-elevation action eventually produces a bite force at the point of resistance, i.e., the food bolus at the incisors, as well as a "joint reaction force" at the articular eminence.

Conventional wisdom holds that forces produced at the tooth row during mastication are absorbed along three "stress trajectories," the canine, maxillary, and pterygoid pillars of the face and cranium, and along a separate set of trajectories in the mandible. The notion of these trajectories has prevailed essentially unchanged since they were first articulated more than 50 years ago by Sicher and Tandler. However, there is now an emerging body of experimental evidence showing that most of the forces generated at the tooth row are absorbed primarily by the food being masticated and secondarily by soft and hard tissues that make up the tooth–peridontal membrane–alveolus complex. Consequently, only relatively small forces are actually conducted from the dentition to the facial skeleton. Moreover, there are indications that other forces acting on the facial skeleton as a consequence of contraction of the jaw elevator muscles (described below) may be of much greater significance than

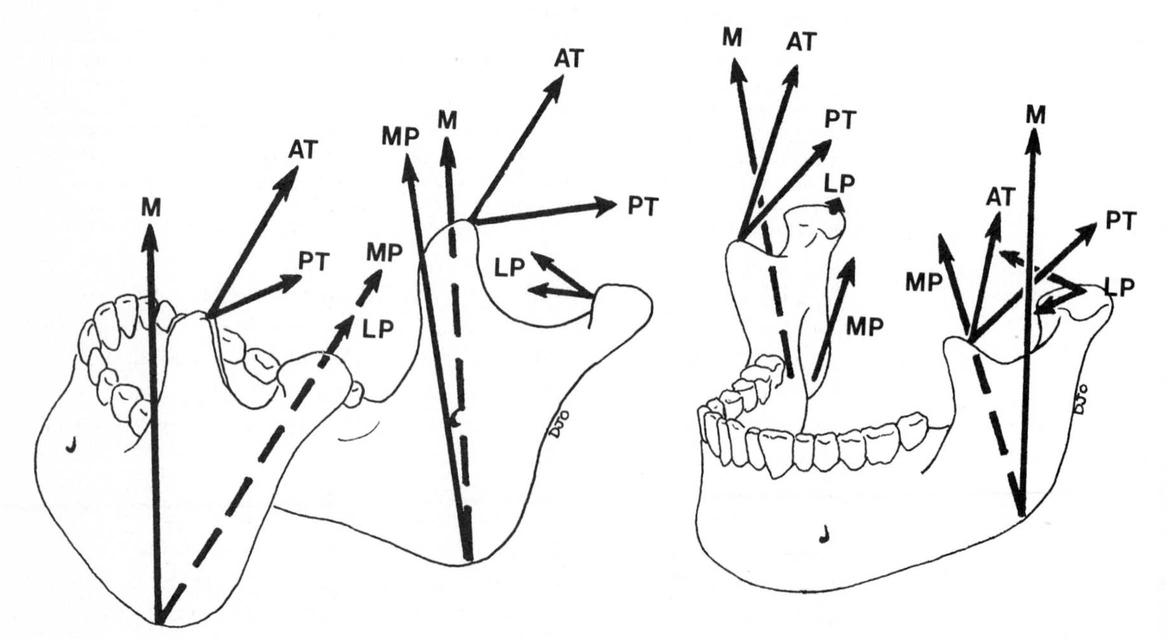

FIGURE 11–6. Contraction of the masseter, medial pterygoid, and temporalis muscles basically cause the mandible to rotate around the axis formed by its bilateral jaw joint and, thereby, to become elevated. The lateral pterygoid muscles act to modify these elevational/rotational actions by affecting the position of the mandible in anterior-posterior terms. The basic directions in which the mandible is caused to move as a consequence of contraction of these muscles, depicted from two different views, result from the lines of action that extend from the origins to the insertions of each muscle, respectively. *M*, masseter; *AT*, anterior temporalis; *PT*, posterior temporalis; *MP*, medial pterygoid; *LP*, lateral pterygoid.

the dentally transmitted forces (Oyen, 1989a and 1989b). Although these findings are subject to further experimental verification, they are consistent with the dynamics of a truly effective masticatory system in which forces generated at the occlusal surface are directed into and absorbed by the food bolus being masticated, rather than being ineffectively directed elsewhere, such as occurs in the Sicher model of pillars and buttresses. In this view, the bite forces are dissipated within the immediate circumdental region. Therefore, the only time in which the traditional buttress notion would apply would be during unopposed occlusion, such as bruxism or swallowing. In that the former behavior is abnormal and therefore relatively infrequent in its occurrence, and the latter involves relatively low forces, it seems unlikely that either of these plays a major role in normal craniofacial skeletal morphogenesis.

The more immediate and obvious effects of the elevator muscles during mastication are those just described. There are other effects, each of which becomes intensified as the bite forces increase, that must also be recognized. Some of these effects are a consequence of spatial arrangements of the jaw muscles relative to their points of origin and insertion. For example, in addition to being superior to its area of insertion on the angle of the mandible, the origins of each masseter muscle are also lateral and anterior. Consequently, when contracting, each masseter will produce a force that will want to move the angle of the mandible laterally. Because of its origins and insertions, the medial pterygoid muscle produces forces that would torque the angle of the mandible medially while elevating the jaw. In the same fashion, the lateral pterygoid muscles produce forces that draw the condylar head medially as well as anteriorly; the anterior component of the temporalis draws the coronoid process somewhat laterally while elevating it; and the posterior temporalis draws the mandible posteriorly while contributing to its elevational rotation around the fulcrum of the temporomandibular joint (Fig. 11–7).

Just as forces produced by the jaw muscles have an effect on the mandible, they also must have an effect on the bones from which they originate, especially when meeting resistance. In contrast to the knowledge we have about the effects of the jaw muscles on the actions of the mandible, surprisingly little has been written about effects brought about by the cranial attachment of these muscles other than observations concerning the formation of lines and crests along the region of musculature origin. If the muscles act in the manner that has just been described for the mandible, then it logically follows that the contraction of the masseter must pull the zygoma in a medial, inferior direction; the medial pterygoid must pull the medial aspect of the lateral pterygoid plate and the maxillary tuberosity in a lateral-inferior direction; and the lateral pterygoids must draw the roof of the infratemporal fossa and the lateral aspect of the pterygoid plates in a posterior direction. It follows, too, that the posterior temporalis must draw bones of the vault to which it is attached in an anterior, inferior, and somewhat lateral direction; the anterior temporalis must exert forces that draw the parietofrontal aspect of the vault inferiorly and laterally; and the fibers that attach to the zygoma in the anterior of the infratemporal fossa must draw this bone in a posterior, inferior direction (Fig. 11–8). Some of the consequences of the effects of these muscle actions on the growing skull and mandible are considered below.

GROWTH RESPONSES

The actions and effects produced by use of the orofacial muscles vary over an individual's life span. What follows is an overview of craniofacial form and

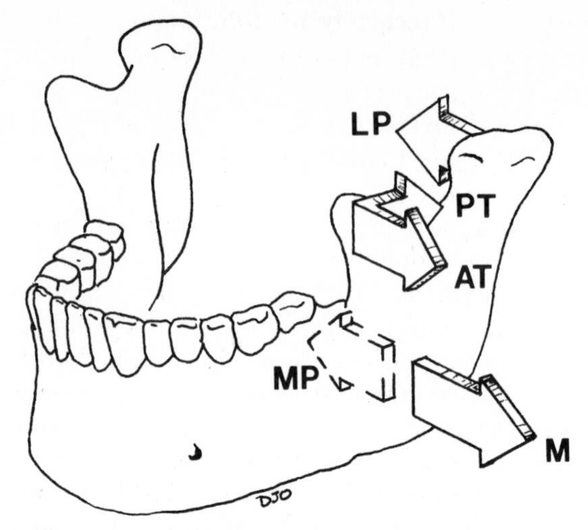

FIGURE 11–7. The origins of the masseter are lateral and anterior, as well as superior, to its insertions on the mandible. Consequently, contraction of this muscle will tend to move the mandible laterally and anteriorly as well as superiorly. Because the mandible is fused and reinforced anteriorly and functions as a single unit, lateral displacement of the jaw by contraction of the masseter on one side is balanced by simultaneous contraction of the contralateral masseter. Simultaneous contraction of the masseters in this fashion will have met the theoretical effect of flexing the sides of the mandible **outward**—i.e., bilateral lateral displacement. Forces acting in the region of the chin as a consequence of the lateral component of masseteric contraction are not countered solely by bony reinforcement in this area. Because of their positions relative to the mandible, the medial and lateral pterygoids both exert medially directed forces to the mandible. Bilateral contraction of these muscles would cause the sides of the mandible to flex **inward** unless countered by other musculoskeletal adaptations. *AT*, anterior temporalis; *LP*, lateral pterygoid; *M*, masseter; *MP*, medial pterygoid; *PT*, posterior temporalis.

function at different phases of life, couched in terms of these actions and effects. While an emphasis is placed on the role played by the jaw elevators, this account also includes consideration of nonmasticatory factors. As will be seen, the overall picture that emerges is one in which the face is continuously changing, passing from one growth phase into another in a virtually seamless fashion in response to a variety of stimuli that also seem to be perpetually changing. The phases to be considered extend from the late fetal stage through partially edentulous adulthood.

Third Trimester Fetus

In the third trimester (weeks 25 to 36) the human face is fully formed and easily recognizable. Because birth is imminent, these last few weeks *in utero* are marked by the attainment of **functional maturation** of those structures that will be relied upon to meet the altered needs that will prevail after birth. Thus, the anatomy of the fetal head is a compromise among the constraints imposed by the uterine environment and birth canal, growth of the postcranial structures, and the specific functional needs served by the head that must be met at birth, which include breathing and eating.

The dominant feature of the fetus is its head, the largest components of which are the cranial vault and the enclosed brain. Shortly before birth, the brain has attained some 20 to 25 per cent of its adult size, and the cranial vault at this age is about 10 times larger than the fetal face. (By adulthood the cranial vault will be only twice as large as the face.) In contrast to the precocious

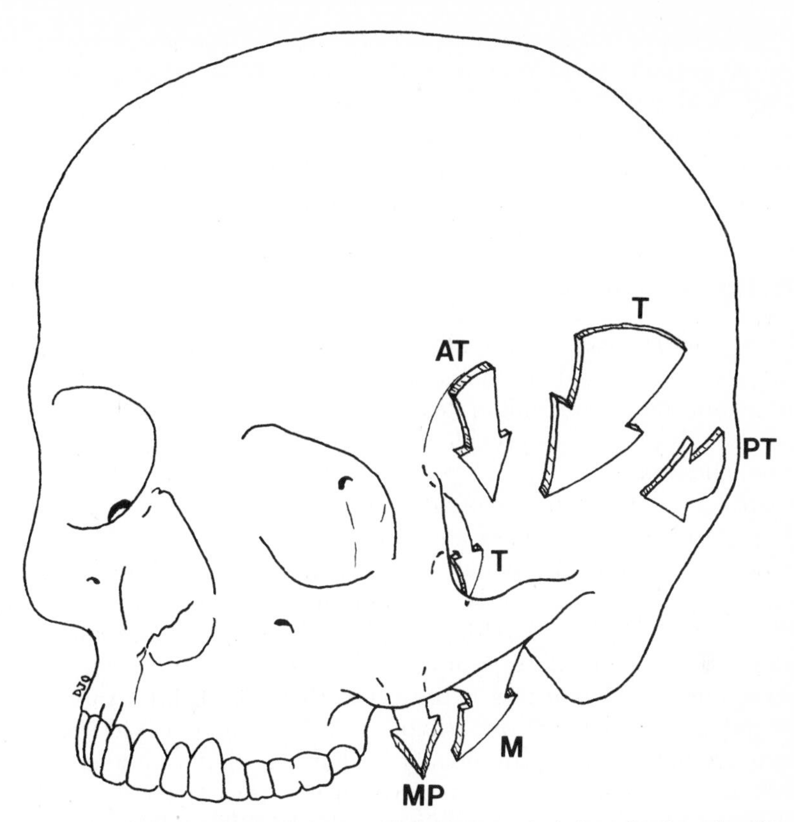

FIGURE 11–8. In addition to causing the mandible to rise up and close, the jaw elevator muscles also have effects on the bones from which they originate. One of these effects is the attempt to cause displacement of the bony area to which the muscle is attached. The direction any such potential displacement will follow must reflect the positions of the origins and insertions of the muscles in question relative to each other. Thus, contraction of the masseter will try to draw the zygomatic arch medially and posteriorly, and contraction of the temporalis will try to draw segments of the zygomatic, frontal, sphenoid, and parietal bones laterally and inferiorly, and so on, while these contractions simultaneously elevate the mandible. The specific effects of these displacement efforts on the cranium and face have never been fully analyzed and documented. *M*, masseter; *T*, temporalis; *AT*, anterior temporalis; *PT*, posterior temporalis; *MP*, medial pterygoid.

neurocranium and its contents, in the third trimester the circulatory system has achieved only about 5 per cent of its ultimate size, and the musculoskeletal system has attained only 3 per cent or so of its adult size.

Unlike changes that occur elsewhere in the body—e.g., the cardiopulmonary system—precocious fetal brain growth does not occur in order to more effectively serve immediate uterine or imminent birth-related needs, nor does brain growth seem to be functionally related to size changes in the body. Rather, this early and extensive brain growth seems to be an essential adaptation that enables the infant to eventually develop and master unique human traits such as language and abstract thought. Thus, the large fetal head is a **pre-**adaptation for **subsequent** needs that are manifest some time after birth.

The size of the fetal brain and its bony enclosure within the restricted womb limits the amount of space available for growth of the rest of the body, including the face. The size and shape of the birth canal places further constraints on both the neurocranium and the face: The skull must be small and flexible enough to pass through the canal. Moreover, because in the normal birthing process the crown of the fetal head is the first to enter the canal, the size and position of the face must be such that the face does not inhibit the cranial molding that is necessary for effective passage through the canal.

Cranial flexibility is made possible because of the fibrous interconnections between the partially formed bones of the vault. While brain growth, uterine constraints, and birthing may set some upper limits on facial growth in the fetus, the size of the face is not wholly determined by these factors. On the contrary, at birth the face must be sufficiently mature **in functional terms** for effective breathing and eating to take place.

Neonate to Deciduous Dentition: The Premasticatory Face

The newborn head is basically constructed to facilitate three things: respiration, ingestion, and continued expansion of the brain. Significant inhibition of any of these can result in death or catastrophic impairment of growth. Accordingly, it is possible to provide a meaningful account of the neonate face in terms of these factors.

At birth the airway must be of adequate size to allow for the unobstructed passage of a sufficient volume of air to meet physiologic needs. Given human infant feeding behaviors, the airway must also be separated from the oral cavity by the palate, thereby allowing both breathing and feeding to occur without interfering with each other. With the exception of some neuromuscular controls, the anatomy of the neonate respiratory tract is essentially a smaller, but wholly mature version of a system that is preserved throughout life. Although there may be changes in the size and shape of its component with growth, the functional anatomy of the respiratory tract does not alter with age.

While the oral anatomy of the newborn is sufficiently well developed to allow effective food ingestion immediately, the suckling behavior seen in the neonate is transient and specialized and does not persist into adulthood. (See Chapter 10 for a complete discussion of the neuromusculature control and actions of suckling behavior.) Suckling is carried out essentially with the use of three anatomic components: the muscles of facial expression (which underlie the cheeks and lips), the tongue, and the palate. This behavior is under genetic, that is, unlearned, neural control. The organization and actions of the facial expression muscles are such that they enable the infant to engage in effective feeding behavior without placing undue loads on any area of the relatively immature facial skeleton. Additionally, to further ensure effective feeding, the muscles of facial expression and the tongue are used in combination with the relatively large and flat palate in a manner that diminishes the likelihood of the infant being able to generate forces or pressures of sufficient magnitude to cause discomfort to the nursing mother. The structural-functional arrangements that so effectively facilitate ingestion in the neonate, however, are transient adaptations that are modified as changes take place elsewhere in the head and face, most notably in the musculoskeletal and dental components of the masticatory system.

In keeping with a pattern of postnatal growth that enables the human infant to attain 60 per cent of its adult brain size within 2 years and 90 per cent of its size within 7 years, the neurocranium predominates over the face in the newborn (see Chapter 16). The fibrous interconnections and the partially formed bones of the cranial vault that facilitated molding and passage of the head through the birth canal enable postnatal growth of the skull to keep pace with the still expanding brain. Although rapid brain growth is not essential to the immediate survival of the newborn, effects of this growth extend beyond neurocranium to the face and directly affect the masticatory system.

The middle cranial fossae are critical areas of brain growth (see Chapters 3, 5, and 6). The lateral boundaries of each fossa are formed by the greater wing of the sphenoid and the frontal, parietal, and temporal bones. The region and bones just identified are involved in masticatory function because they provide anchorage for the jaw elevator muscles, which at birth are small and poorly developed. Additionally, part of the temporal bone serves as the cranial articulation for the jaw joint. Expansion of the middle cranial fossa affects the masticatory system in several ways. For example, lateral expansion of the fossa through lateral displacement of the temporal bone also results in the lateral displacement of the cranial part of the jaw joint.

This displacement, in turn, necessitates growth changes, i.e., widening of the mandible in order to keep pace with the expanding basicranium. (See discussion of the V principle, Chapter 2.) According to simple biomechanics, widening of the jaw in this manner increases the possibility of unfavorable bending moments and/or torsion occurring along its body, especially in the region of the symphysis. Moreover, the likelihood and intensity of these potentially unfavorable effects will increase if asymmetrical loads are placed on the jaw. Although the infantile jaw has the obvious growth potential to keep pace with the expanding cranium, with its unfused symphysis, it is not well suited to resist the bending moments and torque just described. This potential **structural** shortcoming in the jaw, brought about in part by brain growth, is functionally compensated for in the neonate because the forces that act on the jaw during suckling are relatively low and symmetrical. In other words, oral behaviors of the nursing infant diminish the likelihood of unfavorable biomechanical forces acting on the relatively small and weak mandible. The persistence of suckling behavior up to and sometimes past the eruption of the deciduous incisors provides the mandible with the time to undergo structural modifications that enable it to better resist bending and torque. Thus, it is not mere coincidence that sees the mandibular symphysis complete its fusion at about the same time that the deciduous molars erupt.

Anterolateral expansion of the middle cranial fossa through growth along the sphenotemporal suture occurs simultaneously with the lateral displacement just described. In this instance the bones that articulate with the greater wing of the sphenoid experience varying degrees of displacement along an anteroposterior axis. Expansion of the sphenoid and displacement of the frontal, zygomatic, and temporal bones have the following effects on the masticatory system.

First, the lever arm length of the masseter, medial pterygoid, and temporalis muscles are each extended because their anterior-most areas of attachment (origins) are moved away from the jaw joint. Functionally, changes of this sort enable the infant to exert relatively larger jaw-elevating forces without requiring any increase in size of the jaw muscles. Elongation of the lever arm length in the manner described, coupled with a corresponding increase in load arm length, i.e., lengthening of the upper and lower jaws, allows for growth of the masticatory system while preserving its relative effectiveness.

A second effect of endocranial expansion on the masticatory system occurs because growth of the sphenoid and the bones with which it articulates creates a larger surface area for attachment of the origins of the temporalis and medial pterygoid muscles. Growth along the sutures of the greater wing of the sphenoid results in the lengthening of the zygomatic arch through the anterior displacement of the zygoma and the posterior displacement of the temporal bone relative to each other. In addition to increasing lever arm lengths as described earlier, this change also provides a larger area of attachment for the masseteric muscular origins along the zygomatic arch. As will be shown later, functional maturation of the masseter necessitates a modification in the shape of the elongated zygomatic arch.

Sutural growth along the boundaries of the sphenoid is one of several factors that contribute to endocranial expansion and displacement of the maxillary segment of the face (see Chapter 3). An important consequence of the maxillary displacement that results from expansion of the middle and anterior cranial fossae are the changes that are simultaneously brought about on the mandible and jaw elevator muscle function. Anterior displacement of the maxillae does several things, in addition to perhaps "carrying" the mandible forward with it: first, anterior displacement moves the entire dental arcade away from the fulcrum, theoretically increasing the load arm length; second, this movement carries the anterior-most points of attachment of the jaw elevator muscles away from the fulcrum, theoretically extending the lever arm; and third, forward displacement creates space in the posterior region of the dental arch, allowing for subsequent development of the maxillary tuberosity and the growth and eruption of any teeth therein. Simultaneous descent of the maxillae and the accompanying mandible occurs in such a way that the lines of the jaw elevator muscles, especially the masseter and medial pterygoids, move further away from the fulcrum, thereby contributing to the theoretical increases in lever arm length described immediately above. This occurs even though the angle of the mandible decreases, i.e., closes up, moving the line of the ramus more directly under the jaw joint (Fig. 11–9). In addition to working a change in lever arm length, this posterior displacement of the ramus, which ironically occurs in association with the anterior displacment of the maxillae, creates space in the mandibular dental arcade for the development and eruption of the molars. Because of their new position more directly inferior to the jaw joint, the load arm length to these posterior teeth is less than it would have been if the angle of the mandible had not closed up.

The changes described show how brain growth theoretically affects masticatory function, especially in an infant, when the brain is still growing very fast.

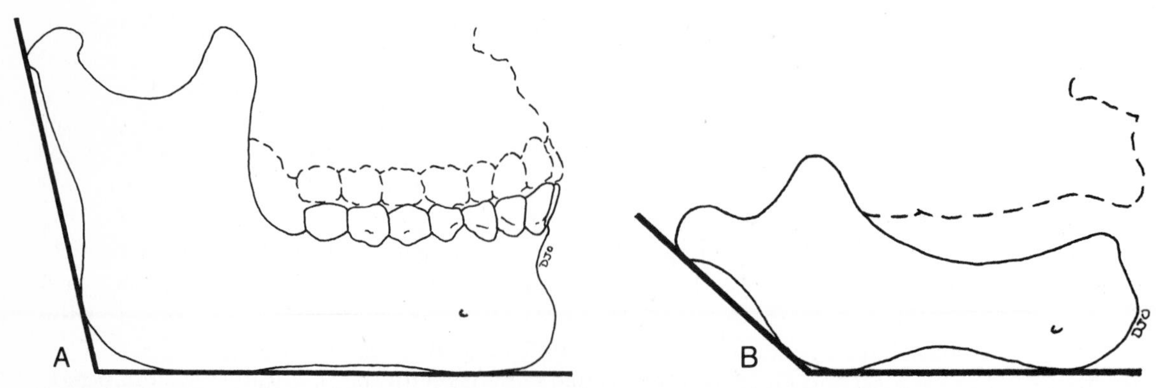

FIGURE 11–9. *A,* The angle of the mandible is formed by the intersection of a line drawn along the inferior border of the mandible with a second line drawn along the posterior border of the ramus. When the deciduous teeth are erupting and coming into occlusion, this angle is relatively open and lies markedly anterior to a line dropped perpendicularly from the functional axis of the jaw joint. Because of this arrangement, the mandibular and maxillary teeth are displaced anteriorly relative to the jaw joint. *B,* As the maxillae descend, the mandibular rami increase in length and remodel in such a fashion that each mandibular angle closes up and drifts posteriorly. These changes have the net effect of potentially shortening the load arm length (by allowing teeth to be positioned more posteriorly) and actually lengthening the lever arm length.

The effects that have been considered relate to potential alterations in mechanical efficiency, size changes in the jaw elevator muscles, and some of the consequences to the lower jaw. When considering these things, it must be borne in mind that they take place when the critical feeding requirements of the infant are being served almost exclusively by the muscles of facial expression, the palate, and the tongue. Thus, even though the effects that have been described are preadaptive for enhanced masticatory effectiveness, their initial appearance is attributable to neurocranial expansion, not functional needs associated with ingestion. Perhaps the simplest verification for this interpretation is provided by the jaw elevator muscles themselves, which are very poorly developed at birth and do not really begin to achieve functional maturation until the deciduous teeth have erupted into occlusion. Also, during this time period the mandibular body is a peculiarly bulbous-appearing bony tube enclosing the developing crowns of the deciduous teeth; and although well suited to the purpose of housing the teeth, the jaw is of dubious value as a force-generating or force-resisting structure at this stage. Although neurocranial expansion provides an early opportunity for growth of the jaw apparatus, the suitability of the suckling apparatus enables the mandible, the maxillae, their enclosed teeth, and the jaw elevator muscles to grow and develop in a manner that seems to better ensure their effectiveness later in life, when they are eventually called upon to generate and resist masticatory forces.

Deciduous and Permanent Dentition: Masticatory Maturation

While it lacks the drama and closely defined time dimensions that make birth such a significant event in the life of an individual, maturation of the masticatory system is the premier event *sine qua non* in the existence of the face. Just as birth is a process, so, too, is maturation of the masticatory system, only in the latter the changes that occur and the manner in which they take place are drawn out over a long period of time. This maturation process never ceases, in the sense that our own personal appearance never stabilizes, to the chagrin of those who seem to prefer a look of youthful exuberance in that most public aspect of anatomy, their faces.

Functional maturation of the masticatory system, and thereby the face, begins when the deciduous maxillary and mandibular incisors have erupted sufficiently into the oral cavity so that they can exert force on an object, such as the probing finger of an inquisitive parent. In a sense, this marks the "birth" of the masticatory system, just as actual birth marks the onset of functional maturation of the respiratory system in an infant. Just as **structural** changes must have taken place before respiratory **function** can occur, so, too, must similar changes have occurred in the masticatory system. Because of size and growth constraints associated with the womb, birth canal, and brain growth (described earlier), structural conditions in the masticatory apparatus at the time of birth are not sufficiently advanced for masticatory function to take place. Through adaptations described in the previous discussion of the neonate, additional time for growth of the masticatory components occurs while suckling takes place. Structural changes that occur during this prolonged period of growth preceding mastication are as follows.

The alveolar ridges of the upper and lower jaws develop sufficiently to hold the now fully formed crowns, whose roots have sufficiently formed to allow the teeth to erupt through the alveolus and gingiva. Alteration of the alveolar margin of the maxillae in this fashion decreases the area of flat surface available on the palate and provides a restricted area of focus for forces generated by elevating the mandible against the maxillae. Formation of the alveolar ridge on the mandible provides space for the deciduous incisors superior to the mandibular body, thereby allowing the bone to thicken in the region of the symphysis, which begins to fuse at this time.

Initially, relatively few other noticeable changes in the skeleton occur early in the maturation process. This is so because in contrast to innate suckling behavior, mastication is acquired behavior in which the infant, through self-exploration and oral experimentation, learns of the existence of the masticatory system and then begins to master its use. Only when this learning process has begun do the jaw elevator muscles significantly begin to develop and subsequent structural-functional changes begin to take place in the facial skeleton.

One of the earliest changes to occur as a consequence of the discovery of the jaw elevator muscles is the substitution of the "mature" swallowing pattern for the infantile pattern (see Chapter 10). In contrast to the infantile pattern, mature swallowing relies on the jaw elevator muscles to position the mandible. This simple shift places demands on the elevator muscles, which, up to this stage, had no significant functional involvement. In response to these novel functional demands, the muscles begin to hypertrophy. Use and growth of these muscles, in turn, place new demands on the bones to which the muscles are attached and on which they apply forces. Now, in a clear stimulus-response fashion, specific changes can be observed in the musculoskeletal system in association with masticatory function.

Whether they occur in the infant with erupting milk teeth or in a sub-adult with occluding second molars, changes wrought in the facial skeleton through alterations in size or function of the jaw elevator muscles are directed at enhancing biomechanical effectiveness and/or diminishing potentially adverse effects caused by the musculoskeletal unit in question. For example, prior to involvement of the jaw elevator muscles in ingestion, swallowing, or mastication, the zygomatic arch consists of two spinally bony processes that extend essentially in a straight line from their zygomatic and temporal bones; at their point of juncture the bones overlap and are joined by a loosely constructed fibrous joint. By the time the deciduous incisors are functionally occluded, not only has the frontozygomatic suture become tighter and more clearly interdigitated, but when viewed laterally the entire bony strut has begun to assume a distinct arched shape with the inferior margin showing a clear concavity. As subsequent teeth erupt and come into occlusion, the zygomatic **arch** progressively more closely resembles its namesake. When viewed superiorly, i.e., peering down into the infratemporal fossa, in the growth sequence just described it becomes readily apparent that the zygomatic arch also develops a distinct, laterally directed convexity, i.e., it bows outward.

The double convexity in the zygomatic arch that has just been described cannot be easily accounted for in any other than structural functional terms. The shape the zygomatic arch assumes is ideally suited to resist forces that are applied to this rather delicate bony structure as a consequence of forceful contraction of the masseter muscle (see Fig. 11–8). Moreover, this structural adaptation occurs in such a way that it enables the arch to maintain effective function while simultaneously undergoing an estimated fivefold increase in its length, largely as a consequence of the relative displacment of the temporal and zygomatic bones through neurocranial expansion, as described earlier.

Growth of the zygomatic arch does not seem to be wholly governed by neurocranial expansion and needs imposed by the masseter muscles. There are good reasons for thinking that the temporalis muscle also influences this bony component of the face. With increasing use of the mandible, the temporalis muscle grows in cross-sectional area, and its origins fan out over the side of the skull, covering portions of the greater wing of the sphenoid, temporal, parietal, frontal, and zygomatic bones. As this happens, the sutural junctions between these bones become better established and provide a firmer base for temporalis attachment but still allowing for cranial expansion through sutural growth. By the time the second deciduous molars have come into occlusion, the temporalis has grown and become sufficiently specialized that evidence of compartmentalization of the muscle is evident within the infratemporal fossa. There, the surface of the greater wing of the sphenoid shows distinct concavity whose superior-posterior-to-inferior-anterior orientation clearly delineates the line of action of the temporalis muscle as it descends to its attachment on the coronoid process of the mandible. In many adults this bony compartmentalization is very pronounced; and in almost every instance, this cranial trait is accompanied by exceptionally well developed coronoid processes. Often visible on the posterior surface of the zygoma, within the intratemporal fossa, are distinct bony spicules, further evidence of intensified use of the temporalis muscle. Although it has never been verified in the laboratory, it seems highly likely that the diameter of the infratemporal fossa, and thereby the size and shape of its lateral boundary, the zygomatic arch, is affected to a significant degree by the use and hypertrophy of the temporalis muscle.

Continuing within the infratemporal fossa, by the time the deciduous cheek teeth have erupted, the infratemporal crests and the plates of the pterygoid process have developed to a significant degree. By the time the second permanent molars are in occlusion, the pterygoid plates have more than quadrupled in size, and distinct fossae are discernible between the infratemporal crests, clear testimony to the expanded growth and use of the medial pterygoid muscles.

To accommodate the space needs of normal development and eruption of the dentition, the arcs of the maxillae and mandible must increase in length. When this occurs in the suckling infant, the masticatory load arms inevitably become longer because the teeth and the incipient temporomandibular joint are essentially in the same plane (see Figs. 11–4 and 11–5). Unless corresponding changes occur in the masticatory lever arm length—e.g., anterior displacement of the line of action of the jaw elevator muscle—lengthening of the upper and lower jaws will result in a theoretical reduction of mechanical efficiency of the jaw system. If an optimal level of bite force is to be maintained under these conditions, an increase in the cross-sectional area of the jaw elevator muscles must take place. Although this situation remains hypothetical in neonates, who do not rely on the masticatory apparatus for any functions, it does have serious implications with regard to subsequent growth: if the spatial relationships just described—i.e., the jaw joint and the occlusal plane in the same plane—were to prevail beyond the suckling phase, the development and eruption of every deciduous and permanent cheek tooth would inevitably result in a lengthening of the mastication load arm. Again, unless compensatory changes in jaw position and/or position of the elevator muscles were to occur, the only way effective mastication could be maintained would be through significant hypertrophication of the elevator muscles. A variation of this situation actually occurs in the olive baboon (Fig. 11–10), in which extensive prognathism offsets any other compen-

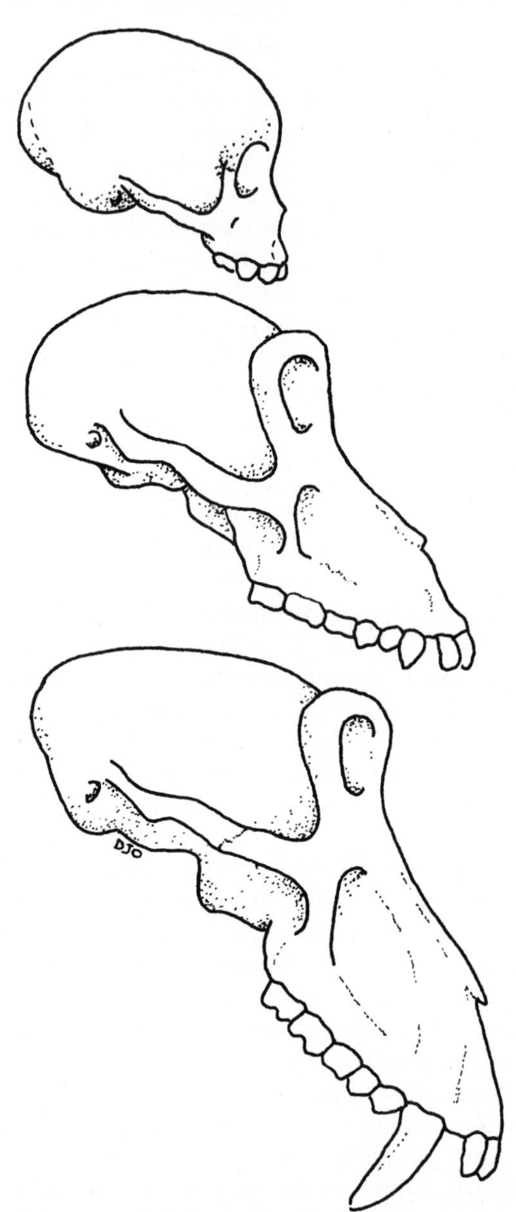

FIGURE 11–10. Olive baboons (*Papio cynocephalus anubis*) experience significant inferior displacement of the maxillae during facial growth. Nonetheless, because of the size of the teeth, the shape of the dental arcade (parallel tooth rows), and respiratory demands placed on the nasopharynx, this species still experiences significant lower facial prognathism that results in relatively unfavorable lever arm–load arm relationships in their masticatory apparatus. As a result, the jaw elevator muscles undergo extensive hypertrophication in order to maintain minimal optimal levels of bite force, and the craniofacial skeleton responds accordingly with striking crests and ridges.

satory changes to the extent that the jaw elevator muscles must undergo enormous changes in cross-sectional area (see Oyen et al., 1979).

Humans are not characterized by extreme lower facial prognathism coupled with massive jaw elevator muscles, even though this most certainly was the case with regard to our fossil antecedents. This is so largely because of the changes in the position of the maxillae, downward and forward, and accompanying alterations of the mandible, i.e., lengthening of each ramus and closing of its angle. Through the remodeling process described elsewhere in this text by Enlow, these shape changes create space at the posterior borders of the maxillae and mandible that not only allow for the formation and eruption of the cheek teeth, but may actually position the posterior-most tooth relatively closer to the jaw joint and/or lines of action of the jaw elevator muscles.

This specific growth adaptation is of particular significance in the maturation of the masticatory process. Because of the anterior displacement that occurs, eruption of the deciduous molars and the first permanent molars in most persons results in a decrease in mechanical efficiency of the jaw apparatus because of inevitable lengthening of the masticatory load arm that ensues. This alteration in load arm length is compensated for, in part, by anterior displacement of the zygomatic root (key ridge) that has the net effect of lengthening the lower arm. Full restoration of bite force at the incisors or at the second milk molars, however, requires an increase in the cross-sectional area of the jaw elevator muscles. Attainment of occlusion with the first permanent molars, however, requires that the dynamics of facial growth and functions alter somewhat. From this time onward, anterior displacement of the teeth relative to the base of the zygoma on the maxilla becomes negligible. Subsequent eruption of the second and third permanent molars are placed relatively **closer** to the lines of action of the jaw elevator muscles. Depending on the extent of further displacement of the entire maxillary complex, the second or third molars may actually be positioned **closer** to the jaw joint. All of this means that in normal persons, after attainment of occlusion of the first permanent molars, subsequent increases observed in cross-sectional areas of the jaw elevator muscles cannot be attributed to decreases in masticatory efficiency brought about by a lengthening of the masticatory load arm. This state of affairs does not indicate that stasis has been reached in the maturation of the masticatory system. On the contrary, the introduction of the possibility of long-lasting increases in mechanical efficiency increases the potential for changes in the musculoskeletal components of the system, because under such conditions it would be conceivable for muscle size actually to diminish without reducing bite force–producing capabilities.

Another factor that ensures potential instability in the masticatory system derives from the fact that when teeth are added to the tooth row, the interdental contact between each successive pair becomes greater because the human dental arcade is parabolic in shape. Thus, beginning with the lateral incisors and working posteriorly, the bite between any two opposing teeth—e.g., the canine, the first deciduous molars, the second permanent molars—will produce progressively larger amounts of torque on the maxillae and mandible because of the asymmetrical loading that occurs. It can be argued, quite persuasively, I think, that the thickening up of the zygomatic process of the maxillae and its zygomatic counterpart is an adaptive response that enables that part of the face to resist the twisting that results when forceful contraction of the masseter muscle on the biting side seems to want to draw the zygomatic arch down and in. The effects of this asymmetrical loading on the mandible are described below.

The changes described in the cranium and the jaw elevator muscles as teeth develop, erupt, and come into occlusion have their mandibular counterparts. By the time the deciduous incisors are functional, the mandibular symphysis has fused. Around this time a striking, but rather poorly understood, phenonemon begins to emerge: in profile the mandible shifts from convex on its anterior surface to concave, and the human chin becomes identifiable. Prevailing thought holds that the fusion of the mandibular symphysis and the emergence of the chin are both adaptations that enable the mandible to effectively resist the torsional force that occurs in this region as a consequence of asymmetrical loads. As already pointed out, the frequency and intensity of asymmetrical loading increase as additional teeth erupt into the parabolic dental arcade.

Another part of the mandible, the condylar neck and head, shows evidence of responding to increasing masticatory loads, especially of the asymmetrical type. To cope with these forces, the condylar neck thickens and the surface area of the condylar head increases considerably, probably 10-fold between the time the deciduous molars are occluding and when the third permanent molars come into occlusion.

As the temporalis, masseter, and medial pterygoid muscles increase in use and cross-sectional area, discernible changes are seen in the mandible. While bony irregularities mark the areas of insertion of the medial pterygoid and masseter muscles on the angle of the mandible, the gonial flare and antigonial notch that characterize attachment of the masseter stand out in a striking manner. The muscle that most affects its area of insertion on the mandible, however, is the temporalis. As one might expect of a muscle whose bony origins cover such an area, the coronoid process raised by the temporalis grows to such an extent that by the time adult muscle size is attained and all of the teeth are in occlusion, the process extends well above the level of the condyloid process. To put this in perspective, it is relatively easy to find individuals in whom the height of the coronoid process, measured from the border of the mandibular notch, exceeds the total ramal height of the mandible measured in a child with complete deciduous dentition.

NONMASTICATORY FACTORS THAT AFFECT THE FACE

In this discussion much has been made of the role played by masticatory function during facial growth and development. Even though the point has been emphasized that functional maturation of the masticatory system is a key process influencing the face, it has **not** been argued that the face can be wholly understood solely in terms of masticatory function. In fact, considerable evidence has been provided showing the enormous influence brain growth has on facial growth and development. There are other factors that should be considered, some of which follow.

The eyes are a striking feature in the infant or child, probably one of the most dominant features of the child's face. Given their precocious development, size, placement in the face, and functional significance, the eyes play an important part in the growth and development of the face. Although we have much left to learn, we are quite knowledgeable about the physiology of vision. Surprisingly, however, no clear consensus exists about the structural-functional interrelationships between the anatomic housing for the eyes, i.e., the orbits, and the rest of the face. Other than a general agreement that the bony orbits do provide protection for the eyes, there is considerable disagreement, particularly about the structural-functional significance of superior and lateral boundaries of the orbits. On the basis of my own research in this area, I have reached a tentative conclusion that in man the circumorbital region, especially the superior and lateral margins, is affected by forces generated in the facial skeletons as a consequence of masticatory function. Moreover, while offering protection along the lateral border, the joined segments of the zygomatic, frontal, and sphenoid bones in this region function primarily to enhance effective contraction of the temporalis and masseter muscles, whose origins are in this general area. The lateral border and wall probably do not function in any significant fashion as a pillar or buttress, resisting compressive forces conducted to this area from the dentition (Sicher and Tandler, 1928).

Breathing is one function served by the face that takes precedence over ingestion in the newborn. With growth its importance does not decrease. In fact, significant anatomic changes occur in the growing child in order to ensure continous, effective breathing, and the face is the site for some of these changes.

The maxillae provide a bony enclosure for most of the facial component of the respiratory system, the nasal airway. As pointed out elsewhere by Enlow in this text, there is persuasive evidence that growth of the maxillae and changes in the cranial base are linked with the fulfillment of physiologic respiratory needs. Displacement of the maxillae, as described earlier in this chapter, and its lateral expansion through growth and remodeling ensue in order to increase the volume and to maintain the favorable anatomic position of the nasopharyngeal airway.

It is well known that respiratory needs are closely related to general body physiology, and thereby to general body growth. Through this relationship, an acceleration in body growth—e.g., the pubertal growth spurt—can necessitate a substantial increase in respiratory demand. If these respiratory demands are of sufficient magnitude and duration, structural accommodations such as an expansion of the nasopharyngeal airway may be called for. In this manner, aspects of masticatory maturation described in terms of changes in the maxillae can actually be traced to physiologic needs associated with respiration. In the example cited, the pubertal growth spurt, even hormonal secretory activity, is linked to growth changes in the face and masticatory system. This example, as well as the discussion just completed, prompts due caution against the overzealous application of a single factor when trying to account for a phenomenon as complex as facial growth.

SUMMARY

The stated purpose of this chapter has been to tell a story about facial growth and development in such a way that the reader will come away with some facts and some ways of thinking about and making sense of the changes that occur in the growing face. An illustrative, rather than exhaustive, set of traits have been described and analyzed in terms of their relationships with the masticatory system. In a few instances nonmasticatory factors acting on the face and the jaw system have also been considered, albeit briefly.

A real attempt has been made to characterize the true potential for change possessed by the musculoskeletal components of craniofacial skeleton, and especially the masticatory apparatus. If anything, by focusing so much on what happens when teeth are **added** to the dental arcade, this potential has been understated. A more complete picture of this capacity for adaptation could be obtained by applying notions about level arm and load arm length, torsional effects of asymmetrical biting, and so forth, in analyses of the conditions that prevail when teeth are **removed** from the tooth row. Similar studies could be undertaken to evaluate the effects of other efforts to manipulate components of the masticatory system—e.g., prosthodontics restorations, orthodontic tooth movements, and maxillofacial alteration by oral surgeons.

Only when these kinds of ideas have been successfully carried out on a broad scale will we ever truly begin to understand **why** the face grows as it does.

Acknowledgments

Many of the notions expressed in this chapter are outgrowths from conversations and arguments I have had over the years with Drs. D. H. Enlow, H. B. Sarles, and A. C. Walker, to each of whom I will be forever indebted. All of the drawings are by Derek J. Oyen. The research on which this chapter is based has been supported by the National Science Foundation.

CHAPTER TWELVE

Prenatal Facial Growth and Development

Part 1

Webster delimits the face as the front part of the head comprising the nose, cheeks, jaws, mouth, forehead (although not actually part of the true face or viscerocranium), and the eyes. However, a 1-month-old embryo has no real face. But the key primordia have already begun to gather, and these slight swellings, depressions, and thickenings are rapidly to undergo a series of mergers, rearrangements, and enlargements that will transform them, as if by sleight of hand, from a cluster of separate masses into a **face.** It is a great story.

Concept 1

The "head" of a 4-week-old human embryo is mostly just a brain covered by a thin sheet of ectoderm and mesoderm. Where the mouth will be is marked by a tiny depression, the **stomodeum** (Figs. 12–1, 12–2, and 12–3). The eyes have already begun to form by a thickening of the surface ectoderm (the future lens), which meets an outpouching from the brain (the future retina). The eyes are still located at the sides of the head, however, as in a fish. As the brain continues to grow and expand, the eyes are rotated toward each other and toward the midline of what is soon to become a face. Does this not greatly reduce the intervening span between the right and left eyes? Yes, but in a relative sense. **Everything** is increasing in size, including the interorbital dimension. The eyes are actually moving farther apart; but because other parts of the head are enlarging even more, the proportionate size of the interorbital area is becoming decreased. When illustrating the process of facial growth, it has always been traditional to show all of the stages as equal in size. Keep in mind, however, that there is actually considerable overall enlargement involved. These changes proceed very swiftly in the early stages.

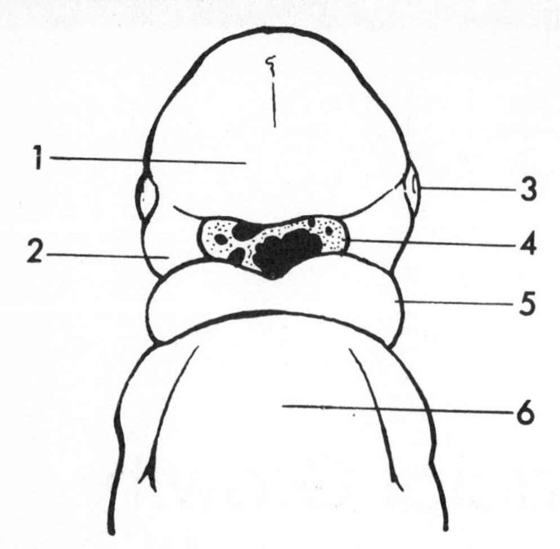

FIGURE 12–1. Human head region at about 4 weeks. *1,* Forebrain. *2,* Maxillary region. *3,* Optic vesicle. *4,* Stomodeal plate (already rupturing). *5,* Mandibular (first) arch. *6,* Cardiac prominence. (Modified from Patten, B. M.: *Human Embryology,* 3rd Ed. New York, McGraw-Hill, 1968.)

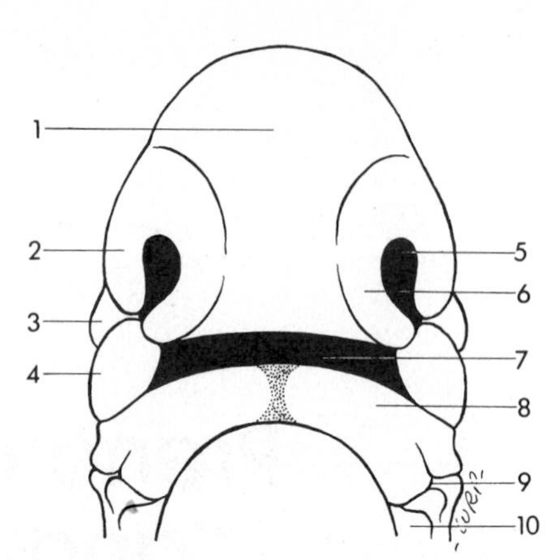

FIGURE 12–2. Face at about 5 weeks. *1,* Frontal prominence. *2,* Lateral nasal swelling. *3,* Eye. *4,* Maxillary swelling. *5,* Nasal pit. *6,* Medial nasal swelling. *7,* Stomodeum. *8,* Mandibular swelling. *9,* Hyomandibular cleft. *10,* Hyoid arch.

Concept 2

As the whole head markedly expands, the membrane that covers the stomodeum does not keep pace with it. This thin sheet quickly breaks through, and the **pharynx** becomes opened to the outside (Fig. 12–4). Everything in front will become the face, and **this** is what is now going to develop. The stomodeum of the embryo is about where the tonsils of the adult are located, deep within the face; so one can see that considerable growth and development have yet to occur in front of it.

The mammalian pharynx is the homologue of the region that develops into the branchial chamber and the gill system of fishes. The human pharyngeal pouches and clefts, however, did not "evolve from gills." More correctly, the

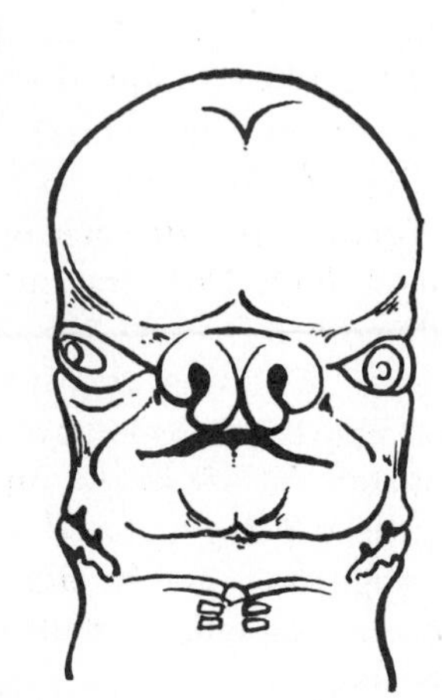

FIGURE 12–3. Face at about 7 weeks.

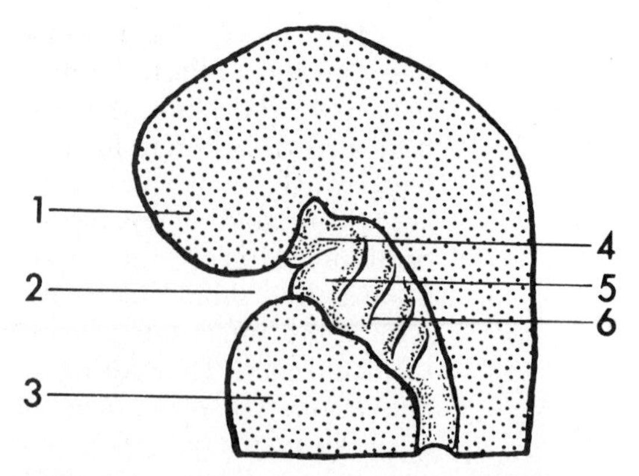

FIGURE 12–4. Internal view of pharyngeal region. *1,* Forebrain. *2,* Stomodeum. *3,* Cardiac prominence. *4,* Maxillary process. *5,* Mandibular process. *6,* Pouch between second and third arches. (Modified from Langman, J.: *Medical Embryology.* Baltimore, Williams & Wilkins, 1969.)

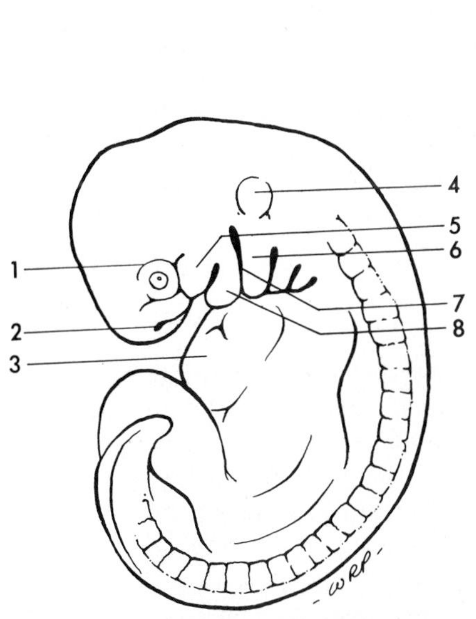

FIGURE 12–5. Human embryo at about 5 weeks. *1,* Eye. *2,* Nasal pit. *3,* Cardiac prominence. *4,* Auditory vesicle. *5,* Maxillary process. *6,* Hyoid arch. *7,* Hyomandibular cleft. *8,* Mandibular arch. (Modified from Patten, B. M.: *Human Embryology,* 3rd Ed. New York, McGraw-Hill, 1968.)

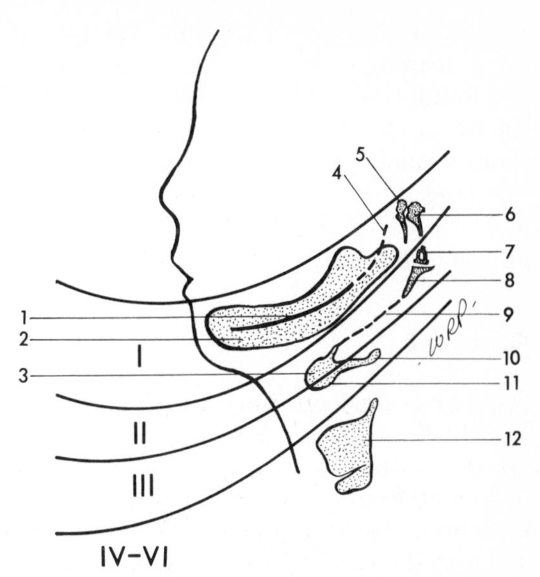

FIGURE 12–6. Pharyngeal arch derivatives (I to VI). *1,* Meckel's cartilage. *2,* Intramembranous bone developing around Meckel's cartilage. *3,* Superior part of body and lesser horn of hyoid. *4,* Sphenomandibular ligament. *5,* Malleus. *6,* Incus. *7,* Stapes. *8,* Styloid process. *9,* Stylohyoid ligament. *10,* Greater horn of hyoid bone. *11,* Inferior part of hyoid body. *12,* Laryngeal cartilages.

primordia that developed into the fish's branchial system were phylogenetically converted to develop into **other** structures instead of gills. This is where many of the parts of the face come in.

Concept 3

The pharynx is the anterior-most segment of the endodermally lined embryonic gut (Fig. 12–5). Its lumen is bounded on the right and left sides by the **pharyngeal arches** (also called visceral, branchial, and, inappropriately, gill arches). Between the arches are the pharyngeal **clefts** on the outside and the **pouches** (see Fig. 12–4) on the inside. Where each cleft meets its pouch, a mesodermally reinforced contact between ectoderm and endoderm occurs. All these arches and some of the clefts and pouches give rise to **specific** adult structures in the face, in other areas of the head, and in the neck. An important point is that many of the fundamental embryonic relationships among them are retained in adult anatomy. The tissues in each arch develop into specific muscles, bones, and cartilages, and the arrangement in the adult is carried forward from the pattern that exists in the embryo. Remember this: each arch has a specific cranial nerve, and each nerve thereby supplies the other structures that are derived from that particular arch.

Concept 4

The **first** pharyngeal arch gives rise to the tissues that will eventually become the mandible and its muscles (Fig. 12–6). It is thus called the **mandibular arch.** A bud develops from it to become the "maxillary swelling," and this is the anlage (that is, primordium) for part of the maxillary arch that is soon to begin forming. The specific cranial nerve to the first arch is the mandibular (V), and it thus innervates the various **muscles of mastication.** The cartilage of the first arch (Meckel's cartilage) serves as the anlage for two of the ear ossicles (malleus and incus). This cartilage does not develop into the mandible itself. The bone of the lower jaws forms intramembranously **around** Meckel's cartilage, and the cartilaginous condyle develops from a separate secondary cartilage that appears later.

Concept 5

The **second** pharyngeal arch is called, appropriately, the **hyoid arch** (see Fig. 12–6). It forms the cartilaginous model from which part of the hyoid apparatus develops and also the third ear ossicle (the stapes). The mesenchyme of this arch gives rise to the stylohyoid muscle and forms all of the various **muscles of facial expression.** These developing, sheet-like muscles spread up and over the face like a superficial sleeve. They are "cutaneous" muscles located in the deep part of the facial skin, and they are much more highly "developed" in man and some primates than in other mammals. This gives our face the capacity for a characteristic repertoire of changing expressions. The specific cranial nerve to the second pharyngeal arch is the **facial** (VII). You can always figure out which nerve goes to which muscle in the face and neck by remembering the simple relationships in Concepts 4, 5, and 6.

Concept 6

The third, fourth, and sixth pharyngeal arches (the fifth drops out) give rise to the remainder of the hyoid apparatus, the laryngeal cartilages, and the muscles of the larynx. The nerves to these arches are the glossopharyngeal (third arch) and the vagus (fourth and sixth arches). In addition, the parathyroids and thymus develop from third and fourth arch tissue.

Concept 7

The main body of the **tongue** develops from the right and left first (mandibular) arches, where they join in the floor of the pharynx by a merger of the paired **lingual swellings** (Fig. 12–7). The mucosal covering has sensory innervation, quite naturally, by the fifth cranial nerve (with a branch joining it from the nerve of the adjacent second arch). The root of the tongue develops from third and fourth arch tissue, and its sensory innervation is thereby provided by the glossopharyngeal and vagus nerves. The primordium for the **thyroid gland** forms a deep diverticulum from the lining, endodermal epithelium into the floor of the pharynx exactly between the first and second arches. Later, it becomes relocated into the neck along with the parathyroids.

Concept 8

Note that the **auditory vesicle,** which was formed by an invagination of the surface ectoderm, is positioned close to the second pharyngeal arch (Fig. 12–8). It is developing into the **inner ear** apparatus (semicircular canals and

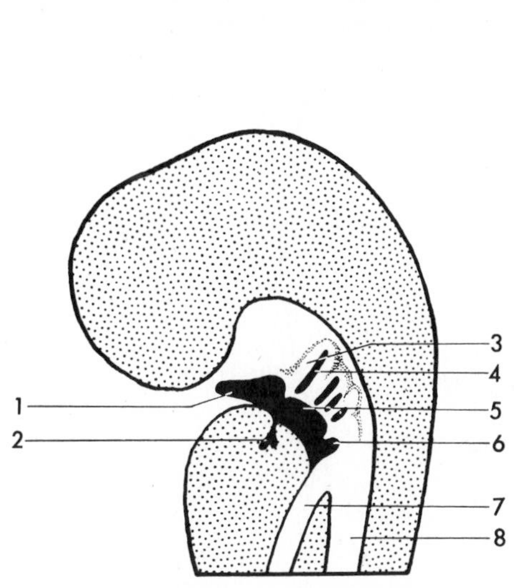

FIGURE 12–7. *1*, Body of tongue (lateral lingual swellings and tuberculum impar). *2*, Thyroid diverticulum. *3*, Mandibular arch. *4*, Pouch between first and second arches. *5*, Root of tongue (copula). *6*, Arytenoid swellings. *7*, Trachea. *8*, Esophagus.

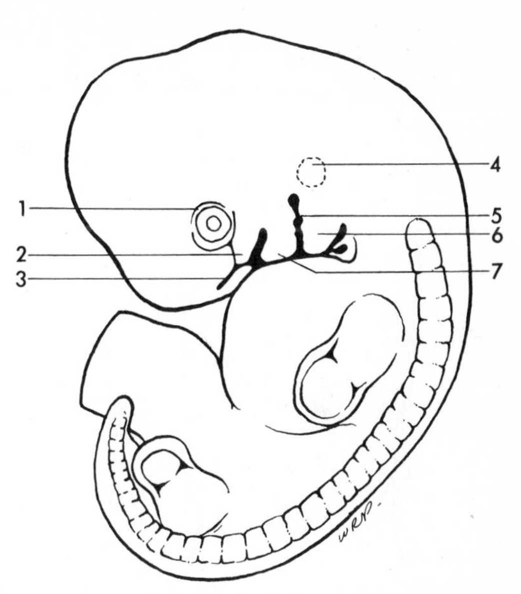

FIGURE 12–8. The face at about 6 weeks. *1*, Eye. *2*, Maxillary process. *3*, Nasal pit. *4*, Auditory vesicle. *5*, Hyomandibular cleft. *6*, Hyoid arch. *7*, Mandibular process of first arch. (Modified from Patten, B. M.: *Human Embryology*, 3rd Ed. New York, McGraw-Hill, 1968.)

cochlea). The **external** ear is being formed by the superficial tissue surrounding the first pharyngeal cleft, and the cleft itself is soon to become the external auditory canal. The **middle** ear chamber, which will contain the auditory ossicles, is formed by an expansion of the first pharyngeal **pouch,** and the ossicles develop from the cartilages of the first and second **arches,** which are, conveniently, right there.

Concept 9

Note that the first pharyngeal arch has given rise to the two sizable pairs of **mandibular** and **maxillary** swellings (Figs. 12–9 and 12–10). Below the

FIGURE 12–9. The face at about 5 weeks. *1*, Frontal prominence. *2*, Eye. *3*, Medial nasal swelling. *4*, Mandibular swelling. *5*, Lateral nasal swelling. *6*, Nasal pit. *7*, Nasolacrimal groove. *8*, Maxillary swelling. (Modified from Langman, J.: *Medical Embryology*. Baltimore, Williams & Wilkins, 1969.)

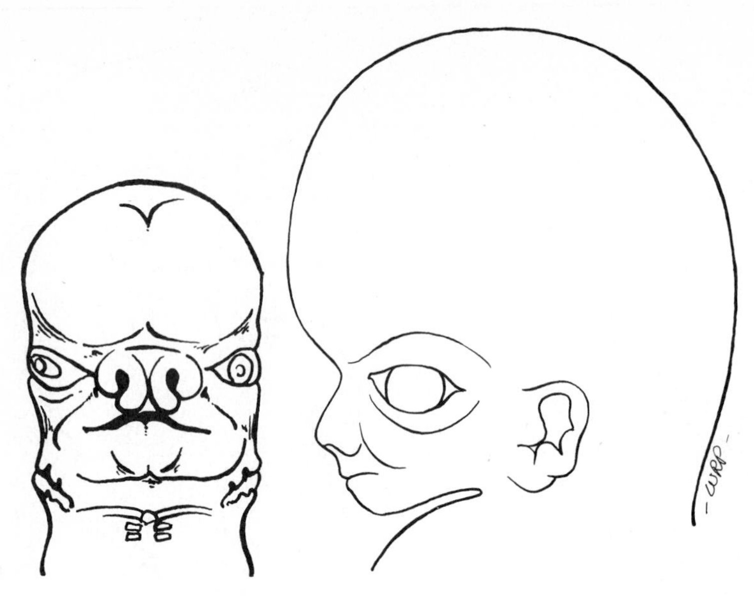

FIGURE 12–10. Face at 7 and about 8 to 9 weeks.

"forehead" is a pair of U-shaped swellings, the **nasal primordia.** At this stage, the embryo is 5 weeks old, but in only about 2 more weeks a fast-moving sequence of changes takes place, and the result is a recognizable face. On each side, the maxillary swellings join with the **medial** limbs of the nasal swellings, and this composite forms the closed maxillary arch.* The middle portion is the "premaxillary" segment that will later house the incisors. It also gives rise to the **philtrum** (cupid's bow) of the upper lip. Above it, the medial limbs merge to form, in addition, the middle part of the nose; the lateral limbs become the nasal wings. Bone is beginning to form in the maxillary and mandibular arches, the eyes are continuing to be displaced into more forward-pointing positions by the enlarging brain, the ear lobes are forming, and there is now a **face.** Within the face, a nasal septum has formed, "shelves" from the right and left maxillae develop and fuse at the midline to form a palate, and the oral and the paired nasal chambers thereby all become partitioned from one another. This will make possible, after birth, breathing and eating at the same time.

Concept 10

In short order, centers of ossification appear for most of the other major bony parts of the face and cranium, some intramembranously and some endochondrally (Fig. 12–11). The bone tissue of each center spreads until the definitive shape of that bone is attained. **Then** the bone begins to "remodel" as it grows. This remodeling process first starts at around 14 weeks for most of the various separate bones and their parts. In Chapter 3, the facial growth and remodeling process in the growing child was explained. The various patterns of resorption and deposition on the surfaces of all the bones were described, and the different growth movements of each bone were accounted for. Is the

*In normal development, this merger involves a closure of the **furrow,** not an actual cleft, between these parts.

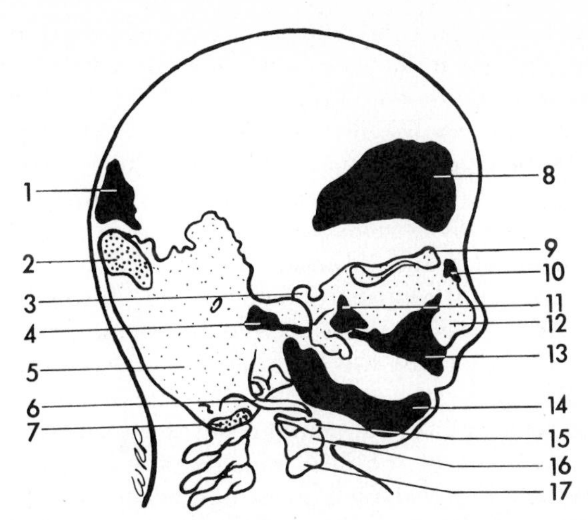

FIGURE 12–11. Developing skull at about 9 weeks. *1,* Occipital bone (interparietal part). *2,* Supraoccipital. *3,* Dorsum sellae (still cartilaginous). *4,* Squamous part of temporal. *5,* Cartilage. *6,* Styloid process. *7,* Occipital (basal part). *8,* Frontal bone. *9,* Crista galli (still cartilaginous). *10,* Nasal bone. *11,* Malar. *12,* Cartilage of nasal capsule. *13,* Maxilla. *14,* Mandible (surrounding Meckel's cartilage.) *15,* Hyoid. *16,* Thyroid cartilage. *17,* Cricoid cartilage. Endochondral sites of ossification are shown by dark stippling, and intramembranous sites are shown in black. (Modified from Patten, B. M.: *Human Embryology,* 3rd Ed. New York, McGraw-Hill, 1968.)

growth and remodeling process for the **fetal** face and cranium the same? In general, yes. The principal differences are in the anterior parts of the upper and lower jaws and the zygoma. In the face of the child after the primary dentition begins to erupt, these surfaces characteristically become **resorptive** (see Fig. 3–131). Throughout fetal life and the early part of childhood, however, they remain depository. The reason is that the bony maxillary and mandibular arches must expand anteriorly to accommodate the development of the primary dentition and the buds for permanent teeth. The bony arches thus grow **forward** as well as posteriorly. The maxillary arch, at the same time, is also growing downward as the nasal chambers expand. As the primary teeth begin to erupt, however, the outer surfaces of the forward part of both the maxilla and mandible become resorptive. Subsequent lengthening of the bony arch then proceeds only posteriorly. The characteristic postnatal resorptive fields in the anterior parts of the arches develop in conjunction with the continuing **vertical** growth of the maxilla and mandible, as explained in Chapter 3. (See Kurihara and Enlow, 1980*a,* for further details.)

Concept 11

A "priority plan" exists during the prenatal growth and development of the face and the body in general. Some specific organs and anatomic parts are given premium status in earlier timing and/or greater growth rate; some other parts receive relative second-order attention until later. This is determined, in general, by the urgency and extent of a given part's functional role in the early physiologic activities of the developing fetus. Certain anatomic components, such as the cardiovascular system and some parts of the nervous system, are essential to fetal life. Other components, including the lungs and the nasal and oral parts of the face, are not utilized as such during prenatal existence, and completeness of development is somewhat deferred relative to the more functionally active fetal organs and tissues. Such growth-delayed parts, however, must achieve readiness for immediate function at the time of birth. Airway size must be adequate to accommodate lung size, which, in turn, must be sufficient to meet the functional needs for the size of the body at the time. Neonatal oral functions must also be ready to respond, virtually instantaneously, upon birth. Thereafter, as described throughout earlier chapters, differential extents and rates of growth maturation occur among the different body (and facial) regions and parts. The neonatal brain, calvaria, basicranium, and eyes are relatively large in comparison with the proportionately much smaller face. However, as general body size progressively increases, the lungs enlarge to match, and the nasal part of the face (not just the external nose) correspondingly begins to increase significantly in height and length. The dentition begins to emerge, chewing begins to replace suckling, neural reflexes change, swallowing patterns change, and the oral part of the face, with its rapidly enlarging masticatory muscles and growing and developing jaw bones, keeps pace.

Part 2

A 1-month-old embryo has no "face" as such. The stage, however, is set. The primordia that will now rapidly develop into the jaws, nose, eyes, ears, and mouth and the many deep structures located within these parts have already begun to form.

Below the bulging **frontal eminence,** an ectodermally lined surface depression marks the developing site of the future mouth. This shallow pit, the **stomodeum,** is separated from the foregut by a thin ectoderm-endoderm floor, the **buccopharyngeal membrane** (Fig. 12–12). This membrane is already beginning to rupture and disappear. The structures around the stomodeum grow and enlarge at a rapid rate. The membrane itself, however, does not continue to grow; so it becomes broken through as massive expansion and separation of the structural parts around it take place. To appreciate how much facial growth is going to occur, realize that the location of the buccopharyngeal membrane in the 1-month-old embryo is at the level of the tonsils in the adult. An enormous amount of facial expansion thus will occur in front of the stomodeum. On the other side of this opening is the endodermally lined pharynx. The pharynx is that part of the foregut characterized by the **pharyngeal** (visceral, branchial) **arches** (Fig. 12–13). Within the pharynx, a **pharyngeal pouch** lies between the arches, and on the outside a **pharyngeal cleft** occurs between the arches. The ectoderm-endoderm contact between each cleft and pouch is termed the **branchial membrane.** All these various pharyngeal parts are major participants in the subsequent formation of many component structures in the head and neck.*

Each right and left pharyngeal arch has a specific nerve, a specific artery (aortic arch), and mesenchyme that develops into specific muscles and specific embryonic cartilages (Fig. 12–14). Certain bones are associated with specific

*Migrating cranial neural crest cells contribute extensively to the early primordia of many tissues developing in the face and the pharyngeal region (see M.C. Johnston et al., 1973).

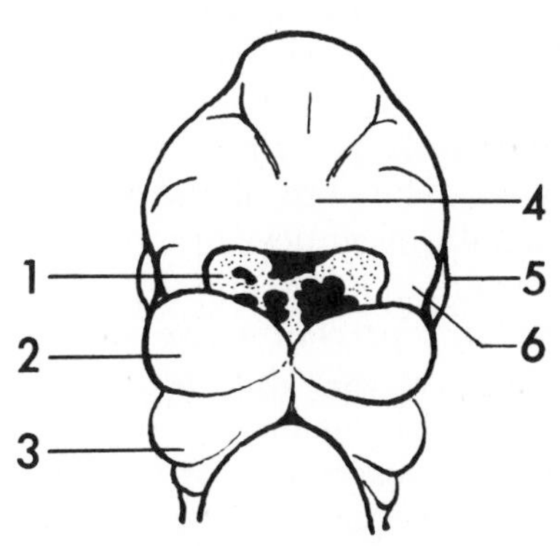

FIGURE 12–12. Human face at about 4 weeks. *1,* Stomodeal plate (buccopharyngeal membrane). *2,* Mandibular arch (swelling or process). *3,* Hyoid arch. *4,* Frontal eminence (or prominence). *5,* Optic vesicle. *6,* Region where the maxillary process (or "swelling") of the first arch is just beginning to form. (Modified from Patten, B. M.: *Human Embryology,* 3rd Ed. New York, McGraw-Hill, 1968.)

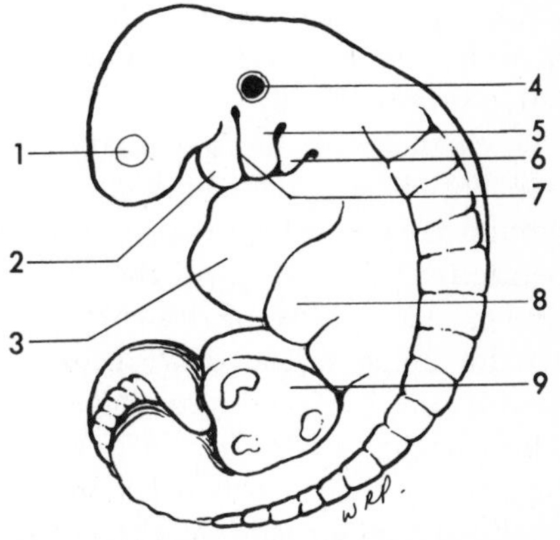

FIGURE 12–13. Human embryo at about 4 weeks. *1,* Optic vesicle. *2,* Mandibular arch (process or swelling). *3,* Cardiac prominence. *4,* Auditory (otic) vesicle. *5,* Hyoid arch. *6,* Third arch. *7,* Hyomandibular cleft. *8,* Hepatic prominence. *9,* Primitive umbilical cord. (Modified from Patten, B. M.: *Human Embryology,* 3rd Ed. New York, McGraw-Hill, 1968.)

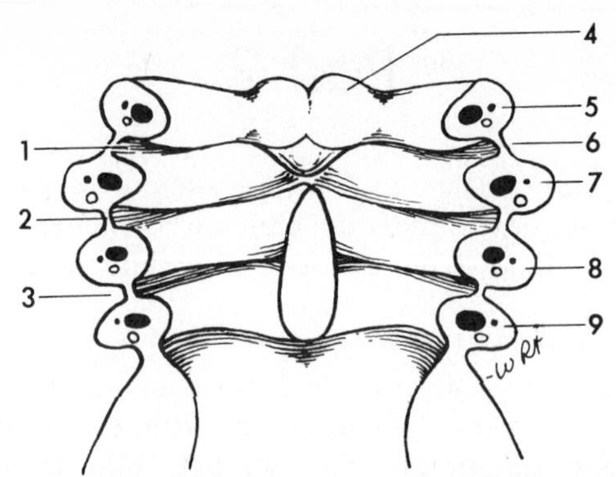

FIGURE 12–14. Internal view of pharyngeal floor and cut arches. *1,* First pharyngeal pouch between first and second arches (to become middle ear chamber). *2,* Branchial membrane. *3,* Pharyngeal cleft. *4,* Region that will develop into the anterior two thirds (body) of tongue. *5,* First (mandibular) arch containing its specific cartilage, cranial nerve, and aortic arch. The pharyngeal arch is also filled with branchiomeric mesenchyme. *6,* First pharyngeal cleft (hyomandibular) to become external ear canal. *7,* Second (hyoid) pharyngeal arch. *8,* Third pharyngeal arch with its own cartilage, aortic arch, cranial nerve, and branchiomeric mesenchyme. *9,* Fourth pharyngeal arch. (Modified from Moore, K. L.: *Before We Are Born: Basic Embryology and Birth Defects.* Philadelphia, W. B. Saunders, 1974.)

pharyngeal arches. This is a basic and important concept because if one can understand the simple embryonic relationships involved, understanding the exeedingly complex adult anatomy is so much easier. The **muscles** that develop in relation to each arch associate directly with the **bones** forming in that arch and are innervated by the resident cranial **nerve** of the same arch. Some of the embryonic pharyngeal pouches and clefts also establish the characteristic anatomic relationships for their adult derivatives. All of this has a logical, systematic, readily recognizable developmental rationale in the embryo. Remembering these specific prenatal relationships, the far less fathomable plan for the adult morphology makes sense.

In the human embryo, there are five pairs of pharyngeal arches. The first is the right and left **mandibular** arch. A bud develops from each first arch to form the paired **maxillary processes.** Both the mandibular and maxillary primordia are thus of first arch origin. The second pharyngeal arch is the **hyoid** arch (see Fig. 12–13). The remaining arches are identified by their respective numbers only.

The cartilage of the first pharyngeal arch is **Meckel's** cartilage, right and left (see Fig. 12–6). It occupies a location that will later be the core of the mandibular corpus, which forms around it. The bony mandible itself develops, independently, directly from the embryonic connective tissue that surrounds Meckel's cartilage. Most of this cartilage actually disappears, but parts of it give rise to the anlagen for two ear ossicles (the malleus and incus), and the perichondrium of Meckel's cartilage forms the sphenomandibular ligament.

The cartilage of the hyoid (second) arch is **Reichert's** cartilage. It forms the third of the three ear ossicles on each side, the stapes. The remainder gives rise to the styloid process of the cranium, the stylohyoid ligament, the lesser horn of the hyoid bone, and a portion of the hyoid body (see Fig. 12–6).

Muscles form from the mesenchyme of the arches. This mesenchyme is termed **branchiomeric** (Gr. *branchia,* gills; Gr. *meros,* segment), in contrast to mesenchyme of somite origin elsewhere in the body. From the branchiomeric mesenchyme of the first arch, the muscles of mastication, the anterior belly of the digastric, the tensor palatini, the mylohyoid, and the tensor tympani muscles all develop. From the branchiomeric mesenchyme of the second arch develop the muscles of facial expression, the stylohyoid, the stapedius, the posterior belly of the digastric, and auricular muscles. The specific cranial nerves (Fig. 12–15) that supply the first arch are the mandibular and maxillary branches of the trigeminal nerve (V). The specific cranial nerve for the second arch is the facial nerve (VII). Thus, the muscles of the first arch (muscles of mastication, and so forth) are innervated by the mandibular division of V, regardless of the place in which each muscle finally becomes located later in development. The muscles of facial expression formed by the branchiomeric mesenchyme of the second pharyngeal arch correspondingly are all innervated by the facial nerve.

Note the gathering of many structures related to the ear in and around the first and second pharyngeal arches (Fig. 12–13). The **auditory placode** differentiates early as a surface thickening of the ectoderm just above and behind the first pharyngeal cleft. This placode rapidly invaginates to form the auditory (otic) vesicle, which then differentiates into the structures of the inner ear (semicircular canals, cochlea). The first pharyngeal **cleft** (between the first and second arches) forms the external auditory meatus and outer ear canal, and the branchial membrane between the cleft and pouch undergoes remodeling changes to participate in the formation of the tympanic membrane (Fig. 12–16). The first pharyngeal **pouch** becomes expanded into the middle ear chamber, and it also forms the auditory (eustachian) tube that retains continuity between the middle ear and the pharynx. The ear ossicles, developing from the cartilages of the first and second arches, lie conveniently next to this area and soon become enveloped within the expanding first pharyngeal pouch (middle ear chamber). They function as the bridge between the tympanic membrane and the inner ear. The auricle of each external ear develops from the surface swellings around the first pharyngeal cleft, and the bumps already present on these embryonic primordia form the characteristic hillocks of the adult ear lobe (see Fig. 12–10).

The cartilage of the **third** pharyngeal arch produces the greater horn of the hyoid bone and part of the body (see Fig. 12–6). The single muscle that

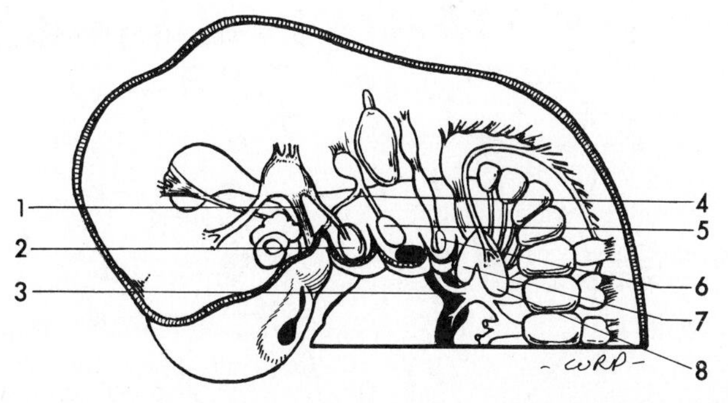

FIGURE 12–15. Five-week embryo showing cranial nerve distribution and developing muscle origins (shown schematically) of the face. *1*, Maxillary division of the trigeminal (no skeletal muscle innervation). *2*, Muscles of mastication developing from branchiomeric mesenchyme of the mandibular (first) arch supplied by the mandibular division of V. *3*, Hypoglossal nerve to the intrinsic muscles of the tongue. *4*, Chorda tympani of the facial nerve leaving second pharyngeal arch to enter tongue and provide sensory (gustatory) innervation. *5*, Facial muscles developing from hyoid arch mesenchyme, supplied by the facial nerve (VII). *6*, Stylopharyngeus muscle of third arch origin, supplied by IX. *7*, Pharyngeal muscles, supplied by X. *8*, Trapezius and sternocleidomastoid muscles, supplied by the spinal accessory nerve. (Modified from Patten, B. M.: *Human Embryology*, 3rd Ed. New York, McGraw-Hill, 1968.)

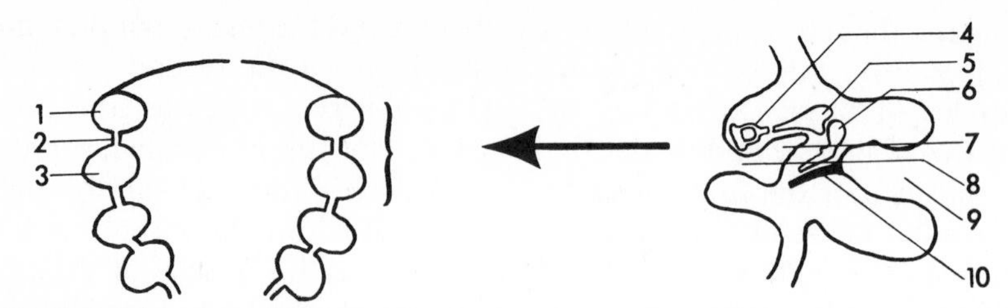

FIGURE 12–16. Developing ear region. *1,* Mandibular arch. *2,* Branchial membrane between the cleft on the outside and pouch on the inside of the pharynx. *3,* Hyoid arch. *4,* Stapes. *5,* Incus. *6,* Malleus. *7,* Middle ear chamber to expand as tympanic cavity surrounding auditory ossicles. *8,* Auditory (eustachian) tube. *9,* External ear canal. *10,* Anlage for the tympanic membrane.

develops from the third arch branchiomeric mesenchyme is the stylopharyngeus. The specific cranial nerve entering the third arch is the **glossopharyngeal.** It therefore supplies the muscle that develops from this arch (see Fig. 12–15). The cartilages in the remainder of the arches form into the thyroid, cricoid, and arytenoid components of the larynx. From fourth arch branchiomeric mesenchyme develop the cricothyroid and pharyngeal constrictor muscles. The specific nerve of the fourth pharyngeal arch is the superior laryngeal branch of the **vagus.** The intrinsic laryngeal muscles develop from the sixth arch and are innervated by that arch's nerve, which is the recurrent laryngeal branch of the vagus.

In each second pharyngeal **pouch,** the lining endoderm and underlying mesenchyme proliferate to form the paired **palatine tonsils** (Fig. 12–17). From the lining of the third pouch develops **parathyroid III** (so called bcause of its third arch origin). This will form the "inferior" parathyroid because it later descends to a level below parathyroid IV. The thymus also develops from the lining of the third pharyngeal arch. Parathyroid IV (the "superior" parathyroid) develops from the fourth pouch.

In the floor of the pharynx, the first (mandibular) arches form rapidly growing **lingual swellings** (Fig. 12–18). A smaller midline swelling, the **tuber-**

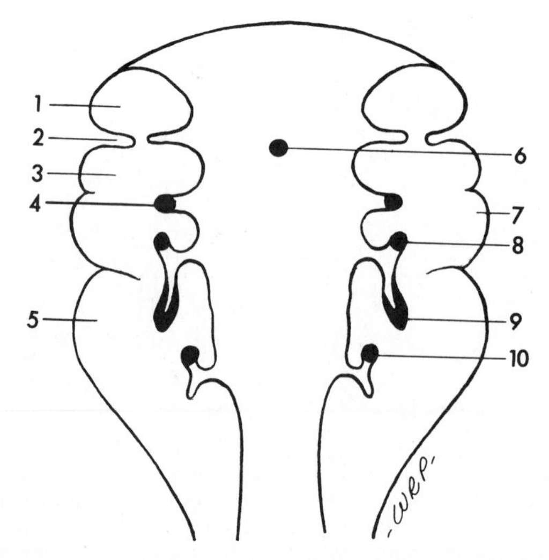

FIGURE 12–17. Pharyngeal pouch derivatives. *1,* Mandibular arch. *2,* Hyomandibular cleft. *3,* Hyoid arch. *4,* Palatine tonsils. *5,* Fourth arch. *6,* Thyroid diverticulum. *7,* Third arch. *8,* Parathyroid III. *9,* Thymus. *10,* Parathyroid IV.

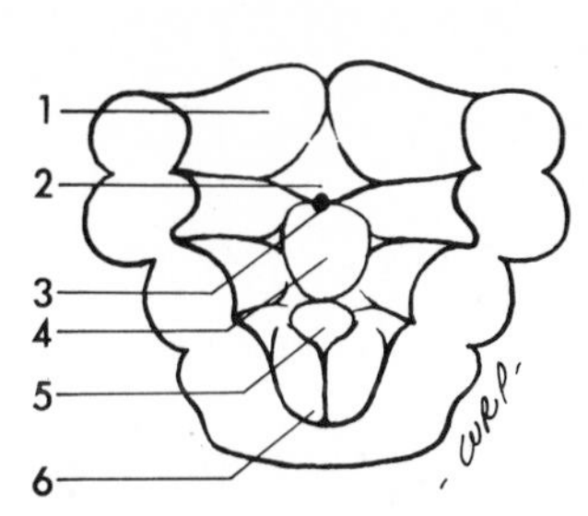

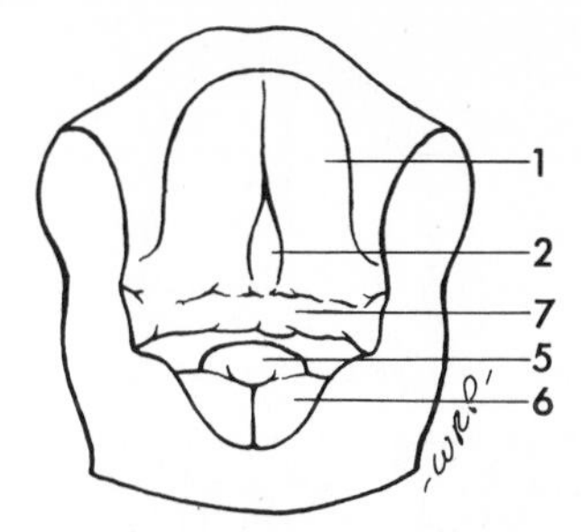

FIGURE 12–18. Developing tongue at 6 and 8 weeks. *1*, Lateral lingual swelling. *2*, Tuberculum impar. *3*, Foramen caecum. *4*, Copula. *5*, Epiglottis. *6*, Arytenoid swellings. *7*, Root of tongue.

culum impar, is also present, and these three structures develop into the mucosal covering for the anterior two thirds, or body, of the tongue. Since the mandibular nerve supplies first arch tissue, it therefore provides the sensory (tactile) innervation for the mucosa of the body of the tongue. The chorda tympani, which is a branch of VII that jumps from the second to the first arch by crossing through the branchial (tympanic) membrane to join the mandibular nerve (lingual branch), provides gustatory innervation for the tongue's mucosa.

At the root of the midventral parts of the second, third, and fourth pharyngeal arches, another prominent swelling occurs, the **copula.** This general region develops into the posterior third (root) of the tongue. The cranial nerves supplying the third and fourth arches are the glossopharyngeal and vagus, and these are thus the sensory nerves that innervate the mucosa of the root of the tongue. The core of the tongue is occupied by its "intrinsic" muscles. These originate from a more caudal region (probably from occipital somatic mesoderm) and grow into the expanding mucosal covering for the tongue being formed by the floor of the pharynx (described above). The motor innervation to these muscles is provided by the paired hypoglossal (XII) nerves. They are carried along with the intrinsic muscles as they migrate anteriorly into the body of the tongue.

Anatomically, the body of the tongue is separated from the root by a V-shaped sulcus (the **sulcus terminalis,** Fig. 12–19). This marks the approximate line between the derivatives of first arch origin and those from the arches behind the first arch. At the midline in this developing groove, between the tuberculum impar and the copula, the thyroid primordium develops as an epithelial diverticulum into the pharyngeal floor (see Figs. 12–7 and 12–17). It then separates from the mucosal lining and migrates caudally. The point of invagination, however, remains as a permanent pit termed the **foramen caecum** (see Fig. 12–18). It is located at the apex of the V and is a landmark that identifies the adult position of the embryonic boundary between the first and second arches. As for most glandular tissues, the thyroid is thus of epithelial origin; and because the primordium develops from the pharyngeal lining, it is of endodermal derivation.

By the time an embryo is about 5 weeks old, the first pharyngeal arch has formed recognizable **maxillary** and **mandibular swellings** (Figs. 12–20 to 12–23). Just above the stomodeum, the paired, laterally located **nasal placodes** have already formed by thickenings of the surface ectoderm, and horseshoe-like ridges **(nasal swellings)** have developed around them to form deepening **nasal pits.** The floor of each pit is termed the oronasal membrane, but it is a

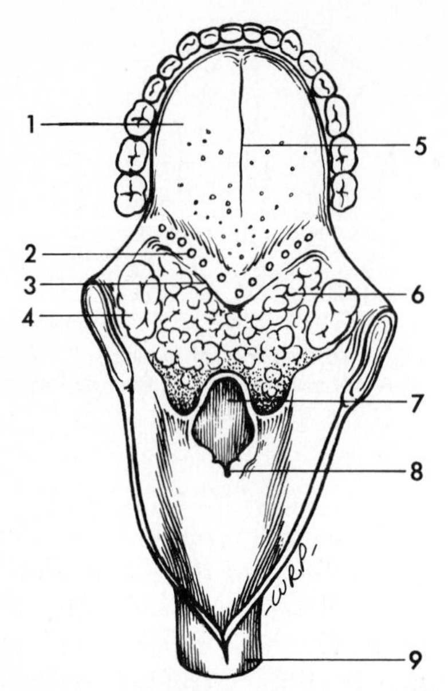

FIGURE 12–19. Adult tongue. *1*, Body, *2*, Circumvallate papilla. *3*, Sulcus terminalis. *4*, Palatine tonsil. *5*, Median sulcus. *6*, Lingual tonsil in root of tongue. *7*, Epiglottis. *8*, Interarytenoid notch. *9*, Esophagus.

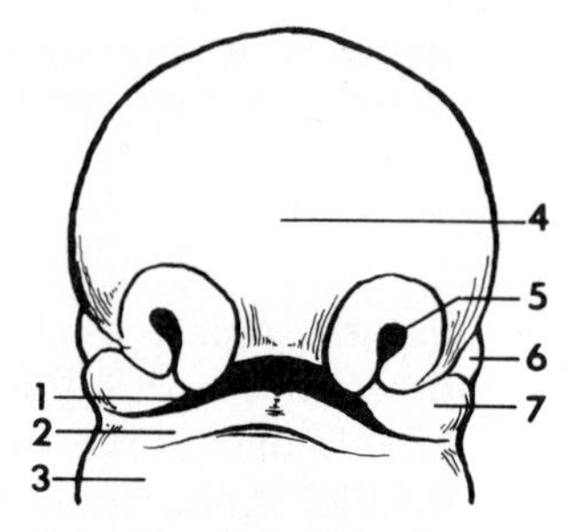

FIGURE 12–20. Facial region at about 5 weeks. *1*, Stomodeum. *2*, Mandibular swelling. *3*, Hyoid arch. *4*, Frontal prominence. *5*, Nasal pit. *6*, Optic vesicle. *7*, Maxillary swelling. (Adapted from Langman, J.: *Medical Embryology.* Baltimore, Williams & Wilkins, 1969.)

transient structure that soon breaks through, thus opening the nasal pits directly into the oral cavity. At the same time, the semicircular nasal swellings continue to enlarge. Each swelling is composed of a lateral and medial limb. The expanding **medial** limbs merge at the midline to form the primordium that will differentiate into the middle part of the nose, the philtrum of the lip, the "incisor" part of the maxilla (premaxilla), and the small primary palate.

The rapidly growing **lateral** limbs of each nasal swelling form the alae of the nose. While these changes occur, the maxillary swellings are also enlarging; and they subsequently merge with the medial limbs of the nasal swellings. The furrow between them (not a complete cleft in normal development) disappears; and a closed, U-shaped arch is thereby formed. The medial limbs form the

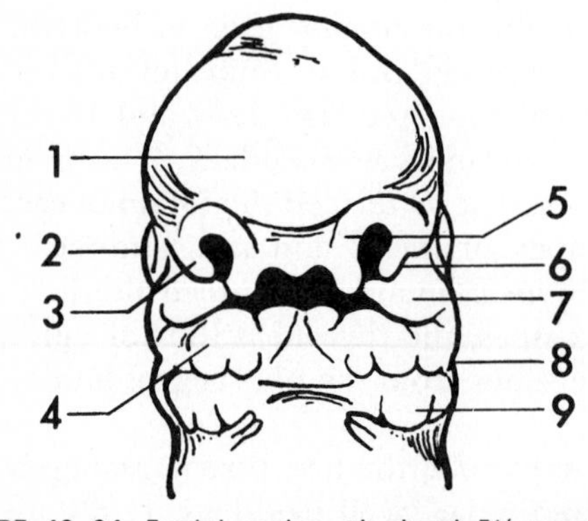

FIGURE 12–21. Facial region at about 5½ weeks. *1*, Forebrain. *2*, Optic vesicle. *3*, Lateral nasal swelling. *4*, Mandibular process. *5*, Medial nasal swelling. *6*, Nasolacrimal groove. *7*, Maxillary process. *8*, Hyomandibular cleft. *9*, Hyoid arch. (Modified from Patten, B. M.: *Human Embryology*, 3rd Ed. New York, McGraw-Hill, 1968.)

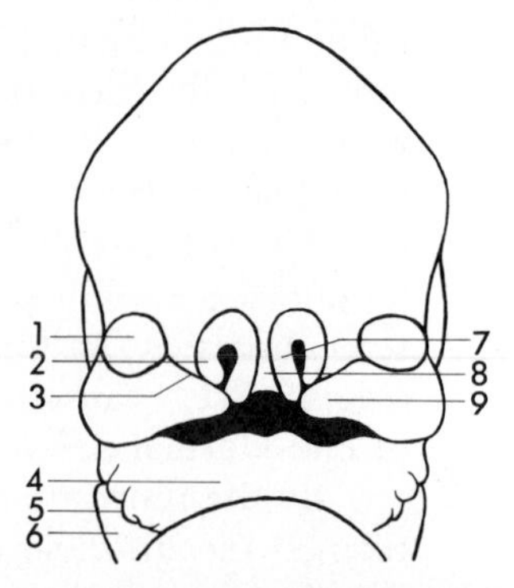

FIGURE 12–22. The face between 6 and 7 weeks. *1*, Eye. *2*, Lateral nasal swelling. *3*, Nasolacrimal groove. *4*, Mandibular swelling. *5*, Hyomandibular cleft. *6*, Hyoid arch. *7*, Medial nasal swelling. *8*, Midline nasal region where nasal septum is forming. *9*, Maxillary swelling.

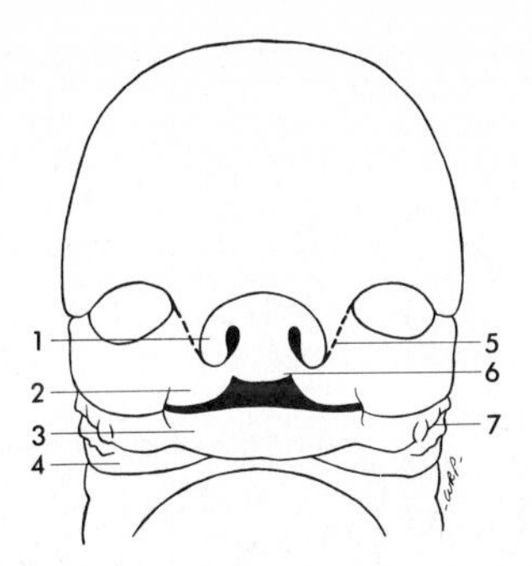

FIGURE 12–23. The face between 7 and 8 weeks. *1*, Nasolateral process. *2*, Maxillary process. *3*, Mandible. *4*, Hyoid arch (note formation of external ear lobes). *5*, Merger line of nasolacrimal groove. *6*, Philtrum. *7*, Hyomandibular cleft.

middle span of both the maxillary arch and the upper lip. The canine, premolar, molar, and lateral lip parts of the upper arch develop from the maxillary processes, and the incisor and medial lip (philtrum) parts develop from the medial nasal swellings. **These** are some of the lines of merger that can be involved in cleft lip and jaw. Sometimes, developmental variations are encountered in which the blastema of a tooth is caught on the "wrong" side of a cleft; this always causes excitement because that is not the way it is supposed to be.

An oblique groove is present between the maxillary swelling and the lateral limb of the nasal swelling. This is the **nasolacrimal groove,** which will soon close, but the line of merger establishes a developmental pathway for the later formation of the nasolacrimal duct. If this merger fails, a permanent facial cleft or fissure results. The superficial tissues in lateral areas of the maxillary process fuse with the mandibular process to form the cheek. **Epithelial pearls** often occur along such lines of mucosal and cutaneous fusion. These are small islands of epithelial cells that were "programmed" to form but were caught up in the fusion process. **Fordyce's spots,** which are remnants of cutaneous sebaceous glands, can similarly be found in the adult buccal mucosa, for the same reason, along lines of fusion.

The growing right and left mandibular swellings join at the midline to form the lower jaw and lip. A cartilaginous interface forms at this junction.*

The frontal prominence forms the forehead and a vertical zone of tissue between the merging medial nasal swellings. Here, the midline **nasal septum** is formed, which is believed by some to function as a pacemaker in later fetal development when its core becomes cartilaginous.

To date, all these facial changes are occurring at about the same time and have proceeded **rapidly** from about the fourth to the sixth week of embryonic development. The paired **palatal shelves** are now forming from each side of the maxillary arch (Figs. 12–24 to 12–29). The oral cavity is still relatively small, however, and the sizable tongue remains interposed between the right and left

---

*Although a limited amount of endochondral ossification will later occur here, the two halves of the mandible fuse completely after birth, unlike the permanently separate lower jaw halves of nonprimates. Except in the "secondary cartilage" of the mandibular condyle (and to a much lesser extent in the mandibular symphysis and a small cartilage on the coronoid process), the greater part of mandibular ossification is by the intramembranous mechanism. Meckel's cartilage does not participate in the endochondral ossification process (except for spots here and there) and disappears except for its contribution to the ear ossicles and ligaments. The maxilla is entirely of intramembranous origin.

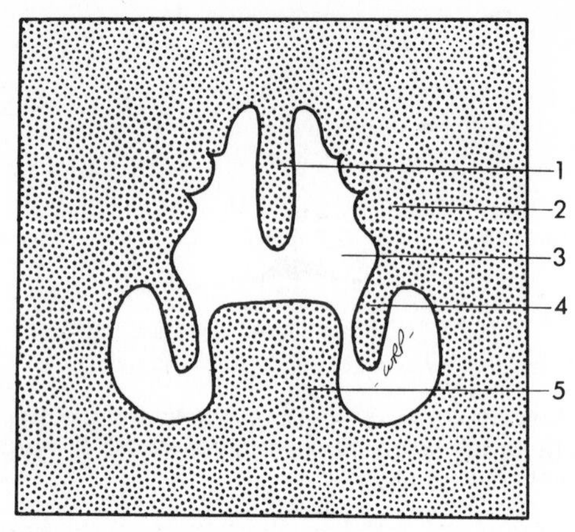

FIGURE 12–24. Frontal section through the oronasal region in a 6½-week embryo. *1,* Nasal septum. *2,* Lateral nasal wall. *3,* Nasal chamber. *4,* Palatal shelf. *5,* Tongue. (Adapted from Langman, J.: *Medical Embryology.* Baltimore, Williams & Wilkins, 1969.)

shelves. The early shelves necessarily enlarge downward in an obliquely vertical manner because of this. However, the inferior expansion of the whole lower part of the face carries the tongue downward. The oral cavity increases greatly in size. The paired nasal chambers are still continuous with the oral cavity (right and left nasal cavities are present because the nasal septum developing downward from the frontal prominence has sustained the original paired nature of the nasal primordia). The oral and nasal cavities at this stage are separated from each other only in the anterior-most region by the tiny, unpaired **primary palate** (median palatine process). The latter was formed by the fusion

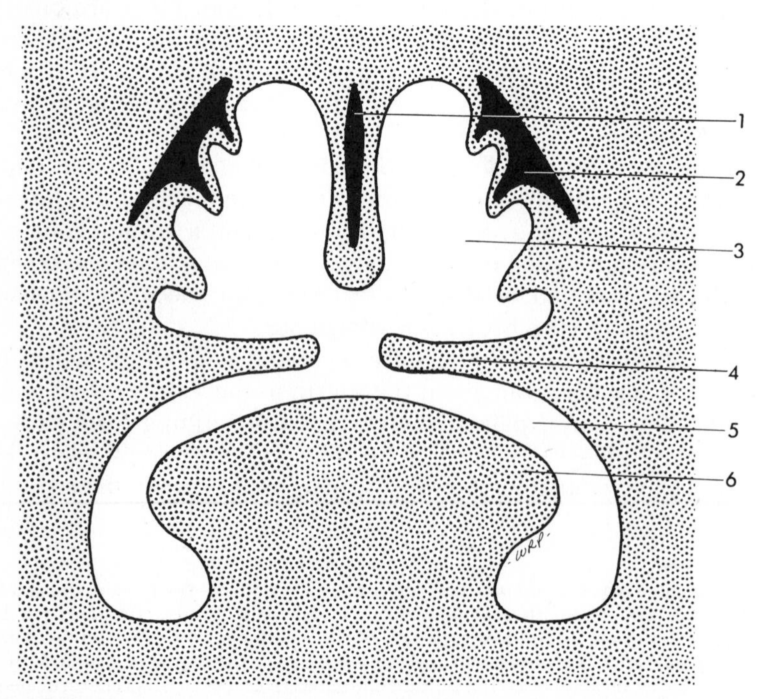

FIGURE 12–25. Frontal section through the oronasal region in a 7½-week embryo. *1,* Cartilage of the nasal septum. *2,* Cartilage of the nasal conchae. *3,* Nasal chamber. *4,* Palatal shelf. *5,* Oral cavity. *6,* Tongue. (Adapted from Langman, J.: *Medical Embryology.* Baltimore, Williams & Wilkins, 1969.)

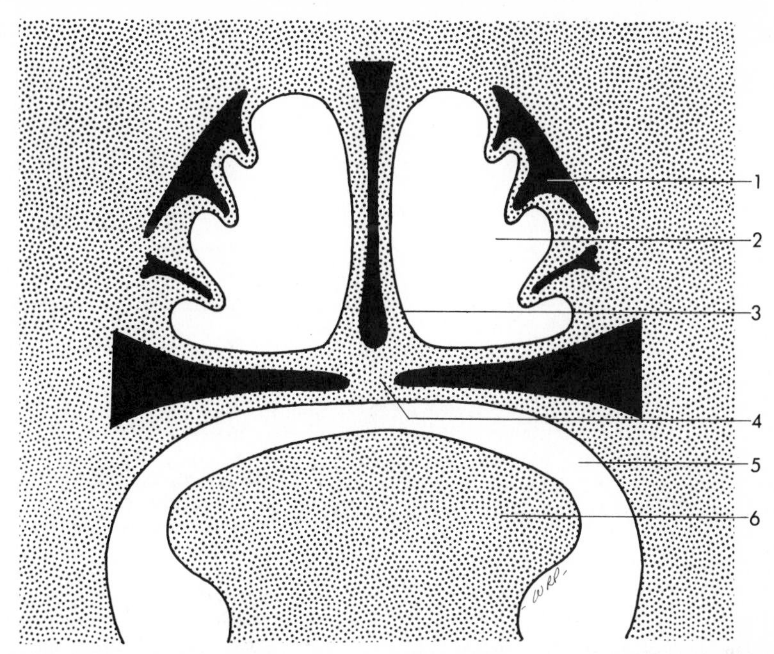

FIGURE 12–26. Frontal section through the oronasal region of a 10-week embryo. *1,* Nasal conchae. *2,* Nasal chamber. *3,* Nasal septum. *4,* Palatal shelves, fused at midline and fused with nasal septum. The intramembranous bone of the palatal shelves (from the maxilla) is beginning to form. *5,* Oral cavity. *6,* Tongue. (Adapted from Langman, J.: *Medical Embryology.* Baltimore, Williams & Wilkins, 1969.)

of the nasomedial ("premaxillary") processes. The whole lower part of the developing face, including the tongue and the floor of the oral cavity, now becomes displaced inferiorly to a greater extent than the palatal shelves are descending, so that the newly formed shelves of the maxilla are free to expand **medially** as well. They come together and soon fuse along the midline (the

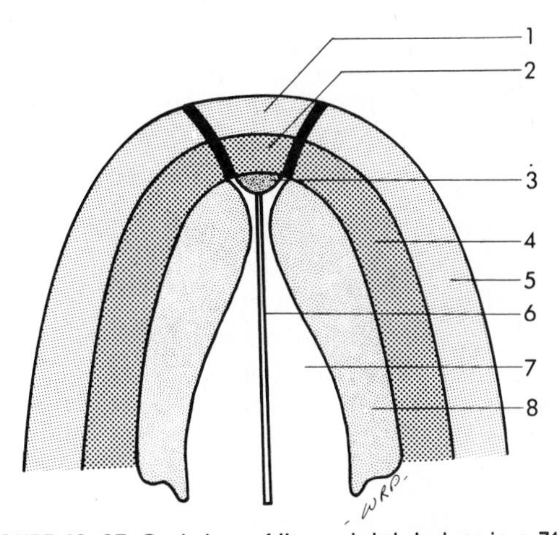

FIGURE 12–27. Oral view of the palatal shelves in a 7½-week embryo. *1,* Philtrum of upper lip. *2,* "Premaxillary" segment from medial nasal processes. *3,* Primary palate. *4,* Upper arch (part derived from maxillary swellings). *5,* Cheek. *6,* Nasal septum. *7,* Open oral and nasal cavities. *8,* Palatal shelves. In this stage, the philtrum and premaxillary segment have already merged with the maxillary swellings.

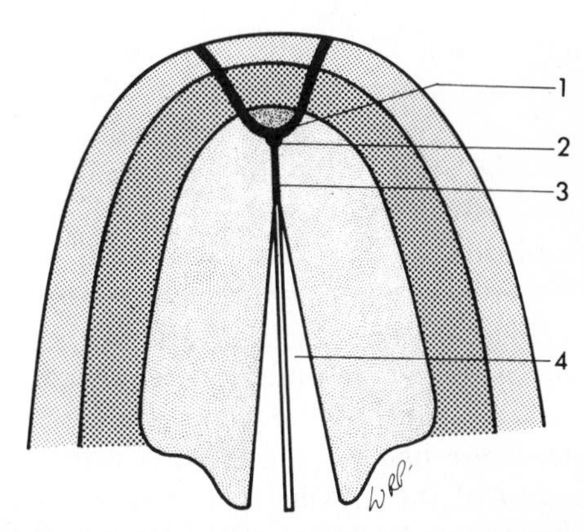

FIGURE 12–28. Oral view of palate showing beginning of fusion. *1,* Merger of midline primary palate with bilateral secondary palatal shelves. *2,* Incisive foramen. *3,* Palatal raphe (midline fusion). *4,* Open nasal and oral chambers.

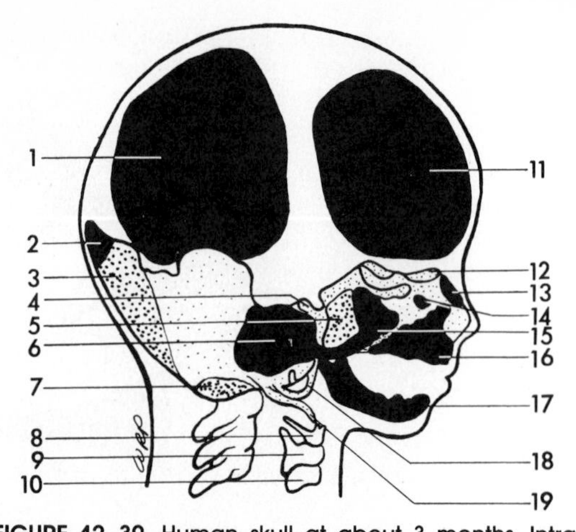

FIGURE 12–30. Human skull at about 3 months. Intramembranous bones are shown in black. Cartilage is represented by light stippling, and bones developing by endochondral ossification are indicated by darker stippling. Approximate time of appearance for each bone is indicated in parentheses. *1,* Parietal bone (10 weeks). *2,* Interparietal bone (8 weeks). *3,* Supraoccipital (8 weeks). *4,* Dorsum sellae (still cartilaginous). *5,* Temporal wing of sphenoid (2 to 3 months; the basisphenoid appears at 12 to 13 weeks, orbitosphenoid at 12 weeks, and presphenoid at 5 months). *6,* Squamous part of temporal bone (2 to 3 months). *7,* Basioccipital (2 to 3 months). *8,* Hyoid (still cartilaginous). *9,* Thyroid (still cartilaginous). *10,* Cricoid (still cartilaginous). *11,* Frontal bone (7½ weeks). *12,* Crista galli, still cartilaginous (inferiorly, the middle concha begins ossification at 16 weeks, the superior and inferior conchae at 18 weeks; the perpendicular plate of ethmoid begins ossification during the first postnatal year, the cribriform plate during the second postnatal year, the vomer at 8 fetal weeks). *13,* Nasal bone (8 weeks). *14,* Lacrimal bone (8½ weeks). *15,* Malar (8 weeks). *16,* Maxilla (end of 6th week; premaxilla, 7 weeks). *17,* Mandible (6 to 8 weeks). *18,* Tympanic ring (begins at 9 weeks, with complete ring at 12 weeks; petrous bone, 5 to 6 months). *19,* Styloid process, still cartilaginous. (Modified from Patten, B. M.: *Human Embryology,* 3rd Ed. New York, McGraw-Hill, 1968.)

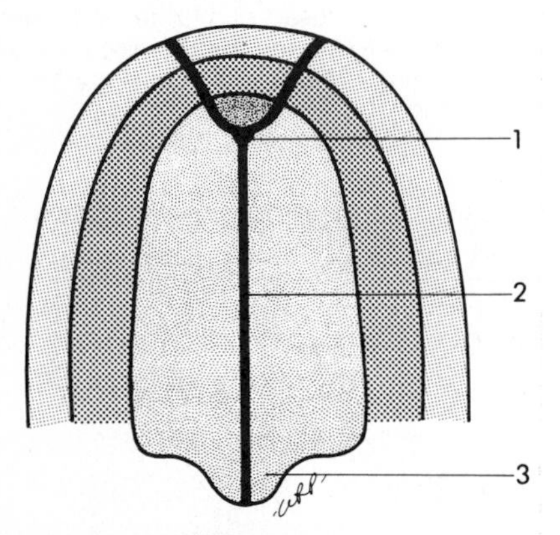

FIGURE 12–29. Full length palatal fusion. *1,* Incisive foramen. *2,* Palatal raphe. *3,* Uvula.

palatal raphe). The shelves "swing upward" in order to contact each other, but some of this apparent upward rotation is produced by differential growth. The shelves are growing and, especially, are becoming displaced inferiorly. By differential growth, the shelves expand downward as well as toward each other, but the **entire** nasal chamber on each side also expands laterally and inferiorly. Some of this apparent "upward swing" is relative and is a consequence of actual inferomedial growth. This is a process in which the different parts in the two nasal chambers and the oral cavity all grow at differential rates and to different extents as the whole midfacial region rapidly increases in size.

The merger of the right and left palatal shelves forms the **secondary palate.** Bone tissue soon appears within it. This part of the palate is a direct extension of the maxilla from which it develops. The original primary palate, formed from the nasomedial (premaxillary) processes, is retained as a small median, unpaired, triangular segment of the palatal complex in the anterior region just ahead of the incisive foramen, a landmark that identifies the midline boundary between the primary and secondary parts of the palate (see Fig. 12–

29). The separate palatine bones and their posterior contribution to the palatal complex do not develop until somewhat later. In the meantime, the nasal septum has merged with the superior surface of the palate. The two nasal chambers are now completely separated, and both have been closed off from the oral cavity along the length of the palate. While these various changes take place, the nasal conchae are already developing as medial and inferior growing processes from the lateral walls of each nasal chamber.

COMPARISON OF PRENATAL AND POSTNATAL GROWTH PROCESSES IN THE FACE AND CRANIUM

The process of "remodeling," involving periosteal resorptive surfaces, first begins in the fetus at about 10 weeks in two principal locations: on lining surfaces of the bone around tooth buds and on the endocranial surface of the frontal bone. The major remodeling throughout the remainder of the early facial skeleton begins at about 14 weeks. Before this time the bones enlarge in all directions from their respective ossification centers. Remodeling, as a process that accompanies growth, starts when the definitive form of each of the individual bones of the face and cranium is attained (Fig. 12–30).

Nasomaxillary Complex

The anterior part of the maxilla in both the fetus and the child is depository on lingual surfaces and resorptive on nasal lining surfaces. A major difference exists, however, on the anterior-most (labial) surface (Figs. 12–31 to 12–33). Here it is depository in the fetus but characteristically becomes resorptive after the first few years following birth. During the fetal period, the exterior surface of the entire maxilla, including its anterior part,* remains depository to provide for increasing arch length in conjunction with the development of the tooth buds and their subsequent enlargement. Resorption occurs on all alveolar lining surfaces surrounding each of the tooth buds (Figs. 12–34 and 12–35). The fetal maxillary arch thus lengthens horizontally in both posterior **and** anterior directions, in contrast to the largely posterior mode of elongation in the later periods of childhood growth.

During the primary dentition period, anterior expansion of the maxillary arch no longer takes place except for small amounts at the margins along the alveolar crest. Not only have all the tooth buds been formed, but the deciduous teeth have erupted and are beginning to be shed, making way for the permanent teeth. The anterior surfaces subsequently become resorptive as a part of the growth and remodeling process that continues to produce the **downward** growth movement of the maxillary arch and palate.†

The posterior and infraorbital surfaces of the maxilla proper are depository in both prenatal and postnatal life (Figs. 12–31 to 12–33). The process of posterior deposition on the maxillary tuberosity progressively increases the

---

*An old game among facial researchers is arguing whether or not a premaxilla exists as a separate bone in man and how many ossification centers are involved.

†During postnatal growth, the vertical length of the maxilla increases more than its width. Upper jaw alveolar width and bigonial width increase about twofold beyond the neonatal size. Facial height increases threefold relative to the orbits. Mandibular corpus height increases about 2.5-fold, ramus height 3.5 fold, and ramus depth about 1.5-fold. The maxillary sinus increases in diameter from about 5 to 6 mm at birth to 12 to 14 mm at 5 to 6 years to 20 to 26 mm in the adult.

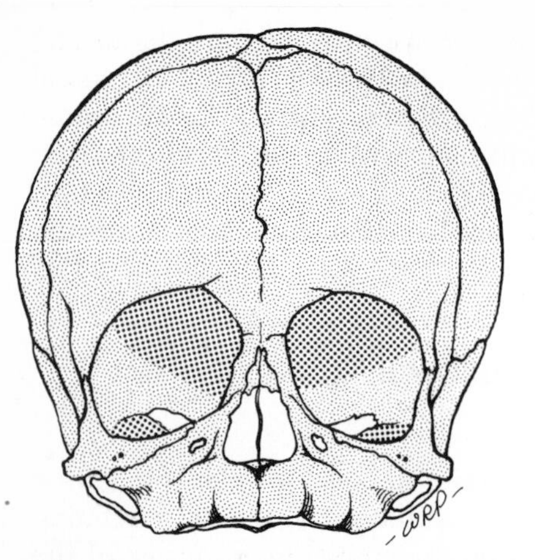

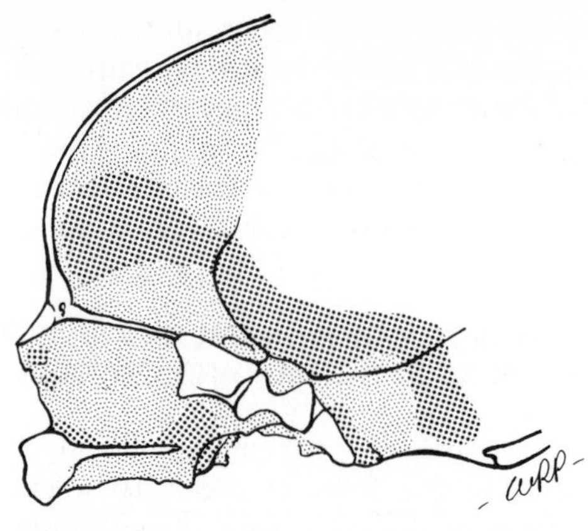

FIGURE 12–31. Resorptive (darkly stippled) and depository (lightly stippled) patterns of growth in a 20-week fetus. See text for descriptions. (From Mauser, C., D. H. Enlow, D. O. Overman, and R. McCafferty: Growth and remodeling of the human fetal face and cranium. In: *Determinants of Mandibular Form and Growth.* Ed. by J. A. McNamara. Center for Human Growth and Development, University of Michigan, Monograph 5, Craniofacial Growth Series, 1975.)

maxilla in horizontal length. Deposition on the orbital floor in the fetal skull keeps it in a constant positional relationship with the eyeball, just as in the growing child. The eyeball enlarges in volume at a decreasing rate after the fourth to fifth fetal months. Its volume increases by over 100 per cent before the fifth month, by 50 per cent during the sixth and seventh months, and only by 23 to 30 per cent in the eighth and ninth months. Remodeling of the orbital floor takes place because the entire maxilla, including the orbit, is displaced in a progressively inferior direction in relation to continuing new bone growth at the frontomaxillary suture. At the same time, deposition on the orbital floor serves to carry it superiorly, thereby maintaining it in a constant position relative

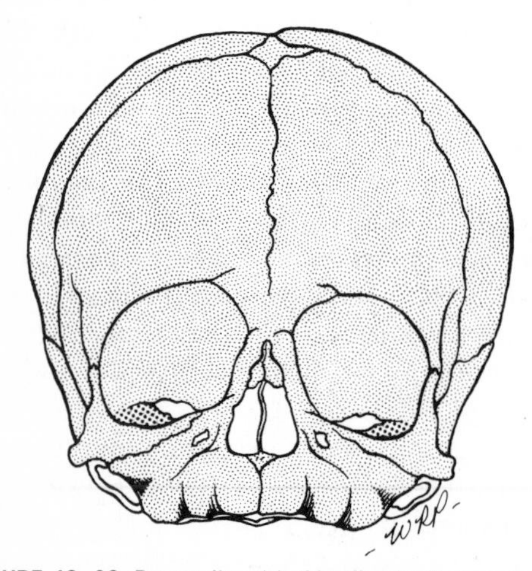

FIGURE 12–32. Resorptive (darkly stippled) and depository (lightly stippled) growth patterns in a term fetus. (From Mauser, C., D. H. Enlow, D. O. Overman, and R. McCafferty: Growth and remodeling of the human fetal face and cranium. In: *Determinants of Mandibular Form and Growth.* Ed. by J. A. McNamara. Center for Human Growth and Development, University of Michigan, Monograph 5, Craniofacial Growth Series, 1975.)

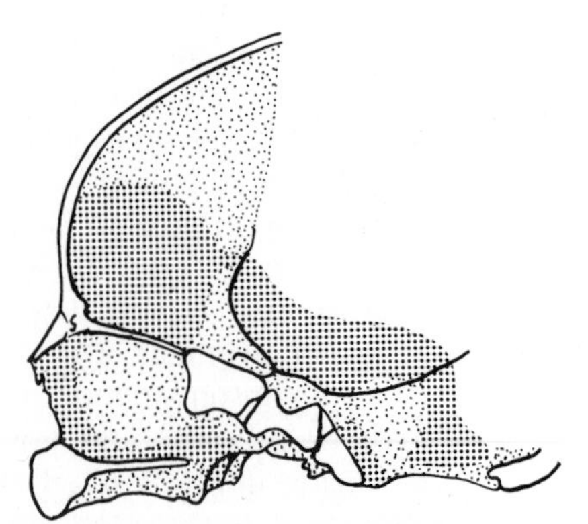

FIGURE 12–33. Resorptive (darkly stippled) and depository (lightly stippled) growth patterns in a term fetus. (From Mauser, C., D. H. Enlow, D. O. Overman, and R. McCafferty: Growth and remodeling of the human fetal face and cranium. In: *Determinants of Mandibular Form and Growth.* Ed. by J. A. McNamara. Center for Human Growth and Development, University of Michigan, Monograph 5, Craniofacial Growth Series, 1975.)

FIGURE 12–34. Resorptive (−) and depository (+) fields in a frontal section through the face of a 26-week fetus. *M,* Maxilla; *Z,* zygomatic bone; *TB,* tooth bud; *V,* vomer; *CG,* crista galli; *SNC, MNC, INC,* superior, middle, and inferior nasal conchae; *N,* inferior orbital nerve; *NS,* nasal septum. (From Mauser, C., D. H. Enlow, D. O. Overman, and R. McCafferty: Growth and remodeling of the human fetal face and cranium. In: *Determinants of Mandibular Form and Growth.* Ed. by J. A. McNamara. Center for Human Growth and Development, University of Michigan, Monograph 5, Craniofacial Growth Series, 1975.)

to the eyeball. The infraorbital canal is also moving upward during maxillary growth by resorption superior to and deposition inferior to the infraorbital nerve. These processes maintain a constant relationship between the nerve and the orbital floor, along which the infraorbital nerve passes before entering the infraorbital canal.

The external surface of the frontal process of the maxilla is depository during both prenatal and postnatal facial development (Fig. 12–35). The

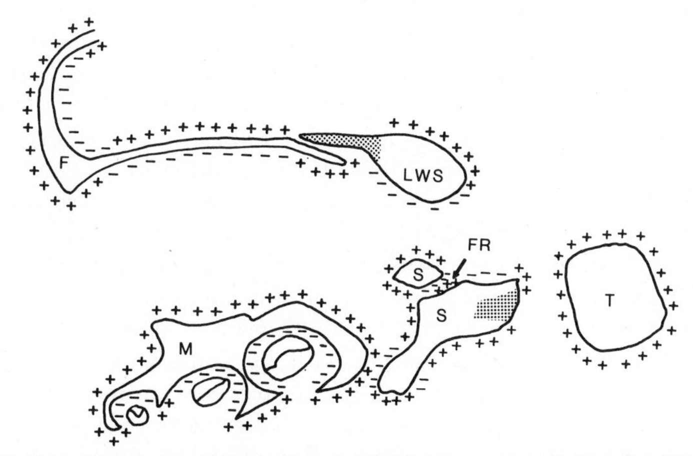

FIGURE 12–35. Resorptive (−) and depository (+) fields in a parasagittal section through the 26-week fetal skull. *M,* Maxilla (with tooth buds); *S,* sphenoid; *FR,* foramen rotundum; *T,* temporal bone; *LWS,* lesser wing of sphenoid; *F,* frontal bone, showing the pattern found until the last trimester. (After Mauser, C., D. H. Enlow, D. O. Overman, and R. McCafferty: Growth and remodeling of the human fetal face and cranium. In: *Determinants of Mandibular Form and Growth.* Ed. by J. A. McNamara. Center for Human Growth and Development, University of Michigan, Monograph 5, Craniofacial Growth Series, 1975.)

contralateral nasal side is mostly depository, with some resorption at older fetal ages, but it is entirely resorptive postnatally. This area in the rapidly growing young child is characterized by a massive lateral expansion of the lateral nasal walls, including the ethmoidal plates and sinuses. The latter part of fetal life appears to be a transitory period in which these surfaces are just beginning a major lateral expansive movement.

In both the fetal and postnatal periods, the nasal side of the palate (including the palatine bone) is resorptive except along the midline, and the oral surface is depository. This provides for an inferior growth movement of the palate and a vertical enlargement of the nasal chambers (Fig. 12–34).

The mucosal surface of the vertical plate of the palatine bone is resorptive, and the opposite lateral surface is depository in both the prenatal and postnatal periods. This provides for the expansion of this part of the nasal chamber in width.

At 18½ weeks, the **vomer** appears as a bone U-shaped anteriorly and Y-shaped posteriorly (Fig. 12–34). Its sloping anterior margin forms a trough for the inferior edge of the proportionately sizable cartilaginous nasal septum. Remodeling of the vomer has already begun at this age. The anterior part is depository inferiorly in the suture between it and the palate. It is also depository laterally and superiorly in the area supporting the cartilaginous nasal septum but is resorptive within the trough. The Y-shaped posterior part of the vomer is depository inferiorly on the surface adjacent to the suture and also superiorly where it abuts the cartilaginous nasal septum. The vomer in this posterior region is resorptive on the inferolateral surface, depository on the superolateral surface, and resorptive on the medial surface adjacent to the area of lateral deposition.

The cartilages in the **inferior nasal conchae** begin endochondral ossification between 15 and 17 weeks. At this age, the cartilage is hypertrophic, and endochondral ossification is in an early stage. By 22 weeks ossification is well under way, and by 26 weeks remodeling has become established, the bone surfaces being resorptive inferiorly and depository superiorly. Later stages show the opposite, with resorption superiorly and deposition inferiorly.

The **superior and middle nasal conchae** ossify somewhat later than the inferior conchae, not beginning until 17 weeks. By 22 weeks ossification is established, and by 26 weeks remodeling is taking place in both of these ethmoidal labyrinths. The middle concha is depository superiorly and resorptive inferiorly. The superior concha is depository inferiorly and resorptive superiorly on the anterior and posterior thirds but depository on the middle third.

The **nasal bones** in both the fetal and postnatal periods are resorptive on the mucosal surface and depository on the external side. The bony bridge and roof of the external nasal protuberance are thus moved anteriorly.

The **lacrimal bone** in the fetus is depository only on the surface adjacent to the nasolacrimal canal. In the adult, however, it is depository on its superolateral and inferomedial surfaces, with the contralateral surfaces showing resorption. In postnatal development, the sutures surrounding this small bone function to adjust to the changing relative positions of the other bones with which it articulates (frontal, maxillary, and ethmoid) by providing sutural "slippage" as they all grow at different rates around it. These facial and cranial changes are not nearly as marked in the fetal skull, however, and the remodeling pattern that characterizes the lacrimal bone in the postnatal skull does not appear until later in childhood when marked vertical midfacial growth and ethmoidal expansion begin relative to a more stable orbit.

The external surfaces of the **zygomatic bone** are primarily depository in fetal life. Resorption does occur, however, on the infraorbital margin where the zygoma overlaps the maxilla. In postnatal development, both the anterior and orbital surfaces are resorptive. Transition from the fetal to the postnatal pattern occurs at some time during the primary dentition period when the premaxillary region ceases to grow in a direct anterior direction. Some resorption is already present on the anterior surface of the zygoma and always on the orbital surface, even during the fetal period. The postnatal change in pattern relates to the posterior relocation of the zygoma in conjunction with the posterior mode of maxillary lengthening in later childhood.

Calvaria and Cranial Base

The squama of the frontal bone in the forehead region has the same pattern of remodeling and growth in the prenatal and postnatal periods (Fig. 12–35). The ectocranial surface is depository, and the endocranial side is resorptive (except along the midline crest). In the region above the frontal eminence, a reversal line occurs, and superiorly the squama becomes depository on both the intracranial and ectocranial sides. This same pattern continues throughout the postnatal growth period.

Deep to the supraorbital rim in the postnatal skull, the entire dural surface is resorptive. On the orbital side beneath the rim, the surface is depository medially and resorptive laterally. The roof of the orbit is entirely resorptive on the dural side and depository on the orbital side. During prenatal growth the medial half of the rim of the orbit, however, is characteristically resorptive, and the lateral half is depository on the intraorbital side (Fig. 12–35). On the dural surface the pattern is reversed, with deposition on the medial half and resorption on the lateral half. The roof of the orbit is resorptive on the intraorbital side and depository on the dural side (floor of the anterior cranial fossa) in early fetal life. In the last trimester, complete reversals of the fetal pattern occur, however. By 39 weeks the roof of the orbit exhibits the characteristic postnatal pattern.

The difference between the prenatal and postnatal patterns of growth and remodeling in the thin-layered roof of the orbit and floor of the anterior cranial fossa appears to be based on differential growth rates by the eyeball and the cerebral hemispheres. As previously mentioned, the relative rate of eyeball growth is much faster before 24 weeks' gestation than after. The frontal lobes of the brain "lag" in their growth until 25 or 26 weeks. The frontal lobes are only 4.5 per cent of their adult size when the fetus is 6 months old and about 11 per cent at birth. After birth a rapid spurt of growth takes place, until about 47 per cent of the adult size is attained at 11½ months and 93 per cent at 7 years.

As the eyeballs pass their stage of maximum expansion and the frontal lobes begin theirs, a reversal in the growth pattern of these parts then occurs to accommodate the changes. Slight variations in the reversal times during this period probably relate to individual variations in the differential growth rates of the different soft tissues involved. The change in the remodeling pattern coincides with the period when the growth of the eyeballs is just slowing and the subsequent spurt of the frontal lobe growth is just beginning.

It is usually stated that the medial part of the anterior cranial base (the mesethmoid) does not begin to ossify until after birth. However, ossification can be noted in some 33- to 39-week-old specimens. When seen, the superior surface of the thin bony plate is resorptive, and the inferior surface is depository. This is the same pattern characterizing the bone in the postnatal skull, and it functions to lower the mesethmoid in conjunction with the inferiorly directed growth and displacement of the remainder of the anterior cranial floor.

The prenatal and postnatal growth and remodeling patterns of the greater wing of the sphenoid are essentially the same (Fig. 12–36). The bone is characteristically resorptive endocranially and depository ectocranially at all fetal ages after its early form has been attained (that is, after 14 weeks). This accommodates the expansion of the temporal lobes; these lobes develop earlier in prenatal life than do the frontal lobes.

In fetal growth, the maxillary nerve and the foramen rotundum through which it passes both move forward. The foramen undergoes this remodeling movement by resorption anterior and deposition posterior to the nerve. The nerve and its canal thereby keep pace with the anterior growth expansion of the entire facial complex.

In the postnatal period, the pterygoid plates are depository on all sides at their base where they connect with the rest of the sphenoid. Inferiorly, however, most of the periosteal surfaces are resorptive, except posteriorly, in the trough, where deposition occurs. The fetal pattern in the last trimester is similar to the postnatal pattern, most surfaces being depository but with some resorption on the anteroinferior surfaces. A variable amount of resorption is also noted on the posterior surface, which enlarges the pterygoid fossa. This is a common variation in the postnatal skull until the forward-growing plates make contact with the maxillary tuberosity, at which time downward growth and fossa expansion can only take place by deposition on the lining surface of the downward-facing fossa itself, with resorption on the anterior side.

The postnatal presphenoid is largely resorptive on the endocranial surface. However, in the fetus the lesser wing is resorptive only on its posterior and inferior edges where it borders the superior orbital fissure. It is depository superiorly where this bone forms the posterior part of the anterior cranial floor. Resorption also occurs anterior to the optic nerve and in the chiasmatic

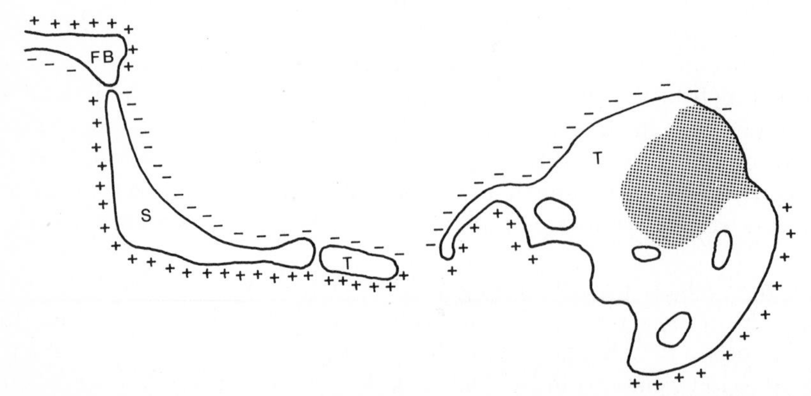

FIGURE 12–36. Resorptive (−) and depository (+) fields in a parasagittal section (lateral to maxillary arch) in a 26-week fetal skull. *FB,* Frontal bone, showing pattern found until reversals occur in last trimester; *S,* sphenoid; *T,* temporal. Stippled area is still cartilaginous. (After Mauser, C., D. H. Enlow, C. O. Overman, and R. McCafferty: Growth and remodeling of the human fetal face and cranium. In: *Determinants of Mandibular Form and Growth.* Ed. by J. A. McNamara. Center for Human Growth and Development, University of Michigan, Monograph 5, Craniofacial Growth Series, 1975.)

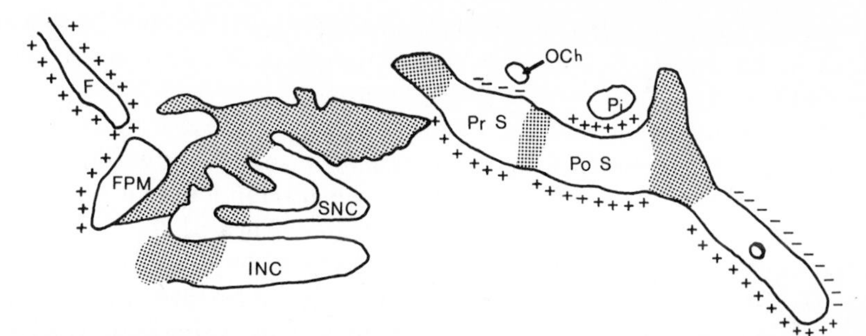

FIGURE 12–37. Resorptive (−) and depository (+) fields in a midsagittal section of a 26-week fetus. *F,* Frontal bone (the only area in which the endocranial surface of the forehead is depository is along the midline crest, as seen here); *FPM,* frontal process of maxilla; *SNC* and *INC,* superior and inferior nasal conchae (endochondral ossification has started in these areas but is not shown here); *OCH,* optic chiasma; *Pr S,* presphenoid; *Po S,* postsphenoid; *Pi,* pituitary; *O,* occipital. Stippled areas are still cartilaginous. (After Mauser, C., D. H. Enlow, D. O. Overman, and R. McCafferty: Growth and remodeling of the human fetal face and cranium. In: *Determinants of Mandibular Form and Growth.* Ed. by J. A. McNamara. Center for Human Growth and Development, University of Michigan, Monograph 5, Craniofacial Growth Series, 1975.)

sulcus. These patterns coincide with the early remodeling pattern of the orbital roof at the time when the eyeball is expanding more rapidly than the frontal lobes. The optic nerve is moving forward to keep pace with the orbit, which also moves forward with the growth of the rest of the face and anterior cranial floor. The dural surface of the presphenoid becomes increasingly resorptive endocranially during a later period, coinciding with the massive growth on the frontal lobes. In the young child, this surface area of the bone usually becomes completely resorptive.

The cortical lining for most of the sella turcica is usually depository during both the prenatal and postnatal growth periods (Fig. 12–37). The superior part of the posterior wall lining the fossa, however, is typically resorptive prenatally as well as postnatally. Variations in reversal line placement exist in the fetus just as they do in the growing child. The floor as well as the anterior and posterior lining walls is variable in growth pattern because of the markedly differential rates of growth of the many separate soft tissue parts in this general region. (See Chapter 3.)

The petrous part of the temporal bone follows essentially the same pattern of growth during both the prenatal and postnatal periods (Fig. 12–36). This involves deposition on the medial surface of the petrous ridge, with resorption on the far lateral portion where it grades into the side of the calvaria. The petrous bone is characteristically depository ectocranially; resorption occurs within the jugular fossa in both age groups.

As in the postnatal period, the fetal basioccipital is resorptive endocranially and depository ectocranially (Fig. 12–37). The lateral part of the occipital complex also follows this pattern. Resorption is common on the surface facing the jugular fossa. Cartilage remains in the occipital condyle at birth.

The overall **length** of the whole braincase at birth is about 63 per cent of its total growth. By the end of the first year, 82 per cent of growth is complete; by 3 years, 89 per cent; by 5 years, 91 per cent; and by 15 years, 98 per cent. The anterior part of the cranial floor (basion-nasion) shows about 56 per cent of adult growth by birth and 70 per cent at 2 years of age. In the newborn, the **width** of the skull base has grown to about 100 mm. By the sixth postnatal month, 50 mm have been added; and by the first year, 20 mm more growth occurs. Thereafter, the rate declines, until only about 0.5 mm per year is added from 3 to 14 years. At birth, the **weight** of the brain is about half that of the adult brain. By the third year, the brain's weight is 80 per cent of the adult brain weight; and at 5 to 8 years, 90 per cent.

In the newborn, the four parts of the occipital bone are still separated by synchrondroses but are fused by about the fifth year (with a range of about ±2 years). At birth, the growth activity of the intersphenoidal synchondrosis has largely ceased. The sutures among the three parts of the temporal bone have undergone partial fusion by about 2 to 4 years. The closure time (or more importantly, the time of growth cessation) of the sphenoethmoidal synchondrosis is not known with certainty, and estimates range from 5 to over 20 years. The early cartilaginous junction can later become membranous, and any subsequent growth proceeds intramembranously. The growth of the spheno-occipital synchondrosis ceases around 15 years, and closure has occurred by about 20 years. The metopic suture usually fuses at about the second year, and the maxillary-premaxillary sutures close during the first or second year. The right and left halves of the mandible fuse during the first postnatal year. The sphenopetrosal and petro-occipital synchondroses may persist in the adult.

At birth, the bones of the calvaria are not yet fully united, and the surfaces of the bones are smooth. The bony cortices are thin and single-layered (lacking a diploë). Six fontanelles are present in the newborn. The posterior closes with bone around the time of birth; the anterior, at about the first year; the anterolaterals, at about 15 months; and the posterolaterals, by 18 months. In the adult, the suture lines are much more jagged; and by the sixth year the bones have become three-layered with a diploë.

In the newborn, the external acoustic meatus characteristically faces downward, but it is already nearly adult-sized. The tympanic bone is still a ring. The mastoid process and pneumatic cells are not present at birth.

The Auditory Ossicles (Fig. 12–38)

All three pairs of the fetal ossicles are fully formed in cartilage by about 8½ weeks. The intramembranous bone of the mandible has already begun to form around Meckel's cartilage, which is still connected with the formative malleus. The external canal and the auditory tube are approaching the developing ossicles from the lateral (auricular) and medial (pharyngeal) sides, respectively. Ossification begins first in the malleus and incus at 16 weeks and follows within 2 weeks in the stapes.

There is one notable difference between the growth and development of the ear ossicles and those of all other bones. They do not undergo extensive remodeling. Their linear dimensions are, for the most part, already established by the cartilaginous anlagen the precede them. Because these tiny bones do not grow in length, the factor of "relocation" is not involved, and the remodeling processes of reshaping and resizing thus do not take place.

The **malleus** is identifiable as a mesenchymal mass in the 10-mm embryo, and this has developed into cartilage in the 28-mm embryo. As Meckel's cartilage begins to undergo degenerative changes anticipating replacement by ligamentous tissue, a center of ossification appears on the cartilaginous manubrium of the malleus at about 15 weeks. This is "perichondral" bone. That is, it develops intramembranously within the connective tissue membrane (the former perichondrium) enclosing the mass of cartilage. Ossification spreads until the entire surface of the element becomes covered with a thin, bony shell (except at

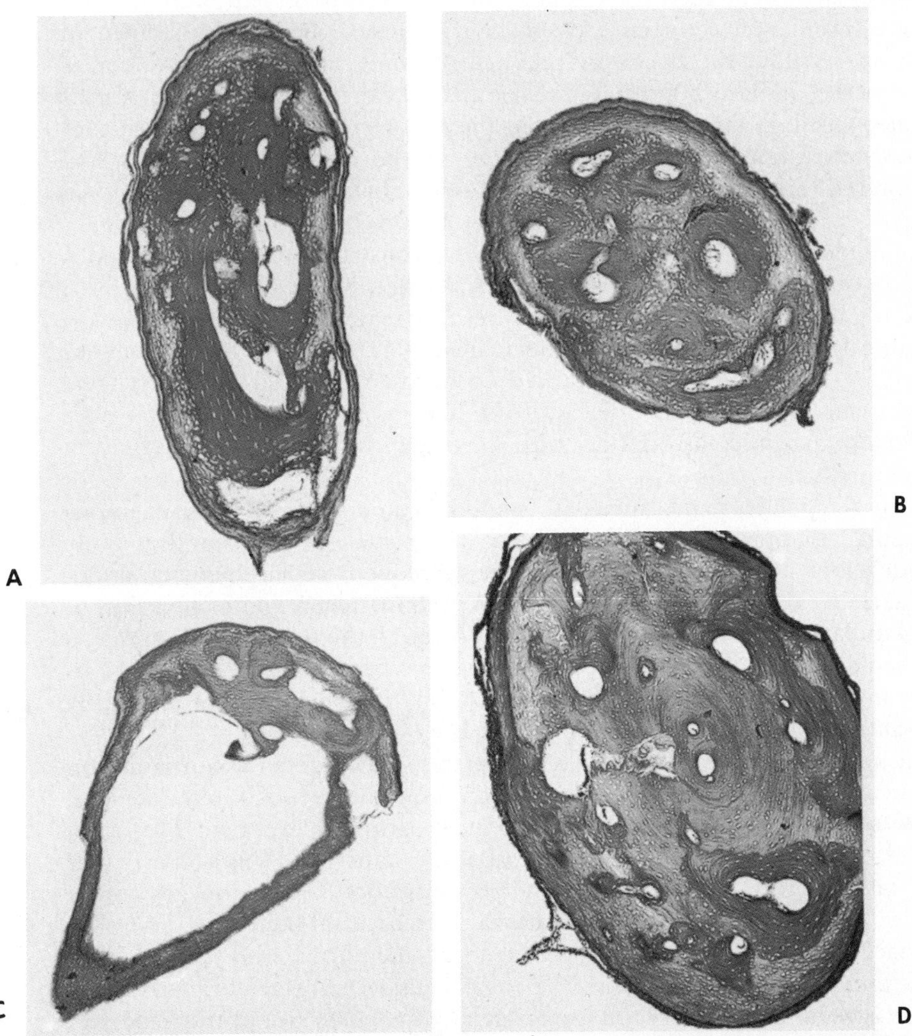

FIGURE 12–38. *A,* Malleus, newborn. This section, from near the head, shows a thin periphery of cartilage enclosing the core of trabecular bone. The latter has completely replaced the endochondral bone, which originally occupied the medulla. *B,* Malleus, newborn. Near the manubrium, a very thin crust of the original cartilage remains. The core of endochondral bone has been largely replaced by several irregular, haversian-like structures that have been deposited within the medullary resorption spaces. Extensive haversian reconstruction does not occur, however. *C,* Stapes, newborn. The obturator foramen in this section is bounded by two crura and the somewhat thicker base (at top). Each crus is composed of perichondral (intramembranous) bone. Although only a portion of the cortex is seen in this section, the crura are hollowed along their medial obturator face, and the cartilaginous core has become completely removed and replaced by mucosal tissue. Note that the external surface of the base retains a thin cartilaginous cover. The perichondral cortex adjacent to the obturator foramen has already been resorbed, and the original endochondral bone in this particular section has been entirely removed. *D,* Incus, 3-month infant. The bone is enclosed by a thin peripheral layer of perichondral (intramembranous) bone tissue. The original endochondral bone of the core has been largely replaced by irregular trabecular bone that has undergone compaction. A few haversian-like structures were produced by deposition of lamellar bone within medullary spaces. Scattered calcified cartilage matrix spicules remain in the medulla. (From Enlow, D. H.: *The Human Face.* New York, Harper & Row, 1968, p. 224.)

articular facets and points of ligament attachment). The core is still cartilaginous, but the chondrocytes now undergo hypertrophy, and the matrix calcifies. Vascular buds invade the cartilage, and partial endochondral bone replacement then takes place. Small medullary spaces form by resorption and become filled with undifferentiated connective tissue, and endosteal bone is laid down in these spaces. Within the perichondral shell of bone, the medullary region is thus composed of endochondral bone, compacted cancellous bone, and islands of calcified cartilage matrix remnants. At the tip of the manubrium, perichondral bone does not develop, but the core of the cartilage receives partial endochondral replacement. It remains surrounded by a covering of the original cartilage.

In the medulla of the malleus, trabecular reconstruction takes place and is said to continue, albeit very slowly, through old age. Major changes, however, do not occur, and extensive haversian reconstruction does not take place. Unlike the stapes, the medulla of the malleus (and also the incus) remains occupied by bone and is not hollowed to become replaced by pharyngeal mucosal tissue. The anterior process of the malleus is the only part that does not develop in direct association with cartilage. It forms in the adjacent connective tissue.

As in the malleus, the **incus** is formed in and around a cartilaginous prototype that approximates adult linear dimensions. A perichondral bony cortex develops around this cartilage, and subsequent replacement by endochondral trabeculae then occurs. Trabecular reconstruction and compaction of some medullary spaces take place very slowly through the years. Small remnants of calcified cartilage, however, may still survive in the aged adult. These various changes involve internal **reconstruction,** not the "remodeling" that ordinarily accompanies the growth process in all other bones.

The cartilage primordium of the **stapes** also has a general configuration similar to the adult form. The stirrup-shaped stapes has a broad, platelike base with two projecting crura encircling the central obturator foramen. The crura converge at the other end to form the head of the stapes, and this is in articular contact with the formative incus. A primary ossification center appears on the surface of the basal plate between the crura. Perichondral bone then forms on the surface of the cartilage within the covering membrane and spreads over the base and onto the crura to form a cortical crust of intramembranous bone. Cartilage is retained, however, on the articular contact surface of the head and also on the inferior side of the base. In both the base and head, the core of cartilage undergoes hypertrophy and calcification, and endochondral bone replacement follows.

Major reconstruction changes now take place on all the medial bone surfaces lining the periphery of the obturator foramen (that is, the two lateral crura and the base and head on the bottom and top). The thin shell of perichondral bone undergoes resorptive removal on these particular surfaces. In the head and base, this exposes the underlying core of endochondral and trabecular bone, which is undergoing compaction to form the irregular, "convoluted" type of bone tissue. Scattered spicules of calcified cartilage survive. The basal plate and the head become thinned as a consequence of the removal of the perichondral bone shell.

In the two crura, unlike the base and head of the stapes, endochondral bone replacement of the cartilaginous core is negligible. As the crust of perichondral bone on the medial surface of each crus is removed, the resorptive process continues on into the cartilage and entirely removes it. The result is a hollowing of the paired crura on their obturator (medial) sides to form trough-shaped processes connecting the base with the head. The hollowed basin of each crus becomes lined by a mucosa of pharyngeal origin. Thus, the crura are composed only of a thin, U-shaped cortex made up of perichondral bone tissue. Original cartilage is retained on the articulating surfaces of the base and head. As in the other ossicles, these cartilage plates do not function as epiphyseal growth centers, and linear growth increases do not occur. Remodeling changes associated with growth are thus not involved. Haversian reconstruction is also negligible in the stapes, and in contrast to what is observed in the malleus and incus, trabecular reconstruction is all but lacking. The histologic structure of the stapes remains essentially unchanged throughout life.

The Mandible

The beginning fetal mandible, as in the earliest growth stages of the other bones of the skull, initially has outside surfaces that are entirely depository in character. At about 10 weeks, however, resorption begins around the rapidly expanding tooth buds and is present thereafter. By 13 weeks, distinct resorptive fields are becoming established on the buccal side of the coronoid process, on the lingual side of the ramus, and on the lingual side of the posterior part of the corpus. The anterior edge of the ramus is already resorptive, and the posterior border is depository. In some specimens, however, the anterior margin along the tip of the coronoid process shows deposition, suggesting a "rotation" to a more upright position (see Chapter 3). By 26 weeks, the basic growth and remodeling pattern that continues on into postnatal development is seen except, notably, in the incisor region (Fig. 12–39). In the fetal and early postnatal mandible, the entire labial side of the anterior part of the corpus is depository. As in the fetal and young postnatal maxilla, the fetal mandibular corpus grows and lengthens **mesially** as well as distally in conjunction with the establishment of the primary dentition. The lingual side of the fetal corpus in the incisor region is resorptive after about the fifteenth week in most (but not all) mandibles. This contributes to a forward growth movement of the entire incisor

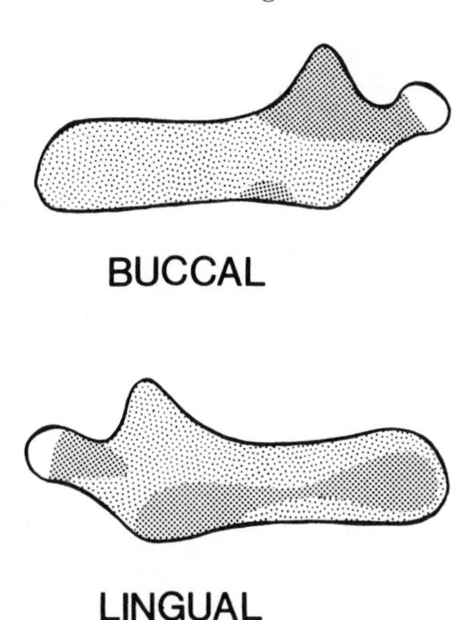

FIGURE 12–39. Mandible in the last trimester of fetal development. Dark stippling represents resorptive fields, and light stippling indicates depository fields.

region of the corpus. During the deciduous dentition period of childhood growth, however, the alveolar bone on the labial side in the forward part of the arch undergoes a reversal to become **resorptive,** and the lingual side becomes uniformly depository. This change occurs in conjunction with the unique lingual direction of incisor movement in the child's mandible. From this time, the chin begins to take on a progressively more prominent form; the mental protuberance continues to grow anteriorly, while the alveolar bone above it moves posteriorly until the lower permanent incisors reach their definitive positions. (See Kurihara and Enlow, 1980a, for more details.)

CHAPTER THIRTEEN

Anomalies, Syndromes, and Dysmorphic Growth and Development

M. Michael Cohen, Jr., D.M.D., Ph.D.

Patients with anomalies are the subject of a field variously known as syndromology, dysmorphology, medical genetics, and teratology. None of these terms reflects what the whole field is about, and although each term has its own advocates, each term also has its limitations. The word "syndromology," derived from Greek, refers to the study of things that run together. Thus, syndromologists deal with children who have birth defects that run together—children with multiple anomalies. However, syndromologists also deal with children who have single anomalies. The word "dysmorphology" means "abnormal morphology." Dysmorphologists deal with children with one or several anomalies, but frequently deal with genetic disorders as well. "Medical geneticists" obviously deal with genetic disorders, but they often deal with environmentally caused defects or even with anomalies whose causes are not known. Finally, "teratology," derived from the Greek, means "the study of monsters." Some clinicians object to using the term in referring to human beings. Teratologists who see patients sometimes do experimental studies on animals as well. Many teratologists concern themselves with studying anomalies in animals exclusively.

We now consider some principles that provide a useful context in which to study anomalies.

ANOMALIES: MALFORMATIONS, DEFORMATIONS, AND DISRUPTIONS

Anomalies are the building blocks of various syndromes. There are three types: malformations, deformations, and disruptions. Most anomalies observed at birth can be sorted into one of these categories. This is practical because the implications of each category are different.

A **malformation** may be defined as a morphologic defect of an organ, part of an organ, or a larger area of the body, resulting from intrinsically abnormal development. Approximately 3 per cent of newborns have significant malformations, and approximately 1 per cent have multiple malformations known as syndromes. The most common class of malformations is incomplete morphogenesis, in which a developmental arrest occurs, as in a **cleft palate** (Fig. 13–1). In normal closure, the palatal shelves fuse. If they become arrested during embryogenesis by some factor, a V-shaped cleft palate will result. Another type of cleft palate is known as **Robin anomaly,** a U-shaped cleft of the palate. If the mandible is hypoplastic during early development, the tongue cannot descend and remains wedged between the palatal shelves. As the palatal shelves attempt to fuse, the tongue obstructs the process, resulting in a U-shaped palatal defect.

Malformations may be minimally or maximally expressed. For example, bifid uvula is a minimal expression of a cleft palate. Malformations are also nonspecific. Each may occur as an isolated defect; each may also occur as a part of various syndromes. Cleft palate, for example, may occur alone or as a component of a great many syndromes of multiple anomalies. For example, **Stickler's syndrome** is an autosomal dominant condition characterized by myopia, retinal detachment, and abnormalities of bones and joints. Cleft palate, especially Robin anomaly, may be a feature of some, but not all, cases. Because malformations occur with various frequencies in different syndromes, they are facultative, rather than obligatory; they may or may not be present in a given example of a condition in which they are known to be features. Because anomalies are both nonspecific and facultative for various disorders, syndrome diagnosis is made not from any one anomaly but from the overall pattern of anomalies.

A **deformation** may be defined as an abnormal form or position of a part of the body caused by nondisruptive mechanical forces. Approximately 2 per

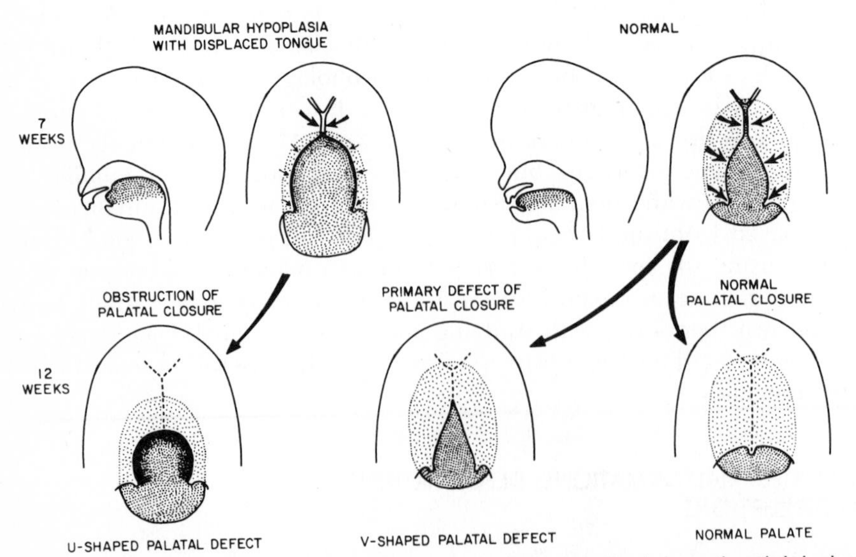

FIGURE 13–1. In the development of ordinary cleft palate, arrest of normal palatal closure produces a V-shaped palatal defect (right). If the mandible is hypoplastic, the tongue fails to descend and obstructs palatal closure, which results in a U-shaped palatal defect (left). (From Hanson, J., and D. W. Smith: U-shaped palatal defect in the Robin anomalad: Developmental and clinical relevance. J. Pediatr. 87:30, 1975.)

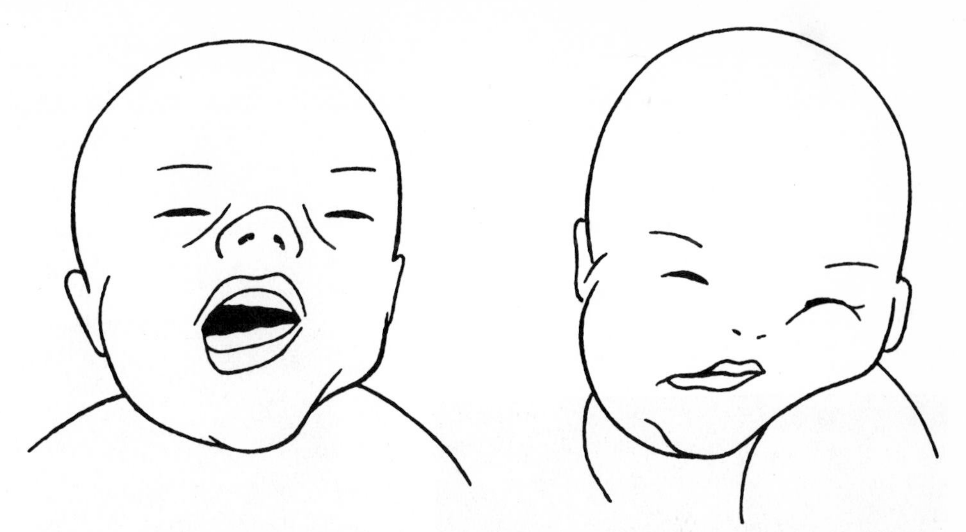

FIGURE 13–2. Mandibular deformation resulting from sharply lateroflexed position of head *in utero* with the shoulder pressed against the mandible for a long period of time. (From Cohen, M. M., Jr.: *The Child with Multiple Birth Defects.* New York, Raven Press, 1982. Courtesy of Mead Johnson Nutritionals.)

cent of newborns have deformations. Important examples include clubfoot, congenital hip dislocation, and congenital postural scoliosis. Figure 13–2 illustrates a **deformation of the mandible** in a newborn infant. For the first few days after birth, infants with deformities can usually be folded into the atypical postures that they held during intrauterine life. If a newborn with the type of mandibular deformity shown is folded into its "position of comfort," the sharply lateroflexed position of the head demonstrates that the deformity resulted *in utero* from the shoulder being pressed against the mandible for a long period of time.

Deformations arise most often during later fetal life. Because the most common cause is intrauterine molding by mechanical forces, the musculoskeletal system is usually affected. The most important factor contributing to deformation is lack of fetal movement, whatever the cause. Deformations occur with higher frequency during first pregnancies than in other pregnancies because of unstretched uterine and abdominal musculature.

Earlier, Robin anomaly was discussed as a malformation, but it may also occur as a deformation. What is the difference? In a deformation, the mandible is not intrinsically malformed by hypoplasia. Rather, it is constrained from growth by intrauterine crowding with the mandible compressed up against the chest during intrauterine life. The small mandible, resulting from constraint, does not allow the tongue to descend, and a U-shaped palate results (Figs. 13–3 and 13–4). Because deformations tend to self-correct once the fetus is out of the intrauterine constraining environment, catch-up growth of the mandible occurs as the child grows (Fig. 13–5). No catch-up growth will occur when the Robin anomaly is based on intrinsic mandibular hypoplasia. Thus, Robin anomaly can be either a malformation or a deformation, depending upon how it arises.

We are now in a position to contrast malformations with deformations in general. The characteristic features of malformations and deformations are compared in Table 13–1. Malformations tend to arise during the embryonic period (first 8 weeks in intrauterine life) at the time of organogenesis and are

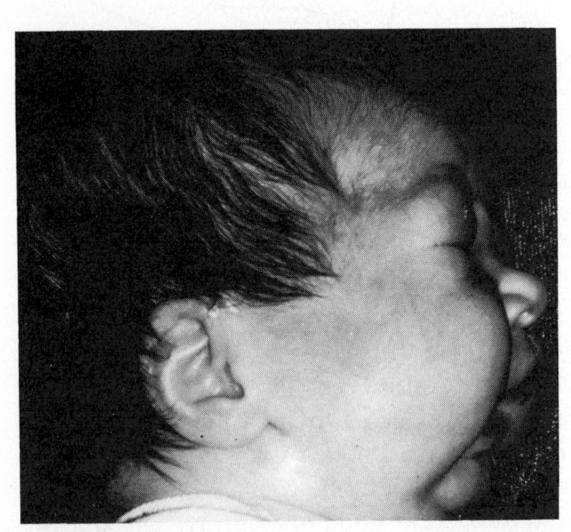

FIGURE 13–3. Infant with Robin anomaly showing micrognathia. (From Cohen, M. M., Jr.: *The Child with Multiple Birth Defects.* New York, Raven Press, 1982.)

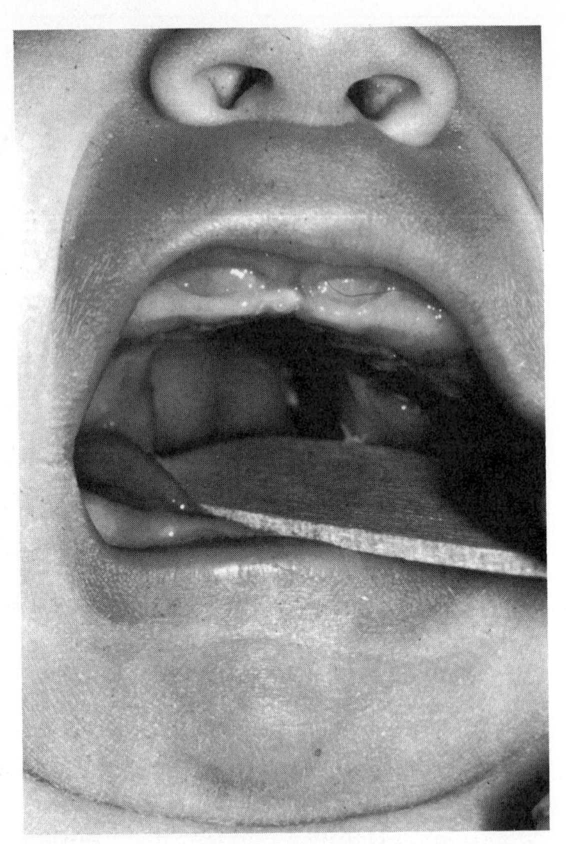

FIGURE 13–4. Same infant as shown in Figure 13–3. Note the U-shaped cleft palate. (From Cohen, M. M., Jr.: *The Child with Multiple Birth Defects.* New York, Raven Press, 1982.)

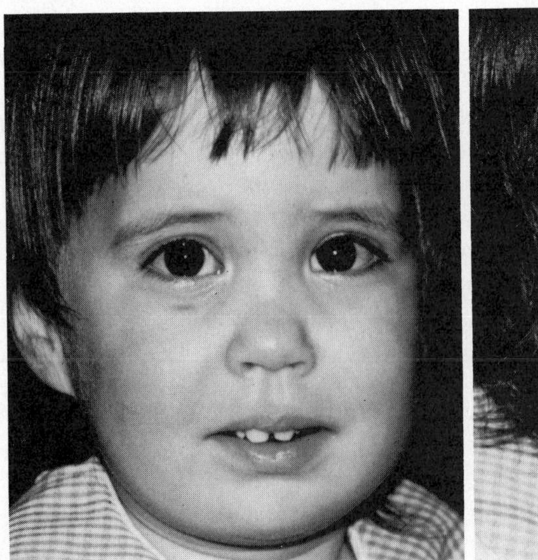

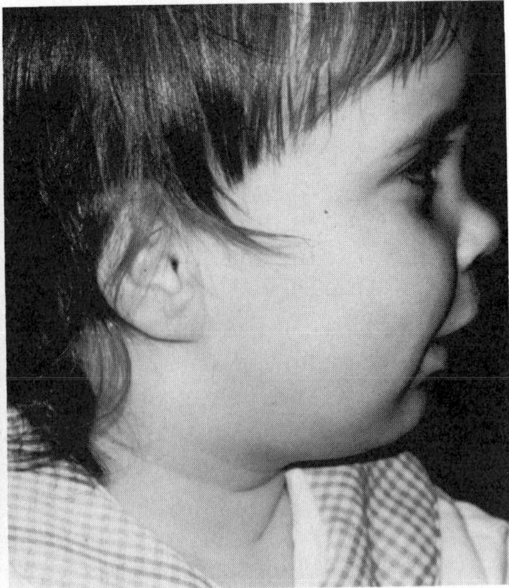

FIGURE 13–5. Same patient as shown in Figures 13–3 and 13–4, showing mandibular catch-up growth in childhood. (From Cohen, M. M., Jr: *The Child with Multiple Birth Defects.* New York, Raven Press, 1982.)

Table 13–1. COMPARISON OF MALFORMATIONS AND DEFORMATIONS

	Malformation	Deformation
Time of occurrence	Embryonic period	Fetal period
Level of disturbance	Organ	Region
Perinatal mortality	+	–
Spontaneous correction	–	+
Correction by posture	–	+

From M. M. Cohen, Jr.: *The Child with Multiple Birth Defects.* New York, Raven Press, 1982.

primarily errors of morphogenesis. They affect formation of organ structure (or field structure, as in the case of orofacial clefting). Deformations, on the other hand, tend to arise during the fetal period (the rest of the gestation following the embryonic period) and are changes in the shape of previously normal parts. Thus, deformations tend to affect intact regions. A clubfoot is not an organ or field defect but a regional defect, because the limbs have already formed. With simple clubfoot, for example, five digits and the proper number of phalanges and metatarsals are present. The whole foot region is deformed. This is not true of malformations such as syndactyly (fusion of digits) or polydactyly (extra digits).

Some degree of perinatal mortality accompanies every statistical survey of malformations, because of the high frequency of central nervous system and cardiovascular malformations. In contrast, the perinatal mortality tends to be low in surveys of deformations. Mandibular asymmetry, clubfoot, and other deformities are compatible with life.

Finally, spontaneous correction or correction by posturing is possible in many deformations. Many deformations correct spontaneously after birth. That self-correction occurs so commonly is not surprising, because after birth, the infant is no longer subject to intrauterine constraining forces. The degree to which self-correction is possible depends on how long during fetal life the constraining forces were acting and on the severity of the deformation. Postural correction is feasible in many cases of scoliosis, congenital hip dislocation, and clubfoot. In contrast, spontaneous correction of malformations is rare, except for small septal defects of the heart, and correction by posturing is impossible.

A **disruption** is a morphologic defect of an organ, part of an organ, or a larger region of the body, resulting from a breakdown of, or interference with, originally normal development. An amputation of a digit *in utero* caused by an amniotic band serves as an example. Amniotic bands arise when a tear in the amnion occurs and the amniochorion "heals" itself. In the process, it sometimes produces string-like bands in which the fetus may become entrapped. The bands become entangled around digits or limbs, cut off the blood supply, and lead to intrauterine necrosis and amputation of digits or limbs. Occasionally, amniotic bands may become entangled in the nostrils, the mouth, and, for some reason that is not understood, the anterior part of the head and face. **Amniotic bands can produce a dramatic disruption of craniofacial structures** consisting of asymmetrical encephaloceles (brain tissue protrusions through a skull defect), bizarre facial clefts, and limb amputations (Fig. 13–6). It will be observed that the condition is suggestive of a malformation. In fact, in the older German literature, these were referred to as secondary malformations. How can we know whether an anomaly is a malformation or a disruption? If the anomaly in question **looks like** a malformation but makes no embryonic sense, it is probably a disruption. Figure 13–6 shows a trifid nostril that cannot be explained embryologically. Hence, it is a disruption. Notice also the amputated fingers.

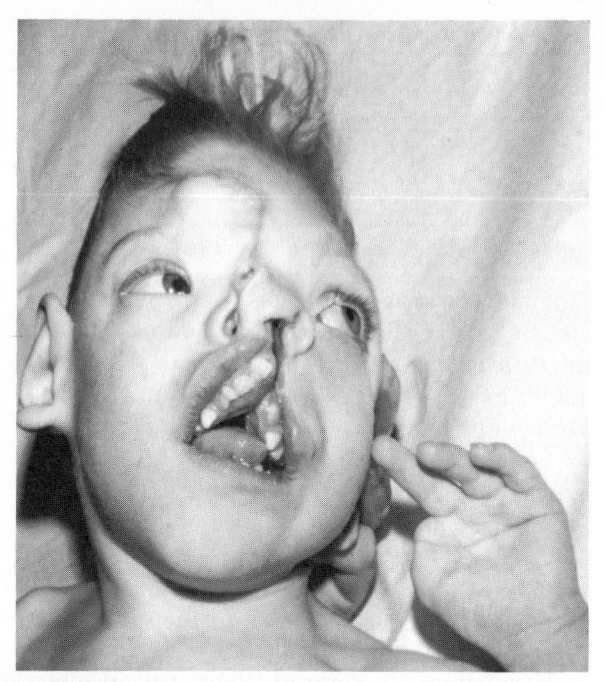

FIGURE 13–6. Bizarre facial clefting and asymmetrical encephaloceles, the result of amniotic bands—a disruptive process. Note the trifid nostril, which cannot be explained embryologically. Note also amputation of the thumb and index finger. (From Cohen, M. M., Jr.: *The Child with Multiple Birth Defects.* New York, Raven Press, 1982.)

Disruptions tend to be sporadic occurrences in contrast to malformations and deformations, both of which may recur in some instances.* No two cases of a given disruption are alike. Examples of a given disruption tend to be more variable than those of a given malformation or deformation. Finally, disruptions can be interpreted mistakenly as embryonic malformations, when, in fact, they arise later during intrauterine life. For example, an open calvarial defect resulting from disruptive amniotic bands may be mistaken for anencephaly, a common malformation with a risk of recurrence.

SYNDROMES

A **malformation syndrome** may be defined as several malformations thought to be pathogenetically related in the same individual. A true malformation syndrome is characterized by **embryonic pleiotropy** in which a pattern of developmentally unrelated malformations occurs; that is, the malformations that make up the syndrome occur in embryonically noncontiguous areas. They are not related to one another at the descriptive embryonic level; but at a more basic level, the malformations have a common cause or are presumed to have a common cause and are thus pathogenetically related. Malformation syndromes lack biochemical definition. The highest stage of malformation syndrome delineation is a known genesis syndrome of the pedigree or chromosomal type. The causes of many malformation syndromes, however, remain unknown.

Many people characterize syndromologists as "rare bird watchers" (Fig. 13–7). This is a misconception. Although many syndromes are individually

---

*Malformations, in a small percentage of cases, may follow mendelian inheritance. Deformations sometimes may have a (lower) multifactorial recurrence risk.

FIGURE 13–7. Some people think of syndromologists as rare-bird watchers. This is a misconception. Although many syndromes are individually rare, in the aggregate they make up a significant portion of medicine. Furthermore, syndromologists are involved in patient care, which is an active, not a passive, endeavor. (From Cohen, M. M., Jr.: *The Child with Multiple Birth Defects.* New York, Raven Press, 1982.)

rare, in the aggregate, they make up a significant portion of medicine. Furthermore, some syndromes are relatively common, such as trisomy 21 syndrome. Finally, syndromologists are actively involved in patient care, which is an active endeavor. Bird watching, on the other hand, is a passive endeavor.

Trisomy 13 syndrome—a well-known chromosomal disorder in which an extra chromosome 13 is present—is characterized by variable features frequently including cleft lip and cleft palate (Fig. 13–8), microphthalmia (small eyes), microcephaly with a severe brain malformation known as holoprosencephaly, congenital heart defects, polydactyly (extra digits), and a host of other anomalies.

Apert syndrome (Figs. 13–9 to 13–15) is a single gene disorder characterized by premature fusion of cranial sutures, bizarre craniofacial appearance, highly arched palate, syndactyly (fusion of digits), and various other abnormalities, occasionally including congenital heart defects. Although the syndrome has autosomal dominant inheritance, the bizarre physical appearance and the mental retardation, which accompanies many cases, reduce the genetic fitness of affected individuals—that is, they are unlikely to marry and have children. Thus, most instances of Apert syndrome represent fresh mutations. Affected families are seen very infrequently, but enough have been reported to indicate autosomal dominant inheritance.

DYSMORPHIC GROWTH AND DEVELOPMENT

Many malformations and malformation syndromes exhibit dysmorphic growth and development. Several examples follow.

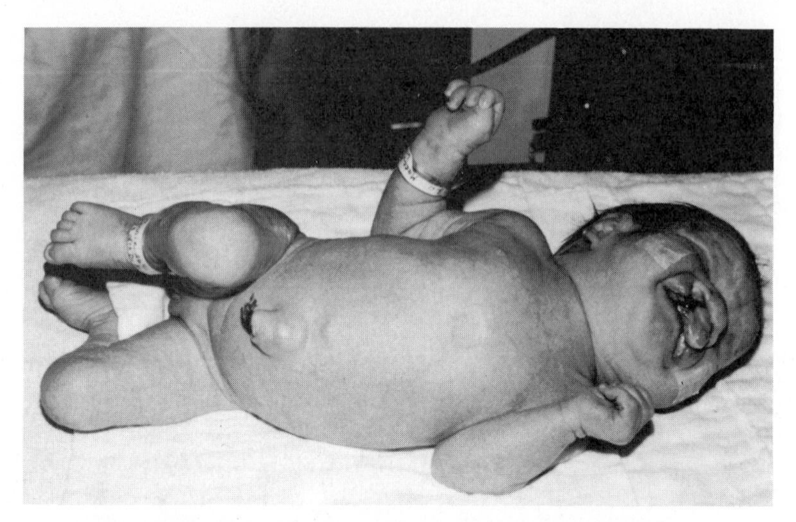

FIGURE 13–8. Trisomy 13 syndrome. Note bilateral cleft lip. Patient also has microcephaly, a severe brain malformation, small eyes, a congenital heart defect, and an umbilical hernia.

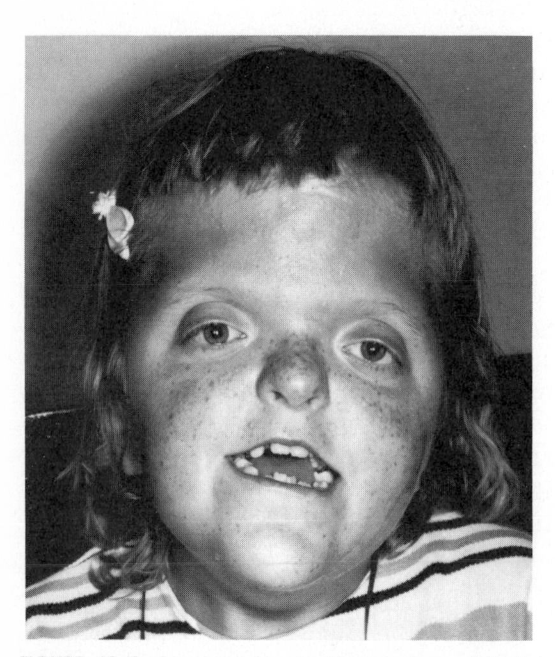

FIGURE 13–9. Apert syndrome. Abnormal craniofacial appearance.

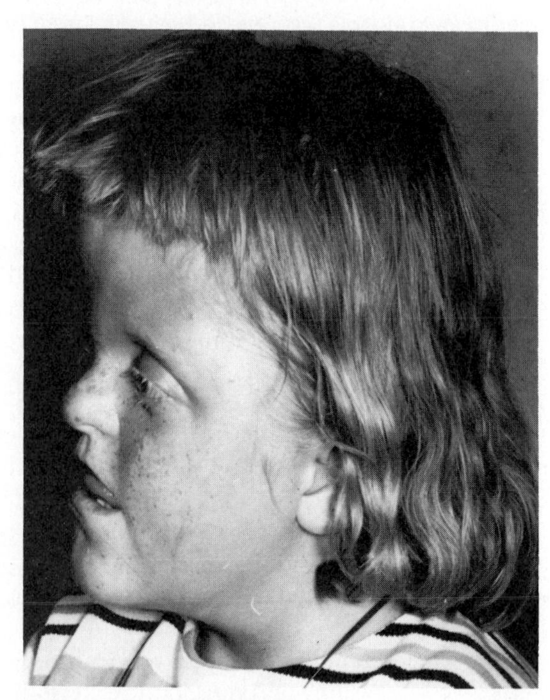

FIGURE 13–10. Apert syndrome. Abnormal craniofacial appearance.

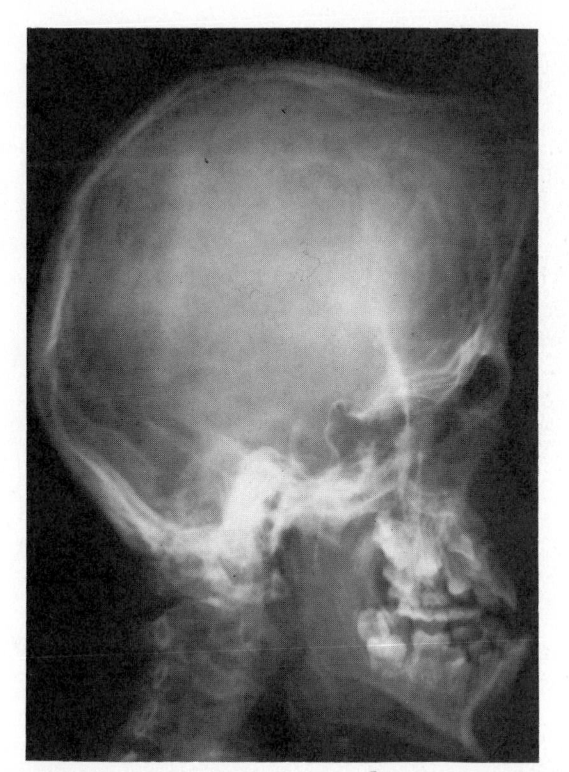

FIGURE 13–11. Lateral skull radiograph showing grossly abnormal calvaria, cranial base, and maxillary configuration. (From Cohen, M. M., Jr.: *Craniosynostosis: Diagnosis, Evaluation, and Management.* New York, Raven Press, 1986.)

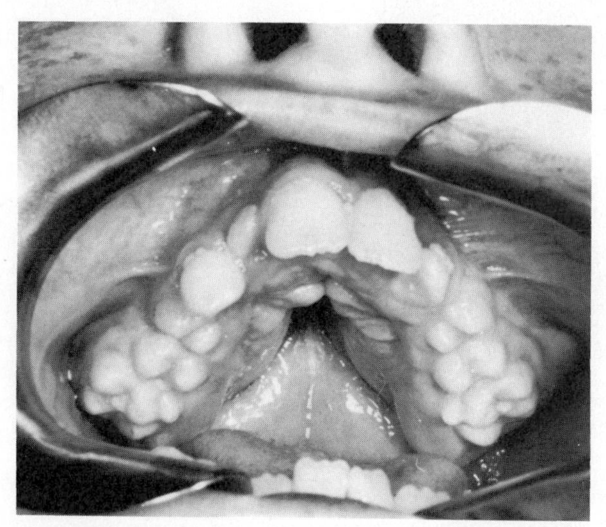

FIGURE 13–12. Apert syndrome. Highly arched palate.

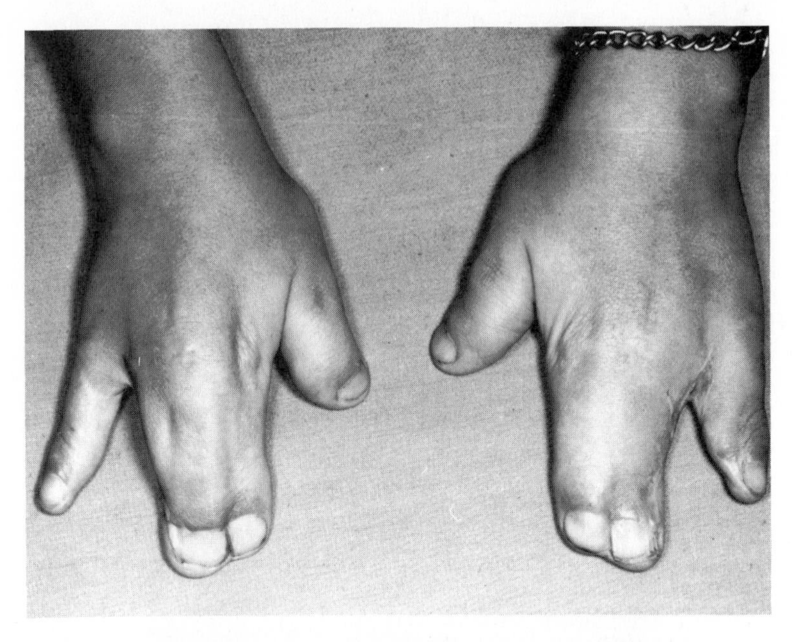

FIGURE 13–13. Apert syndrome. Characteristic type of syndactyly of the hands.

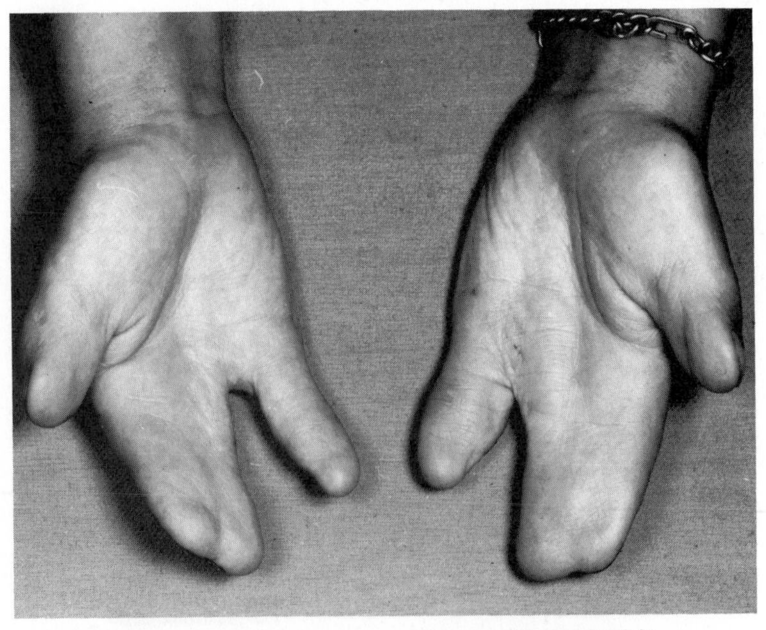

FIGURE 13–14. Apert syndrome. Characteristic syndactyly.

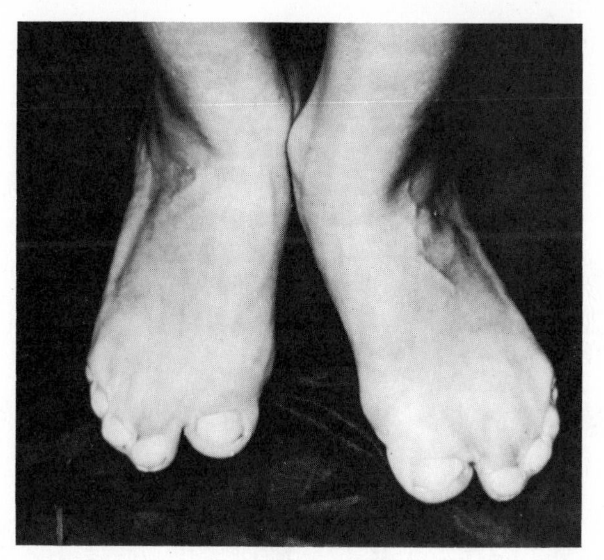

FIGURE 13–15. Apert syndrome. Characteristic syndactyly of the feet.

Figures 13–16 and 13–17 illustrate a patient with **mandibulofacial dysostosis.** The clinical appearance is characterized by down-slanting palpebral fissures and malar deficiency. A hearing aid will also be observed. Such patients have various malformations of the ear ossicles in addition to external ear malformations and a small mandible with a concave lower mandibular border.

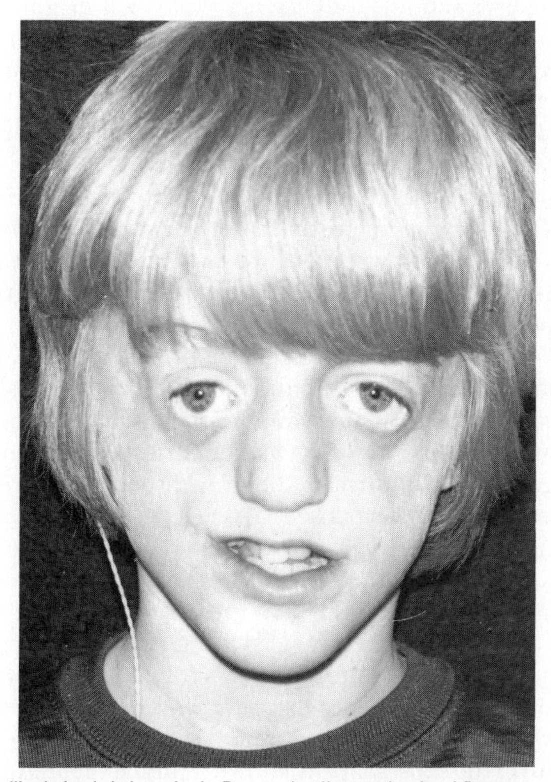

FIGURE 13–16. Mandibulofacial dysostosis. Down-slanting palpebral fissures, malar deficiency, and hearing aid. (From Cohen, M. M., Jr.: *The Child with Multiple Birth Defects.* New York, Raven Press, 1982.)

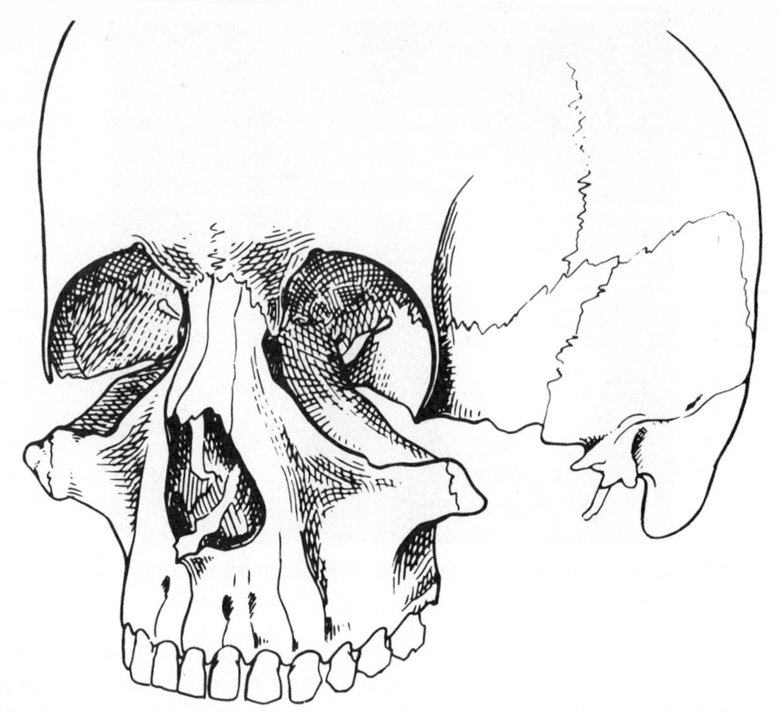

FIGURE 13–17. Mandibulofacial dysostosis. Bony defect of ovoid-shaped orbits with absence of the zygomatic arches. (From Tessier, P.: Anatomical classification of facial, cranio-facial and latero-facial clefts. J. Maxillofacial Surg. 4:69, 1976.)

The down-slanting eyes and malar deficiency trace their origin to an underlying bony defect characterized by absence of the zygomatic arches and ovoid-shaped orbits. The term "dysostosis" refers to a malformation of individual bones either singly or in combination, but it is not a generalized skeletal disorder. Thus, mandibulofacial dysostosis is limited to facial structures. Autosomal dominant inheritance is characteristic.

Craniosynostosis is a condition resulting from premature fusion of the cranial sutures. This is a characteristic feature of Apert syndrome, which has already been discussed. In this section, we turn to a consideration of simple, nonsyndromic craniosynostosis. Head shape depends on which sutures are prematurely synostosed, the order in which they synostose, and the timing at which they synostose. Craniosynostosis may be prenatal or perinatal in onset or may occur later during infancy or childhood. The earlier synostosis occurs, the more dramatic is the effect on subsequent cranial growth and development. The later synostosis occurs, the less is the effect on cranial growth and development. Figure 13–18 illustrates the growth pattern of the calvaria when various sutures are prematurely closed. As a general rule, growth restriction occurs at right angles to the fused suture with compensatory expansion in the same direction as the fused suture. **Normocephaly** is the normal head shape. If the sagittal suture is prematurely synostosed, the calvaria is restricted in its lateral growth and compensates by permitting more passive growth to occur at the coronal and lambdoidal sutures. The skull shape is known as **dolichocephaly.** If the coronal suture is prematurely synostosed, growth is arrested in an anteroposterior direction, compensatory growth occurring laterally at the patent sagittal suture; the skull shape is known as **brachycephaly. Plagiocephaly** is an asymmetric skull shape. This may be produced by unilateral closure of the coronal suture or unilateral closure of the lambdoidal suture. Finally, if the metopic (interfrontal) suture closes prematurely, a triangular calvaria results, known as **trigonocephaly.**

Craniosynostosis can be viewed from two different perspectives—anatomic and genetic. When the main interest is clinical description, growth and development, or surgical management of craniosynostosis, the perspective is anatomic, as shown in Figure 13–18. In this context, which particular suture is synostosed is of primary importance. From a genetic perspective, however, classifying various syndromes on the basis of which particular suture is synostosed can be very misleading. Because patients with the same genetic condition may have fusion of different sutures, which specific suture is synostosed is of secondary importance in this context. For example, in Figure 13–19, an affected father and daughter are shown. The father has sagittal synostosis, and his daughter has unilateral coronal synostosis. Thus, they have different cranial configurations, and yet both have exactly the same genetic disorder. The clinical geneticist's main concern is with the overall pattern of anomalies (from which the diagnosis of a specific syndrome follows) and with which family members are affected (from which genetic counseling follows). Thus, from a genetic perspective, craniosynostosis and syndromes with craniosynostosis should not be classified on the basis of which sutures are synostosed. It is important to emphasize that although anatomic and genetic means of classifying craniosyn-

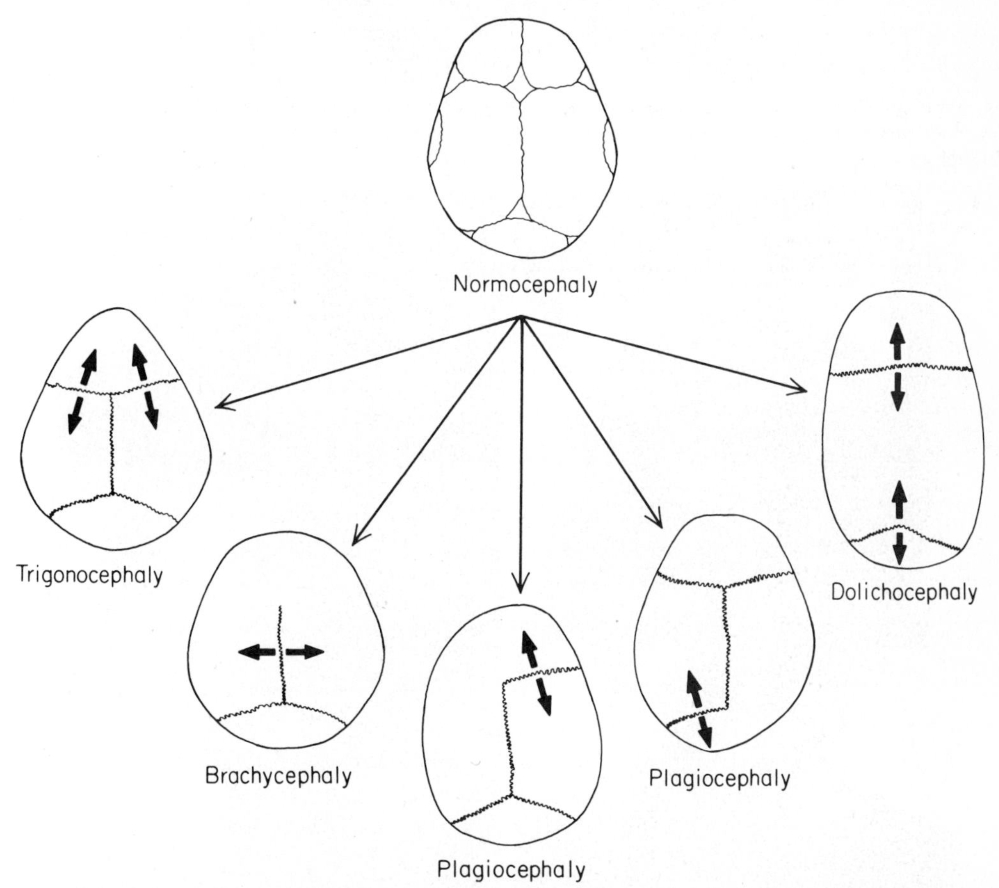

FIGURE 13–18. With premature craniosynostosis, growth restriction occurs at right angles to the fused suture with compensatory expansion in the same direction as the fused suture. (From Cohen, M. M., Jr.: *Craniosynostosis: Diagnosis, Evaluation, and Management.* New York, Raven Press, 1986.)

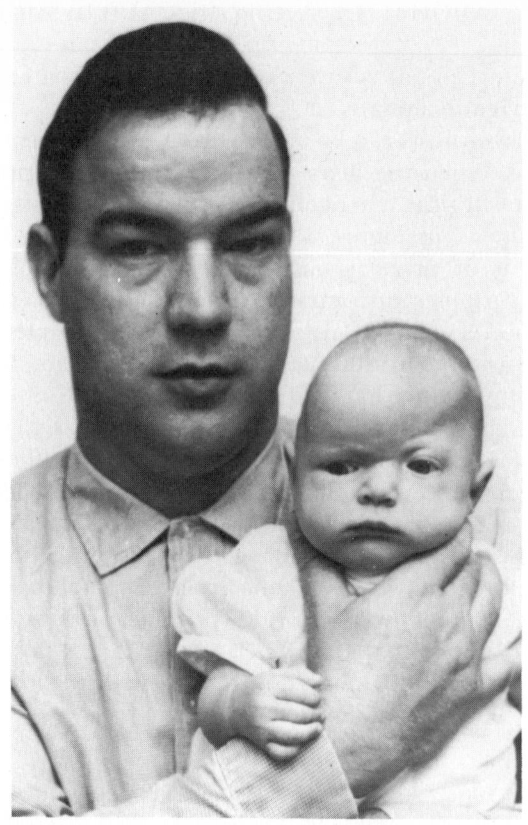

FIGURE 13–19. Autosomal dominantly inherited craniosynostosis. Father has sagittal synostosis. Infant has unilateral coronal involvement. (From Anderson, F. M., and L. Geiger: Craniosynostosis: A survey of 204 cases. J. Neurosurg. 22:229, 1965.)

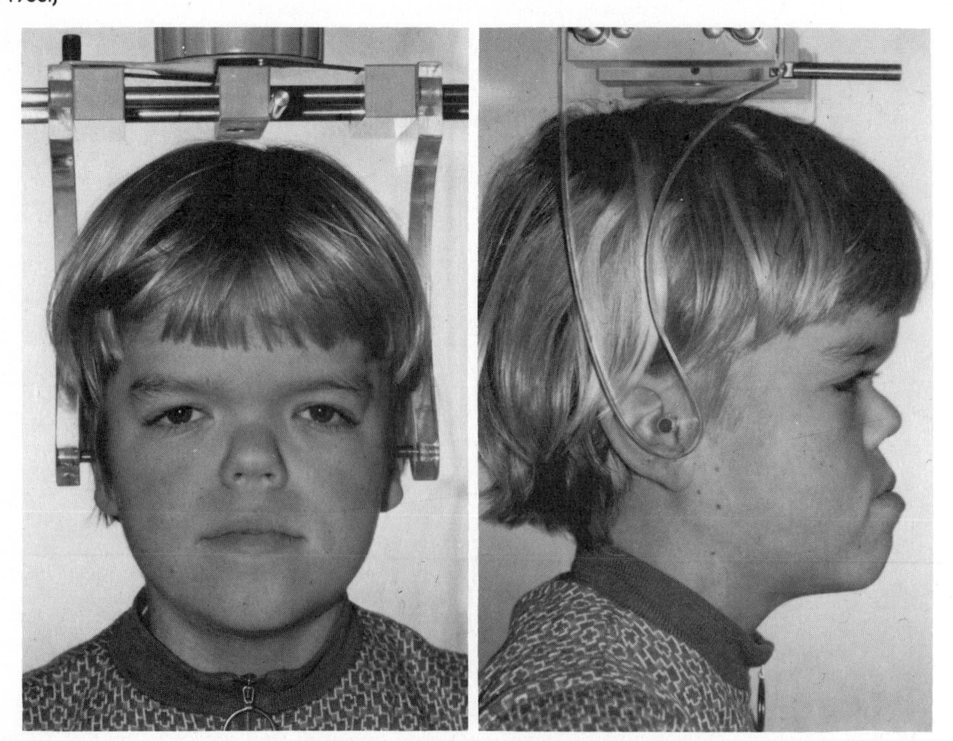

FIGURE 13–20. Achondroplasia. Note midface deficiency. (From Cohen, M. M., Jr.: Malformation syndromes. In: *Surgical Correction of Dentofacial Deformities,* Vol. 1. Ed. by W. H. Bell, W. R. Proffit, and R. P. White. Philadelphia, W. B. Saunders, 1980.)

ostosis may cross-cut each other, both perspectives are equally valid and depend on context.

A patient with **achondroplasia** is shown in Figure 13–20. The condition is characterized by an unusual craniofacial configuration and disproportionate short stature. Autosomal dominant inheritance is characteristic, although most cases represent fresh mutations. Significant craniofacial findings include enlarged calvaria, frontal bossing, large frontal sinuses, occipital prominence, normal anterior cranial base length, strikingly shortened posterior cranial base length, an acute cranial base angle, a short nasal bone that is deformed and depressed, short upper facial height, recessed maxillae, posterior tilt of the nasal floor, and a prognathic mandible that is anteriorly displaced but of normal size and with a normal gonial angle and a high coronoid process. In achondroplasia, bone that is preformed in cartilage is affected. Membrane bones are not disturbed, except secondarily because both cartilage bones and membrane bones interarticulate in the skull. The craniofacial appearance can be viewed as the effect of abnormal endochondral bone formation on the development of the skull as a whole. The mandible is normal in length because growth at the condylar cartilage is appositional; in achondroplasia, only interstitially growing cartilage is affected.

CHAPTER FOURTEEN

Cephalometrics

William W. Merow, D.D.S.
B. Holly Broadbent, Jr., D.D.S.

Cephalometric radiography, or "cephalometrics," as it is more frequently called, is a technique employing oriented radiographs for the purpose of making head measurements. It has found wide use in growth research and in orthodontic diagnosis and treatment evaluation. The principles of cephalometrics are patterned closely after the science of craniometry, which has long been used in anthropology in the quantitative study of the skull.

Craniometry was initially applied in measuring dried skulls. Standard landmarks and measurements were developed, and much useful information was obtained. The technique, however, had the obvious shortcoming of being a one-time-only evaluation of a static subject. Serial study of growth changes was obviously not possible. When the technique was transferred to living subjects for growth measurement, accuracy was lost because landmarks were obscured, the soft tissue covering was of varying thickness, and there was no access to deeper structures.

In 1931 the first comprehensive publication on the basic technique of cephalometric radiography was introduced by Dr. B. Holly Broadbent, Sr. From the early 1920s, Broadbent had been associated with Dr. T. Wingate Todd of the Anatomy Department of Western Reserve University of Cleveland, Ohio. Todd had been brought from Manchester, England, to head the Brush Inquiry, a study conceived and financed by the famous inventor Charles F. Brush, who was interested in "normal" human development growth. The Brush Inquiry included, among a very comprehensive group of tests, the concept of the study of the skeletal development of youngsters through serial or longitudinal radiographs of developing epiphyses throughout the entire body.

At that time there was no method for accurately positioning the child's head so that cephalometric radiographs could be compared in a serial study. Dr. Broadbent, through his association with Todd, and with the assistance of an excellent machinist and arduous experimentation, transformed the Todd craniostat into the first cephalometer (Fig. 14–1) in 1925. Thus, cephalometric

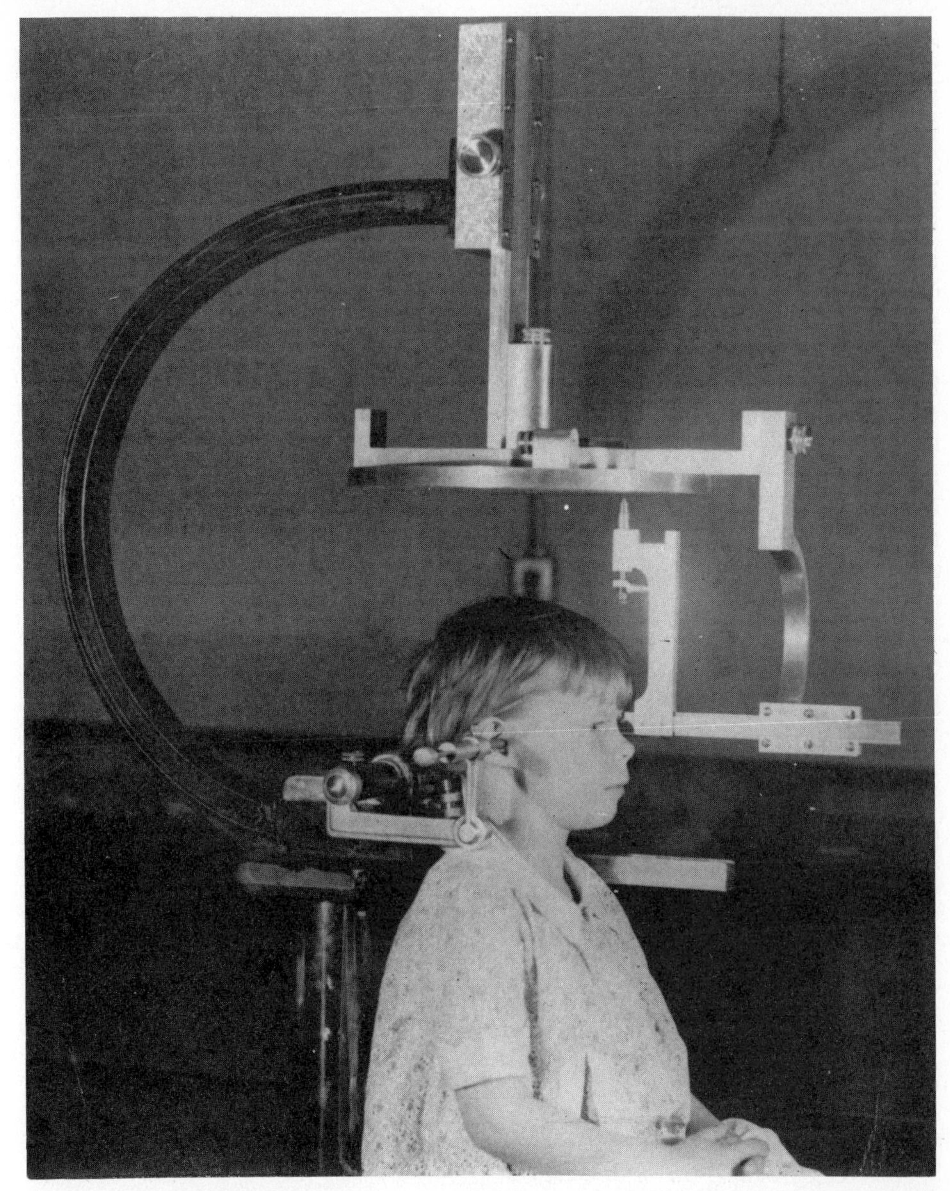

FIGURE 14–1. First Broadbent cephalometer, designed to hold the living head in a manner similar to that used by the radiographic craniometer. (From Broadbent, B. H., Sr., B. H. Broadbent, Jr., and W. Golden: *Bolton Standards of Dentofacial Developmental Growth.* St. Louis, C. V. Mosby, 1975.)

radiography became a means for producing serial lateral and frontal cephalograms, with the subject in a fixed and reproducible position.

Concurrent with the Brush Inquiry, but as an autonomous research project, the Bolton Study was initiated and supported through the interest of Mrs. Frances P. Bolton and her son Charles. This study investigated the craniofacial and dental development of the Brush Inquiry group as well as others brought in by interested clinicians. The Broadbent-Bolton cephalometer was used in collecting cephalometric records, many of them comprehensive longitudinal studies, on over 5000 children. The Brush Inquiry and Bolton Study records and data are housed at Case Western Reserve University in Cleveland as the Bolton-Brush Growth Study Center.

Simply stated, cephalometrics involves making measurements from lateral and frontal head radiographs taken with the head held in a fixed position in a cephalometer (Figs. 14–2 and 14–3). The head is held in this position by means of ear rods, which are aligned on the central axis of radiation from the x-ray tube. Thus, for a profile view, the sagittal plane of the head is at right angles to the direction of the x-rays, the film cassette placed as close as possible to the left side of the face. For a frontal or posteroanterior (PA) view, the frontal plane of the head is perpendicular to the x-ray beam, the film cassette placed as close as possible to the face. A standard distance of 60 inches from the source of radiation to the midsagittal plane or to the porionic axis is maintained. Vernier scales for measuring the film distances from the midsagittal plane or porionic axis for varying head sizes permit accurate correction for enlargement.

Standardization of the technique is necessary to minimize error when serial radiographs of the same individual are taken at different times and to permit universal use of cephalometric data obtained from many different sources. Possible errors in the technique include (1) a lack of perpendicularity of the x-ray beam to the subject's midsagittal plane and the film surface and (2) not placing the film in the closest approximation to the head and face for minimizing enlargement (Fig. 14–4).

Obviously, norms and standards developed from the Broadbent-Bolton technique, where the enlargement factor is minimized, may not be compatible with those derived from a fixed cassette relationship, which may vary from 12

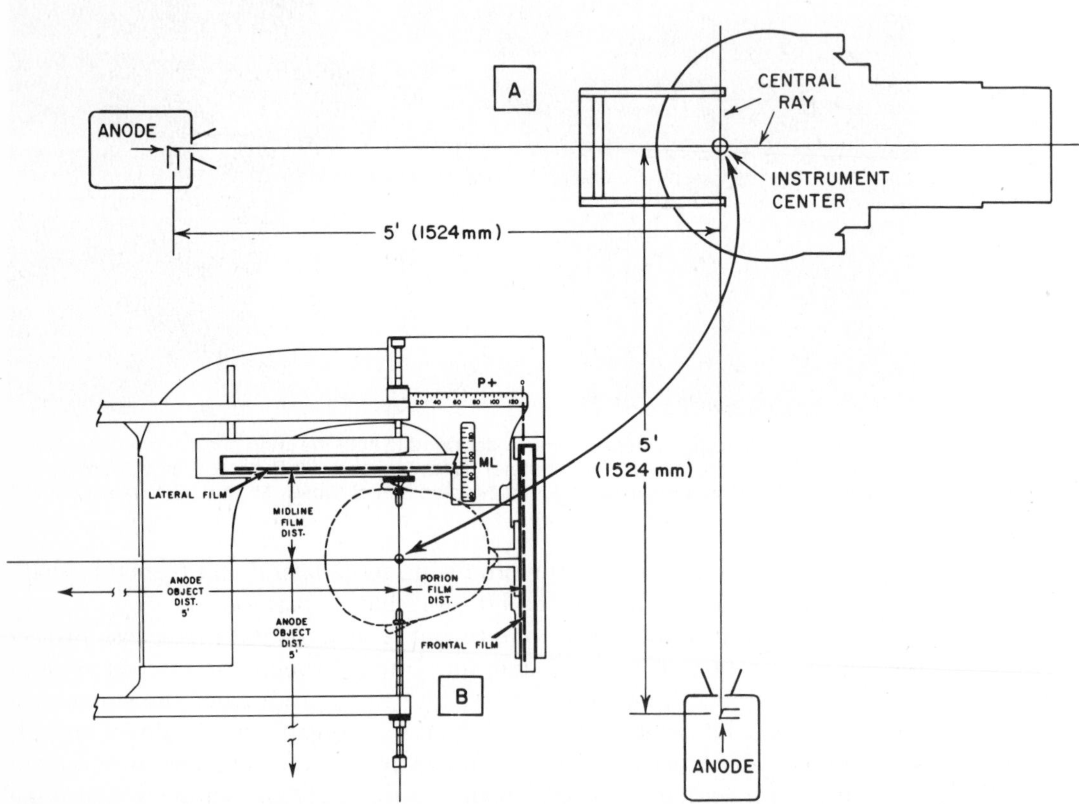

FIGURE 14–2. Bolton cephalometric floor plan. *A,* Relation of the anodes, the central rays, and the instrument center. *B,* Relation of the instrument center to the lateral and frontal films and the ML (midline-lateral film distance) and P+ (porion film distance) scales. (From Broadbent, B. H., Sr., B. H. Broadbent, Jr., and W. Golden: *Bolton Standards of Dentofacial Developmental Growth.* St. Louis, C. V. Mosby, 1975.)

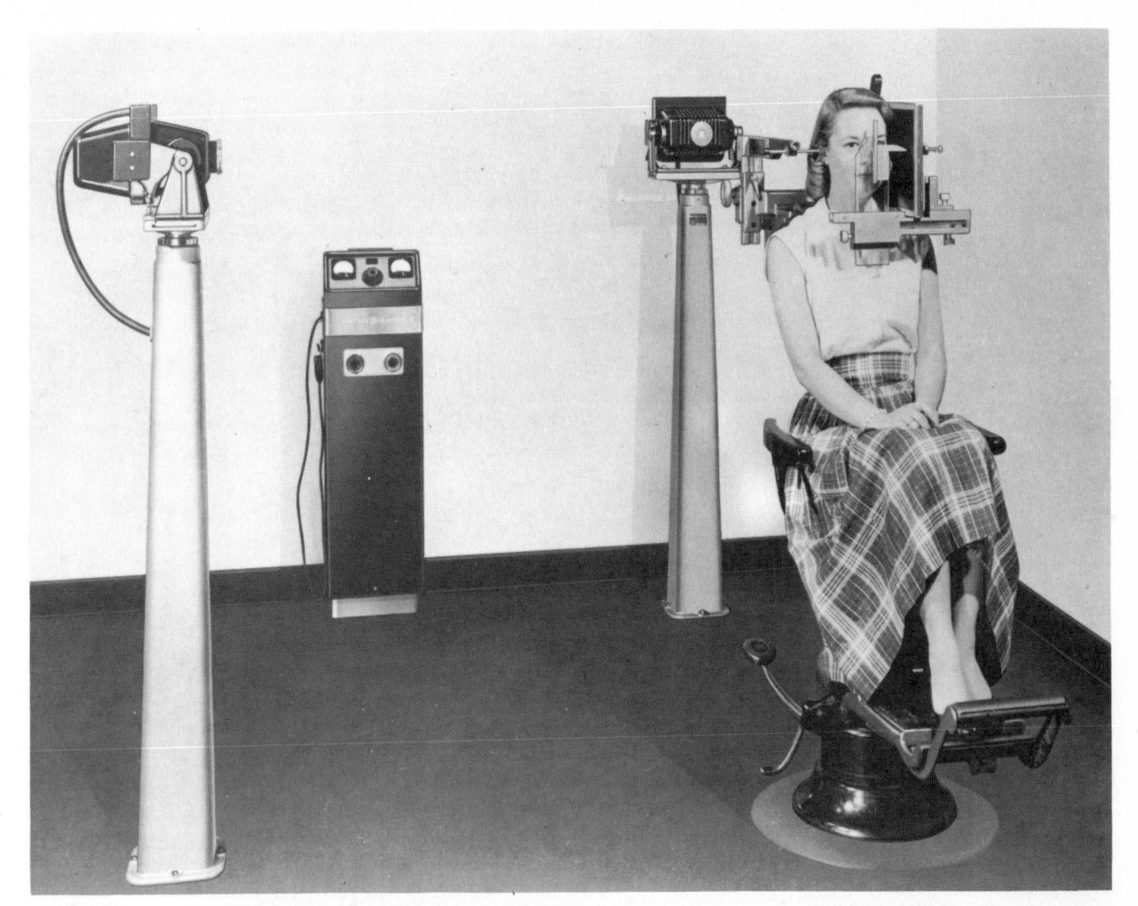

FIGURE 14–3. Broadbent-Bolton radiographic cephalometer as it was used by the Bolton Study, Case Western Reserve University, Cleveland, Ohio. (From Broadbent, B. H., Sr., B. H. Broadbent, Jr., and W. Golden: *Bolton Standards of Dentofacial Developmental Growth.* St. Louis, C. V. Mosby, 1975.)

cm to 18 cm distance from midsagittal plane to film surface, thus causing a disparity in the enlargement factor.

The safety features of the cephalometric technique include the use of the 90-kv peak to minimize softer x-rays, which are absorbed by the patient and are ineffective in contributing to the resulting picture. Additionally, the beam is filtered to remove as many of these softer x-rays as possible. The film is a double-emulsion film sandwiched in a cassette with compatible intensifying screens, which greatly reduce the radiation necessary for a satisfactory radiograph. Patient and operating personnel protection is, of course, a prime consideration.

Figures 14–5 and 14–6 are a representative pair of cephalometric radiographs that have been taken with the standardized technique. In the lateral x-ray, a soft tissue filter is placed either in front of the patient, or against the x-ray film cassette, in order to decrease the radiation in this area, and to give an enhanced soft tissue outline.

ORIENTATION

In the Broadbent technique, as was noted in the diagrammatic representation of the head holder, a recording of the distance of the lateral film from

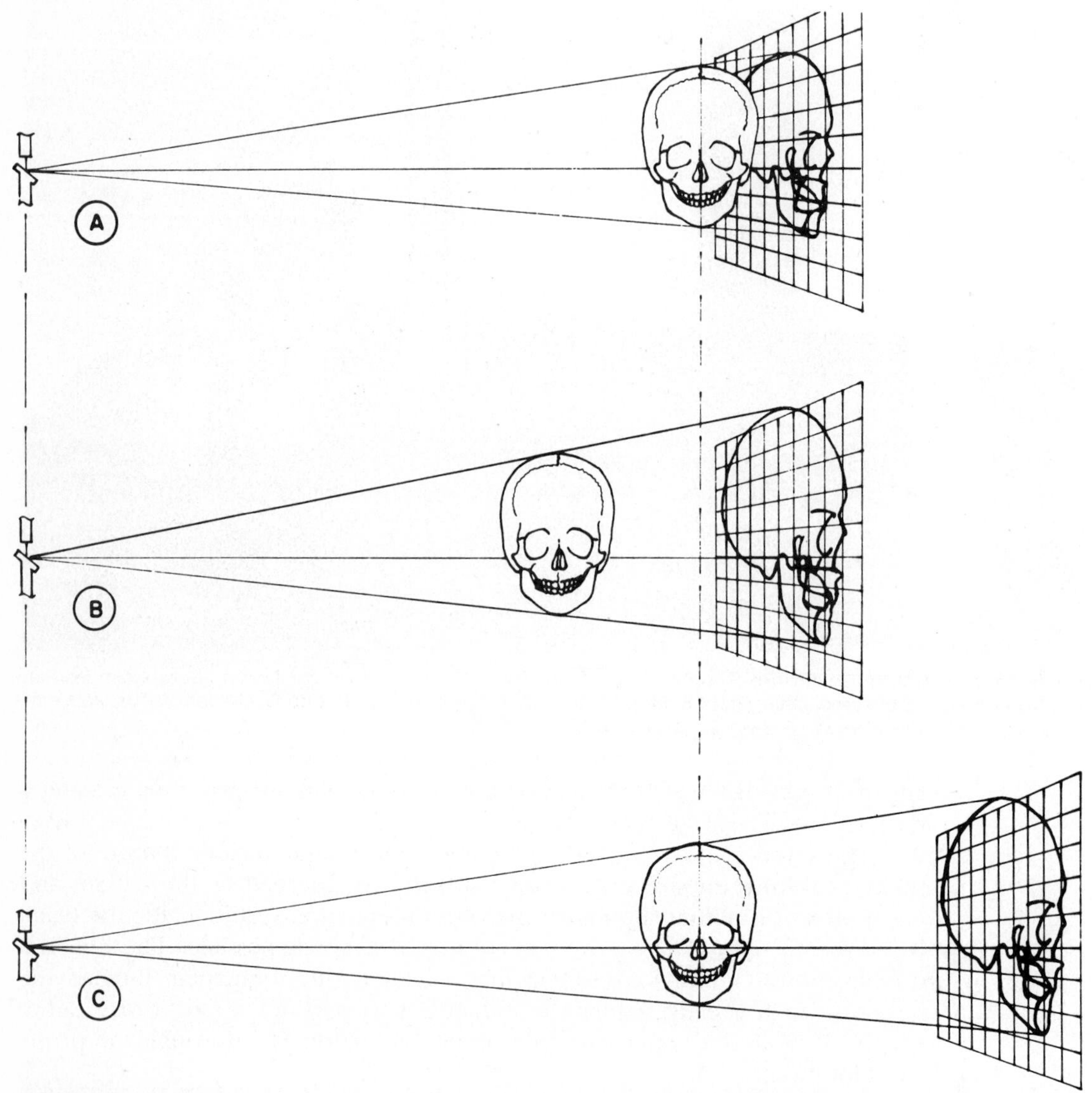

FIGURE 14–4. Enlargement of shadow. *A*, Skull distant from the anode and close to the film produces minimal enlargement and a sharp shadow. *B*, Skull situated closer to the anode and more distant from the film produces unnecessary enlargement and less sharpness of the shadow. *C*, Skull far from the anode and also distant from the film produces unnecessary enlargement and less sharpness of the shadow. (From Broadbent, B. H., Sr., B. H. Broadbent, Jr., and W. Golden: *Bolton Standards of Dentofacial Developmental Growth.* St. Louis, C. V. Mosby, 1975.)

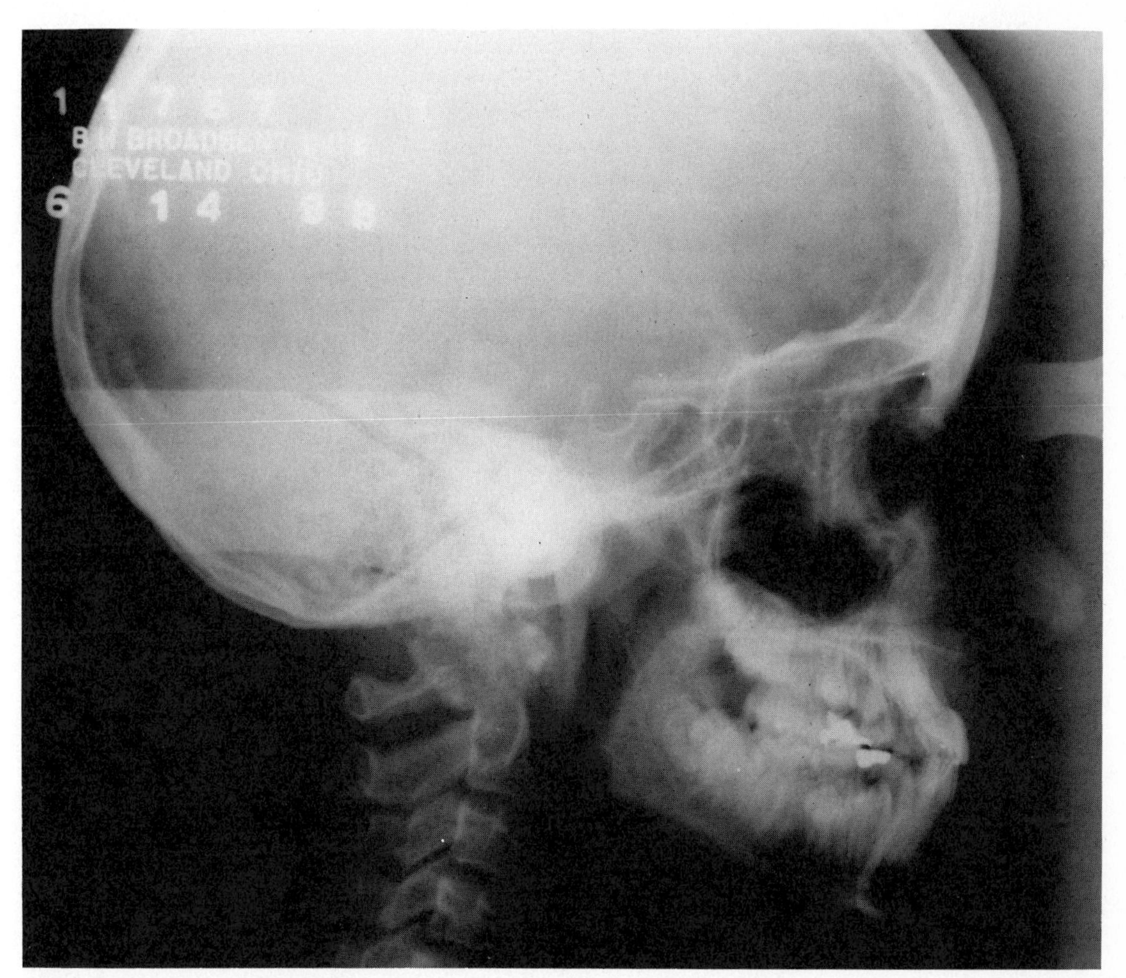

FIGURE 14–5. Lateral cephalometric radiograph. (From the Bolton Study, Bolton-Brush Growth Study Center, Case Western Reserve University, Cleveland, Ohio.)

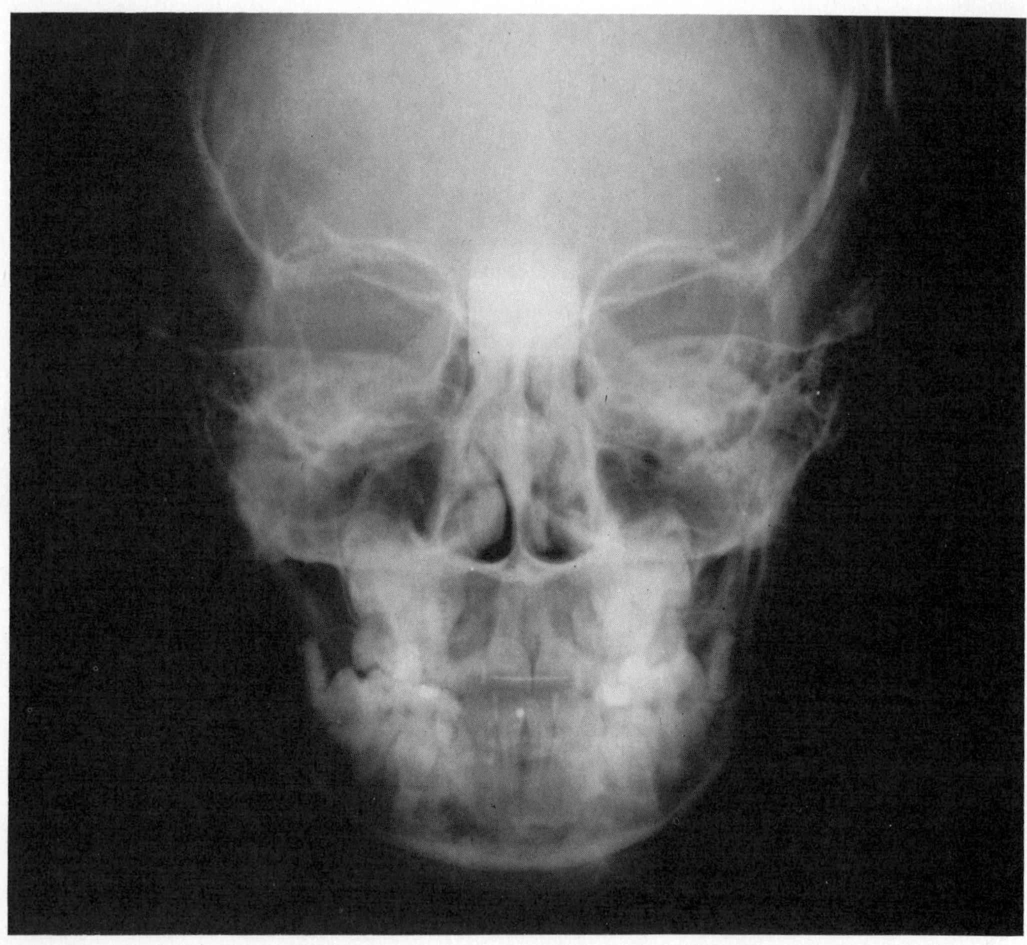

FIGURE 14–6. Frontal cephalometric radiograph. (From the Bolton Study, Bolton-Brush Growth Study Center, Case Western Reserve University, Cleveland, Ohio.)

the midsagittal plane and also the distance of the frontal film surface from the porionic axis allows for direct orientation of the frontal to the lateral for transfer of right and left structures, peripheral to the midline, from the lateral x-ray to the frontal, and the reverse (Fig. 14–7). This orientation is of significant assistance, not only in discerning right and left structures, but also where correction might be necessary for a frontal radiograph in which the head has been tilted down or up from Frankfort relation.

Most contemporary cephalometers used in orthodontic offices incorporate the basic elements of roentgenographic cephalometry but utilize only one x-ray source, with the associated ability to rotate the head holder 90 degrees to take a complementary frontal view. The wall-mounted arrangement (Fig. 14–8) allows for conservation of space and is very utilitarian, because the x-ray tube may also be used for periapical radiographs. Cephalometers that have provision for taking panoramic radiographs are also available for clinical use.

TRACING THE X-RAYS

Because of the need for measuring angles, planes, and linear dimensions on the radiographs, as well as comparing morphologic features, the historic technique of tracing the x-rays on 0.003 matte acetate paper is widely used so that the tracings of serial records may be superimposed and viewed on a transilluminated screen. The Broadbent-Bolton technique is that of averaging right and left structures in order to produce a tracing that represents these structures as they might be projected on the midsagittal plane. Figure 14–9 indicates the tracing technique of drawing between points of the right and left structures, rather than just between lines, in order to make an accurate average of the two images.

Figure 14–10 is a diagram of comprehensive lateral and frontal tracings and also includes many of the landmarks used in the cephalometric analyses to be described. A glossary of definitions is included at the end of this chapter.

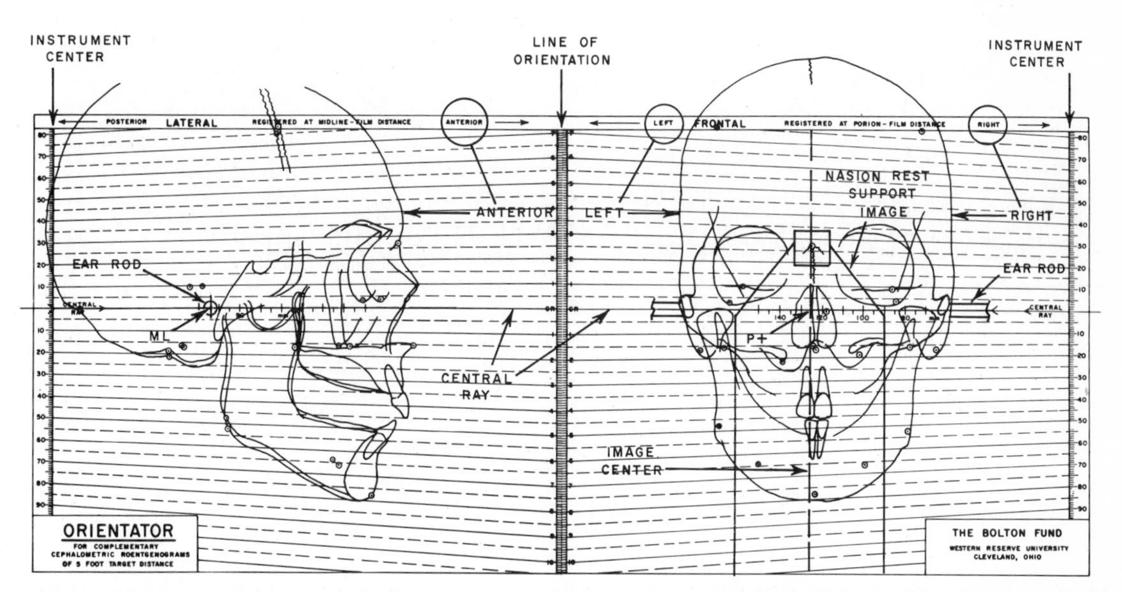

FIGURE 14–7. Bolton Orientator with lateral and frontal tracing in position. (From Broadbent, B. H., Sr., B. H. Broadbent, Jr., and W. Golden: *Bolton Standards of Dentofacial Developmental Growth.* St. Louis, C. V. Mosby, 1975.)

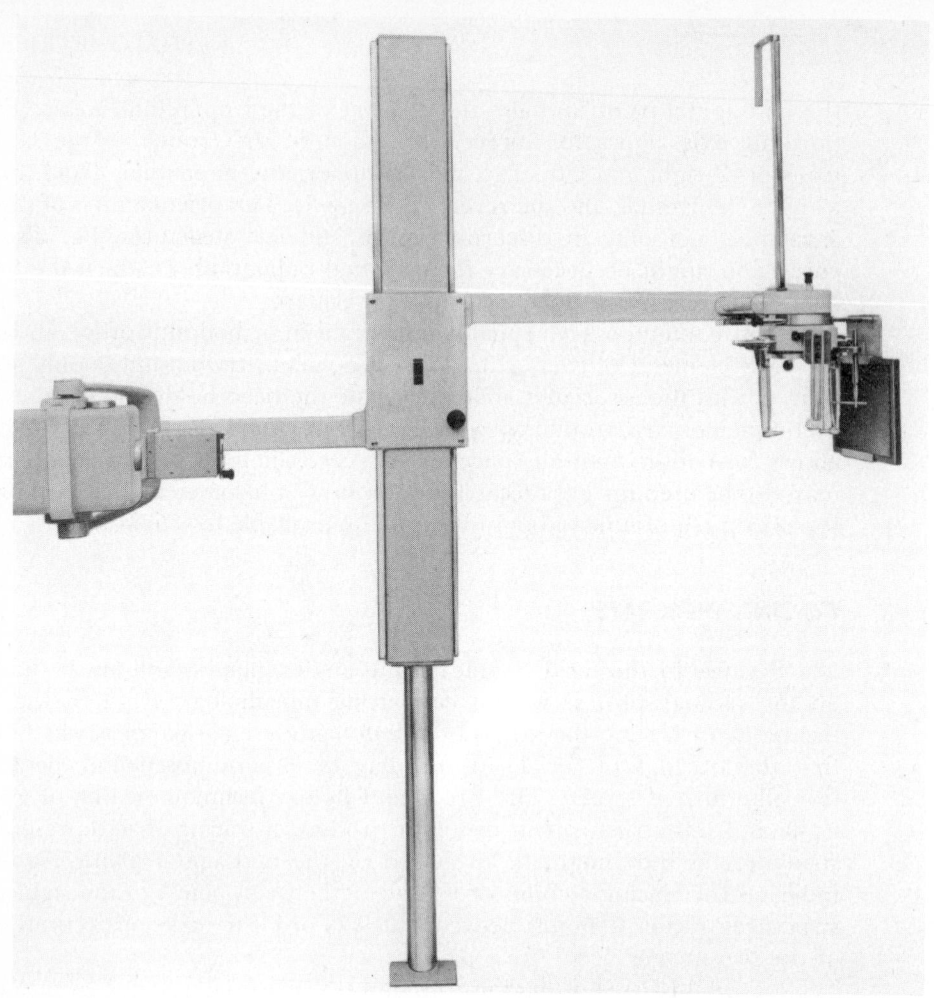

FIGURE 14–8. Contemporary wall-mounted cephalometer. (From the Bolton Study, Bolton-Brush Growth Study Center, Case Western Reserve University, Cleveland, Ohio.)

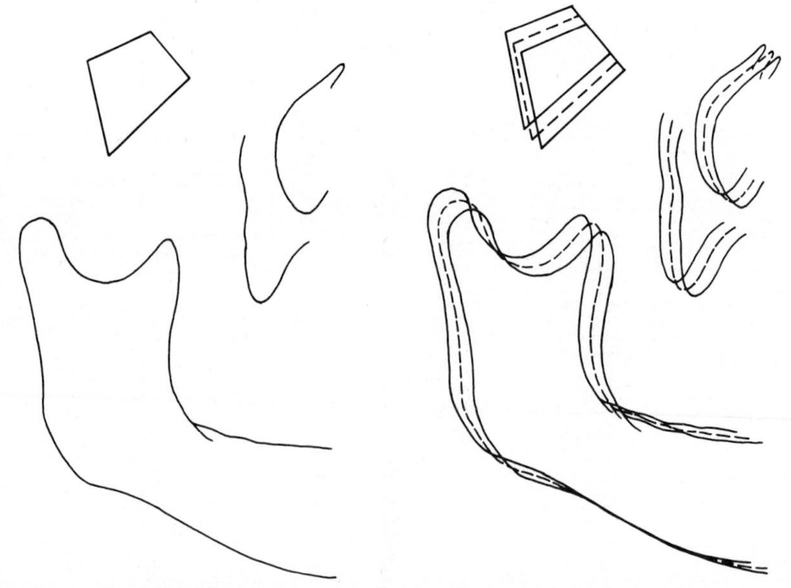

FIGURE 14–9. Averaging method. Note that the average (broken line) maintains the contours of the surfaces being averaged. The average depicts identical facets of right and left structures and not simply adjacent structures. (From Broadbent, B. H., Sr., B. H. Broadbent, Jr., and W. Golden: *Bolton Standards of Dentofacial Developmental Growth.* St. Louis, C. V. Mosby, 1975.)

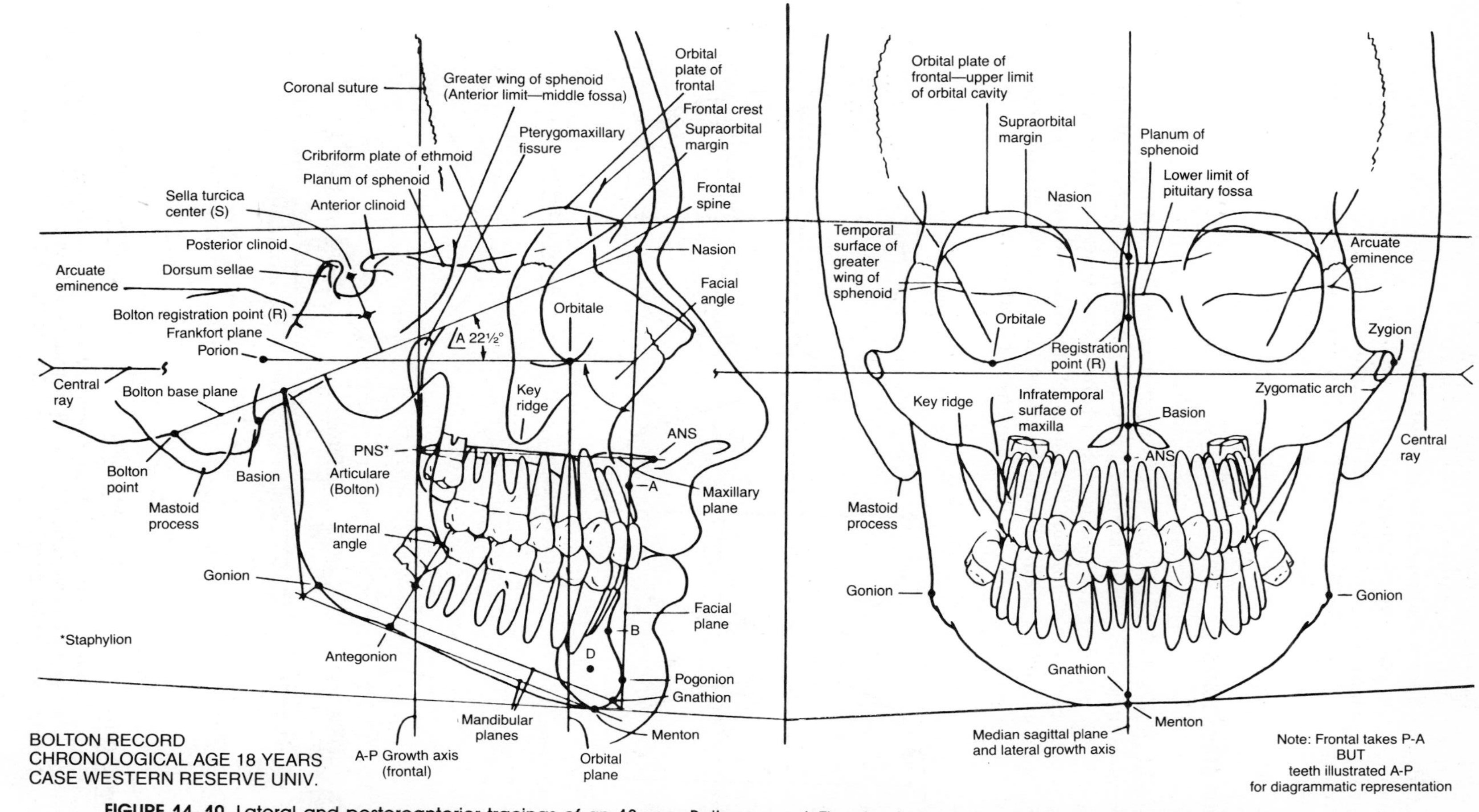

FIGURE 14–10. Lateral and posteroanterior tracings of an 18-year Bolton record. The structures, lines, and points that are frequently traced have been labeled. (From Broadbent, B. H., Sr., B. H. Broadbent, Jr., and W. Golden: *Bolton Standards of Dentofacial Developmental Growth.* St. Louis, C. V. Mosby, 1975.)

In making the tracing, of course, the individual investigator or diagnostician may elect to trace whatever features are of interest.

Although the lateral cephalogram is most routinely used in orthodontic practice today, a complementary frontal picture can be of inestimable value in understanding facial proportions (Fig. 14–11), asymmetries that may be present, and the positions of specific dental units such as the incisors, cuspids, and molar series.

SOFT TISSUE STRUCTURES

Figure 14–12 is a drawing of the soft tissue structures that may be observed in the lateral cephalogram. They are of significant importance to the cephalometrician in understanding the relationship of these structures to the skeletal elements as well as to the dentition. Diagnosis of an enlarged adenoid mass, turbinates, tongue and lip position and thickness, and so forth, are of great value in the process of diagnosis, treatment planning, and prognosis.

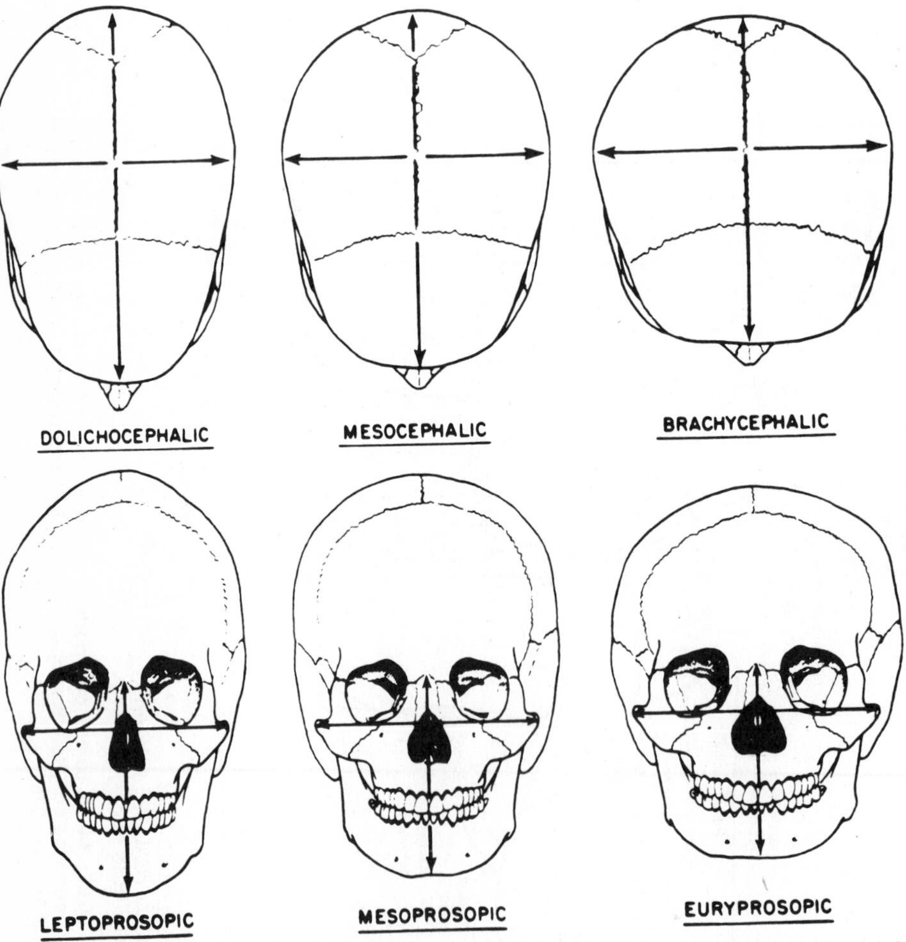

FIGURE 14–11. Cephalic and facial indices. Three distinct facial types as related to the height, width, and depth of the skull. (From Broadbent, B. H., Sr., B. H. Broadbent, Jr., and W. Golden: *Bolton Standards of Dentofacial Developmental Growth.* St. Louis, C. V. Mosby, 1975.)

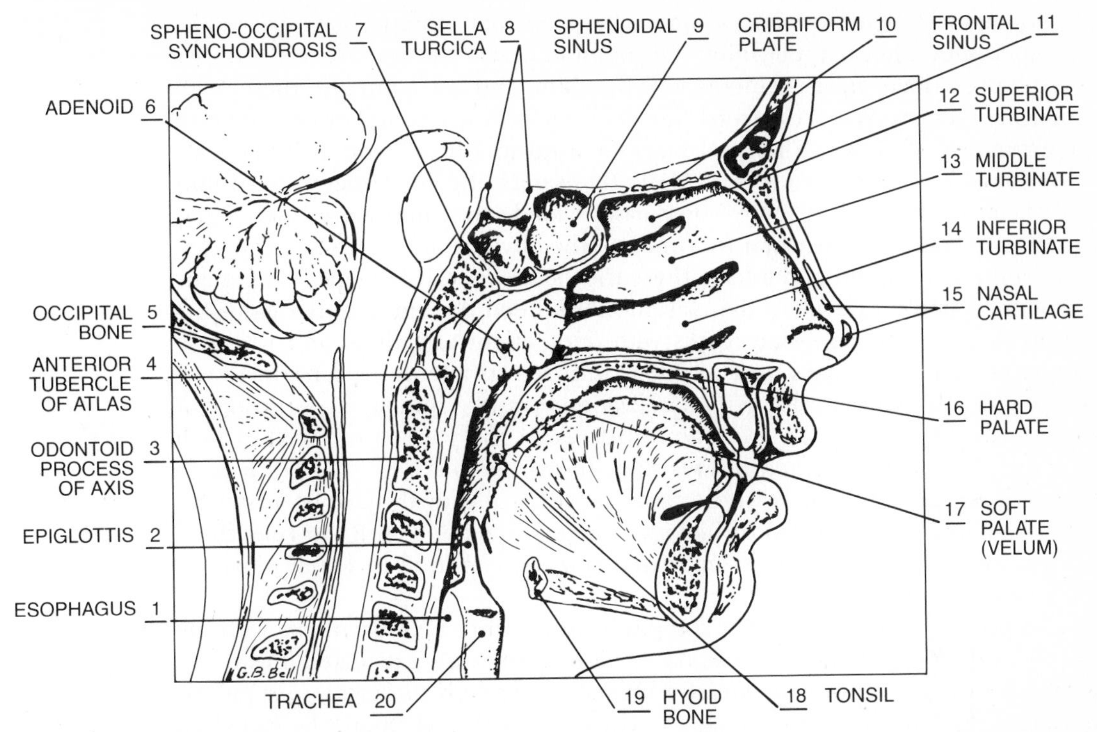

FIGURE 14–12. Soft tissue structures. Midsagittal section of the head, indicating the pertinent soft and hard tissue structures. (From the Bolton Study, Bolton-Brush Growth Study Center, Case Western Reserve University, Cleveland, Ohio.)

NORMS, VARIABILITY, COMPARISON

Before entering into a description and discussion of the various cephalometric measurements and some of the more frequently used analyses, it seems appropriate to attempt to define a concept of how and for what purpose we are to use this technique.

It has already been stated that measurements made on oriented headfilms provide useful information in both research and clinical practice. In the research application three primary efforts have been made: first, the accumulation of data related to craniofacial growth changes; second, the establishment of statistical norms for numerous cranial and dentofacial dimensions; and third, the evaluation of response to various treatment procedures. All these approaches have taken into account the obvious variables of age, sex, and racial and ethnic background. In growth research much has been learned of the morphologic and dimensional patterns of skull growth. However, the information is limited to just those aspects: change in shape and size. Although it is true that changes in rate, direction, and pattern of growth have been recorded, the cephalometric technique does not locate the sites of growth or measure the contribution of growth sites. Realistically, the single film provides only a static evaluation of that individual's size and shape at that point in time. A subsequent film on the same individual permits an evaluation of size and shape changes that occurred in the interval between films but does not tell exactly where the growth occurred.

In addition, and not covered in this overview, is the importance of the individual's biologic age (skeletal age), because either delayed or accelerated development is of utmost importance when "normal" comparisons are made. In orthodontic clinical application, the common practice is to make any number of the prescribed measurements on the film and to compare these with established norms. When the word "normal" is used, the researcher and clinician alike may ask, "What is a normal face—or a normal head?" or they may ask, "Normal with reference to what?" Many thousands of cephalometric headfilms have been traced and as many comparisons made in a massive effort to find this elusive "norm." But when we combine the obvious differences in age, sex, race, and environment and add to these the vicissitudes of biologic and genetic variation, it becomes evident that variation is indeed the name of the game.

Realistically, the clinician works with one patient at a time; and when he compares that individual's measurement with statistically derived norms, he is comparing his patient with a generalization drawn from a large group of individuals. Or he may be attempting to forecast the pattern of facial growth of a young patient by comparing his static measurements with those of a large group of similar individuals whose patterns of facial growth have been statistically evaluated. In the world of biology, where variability is still the rule, these efforts, although meeting with increased success, continue to entail a certain amount of hazard.

In practice, the orthodontist compares his patient's measurements with the norms and notes areas of deviation. These norms are calculated means or averages of many equivalent measurements. Along with the mean, a standard deviation (SD) is usually calculated. In clinical use, this SD might be called an acceptable range of variability. In other words, when there is variation within one SD, treatment by conventional orthodontic means should yield a good result. As the degree of deviation increases, the relative success of treatment decreases. This may appear to be an oversimplification, but essentially it means that, to some extent, the treatment is the victim of the individual pattern presented.

On the positive side, cephalometric-aided research has identified growth patterns and predictable responses to certain treatment procedures, either of which may assist the treatment planning process. The extent to which deviations from the norm influence the evaluation of the individual patient is a judgment made by the clinician, which he incorporates with all the other diagnostic information he has accumulated.

LINES, PLANES, ANGLES

In an effort to provide the beginning student with a basic understanding of cephalometric evaluation, several of the points, lines, planes, and angles that are common to several of the analyses will be diagrammed and discussed separately.

An example of a lateral tracing is shown in Figure 14–13. Definitions of the various landmarks and points used in most contemporary analyses are provided in the Glossary on page 391. The lateral tracing should include the soft tissue profile, the bony profile, the outline of the mandible, the posterior outline of the brain case, the odontoid process of the axis, the anterior lip of the foramen magnum, clivus, planum, and sella turcica outline, the roof of the orbit, the cribriform plate, the lateral and lower borders of the orbit, the outline of the pterygomaxillary fissure, the floor of the nose, the roof of the palate, the soft palate, the root of the tongue, the posterior pharyngeal wall, and the

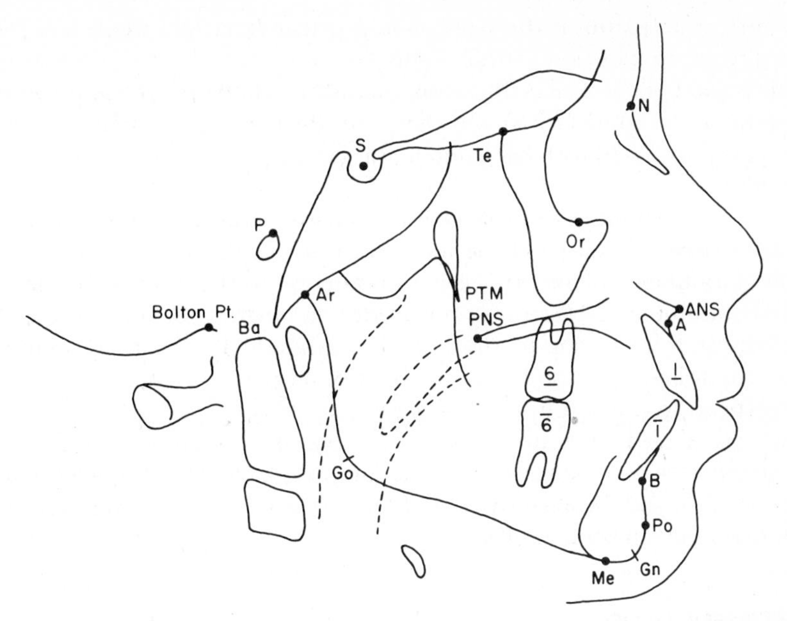

FIGURE 14–13. Lateral tracing with standard cephalometric landmarks. See Glossary of Terms for definitions.

body of the hyoid bone. Minimum traced teeth should include the first permanent molars and the most anterior incisors.

Figure 14–14 illustrates the most frequently used horizontal planes and lines. The sella-nasion (SN) is drawn from the selected point sella to nasion. It is described as representing the anteroposterior extent of the anterior cranial base and serves as a reference line when one is relating facial structures to the cranial base.

The Frankfort horizontal plane (FH) is drawn tangent to the superior outline of porion (P) and extends through orbitale (Or). It is widely accepted

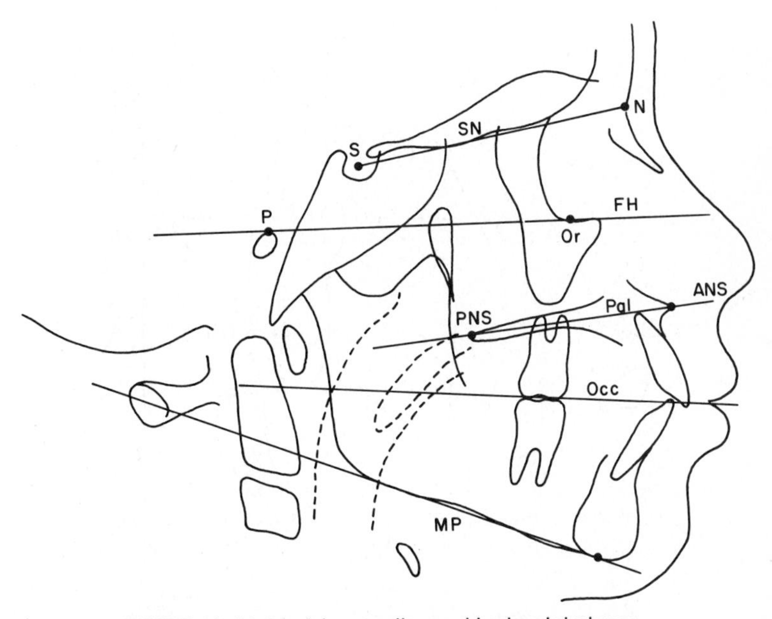

FIGURE 14–14. Most frequently used horizontal planes.

as **the** horizontal plane of the head. Some researchers feel there is a postural significance in the position of this plane.

The palatal plane (Pal) is drawn by extending a line through and connecting the anterior nasal spine (ANS) and the posterior nasal spine (PNS). By relating palatal plane to Frankfort horizontal, the postural tilt of the maxilla can be measured.

The occlusal plane (Occ) bisects the incisor overbite (or open bite) and passes over the distal cusps of the most posterior teeth in occlusion.

The mandibular plane (MP) is drawn tangent to the inferior border of the symphysis outline and extending posteriorly is tangent to the inferior border of the mandible posterior to the antegonial notch. Relating the mandibular plane to the sella-nasion or Frankfort horizontal plane provides an assessment of the vertical proportion in the lower face.

The plane and lines just described serve as reference planes for other measurements or may be related to one another, as will be discussed later. All tracings used in the illustrations are of the same patient and will also be used later in discussing several analyses.

SKELETAL ASSESSMENT

Profile assessment includes determining the anteroposterior position of the chin, the maxilla, the anterior teeth, and the soft tissue. The facial angle is used in determining anteroposterior chin position (Fig. 14–15). It is the angle between the Frankfort horizontal plane and the facial plane. The facial plane is a line drawn from nasion (N) through pogonion (Pog). The mean value for this angle is 87.8 degrees, with a range from 82 to 95 degrees. Values larger than these would indicate a lower face prognathism and Class III malocclusion, whereas smaller values would be associated with a retrognathic mandible and Class II malocclusion. The tracing used in Figure 14–15 shows a facial angle of 90 degrees, which indicates an acceptable anteroposterior position of the chin.

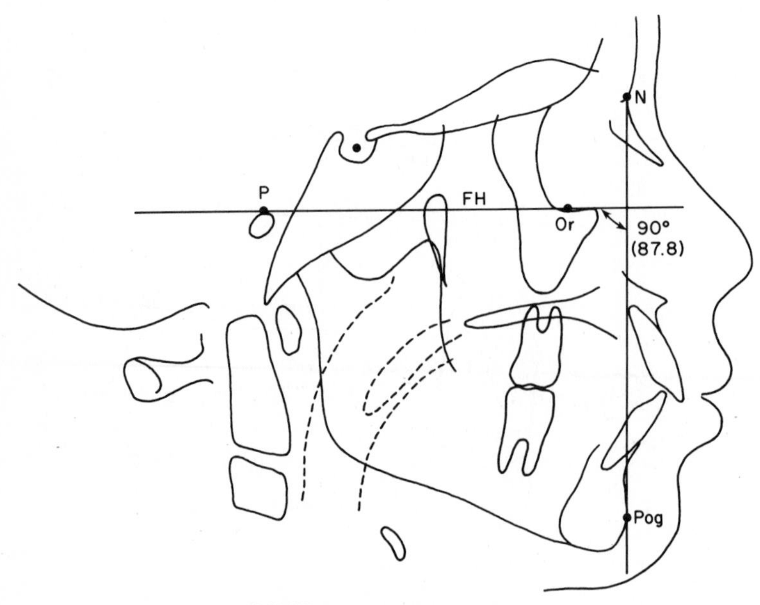

FIGURE 14–15. Facial angle.

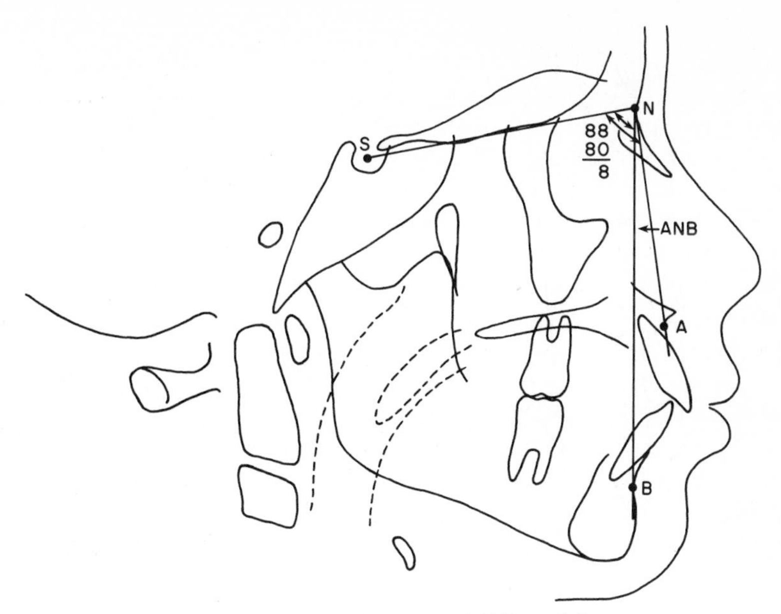

FIGURE 14–16. SNA, SNB, and ANB angles.

The maxilla and mandible may be related to each other anteroposteriorly by the SNA-SNB angles (Fig. 14–16). The angles are read between the SN and the NA and NB lines, respectively. Although points A and B, by definition, appear to represent maxillary and mandibular basal structures, respectively, some authors question their validity on the basis that they may be influenced by incisor tooth movement during treatment. In any event, their mean values, 82 and 80 degrees, respectively (at ages 12 to 14), are used to assess the anteroposterior position of the maxilla and mandible with respect to the anterior cranial base. Of perhaps more interest to the clinician is the difference between the angles—the ANB angle. The mean value of the ANB angle is 2 degrees, and significant deviations from this mean indicate an anteroposterior discrepancy of those basal structures that support the dentition. A high ANB angle indicates a maxilla that is forward, a retrognathic mandible, or a combination of these deviations (Fig. 14–16). The tracing used in this illustration shows a high ANB angle (8 degrees), but the SNB angle is normal (80 degrees). It must be assumed, then, that the maxilla, or point A at least, is positioned anteriorly. Large ANB deviations in either direction indicate to the clinician that the problem to be treated has an orthopedic significance, and it may not be responsive to treatment by tooth movement alone.

Anteroposterior variations in facial profile may be assessed by the angle of convexity (Fig. 14–17). The angle of convexity provides information similar to that from the ANB angle, but in this case it takes into account the influence of the "chin button" or prominence of pogonion. The mean value of the angle of convexity is 0 degrees, with a range of −8.5 to +10 degrees. In the tracing used in the illustration, the high angle of 17 degrees indicates a very convex facial profile and agrees with the information previously obtained from the SNA-SNB angles.

Another means of measuring convexity is by the relationship of point A to the facial plane (Fig. 14–18). The measurement is made horizontally from point A to the facial plane and recorded in millimeters. The mean value is 0, with a

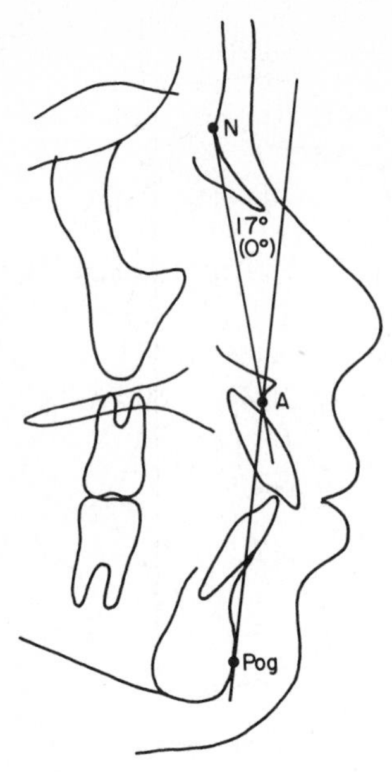

FIGURE 14–17. Angle of convexity.

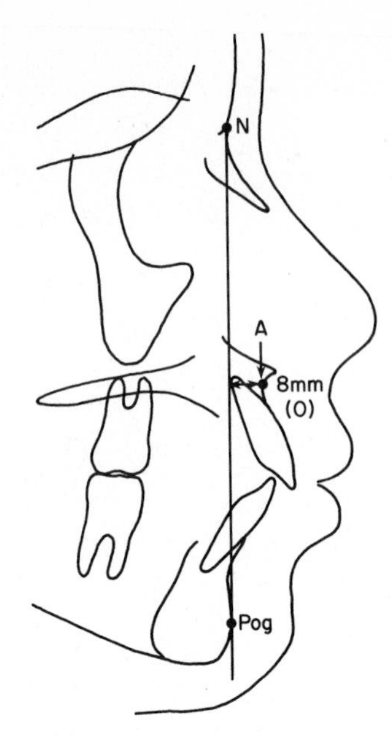

FIGURE 14–18. Point A to facial plane.

range of −3 mm to +4 mm. Deviations greater than 5 mm in front of the facial plane or 3 mm behind it are suggestive of an orthopedic problem in anteroposterior skeletal relation. In the illustration, the deviation is 8 mm in front of the facial plane, agreeing with previous assessments of profile convexity.

The measurements outlined above are primarily related to anteroposterior assessment of skeletal profile. The mandibular plane angle provides a means of assessing vertical relation and the morphology of the lower third of the face. The mandibular plane angle may be measured in relation to the Frankfort horizontal plane (FMA) or in relation to the SN line (SN-MP). The mean FMA is 21.9 degrees (Fig. 14–19), and the mean SN-MP angle is 33 degrees. High mandibular plane angles may result from a short ramus, obtuse gonial angle,

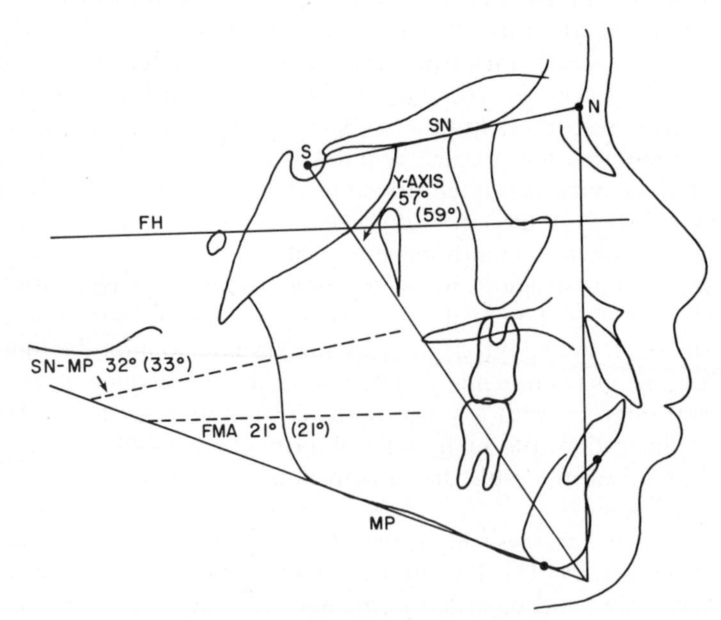

FIGURE 14–19. Mandibular plane angle.

high position of the glenoid fossa, long anterior face height, or any combination of these. High mandibular plane angles are frequently associated with anterior open bites and vertically growing facial patterns. Conversely, low or flat mandibular plane angles, frequently associated with deep anterior overbite and horizontal mandibular growth patterns, are the result of a long ramus, acute gonial angle, short anterior face height, or any combination of these.

Figures 14–20 and 14–21 illustrate tracings showing high and low mandibular plane angles, respectively, and the differences in facial pattern associated with them. Note the differences in the ratio of facial height to depth and in the ratio of posterior face height to anterior face height.

The Y-axis, drawn from the sella point through gnathion and related angularly to the Frankfort horizontal plane, is judged to be indicative of the direction the mandibular symphysis will follow in future growth. The mean value of the Y-axis is 59 degrees. The tracing of the sample patient (Fig. 14–19) shows a measurement of 57 degrees. Angles significantly larger than the mean indicate a greater proportion of vertical growth at the symphysis, whereas smaller angles indicate a relatively more forward pattern. Figure 14–20 shows a 66-degree Y-axis and Figure 14–21 a 52 degree measurement. These measurements are consistent with the respective vertical and horizontal growth patterns of these two patients.

DENTAL ASSESSMENT

The dental assessment is made by a combination of several measurements, both angular and linear, primarily involving the incisors. Anteroposteriorly, the incisor crowns may be related to the facial plane (Fig. 14–22), the ideal position of the lower incisor crown being right on the plane or within an acceptable range of −2 mm to +3 mm. The illustration (Fig. 14–22) would be described

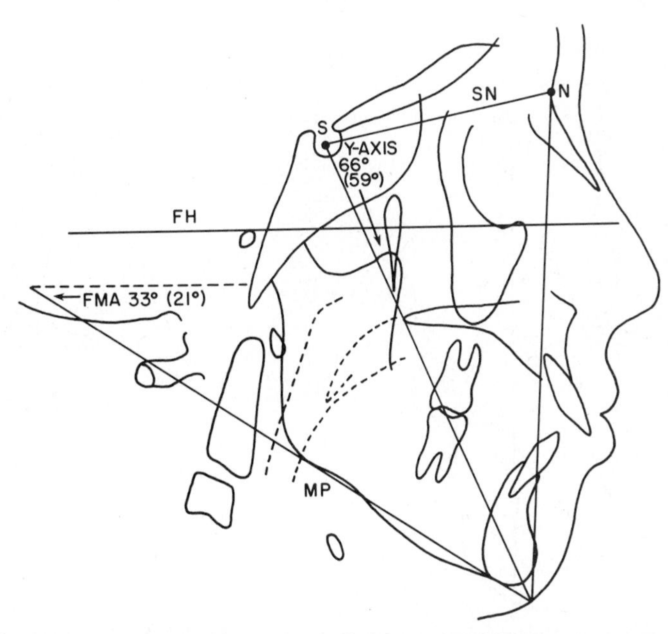

FIGURE 14–20. Steep mandibular plane angle. Note the disproportion of anterior to posterior face height and shallow face depth.

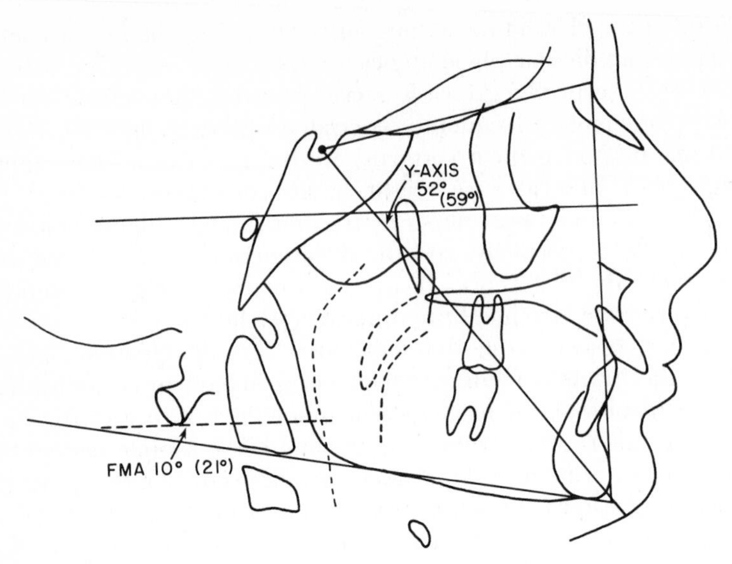

FIGURE 14–21. Low mandibular plane angle. Note the relative decrease of anterior face height and increased depth of face.

as protrusive, the maxillary incisor being 15 mm in front of the facial plane and the mandibular incisor 10 mm forward.

The incisors may be related anteroposteriorly in the same general manner with reference to the A-Po line (Fig. 14–23). Again, the ideal for the lower incisor is right on the line, with an acceptable range of −2 mm to +3 mm. It must be pointed out that in this instance the reference line (A-Po) is related to the maxillary denture base and the chin, rather than to the facial plane, and will thus vary with anteroposterior deviation between maxilla and mandible. In the patient illustrated in Figure 14–23, the maxilla is markedly in front of the

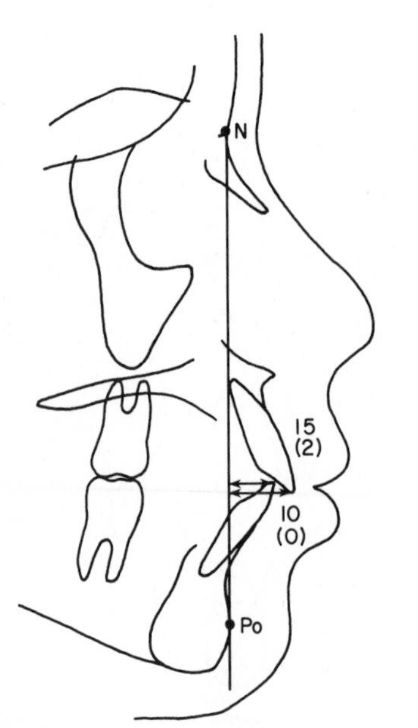

FIGURE 14–22. Upper and lower central incisors to the facial plane.

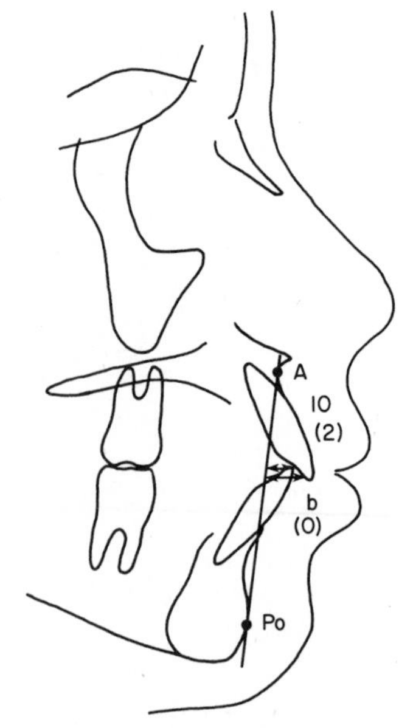

FIGURE 14–23. Upper and lower central incisors to the A-Po line.

mandible, and the lower incisor is 6 mm in front of the A-Po line, compared with the 10 mm relationship to the facial plane.

The most commonly used angular measurements are the interincisal angle, the long axis of the lower incisor to the mandibular plane, and the long axis of the upper incisor to the Frankfort horizontal plane. Figure 14–24 illustrates these measurements. Various authors place the mean interincisal angle between 125 and 135 degrees. Larger angles result from very upright incisors and are frequently associated with deep overbite. Small angles occur in dental protrusion, as illustrated in Figure 14–24.

The ideal lower incisor to mandibular plane angle is 90 degrees and would be expected to occur with an average mandibular plane angle (FMA). Some variations in lower incisor posture according to mandibular posture are mentioned later in the discussion of the Tweed analysis.

The upper incisor angulation of the Frankfort horizontal plane is also illustrated in Figure 14–24 and has a mean value of 110 degrees. Larger angles usually indicate maxillary incisor protrusion.

SOFT TISSUE ASSESSMENT

Perhaps the most commonly used evaluation of soft tissue profile is the relationship of the lips to the esthetic plane (Ricketts), as shown in Figure 14–25. Although this is primarily an esthetic evaluation, it is based on the fact that lip posture is influenced directly by the anteroposterior position of the teeth behind the lips. As a criterion, the lower lip position should be within a range of 2 mm behind the E-plane (esthetic plane) to just touching it. In the patient shown in the illustration, both lips are well in front of the plane and reflect the extent of dental protrusion already documented.

The measurements just described are but a few of those used in cephalometrics and are illustrated here not because they are more significant than

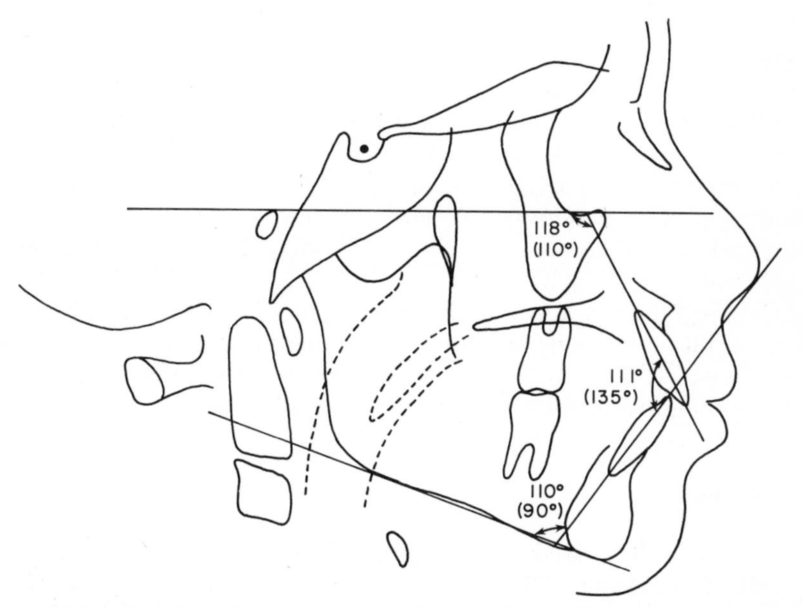

FIGURE 14–24. Interincisal angle, lower incisor to the mandibular plane, and upper incisor to the Frankfort horizontal plane.

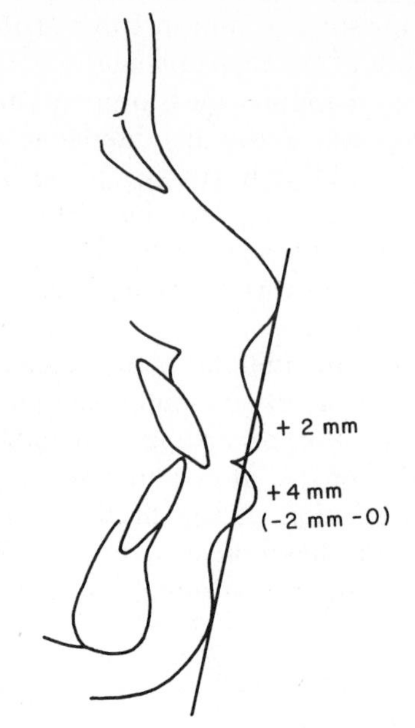

FIGURE 14–25. Esthetic plane (Ricketts' analysis).

others, but because they are representative of the type of measurement made and provide a good initial understanding of this type of evaluation. The following pages will illustrate several of the more commonly used analyses incorporating these and similar measurements. Again, there is no suggestion that these are the most significant or the most accurate, but they are selected as being representative of a cross section of contemporary cephalometric thought. Each analysis is based on a background of research and hypothesis, and its author presents it as an attempt to provide a practical clinical application of the research effort.

ANALYSES

In an effort to illustrate the techniques of cephalometrics analysis, seven different analyses are presented. Some of these are "total" in the sense that they make an effort at "total facial analysis." Others are limited to emphasis on a particular area or dimension. A single patient's lateral tracing is used to illustrate each analysis, and comments are made pertaining to interpretation when appropriate.

The subject used is a Caucasian male, aged 13 years, 7 months, in good health and with no previous orthodontic treatment. An earlier headfilm, taken 3 years previously, is used as a final illustration to demonstrate the technique of superimposing tracings for evaluation of growth changes.

Downs' Analysis

Downs' analysis is based on a sample of 20 children, aged 12 to 17 years, with excellent occlusions. The diagram (Fig. 14–26) consists of the lines Na-Pog, Na-A, A-B, A-Pog, S-Gn, the occlusal plane, the mandibular plane, the long axis of upper and lower incisors, and the Frankfort horizontal plane.

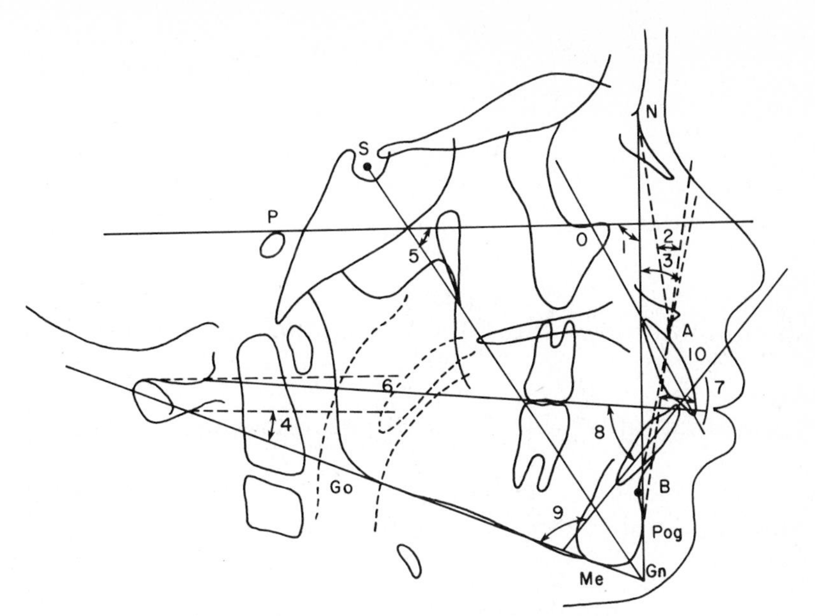

FIGURE 14–26. Tracing for Downs' analysis.

An addition to Downs' analysis is a chart by Voorhies and Adams providing a graphic portrayal of the ten measurements in the analysis (Fig. 14–27). The line of small arrows down the center of the diagram identifies the mean figure for each measurement, and the extent of the polygon outlines the range of each measurement. The dotted line in the illustration is the plot of the measurements of the subject used in this discussion.

The top half of the diagram charts those measurements related to skeletal configuration, whereas the lower half shows denture relationships. The interpretation for the patient illustrated shows that the chin is well situated anteroposteriorly, and the mandibular plane and Y-axis measurements forecast continued normal downward and foward growth in this area. The high angle of convexity and A-B plane angle confirm the midface convexity. If the chin position is normal, the convexity must be due to midface or maxillary prominence.

The lower half of the diagram charts the marked bi-dental protrusion of this patient, as indicated by the deviations in the interincisal angle, the angle of the lower incisor to the occlusal plane, and the angle of the lower incisor to the mandibular plane.

The anterior prominence of point A, which moves the A-Po line forward at its upper end, masks the protrusion of the upper incisor.

In summary, this patient would be described as having normal skeletal arrangement, except for the midface convexity, and a superimposed bi-dental protrusion.

Downs' analysis might be described as being profile-oriented. The primary reference plane is the Frankfort horizontal. Vertical assessment is only with the mandibular plane and Y-axis.

Steiner's Analysis

The Steiner analysis is actually a composite of measurements from several other sources (Margolis, Thompson, Riedel, Wylie, and Downs). It is based

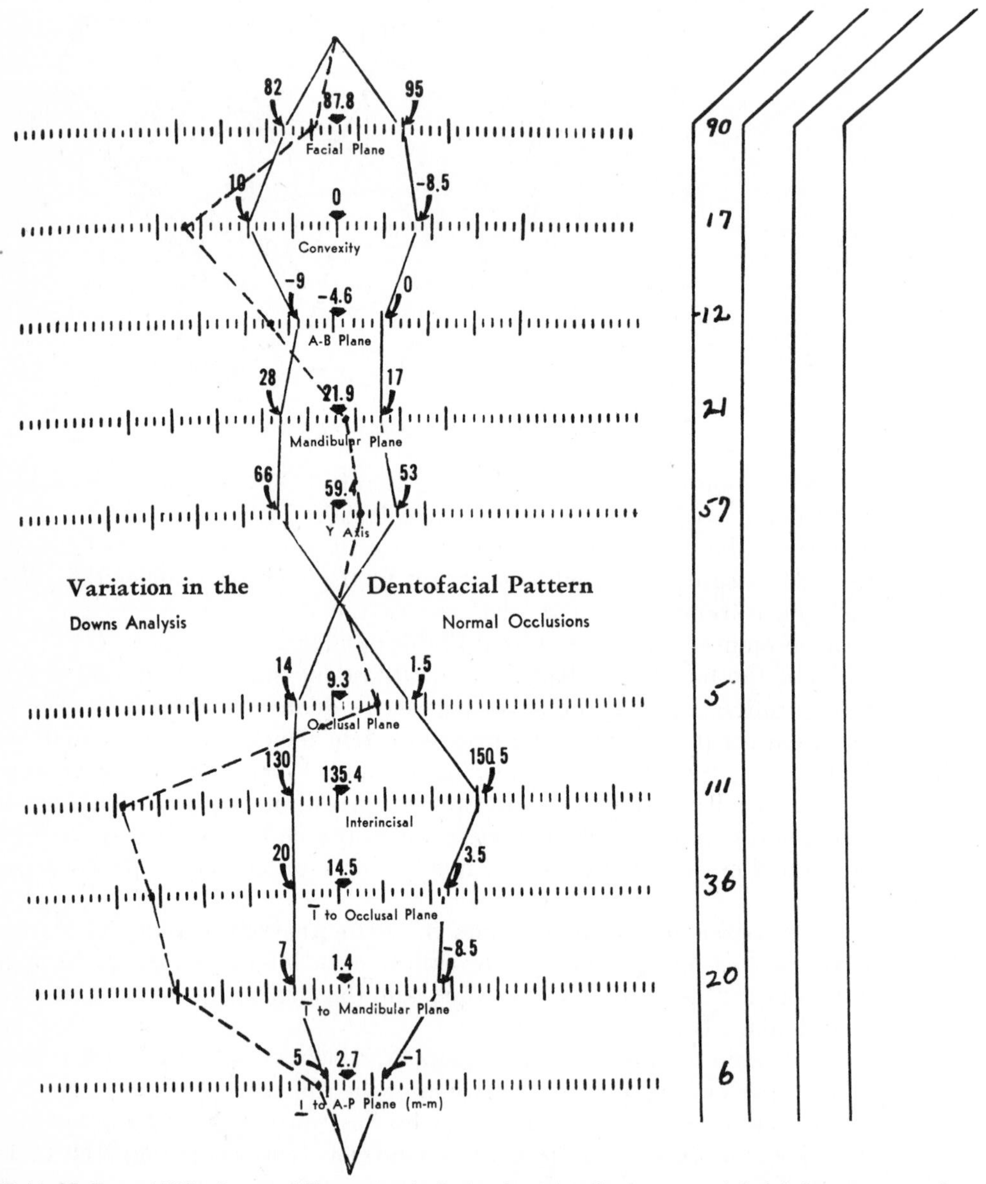

FIGURE 14–27. Downs' "Wigglegram." The upper half of polygon outlines range of skeletal measurements; and the lower half, the range of dental measurements. The dotted line shows the relation of the sample patient's measurements to the acceptable ranges. (Adapted from Voorhies, J. W., and J. W. Adams: Polygonic interpretations of cephalometric findings. Angle Orthod., 21:194, 1951.)

primarily on a single plane of reference—the S-Na line—and does not take into account variations in the length or cant of this reference plane. A particular feature of the analysis is the linear as well as angular relation of the incisors to reference lines (Na-A and Na-B).

Lines to be drawn are S-Na, Na-A, Na-B, Go-Gn, the occlusal plane, and the long axis of the upper and lower incisors (Fig. 14–28). The Steiner analysis form (Fig. 14–29) lists the reference norms and the measurements made from the subject.

The appraisal of this patient (Fig. 14–29) reveals the chin well related anteroposteriorly, the maxilla forward, and a high ANB angle, confirming the basal discrepancy. The upper incisor is well related linearly and angularly to its base, but the lower incisor is markedly procumbent by both forms of measurement. The low interincisal angle is consistent with the bi-dental protrusion.

Summary appraisal of the patient shows normal chin position, forward maxillae, and bi-dental protrusion. The analysis does show that although the lower incisor posture can be corrected by simple uprighting movement, it would be necessary to move the maxillary incisor bodily to preserve its normal angular posture.

An additional feature of the Steiner analysis is the provision made on the form (Fig. 14–29) for a treatment planning procedure. It is not within the scope of this chapter to go through this procedure in detail, but briefly it is a means of synthesizing treatment objectives, taking into account the original cephalometric data, the arch length discrepancy, and selected treatment objectives.

The Steiner analysis is also profile-oriented and provides excellent visualization of incisor position and anterior facial profile detail. With the addition of its treatment planning rationale, it has enjoyed wide clinical use in the field of orthodontics.

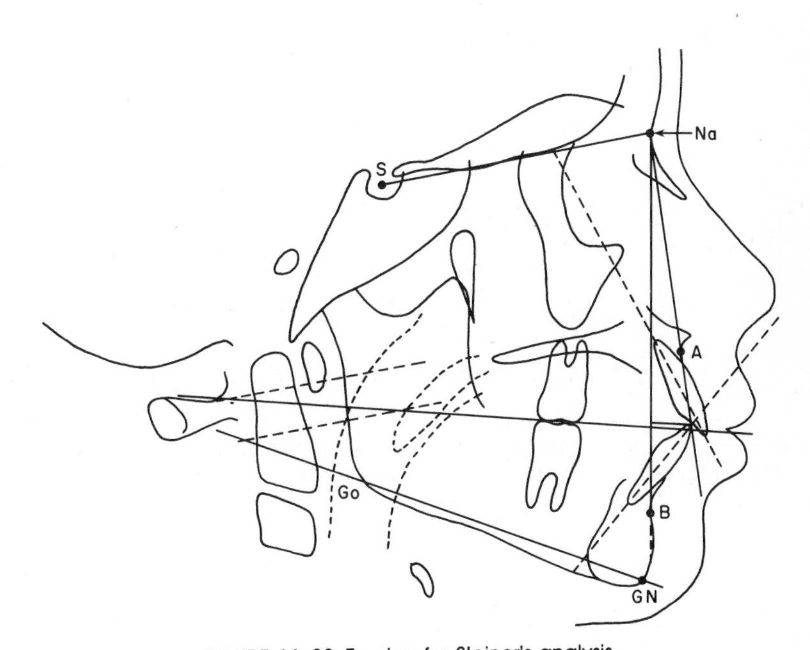

FIGURE 14–28. Tracing for Steiner's analysis.

STEINER ANALYSIS

		Ref. Norm.						
SNA	(angle)	82°	87					
SNB	(angle)	80°	79					
ANB	(angle)	2°	8					
SND	(angle)	76° or 77°						
$\underline{1}$ to NA	(mm)	4	4					
$\underline{1}$ to NA	(angle)	22°	20					
$\overline{1}$ to NB	(mm)	4	11					
$\overline{1}$ to NB	(angle)	25°	40					
Po to NB	(mm)	not established	1					
Po & $\overline{1}$ to NB	(Difference)		10					
$\underline{1}$ to $\overline{1}$	(angle)	131°	111					
Occl to SN	(angle)	14°	14					
GoGn to SN	(angle)	32°	30					
Arch length discrepancy			-3					

(mm)	+	−
Correcting Arch Form Moves $\overline{1}$		

LOWER ARCH	+	−
Discrepancy		
Expansion		
Relocation $\overline{1}$		
Relocation $\overline{6}$		
$\overline{E}$ Space		
Intermaxillary		
Extraction		
Total Net		

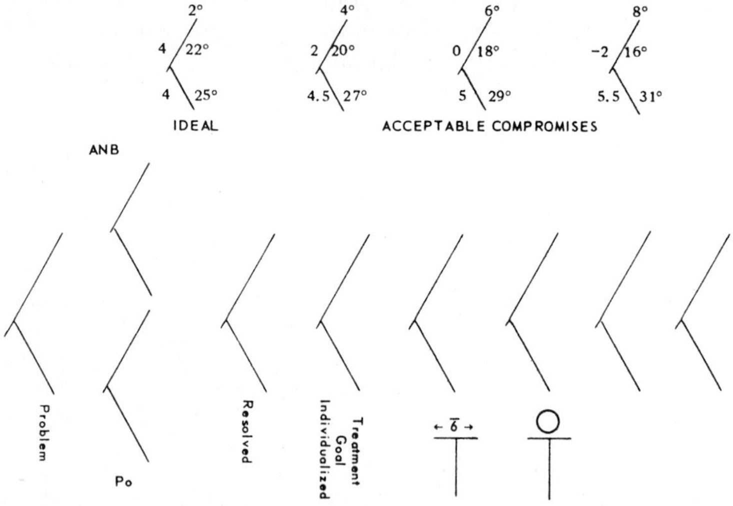

2° 4 22° 4 25° IDEAL

4° 2 20° 4.5 27°

6° 0 18° 5 29°

8° −2 16° 5.5 31°

ACCEPTABLE COMPROMISES

ANB

Po

Problem

Resolved

Treatment Goal Individualized

← $\overline{6}$ →

* *These estimates are useful as guides but they must be modified for individuals.*

Dental Corporation of America • P.O. Box 1011 • Washington, D. C. 20013

FIGURE 14–29. Recording and treatment planning form for Steiner's analysis, with measurements of sample patient, R.B. (Courtesy of Dental Corporation of America, Washington, D.C.)

Wylie's Analysis

In the following paragraphs, only the "assessment of anteroposterior dysplasia" as presented by Wylie in 1947 will be described. The averages were derived from a sample group of an equal number of males and females with an average age of 11 years, 6 months.

All measurements except mandible length are made parallel to the Frankfort horizontal plane from projections to the following points: posterior border of the condyle, sella, pterygomaxillary (PTM fissure), upper first molar, and the anterior nasal spine. The mandibular plane is drawn and perpendicular projections made to the posterior border of the condyle and pogonion for mandibular length.

Figure 14–30 shows the diagram and measurements on the sample patient. Comparison with the means is shown in Table 14–1.

Basically, the analysis provides a means of evaluating anteroposterior size and position of the maxilla and mandible. For maxillary measurements below the norm, the difference is put in the prognathic column; and for values above the norm, in the orthognathic column. For mandibular measurements above the norm, the difference is put in the prognathic column; and when the measurements are below normal, in the orthognathic column.

In the appraisal of this patient, there is a net difference, indicating that the mandible is 5 mm smaller than the maxilla. Actually, the individual measurements show that this is a very deep face with greater depth in the maxilla than in the mandible. This analysis is particularly useful in evaluating Class III skeletal patterns.

Tweed's Analysis

It should be stated at the outset that the Tweed analysis is not a total facial analysis. Dr. Tweed did not intend it to be a total analysis, and it is unfortunate

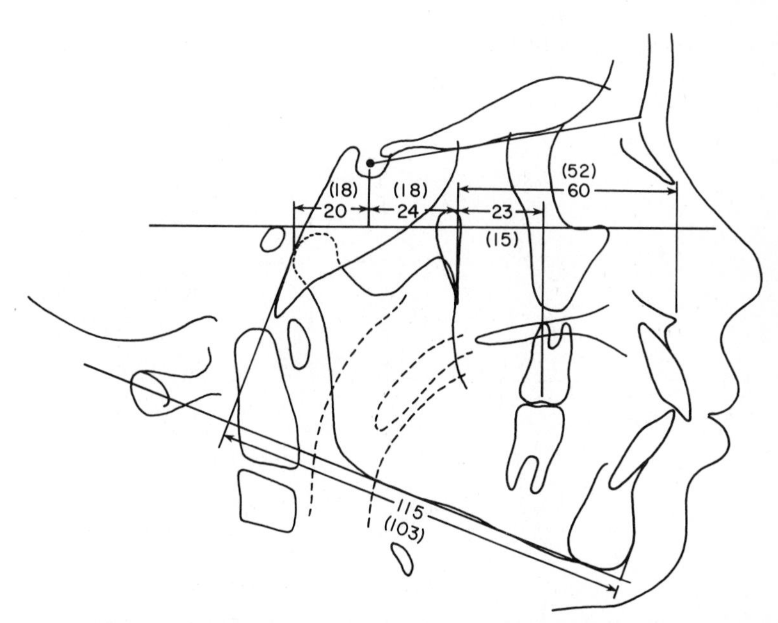

FIGURE 14–30. Tracing for Wylie's assessment of anteroposterior dysplasia. Mean values for each measurement are shown in parentheses.

Table 14–1. PATIENT (see Fig. 14–30) COMPARED WITH SAMPLE MEANS

Dimension	Male	Female	R.B.	Orthognathic	Prognathic
Glenoid fossa to S	18	17	20	2	
S to PTM	18	17	24	6	
PTM to ANS	52	52	60	8	
PTM to 6	15	16	23	1	
Mandibular length	103	101	115		12
Total				17	12

that some practitioners have used it in this manner. The analysis is based primarily on the deflection of the mandible, as measured by the Frankfort-mandibular plane angle, and the posture of the lower incisor. The objectives of the analysis appear to be twofold. The first is to determine the position the lower incisor should occupy at the end of treatment. Predetermination of this relationship provides useful treatment planning information, especially with regard to the extraction decision. Second, Dr. Tweed established a prognosis on the treatment result, based on the configuration of the triangle.

Basically, the analysis consists of the so-called Tweed triangle, formed by the Frankfort horizontal plane, the mandibular plane, and the long axis of the lower incisor (Fig. 14–31). The three angles thus formed are the Frankfort-mandibular plane (FMA), lower incisor to mandibular plane (IMPA), and lower incisor to Frankfort horizontal (FMIA). The basis is the FMA, as the following norms and prognoses indicate:

1. FMA 16° to 28°: prognosis good
   at 16°, IMPA should be 90° + 5° = 95°
   at 22°, IMPA should be 90°
   at 28°, IMPA should be 90° − 5° = 85°
   Approximately 60% of malocclusions have FMA between 16° and 28°
2. FMA 28° to 35°: prognosis fair
   at 28°, IMPA should be 90° − 5° = 85°, extractions necessary in majority of cases.
   at 35° IMPA should be 80° to 85°

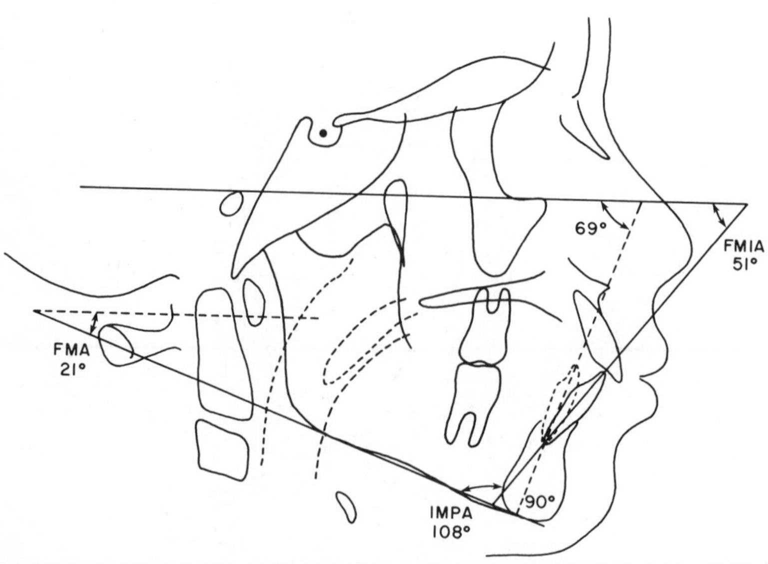

FIGURE 14–31. Tweed triangle. The dotted line intersecting mandibular plane is parallel to Frankfort horizontal. The dotted outline of the lower incisor represents the ideal position of this tooth.

3. FMA above 35°: prognosis bad, extraction frequently complicates problem

Sometime after introducing the original figures shown above, Tweed began stressing the importance of the FMIA, recommending that it be maintained at 65 to 70 degrees.

In the sample tracing (Fig. 14–31), the FMA is 21 degrees, FMIA 51 degrees, and IMPA 108 degrees. According to the analysis, with the FMA of 21 degrees, the IMPA should be 90 degrees, as shown by the dotted line on the tracing. With this change, the FMIA would be 69 degrees, which is within the recommended range. To attempt to achieve this relationship would obviously necessitate removal of dental units.

The Tweed analysis is primarily for clinical treatment planning by establishing a position the lower incisor should occupy, with provision made for variations in mandibular position.

Ricketts' Analysis

Ricketts' analysis has progressed through a series of modifications and has now burgeoned into a detailed evaluation of craniofacial and dental morphology. It has been adapted to a computer-based diagnostic and treatment forecasting service. Continued refinements of growth and treatment forecasting aspects may be anticipated as additional data are accumulated in the computer. Because this chapter is providing only an overview of several of the lateral or profile analyses, only the lateral summary analysis will be presented.

The lines traced are the Frankfort horizontal, facial, occlusal, mandibular, and esthetic (tip of nose to tip of chin) planes, N-Ba, Pt-vertical (tangent to the posterior outline of the PTM and perpendicular to the Frankfort), the facial axis (upper margin of foramen rotundum to gnathion), and the long axis of the incisors.

From this tracing (Fig. 14–32), eight relations are measured to provide an

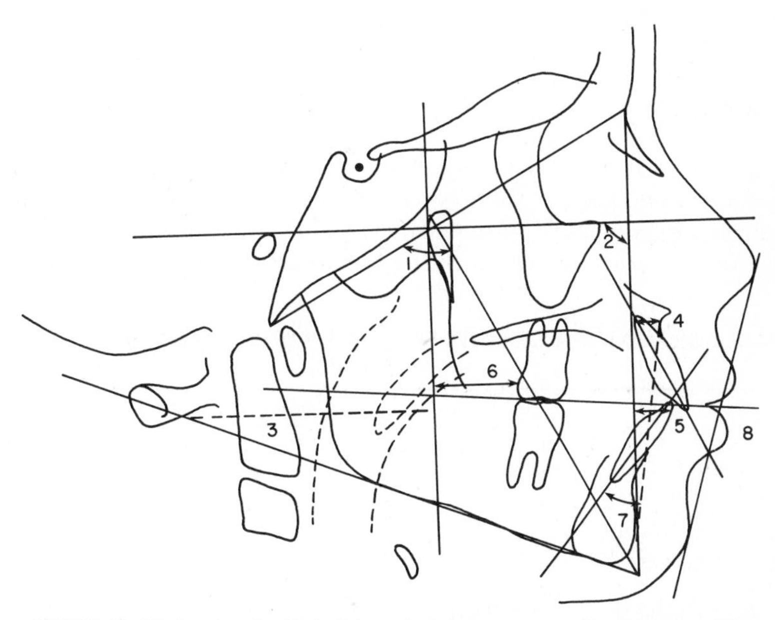

FIGURE 14–32. Tracing for Ricketts' analysis (summary profile analysis only).

overall appraisal of the case. The means and age changes from age 9 are shown in Table 14–2. The eight measurements are as follows:

1. Facial axis: the angle between the Ba-Na plane and the line from the foramen rotundum to gnathion. This gives the direction of the growth of the chin and is a modification of the Y-axis from Downs' analysis.
2. Facial depth: the angle between the Frankfort plane and the facial plane.
3. Mandibular plane to Frankfort plane angle.
4. Convexity: the horizontal distance between point A and the facial plane.
5. Lower incisor to A-Po: locates the lower teeth anteroposteriorly in the mandible.
6. Upper molar position: the horizontal distance from PTM to the distal surface of the upper first molar.
7. Lower incisor inclination: the angle between the lower incisor axis and the A-Po line. This is a refinement of the lower incisor to NB line (Steiner), which takes into account basal relationships.
8. Esthetic plane: lower lip relation anteroposteriorly to the esthetic plane.

This portion of the analysis provides an excellent initial survey of the case under study. For the deeper, more sophisticated analysis and the mechanism of growth forecasting, reference to the syllabus *An Orthodontic Philosophy* by Carl Gugino is recommended.

For an evaluation of the patient R.B., the measurements (see Table 14–2) show that the skeletal configuration (facial axis, facial depth, and mandibular plane) is essentially normal, taking into account the age difference between R.B. and the 9-year-old norms. However, the last five measurements are all indicative of the midface convexity and dental protrusion evident in this patient.

This basic summary analysis is somewhat similar to Downs' analysis, except that convexity is measured by direct linear relation of point A to the facial plane. Although Ricketts uses Frankfort as the horizontal reference plane, it should be pointed out that anatomic porion is used instead of "machine" porion, as described earlier. Also, the facial axis, which replaced the Y-axis, is quite different, as described.

Bjork's Analysis

Bjork has been an outstanding researcher in the field of cephalometrics, and his work "The Face in Profile" is certainly recommended reading for those interested in cephalometric studies. His research was based on a study of 322 Swedish boys 12 years of age and 281 conscripts 21 to 23 years of age and

Table 14–2. AGE AND MEAN CHANGES FROM AGE 9 YEARS USED IN RICKETTS' ANALYSIS (see Fig. 14–32)

	Mean	For 9 Years Old + Change	R.B.
1. Facial axis	90° ± 3°	No change	90°
2. Facial depth	86° ± 3°	− 1°/3 yrs	90°
3. Mandibular plane	26° ± 6°	− 1°/3 yrs	21°
4. Convexity of point A	2 ± 2 mm	− 1 mm/3 yrs	8 mm
5. Te to A-Po	+ 1 ± 2 mm	No change	10 mm
6. Upper molar to PTV	Age + 3 ± 2 mm	1 mm/yr	23 mm
7. T to A-Po	22° ± 4°	No change	32°
8. Lower lip to E-plane	− 2 ± 2 mm	Decrease	4 mm

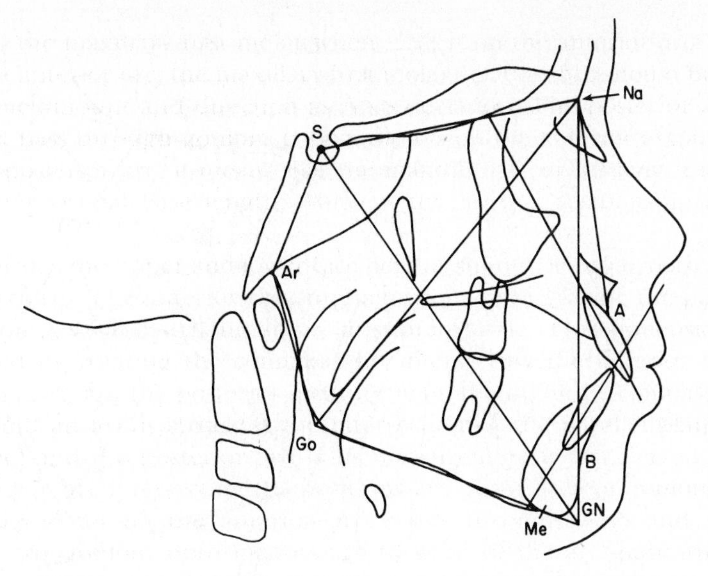

FIGURE 14–33. Tracing for Bjork's analysis (as adapted by Jarabak).

included almost 90 different measurements. Because it is not within the scope of this brief overview to attempt to abstract his massive work, only the principal portions of Bjork's analysis, as adapted and modified by Jarabak, will be presented.

The profile analysis is similar to Steiner's in that S-Na is the reference line, and SNA-SNB, along with Go-Gn, provides basic skeletal evaluation. The incisor axis and incisor to A-Po relate the dentition to the skeletal base.

Lines to be drawn are S-Na, S-Ar, Ar-Go, Go-Gn, Na-Pog, S-Gn, Na-Go, Na-A, Na-B, A-Po, the occlusal plane, and the long axis of the incisors (Fig. 14–33).

A feature of the analysis is the use of the polygon N-S-Ar-Go-Gn to assess anterior and posterior face height relationships and to predict direction of growth change in the lower face. The basis of this approach is the relationship of three angles—saddle angle (Na-S-Ar), articular angle (S-Ar-Go), and gonial angle (Ar-Go-Me)—and the lengths of the sides of the polygon.

In the example illustrated (see Fig. 14–33), a lateral film of R.B. taken in 1971 was traced and measured. Another film, taken in 1974 (Fig. 14–34), was also measured; the values, together with the means, for each are shown in Table 14–3.

Briefly, at age 11 the anterior cranial base (S-Na) should equal mandibular body length (Go-Me). An ideal ratio of posterior cranial base length (S-Ar) to ramus height (Ar-Go) is 3:4. If the sum of the three angles described earlier is greater than 396 degrees, there would be a tendency toward "clockwise" growth change in the mandible. The reverse (counterclockwise) would be the case with angles with a sum less than 396 degrees. A ratio of posterior face height (S-Go) to the anterior face height (Na-Me) of 56 to 62 per cent would indicate a clockwise pattern of growth change in the mandible, whereas a ratio of 65 to 80 per cent would indicate counterclockwise change. Clockwise change means that anterior face height is increasing more rapidly than posterior face height and would be associated with downward and backward growth change at the symphysis and anterior open bite tendency. Counterclockwise change indicates

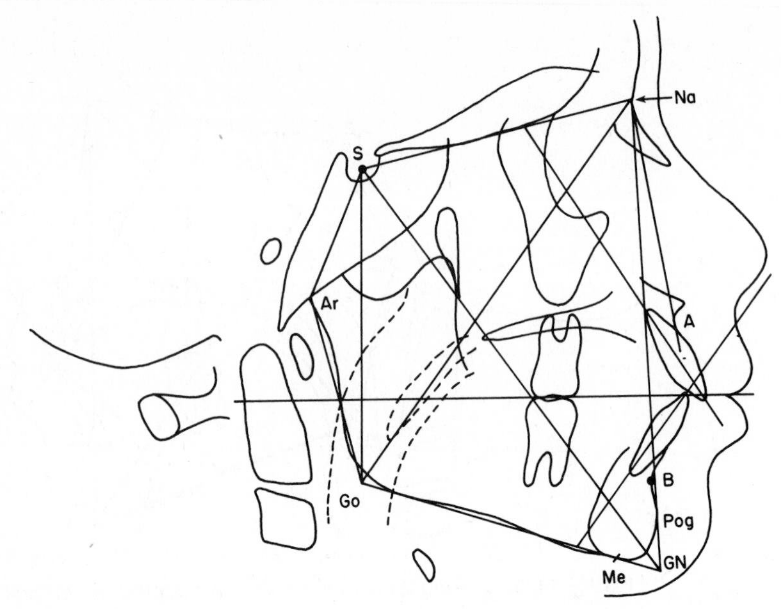

FIGURE 14–34. Bjork's analysis tracing on the patient shown in Figure 14–33 after 3 years' growth change.

Table 14–3. PATIENT (R.B.) VALUES AT TWO AGES COMPARED WITH BJORK MEANS (see Figs. 14–33 and 14–34)

Measurement	Average	R.B.	
		8/3/71	*6/6/74*
Saddle angle	123° ± 5° (Bjork)	123°	126°
Articular angle	143° ± 6° (Bjork)	141°	142°
Gonial angle	130° ± 7° (Bjork)	127°	122°
Sum	396° (Bjork)	391°	390°
Anterior cranial base length	71 ± 3 mm (Bjork)	71.5 mm	74 mm
Posterior cranial base length	32 ± 3 mm (Bjork)	35 mm	36 mm
Gonial angle			
Upper	52–55°	56°	53°
Lower	70–75°	71°	69°
Ramus height	44 ± 5 mm (Bjork)	43 mm	49 mm
Body length	71 ± 5 mm (Bjork)	68 mm	74 mm
Mandibular body–anterior cranial base ratio	1:1	1:1	1:1
SNA	80°	86°	87°
SNB	78°	80°	80°
ANB	2°	6°	7°
SN-MP		31°	31°
Y-axis		57°	57°
Anterior face height		114 mm	122 mm
Posterior face height		74 mm	81 mm
Posterior face–anterior face ratio		65%	66%
56–62% clockwise			
65–80% counterclockwise			
Facial angle (SN-Po)		86°	87°
Denture			
Occlusal P1-M-P1		13°	15°
1 to M-P1	90° ± 3°	100.5°	110°
1 to SN	102° ± 2°	106°	108°
1 to facial plane	5 ± 2 mm	14 mm	15 mm
1 to facial plane	−2 ± 2 mm	9 mm	10 mm
1 to *1*		121°	112°

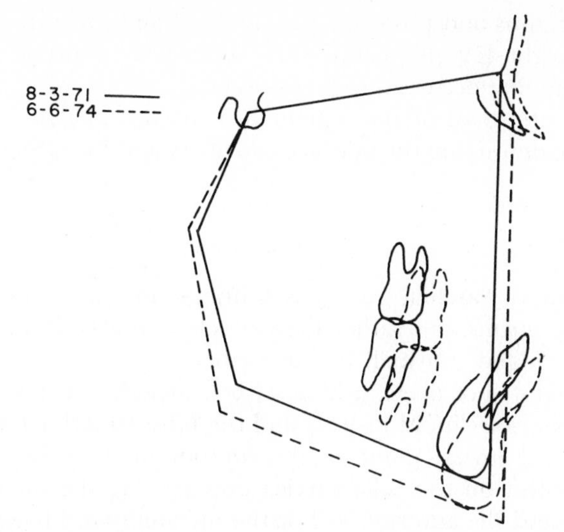

FIGURE 14–35. Bjork's polygons from tracings in Figures 14–33 and 14–34 superimposed to show growth change from age 10 years, 9 months to age 13 years, 7 months. The SN lines are superimposed, registered at sella.

more rapidly increasing posterior face height, forward growth of the chin, and anterior deep bite tendency.

In the example illustrated, the sum of the posterior angles (390 degrees) and the 65 per cent ratio of posterior to anterior face height suggest a closing, or counterclockwise, pattern. However, an examination of individual components shows the ramus height to be short in its ratio to posterior cranial base length, which would counteract this tendency. The prediction would be a normal downward and forward growth pattern with the mandibular plane dropping in a parallel manner, which is what occurred (Fig. 14–35). The mandibular corpus length, which was 3.5 mm shorter than the anterior cranial base length, grew to the normal 1:1 ratio.

Figure 14–36 shows a female patient, aged 10, who was observed for 1 year of growth change. In this example, the high posterior angle sum (401

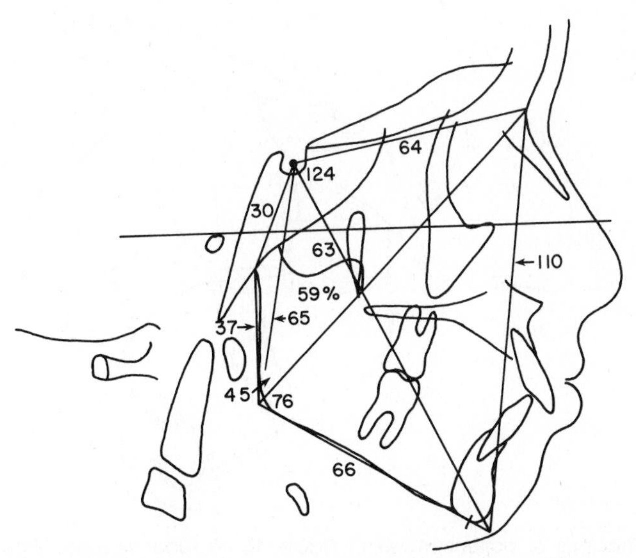

FIGURE 14–36. Tracing of Bjork's polygon on a female patient at age 10 years, 7 months. Data indicate a clockwise pattern of growth change in the lower face.

degrees), short ramus and posterior cranial base, and low ratio of posterior to anterior face height (59 per cent) were combined to predict the clockwise pattern of change that occurred (Fig. 14–37).

The overall appraisal of the patient R.B. by this analysis agrees with the other analyses in identifying the midface convexity and the bi-dental protrusion.

Sassouni's Analysis

The Sassouni, or "archial," analysis is unique in that it does not employ a set of established norms, but rather defines relationships within the individual pattern that are judged "normal" or "abnormal."

Lines and points not previously described include the supraorbital plane, tangent to the anterior clinoid process and the most superior point on the roof of the orbit; Si, the lowest point on the contour of the sella turcica; Sp, the most posterior point on the sella turcica outline; Te, the intersection of the cribriform plate and the anterior wall of the infratemporal fossa; and point O, the center of the convergence where the four horizontal planes tend to intersect.

A plane is drawn parallel to the supraorbital plane, tangent to Si, and the occlusal, palatal, and mandibular planes are drawn. These four planes should converge toward a point. In the event that only three of the planes converge toward a point, the fourth plane is divergent from the facial pattern. If only two of the planes converge, as in Figure 14–38, the junction of cranial base plane and mandibular plane is used as point O. From point O, four arcs are drawn: from nasion, from point B, from Te, and from Sp. These arcs are called the anterior, basal, midface, and posterior arcs, respectively.

General evaluation of the planes shows that the more they are parallel, the greater is the tendency toward skeletal deep bite; and the more they are steep to each other, the greater is the tendency toward skeletal open bite.

The anterior arc from nasion should pass through ANS, the tip of the maxillary incisor, and the pogonion. The basal arch from point A should pass through point B. The midfacial arc from Te should pass tangent to the mesial

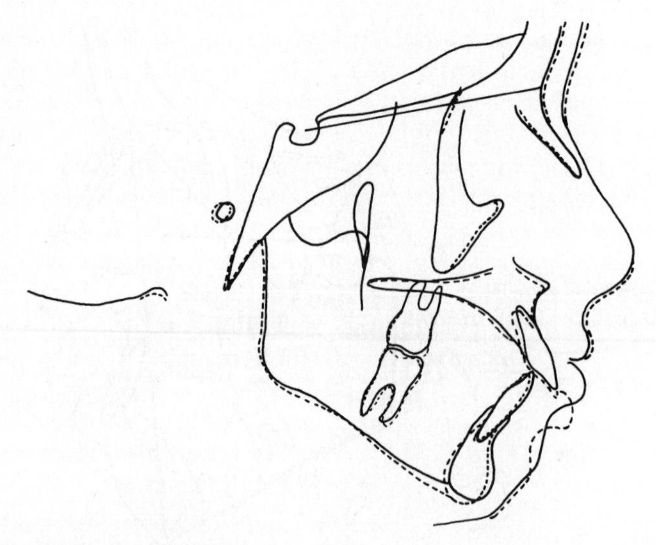

FIGURE 14–37. Tracings of patient shown in Figure 14–36 superimposed, showing 1 year of clockwise growth change in the mandible, with the symphysis dropping down and back.

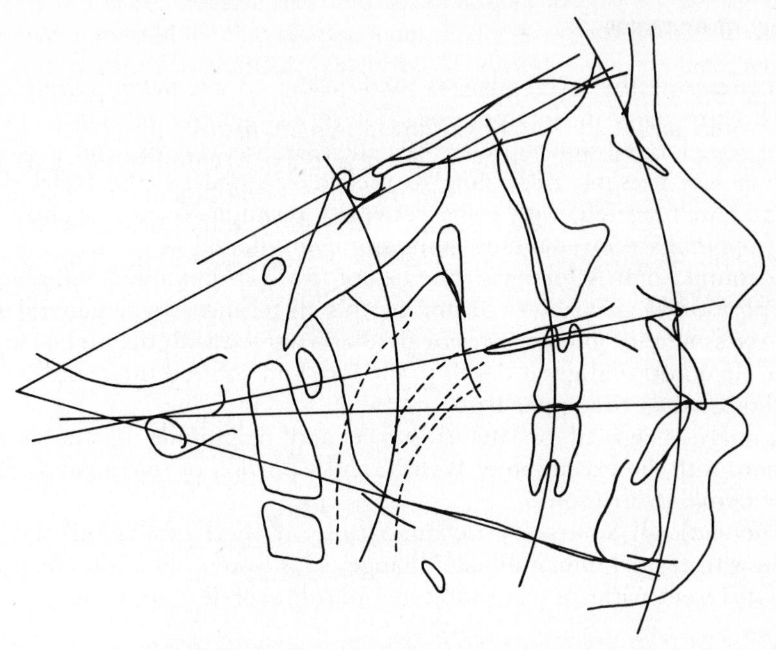

FIGURE 14–38. Tracing for archial analysis (Sassouni's).

surface of the maxillary first molar when ANS is on the anterior arc. If ANS is not on the anterior arc, the maxillary first molar relationship should be adjusted in the same amount and direction as ANS deviation. The posterior arch from Sp should pass through gonion. If pogonion is on the anterior arc and gonion is on the posterior arc, it means that the mandibular corpus length is equal to the anterior cranial base length, which is the normal relationship at age 12 years.

Vertically, the upper and lower face heights should be equal, both anteriorly and posteriorly. The anterior measurement is made by placing the point of the compass on ANS and striking an arc at supraorbitale. That dimension is then transferred by rotating the compass and intersecting the anterior arc at the symphysis area. For the posterior measurement, the tip of the compass is placed at PNS, and an arc is struck at the intersection of the parallel plane (cranial base plane) and the posterior arc. This measurement is transferred inferiorly by striking an arc intersecting the posterior arc in the area of gonion.

In Figure 14–38 the anterior arc passes through ANS and is slightly forward of pogonion. Both incisors are forward of the arc, indicating incisor protrusion. Point B is behind the basal arc, indicating a basal discrepancy.

The mesial contour of the maxillary first molar is forward from the midfacial arc, indicating the forward position of the maxillary teeth.

Gonion is behind the posterior arc slightly more than pogonion is behind the anterior arc, but corpus length in relation to anterior cranial base length is essentially normal for the age of this patient.

Vertically, the palatal plane is deviant, appearing elevated in front and depressed posteriorly. This results in anterior lower face height being slightly excessive and posterior lower face height slightly deficient.

In summary, this analysis indicates that the subject's general facial pattern is well coordinated, except for the tilted palatal plane, the protrusion of the dentition, and the mildly retrognathic mandible.

Summary of Analyses

In comparing the seven analyses used on the sample patient, there is little serious disagreement in the appraisals. There was general agreement that the chin was well located anteroposteriorly, the face was convex, and a bi-dental protrusion was present. According to the archial analysis, the facial skeletal profile was well related, with some convexity resulting from a slightly retrognathic mandible, the remaining convexity being dental in nature. The other analyses found point A forward. The extent to which point A is influenced by incisor position is probably a factor in this difference. In a general sense, Wylie's assessment of anteroposterior dysplasia agreed with the archial analysis in that it demonstrated great depth to the face and showed the mandible to be 5 mm shorter than the upper base.

All analyses agreed on the existence and magnitude of the bi-dental protrusion, with the exception of Wylie's, and a portion of that analysis did not measure dental protrusion.

In general, all analyses were quite close in their evaluation of growth direction with regard to mandibular change. This would not always be the case if many different patterns were subjected to this overview approach.

BOLTON STANDARDS

Many different types of overlay cephalometric appraisals have been conceived through the years, from grids to polygonal patterns, to various transparencies, and so forth. Each has made a contribution to our ever-increasing knowledge of "normal" dentofacial growth and development.

Probably the most comprehensive of these to date are the Bolton Standards of Dentofacial Development Growth. These standards are optimum serial facial patterns that include annual lateral transparencies from 1 to 18 years with coordinated frontal standards from 3 to 18 years for both males and females (Fig. 14–39).

Published by Broadbent, Broadbent, and Golden in 1975, they provide a comprehensive method of analysis, because the soft tissue profile as well as the skeletal elements of the cranium, face, and dentition are provided for morphologic comparison as well as cephalometric measurement.

Because there is no dichotomy of size differentiation between the male and female groups, an average of the two sexes is used as the "standard" for each age. There is, of course, the need to consider sexual dimorphism in the peripheral skeletal morphology (Figs. 14–40 and Table 14–4), but the basic difference in size takes place after puberty, with the males growing for a longer period and to a larger size, as a group, than the females (Fig. 14–41).

It must be recognized that in the averaging tracing technique employed for arriving at single annual "norms," the individual circumpubertal growth spurts of the subjects, which occur at differing ages, are blended out. The Bolton Standards may be superimposed over the cephalogram directly or, for greater ease of interpretation, over the traditional tracing of the radiograph.

There is no set method of superimpositioning, and the diagnostician is free to use whatever landmarks suit the case at hand. Usually the four basic areas of morphology—cranial base, skeletal maxillary base, mandibular base, and soft tissue—are compared with appropriate chronologic standards; and then specific relationships of these bases and the dentition are assessed (Fig. 14–42).

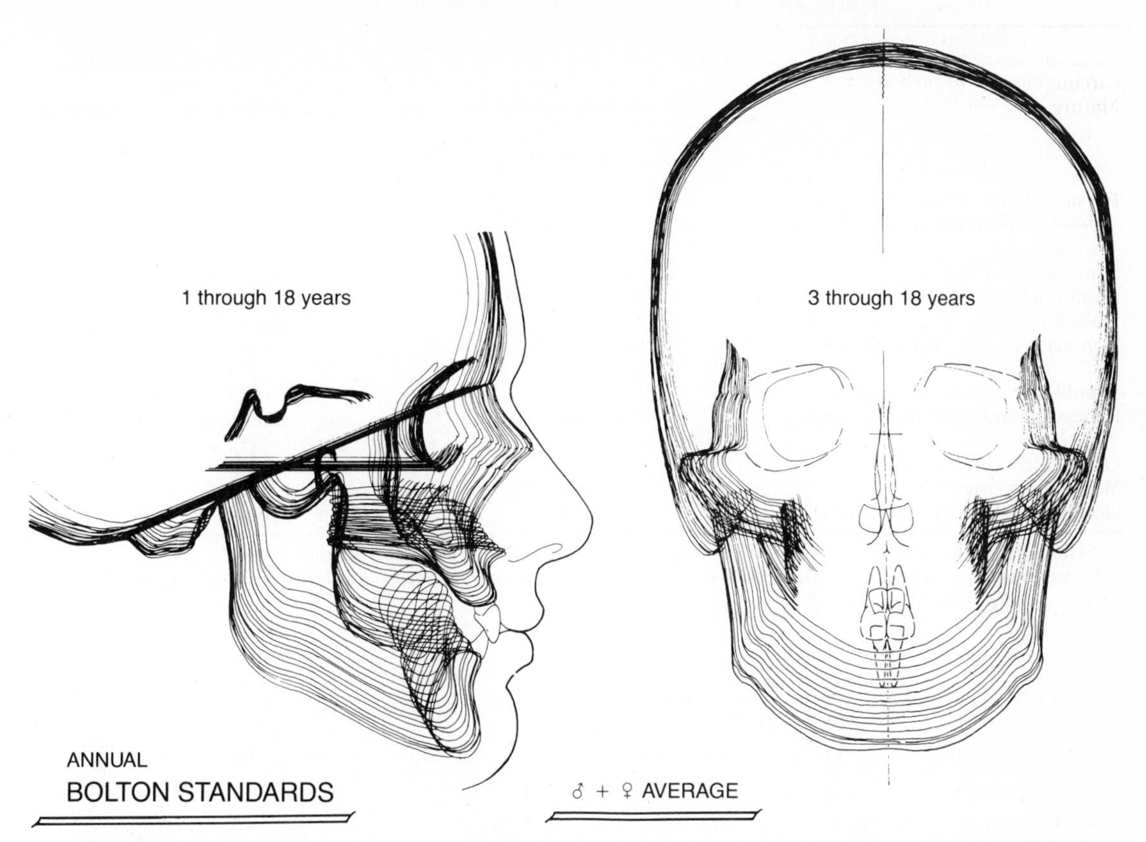

FIGURE 14–39. Annual Bolton Standards. Both lateral and frontal views superimposed in Bolton relation. (From Broadbent, B. H., Sr., B. H. Broadbent, Jr., and W. Golden: *Bolton Standards of Dentofacial Developmental Growth.* St. Louis, C. V. Mosby, 1975.)

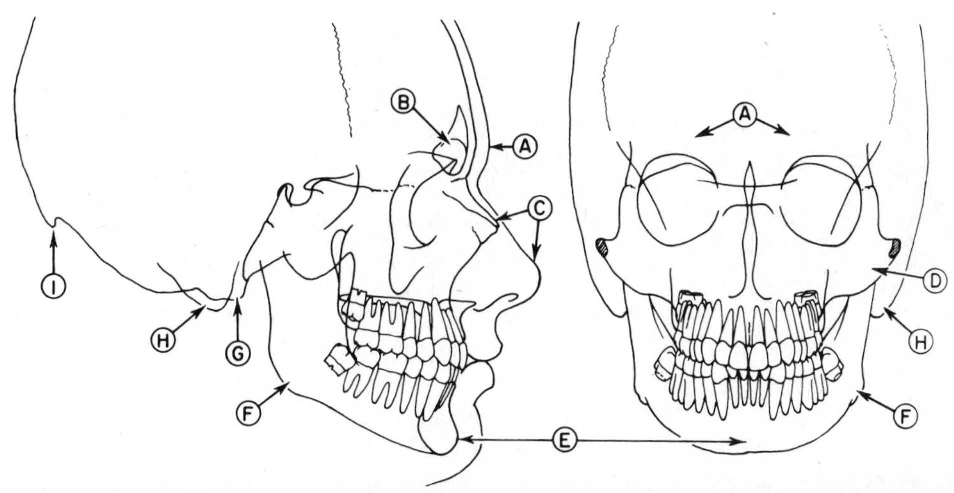

FIGURE 14–40. Sexual dimorphism in craniofacial development. The diagram indicates the basic areas of differentiation that occur between males and females in the adolescent years. (From Broadbent, B. H., Sr., B. H. Broadbent, Jr., and W. Golden: *Bolton Standards of Dentofacial Developmental Growth.* St. Louis, C. V. Mosby, 1975.)

Table 14–4. SEXUAL DIMORPHISM IN CRANIOFACIAL PATTERNS: THREE AREAS OF COMPARISON OF MALES AND FEMALES THAT INDICATE THEIR DISSIMILARITIES*

	Females	Males
Circumpubertal growth spurt	10 to 12 years	12 to 14 years
Mature size	Growth plateaus at approximately 14 years Moderate increase to 16 years	Active to approximately 18 years
Physical characteristics (differences develop in middle to late adolescence)		
Supraorbital ridges *(A)*	Virtually absent	Well developed
Frontal sinuses *(B)*	Small	Large
Nose *(C)*	Moderate	More massive
Zygomatic prominences (cheekbones) *(D)*	Small	Large
Mandibular symphysis (pogonion) *(E)*	Rounded	Prominent
External mandibular angle (gonion) *(F)*	Rounded	Prominent lipping
Occipital condyles *(G)*	Small	Large
Mastoid processes *(H)*	Small and delicate	Large
Occipital protuberance (inion) *(I)*	Insignificant	Prominent

*These dissimilarities are not significantly related to skeletal balance or malocclusion.

From Broadbent, B. H., Sr., B. H. Broadbent, Jr., and W. Golden: *Bolton Standards of Dentofacial Developmental Growth.* St. Louis, C. V. Mosby, 1975.

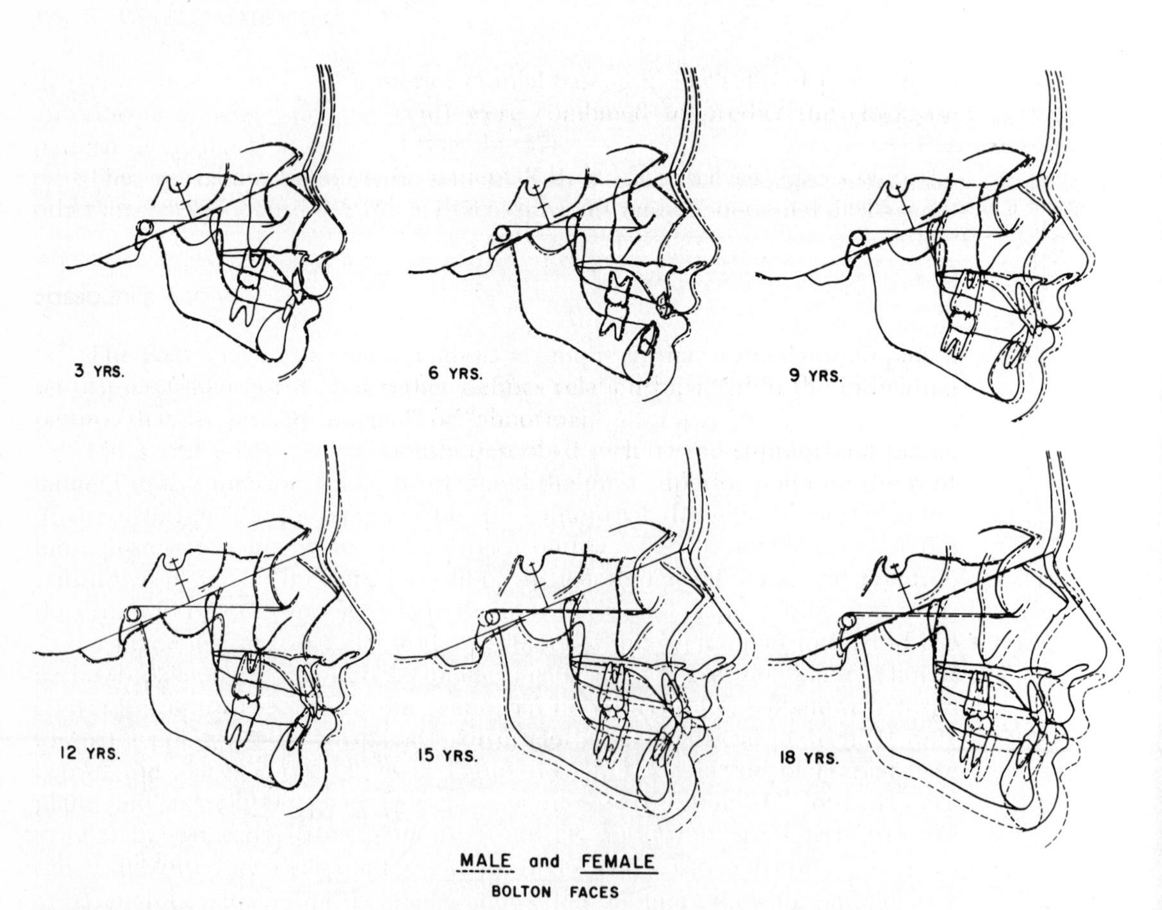

FIGURE 14–41. Average male and female Bolton faces superimposed in Bolton relation. The similar morphologic pattern is demonstrated with the first significant change in size occurring between 12 and 15 years at approximately 14 years. (From Broadbent, B. H., Sr., B. H. Broadbent, Jr., and W. Golden: *Bolton Standards of Dentofacial Developmental Growth.* St. Louis, C. V. Mosby, 1975.)

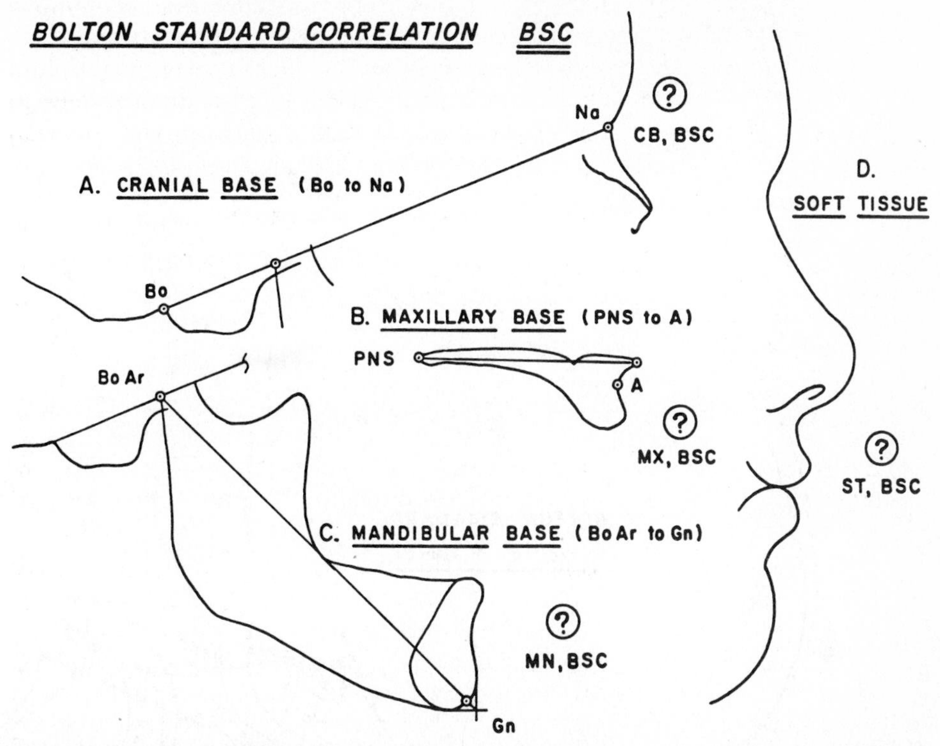

FIGURE 14–42. Bolton Standard Correlation (BSC). Diagrammatic representation of the landmarks employed when the composite standards are used for assigning Bolton Standard age levels to the three skeletal components of the craniofacial complex. (From Broadbent, B. H., Sr., B. H. Broadbent, Jr., and W. Golden: *Bolton Standards of Dentofacial Developmental Growth.* St. Louis, C. V. Mosby, 1975.)

An example of the use of the standards in analyzing the lateral cephalogram of a 10-year-old Angle Class II, Division I, female is shown in Figure 14–43. The superpositioning is arbitrarily done in this instance on the Bolton plane at nasion.

The frontal example is that of a case with a significant mandibular asymmetry. The frontal standard is superimposed on a "best fit" basis on the orbits and upper face to most clearly show the lower face deviations (Fig. 14–44).

It goes almost without saying that standards of the Broadbent-Bolton Standards can only be used, with accuracy, in analyzing cephalograms taken by a similar (i.e., minimal enlargement) technique.

SUPERIMPOSING FOR GROWTH CHANGE EVALUATION

One of the most useful procedures to the clinician is acquiring headfilms of individual patients at periodic intervals and comparing them for obtaining a general view of growth changes. As mentioned at the beginning of this chapter, this technique will not locate the sites of growth, but will provide a quantitative directional appraisal of changes that have occurred.

There are numerous planes and landmarks that may be superimposed for viewing growth change. Probably the most common is to superimpose two serial tracings with point registration at sella and the S-Na lines superimposed one over the other. This provides a composite view of growth change during the

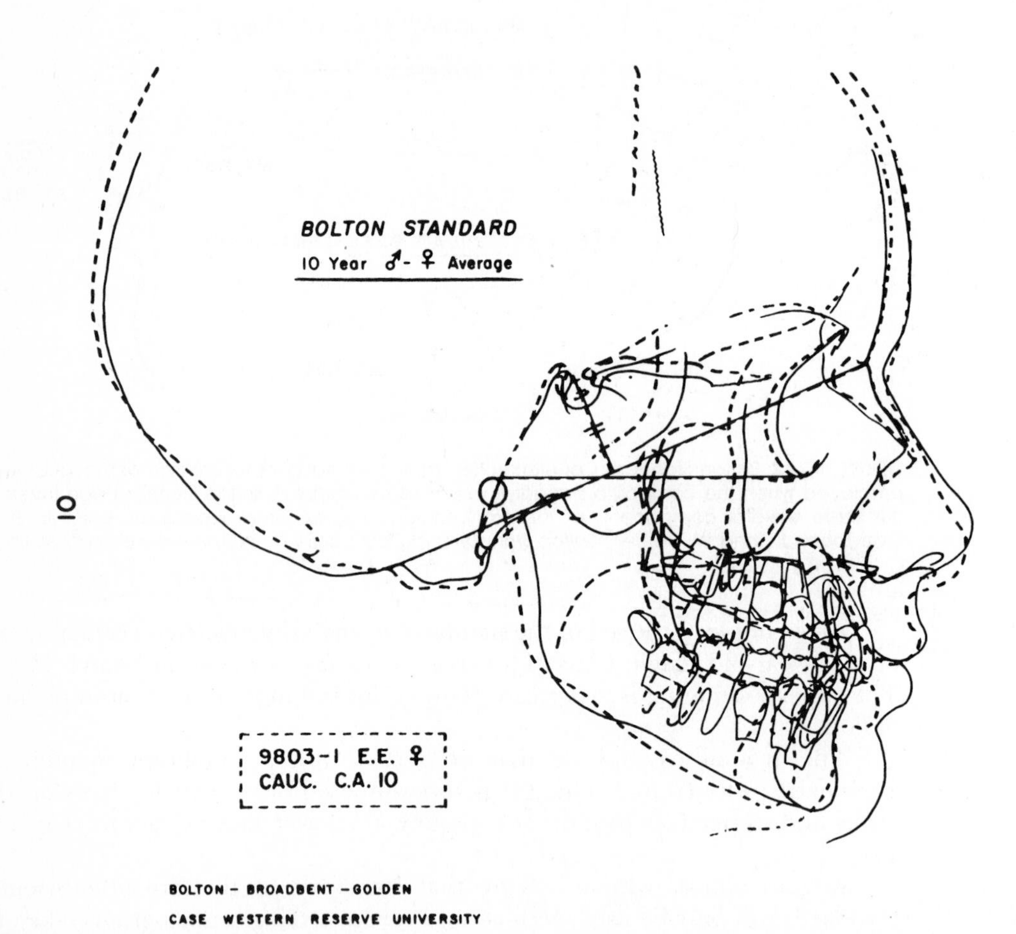

FIGURE 14–43. Ten-year-old Bolton Standard (superimposed on the Bolton plane at nasion) to indicate the structural variables in a typical Class II, Division 1 case. (From the Bolton Study, Bolton-Brush Growth Study Center, Case Western Reserve University, Cleveland, Ohio.)

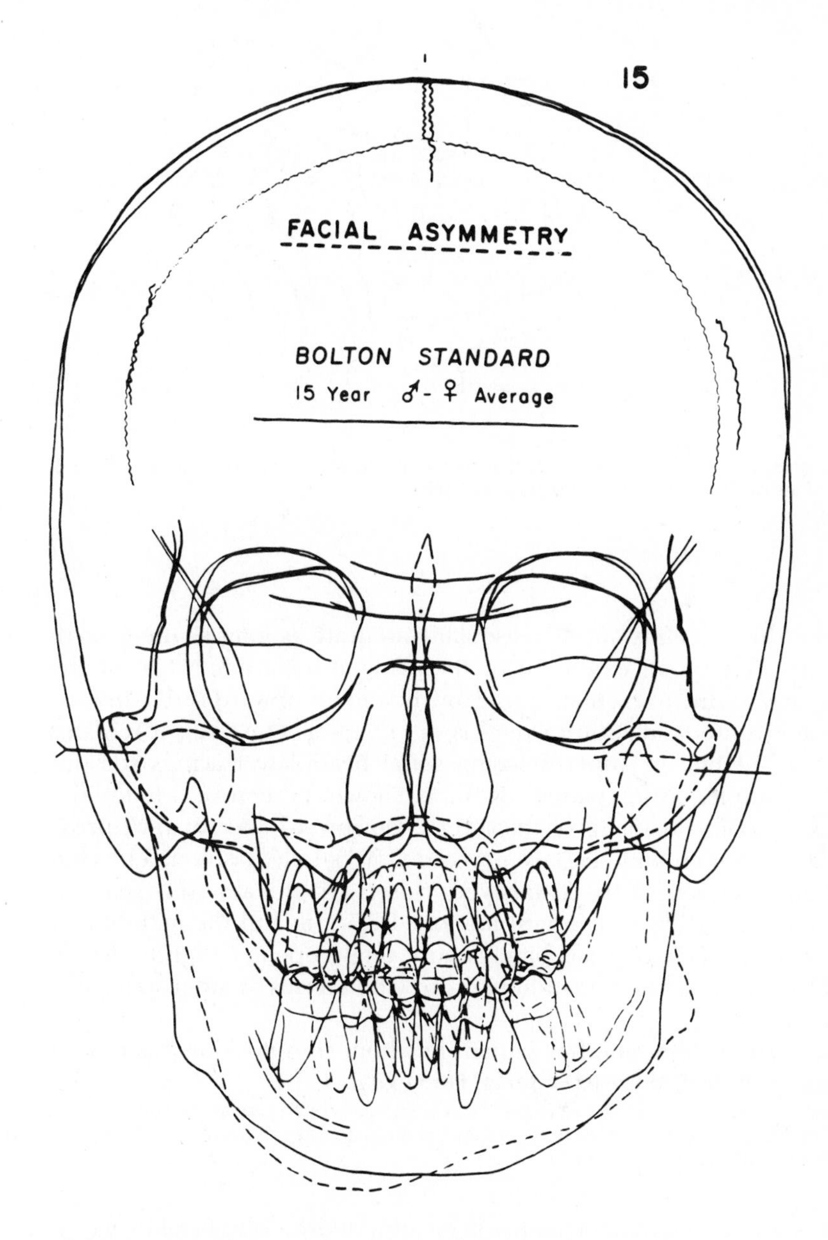

FIGURE 14–44. Fifteen-year-old frontal Bolton Standard (superimposed on the midsagittal plane and orbits) to indicate facial asymmetry in a typical case. (From the Bolton Study, Bolton-Brush Growth Study Center, Case Western Reserve University, Cleveland, Ohio.)

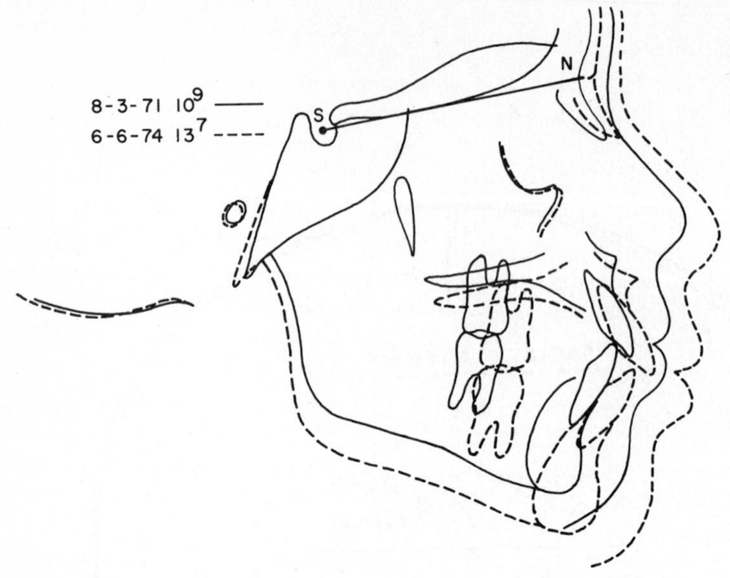

FIGURE 14–45. Tracings of sample patient, R.B., superimposed to show growth change over a 3-year period. SN lines are superimposed, registered at sella.

period between the two films and is reasonably accurate as long as the growth change at nasion follows an extension of the original S-Na line. Most of the change in this area is due to frontal sinus growth, and an upward or downward migration of the frontonasal suture would result in error. However, it remains a most common method of superimposing serial headplate tracings for this purpose. An example, using patient R.B., is shown in Figure 14–45 and illustrates an essentially normal growth pattern. Downward and forward progress of both maxilla and mandible have proceeded in parallel fashion. The chin has followed an extension of the Y-axis very closely. In the almost 3 years of growth, the anterior cranial base length increased 2.5 mm and the mandibular body length, 6.0 mm. The cranial base flexure angle opened slightly, which counteracted to some extent the mandibular growth in terms of anteroposterior chin position.

Figure 14–37 illustrates another patient in whom chin position has moved downward and backward, increasing facial convexity.

SUMMARY

This discussion of cephalometrics has been admittedly a very brief overview of the technique and principles used in this form of craniofacial evaluation. The technique of acquiring headfilms and tracing and measuring them is straightforward and has demonstrated good accuracy. Over the past 60-odd years, various measurements and systems of analysis have been developed and currently are widely used, especially in the orthodontic field. Cephalometric research and its clinical application have centered on facial pattern evaluation, growth prediction, and treatment evaluation, all of which are external morphologic appraisals. The widespread clinical interest in cephalometrics has undoubtedly resulted in the development of measurements and analyses that help the clinician evaluate areas of primary interest, with emphasis on locating and quantifying external positional deviations, rather than pinpointing sites of growth change or treatment response. The immediate concern of the clinician was, "What is the problem? Where is it? How severe is it? What can I do to correct it?" Cephalometrics has helped answer these questions, but it has not added significantly to an understanding of the true nature of the deviation, how it got that way, and by what mechanism it will respond to treatment.

In defense of cephalometrics, it must be stated that many of the landmarks and lines were carry-overs from preradiographic craniometric techniques, and although they bore little relationship to sites of growth and/or treatment change, they were easy to locate and lent themselves well to the kind of geometric evaluation in current use.

Deviation was recognized by comparison of patient measurements with statistically derived means obtained by study of selected population samples. Perhaps it is an overstatement to say that this procedure appears to have the potential error of not recognizing biologic variability, and there is an implication of trying to make everybody look alike.

Gross evaluation of growth and treatment change, which was illustrated earlier, is of limited value in this context because it records only what has happened without divulging the true nature and location of the change. This is not to say that cephalometrics is not a useful tool for the clinician and researcher alike, because it is quite likely that its present mode of use will continue to be an important part of craniofacial research and clinical evaluation.

More recently, growth and development research has shifted emphasis to a more complete understanding of individual patterns of growth change and of what is happening in this complex anatomic area to cause these patterns to develop. The principal author of this text, as a counterpoint to the system and philosophy of analysis just described, visualizes use of the cephalometric principle and technique for a quite different purpose. Although the cephalometric study to be described by Dr. Enlow in subsequent pages may not be designed for routine clinical use, an understanding of it may help to close the "gap" between the questions, "What is the problem?" and, "How did it get that way?"

COUNTERPART ANALYSIS (OF ENLOW)

This is a method in which the various facial and cranial parts are compared with each other to see, simply, how they fit. The individual is measured against himself, rather than compared with population standards and norms. Most conventional methods of analysis and cephalometric growth studies are intended essentially to determine **what** a particular growth or form pattern is. This procedure was developed to explain **how** such a pattern was produced in any given person. The ANB angle, for example, tells one the nature of the positional relationship between the anterior part of the upper and lower arches and provides an index with which one can gauge the extent of malocclusions. The counterpart procedure is intended to account for the composite of the anatomic and morphogenetic factors that **produced** the particular ANB angle (and other measurements) found in a given person.

Most conventional cephalometric planes and angles are not intended to coincide with or indicate actual sites and fields of growth and remodeling, and they are thus not appropriate for the essentially anatomic purposes just described. Because most standard planes and angles do not represent the patterns and distribution of growth fields, comparisons of the individual with population standards are required; there is usually no other basis for interpretation, owing to the nature of the planes themselves. However, if planes are constructed so that the activities of the growth and remodeling fields are in fact directly represented, a built-in and morphologically natural set of "standards" is identifiable that allows meaningful evaluation of overall craniofacial form and pattern without population comparisons.

The analysis is based on the **counterpart principle**. This is the actual design basis upon which the face is constructed and which underlies the plan of its intrinsic growth process. The counterpart concept was described in Chapters 3 and 5, and it was used as the working basis for explaining how the face grows. The counterpart analysis is, in effect, the same. It shows **where** imbalances exist, **how much** is involved, and what the **effects** are. Refer to pages 59 to 61 and 171 to review just what a "counterpart" is.

In Figure 14–46, construction lines have been drawn on a headfilm tracing to represent several key fields and sites of growth. These include the maxillary tuberosity, the mandibular condyle (using articulare for convenience, rather than condylion), the ramus-corpus junction, the posterior border of the ramus, the anterior surfaces of both the maxillary and mandibular bony arches, the occlusal plane, and the junction between the middle and anterior cranial fossae (the anterior-most extent of the great wings of the sphenoid where they cross the cranial floor). Other planes may be added to represent other major growth areas, if desired, such as the zygomatic arch, the palate, the olfactory plane, and the anterior-vertical plane of the midface.

Note that the **PM** vertical plane is represented. This is the important boundary that separates the anterior cranial fossa and nasomaxillary complex from the middle cranial fossa and pharynx. The ramus relates to the latter and the corpus to the former (see page 171).

Two basic factors are important in evaluating the role of any bone or part of a bone in a composite assembly of several different bones. The first is the bone's **size** (horizontal and vertical), and the second is its **alignment** (rotational position). In this analysis, both must be considered. The reason is that the nature of the alignment of any bone affects the **expression** of its various dimensions, as explained in Chapter 6. The determination of a bone's dimension alone is not enough (and can be misleading); its alignment must also be known for one to see just how this factor affects its actual dimensions. In the counterpart analysis, both are determined for all the various bony parts and counterparts.

The rationale, in brief, is that the vertical and/or horizontal size of one given part is compared with that of its specific counterpart(s). If they exactly match, or nearly so, a dimensional "balance" exists between them. If one or the other is long or short, however, the resulting imbalance can cause either **protrusion** or **retrusion** of the part of the face involved and thereby affect the profile, either directly or indirectly (Fig. 14–46). The various parts and counterparts are then checked for their alignment to see if each, independently, has a protrusive or a retrusive effect, regardless of the nature of the dimensions. Then all the regional part-counterpart relationships are added up to see how the sum of all underlies the face of any given individual. This may be done on a single headfilm tracing at any age, or serial headfilms can be used for determining the progressive effects of age changes or of treatment results.

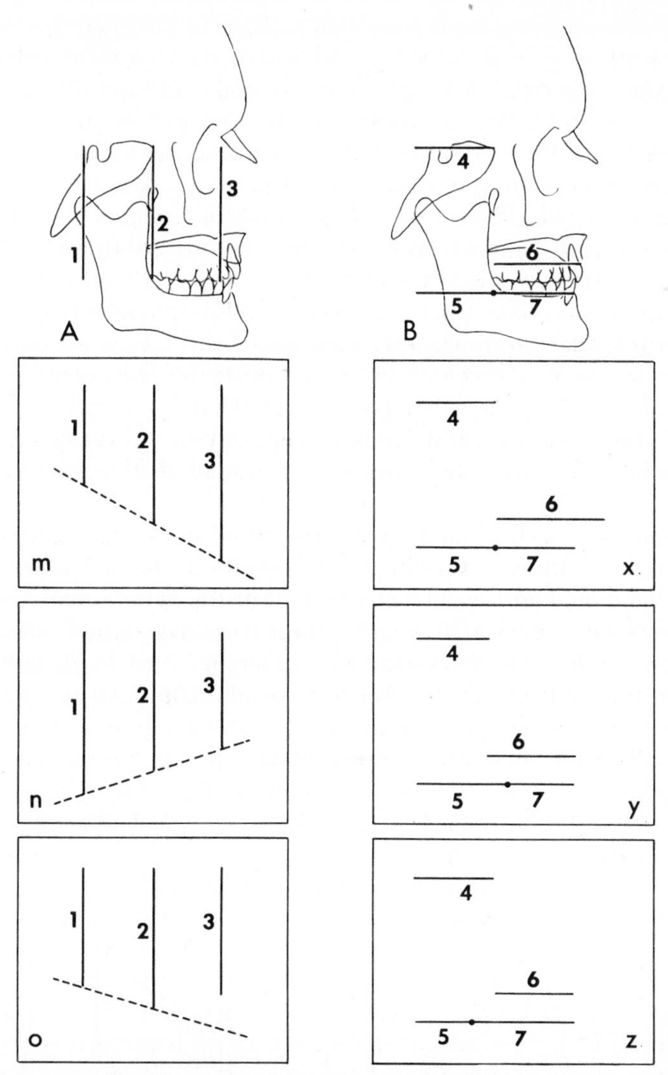

FIGURE 14–46. *A,* Three vertical architectural counterparts: the cranial floor–ramus vertical *(1);* the posterior nasomaxilla *(2);* and the anterior nasomaxilla *(3).* They are all in exact dimensional balance, and the functional occlusal plane therefore coincides with the neutral occlusal axis, which is perpendicular to these three vertical planes. If vertical dimensional imbalance occurs, however, downward occlusal rotation *(m),* upward occlusal rotation *(n),* or open bite *(o)* necessarily results. *B,* Four horizontal architectural counterparts: the middle cranial fossa *(4),* the ramus *(5),* the maxilla *(6),* and the mandibular corpus *(7).* The anterior cranial fossa is also involved but is not included in this diagram. If all these horizontal counterparts are balanced, as in *B,* their effective dimensions are very close to an exact match and the bones "fit" relative to each other. Diagram *x* shows a maxillary arch that is excessive relative to a disproportionately smaller corpus. The other segments *(4* and *5)* are balanced. The result is maxillary protrusion. Diagram *y* demonstrates a similar maxillary-mandibular imbalance, but the ramus, although actually out of balance to the cranial base, serves to provide dimensional compensation, so that aggregate balance is achieved in the overall composite of parts. Diagram *z* illustrates a "long" middle cranial fossa that is not dimensionally in balance with a "short" ramus. However, aggregate balance results because an imbalance also exists between the corpus and the maxilla, so that the sum of all their dimensions has been balanced. It is apparent that many other similar combinations are possible among the composites of these architectural counterparts which will produce either a balanced or imbalanced face. **Note**: The important factor of **alignment** is not included in these figures. (From Enlow, D. H., R. E. Moyers, W. S. Hunter, and J. A. McNamara, Jr.: A procedure for the analysis of intrinsic facial form and growth. Am. J. Orthod., 56:6, 1969.)

Figure 14–47 shows a Class II individual in whom major variations and imbalances are present for the different horizontal and vertical dimensions and for the alignment relationships (compare with the Class III individual described in the next paragraph). Note that (1) the mandibular corpus is short **relative** to its counterpart, the maxillary bony arch (both skeletally and dentally in this individual); (2) the corpus is aligned (rotated) upward (that is, the "gonial angle" is more closed); (3) the middle cranial fossa is aligned obliquely more forward (the dashed lines represent "neutral" alignment positions); (4) the ramus is aligned more backward; and (5) the nasomaxillary complex is vertically long (resulting in a downward and backward ramus rotation). **All** these features are either mandibular retrusive or maxillary protrusive, and they have combined to produce the multifactorial basis for a Class II malocclusion and retrognathic profile. Note, however, that the horizontal breadth of the ramus **exceeds** its counterpart, the horizontal (not oblique) dimension of the middle cranial fossa. This is a compensatory feature that has partially offset the aggregate effects produced by the other features and thereby reduced the severity of the malocclusion. If desired, the actual amounts of each and all of these effects can be measured.

Figure 14–48 shows an individual in whom the dimensions and alignments combine to produce the composite, multifactorial basis for a Class III malocclusion. Note that (1) the dental **and** skeletal dimensions of the mandibular corpus in this individual exceed maxillary arch length; (2) the corpus is aligned (rotated) markedly downward; (3) the middle cranial fossa is aligned backward; and (4) the ramus is aligned forward. These relationships are all mandibular protrusive or maxillary retrusive, and they combine to produce the prognathic face and Class III malocclusion. The horizontal breadth of the ramus, however, is less

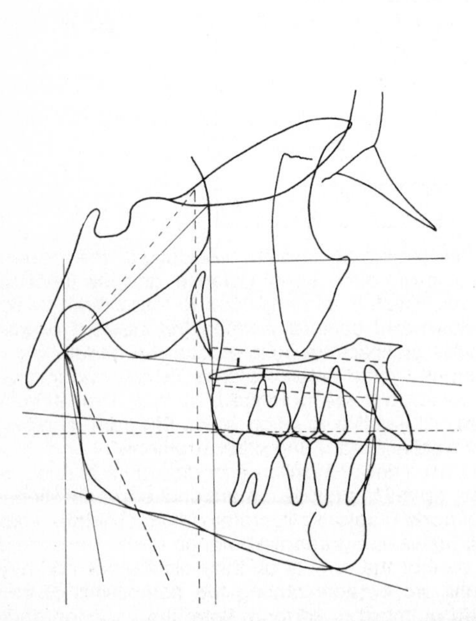

FIGURE 14–47. Headfilm tracing of a Class II patient. Construction lines have been added for the counterpart analysis. (From Enlow, D. H., T. Kuroda, and A. B. Lewis: The morphological and morphogenetic basis for craniofacial form and pattern. Angle Orthod., 41:161, 1971.)

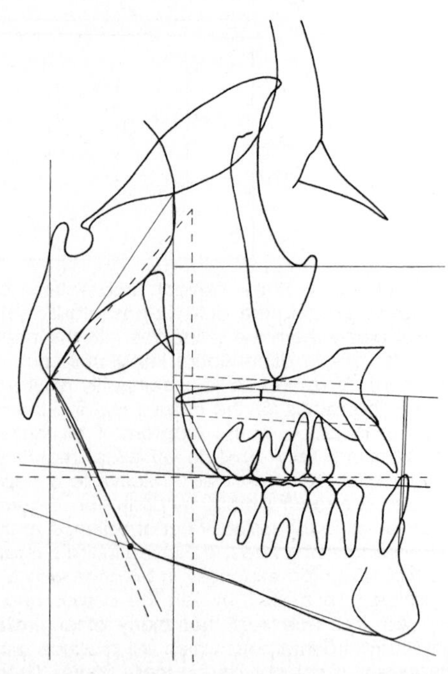

FIGURE 14–48. Headfilm tracing of a Class III patient. Compare with Figure 14–47. (From Enlow, D. H., T. Kuroda, and A. B. Lewis. The morphological and morphogenetic basis for craniofacial form and pattern. Angle Orthod., 41:161, 1971.)

than that of its counterpart, the middle cranial fossa. This is a compensatory feature that, in this particular individual, has partially offset the composite effects of the other relationships and thereby reduced the extent of the malocclusion.

For a more detailed description of the construction lines used, how to determine the actual dimensions, and how to establish the "neutral" alignment planes for any given individual, see Enlow et al. (1971).

The "counterpart analysis" is not intended as a routine clinical tool for everyday office use in diagnosis and treatment planning. It is not needed for this, because the rationale for treatment procedures, at least today, is not usually based on corrections of the actual underlying causes of malocclusions and other kinds of facial and cranial dysplasias. The counterpart analysis is useful, however, in determining what treatment has done in terms of the specific anatomic and developmental changes that have been brought about, more so than most types of analyses, because the others deal more with correlative geometry than with morphologic and morphogenetic relationships. Actually, the immediate payoff for the counterpart analysis has already been largely achieved. It has pointed out more clearly the multifactorial basis for malocclusions and just what some of the specific anatomic and developmental factors are. It has shown how a number of compensatory features participate. It has explained how and why population groups have either Class II or Class III tendencies. Except for such specific types of research studies, however, the counterpart analysis is inappropriate as a routine clinical method. When the intrinsic control processes of facial growth become better understood, when the control processes themselves can be controlled, and when treatment procedures can **then** become based on the real causative factors that underlie structural imbalances, then cephalometric analyses utilizing genuine anatomic and developmental relationships will become increasingly more relevant. The counterpart analysis itself, of course, is far from complete and is only a beginning, but it is a concept. It is also very useful in understanding the rationale for the basic plan of normal facial construction as well as malocclusions (as in Chapters 3 and 6) and in explaining and teaching this complex subject to students in a way that is relatively easy to understand.

GLOSSARY OF TERMS

The terms to be defined are primarily related to landmarks used in roentgenographic cephalometry. The definitions used are those most commonly found in the craniometric and orthodontic literature.

A: See Subspinale.

Antegonion: The highest point of the notch or concavity of the lower border of the ramus where it joins the body of the mandible.

ANS, Anterior nasal spine: A sharp median process formed by the forward prolongation of the two maxillae at the lower margin of the anterior aperture of the nose.

Anteroposterior (AP) or frontal growth axis of the head and face: A transverse zone delineated by a plane through the coronal suture above, passing down through the pterygomaxillary fissure near the posterior termination of the hard palate, along the anterior border of the ascending rami, and through the junction of the horizontal and vertical components of the mandible. It marks the division of the anterior from the posterior component of craniofacial developmental growth when lateral tracings are oriented in Bolton relation.

Ar, Articulare: Bjork—The intersection of the image of the posterior border of the ramus with the base of the occipital bone. Bolton—The point of intersection, in lateral aspect, of the posterior border of the condyle of the mandible with the Bolton plane.

B: See Supramentale.

Ba, Basion: The point where the median sagittal plane of the skull intersects the lowest point on the anterior margin of the foramen magnum.

Bolton plane: A line joining the Bolton point and nasion on the lateral cephalogram.

Bo, Bolton point: A point in space about the center of the foramen magnum that is located on the lateral cephalogram by the highest point in the profile image of the postcondylar notches of the occipital bone.

BSC, Bolton standard correlation:

CB, Bolton cranial base: A line from Bolton articulare to nasion.

MX, Bolton maxillary base: A line from PNS to ANS.

MN, Bolton mandibular base: A line from Bolton articulare to gnathion.

Bregma: The point on the skull corresponding to the junction of the coronal and sagittal sutures.

Cephalogram or Radiograph: A generally accepted term describing a standardized roentgenographic (x-ray) picture of the head.

Cephalometer (roentgenographic cephalometer): In craniometry, an instrument for measuring the head. A cephalometer (device for holding the head) combined with roentgenographic equipment for the production of standardized complementary lateral and frontal radiographs used for measuring developmental growth of the dentition, face, and head.

Cd, Condylion: The most superior point on the head of the condyle.

Convexity, angle of: The angle formed by a line nasion to A and a projection of a line pogonion to A.

Coronal suture: The transverse union of the frontal with the parietal bones.

Craniostat: A device for holding the head for craniometric study.

Dacryon: A point on the inner wall of the orbit at the junction of the frontal and lacrimal bones and maxilla.

Facial angle: The angle formed by the junction of a line connecting nasion and pogonion, FP (facial plane), with the horizontal plane of the head, FH (Frankfort plane).

Facial height:

Total: The distance between nasion and gnathion when projected on a frontal plane.

Lower face: The distance between ANS and gnathion when projected on a frontal plane.

Upper face: The distance between ANS and nasion when projected on a frontal plane.

FP, Facial plane: The line connecting nasion and pogonion on the lateral cephalogram.

FH, Frankfort horizontal plane: A horizontal plane determined by the two poria and left orbitale. It approximates closely the position in which the head is carried during life and is established on the lateral cephalogram by a line joining orbitale with porion as indicated by the top of the ear rod.

FMA angle: The angle formed by the mandibular plane and the Frankfort horizontal plane.

Foramen rotundum: A round opening in the greater wing of the sphenoid bone for the passage of the superior division of the fifth nerve.

FOP, Functional occlusal plane: A horizontal line from the posterior-most occlusal contact of the last fully erupted maxillomandibular molars extending anteriorly to the anterior-most occlusal contact of the fully erupted premolars.

Frontotemporale: A point near the root of the zygomatic process of the frontal bone at the anterior-most point along the curvature of the temporal line.

Gl, Glabella: The most anterior point on the frontal bone.

Gn, Gnathion: The lowest, most anterior midline point on the symphysis of the mandible.

Go, Gonion: The external angle of the mandible, located on the later cephalogram by bisecting the angle formed by tangents to the posterior border of the ramus and the inferior border of the mandible.

Ii, Incisor inferius: The tip of the crown of the most anterior mandibular central incisor.

Is, Incisor superius: The tip of the crown of the most anterior maxillary central incisor.

Id, Infradentale: The most anterior point of the tip of the alveolar process between the mandibular central incisors.

I, Inion: The apex of the external occipital protuberance.

Interincisal angle: The angle formed by the long axis of the lower incisor and the long axis of the upper incisor.

Internal angle of the mandible: Located on the lateral cephalogram by bisecting the angle formed by tangents to the anterior border of the ramus and the superior border (alveolar crests) of the mandible. Note: A line joining the internal angle and antegonion marks the junction of the ramus with the body of the mandible.

Key ridge: The prominent ridge, formed by the malar process, which divides the canine fossa from the infratemporal fossa on the lateral surface of the maxillary bone.

Lateral growth axis: The division between the right and left lateral components of growth (see Median sagittal plane).

MP, Mandibular planes: Variations of definitions include:

A tangent to the lower border of the mandible.

A line joining gonion and gnathion.

A line joining gonion and menton.

A line from menton tangent to the posteroinferior border of the mandible.

Maxillary plane: See Palatal plane.

Median sagittal plane: See Lateral growth axis. The anteroposterior median plane of the cranium and face.

Me, Menton: The most inferior point on the symphysis of the mandible in the median plane. Seen on the lateral cephalogram as the most inferior point on the symphyseal outline.

Metopion: A point in the median line of the forehead between the summits of the frontal eminence.

N, Na, Nasion: The craniometric point where the midsagittal plane intersects the most anterior point of the nasofrontal suture. (The anterior termination of the Bolton plane.)

Normal face: By normal face, we do not mean a face of certain dimensions or particular form (features), but a well-grown face, harmoniously developed, skeletally and dentally, and consistent in developmental progress with its years (Bolton).

O point: Center of convergence area of horizontal planes used in Sassouni's analysis.

Occ, Occlusal plane: A line passing through one-half of the cusp heights of the first permanent molars and one-half of the overbite of the incisors.

Op, Opisthion: The most posterior point of the foramen magnum.

Orbital plane: The frontal (transverse) plane of the head passing through the left orbital point.

Or, Orbitale: In craniometry, the lowest point on the inferior margin of the orbit. The left orbital point is used in conjunction with the poria to orient the skull on the Frankfort horizontal plane.

Pal, Palatal plane, Maxillary plane: A line connecting the tip of the anterior nasal spine with the tip of the posterior nasal spine as recorded in the lateral cephalogram.

PM, Posterior maxillary plane: A vertical line from the averaged intersections of the great wings of the sphenoid and the anterior cranial floor, extending inferiorly to the averaged lower-most points of PTM.

Po, Pog, Pogonion: The most anterior point on the symphysis of the mandible in the median plane.

Points:

A: See Subspinale.

B: See Supramentale.

D: The center of the cross section of the body of the symphysis. It is established by visual inspection.

R: Bolton registration point. The center of the Bolton cranial base; a point midway on the perpendicular erected from the Bolton plane to the center of the sella turcica (S).

P, Porion: Anatomic porion is the outer upper margin of the external auditory canal; machine porion is the uppermost point on the outline of the ear rods of the cephalometer.

Porionic axis: A line drawn between the two poria.

PNS, Posterior nasal spine: A process formed by the united, projecting median ends of the posterior borders of the two palatine bones.

Pr, Prosthion: The most anterior point of the alveolar portion of the premaxilla, between the upper central incisors.

PTM, Pterygomaxillary fissure: In the lateral cephalogram, an inverted, elongated, teardrop shaped area formed by the divergence of the maxilla from the pterygoid process of the sphenoid. The posterior nasal spine and staphylion are generally located beneath the lower pointed end of this area.

Pt-vertical: A vertical line tangent to the posterior contour of PTM perpendicular to FH.

S, Sella turcica (Turkish saddle): The hypophyseal or pituitary fossa of the sphenoid bone lodging the pituitary body. The landmark S is the center of sella as seen in the lateral cephalogram and located by inspection.

SE, Sphenoethmoidal suture: The most superior point of the suture.

Si: The most inferior point on the lower contour of the sella turcica.

SN, Sella-nasion plane: The plane formed by connecting a line from sella to nasion.

Skeletal age: The maturational age of an individual as determined by the analysis of bone age indicated by the hand-wrist x-ray (biologic age).

SO, Spheno-occipital synchondrosis: The most superior point of the junction between the sphenoid and occipital bones.

Sp: The most posterior point on the posterior contour of the sella turcica.

Supraorbital plane: A line tangent to the anterior clinoid process and the most superior point on the roof of the orbit, as seen on the lateral cephalogram.

Sta, Staphylion: The point in the median line (interpalatal suture) of the posterior part of the hard palate where it is crossed by a line drawn tangent to the curves of the posterior margins of the palate. (In the lateral cephalogram, the posterior curved margins of the hard palate frequently may be seen more clearly than the posterior nasal spine.)

SOr, Supraorbitale: The uppermost point of the orbital ridge on the lateral cephalogram, it can be located at the junction of the roof of the orbit and the lateral contour of the orbital ridge.

Subspinale (Point A): That point in the median sagittal plane where the lower front edge of the anterior nasal spine meets the front wall of the maxillary alveolar process (Downs' point A).

Supramentale (Point B): The deepest midline point on the mandible between infradentale and Pogonion (Downs' point B).

Te, Temporale: The intersection of the shadows of the ethmoid and the anterior wall of the infratemporal fossa.

Vertex: The most superior point on the cranial vault.

Y-axis: The line joining the sella turcica (S) center and gnathion.

Zygion: The point on the zygoma on either side, at the extremity of the bizygomatic diameter.

CHAPTER FIFTEEN

General Body Growth and Development

Robert Cederquist, L.D.S., D.D.S., M.S., M.A., Ph.D.

There are several reasons for the importance of the availability and use of data on the somatic growth of normal children. First, the determination of whether the growth of a particular individual deviates from what may be considered normal variation must be based on a comparison with healthy children. Second, in order to make accurate descriptions of a growth aberration, correspondingly precise information about the normal state must be available. Third, changes in patterns of growth that occur over time within representative samples of a population are valuable indicators of changes in the general health and nutritional status of that particular population. Fourth, it would not be possible to design and conduct investigations regarding control mechanisms of growth if no precise data were available describing the resultant somatic effects, i.e., the growth changes themselves that are observed. Thus, and as stated by Marshall (1977), whether growth is dealt with from a clinical point of view or as a tool in public health or in scientific investigations, well-planned studies of normal children are of crucial importance.

COLLECTION OF GROWTH DATA

Standards of growth should be representative of the group of individuals with whom they are to be compared. There are two basic but quite different rationales in designing growth studies: (1) to describe in detail the growth of individual children and (2) to construct growth standards to indicate the variation around the mean values of a particular measurement at various ages. In the first instance the collection of longitudinal data is required, whereas cross-sectional data are well suited for the latter. The distinction between these two types of investigations is very important. The appropriate way to study the growth of any individual child is to measure it repeatedly. The collection of repeated measurements on the same child, or group of children, at each age is called a **longitudinal study.** On the other hand, a **cross-sectional study** is one in which different children are measured at each age; i.e., no individual is measured more than once. This is the most effective method of estimating the mean value, including the variation around this mean, of any given measurement, e.g., stature, of children in different age groups. Thus, population standards are correctly based on cross-sectional material because a longitudinal study is less satisfactory for estimating the population mean at successive ages (Healy, 1986). The samples at each age in a longitudinal study are not independent of each other because they represent the same individuals.

A longitudinal study may span only a few years or it may extend from birth to adulthood, children being examined every year up to age 20 or beyond. It is usually not possible to maintain the same sample of children within a growth study for extended periods of time (Tanner, 1978a). Some children will leave the study during its course. On the other hand, the investigators may decide to add individuals while the study is going on. This type of data collection is called mixed longitudinal, and is the most practical sampling technique for describing longitudinal growth patterns. A special design of a mixed longitudinal study is one in which a series of short-term pure longitudinal studies overlap, i.e., the grouping may be from birth to 6, from 5 to 11, from 10 to 16, and from 15 to 20 years of age (Tanner, 1986a). Special analytic techniques are needed to extract accurate information from such a study.

The main advantage of longitudinal studies is that they reveal accurately the growth patterns of individuals; i.e., they permit the study of changes in growth rate and of the sequence of timing of events during the growth period. A growth curve obtained by plotting the means at each age, as obtained by cross-sectional data, will not describe correctly the shape of the curve of any individual. This error is not noticeable in a significant way until the age of 9 years. However, beyond that age a major bias is introduced when cross-sectional data are utilized for growth evaluation of individuals (Tanner, 1986b). The reason is that the growth acceleration at puberty occurs early in some children and late in others. It would be quite misleading to represent an average growing child by averaging the values at each successive age. Such a procedure would not take into account the difference in timing between growth curves of different individuals; i.e., it would disregard the variation in tempo of growth (Tanner, 1987). Figure 15–1A shows five growth curves during the pubertal growth spurt from which a mean curve has been constructed by averaging the values at each age. This curve erroneously suggests that the growth spurt is more prolonged in time and is flatter than any of the individual curves. It is thus a poor representation of an average (i.e., typical) child. On the other hand, one may calculate the mean age at which the peak of the pubertal growth spurt occurs for the five individuals and then redraw the individual curves accordingly. This will generate a mean curve that better depicts the growth of a "typical" individual. Such realignment of curves is shown in Figure 15–1B. They have been plotted against a time scale (the abscissa) of a developmental index, rather than chronologic age. A graph like this can only be constructed with longitudinal data.

DISTANCE AND VELOCITY CURVES

By measuring stature of an individual annually over a long period of time, e.g., from birth to 20 years of age, a series of measurements will be collected

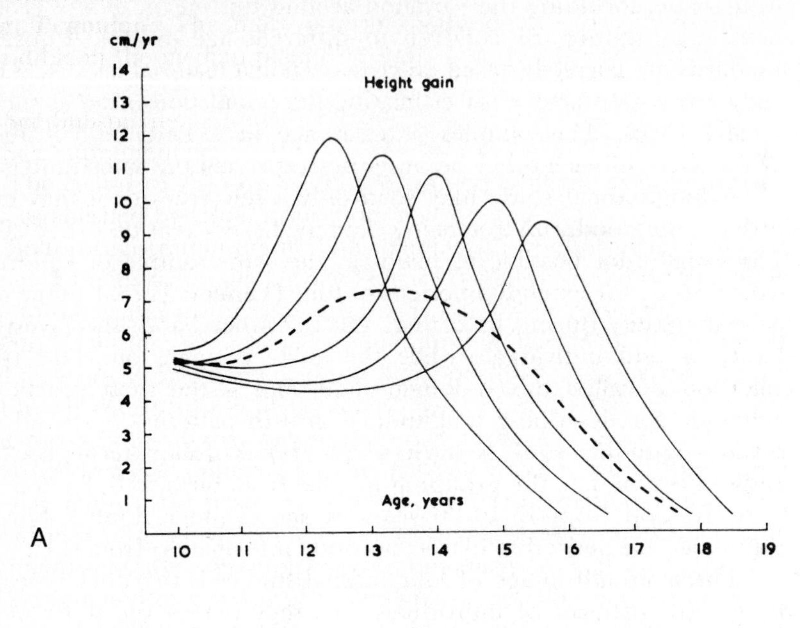

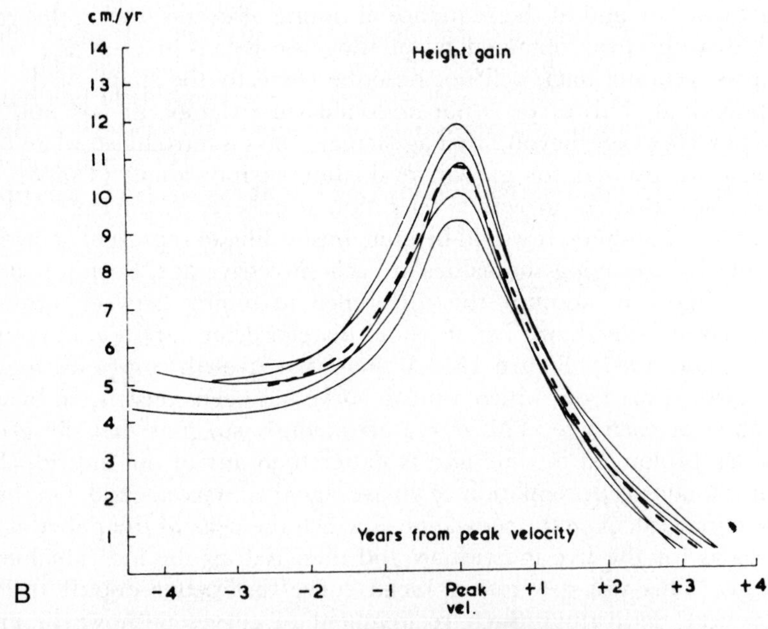

FIGURE 15–1. *A,* Five growth curves (solid lines) during the pubertal growth spurt from which a mean curve (broken line) has been constructed by averaging the values of each age. *B,* The five curves have been plotted according to the mean age at peak height velocity. This will produce a mean curve (broken line) that more closely represents an "average growing" individual. (From Tanner, J. M., R. H Whitehouse, and M. Takaishi: Standards from birth and maturity for height, weight, height velocity, and weight velocity: British children, 1965. Arch. Dis. Child., 41:454, 613, 1966.)

that may be used to illustrate growth in one of two ways. First, stature may be plotted against age to indicate height achieved at each age. This type of curve is called a distance curve (Tanner, 1955) or an achievement curve (Rallison, 1986). Figure 15–2 is a distance curve, i.e., a height-for-age curve. Second, if one wishes to know with what speed an individual has been growing between successive measurements, a distance curve is not adequate. To analyze this, one must calculate growth velocity (also called growth rate) which refers to the increase in a given measurement, in this instance stature, during a certain time period. A velocity curve is shown in Figure 15–3. Although height velocity is expressed in centimeters per year, it may be calculated over any period of time. However, for a velocity value to be reliable for clinical use, it should span a period of close to 1 year due to variation in growth rate during the course of a year (Marshall, 1975).

For example, for the calculation of growth velocity, assume that an individual is measured on his twelfth birthday and again 3 months later. Let us assume that measurements reveal an increase in stature of 2 cm during this time. The velocity is then 8.0 cm/yr (2.0 cm divided by 0.25 [yr]). In growth charts age is usually indicated by years and tenths of years, and not by years and months. Thus, in plotting a measurement of a subject, age should be expressed as decimal age. This becomes a simple operation with the use of a conversion table for calendar dates to decimal dates (Table 15–1).

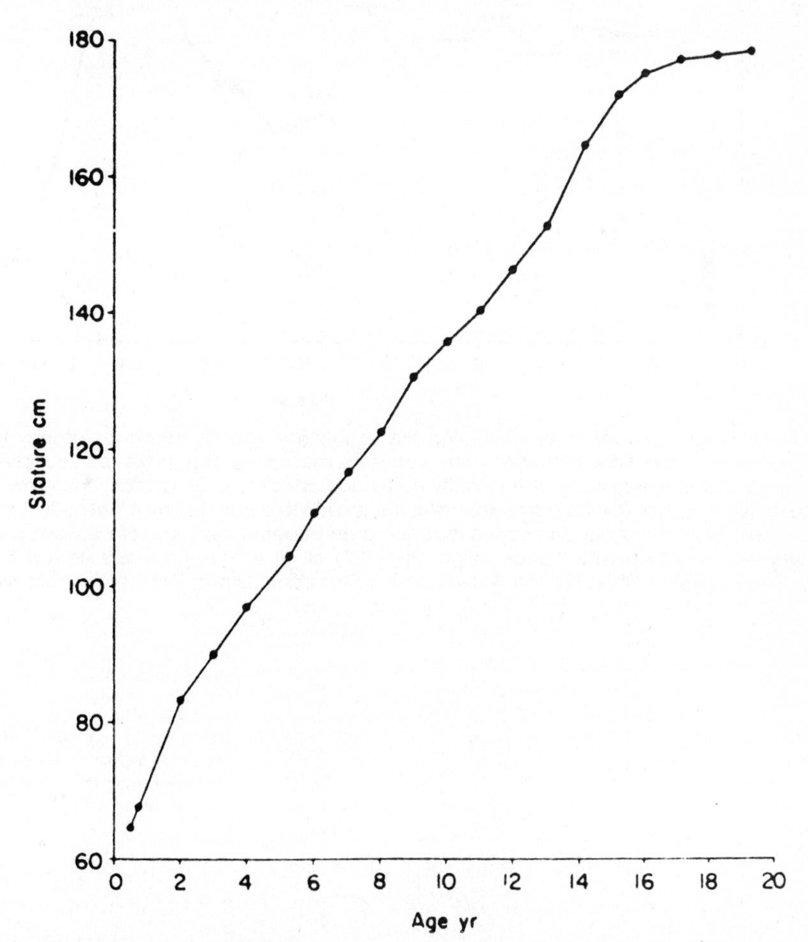

FIGURE 15–2. Distance curve based on repeated height measurements of the same individual. (From Marshall, W. A.: *Human Growth and Its Disorders.* London, Academic Press, 1977.)

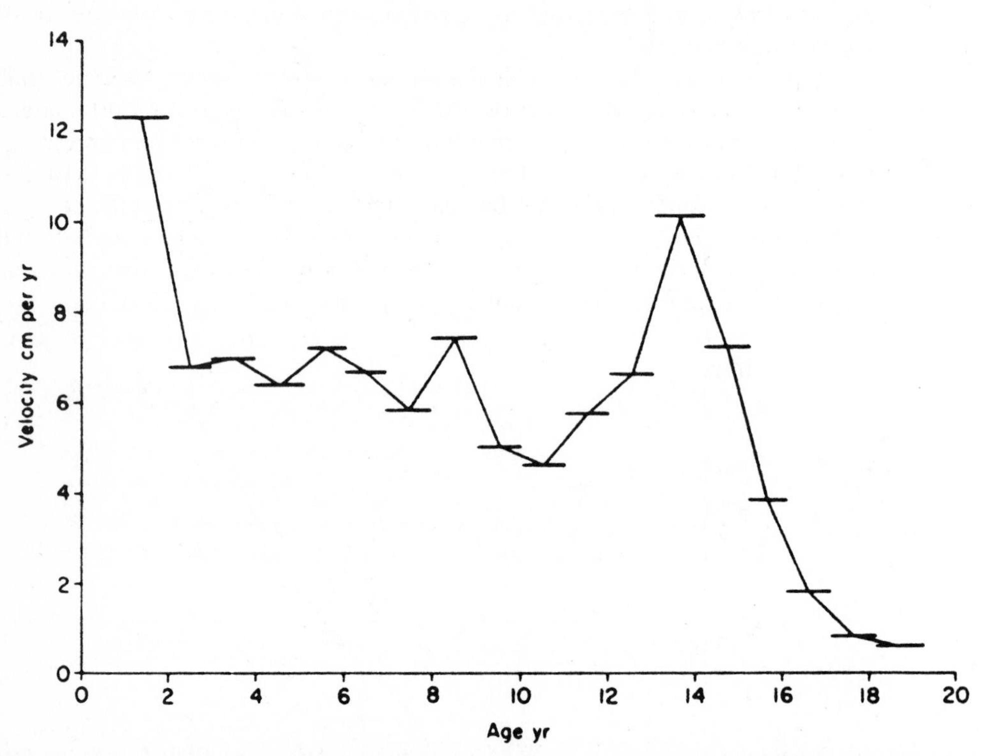

FIGURE 15–3. Height velocity curve illustrating the variation in growth rate from infancy to early adulthood. The horizontal lines represent time spans between successive recordings. The growth curve is constructed by connecting the midpoints of each time span. The velocity curve describes more graphically the growth of the individual than does the distance curve. For this particular child, the growth rate was 12 cm/yr between 1 and 2 years of age. After that, it showed fluctuations but decreased to reach a prepubertal minimum of between 4 and 5 cm/yr. During the spurt there was a growth rate ("peak height velocity") of 10 cm/yr. This was followed by a sharp decrease in velocity. (From Marshall, W. A.: *Human Growth and Its Disorders.* London, Academic Press, 1977.)

Table 15–1. CONVERSION TABLE FOR CALENDAR DATES TO DECIMAL DATES

Date	Jan	Feb	Mar	Apr	May	June	July	Aug	Sept	Oct	Nov	Dec	Date
1	000	085	162	247	329	414	496	581	666	748	833	915	1
2	003	088	164	249	332	416	499	584	668	751	836	918	2
3	005	090	167	252	334	419	501	586	671	753	838	921	3
4	008	093	170	255	337	422	504	589	674	756	841	923	4
5	011	096	173	258	340	425	507	592	677	759	844	926	5
6	014	099	175	260	342	427	510	595	679	762	847	929	6
7	016	101	178	263	345	430	512	597	682	764	849	932	7
8	019	104	181	266	348	433	515	600	685	767	852	934	8
9	022	107	184	268	351	436	518	603	688	770	855	937	9
10	025	110	186	271	353	438	521	605	690	773	858	940	10
11	027	112	189	274	356	441	523	608	693	775	860	942	11
12	030	115	192	277	359	444	526	611	696	778	863	945	12
13	033	118	195	279	362	447	529	614	699	781	866	948	13
14	036	121	197	282	364	449	532	616	701	784	868	951	14
15	038	123	200	285	367	452	534	619	704	786	871	953	15
16	041	126	203	288	370	455	537	622	707	789	874	956	16
17	044	129	205	290	373	458	540	625	710	792	877	959	17
18	047	132	208	293	375	460	542	627	712	795	879	962	18
19	049	134	211	296	278	463	545	630	715	797	882	964	19
20	052	137	214	299	381	466	548	633	718	800	885	967	20
21	055	140	216	301	384	468	551	636	721	803	888	970	21
22	058	142	219	304	386	471	553	638	723	805	890	973	22
23	060	145	222	307	389	474	556	641	726	808	893	975	23
24	063	148	225	310	392	477	559	644	729	811	896	978	24
25	066	151	227	312	395	479	562	647	731	814	899	981	25
26	068	153	230	315	397	482	564	649	734	816	901	984	26
27	071	156	233	318	400	485	567	652	737	819	904	986	27
28	074	159	236	321	403	488	570	655	740	822	907	989	28
29	077		238	323	405	490	573	658	742	825	910	992	29
30	079		241	326	408	493	575	660	745	827	912	995	30
31	082		244		411		578	663		830		997	31

The decimal should be preceded by the year, e.g., January 3, 1974 = 74.005.
From Marshall, W. A.: Growth and secondary sexual development and related abnormalities. Clin. Obstet. Gynecol., 1:593, 1974a.

Example 1 (refer to Table 15–1): A boy with a birthdate of February 10, 1983, was measured on October 20, 1988. His age at that time was 5.690 years (88.800 − 83.110). The advantage of this technique will be apparent when growth velocities are calculated. **Example 2:** A girl is first measured on April 12 and then the measurement is repeated on December 17. The growth velocity is obtained by dividing the change in the measurement—assume it is 2.5 cm—by the time span—i.e., 0.682 year (0.959 − 0.277), which is 3.7 (3.666) cm/yr. Notice that the result will be directly expressed in centimeters per year.

Both distance and velocity curves are important. They yield different information. The **distance curve** shows the actual size of a particular parameter, and the **velocity curve** indicates the rate of change in that parameter.

MEASUREMENTS FOR GROWTH STANDARDS

The most common measurements for monitoring growth in the clinic are weight, height, and skinfold thickness. In some disorders of growth the relationship between lower limb length and upper body length is altered. In such cases sitting height, when compared with total height, gives valuable information.

By far the most important measurement is height. Up to the age of 2 years supine length is used as stature measurement. It is taken with the child positioned horizontally on a measuring board. This measurement averages about 1 cm more than that when the child is standing up (Tanner, 1978a). This difference in measuring technique is the cause for the interruption in the distance curve observed at age 2 in growth charts (Tanner, 1978b). Both supine length and stature are highly reliable measurements if proper standardization of technique is exercised. Two examiners should not differ more than 4 mm in 95 per cent of occasions (Tanner, 1978a). All too often the error is inherent in the measuring device. The flimsy angle arms attached to many of the weighing scales available on the market are generally not acceptable because they lack rigidity, which causes the vertical rod to bend. Frequently the horizontal arm does not remain consistently at a right angle to the vertical rod. The error may well exceed 1.5 cm with this type of nonrigid instrument even with proper technique of positioning the individual during the measuring procedure (Marshall, 1977). A good instrument for measuring stature is the Harpenden stadiometer, shown in Figure 15–4. This instrument has a rigid vertical backboard and a counterbalanced horizontal headboard that moves vertically on ball-bearing rollers. It has a digital read-out in millimeters with a recording range of 600 to 2000 mm.

So that measuring errors are kept to a minimum, it is essential to use a standardized position of the subject to be measured. The child must be standing with back and heels against the backboard of the instrument. Shoes are removed.

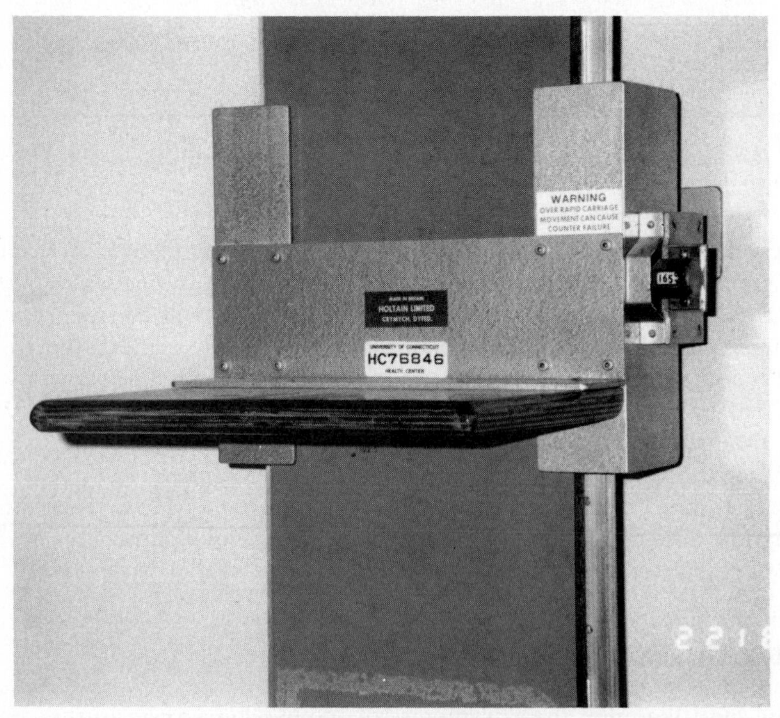

FIGURE 15–4. The Harpenden stadiometer for measuring stature. This instrument is of rigid construction. It has a counterbalanced horizontal headboard with digital read-out. (Courtesy C. J. Burstone, Department of Orthodontics, University of Connecticut Health Center.)

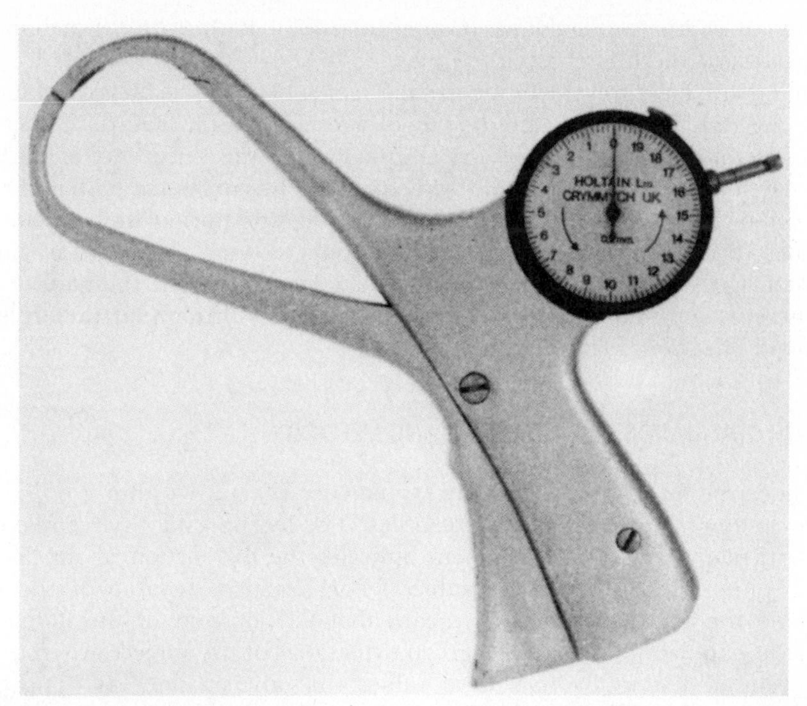

FIGURE 15–5. The Tanner-Whitehouse skinfold dial caliper. This instrument is used for measuring the amount of subcutaneous fat. Both caliper and stadiometer (Fig. 15–4) are available from Seritex, Inc. (450 Barell Ave., Carlstadt, NJ 07072).

The head is oriented to the Frankfort horizontal. The subject is told to stretch, but without lifting the heels off the floor. During the recording the examiner is standing in front of the subject and exerting gentle, upward pressure on the mastoid processes to further encourage stretching. Both height and supine length should be recorded to the last completed millimeter.

Weight, although frequently used as a measure of growth, is not a totally adequate measurement. As long as some basic rules regarding standardization are followed, it is an easy measurement to take, which explains its common use. However, it is important to realize that weight is an expression of a composite of growth parameters. For example, a child may have normal rate of growth which is not reflected in weight measurements if the individual is losing subcutaneous fat simultaneously. Similarly, a child with an exceedingly low velocity of growth may show adequate weight gain due to increase in subcutaneous fat. Weight may be a good indicator of degree of obesity if it is evaluated relative to stature. However, loss or gain of fat is best measured by skinfold thickness. Such measurements are taken with a skinfold caliper (Fig. 15–5). To facilitate standardization of measuring technique, such calipers, by convention, exert constant pressure of 10 g/mm^2 regardless of the degree of jaw separation. Limb fat and trunk fat follow slightly different growth curves. Thus, one measurement, the triceps skinfold, is taken as a representative record of limb fat; whereas the subscapular skinfold is used as a measurement for trunk fat. The triceps recording is made on the posterior surface of the left upper arm halfway between acromion and olecranon (Cameron, 1986). The subscapular measurement is made below the left scapula at the inferior angle formed by the lateral and medial scapular borders (Cameron, 1986). Like measurements for body weight (Buckler, 1979), skinfold measurements do not have a normal distribution within a population (Cameron, 1986). Rather, the distribution is skewed toward the higher values. Figure 15–6 shows distance curves for both limb and trunk fat. After the first postnatal year subcutaneous fat is continuously decreasing until between 6 and 8 years of age, whereafter it starts to increase. Girls have a little more fat than boys in infancy, and this difference is gradually accentuated during childhood and adolescence. The curves for limb and trunk fat become more divergent during puberty. At this period limb fat in boys decreases (triceps measurement), and there is a slowing down in the gain of trunk fat. Girls at puberty show a temporary leveling off in the gain of limb fat, whereas trunk fat in girls has a steady increase throughout puberty until adulthood (Tanner, 1978b).

GROWTH STANDARDS FOR DISTANCE AND VELOCITY

The variation in a given measurement—e.g., height—within a population is most commonly expressed as percentiles. The location on a percentile chart of a particular subject's measurement indicates the proportion of smaller and greater values than that of the subject. For instance, a value of the third percentile for a given parameter means that 97 per cent of the individuals within the population have values greater than that of the subject, whereas only 3 per cent of the individuals have values that are smaller. As opposed to standard deviation (SD), percentiles can be used for parameters regardless of whether they conform to a normal distribution or not. Thus, the variations in weight and skinfold thickness that do not show normal distributions (supposedly because of the accumulation of excessive subcutaneous fat within modern populations) are most conveniently described in terms of percentiles. On the other hand, parameters like stature and limb length, which follow a normal distribution, can easily be described with either method, i.e., percentiles or SD. For such measurements the value at the fiftieth percentile equals the statistical mean for the population. In growth standards it is common to indicate the ninety-seventh and third percentiles. In a normal distribution these values correspond very closely to ± 2 SD. Figure 15–7 depicts the relationship between percentiles and SD.

The National Center for Health Statistics (NCHS) has published growth charts based on cross-sectional data (Hamill et al., 1977). The data were derived from (1) the Fels Research Institute (collected 1929 to 1975), (2) the NCHS Health Examination Survey (HES) Cycles II (1963 to 1965) and III (1966 to 1970), and (3) the first National Health and Nutrition Examination Survey (NHANES I) with data collected 1971 through 1974. The Fels Research Institute data were used in constructing weight charts for the ages between birth and 3 years. The HES-NHANES I data formed the basis for six charts, all conditioned for sex: two distance charts for height (Fig. 15–8A and B), two distance charts for weight (Fig. 15–8A and B), and two weight-for-stature charts for preadolescent children (Fig. 15–8C and D). The charts were developed after the National Academy of Sciences had recommended that such information be made available for the nutritional assessment of infants and children in the United States. Because of the specific characteristics of the underlying study—e.g., large sample size, a well-nourished population, and cross-sectional design—the set of NCHS charts have been recommended as an international reference to be used by the World Health Organization (WHO) (Waterlow et al., 1977). The charts are now included in various WHO publications on nutritional and growth monitoring (World Health Organization, 1978, 1983, 1986). It should

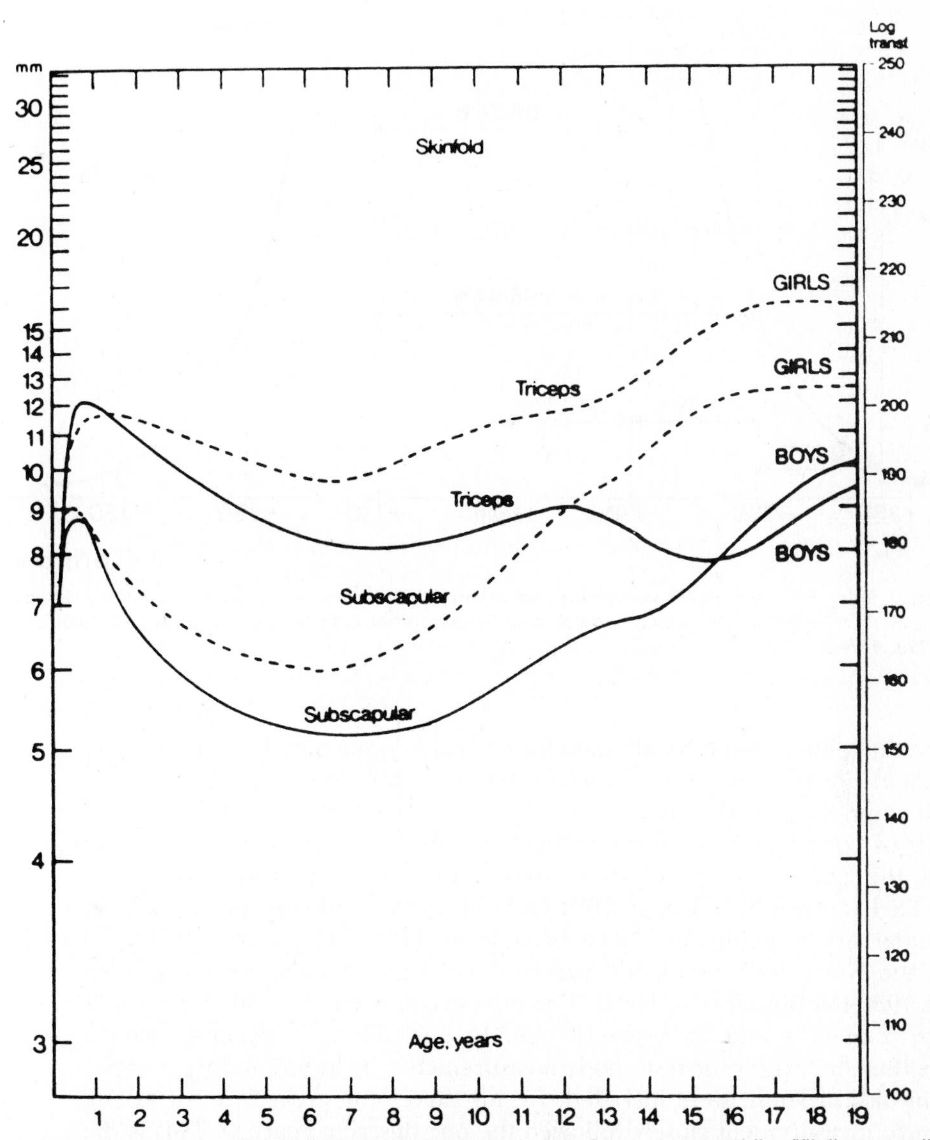

FIGURE 15–6. Distance curves for skinfold thickness (subcutaneous fat). Curves represent the fiftieth percentiles. Two measurements, limb fat (triceps) and trunk fat (subscapular), for girls and boys, respectively, are shown. The scale is in millimeters on the left of the graph and logarithmic transformation on the right. (From Tanner, J. M.: *Foetus into Man: Physical Growth from Conception to Maturity.* Cambridge, Harvard University Press, 1987, original data from Tanner, J. M. T., and R. H. Whitehouse. Arch. Dis. Child., 50:142, 1975.)

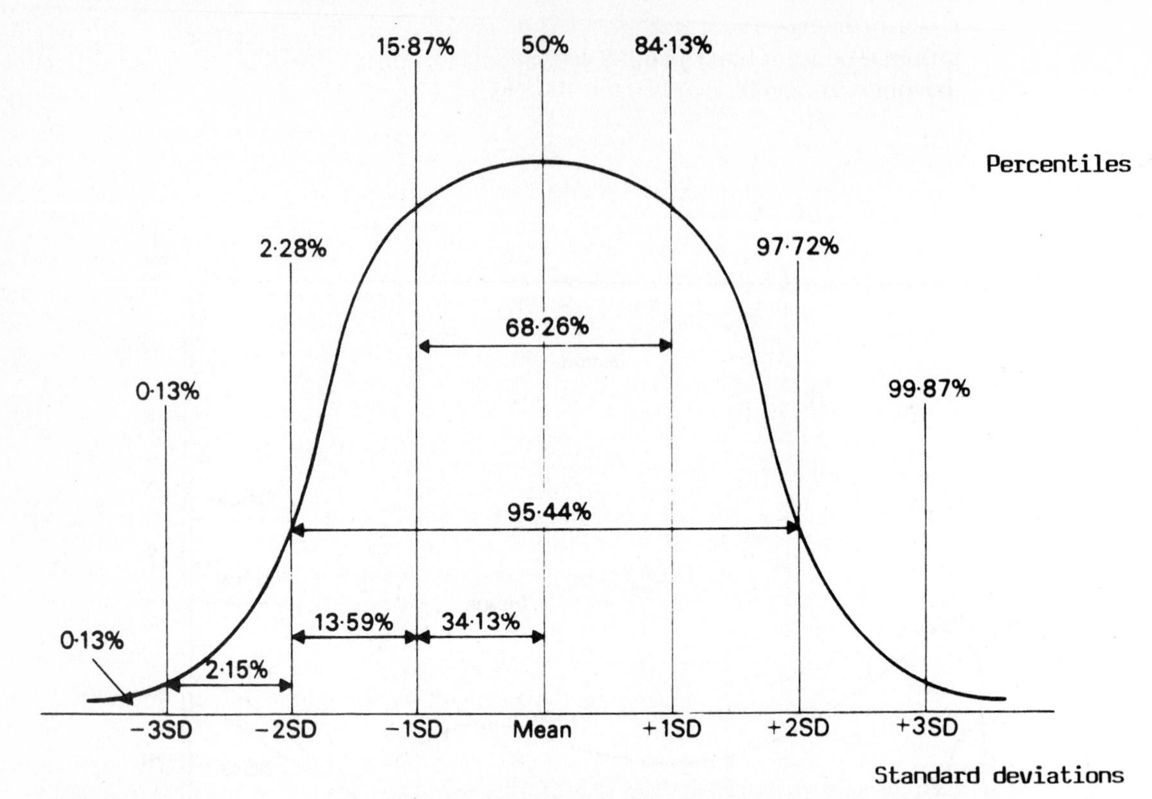

FIGURE 15–7. Relationship between percentiles and standard deviations for values with a normal distribution. (From Buckler, J. M. H.: *A Reference Manual of Growth and Development.* Cambridge, Massachusetts, Blackwell Scientific Publications, 1979.)

be noted that although the NCHS data have been recommended as a reference, they should not be used as a universal standard (Waterlow et al., 1977). The target for optimal growth is not necessarily the same for all populations.

It is of considerable significance that in order to keep their growth data current, the NCHS is renewing their surveys on somatic measurements. Since NHANES I was conducted from 1971 to 1974, newer anthropometric data on the United States population have become available. The second National Health and Nutrition Examination Survey (NHANES II) was undertaken from 1976 to 1980 (McDowell et al, 1981). The sample consisted of 27,801 individuals between 6 months and 74 years of age. In addition to a general medical examination and diagnostic tests, body measurements, including stature, weight, and skinfold thickness, were included. In this survey the procedure for taking the stature measurement closely followed the one described earlier. This is the so-called Tanner-method—i.e., gentle upward pressure is exerted on the subject's mastoid processes. In earlier NCHS surveys this technique was not used (Hamill et al., 1977). This difference in methodology, when related to comparative studies, may be of significance, because it has been shown that diurnal variation in height can be considerable (Strickland and Shearin, 1972), a phenomenon that stretching minimizes. Information on anthropometric data from NHANES II is currently only available in tabular form with estimates of means, SD, and fifth to ninety-fifth percentiles (Najjar and Rowland, 1987).

The NCHS type of growth charts is based on "sizes attained" at a given age and is suitable for comparing body measurements of children at one given

Text continued on page 411

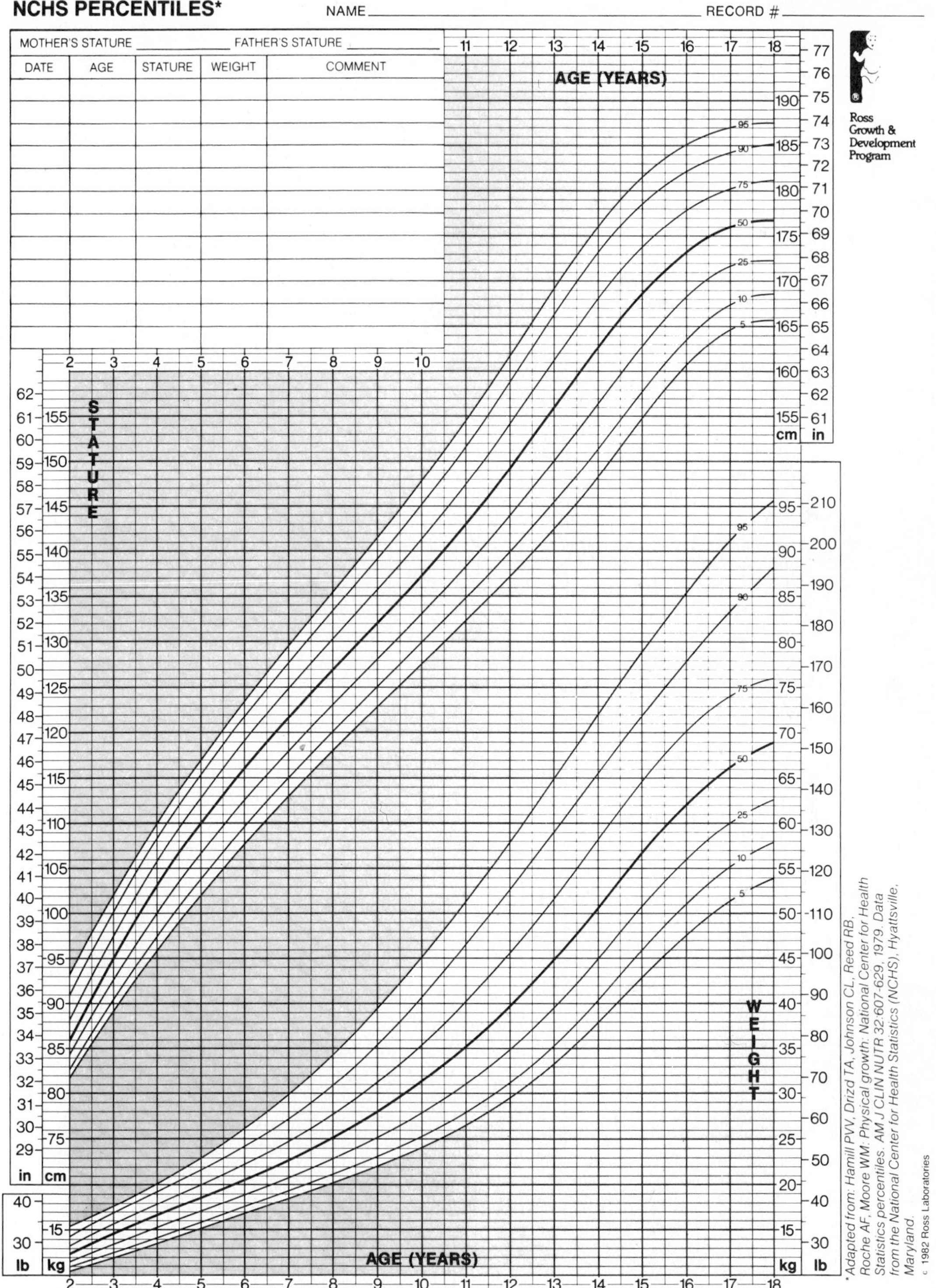

FIGURE 15–8. *A*, Growth chart from National Center for Health Statistics. Cross-sectional data. Stature by age percentiles and weight by age percentiles for boys, 2 to 18 years of age.

Illustration continued on following page

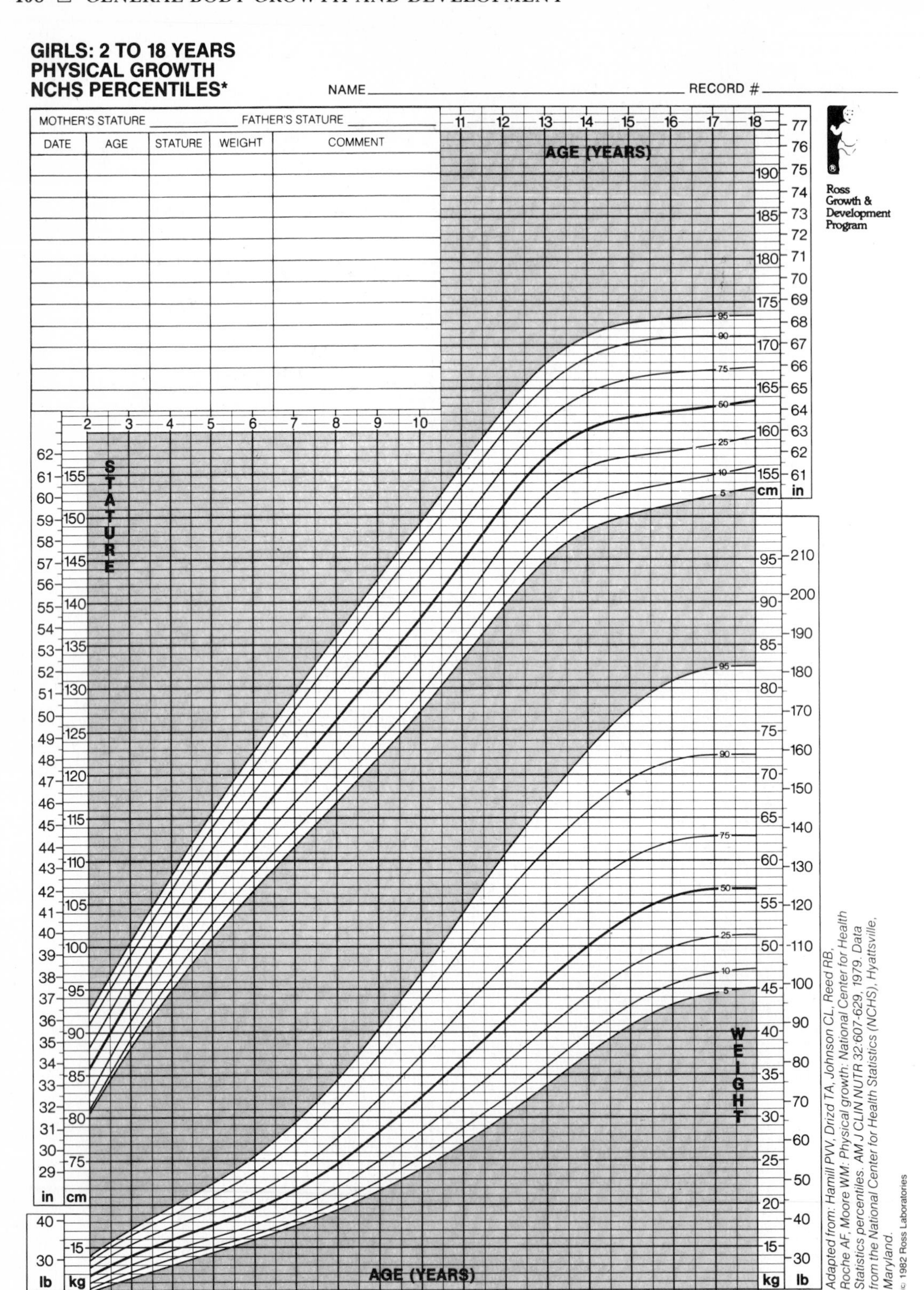

FIGURE 15–8 *Continued B,* Stature by age percentiles and weight by age percentiles for girls, 2 to 18 years of age.

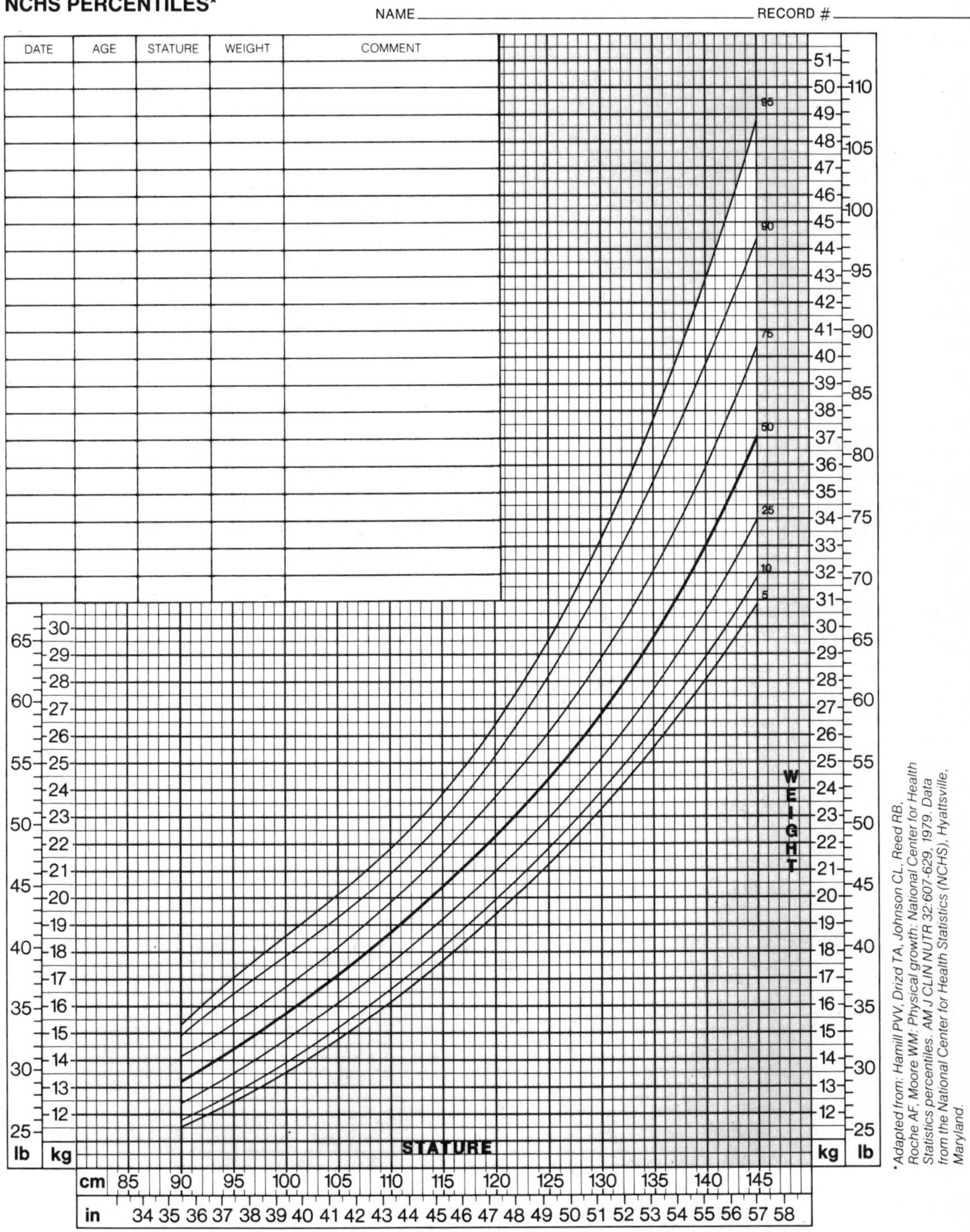

FIGURE 15–8 *Continued C,* Weight by stature percentiles for prepubescent boys.

Illustration continued on following page

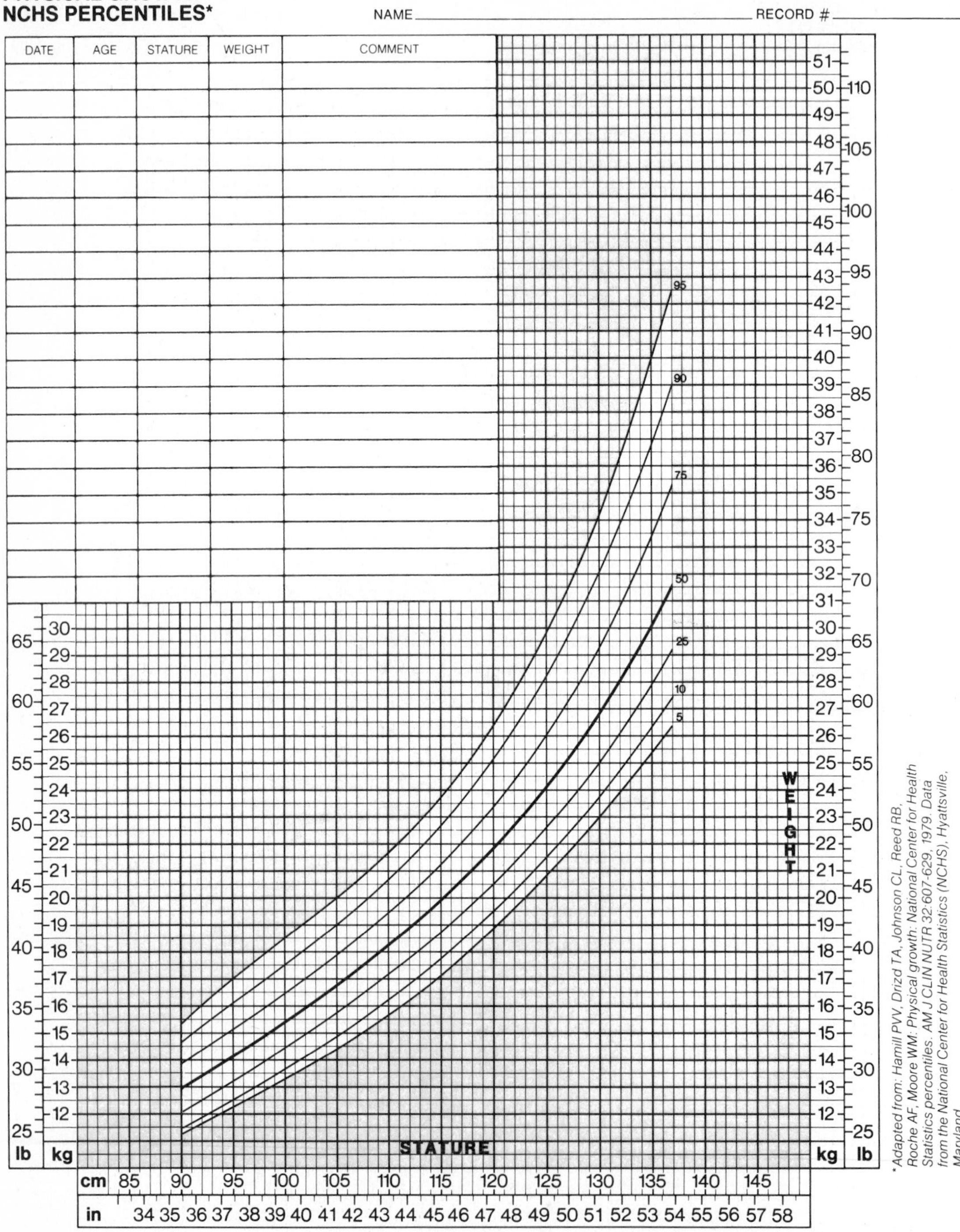

FIGURE 15–8 *Continued D,* Weight by stature percentiles for prepubescent girls. (Original data from Hamill, P. V. V., T. A. Drizd, C. L. Johnston, R. B. Reed, and A. F. Roche: NCHA Growth Curves for Children, Birth–18 years. United States. DHEW Publ. No. (PHS) 78-1650. Vital Health Statistics (11), No. 165, 1977. Reproduced by permission of Ross Laboratories. Columbus, Ohio)

point in time. Allowing for some degree of error, it may also be used in the clinic to chart a preadolescent child up till about age 9. Beyond that age children vary too much in their tempo of growth (Tanner and Davies 1985; Tanner, 1986b), i.e., some chidren are early and others are late maturers. Thus, if all children, regardless of growth tempo, are included, the variation in height at any given age will be inflated, compared with the variation that exists among children of a given tempo. Growth curves that do not account for growth tempo are not entirely suitable for clinical use. Standards that separate early and late maturers from average maturers were originally attempted by Bayley (1956). However, Tanner et al. (1966) refined the construction of such standards and produced so-called longitudinal standards for clinical use. These are well suited for monitoring the growth of individuals, especially during puberty. Such standards are based on relatively small longitudinal studies combined with large-scale cross-sectional data. The longitudinal data give the **shape** of the curve during puberty, and the cross-sectional material is used to generate the pre- and postpubertal percentiles. Thus, Tanner's original British standards of 1965 were later revised to include percentiles for early and late maturers in addition to children who mature at an average age (Tanner and Whitehouse, 1976). This has considerably extended the use of the previous charts. In 1985 Tanner and Davies published longitudinally based, tempo-conditional stature and stature velocity standards for the United States population. These standards are based on the cross-sectional data of the NCHS (Hamill et al., 1977) for the calculation of (1) pre- and postpubertal percentiles and (2) the age at the peak of the pubertal growth spurt. The shape of the curves during puberty is based on the earlier British longitudinal data. Percentiles are indicated for average maturing children as well as early and late maturers. The standards are available as distance and velocity curves for stature for boys and girls separately. Comments asking for clarification regarding the construction of the charts appeared shortly after their publication (Himes and Roche, 1986; Burstein and Rosenfield, 1986). The charts (available from Serono Laboratories, Inc., 280 Pond St., Randolph, MA 02368) are well suited and can be recommended for clinical use with North American children.

THE USE OF GROWTH CHARTS

It is important to make a clear distinction between height-attained charts (distance curves) and height velocity charts (growth rate curves). When a child is repeatedly plotted on a height-attained chart he/she tends, on the average, to remain in the same percentile. However, repeated measurements of growth velocity rarely stay within the same percentile unless they are on or close to the fiftieth percentile. Velocity charts indicate the range of velocities at which normal individuals may fall in any 1 year. An individual who for several years was plotted at the tenth percentile for **stature** would be a normally growing child. However, someone who grew at the tenth percentile for **stature velocity** during several years would show abnormal growth and would be falling steadily further below the average height of his or her age peers. A tenth percentile growth velocity is not abnormal for a limited time. However, it must be compensated by a velocity above the fiftieth percentile; otherwise the child will fall in percentiles on the stature chart. Thus, it seems clear that an essential part of growth assessment is evaluating growth speed. This is an important evaluation even if the child's present stature is within normal limits. Figure 15–9 (Marshall, 1974a) illustrates the importance of velocity recordings. Depicted

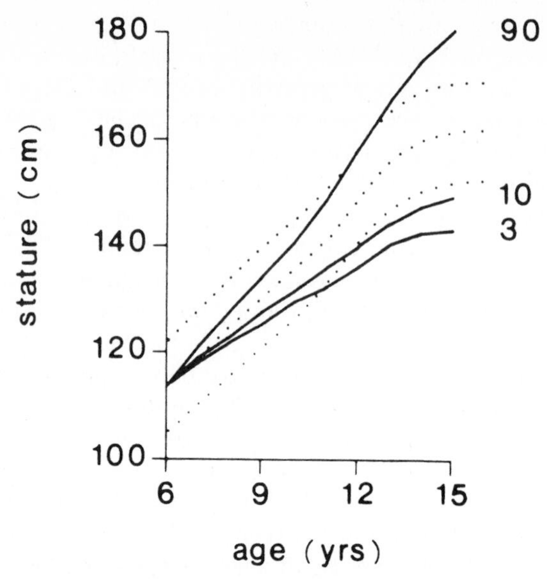

FIGURE 15–9. Distance curves (solid lines) for height of three hypothetical individuals whose growth velocities consistently were at the 3rd, 10th, and 90th percentiles, respectively. The dotted lines on the graph indicate the third, fiftieth, and ninety-seventh percentiles for stature. (From Marshall, W. A.: Growth and secondary sexual development and related abnormalities. Clin. Obstet. Gynecol., 1:553, 1974.)

are the distance curves of three hypothetical individuals who at 6 years of age were plotted at the fiftieth percentile. After age 6 they consistently grew at the ninetieth, tenth, and third percentiles for **velocity,** respectively. Although the velocity for each of the three children is within normal limits for any 1 year, the cumulative effect will result in abnormal growth curves, as the graph indicates.

By studying distance charts it becomes apparent how wide the range of stature is for any single age group. It is in fact much greater than the amount of growth that occurs in an individual child during any 1 year. For boys the average increase in stature between ages 8 and 9 is between 5 and 6 cm. However, the normal limits of variation (i.e., between the fifth and ninety-fifth percentiles) is over 20 cm. Thus, a tall (ninety-fifth percentile) 7-year-old boy is more than 4 cm taller than an average 8-year-old. He is in fact taller than more than 10 per cent of 11-year-old boys. On the other hand, a small but normal 7-year-old is no taller than a tall 4-year-old. These examples show that a child's chronologic age is an imprecise indicator of what the height of the child ought to be.

THE ADOLESCENT GROWTH SPURT

Whole-Year Peak Height Velocity

During puberty the growth velocity curve rises to a maximum and then quickly begins to fall again. The maximum speed (at the height of the curve) is called peak height velocity (PHV). For practical purposes the significant recording is the gain in stature that occurs over the whole year centered on the age of PHV. Tanner and Davies (1985), after having reviewed a number of empirically determined values (Tanner et al., 1966; Lindgren, 1976; Taranger et al., 1976a; Billewicz et al., 1981), determined midrange figures for the United

States population of 9.5 cm/yr for boys and 8.3 cm/yr for girls. These values correspond to the fiftieth percentile for children who mature at an average speed.

Age of Peak Height Velocity

From the observed cross-sectional semiannual values of the National Center for Health Statistics (Hamill et al., 1977) Tanner and Davies (1985) estimated the age at PHV for the United States to be 13.5 years for boys, with an SD of 0.9 year. The corresponding value for girls is 11.5 years, also with an SD of 0.9 year. The ages of 13.5 and 11.5 years equal the fiftieth percentiles for average maturing boys and girls, respectively.

Early and Late Maturers

Tanner and Davies (1985) also determined the fiftieth percentile curves for boys and girls who have their PHV either 2 SD of age early or late. As mentioned earlier, the SD of age at peak height velocity is 0.9 years for both sexes. Thus, 2 SD early maturing boys have their PHV at 11.7 years and 2 SD late maturing boys at 15.3 years. The corresponding ages for girls are 9.7 years and 13.3 years, respectively. It has been well documented that early maturing children, on the average, have a higher PHV than do late maturing ones. The values for PHV (fiftieth percentile) for 2 SD early maturing boys was estimated at 10.3 cm/yr and for 2 SD late maturing boys at 8.5 cm/yr. The corresponding estimates for girls were 9.0 cm/yr and 7.6 cm/yr, respectively. Although they differ in the magnitude of their peak and in the age at which it occurs, early and late maturers do not show significant differences in their final adult height. That is to say, there is no relationship in normal boys and girls between age at PHV or magnitude of PHV and final adult height (Lindgren, 1978; Marshall and Tanner, 1986).

PUBERTY

Puberty is a combination of morphologic and physiologic changes that occur in children during the time of the maturation of testes in boys and ovaries in girls. These changes involve the whole body, but they do not commence at the same age, nor is their duration the same in all children. Except in the term "adolescent growth spurt," the use of the word "adolescence" is usually restricted to psychologic changes that occur simultaneously with the anatomic and physiologic. However, this distinction between "puberty" and "adolescence" is by no means universal. The main somatic indicators of puberty are (1) the adolescent growth spurt; (2) the development of the gonads; (3) the development of the secondary sex characteristics; (4) changes in body composition, i.e., muscle mass and amount and distribution of subcutaneous fat; and (5) an increase in circulatory and respiratory capacities, especially in boys (Marshall and Tanner, 1986).

The pubertal growth spurt occurs, on the average, 2 years earlier in girls than in boys (see Age of Peak Height Velocity). Thus, many girls will begin to grow rapidly, whereas most of the boys of the same age are still growing at their prepubertal rate. The onset of the spurt, i.e., the age at which the curve shows continuous increase, has been estimated by Taranger and Hägg (1980). Their data show an age of 10.04 years (SD, 1.26 years) for girls and 12.08 years (SD, 1.20 years) for boys. These investigators found no significant difference between sexes with regard to the duration of the spurt, being 4.73 and 4.91 years for girls and boys, respectively. However, they found the postpubertal period (defined as the time between the first annual increment below 20 mm following PHV and the first of three consecutive annual increments, each being below 5 mm) to be significantly longer in girls than in boys.

In both European and US data the sex difference in adult stature is about 13 to 15 cm, i.e., men being on the average 13 to 15 cm taller than women. The major portion of this difference (about 10 cm) is accounted for by the 2-year delay in the boys' spurt (Marshall and Tanner, 1986). The remaining, but much smaller, part of the difference (3 to 5 cm) is related to the greater magnitude of the spurt in boys. The variation in timing of the spurt between boys and girls means that boys are about 10 cm taller when they start their spurt, compared with girls when they start theirs. The girls, of course, are 2 years younger at that stage. Until 9 years of age there is little difference in height between boys and girls. Sexual dimorphism in stature is not observed until the beginning of the pubertal spurt.

Both boys and girls show great variation not only in the age at which their secondary sex characters begin to develop but also in the amount of time they take to go through the various stages in the sequence of pubertal changes. In 95 per cent of normal girls breast development may start at any age between 9 and 13 years, with a mean age of 11.5 years (Marshall and Tanner, 1969). The mean time is 4.0 years to pass through the stages of breast development and reach the adult stage. In 95 per cent of normal boys genital development begins at any age between 9.5 and 13.5 years, with a mean age of 11.64 years (Marshall and Tanner, 1970). The mean time is 3.05 years for a boy to go through the stages of genital development to reach the mature stage. The timing of the pubertal growth spurt related to sexual development is quite different in girls and boys. In girls the spurt is an event that occurs in the early stages of breast development. About 40 per cent of girls reach PHV, whereas their breast development is still in the "bud" stage (Marshall and Tanner, 1969). Another 50 per cent experience PHV while breast development is in an early intermediate stage, and all girls have gone through PHV before their breasts are fully developed. Thus, one cannot assume that girls in early stages of development still have PHV in front of them. They may have reached their spurt or they may, in fact, already be slowing down. This is quite different from the situation in boys, of whom 75 per cent reach PHV during a late intermediate stage of genital development and 20 per cent do not reach PHV until after complete genital development has occurred (Marshall and Tanner, 1970). Only 5 per cent of boys experience their PHV during the early stages of sexual development. Thus, in contrast to girls, in whom it occurs early, the peak of the pubertal growth spurt in boys is a late event in the sequence of changes during puberty.

Although there is a distinct difference between the sexes in the timing of the pubertal growth spurt, being about 2 years earlier in girls, there is much less difference in the timing of the secondary sex characters. Available data suggest that in boys genital development begins only a few months to a year after breast development in girls (Marshall and Tanner, 1969, 1970; Taranger et al., 1976a; Largo and Prader, 1983a, 1983b). These data also show that final stages of maturity are reached at the same age in girls and boys (Marshall and

Tanner, 1969, 1970; Taranger et al., 1976a) or less than 10 months apart (Largo and Prader, 1983a, 1983b). Despite these small differences, from a social standpoint girls appear to reach puberty much earlier than boys. Marshall (1977) has pointed out that this is obviously so since early indicators of puberty in girls, e.g., the growth spurt and breast development, are easily noticeable in ordinary social contexts, while early signs of puberty in boys, e.g., genital development, are not. More conspicuous indicators in boys, e.g., the growth spurt and voice change (Hägg and Taranger, 1980) are late events in the sequence of pubertal changes. Thus, from a social standpoint, an average maturing girl becomes an obvious adolescent about 2 years before an average maturing boy. However, it now seems clear that this is a consequence of differences in the **sequence** of pubertal changes, rather than significant age difference at reaching puberty.

In girls menarche, unlike the growth spurt and breast development, is usually a late event during puberty. In over 70 per cent of all girls it occurs in the late stages of breast development (Marshall and Tanner, 1969). Menarche almost always occurs after the peak of the growth spurt. Thus, if one knows that a girl has reached menarche, one can with fair amount of certainty assume that she has passed her highest growth velocity and that her growth is slowing down. In Western European populations the age at menarche is 13.0 to 13.5 years (England, 13.5 years [Marshall and Tanner, 1969]; Switzerland, 13.4 years [Largo and Prader, 1983]; Sweden, 13.0 years [Taranger et al., 1976a]; Norway, 13.2 years [Brundtland and Walløe, 1973]). In the United States girls are slightly younger when they reach menarche: 12.8 years for whites (MacMahon, 1973; Zacharias et al., 1976) and 12.5 years for blacks (MacMahon, 1973). In relation to PHV, menarche occurs, on the average, 1.2 to 1.3 years later (Marshall and Tanner, 1969; Taranger et al., 1976a; Zacharias and Rand, 1983). There is a definite association between age at PHV and age at menarche. It is probably the closest association between any two indicators of skeletal growth and sexual maturation (Marshall and Tanner, 1986). Marshall and Tanner (1969) calculated the correlation coefficient between age at menarche and age of PHV to be 0.91. Menarche is also related to bone maturation. Using a sample of about 70 girls, Marshall (1974b) found that 85 per cent had "bone ages" of between 13 and 14 "years" when they first experienced menstruation.

GROWTH IN DIFFERENT PARTS OF THE BODY

Most skeletal dimensions, internal organs, and soft tissues, like muscles, have growth curves similar to those for stature. There are, however, some variations in the general growth pattern. Those for the growth of subcutaneous fat have been described earlier in this chapter. Other tissues and organs with growth patterns differing from those of height growth are the brain and skull, the reproductive organs, and the lymphoid tissues of tonsils, adenoids, and intestines. Such differential growth patterns are well illustrated in Figure 15–10, which is a frequently reproduced graph by Scammon (1930). It shows growth of various tissues expressed as a percentage of total gain from birth to maturity. The "general" curve exemplifies height growth. The "genital" curve represents the reproductive organs. This curve shows very slow prepubertal growth, whereas during puberty the growth rate is quite dramatic. The "neural" curve has a very characteristic postnatal shape because the brain develops earlier than any other part of the body. Between birth and 18 years of age the

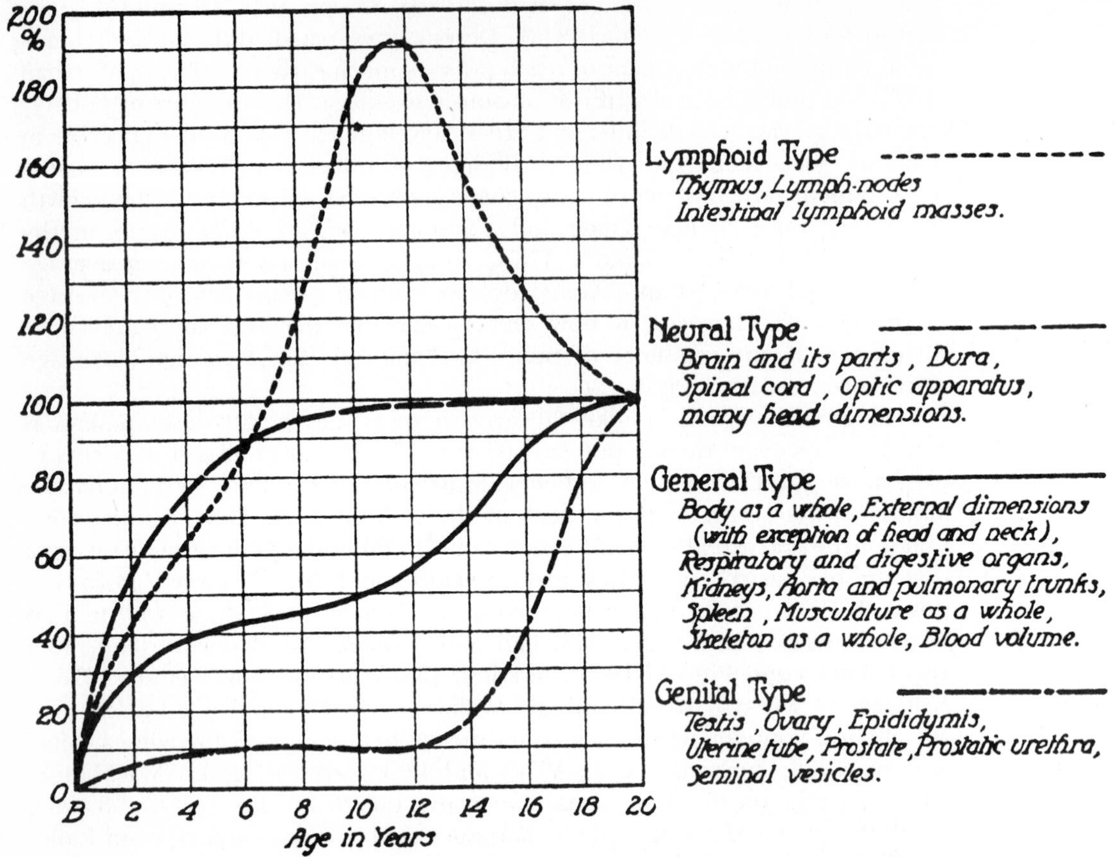

FIGURE 15–10. Postnatal growth of various tissues expressed as a percentage of total gain from birth to maturity. (From Scammon, R. E.: The measurements of the body in childhood. In: *Measurement of Man.* Ed. by J. A. Harris, C. M. Jackson, D. G. Paterson, and R. E. Scammon. Minneapolis, University of Minnesota Press, 1930.)

head circumference has an average increase of 20 cm. However, half of this has already been achieved at around 8 to 9 months of age, and 75 per cent of the total increase has been gained at the age of 2 years. The growth of the eye follows the brain growth curve with a probable, although slight, pubertal spurt. Evidence of such a spurt is the fact that during puberty there is a faster increase in the incidence of myopia than in other age groups (Largo, 1978). The "lymphoid" curve has also a quite different shape relative to the "general" type. As shown in Figure 15–11, this curve reaches its maximum before puberty, whereafter it continually regresses until maturity.

Almost all skeletal dimensions show acceleration during puberty, but the effect is not uniform throughout the skeleton. Some regions of the body go through the growth process faster than others. At any given age during childhood and adolescence the head is more advanced in its progress toward its adult size, compared with the trunk. Within the trunk the shoulder area is further advanced at any given age than the pelvic region. This differential in timing of growth has been characterized as a craniocaudal maturity gradient (Marshall, 1977). In the lower extremities the maturity gradient is directed from the foot toward the proximal end (Cameron et al., 1982). The foot shows a peak velocity about 6 months before the lower leg, which in turn has a peak velocity slightly ahead of the upper leg. In most children the foot reaches its adult size before any other part of the limbs or trunk. This may temporarily result in disproportionately large feet (Marshall and Tanner, 1986). However, proportions are usually restored once other linear body dimensions catch up. Also, in the upper extremities there appears to exist a distal-to-proximal maturity gradient. The forearm has a peak velocity a short time after the hand but about 6 months before the upper arm (Maresh, 1955; Cameron et al., 1982).

It is common practice to measure trunk length as sitting height. This is then subtracted from the measurement of stature to give leg length (also called subischial length). In most adolescents the peak growth velocity in leg length occurs about half a year before that of the trunk. During childhood growth in leg length exceeds that in trunk length. However, during puberty the spurt in sitting height exceeds the gain in leg length (Eveleth, 1978), which increases the effect of the trunk on the overall proportions of the total stature. The growth spurt in stature is dependent upon the patterns of growth in both trunk and legs, and thus PHV in stature is reached when the sum of the velocities of trunk and leg recordings are maximal (Marshall, 1977).

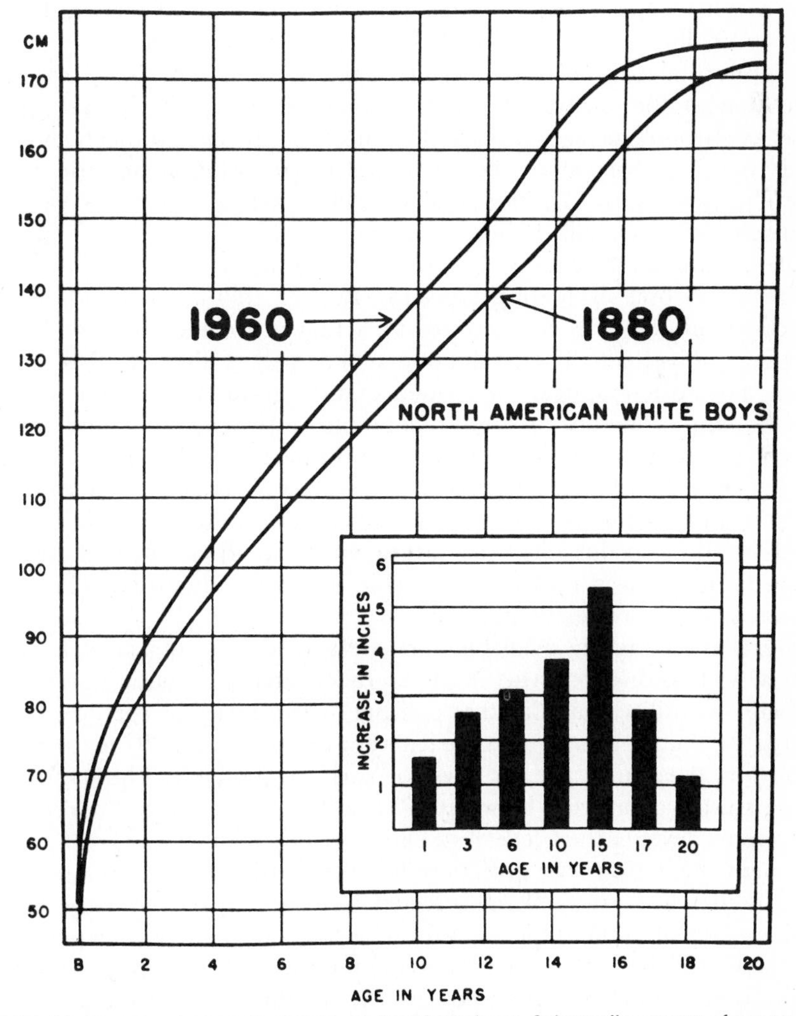

FIGURE 15–11. Secular change in stature of American boys. Schematic curves of mean stature for 1880 and 1960. The bar graph illustrates the differences in increase between various age groups. (From Meredith, H. V.: Change in the stature and body weight of North American boys during the last 80 years. Adv. Child. Dev. Behav., 1:69, 1963.)

SKELETAL MATURATION

As has earlier been discussed, children vary greatly in the speed with which they advance through their postnatal growth changes; i.e., they vary in tempo of growth (Tanner, 1978b). They also show extensive age variation when they complete these changes. Thus, at any given chronologic age children differ in their progress toward adulthood, which is to say that they vary in their physical maturity.

There is no completely satisfactory method of expressing degree of maturity. However, skeletal development is one of the most useful tools in quantifying maturity. Because the centers of ossification of the skeleton undergo a relatively fixed pattern of change in size and shape, these can be identified and described from roentgenograms. Skeletal development fulfills two essential criteria for an adequate index of maturity: (1) that its end result is the same in all normal individuals for whom growth is complete and (2) that it goes through a series of recognizable changes that are common to all children (Marshall, 1977).

It is occasionally suggested that the degree of skeletal maturity (skeletal age) is a reliable predictor of when pubertal changes will occur. However, Marshall (1974b) has demonstrated that such correlations are very low, with one exception, namely, between skeletal age and menarche (see Puberty). He showed that at the various stages of sexual development skeletal age is as variable as chronologic age in both girls and boys. It has also been shown that there is no close correlation between skeletal maturation and peak of the pubertal growth spurt. At the time of PHV skeletal age is as variable as chronologic age (Marshall, 1974b; Houston, 1980).

The assessment of skeletal maturity is based on the recognition of specific maturity indicators on roentgenograms (Roche, 1980). Such indicators are roentgenographically visible bony features that are common to all normally developing individuals. The evaluation makes use of the fact that the bony features most often appear in sequence during the time from infancy to complete maturation. The area most commonly assessed is the hand and wrist, where the carpal bones, the metacarpals, and phalanges, and the distal ends of the radius and the ulna are specific sites used for evaluation. For each of the sites a series of recognizable stages of ossification are determined. The stages reached by the various sites are averaged to yield an overall degree of ossification, which is a measure of a child's skeletal maturity. This may be expressed as bone or skeletal age.

Several methods are available for assessing degree of skeletal maturity. The most well known of these is the so-called atlas technique (Greulich and Pyle, 1959). It utilizes an atlas that consists of two series of roentgenographic standards of the hand and wrist, each representing some 30 maturity levels from birth to the adult stage for boys and girls, respectively. The standards are based on studies conducted between 1931 and 1942 on white children of high socioeconomic level in the Cleveland, Ohio, area. The films which were selected to represent the various age groups reflect the median stage of maturation for each particular age group. Each selected film was assigned a bone age (i.e., skeletal age) equal to the subject's chronologic age at the time of exposure. The use of the Greulich-Pyle atlas is by a matching method. The hand/wrist film of the child to be assessed is compared with the standards. Ossification centers are compared until one standard is found that most closely approximates the child's film. The skeletal age assigned to this standard is the skeletal age of the child. In addition to the standards, the atlas lists maturity indicators of the individual bones and epiphysis with description of their maturational changes. It is thus possible to assign a bone age to each individual bone.

One problem with the atlas method of assessment is that a child's hand/wrist film often does not match any particular standard; i.e., there are differences in the maturation levels of the various ossification centers. Rallison (1986) has suggested that when such discrepancies exist, one should base the evaluation on those epiphyseal centers with the highest statistical "community," i.e., those centers that tend to progress together in sequence. Such ossification centers have been identified and described by Garn and associates (Garn and Rohmann, 1959; Garn et al., 1967c), who argue that these centers have the greatest predictive value in skeletal assessment. The replicability of the atlas method when used by skilled assessors has been studied by Roche et al. (1970, 1974, 1975). Repeated ratings of hand/wrist films by same assessor (intra-assessor error) differed by less than 0.5 year in 95 per cent of the cases, whereas ratings by different assessors (interassessor error) differed by about 0.8 year in 95 per cent of the cases. Because of its highly favorable socioeconomic circumstances, the sample on which the Greulich-Pyle data is based showed vary rapid maturation. It is developmentally advanced even when compared with present-day American children (Tanner et al., 1983b). Another possible drawback has been pointed out by Rallison (1986), who does not consider the atlas method accurate enough for clinical use with subjects under 2 years of age.

An alternative to the atlas system is the Tanner-Whitehouse (TW2) method by Tanner et al. (1975b, 1983b). It is based on assigning numeric scores to the bones of the hand and wrist, depending on their level of maturity. Twenty centers of ossification have been assigned either eight or nine stages, identified by the letters A to I. The bones of the second and fourth digits have been omitted because their maturity levels show high correlation with those of corresponding bones of the other digits. Thus, only redundant information would be provided, were they included (Tanner et al., 1983b; Roche, 1986) (note the reverse argument regarding highly correlated indicators used by Garn in his work on communality indices mentioned in the previous paragraph). Each stage of each bone in the TW2 method has been given a numeric score based on its contribution to the overall maturity. For example, a bone that matures early is given a lower score. The sum of all the individual bone scores constitutes the maturity level of the child. The adult stage is indicated by a score of 100. The TW2 system describes three separate scoring systems. One of these involves only the carpal bones; another includes the radius, ulna, and short bones of the digits (RUS); and a third system is a combination of the two. The most commonly used system is RUS. Once the maturity score has been calculated, it is plotted on a maturity standard, i.e., a chart that gives means and percentiles of maturity scores based on hand/wrist films of children at given ages. Thus, it is possible to determine at what percentile a particular child is with regard to skeletal maturity in comparison with children his or her age. The chart may also be used to convert a child's score into a "bone age." The TW2 method involves less subjectivity and may be considered more reliable than the Greulich-Pyle method (Roche, 1986).

A third method of assessing skeletal maturation from hand/wrist roentgenograms is the mean appearance time (MAT) method (Taranger et al., 1976b, 1976c). The basis for this method is a study of the skeletal development of 212 children from birth to 7 years of age. The key point of this method is the calculation of maturity scores for the various bone stages by using the mean MAT for each stage. Bone stages correspond to those in the TW2 method

(Tanner et al., 1975b). Mean appearance time of the bone stage represents the age when 50 per cent of the population has made the transition into that stage from the preceding one. The MAT method includes the plotting of an individual's "skeletal profile" on a bone stage chart. This provides a graphic representation of how well the various bones conform to the reference mean appearance time. The plot on the chart clearly shows when sequence polymorphism is present, i.e., out of sequence patterns of maturation of the ossification centers. This type of variation in maturation sequencing exists not only between populations but also within populations (Garn et al., 1971).

Of the different methods available for the determination of skeletal maturity, the TW2 system is the one currently recommended by leading investigators for general use (Roche, 1980; Taranger et al., 1987).*

SECULAR CHANGE

In auxology secular change means alteration in growth and maturation patterns that occur within a population over a prolonged period of time, i.e., over decades or more. Such changes involve (1) increased growth velocity, (2) height increase within the various age groups of growing individuals, (3) increased final adult height, and (4) earlier maturation (e.g., decrease in age of menarche in girls).

During the last century children in industrialized countries have been progressively becoming taller and maturing more rapidly. The secular increase in stature is well documented in Europe (Vlastovsky, 1966; Ljung et al., 1974; Taranger et al., 1978; Brundtland et al., 1980; van Wieringen, 1986), in East Asia (Takahashi, 1966; Kimura, 1967; Low et al., 1981; Matsumoto, 1982; Tanner et al., 1982; Lau and Fung, 1987), and in North America (Meredith, 1963, 1976; Stoudt et al., 1965; Damon, 1968; Hammill et al., 1970; Malina, 1972; Leung et al., 1987). The data from the various countries are in good agreement and indicate that the magnitude of the secular increase in stature is greatest during the years of adolescence. The increase is noticeable already in childhood but is progressively getting larger with advancing age up till mid-adolescence (Meredith, 1976). From this point until early adulthood the increase is steadily getting smaller; and measurements of final adult height, although they still show secular increase, are not as dramatically different as are measurements during adolescence. Meredith's (1963) review of North American data on Caucasian boys shows that there has been a height increase of approximately 1 cm/decade, if one counts from the latter part of the last century. Ten-year-old boys have increased in height by about 1.25 cm/decade, and 15-year-olds have increased by 1.65 cm/decade. However, 17-year-old boys have experienced a secular increase of only 0.84 cm/decade; and in young adults, at age 20, the increase has not been more than 0.34 cm/decade. These data are graphically summarized in Figure 15–11, which shows (1) two distance curves for height, based on data collected 80 years apart, and (2) a histogram illustrating the magnitude of the secular height increase within the various age groups.

The secular change in tempo of growth, i.e., the rate of maturation was well documented by Ljung et al. (1974). They compared growth data on

---

*In 1988 a new method for skeletal assessment was published (Roche, et al., 1988). However, it was not available to me for review until after the completion of the manuscript.

Swedish school children from 1883, 1938, and 1968 and demonstrated a progressively lower age at PHV. For girls PHV occurred at age 12.8 in 1883, at age 12.2 in 1938, and at age 11.6 in 1968. For boys the ages were 15.2 years, 14.3 years, and 14.0 years, respectively. Another significant effect of progressively earlier maturation is the continuous decrease in the age of menarche in girls. European data show that from the middle of the last century there has been a steady decline from about 17 years of age to about 13 (Tanner, 1978b). From data collected during this century it is evident that similar trends exist in the United States (Wyshak, 1983). However, several studies indicate that the decrease in age is leveling off or has actually stopped in several places, like Olso and London (Brundtland and Walløe, 1973; Tanner, 1973).

When secular increase in stature is studied, it must be considered together with secular change in maturation. The greater part of the height increase in children can be explained by earlier maturation, i.e., a shorter growth period. A present-day 10-year-old boy has completed a greater percentage of his growth, compared with a 10-year-old a century ago. However, all stature increase cannot be explained this way. There is also a real secular increase in adult height, although not as pronounced as the one observed in children.

After a comparison of stature between several generations of Harvard students, Damon (1968) was one of the first to suggest that the secular trend might be stopping. He found a height increase of 2.6 cm between two generations with mean birth dates of 1858 and 1888, respectively. There was an increase of 1.1 cm between cohorts with mean birth dates of 1888 and 1918 but no difference between cohorts with mean birth dates of 1918 and 1941. In contrast to these data, Damon (1974) found a slight but statistically significant increase in stature between mothers (median birth date, 1920) and daughters (median birth date, 1948) who were all graduates from the same two private colleges. It is currently a debatable issue whether the secular trend is still continuing. This question is very much related to what area of the world and what socioeconomic groups within populations are being studied. Data, reviewed by Roche (1979), are accumulating indicating that it is slowing down or has indeed stopped at least in many of the higher socioeconomic groups in several countries of the developed world. The situation for the population at large in these countries may be somewhat different. Although the increase is quite small, adults in the United States measured for NHANES I (1971 to 1974) were taller than those in NHANES II (1976 to 1980) (Abraham et al., 1979). The increase was slightly over 1.75 cm, the biggest difference in the age group 18 to 24 years, for which the increase was 2.5 cm. Damon's statement of 1974 regarding the question of an ongoing secular trend may still be valid today: "Trends toward larger body size and earlier menarche, documented for well over a century and still continuing in the general population of the United States, may be coming to an end among the well-to-do, for whom body size has nearly stabilized, and age at menarche may have stabilized" (Damon, 1974).

No documented cause has been established for the secular trend, but several factors are considered to be of significance. Malina (1979), in an extensive review of the subject, discussed environmental as well as genetic factors that have probably formed the underlying basis for secular changes. It is obvious that the most frequently mentioned environmental factor is improved nutrition. However, dietary intake interacts with other environmental factors, and that nutrition is part of a complex of factors, all of which are responsible for improved environmental circumstances in general, is currently the preferred view (Malina, 1979; van Wieringen, 1986). This group of factors provides for better socioeconomic conditions and improved health status, as reflected in reduced morbidity and mortality rates. Tanner (1968) has suggested reduction of family size as a significant environmental factor. Small family size is likely to result in improved living conditions, increased food intake, and a decrease in the frequency of disease. All of these effects will have a positive influence on general health.

Outbreeding seems to be the most commonly mentioned genetic factor possibly associated with the secular trend (Tanner, 1968; Malina, 1979). In the last century there has been a dramatic increase in the mobility of people within, as well as between, populations, which has caused a gradual relative decline in the reproduction rate within breeding isolates and a simultaneous increase in the frequency of outbreeding (i.e., parents having distant geographic origins, rather than coming from the same isolate). It has been proposed that this demographic change had an effect on stature by the genetic phenomenon called heterosis (Tanner, 1978b). There are data available that indicate that outbreeding may indeed result in increased stature in man (Hulse, 1957).

Tanner (1968) has stated that it is conceivable that environmental factors are primarily responsible for those secular changes in children that reflect earlier maturation. The trend toward increased adult height, on the other hand, Tanner suggests, may be due to a combination of genetic and environmental factors. Malina (1979) thinks that outbreeding may be a factor, but if it is, the effect is minimal. Others (van Wieringen, 1986), however, have rejected heterosis and other genetic factors as possible causes. What most authors do seem to agree upon is that the dominant factors are conditions related to reduced infant and child morbidity and mortality, i.e., better socioeconomic conditions, including improved nutrition, better living standards, and reduced family size. These are all interrelated circumstances and, taken together, may well be behind the secular trend.

CHAPTER SIXTEEN

Adult Facial Growth

Rolf G. Behrents, D.D.S., M.S., Ph.D.

The changing face of the human clearly reflects time. From the massive alterations initiated in prenatal development, through the growth years of childhood and adolescence, into the mature period of adulthood, and then continuing on into old age, the altering contours characterizing the aging process of human facial form are consistently demonstrated time and time again. Many studies have been conducted to describe and analyze the external changes, and more studies have been conducted to characterize the underlying osseous changes that provide the template for the surface alterations.

Fascination with this continuing example of biologic alteration has traditionally focused on the periods of life when the most rapid and obvious changes were taking place. This is practical and logical, for the information gained is deemed important in understanding the growth process and effecting treatment.

Based on many years of study most of the contemporary textbooks describing the growth process suggest that postnatal growth peaks in mid adolescence and slows dramatically in late adolescence, and that no growth occurs in adulthood. Common dates of cessation revolve around 14 years of age in females and 16 in males. This is considered the norm, but some variation is acknowledged, such that an occasional "late maturer" is a recognized phenomenon.

As a result of these deeply rooted notions about growth cessation, the adult craniofacial skeleton is viewed as a stable, static entity in terms of size and shape change; and later in life, degenerative changes are often described in characterizing old age. Facial wrinkles, sagging tissues, resorbed alveolar bone, loss of teeth, and decreased vertical dimensions of the lower third of the face are often used to caricature old age.

Conclusions drawn from this information, while seemingly logical and satisfactory to the casual observer, however, are not founded on actual adult data but, rather, are based on extrapolations of the adolescent growth pattern. Simply stated, growth is assumed to cease in adulthood because it slows in adolescence. Thus, our common knowledge of growth termination is based on incomplete information. And when the literature is scrutinized, little evidence is found to verify that progressive facial changes noted in adulthood have no underlying skeletal basis. Furthermore, and more important, there is no reason to suspect that biologic alteration is impossible nor even improbable and that growth ceases in the adult years. The purpose of this chapter is to convince the reader that the concept of a termination of craniofacial growth in adolescence is an erroneous belief.

BIOLOGIC PROCESSES IN ADULTHOOD

Contrary to what might be thought in some disciplines, biologic processes are not dormant during adulthood. Much literature is available suggesting that biologic activity occurs or, rather, continues into and during adulthood and that these processes have a direct bearing on alteration of the craniofacial skeleton. Although all tissues are subject to change during adulthood, the present discussion will focus on some of the morphologic changes that occur in bone. A more complete picture of the manifest changes that occur during aging at the cell, tissue, and organ levels may be seen in the works of Andrew (1971), Finch and Hayflick (1977), Sinclair (1978), Kohn (1978), and others.

CHANGES IN BONE

Although in the earliest concepts bone was regarded as static in nature during adulthood, it is now believed that various dynamic changes occur throughout life. Bone, with regard to age, continues to alter qualitatively and quantitatively. In addition to changes in cellular morphologic features, in old age there is a decrease in bone water content, an increase in apatite crystal size, an increase in the volume but a decrease in the weight of the skeleton, a decrease in bone mass, and a decrease in physical density.

With regard to bone turnover, bone remodeling appears to be slower, but not absent, with age. Usually the literature describes a high bone turnover rate in childhood, deposition of new bone being more prevalent than resorption. In young adult bone, the turnover rate is slower, and markedly less bone formation and deposition occurs, but nonetheless both processes still continue. In old bone the delicate balance between deposition and resorption is found to be disturbed, with a gradual increase in resorptive activity, particularly on endosteal bone surfaces. However, little evidence of substantial bone formation changes are evident (i.e., little evidence of a reduction in bone formation with age). This accounts for the more-than-occasional report of active bone formation (even involving the facial bones) and even *increased bone formation* during and after middle age in spite of the fact that more resorptive activity is also apparently taking place.

Although this appears to be the general human condition, considerable individual variation also appears to be the rule. Depending on the age and the bone studied, new lamellar bone may be found in some specimens well into the ninth decade of life, whereas in other specimens such activity is not easily discernible beyond adolescence. In terms of localization, Enlow (1982b) points out that there may be some areas of primary (nonhaversian) periosteal or endosteal bone encountered in old bone, because these deposits may represent remodeling alterations in response to morphologic skeletal changes associated with changes in weight, posture, loss of teeth, and other factors.

Thus, it appears logical to assume that some degree of bone turnover continues well into adulthood and, further, that the phenomenon of appositional growth and functional adaptation may well continue into senility. Such situations, if general for man, would be expected to result in alterations of bones throughout aging that are measurable with common analytic tools. This is indeed the case.

In addition to the continuing replacement of cancellous and cortical bone through microscopic views of resorption and deposition, gross changes are noted with age. Enlow (1982b) points to a change (decrease) in the width of the cortex in advanced age. This is noted in the literature with respect to many of the tubular bones of the appendicular skeleton such as the rib, metacarpal, humerus, and femur. Some trabecular bone loss is also noted, especially in women. As a result, there is an increase in the various canals and cancellous spaces of the aged and cortical "thinning," a condition often termed osteoporosis. The cortical size decrease also has been implicated in accounting for the greater incidence of "spontaneous" fractures noted in the aged. However, while the width of the cortex is decreasing, the external dimensions of the bone are increasing.

This aspect of bone development or bone redistribution involving gross bone changes on both the periosteal and endosteal surfaces has been extensively studied by Garn, his associates, and others. Their studies have demonstrated that up through the fourth decade, continuing bone gain in both sexes is accomplished by both subperiosteal and endosteal deposition, resulting in an increase in cortical thickness and a maximum bone mass between 30 and 40 years of age. Beyond that age, subperiosteal apposition continues throughout life, resulting in increased external dimensions (more for women) and, because of endosteal resorption, an expansion of the medullary cavity (Fig. 16–1). These two effects result in a decrease in cortical thickness with advanced age. This phenonenon is apparently a general human condition and occurs regardless of genetic, socioeconomic, nutritional, or environmental factors.

Because such bone "expansion" has been shown to occur in a variety of situations, including the bones of the hand and arm, leg and foot, and vertebral column, it might be expected that complex physical assessments that collectively evaluate many bones would show increases during adulthood even though the changes noted per bone were small. This has been found to be the case in several important situations. For example, reports of stature assessment are often quite varied in their findings, with growth cessation (maximum height) reported to occur at anywhere from 17 to 45 years of age (which is long after epiphyseal union). Given that height is a gross measure of all the tissue alterations that occur from the top of the head down to the bottoms of the feet, and given that increases in dimensions have been reported for the bones in between, it is not surprising that a large variation in age of attained maximum height has been reported. With the bones growing and with (some) tissues being variously compressed, age of maximum height attainment becomes more a varied individual attainment than a specific human event.

Another example of complex osseous change involves the bones of the hand. It has been shown that during adulthood some of the bones gain in dimension while others show a diminution in size. The phalanges gain in length while the metacarpals lose. However, on the whole, the overall dimensions of

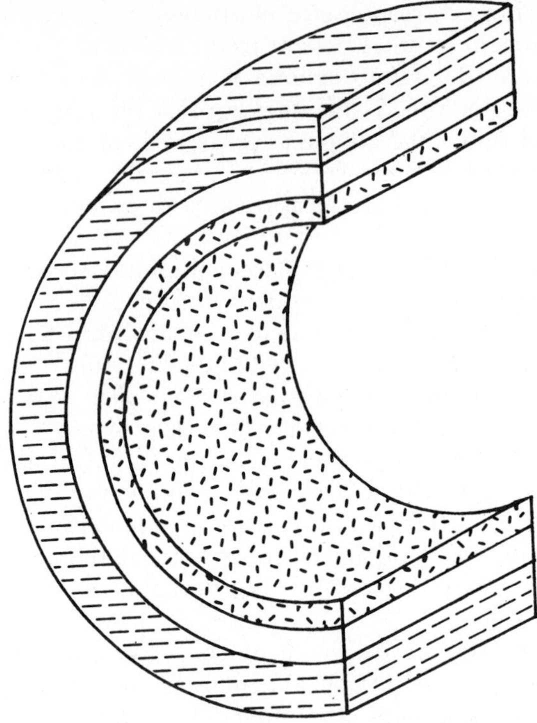

FIGURE 16–1. Schematic drawing of a tubular bone. Bone is deposited subperiosteally (parallel dashes) and endosteally (random dashes) until 30 to 40 years of age. Thereafter, bone is deposited subperiosteally and resorbed endosteally. External bone diameters increase during this process.

each finger (ray) remain approximately the same, the increases canceled by the negative changes. This differential situation, once unrecognized, is another example of the complex osseous changes that may occur during adulthood.

ADULT CRANIOFACIAL GROWTH

More to the point, alterations of the craniofacial skeleton have been reported during adulthood with consistency. As might be predicted, studies conducted before the routine use of roentgenographic cephalometric techniques were only successful in *suggesting* that alteration of the craniofacial skeleton continued into adulthood. Early studies often used cross-sectional material—measurements were performed on dry skulls by the use of craniometric techniques or the external surfaces of living individuals with various anthropometric assessments—or depended on casual observation. As a result, group means usually pointed inconclusively to differences between younger and older groups of individuals. However, several important observations were made. For example, results of early studies suggested that during adulthood the cranium became thicker and the depth, width, and height of the face became greater by several millimeters. Similar alterations of soft tissue structures were also suggested: nose height and breadth increased significantly into later adulthood, as did ear and lip height. On the other hand, contrary views were prevalent questioning whether changes occurred in the craniofacial skeleton at all. This conflict was understandable because study designs and uncontrolled variables (such as the loss of teeth) easily confounded findings and fueled differences of opinion. Furthermore, cross-sectional studies conducted more recently have not overcome past difficulties in interpretation.

With the advent of the cephalometer, precisely controlled longitudinal studies became possible on living, growing individuals. Several decades of investigation using this approach have shown quite conclusively that the craniofacial skeleton continues to "grow" during adulthood. In general, the suggestions pointed to using the cross-sectional approach were corroborated, extended, and detailed with the longitudinal approach and the cephalometric technique. Credit should be given to Buchi, Thompson and Kendrick, Carlsson and Persson, Tallgren, Israel, Forsberg, Susanne, Sarnas and Solow, and Lewis and Roche for their contributions in documenting adult craniofacial change.

A major contribution was provided by Israel in a series of studies. With lateral skull x-rays and cephalograms of adult men and women, his findings were clear. All of the craniofacial measures used demonstrated size increases. Dimensional increases were noted for cranial thickness, upper facial height, sinus size, and other aspects of the craniofacial skeleton. He concluded that his results infer a virtual symmetrical magnification enlargement process that generally amounts to approximately a 4 to 5 per cent increase in size. Regional assessments demonstrated differential change. In this regard, he suggested that the upper face increased by 6 per cent, the frontal sinuses by 9 to 14 per cent, and the mandible by 5 to 7 per cent. Although criticized on technical grounds, the works of Israel clearly demonstrate that an enlargement of the craniofacial skeleton occurs; the amount of change was not as clear.

The investigators who followed Israel have to a great extent validated his findings. However, because of the difficulties attendant to the conduct of longitudinal investigations, many adult studies conducted have suffered in application because of difficulties involving the short age spans studied, abnormal dentition, small samples, and technical limitations. For example, numerous papers describing craniofacial change in the adult are based on longitudinal observations made on fewer than 100 individuals.

A recent study by Behrents (1985), however, appears to have largely overcome the previous limitations and perhaps provides new insight into the nature of the specific morphologic changes within the craniofacial complex associated with adulthood. The sample was drawn from those individuals who had previously participated in the Bolton Study in Cleveland, Ohio. The Bolton Study was an extensive longitudinal investigation initiated in the 1930s and 1940s and was conducted to document and describe craniofacial changes that occurred in healthy, normal children and adolescents. The most unusual feature of the Bolton Study was a new technique developed by the director of the study, Dr. B. Holly Broadbent, Sr.: roentgenographic cephalometrics. Fortunately, the original Bolton Study continued for several decades, and many of the participants were studied continuously until they reached young adulthood. Behrents then recalled, in the early 1980s, 113 of the original participants for new data collection. Together with existing records in the Bolton Study collection, 163 cases spanning the ages 17 to 83 years were used for determining the nature and extent of the craniofacial changes that occurred between young adulthood and later adulthood. Upon recall, the participants were in good health, and most had a full complement of teeth.

The facial changes noted by direct observation during the recall examinations revealed changes consistent with expectations. Typically, faces appeared somewhat larger, with increased size noted especially for the nose and the ears. Size increases were apparent for ear length, ear breadth, and thickening of the lobe. The nose appeared to be broader and longer and had a more downturned tip. The lower third of the face showed evidence of increased dimensions, although lip prominence had lessened (Fig. 16–2).

Consistent with the visual change, radiographic examinations revealed that continuing bone alteration was the norm. This affected not only external osseous architecture but also internal architecture. Sinus size and shape alterations were a consistent finding when the frontal, sphenoidal, and maxillary sinus cavities were assessed by comparing x-rays from young adulthood with those collected during older adulthood (Fig. 16–3). Obliteration of the radiographic images of the sutures, thinning of the parietal bones, and calcification of the falx cerebri were also seen.

Extensive cephalometric data, adjusted for magnification, revealed continuing growth of the craniofacial complex at all age levels. For both male and female subjects, most of the distance measures and most of the angular values demonstrated significant change (Figs. 16–4 and 16–5). Presence of growth during adulthood was the norm; it was not unusual for 95 per cent of the sample to show a dimensional increase for a particular measure during adulthood. It is thus clear that the designation of "late maturer" is really a misnomer, for continuing maturation during adulthood is the norm.

Overall, both size and differential shape changes were noted for various areas of the face (Figs. 16–6 to 16–9). A 2 to 10 per cent increase was the rule, the bones of the cranial base altering least, the facial bones a moderate amount, the frontal sinus more, and the soft tissues most. The changes seen were similar to typical adolescent alterations but of a lesser magnitude and rate. The typical direction of growth varied to a certain extent, depending on the period of adulthood studied. In young adulthood the direction of growth was specific to

Text continued on page 433

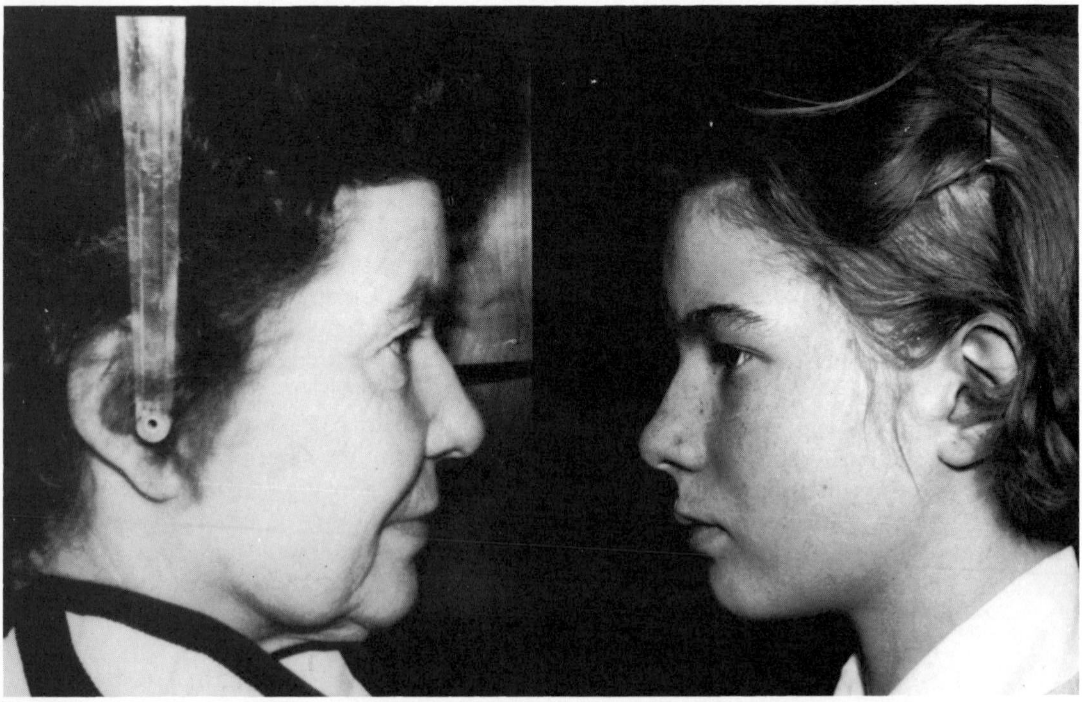

FIGURE 16–2. Facial changes in a female from age 14 to age 60 (From Behrents, R.G.: *Growth in the Aging Craniofacial Skeleton.* Monograph 17, Craniofacial Growth Series. Ann Arbor, University of Michigan, Center for Human Growth and Development, 1985.)

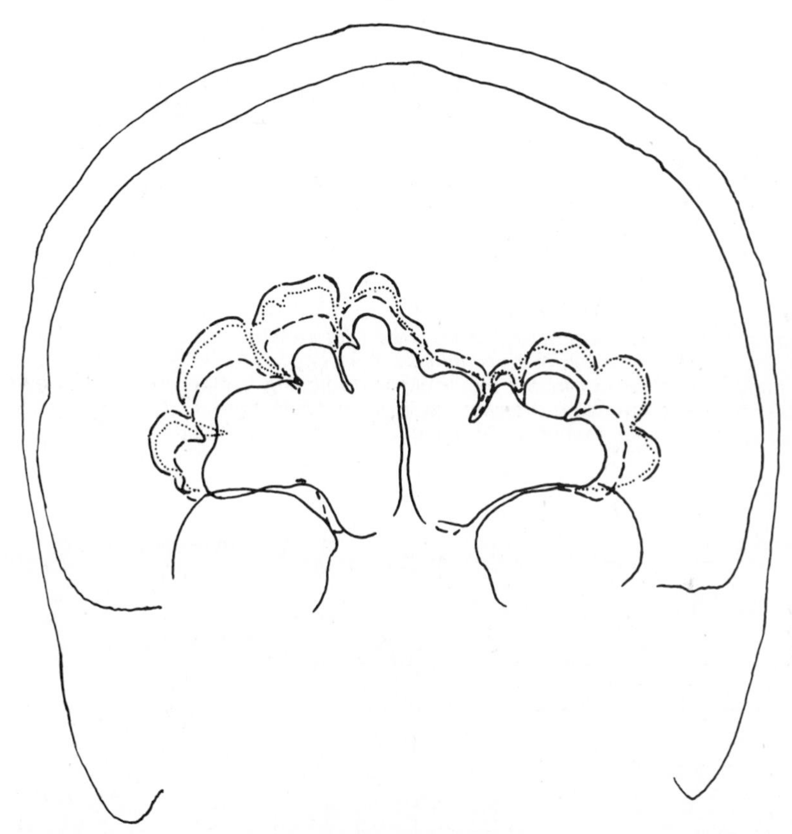

FIGURE 16–3. An example of enlargement of the frontal sinus during adulthood. The solid line indicates the sinus outline at age 17, and the dashed, dotted, and dash-dot lines indicate the change in the sinus outline occurring over progressively older ages (21, 36, and 52 years of age). (From Behrents, R.G.: *Growth in the Aging Craniofacial Skeleton.* Monograph 17, Craniofacial Growth Series. Ann Arbor, University of Michigan, Center for Human Growth and Development, 1985.)

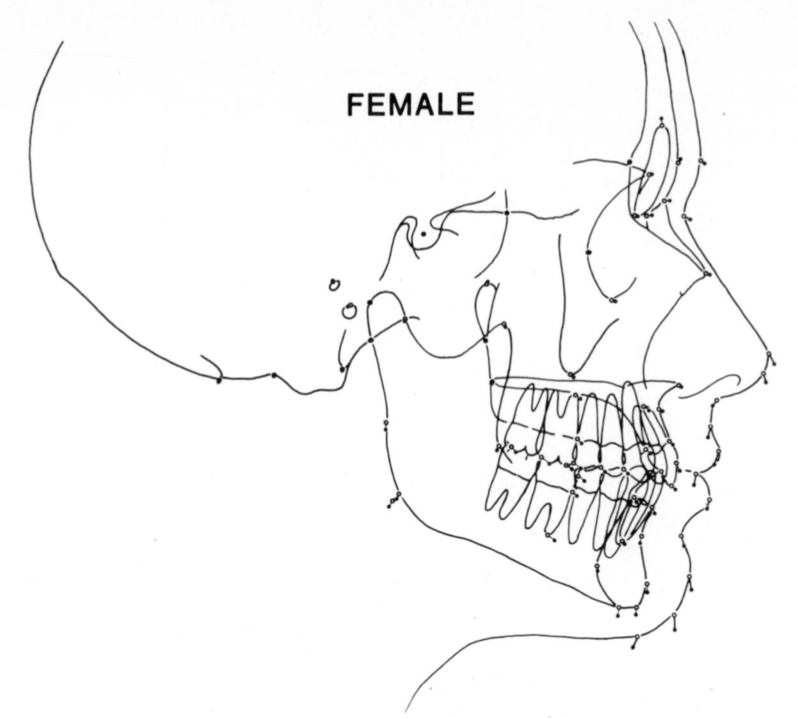

FIGURE 16–4. Mean change in the female during adulthood for skeletal, dental, and soft tissue landmarks. The open circles indicate the mean location of the landmarks in young adulthood while the solid circles denote the mean landmark locations in older adulthood. The background tracing is based on young adulthood anatomy. (From Behrents, R.G.: *Growth in the Aging Craniofacial Skeleton.* Monograph 17, Craniofacial Growth Series. Ann Arbor, University of Michigan, Center for Human Growth and Development, 1985.)

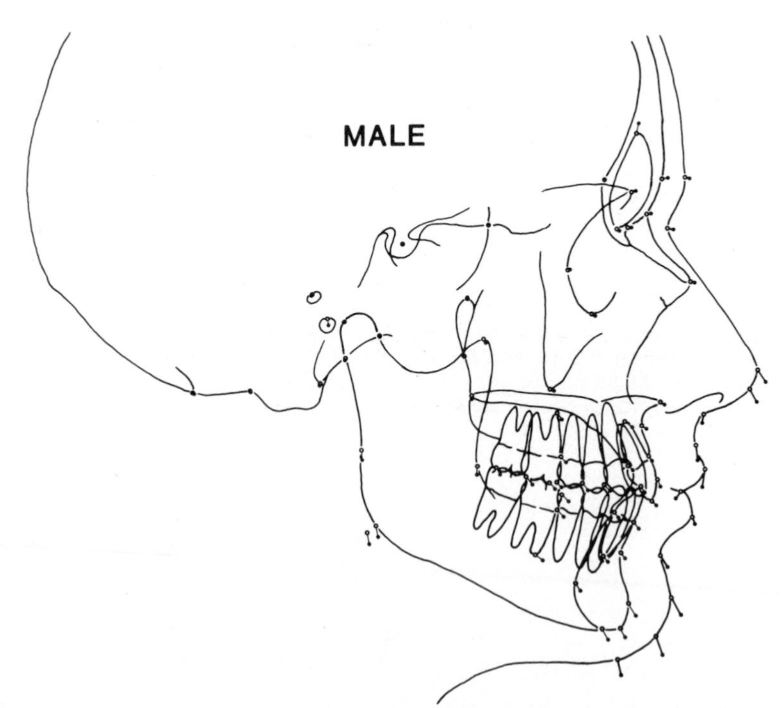

FIGURE 16–5. Mean landmark change in the male during adulthood. Constructed as in Figure 16–4, the mean amount and direction of change of each landmark can be visualized. (From Behrents, R.G.: *Growth in the Aging Craniofacial Skeleton.* Monograph 17, Craniofacial Growth Series. Ann Arbor, University of Michigan, Center for Human Growth and Development, 1985.)

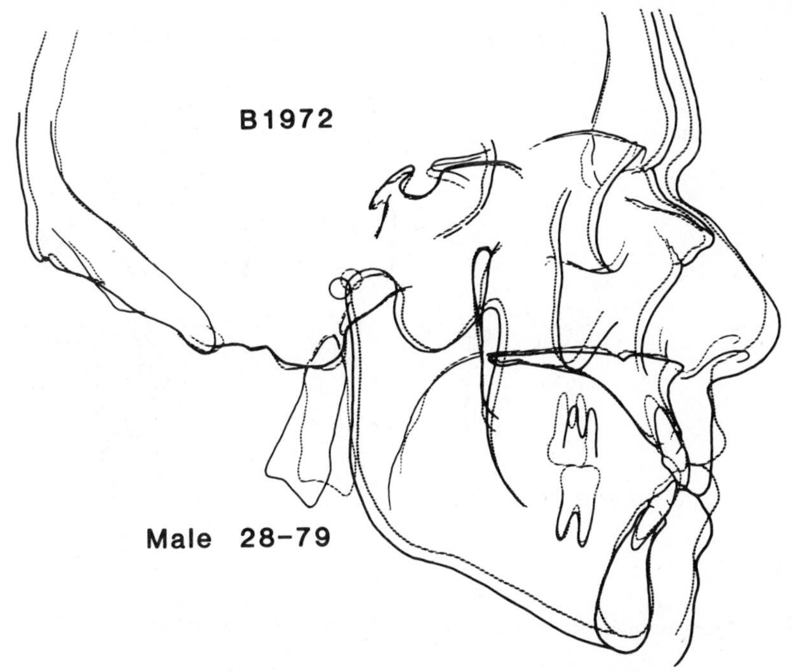

FIGURE 16–6. Superimposed tracing based on registration on sella and orientation along sella–nasion. Male at age 28 (dashed line) and age 79 (solid line).

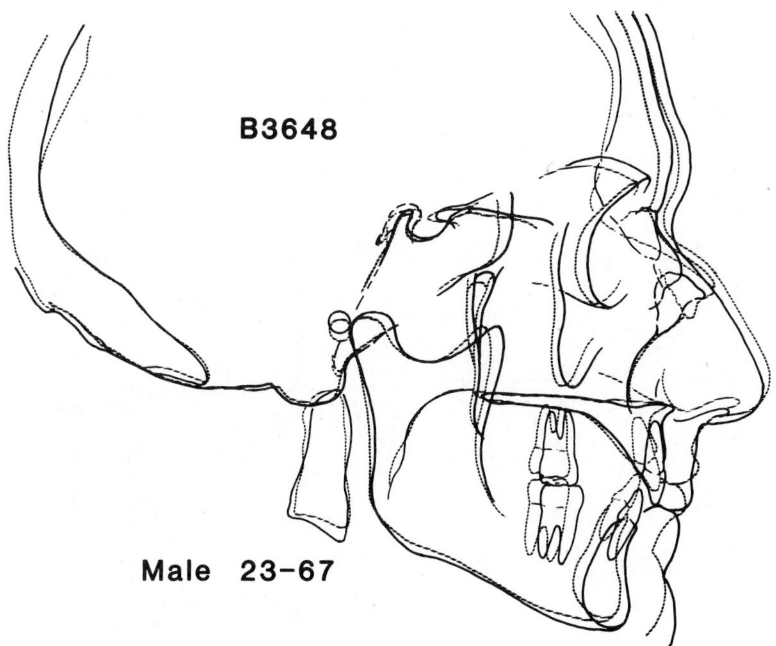

FIGURE 16–7. Superimposition of a male at ages 23 and 67. This individual had also undergone a rhinoplasty in the intervening years.

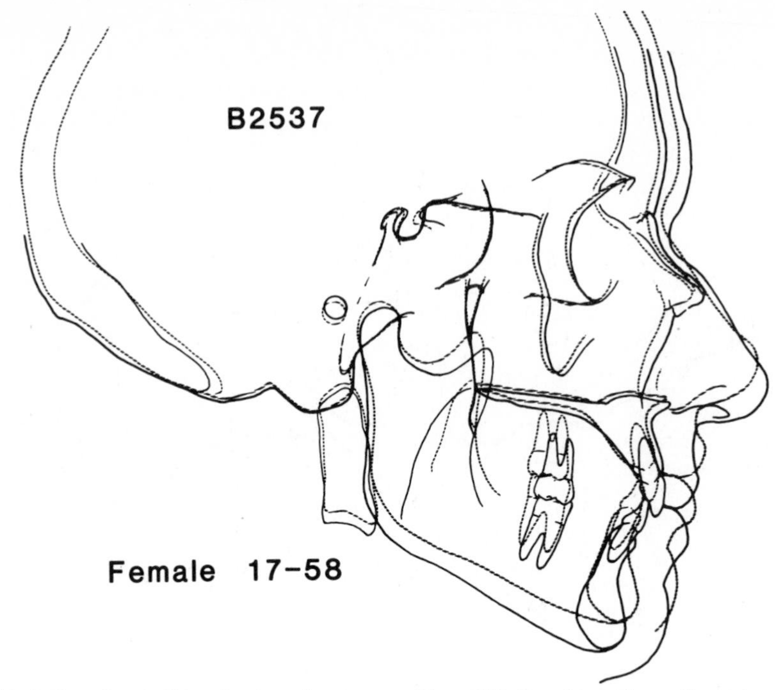

FIGURE 16–8. Superimposition of a female at ages 17 and 58. Considerable vertical change is seen.

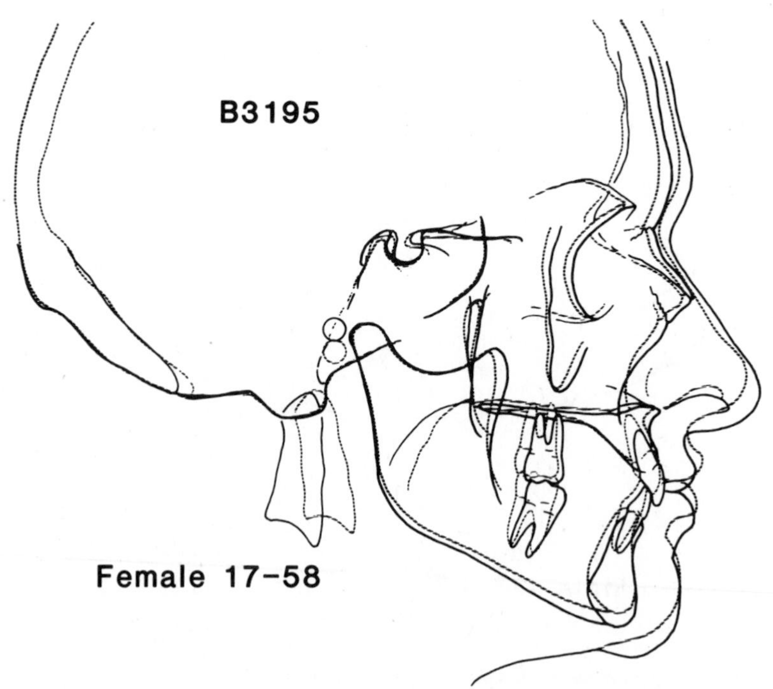

FIGURE 16–9. Superimposition of a female at ages 17 and 58. (From Behrents, R.G.: The consequences of adult craniofacial growth. In: *Orthodontics in an Aging Society*. Ed. By D.S. Carlson. Monograph 22, Craniofacial Growth Series. Ann Arbor, University of Michigan, Center for Human Growth and Development, 1989.)

an individual's growth pattern. In other words, a "horizontal grower" grew horizontally in young adulthood, and a "vertical grower" grew vertically. Later in adulthood, however, vertical dimensional change appeared to predominate. Soft tissue changes were more dramatic than skeletal change but remained related and patterned with the skeletal alteration.

Definite differences in the nature and extent of some changes were found between males and females. Typically, females were smaller than males at comparable ages in young adulthood, and during adulthood females grew less; males were generally 5 to 9 per cent larger. Although the anterior length of the face was increasing in both sexes, vertical change was more characteristic of the female. It appears that this lengthening of the face is occurring by two different means, depending on sex. A forward rotation of the mandible was seen in the male, and a converse rotation in females was common. Regardless of the rotation, vertical dimensions of the face increased with time. Beyond this, dimorphic features characterizing the sexes were especially prominent in the upper face. Differences were noted in the orbital region, the orbit being more upright in the female; the area of the glabella, the male glabellar area being more robust; and the nose, the male nose being larger and longer.

The data further suggest that females undergo a generalized growth deceleration in their teens, but with a reacceleration later (Fig. 16–10). This cyclic process may be related to the fact that most of the females in their 20s and 30s were bearing children. It has been demonstrated that pregnancy does

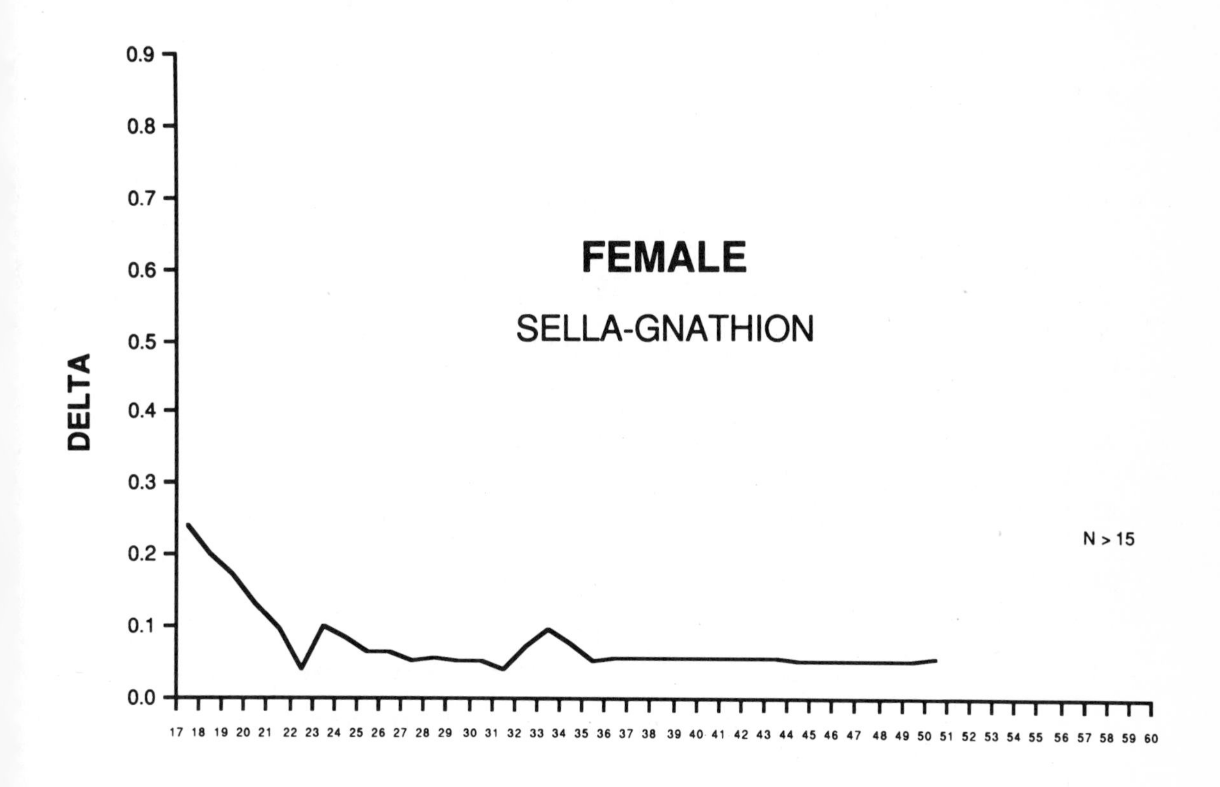

FIGURE 16–10. Growth curve for the distance measure sella–gnathion in the female based on partitioned growth increments. (From Behrents, R.G.: The consequences of adult craniofacial growth. In: *Orthodontics in an Aging Society.* Ed. by D.S. Carlson. Monograph 22, Craniofacial Growth Series. Ann Arbor, University of Michigan, Center for Human Growth and Development, 1989.)

affect bone turnover. During pregnancy periosteal bone formation rates are elevated; and as a result the cross-sectional area of the medullary cavity and the endosteal and periosteal perimeters increase. It is thus quite conceivable that the external dimensions of the facial bones may be affected during the process. Males, on the other hand, show a very regular gradually decelerating pattern of growth over the adult years (Fig. 16–11).

Regional considerations demonstrated that little significant change occurs in the cranial base region except at its extreme extensions. The occipital condylar area tends to be displaced downward and forward with time; and as might be expected, the area around nasion tends to develop in an anterior direction. The endocranial surface of the frontal bone appears to be fairly stable; however, the ectocranial surface of the outer table continues to develop significantly through time (9 per cent enlargement). As might be expected, the frontal sinus increases substantially in size. Positionally, the upper and lower extremes of the sinus move forward and apart during adulthood. As a result of all these changes, the usual measurements designed to measure the length of the cranial base indicate that a small but significant increase occurs over time. Furthermore, the area including the external aspects of porion tended to shift downward relative to the remainder of the cranial base. This might well affect measures using Frankfort horizontal reference schemes. Anatomic porion, on the other hand, remains quite stable with regard to other cranial base structures.

In the midface, differing amounts of activity occur when the anterior and posterior aspects are studied. In the posterior region activity is slight, with only

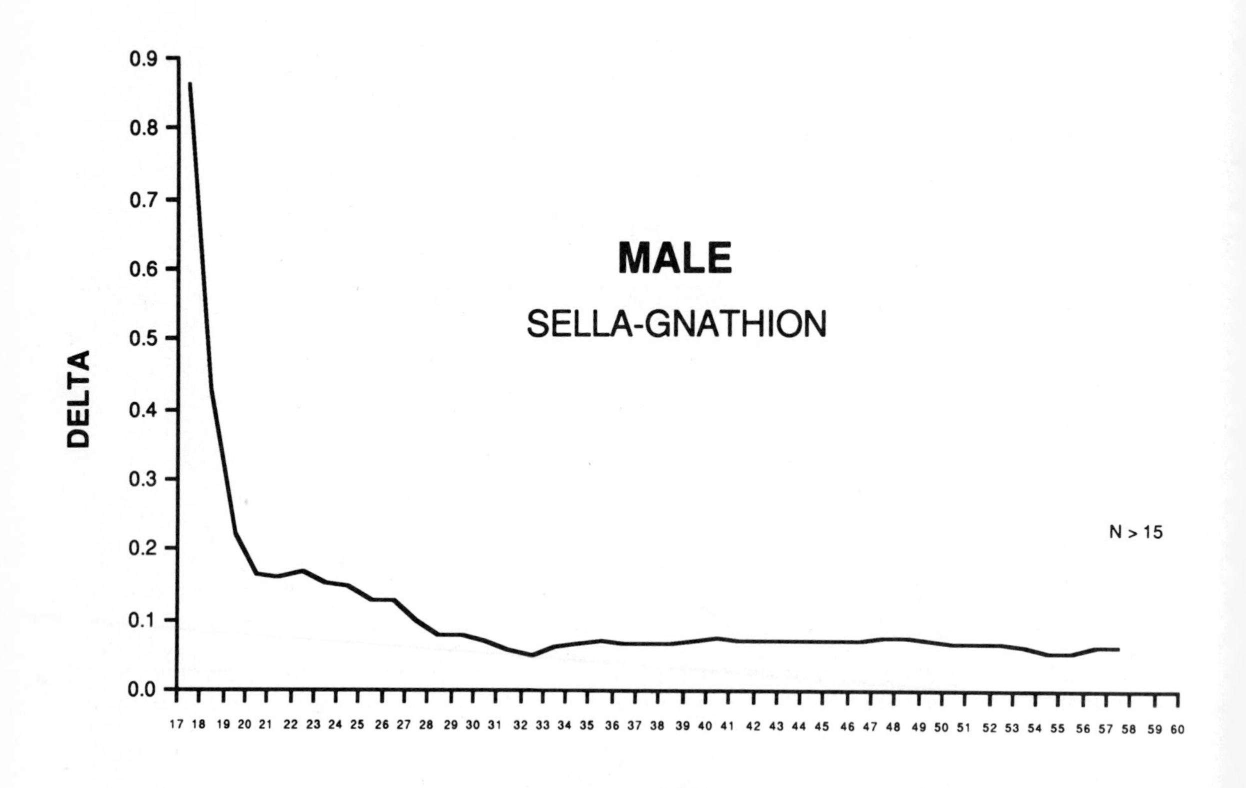

FIGURE 16–11. Growth curve for sella-gnathion in the male. (From Behrents, R.G.: The consequences of adult craniofacial growth. In: *Orthodontics in an Aging Society.* Ed. by D.S. Carlson. Monograph 22, Craniofacial Growth Series. Ann Arbor, University of Michigan, Center for Human Growth and Development, 1989.)

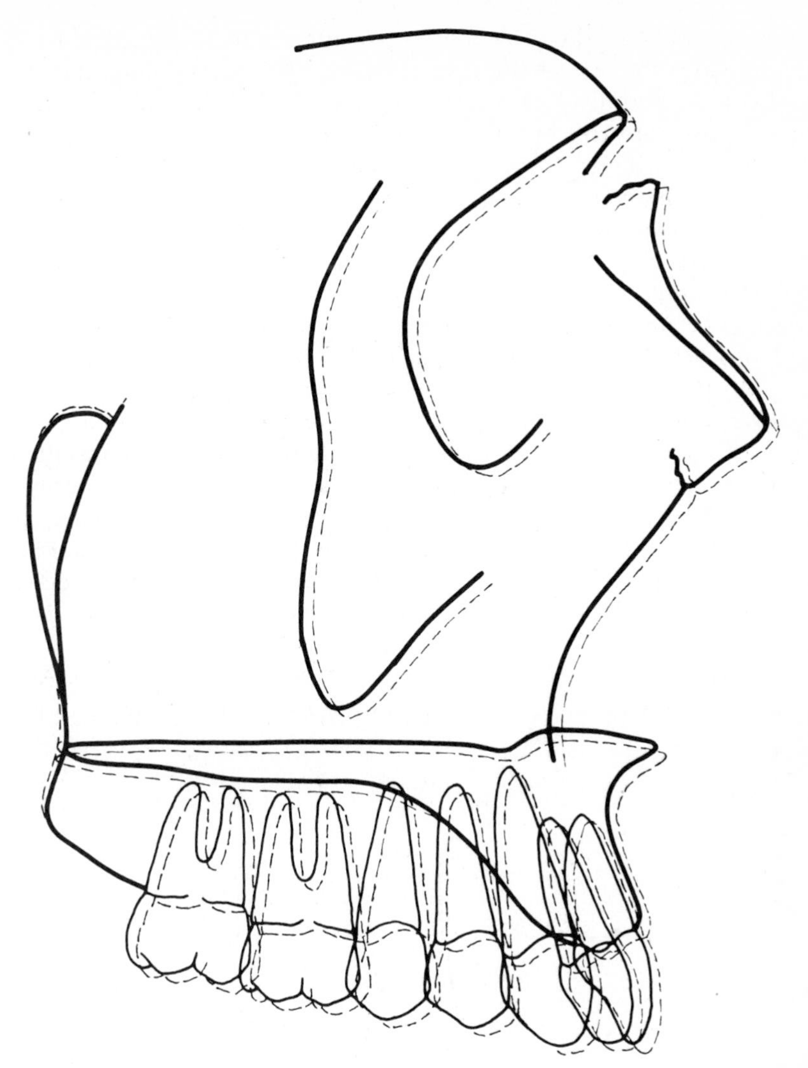

FIGURE 16–12. Schematic diagram indicating the generalized changes occurring in the midface (sella/sella–nasion orientation). Slight differences are noted between males and females but are not represented here. (From Behrents, R.G.: *Growth in the Aging Craniofacial Skeleton.* Monograph 17, Craniofacial Growth Series. Ann Arbor, University of Michigan, Center for Human Growth and Development, 1985.)

a few subtle changes noted; little substantive change is seen in the region of the pterygomaxillary fissure. However, definite change occurs in the palatal region. Apparently the palatal structures continue to relocate posteriorly and inferiorly. Some differences were noted with regard to sexual dimorphism; males showed a greater and more inferior development of this region than females. This latter change is consistent with a general observation characteristic of the male where a consistent posterior counterclockwise development and rotation of structures is seen (Fig. 16–12).

The anterior aspects of the midface showed consistent change in both sexes. Most development was expressed anteriorly, even though vertical displacement was also apparent. As mentioned above, the nasal region continued to develop anteriorly. Thus, the position of nasion and the tip of the nasal bone were located in an anterior position with time. In the female there was a tendency for the tip of the nasal bone to elevate. Likewise, the superior, lateral, and inferior aspects of the orbit moved consistently forward with time. Such a change increases the size of the orbital cavity. Such remodeling and displacement, of course, also involves the zygomatic processes. The anterior aspects of the palate move forward, but also substantially downward, during adulthood. The alveolus also increases in size, vertically. Apparently in the absence of periodontal disease continued development of the alveolus is possible.

It might be expected that the changes seen in the cranial base and midface have an additive effect on the mandible such that considerable change in size and position should occur (Figs. 16–13 and 16–14). This is the case. The chin continues to be displaced in an anterior direction during all ages, but much more of this activity is seen in the male. In the female the mandible comes forward, but not to the extent seen in the upper and lower anterior regions of the midface. Thus, there is a tendency for the female mandible to appear more retruded with age, even though the chin is coming forward. Significant vertical translation of the chin also occurs in both sexes. Thus, the anterior facial dimensions consistently increase. The mean increase for total face height (nasion to menton) was 2.8 mm during adulthood, but in individual cases change on the order of 10 mm (1 cm!) did occur. Compared with changes seen in the upper face, the amount of lower facial change was double (0.9 mm, compared with 1.9 mm).

Of separate, but related interest is the rotation of the mandible. Although this rotation is not extensive, it is apparent the mandible continues to rotate in

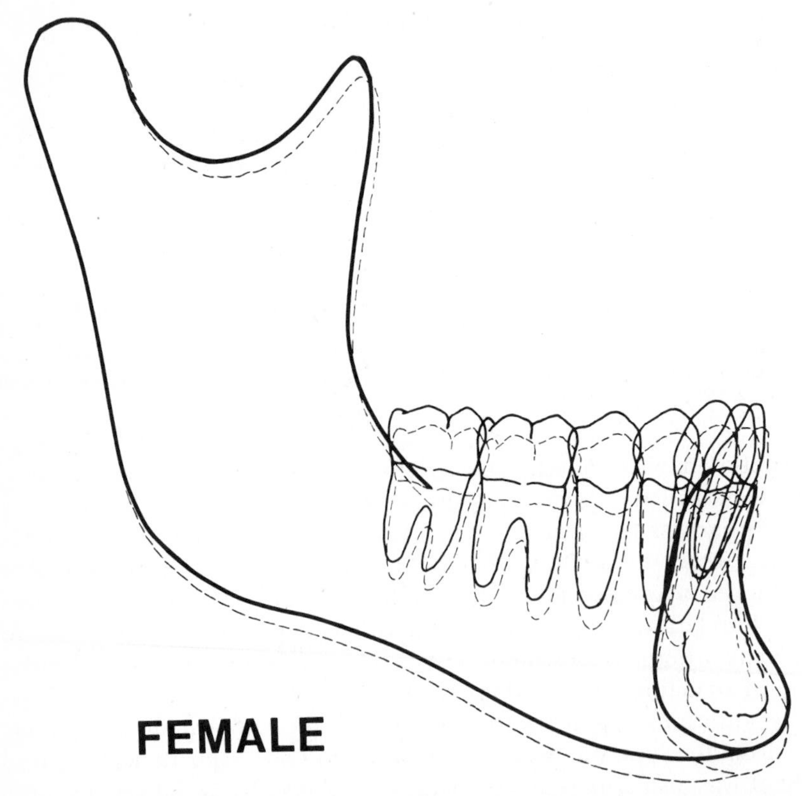

FIGURE 16–13. Schematic diagram of the change noted for the female mandible in relation to a sella/sella–nasion orientation. Vertical and clockwise rotations are seen. (From Behrents, R.G.: *Growth in the Aging Craniofacial Skeleton*. Monograph 17, Craniofacial Growth Series. Ann Arbor, University of Michigan, Center for Human Growth and Development, 1985.)

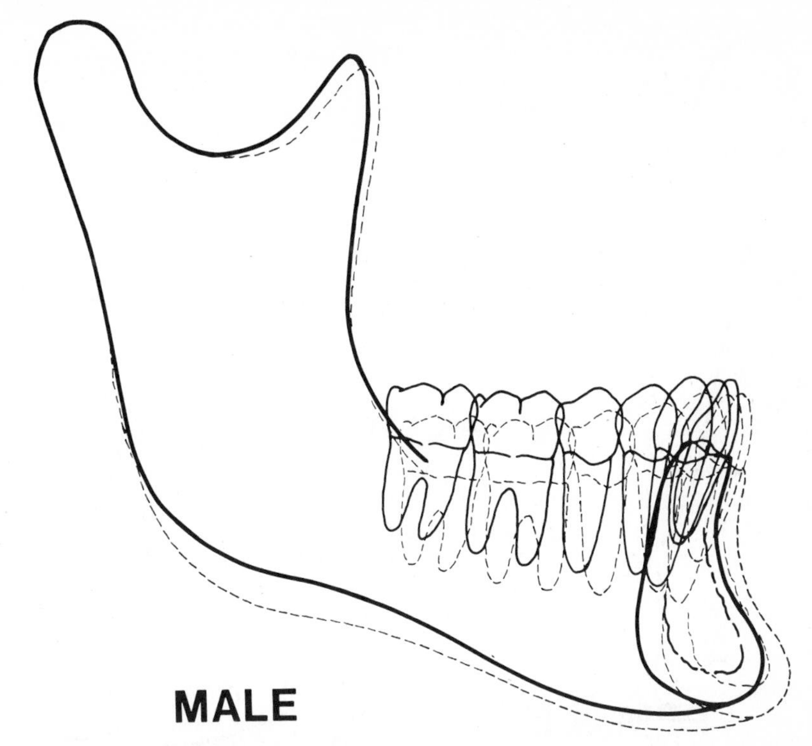

FIGURE 16–14. Schematic diagram of the change noted for the male mandible in relation to a sella/sella–nasion orientation. Counterclockwise and anterior movement is characteristic of the male. (From Behrents, R.G.: *Growth in the Aging Craniofacial Skeleton.* Monograph 17, Craniofacial Growth Series. Ann Arbor, University of Michigan, Center for Human Growth and Development, 1985.)

a counterclockwise fashion in the male and in a clockwise direction in the female. Although both movements are subtle, they tend to effect an elongation of the face, although by differing means. Growth also affects the position of gonion, with that structure being relocated inferiorly and anteriorly in the male and inferiorly and posteriorly in the female. More general posterior development of the mandible is seen in the male consistent with what is occurring in the midface.

Although it might be suggested that many of the changes noted for the mandible are due to growth effects produced elsewhere in the craniofacial skeleton, it is clear that the mandible itself is growing (Figs. 16–15 and 16–16). Distances intended to describe the growth of the regions of the mandible show that the overall length of the mandible, the body, and ramus, and the alveolar regions are all increasing in size. Furthermore, in the male, the angle between the body and the ramus becomes more acute by a small amount with age. The anterior border of the ramus continues to relocate posteriorly with time. This suggests that resorption of the anterior border of the ramus continues in adulthood much like that seen in adolescence. Such activity during adulthood might have some effect on the ability of the third molars to erupt later in life. The posterior border of the ramus appears to be stationary in the female and moving anteriorly in the male. These two effects produce a decrease in the width of the ramus with time.

As might be expected, the dentition reacts to the osseous changes that occur during adulthood (Figs. 16–17 and 16–18). Consistently, for both sexes, the maxillary anterior teeth become more vertically upright during adulthood.

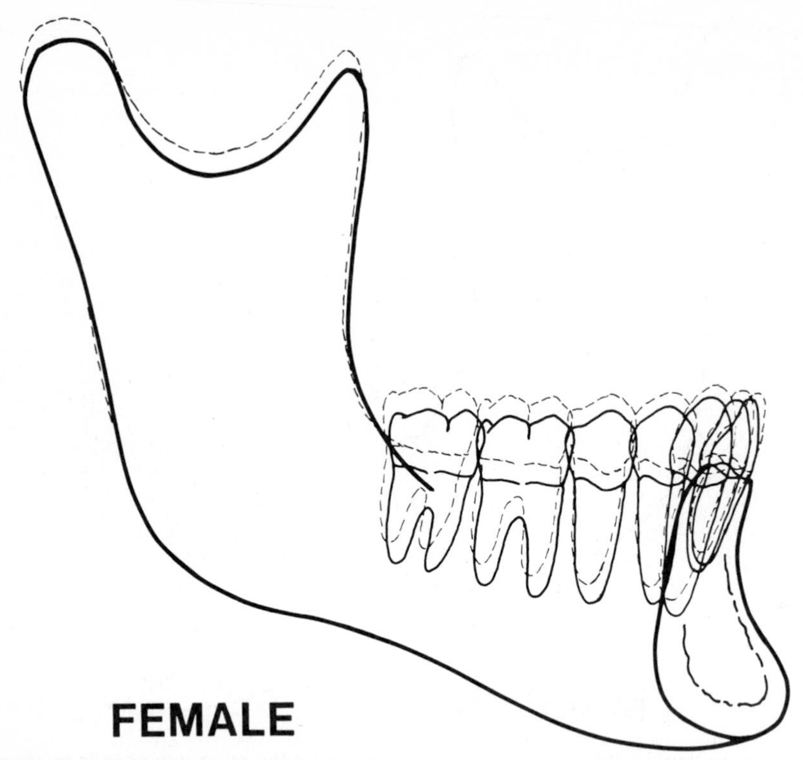

FIGURE 16–15. Schematic diagram generalizing the size and shape change seen in the female mandible during adulthood. According to this regional superimposition there is an apparent increase in the height of the alveolar process. A slight amount of resorption (anterior) and deposition (posterior) is seen in the ramus. (From Behrents, R.G.: *Growth in the Aging Craniofacial Skeleton.* Monograph 17, Craniofacial Growth Series. Ann Arbor, University of Michigan, Center for Human Growth and Development, 1985.)

The lower anterior teeth, however, appear quite stable in their orientation, with only a tendency for proclination in the female. The posterior teeth apparently change their inclination in response to the altered position of the mandible. The axis of the molars shows a significant uprighting in the male and a tendency for being more distally inclined in the female. The lower molar movements complement those of the upper molars: in the male the molar becomes more upright, and in the female there is a tendency for a more mesial inclination. By virtue of these changes, it might be expected that the dentition would appear less prominent in the older adult. With regard to the relationship of the teeth to the bony profile, this is not the case. However, with regard to the relationship of the teeth to the soft tissue profile, the teeth do appear less prominent. Curiously, overbite did not increase with time. However, when it was realized that attrition was commonplace, it is reasonable to surmise that overbite did increase but it was decreased by incisal wear.

The soft tissue mask surrounding the osseous structures undergoes notable alterations with age. The changes are more extensive than those seen within and among the osseous structures, but still the soft tissue changes remain related to the underlying osseous change. Anterior movements of the soft tissues over the glabella, the nasal region, the midface, and the chin are seen. Likewise, the soft tissues also reflect vertical osseous changes as seen over the nasal region, the midface, and the chin. But some differences are noted. The nose grows a great deal in size, and the tip becomes more angled and downturned. The height of the upper lip follows this same developmental course and lengthens to the same extent that the nose grows. Furthermore, because of the dental changes together with the growth of the nose and the anterior movement of the chin, the teeth appear less prominent, the lip area flattened, and the lips located more inferiorly, almost completely covering the upper incisors. Thus, overall, with age there is a straightening and elongation of the profile. Such findings mirror those found in other studies.

In addition to the healthy, dentate, orthodontically untreated sample, small subsets of individuals who had undergone various treatments (orthodontics, multiple extractions, rhinoplasty) were also studied. In these subjects continuing adult growth was demonstrated, but the nature and amount of the alteration were different from those of their untreated counterparts. In general, in young adulthood the group of individuals who had previously received orthodontic treatment were indeed different in craniofacial configuration when compared with untreated individuals. During adulthood they also grew poorly, as they had in the past. For example, persons with a retruded mandible retained this characteristic during adulthood, and in some instances it worsened. Persons who had experienced the loss of many teeth also grew in adulthood, but their growth was again different from that of either the treated or untreated individuals. Typically these cases grew less in adulthood; this was particularly evident in the anterior portions of the face. Such cases also demonstrated loss of vertical dimension with time; almost all of the persons in the untreated sample (most of whom had complete dentition) showed the opposite effect. The loss of teeth obviously alters the growth of the adult craniofacial skeleton.

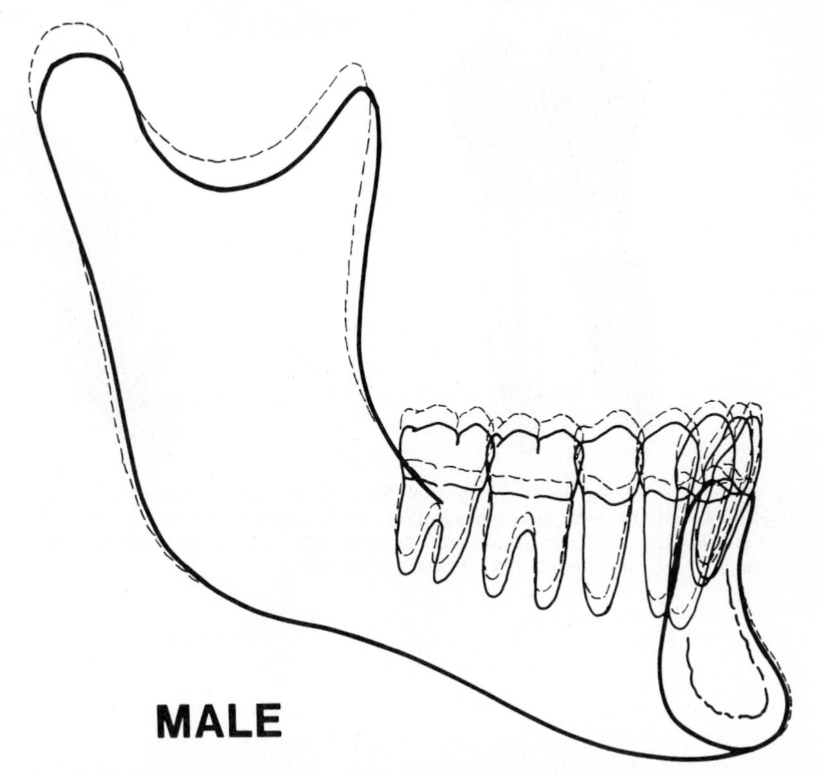

FIGURE 16–16. Schematic diagram generalizing the size and shape change seen in the male mandible during adulthood. In the male, compared with the female, similar and greater changes are noted. (From Behrents, R.G.: *Growth in the Aging Craniofacial Skeleton.* Monograph 17, Craniofacial Growth Series. Ann Arbor, University of Michigan, Center for Human Growth and Development, 1985.)

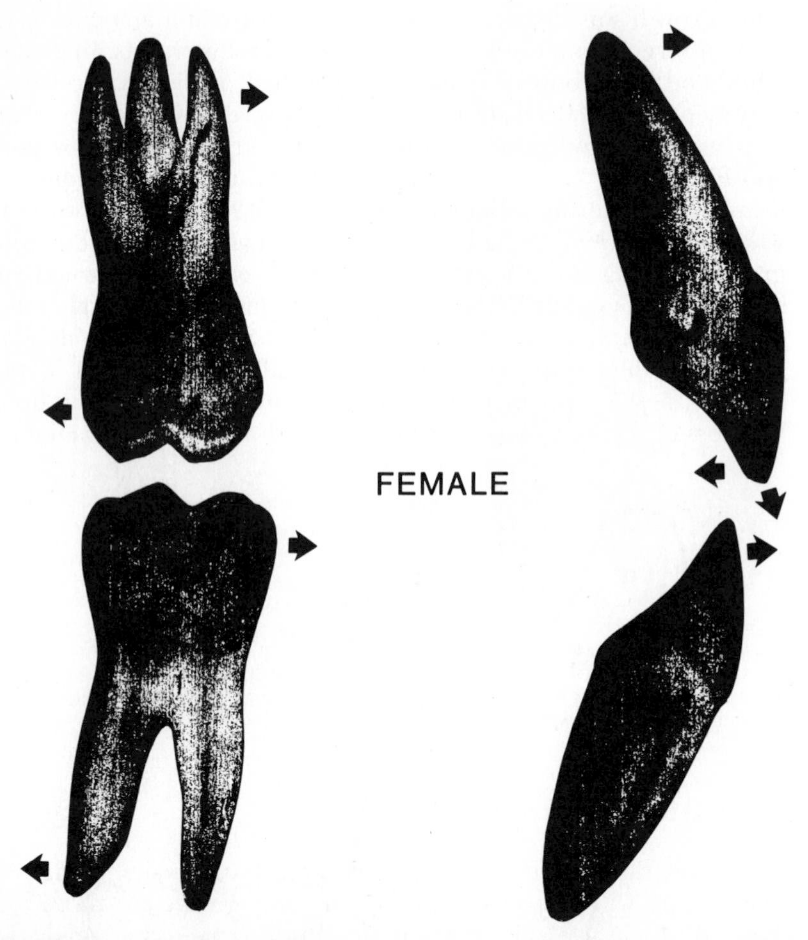

FIGURE 16–17. Diagram of the tooth movements seen in the female during adulthood. Uprighting of the upper anterior teeth is consistently seen, and there is a tendency for the lower anterior teeth to tip forward. (After Behrents, R.G.: *Growth in the Aging Craniofacial Skeleton.* Monograph 17, Craniofacial Growth Series. Ann Arbor, University of Michigan, Center for Human Growth and Development, 1985.)

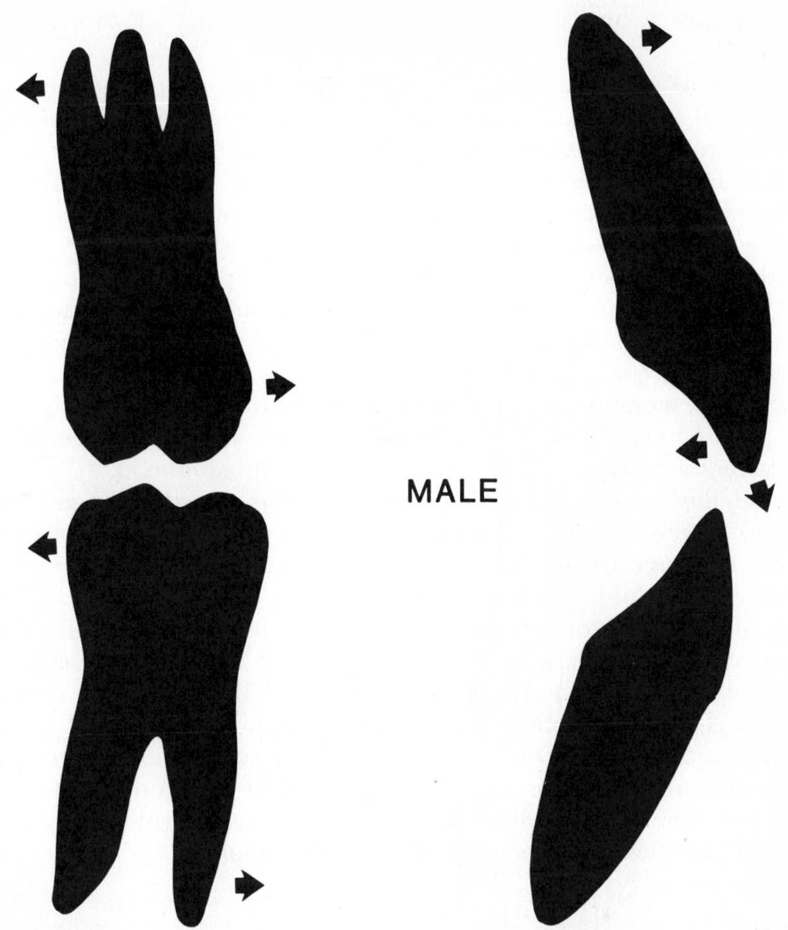

FIGURE 16–18. Diagram of the tooth movements seen in the male during adulthood. Uprighting of the upper anterior teeth is also consistently seen. (After Behrents, R.G.: *Growth in the Aging Craniofacial Skeleton.* Monograph 17, Craniofacial Growth Series. Ann Arbor, University of Michigan, Center for Human Growth and Development, 1985.)

EXPLANATORY MECHANISMS

Given that continued craniofacial growth is a reality, it might be helpful to understand which mechanisms active in adolescence are also present in adulthood. Unfortunately, in terms of mechanisms that can serve to account for the change, we are pressed by the available literature to conclude that few mechanisms remain after late adolescence. For example, it is difficult to imagine that the translation mechanisms conceptualized by Moss and Scott are active during adulthood (with regard to the nasal septum, brain growth, orbit, etc.), and little literature really exists on the topic. In adulthood, the brain apparently becomes smaller, the septal cartilage may continue to grow until the 30s but mainly at its anterior free end. Further, although it has been shown that a condylar contribution to mandibular growth can occur into the early 20s, weak evidence (acromegaly) is present to suggest that the condylar cartilage retains the potential for substantial change during later adulthood. Although the major translatory mechanisms supposedly active during adolescence do not appear to be plausible with regard to adult growth, some other mechanisms present during adolescence may continue into the adult period. For example, considerable literature is available with regard to the patency of sutures during adulthood.

Estimates of suture closure vary widely depending on the suture studied, the manner of study, and the specimens under review. Some studies suggest that the sutures of the cranium close in late adolescence; and those of the face, soon after. However, there is no clear consensus on these issues, and some investigators suggest that cranial and facial sutures remain patent past age 18 and clearly into the 20s and perhaps even into the 30s. Also, there appears to be considerable variation within each region. For example, it has been suggested that the palatine and intermaxillary sutures remain unossified into the 30s and that the frontozygomatic suture remains open until the eighth decade of life. Furthermore, there may be individual variation as well. Therefore, on the basis of the available literature, it is possible that the sutures may in some ways and in some areas participate in the growth activity seen in the adult. Their principal postadolescent contribution, however, occurs early in adulthood.

Remodeling of the facial bones to effect the growth changes seen in adulthood, however, seems a rationale approach to the explanation of adult alteration. Although the mechanism, amounts, and rates of change differ for adolescence and adulthood, remodeling continues throughout life and is the most plausible explanatory mechanism relating to bone surface activity. However, one might question why such activity would be *expected* to result in a change in morphology. In this regard, Enlow (1986) believes that once basic adult form is obtained, unless there is an intrinsic environmental alteration (change in function, change in biomechanical circumstance, loss of teeth) gross (not histologic) remodeling is basically static in effect. Environmental changes, however, are quite likely—indeed, certain. Therefore, a morphologic change is logical, and remodeling seems the probable explanation.

SUMMARY

What is evident from this discourse is that growth is apparently operative at wider age spans than previously thought, and there may be no cessation of growth at all. Recent reports support the view that active growth of the craniofacial complex continues into adulthood, but perhaps only into the 30s. Whether or not growth ceases in middle age or continues is probably an unanswerable question, given the sensitivity of the available recording techniques. Regardless, that adults grow has been confirmed; growth of the craniofacial complex does not cease in adolescence. The amount of growth occurring during adulthood is small, especially when the amount per year is quantified. Nonetheless, the cumulative increments of growth over time cause a modest amount of differential alteration of the craniofacial skeleton. Generally a 2 to 10 per cent enlargement occurs. Characterization of the growth in adulthood suggests that it is an extension of the adolescent change except that in the later ages, a generalized vertical elongation of the facial structures predominates.

CONCLUSIONS AND APPLICATIONS

It is clear that tissues in general and the skeleton in particular are far from quiescent during adulthood and aging. Distinct, sometimes complementary and sometimes contradictory (different from activity during childhood and adolescence) processes are involved. Furthermore, such processes may exert influences that alter a skeletal structure so as to decrease or increase its prominence. Thus it follows that such change could be noted during anthropometric, craniometric, or cephalometric study.

On the basis of the findings presented here, it must be recognized that growth and development of the craniofacial skeleton is a continuing, long-term process that apparently has its periods of exuberance and relative quiescence, but the biologic mechanisms that incite or regulate the changes remain intact and never really terminate. We know that bone is continuously remodeled irrespective of physiologic demands and aging, we know that bone responds to injury and manipulation at all ages, and now we know that the process of growth apparently does not terminate as was once thought and that changes in morphology may continue long after puberty—adults grow.

Considerable discussion has resulted, since the aspects of adult change have been documented and accepted, over what to call such change. Arguments have suggested "adaptation," "maturation," something other than "growth." This issue, although mainly semantic, cannot be resolved. The difficulty lies in our definitions of growth, which are imperfect because of our lack of understanding of its mechanisms and its temporal nuances. For now, the adult alteration noted here is indeed "growth," for it is not unique to the adult (although it has some unique features); it is merely an extension of the changes seen during earlier years. The nature of the differences is best described in terms of degree, not in terms of kind.

Many of the technical procedures that have been developed to improve facial form in the adolescent have been based on our understanding of the nature of man, including growth. One would expect, therefore, that the present information about growth during adulthood might come to influence the clinician's diagnostic and treatment planning activities on patients within that category. So, too, might the present knowledge find application and even misapplication in providing a basis for some treatments. Perhaps more important, present knowledge should direct the clinician to understand that osseous change will continue long after supposedly definitive treatments are rendered; rigid stability of the craniofacial skeleton is an untenable concept.

Of greatest significance, however, is acceptance of the concept that adults retain the capacity to change and may do so. Given that concept, ultimately prediction and control of the processes of growth will become possible.

CHAPTER SEVENTEEN

Facial Growth and Development in the Rhesus Monkey

Primates have become increasingly popular, in recent years, for laboratory studies of the facial growth process. Other animals, of course, have been used quite profitably for basic research, but it is felt by many investigators that the macaque (and some other simian species as well) is particularly appropriate for certain clinically related types of experimental studies. The purpose of this short chapter is to introduce the researcher to the elements of craniofacial growth in the monkey and to outline briefly major similarities and differences in comparison with the human facial growth process.

The basic plan of growth in the monkey's mandible, nasomaxillary complex, and cranium generally parallels that found in the human face and cranium. Although there are certain specific and important differences (Fig. 17–1), as pointed out below, the overall distribution and pattern of resorptive and depository growth fields and the nature of displacement movements are similar. The simian mandible grows predominantly posteriorly, and the ramus lengthens in a superoposterior direction (Figs. 17–2, 17–3, and 17–4). As in man, a continuous sequence of remodeling changes is involved in backward and upward ramus growth and the progressive remodeling conversion from ramus to corpus. The same basic rationale exists for the various regional growth fields in both ramus and corpus. As in man, the monkey's mandible becomes displaced anteriorly and inferiorly as it enlarges. The nasomaxillary complex in the monkey also grows posteriorly as it is simultaneously displaced anteriorly. The maxillary arch grows downward by the same pattern of deposition, resorption, sutural growth, and displacement as in the human face. The floor of the monkey's cranium, similarly, shows remodeling changes resembling those of the human cranial base, although differences in extent are involved in relation to the respective sizes of the brain (see Fig. 17–11).

Because there are differences in the gross topographic shape in several regions of the simian skull, there are corresponding differences in the growth and remodeling of these particular parts. One notable difference exists in the mental region of the mandible (Fig. 17–5). In the human mandible, a prominent

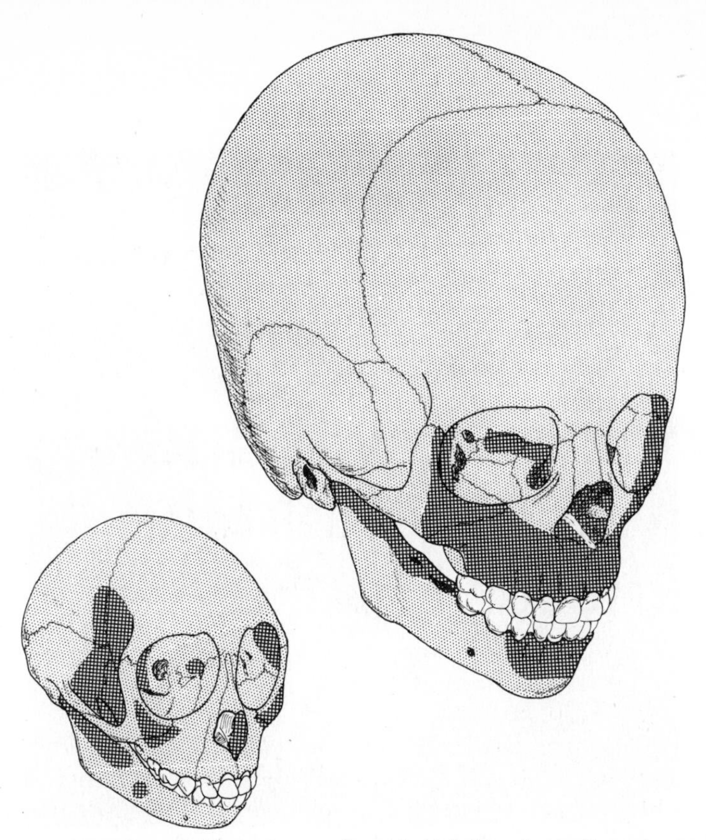

FIGURE 17–1. The various depository (light) and resorptive (dark) fields of growth and remodeling in the rhesus monkey are compared with those in the human face. (From Enlow, D.H.: A comparative study of facial growth in *Homo* and *Macaca*. Am. J. Phys. Anthropol., 24:293, 1966.)

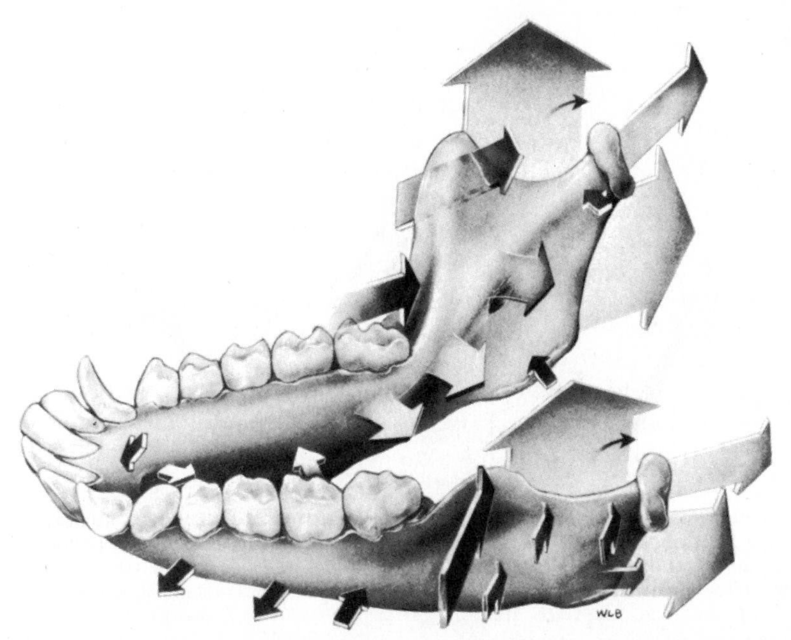

FIGURE 17–2. The growth and remodeling of the rhesus monkey's mandible is shown by resorptive arrows (going into the bone surface) and depository arrows (coming out of the bone surface). Compare with the human mandible in Figure 3–129. (From Enlow, D. H.: *Principles of Bone Remodeling.* Springfield, Ill., Charles C Thomas, 1963.)

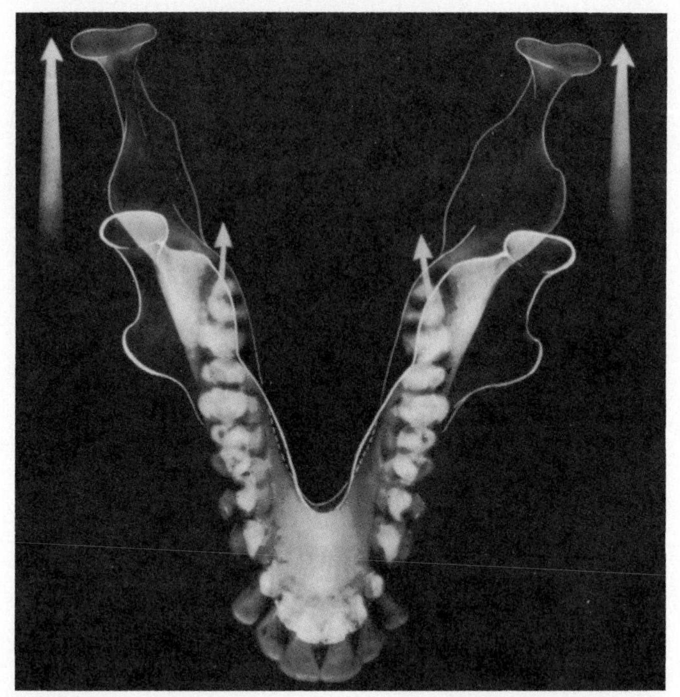

FIGURE 17–3. Superimposed growth stages of the mandible in the rhesus monkey. Posterior growth of the ramus from the lingual surface (principle of the V) is indicated by the small arrows, and the direction of condylar growth is shown by the larger arrows. (From Enlow, D. H.: *Principles of Bone Remodeling.* Springfield, Ill., Charles C Thomas, 1963.)

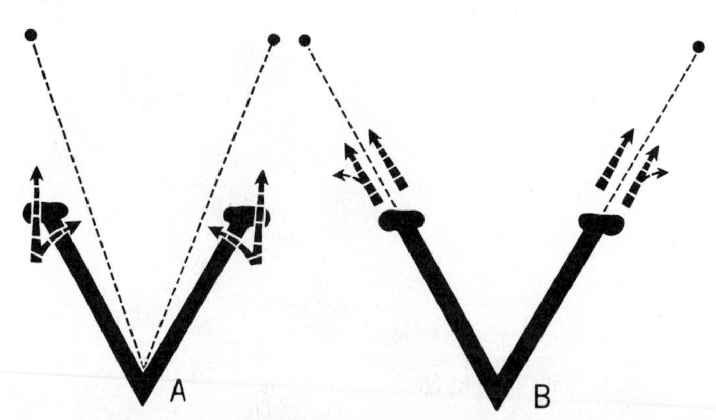

FIGURE 17–4. This diagram illustrates the growth of the monkey's mandible by the V principle. The condyle and ramus do not simply grow backward along the principal axis, as represented by *B.* Rather, the enlarging V involves deposition on the lingual side, with some areas of resorption on the buccal side, because it increases in length more than in width. (From Enlow, D. H.: *Principles of Bone Remodeling.* Springfield, Ill., Charles C Thomas, 1963.)

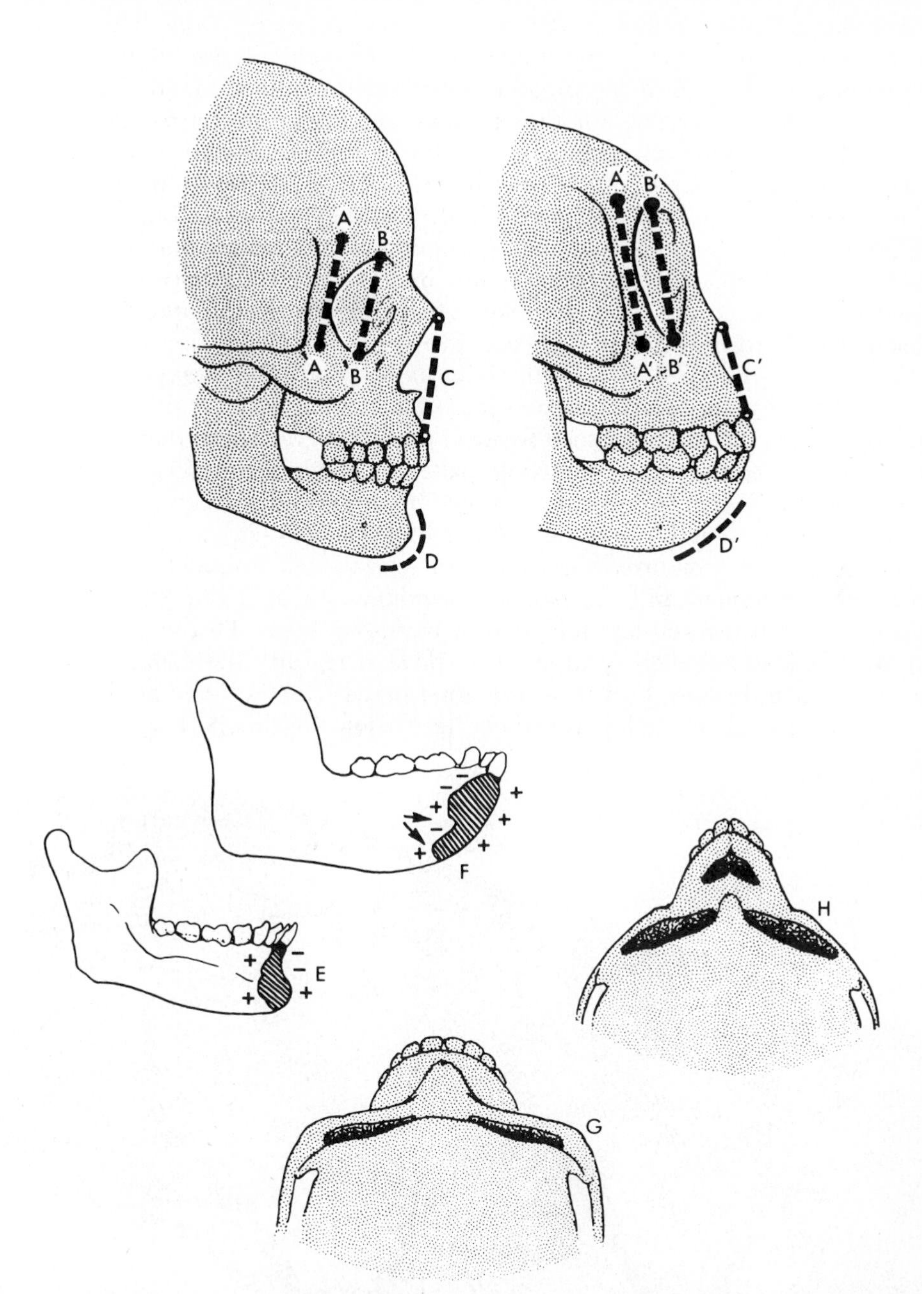

FIGURE 17–5. Because of anatomic differences between the human and monkey facial parts, corresponding differences are involved in their respective growth and remodeling processes. These two sub-adult skulls are oriented so that their occlusal planes are approximately parallel. Compare the size and slope of the frontal prominence, the relative positioning of the upper and lower orbital rims *(B)*, the angle of the lateral orbital rims *(A)*, the relationships between the tip of the nasal bone and the premaxilla *(C)*, the structure of the mandibular mental region *(D)*, the presence of a simian shelf in the monkey's mandible and a chin on the human mandible *(E* and *F)*, and differences in the contour of the malar region *(G* and *H)*. (From Enlow, D. H.: A comparative study of facial growth in *Homo* and *Macaca*. Am. J. Phys. Anthropol., 24:293, 1966.)

chin marks this region, a distinctive feature that characterizes the face of modern man (and also, for reasons yet to be studied, the elephant). This structure is lacking in the monkey as well as in extinct species of *Homo*. The mandible of the monkey, however, has a unique "simian shelf" on the lingual side in the mental region. This is not present in man. In the human mandible, the chin is formed by (1) variable amounts of bone deposition on the mental protuberance, together with (2) surface resorption in the alveolar area just above it. The lower incisors tip lingually, and overjet and overbite result in lieu of end-to-end tooth contact. In the pointed, chinless mandible of the monkey, the entire labial surface of the mental region undergoes progressive anterior bone growth (Fig. 17–6). The incisors do not drift lingually (posteriorly). Rather, the mandibular corpus of the monkey grows forward in the anterior part as well as backward in the posterior part. The result, in conjunction with an equivalent mode of maxillary growth, is a more protrusive muzzle as compared with that of man. In mammalian species having markedly protrusive lower and upper jaws, the extent of such forward growth is even more than in the monkey. This growth and remodeling pattern produces an elongate, angular, chinless mandible in contrast to the short, broad, rounded, regressive, flattened, and chin-tipped arch of the human mandible (Fig. 17–7).

On the lingual side, a contrasting pattern of growth and remodeling is present between the human and the monkey mandibles. In man, the entire lingual surface shows a marked accumulation of periosteal bone. This lingual surface in the monkey, however, is largely resorptive, a feature that complements the forward, protrusive growth in the anterior part of the bony arch. But at the genial crest, a reversal line is present; and below this line part of the

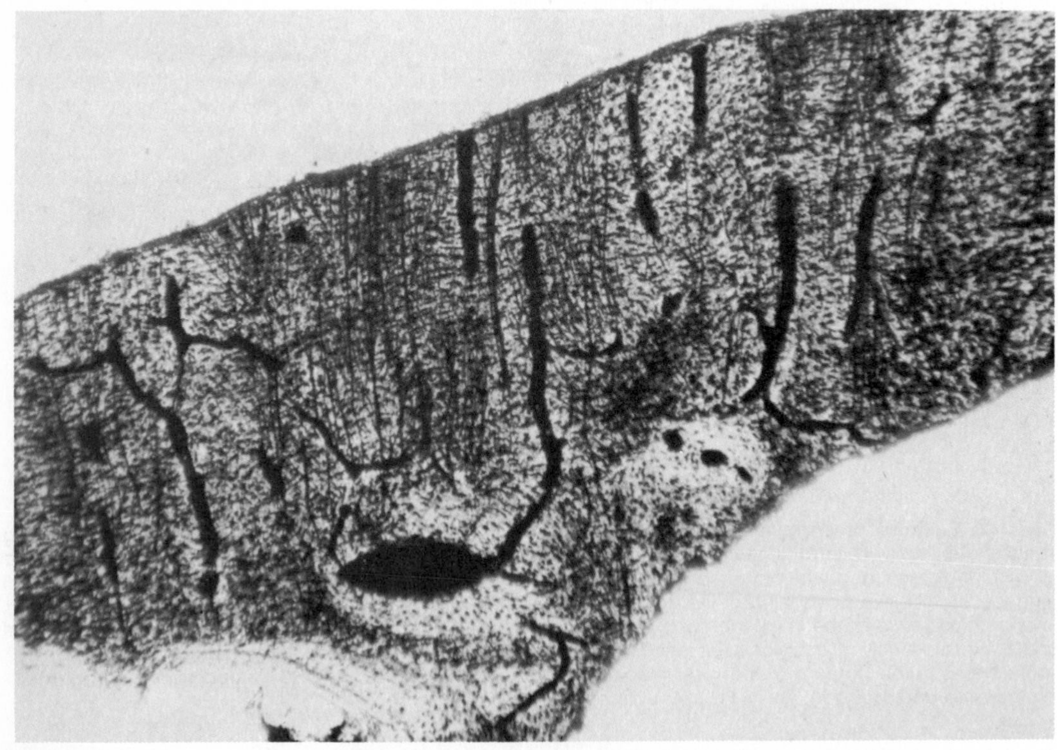

FIGURE 17–6. This cortical section from the monkey's mandible was taken from the labial side of the cuspid region. It shows a depository periosteal surface and a resorptive endosteal surface. Note the attachment fibers embedded in the cortex. (From Enlow, D. H.: *Principles of Bone Remodeling*. Springfield, Ill., Charles C Thomas, 1963.)

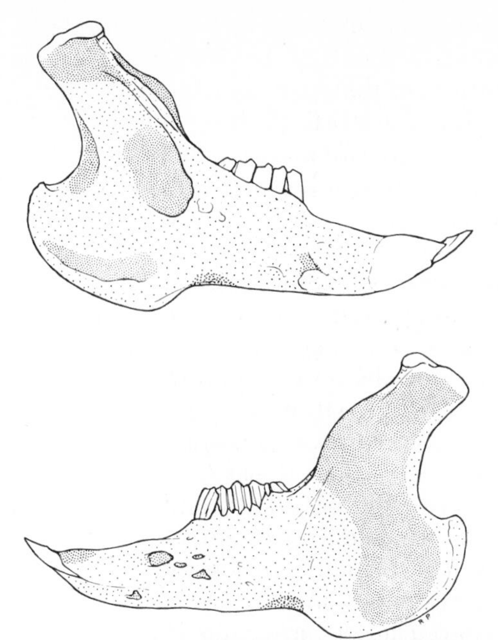

FIGURE 17–7. The depository (light) and resorptive (dark) fields of growth and remodeling in the mandible of the rabbit, a popular laboratory animal in bone research, are shown. Although the overall plan is comparable to that of the human mandible, specific differences exist that relate to corresponding differences in topographic morphology. While growth takes place posteriorly, a major direction of growth also occurs anteriorly in the protrusive face of the rabbit. As the condyle and ramus grow posteriorly, sequential relocations are involved, just as they are in man. The lingual side of the ramus is predominantly resorptive, and the buccal side is largely depository. Note that a coronoid process is poorly represented, as is the lingual tuberosity. The patches of resorption on the lingual surface of the ramus relate to corresponding surface hollows. The resorptive zone on the dorsum of the alveolar bone just behind the incisors is involved in a progressive downward relocation of this area as the incisors move mesially with the lengthening mandible. (Adapted from Bang, S., and D. H. Enlow: Postnatal growth of the rabbit mandible. Arch. Oral Biol., 12:993, 1967.)

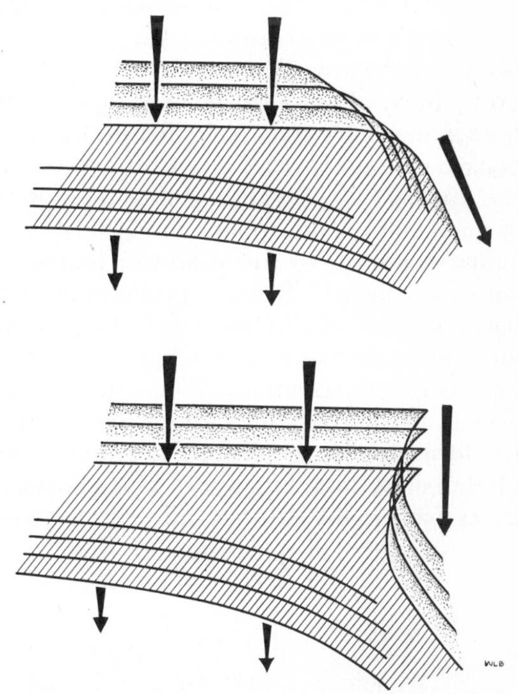

FIGURE 17–8. The monkey's palate (top) grows downward by resorption on the nasal side and deposition on the oral side. The human palate (bottom) has an equivalent mode of growth. Note that the protrusive premaxillary region in the monkey, however, grows **forward** by external deposition, in contrast to the human maxilla, in which this region grows straight downward (after the primary teeth are established) by external resorption. (From Enlow, D. H.: A comparative study of facial growth in *Homo* and *Macaca*. Am. J. Phys. Anthropol., 24:293, 1966.)

lingual surface is depository, not resorptive. This remodeling combination produces the prominent, characteristic **simian shelf** of the monkey's mandible. These contrasting remodeling differences between man and monkey in the forward part of the mandible result in a chin on one side and a simian shelf on the other, respectively (see Fig. 17–5).

Because the relative dimensions of the human brain and its cranial floor are a great deal wider in man, the bi-condylar dimension of the human lower jaw is proportionately more broad. Together with the proportionately shorter arch, the U-shaped human mandibular configuration is characteristically much more rounded. In contrast, the V-shaped mandible of the monkey is proportionately longer, more narrow, and quite angular (see Fig. 17–2.) The massive trihedral eminence on each side of the human mandibular corpus contributes to its rounded form. In the monkey, a distinctive resorptive field, not usually present in man, occurs just anterior to the trihedral region (see Fig. 17–1).

This produces a lingual shift of the corpus and contributes to the more angular shape of the monkey's bony arch (see Fig. 17–2).

A major growth difference exists in the anterior part of the maxillary arch. The outside (labial) surface of the human "muzzle" is characteristically resorptive. This is a unique human facial growth feature associated with the markedly decreased extent of muzzle protrusion and the essentially straight-down growth direction of the maxilla. In the monkey (Figs. 17–1, 17–8, and 17–9), the labial surface of the entire anterior part of the arch, including the separate premaxillary segments, is depository (as it is in other mammalian species as well). It grows **forward** as well as downward and is therefore more protrusive than the human maxillary arch. Note that the external surface of the monkey's premaxillary region is convex, in contrast to the concave nature of the forward part of the human arch. Because of this, the downward mode of growth by the arch in man requires resorption on the labial side of the alveolar cortex (the portion below *A* point). In the **postnatal** human face, the loss of the premaxillary sutures, a lack of forward premaxillary cortical growth, the downward and backward rotation of the whole facial complex, and the reduced extent of anterior displacement all combine to result in a marked regression in the extent of maxillary prognathism. The anterior mode of cartilaginous and soft tissue growth by the overlying nasal region, although reduced in extent, results in the formation of the distinctive human "nose," which protrudes well forward of the short maxillary arch. This provides for **downward**-directed external nares that aim the inflow of air vertically, in the vertically aligned nasal chamber,

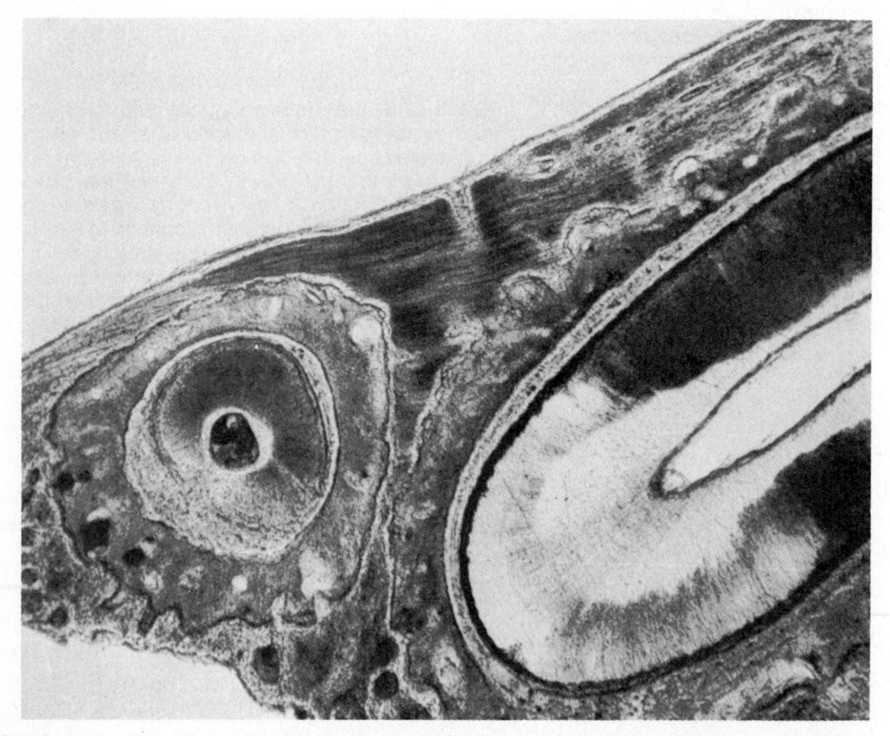

FIGURE 17–9. This photomicrograph shows the external depository nature of the monkey's premaxillary region. The forward alveolar lining surface is resorptive as the teeth drift forward in the protrusively growing muzzle. Compare with the converse pattern in the human maxilla in Figure 3–137. (From Enlow, D. H.: A comparative study of facial growth in *Homo* and *Macaca*. Am. J. Phys. Anthropol., 24:293, 1966.)

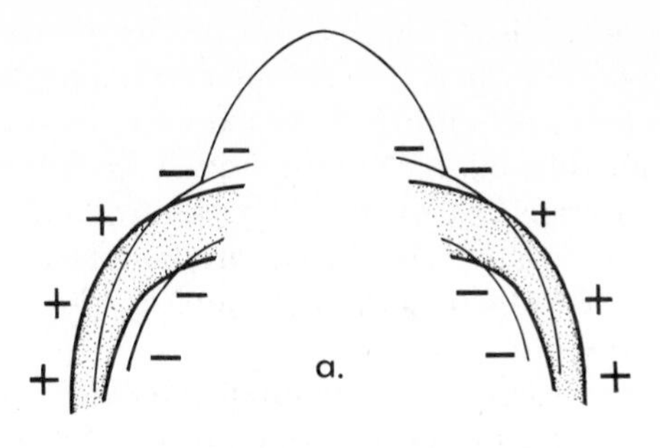

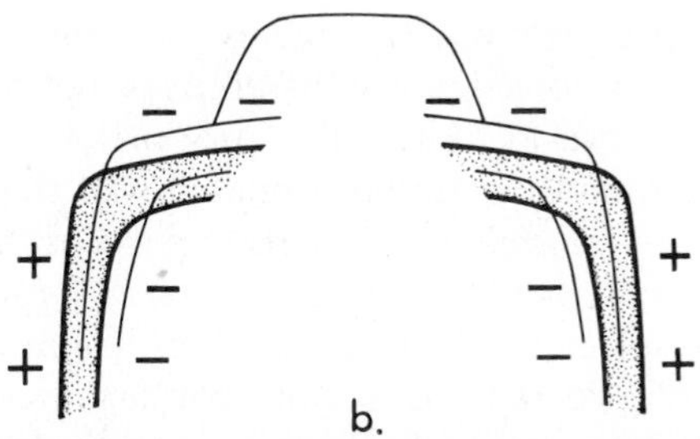

FIGURE 17–10. Remodeling differences in the malar region between the monkey *(a)* and the human *(b)* faces are shown. Note that the anterior surfaces of both are resorptive (−) but that the squared configuration of the human cheekbone involves more extensive surface resorption. The rounded contour of the monkey zygomatic region, in contrast, has a lateral depository surface (+) that extends farther onto the anterior face of the malar. These differences also relate to the lesser extent of backward malar relocation in the more protrusive forward-growing muzzle of the monkey. (After Enlow, D. H.: A comparative study of facial growth in *Homo* and *Macaca*. Am. J. Phys. Anthropol., 24:293, 1966.)

toward the sensory olfactory nerves in the **roof** of the chamber (in comparison to the posterosuperior wall in other species).

In the face of both man and the monkey, the outer surface of the cheekbone region is resorptive. The extent, however, is much less in the monkey (Fig. 17–10). There are two reasons. First, because the maxillary arch in the monkey grows anteriorly as well as posteriorly, the amount of backward relocation needed by the malar protuberance is not as great. In man, the direction of bony maxillary arch growth is posterior (after the primary dentition is established), and the malar region must necessarily grow backward to a greater extent than in the monkey in order to keep proportionate position relative to it. Second, the topographic configuration of the malar region is different in the two species. The more forward-rotated orbits in man produce a distinct squaring of the cheekbone and a flattening of the whole face. The malar region of the monkey is more rounded. Because of this, the lateral drift of the monkey's zygoma requires periosteal deposition that extends farther around onto the anterior-facing surface of the cheek, as shown in Figure 17–10.

Another basic difference occurs in the growth patterns of the lateral orbital areas in the human and the simian skulls. The orbital rims in man are much more vertically disposed because of the larger frontal lobes and upright forehead. The anterior edge of the lateral rim is entirely resorptive, and the postorbital surface is depository. This combination moves the lateral rim backward as the contiguous malar region beneath it grows posteriorly. The backward growth of the malar region together with the forward growth of the forehead and superior orbital rim result in a human lateral orbital rim that is inclined obliquely forward (see Fig. 17–5). In the monkey, conversely, the anterior edge of the lateral orbital rim is depository, and the postorbital side is resorptive. In conjunction with a much lesser extent of frontal lobe and forehead growth, the lateral rim remains inclined obliquely backward. Its inferior part grows forward and laterally, rather than backward and laterally, as in the human face. The upper orbital rim in the human face is rotated so that it protrudes forward of the lower rim. In the monkey, the superior rim remains at a level posterior to the inferior rim.

As in the human calvaria, the individual bones of the skull roof in the monkey enlarge in perimeter by sutural bone growth. Both the inner and outer surfaces are depository except for limited remodeling near the sutures. The bony cortex is usually composed of a single thin layer of lamellar bone. Where diploic spaces exist, the cortex has two tables, and the endosteal surfaces of both are resorptive. Two conspicuous differences between the human and simian calvariae exist, however. First, the proportionately deeper temporal fossa, medial to the zygomatic arch, has a medial wall that is resorptive in the monkey, and this sizable remodeling field covers parts of the temporal, parietal, and frontal bones. Second, the nuchal region has a large resorptive field on the exterior of the monkey skull. This provides an inward rotation of the inferior part of the occipital bone to sustain contact with the slower-growing cerebellum as the whole occipital bone is being displaced outward by the cerebrum.

The occipital part of the clivus is resorptive on the endocranial surface in both species (Fig. 17–11). It therefore moves forward and downward as the clivus lengthens at the spheno-occipital synchondrosis and at the rim of the foramen magnum. Unlike that in man, however, the sphenoidal part of the clivus is always depository in the monkey (this area is variable in man). It flares posteriorly much more than in man and moves markedly upward and backward during growth, together with the posterior lining wall of the sella turcica, which is resorptive. In man, the superoposterior wall of the sella turcica also moves back but apparently not to the same relative extent. The floor of the monkey's

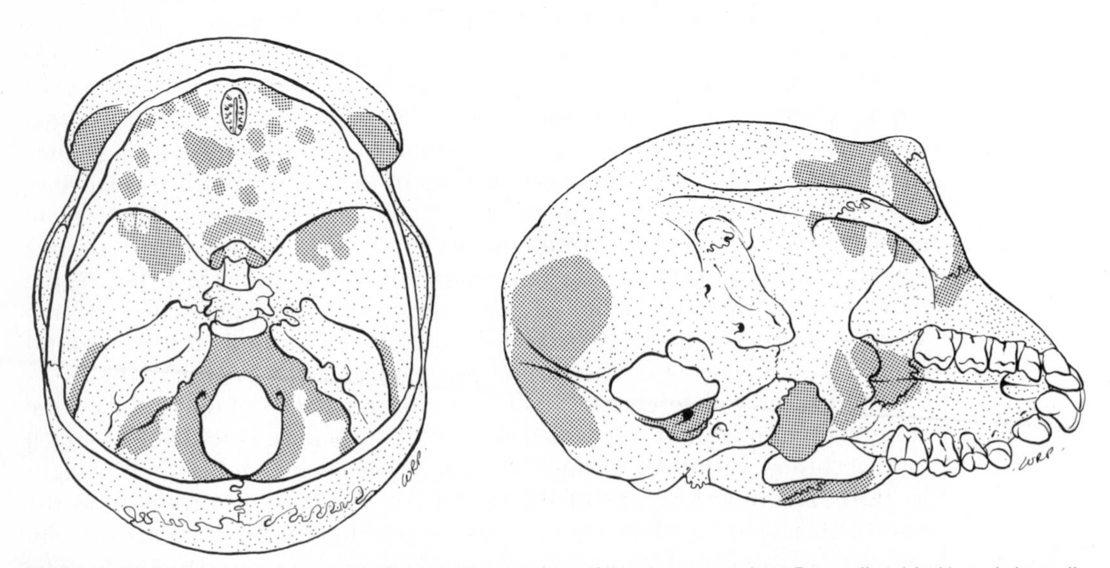

FIGURE 17–11. Growth and remodeling fields in the cranium of the rhesus monkey. Resorptive (dark) and depository (light) surfaces are indicated. See text for descriptions. (Adapted from Duterloo, H.S., and D.H. Enlow: A comparative study of cranial growth in *Homo* and *Macaca*. Am. J. Anat., 127:357, 1970.)

pituitary fossa is depository (see page 100). This important region, however, needs further basic study in both species; the growth process here is complex and relates to differential extents of growth between the midline and lateral parts of the brain, the displacement versus enlargement of the hypophysis, and the extents of remodeling versus the extents of displacement of the various bony parts involved. Differences at various prenatal and postnatal age levels are also likely to be involved because of differentials in growth rates among the parts of the cranial floor and among the lobes of the brain.

In the monkey, a sizable ring of resorption occurs on the floor of the cranium surrounding the endocranial side of the foramen magnum. This serves to lower as well as enlarge its diameter. Unlike the endocranial surface of the human cranial floor in general, only scattered patches of resorption exist in the various endocranial fossae of the monkey. This relates to the much more shallow nature of the cranial fossae, in contrast to the deep endocranial pockets present in man, and to the much lesser extent of cranial base flexure.

The entire superolateral half of the orbital roof in the monkey is resorptive. This same resorptive growth field is present in the human orbit but is more restricted in extent (postnatally). It is confined to that part of the roof beneath the overhanging supraorbital ridges. The roof of the orbit is also the floor of the anterior cranial fossa. In man, the endocranial side is resorptive in conjunction with the massive enlargement of the cerebrum and the flexure of the cranial base (see section on early fetal orbital growth, page 320). The opposite (orbital) side of the bony plate receives deposits as the entire anterior cranial fossa grows downward. The situation in the monkey is different. The anterior cranial floor slopes upward much more, and there is a lesser extent of cranial base flexure. The simian face lies essentially **in front of**, rather than beneath, the floor of the anterior cranial fossa. The surface of the bony floor of the anterior cranial fossa is largely depository, rather than resorptive, in conjunction with a much lesser extent of growth expansion by the brain. Sutural growth provides a proportionately greater amount of the bony enlargement involved. This depository surface is complemented by a resorptive surface on its opposite (orbital) side, a pattern that also provides for orbital expansion in conjunction with growth at the various orbital sutures.

The pterygoid plate in the human skull is typically depository on the surface lining the trough-like pterygoid fossa, and a large part of the external side is resorptive, at least during the latter period of childhood growth, when middle endocranial fossa growth is complete. This combination produces growth enlargement essentially straight downward. In the monkey, the pterygoid plates are characteristically resorptive on the fossa side and depository on the external sides. This produces a downward and also a **forward** growth expansion in conjunction with the more protrusive nasomaxillary complex and the lesser extent of cranial base flexure.

In man, the bony lining of the auditory canal is usually resorptive on the anterior surface and depository on the posterior surface. This results in an anterior drift of the entire canal. The lining of the monkey's auditory canal, in contrast, is entirely resorptive. This enlarges the canal to its definitive size, but a forward drift movement is not involved.

The oral side of the horizontal shelf of the palatine bone in man is depository, and the nasal side is resorptive. This produces the downward growth of this part of the palate in conjunction with corresponding growth movements by the other bony palatal parts. In the monkey, however, the oral side of the palatine bone is often resorptive and the nasal side depository, although variations occur. The remainder of the bony palate is basically similar to that in man with regard to the growth pattern. The meaning of this difference is not known but probably relates to a composite remodeling-displacement rotation of the whole maxilla. In man a similar compensatory function is achieved by variations in the amounts of deposition or resorption in the anterior versus posterior parts of the palate, rather than variations in the actual pattern. (See page 119.) This should be taken into account in using the monkey in cleft palate research. The anterior part of the nasomaxillary complex is not displaced inferiorly to the same extent as the posterior portion (in relation to sutural growth). This has the effect of a counterclockwise "rotation," which is masked because the anterior part of the midface grows downward by direct deposition and resorption (remodeling growth) to a greater extent than the posterior part, as shown by the studies of McNamara. A similar combination exists in the human face as well, but the extent of direct inferior cortical growth in the anterior part of the maxilla in man is notably greater (as shown by Bjork). In many simians and anthropoids, there is a vertical hypoplasia of the anterior midface as an established anatomic feature, a characteristic not shared by the face of man. The posterior part of the anthropoid and simian maxilla appears quite large and "low," but it is actually of proper dimension (relative to the brain) and in its correct anatomic position (relative to the senses, as described in Chapter 5). It is the **anterior** part of the maxilla that is vertically shallow. Anterior open bites in the monkey are common. Moreover, an end-to-end type of incisor contact exists, rather than an overbite, as in the human dentition. This also relates to the absence of a chin on the monkey's mandible. The human chin, which is formed in association with the development of overbite and overjet, may thus relate to the more extreme extent of maxillary rotation caused by frontal lobe expansion and the greater extent of anterior maxillary remodeling compensation. (See also page 122.)

CHAPTER EIGHTEEN

Bone and Cartilage

Part 1

Concept 1

Because of the unique nature of its intercellular matrix, cartilage is a rigid and firm tissue, but it is not hard. Cartilage provides three basic functions. It gives **flexible** support in appropriate anatomic places (the nasal tip, ear lobe, thoracic cage, tracheal rings); it is a **pressure-tolerant** tissue located in specific skeletal areas where direct compression occurs (articular cartilage); and it functions as a "**growth cartilage**" in conjunction with certain enlarging bones (synchondrosis, condylar cartilage, epiphyseal plate). Cartilage is a nonvascular connective tissue that is ordinarily noncalcified (both vascularization and calcification, however, are involved as steps in the replacement process by bone tissue).

Concept 2

Cartilage usually has a perichondrium, but, significantly, it can exist without this covering membrane. Cartilage grows **appositionally** by the activity of its chondrogenic membrane, and it also grows **interstitially** by cell divisions of chondrocytes and by additions to its intercellular matrix. Together with the noncalcified nature of the matrix and the absence of vessels, these various features combine to allow the cartilage to function and to grow in areas of direct pressure. Because it can exist without a covering membrane, it is suited for articular surfaces, synchondroses, and epiphyseal plates. Because it can expand interstitially, cartilage can thereby grow even without a membrane. Because its matrix is noncalcified, unlike bone, cell divisions are possible, and diffusion of nutrients and wastes can take place through the permeable matrix. Because there are no vessels to press closed, either in a covering membrane or within the matrix, cartilage is pressure-adapted and can **grow** where compression exists because of its noncalcified, interstitial expansion features.

Concept 3

Unlike cartilage, which is pressure-related in many of its anatomic locations, bone is **tension-adapted**. Bone **must** have a vascular, osteogenic covering membrane (or other soft tissue), and it can only grow appositionally. Bone cannot grow directly in heavy pressure areas because its growth is dependent upon a sensitive, vascular membrane. Bone needs a membrane in order to support its internal vascular system, which, in turn, is essential because the matrix is calcified and will not allow diffusion of oxygen, nutrients, and metabolic wastes to and from cells. Moreover, bone cannot grow interstitially because of its calcified matrix.

Concept 4

In all areas of skeletal growth, bone grows **intramembranously** in tension areas and **endochondrally** in pressure areas. "Growth cartilages" take part in the latter ossification process. They provide for the **linear** growth of a bone **toward** the direction of pressure. As interstitial cartilage expansion provides pressure-adapted growth on the pressure side of the cartilage plate,* an equal amount of cartilage is removed and replaced by bone on the other side. This allows the bone to lengthen toward its force and weight-bearing articular contacts. The remainder of the bone, including all its cortical plates, grows by membranous ossification in conjunction with the periosteal and endosteal membranes.

Concept 5

The membranes associated with bone (periosteum, sutures, periodontium) have their **own** internal growth and remodeling process. The whole membrane, as a sheet, does not merely move away as new bone is deposited by it. Rather, it undergoes extensive fibrous changes in order to sustain constant connections with the bone by means of collagenous fiber continuity from the membrane into the matrix of the bone. As fibers in the membrane become enclosed within the new bone deposits, the membrane-produced fibers become incorporated as bone fibers. This is accompanied by fibrous remodeling within the membrane to provide continuity between membrane and bone fibers. The membrane **grows** outward, rather than just backing off as bone is laid down by it. The process involves complex linkages among the periosteal collagenous fibers (see Fig. 18–15). The movements of muscle attachments along remodeling surfaces of bones and the insertion of muscles on **resorptive** bone surfaces are also carried out by this membrane remodeling and relinkage process.

Concept 6

The periodontal membrane converts the **pressure** exerted by a tooth during chewing into **tension** on the collagenous fibers attaching the tooth to the pressure-sensitive alveolar bone. This is accomplished by placing the tooth

---

*The cartilage and capsule of the mandibular condyle represent a significant and specialized exception to the interstitial versus appositional modes of osteogenesis. See Chapters 3 and 4.

in a sling of membrane fibers (that is, of the periodontium), so that inward pressure by the tooth caused by biting is thereby translated into a pull on the fibers. The periodontal membrane **also** provides for the mesial, lateral, vertical, and distal drifting and for the eruption, tipping, and rotations of each individual tooth. As the face enlarges, the teeth are **moved** along with the various parts of the growing bones. There are significant amounts of such tooth movements associated with progressive facial growth. The maxillary teeth, for example, move inferiorly for a considerable distance in conjunction with vertical nasal expansion. Everybody is familiar with "mesial" drift, but the movements of teeth during the facial growth process involve **much** more than this. All the various tooth movements are carried out by the periodontal membrane. It is customary to describe drift as a process of resorption on the lead, or "pressure," side and deposition on the trail, or "tension," side of the alveolar socket. This is not, however, simply a two-dimensional process (Fig. 3–120). The **same** deposition and resorption of bone **also** bring about vertical movements of the teeth as the jaws and other parts of the face grow vertically, and it provides for all of the many other growth movements of the teeth as well (Fig. 3–90).

Concept 7

Tooth movements are accomplished by relinkage changes of the collagenous fibers within the periodontal membrane (Fig. 3–122). The membrane does not simply "move" in the direction of the tooth movement; it **grows** in this direction. As parts of the periodontal fibers become enclosed by bone on the depository side of the moving alveolar socket, the fibers within the periodontium are lengthened by the fibroblasts in the membrane, and new fibers are produced as well. This involves continued fibrous remodeling to provide direct continuity between the fibers in the membrane and the attachment fibers embedded in the bone. The remodeling process is one in which slender "precollagenous" fibrils link, relink, lengthen, or shorten to provide continuous, uninterrupted connections between the coarse "mature"collagenous fibers attached to the tooth on one end and the alveolar bone on the other (Figs. 3–121, 3–123, 3–124, and 3–125). This is presumably done in some parts of the membrane by enzymatic removal of ground substance to free the precollagenous fibrils (which are bundled together to form the larger mature fibers, with ground substance as the binding agent), by fibroblast activity to lengthen or shorten them in the intermediate layer, and by additions of ground substance by fibroblasts to bundle them up in some other parts of the membrane. The fibrous membrane thereby grows to new locations along with the drifting teeth and bone. In some instances, an **intermediate plexus** of precollagenous fibrils can be distinguished as a distinct middle layer in the periodontal membrane during active periods of tooth movement. Or linkage changes may be more diffuse throughout most of the membrane without producing a distinguishable, separate intermediate plexus. The remodeling process in the membrane, in either case, continues to occur as long as the tooth actively moves. When bone growth and tooth movements largely cease, this active, dynamic membrane then functions essentially as a ligament. The term "periodontal ligament," rather than "periodontal membrane," is not appropriate during the growth period because of the many different functional activities involved. This remarkable membrane is much more than just a stable ligament. (See page 130.)

Concept 8

Histology textbooks teach (at least at present) that the haversian system (secondary osteon) is the structural unit of bone. This is incorrect. In the bone of the young, growing child, the haversian system is **not** a major structural feature. This old haversian system notion not only has misled students of dentistry and medicine but also has concealed an important concept. There are other, much more widespread kinds of bone in the child's growing skeleton. The concept is this: different functional circumstances exist, and there is a specific type of bone tissue for each circumstance. Some bone types are fast-growing, others slow-growing. Some bone types grow inward, others outward. Some are associated with muscle, tendon, or periodontal attachments; others are not. Some bone types form a thick cortex; others, a thin cortex. Some relate to a dense vascular supply; others, to scant vascularization—and so on. "Haversian systems" could not do all this. These are important points because a basic feature of bone is its developmental versatility and adaptability as a tissue.

Concept 9

Primary vascular bone tissue (Fig. 18–1) is the principal type of periosteal cortical bone in the growing skeleton of the child. The vessels are enclosed within canals as new bone is laid down around each vessel in the osteogenic part of the periosteum. These canals are not surrounded by concentric "haversian" lamellae. If the bone is fast-growing, many vessels and their canals characteristically become enclosed. If it is slower-growing, fewer or even no canals are incorporated within the compact bone substance. **Compacted coarse cancellous** bone (Fig. 18–2) is the principal cortical type formed by the endosteum. Half to two-thirds of all the cortical bone in the body is composed of this important, distinctive structural type. It is formed by the inward growth of the cortex into the medulla (that is, periosteal resorption and endosteal deposition); medullary cancellous bone is converted into cortical compact bone by filling the spaces until they are reduced to vascular canal size. **Fine cancellous** bone (Fig. 18–3) is one of the fastest growing types. It is formed throughout the fetal skeleton and also occurs in rapidly enlarging parts of a postnatal bone. This type of cortical bone tissue is characterized by spaces that are larger than ordinary vascular canals but smaller than the coarse cancelli of the medulla. **Nonlamellar** bone is also a rapid-growing type, and it occurs in conjunction with fine cancellous bone formation, although many "compact" areas of the cortex may also be nonlamellar. **Lamellar** bone is a slower-growing type found throughout most parts of the adult's skeleton and the slower-forming areas of the child's skeleton. The various sections of cortical bone seen in Figures 18–33, 18–34, 18–35 and 18–36, for example, are all composed of lamellar bone tissues. A limited distribution of **haversian** bone is formed in some areas of muscle attachment on resorptive or remodeling surfaces of a bone in the child (Fig. 18–4), but it is not a principal feature of the young skeleton. Most haversian systems develop much later in life and are concerned with the secondary reconstruction of the original primary cortical bone (for various reasons described in Part 2). A great many species, however, do not have **any** haversian systems at any age. **Bundle bone** is characterized by dense inclusions of attachment fibers from the periodontal membrane (see Fig. 3–124). This bone type is formed only on the depository sides of the alveolar socket.

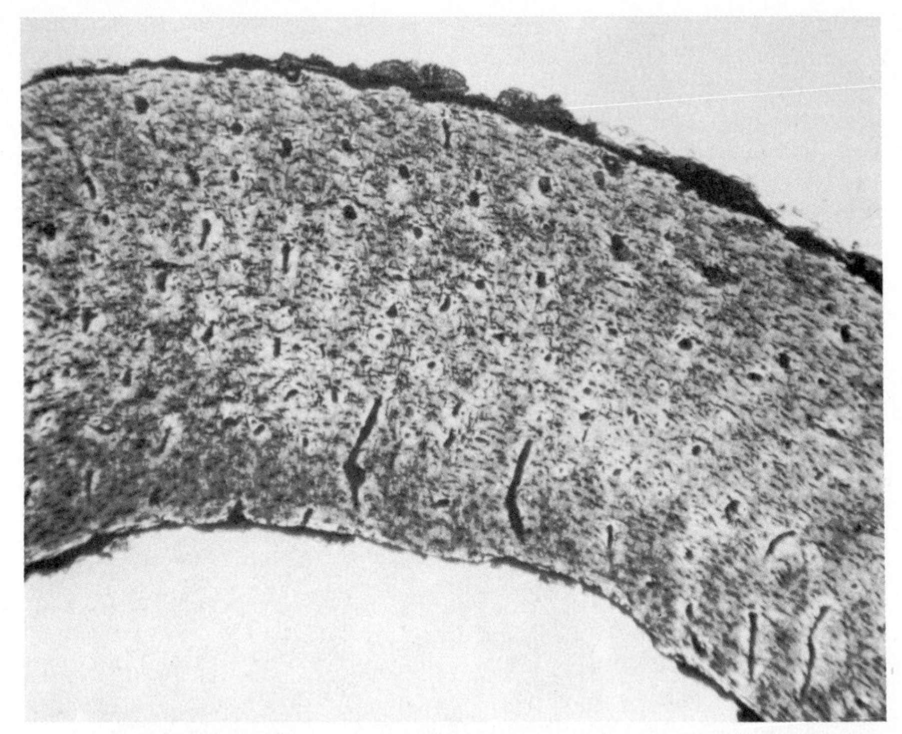

FIGURE 18–1. Primary vascular bone tissue. This is the "standard" type of bone found in the growing human child as well as in most other vertebrates. Note that haversian systems are not present in this transverse section of cortical bone. (From Enlow, D.H.: *Principles of Bone Remodeling.* Springfield, Ill., Charles C Thomas, 1963.)

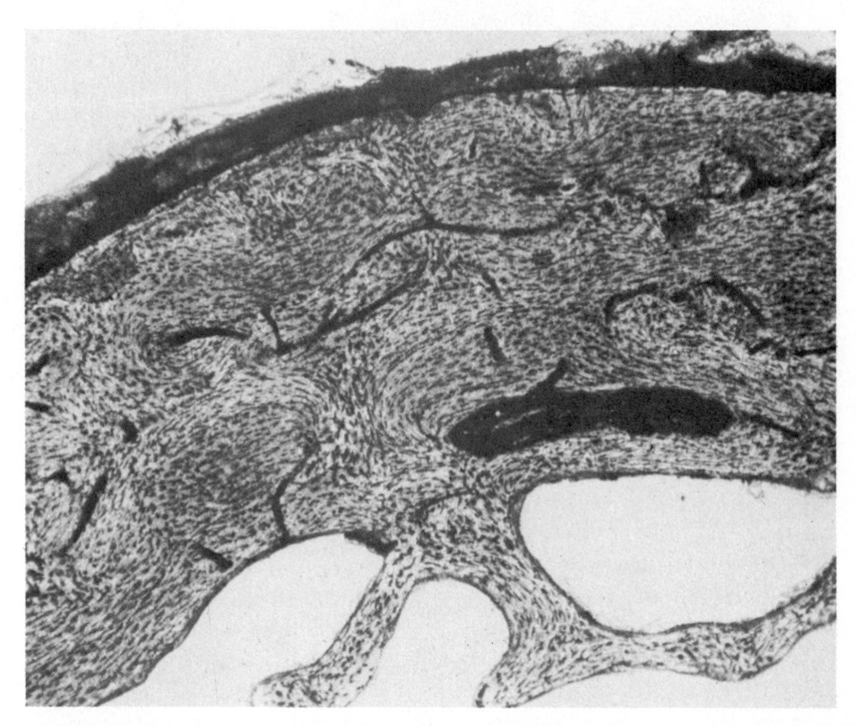

FIGURE 18–2. This section was taken from an inward-growing region of the cortex. It is compacted cancellous bone produced by endosteal deposition and periosteal resorption. The large spaces between the coarse cancellous trabeculae have been filled with lamellar bone. (From Enlow, D.H.: A study of the postnatal growth and remodeling of bone. Am. J. Anat., 110:79, 1962.)

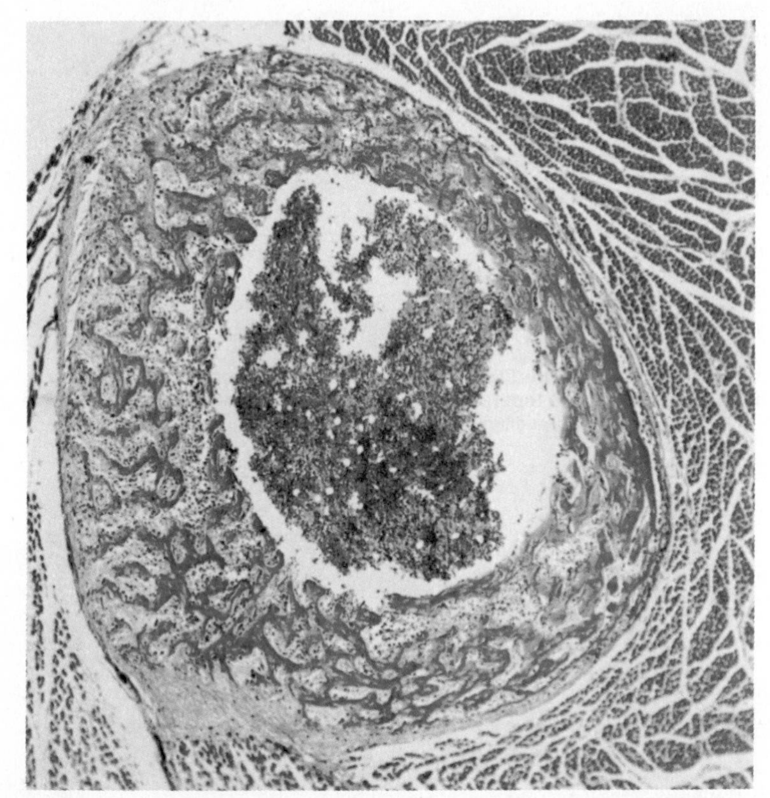

FIGURE 18–3. The cortices in fetal bones are composed of fine cancellous, nonlamellar bone tissue. Note the relatively small connective tissue-filled spaces. Areas of very fast growth in postnatal bones can also be fine cancellous in structure. (From Enlow, D.H.: A study of the postnatal growth and remodeling of bone. Am. J. Anat., 110:79, 1962.)

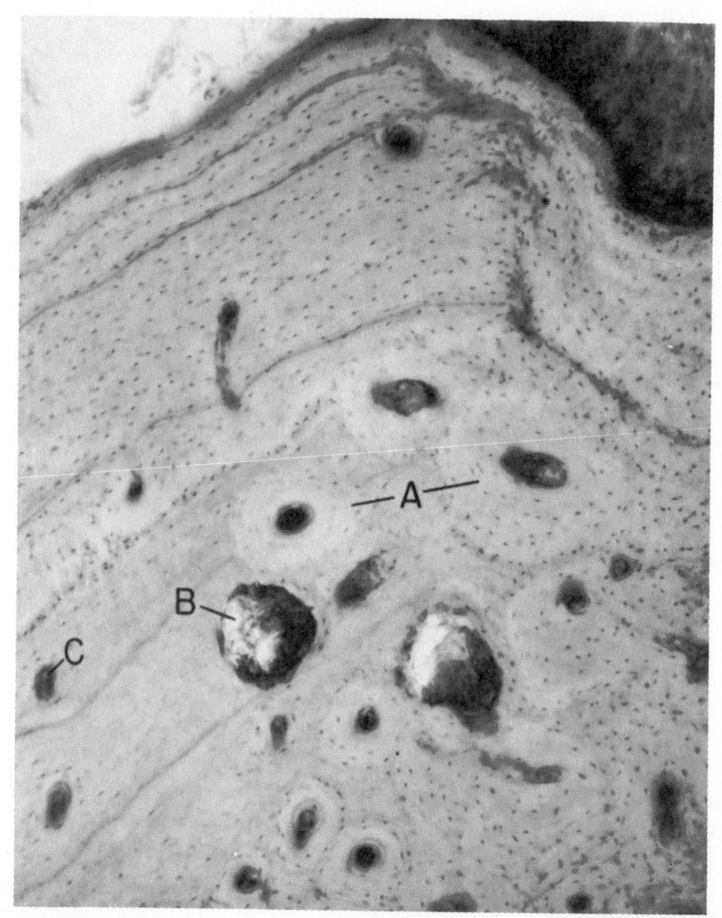

FIGURE 18–4. These secondary osteons are located in a tuberosity that is undergoing a remodeling movement. The shifting of the attached muscle involves the continued formation of haversian systems. A primary vascular canal *(C)* is enlarged into a resorption canal *(B)*, and concentric lamellae subsequently deposited within the resorption spaces result in fully formed secondary osteons *(A)*. (From Enlow, D.H.: Functions of the haversian system. Am. J. Anat., 110:269, 1962.)

The resorptive side is usually composed of compacted coarse cancellous (endosteal) bone or, if the alveolar plate is very thin, of bundle bone formed on the depository side but translated over to the resorptive side as alveolar drift proceeds. **Chondroid** bone (see Fig. 18–31) is found at the apex of the alveolar rim and other rapidly forming areas throughout the skeleton (such as the apex of growing tuberosities where tendons attach). This bone tissue type resembles cartilage because of the large, rounded appearance of its crowded osteocytes surrounded by a nonlamellar, basophilic matrix. Because it undergoes internal metaplasia into **other** bone tissue types, chondroid bone is perhaps the only kind of bone tissue that actually has what might be regarded as an interstitial mode of growth.

Part 2

Haversian systems have long been popularized as the "units" of bone tissue structure, but this is not really the case at all.* It is important to know that **other**, quite basic kinds of bone tissue exist and that they are directly involved in the growth process. Elementary texts adequately describe the two basic modes of bone growth, endochondral and intramembranous, but they do not explain **why** both occur, and this is important. This chapter presents a short supplement to standard histology texts in order to provide this kind of essential and useful information.

CARTILAGE

This tissue, historically, was one of the first to be described histologically (bone is another) because of its stiff, turgid matrix, which can be readily sectioned and studied microscopically. It was a basic tissue utilized by Schwann in his landmark development of the animal cell doctrine. Curiously, it is a relatively poorly known tissue today, compared with most of the others. Many monographs and textbooks have been written on bone, but (at this writing) few such encyclopedic references exist for cartilage compared to an enormous library centered on bone tissue. Numerous symposia have been conducted on the subjects of bone, the lymphocyte, muscle, nervous tissue, and so on; but very few have dealt especially with cartilage. There are no cartilage "societies," as there are for many other tissues and organs. Many internationally recognized authorities are identified with most kinds of tissues; but only a relative handful are concerned with the normal structure, physiology, and growth of cartilage. Considering that the controversial cartilaginous growth sites are presumed by many to represent **the** key growth pacemakers for so many parts of the body, including the face and braincase, this is not as it should be. There are great opportunities here for researchers in a relatively uncrowded but very important area.

Several distinctive structural features relate to cartilage (Figs. 18–5 and 18–6). First these are listed, and then the nature of their interrelationships is explained in terms of the various fundamental functions of cartilage.

1. Cartilage has a stiff but **not hard** intercellular matrix. It provides rigid support, but it is so soft that it can be cut with a fingernail. This feature is based on the exceptionally high content of water-bound ground substance. The rich amount of chondroitin sulfate (Gr. *chondros*, cartilage) in the cartilage matrix is associated with a noncollagenous protein, and this combination has the special property of marked hydrophilia. This gives the turgid, firm character to the matrix. Cartilage develops in locations around the body where **flexible** (not brittle, unyielding) support is appropriate.
2. The matrix of ordinary cartilage is noncalcified.
3. The matrix is nonvascular.
4. Cartilage can grow **both** interstitially and appositionally.

*Another basic point is **why** haversian systems develop. This is explained later.

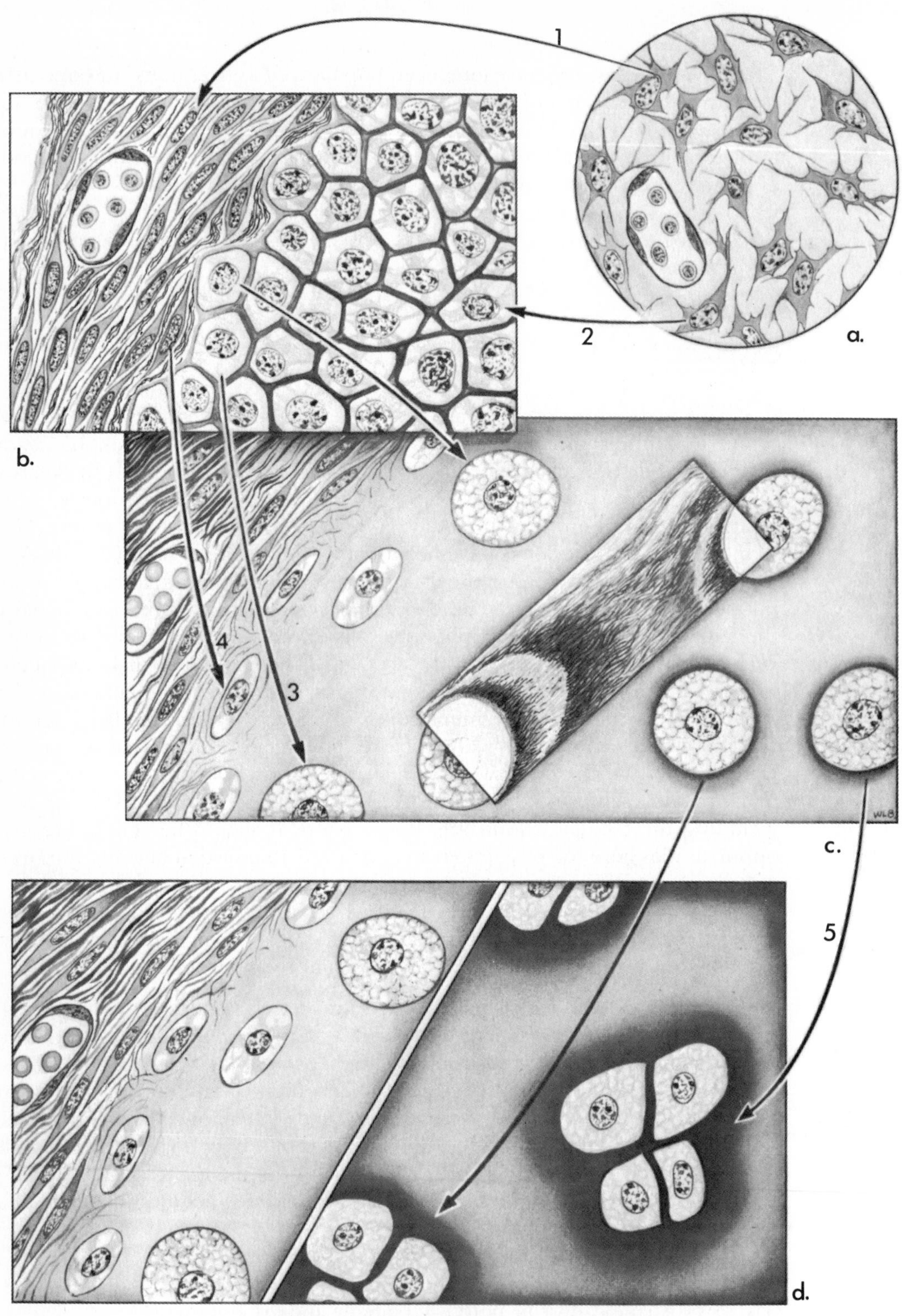

FIGURE 18–5 *See legend on opposite page*

5. Cartilage has an enclosing, vascular membrane, but it can exist without one; it does this in certain specific locations.
6. Cartilage is uniquely pressure-tolerant.

Now, let us combine these various structural and physiologic features in order to relate cartilage to its special roles in the body and particularly in the face.

1. Because the matrix is noncalcified, the matrix is also able to be nonvascular. Nutrients and metabolic wastes diffuse directly through the soft matrix to and from cells. Thus, blood vessels are not required in cartilage as they are in bone, with its hard, impervious matrix.
2. Because the matrix is nonvascular, cartilage is pressure-tolerant (one of several reasons). There are no vessels near the surface to mash closed by compression, in contrast to other kinds of soft tissues, thereby allowing cartilage to operate metabolically because there are no lines of supply to occlude. The water-bound, noncompressible matrix is not badly distorted by the force, and its turgid nature protects the cells within it.
3. Unlike bone, cartilage can function without a covering membrane. This is possible because the matrix is noncalcified (allowing diffusion) and because the matrix is nonvascular. It is thus not dependent upon **surface** blood vessels in an enclosing **surface** membrane, since it does not have vessels within the matrix.
4. Because cartilage can function without a covering membrane, it is especially adaptable to sites involving pressure, as on articular surfaces and the surfaces of epiphyseal plates. If a soft connective tissue membrane were present, its vessels would be closed off by the pressure, and the cells would be subject to anoxia as well as the direct pressure itself. Furthermore, a delicate perichondrial membrane could not withstand abrasive articular movements. In conjunction with synovial secretions, however, the naked surface of the articular cartilage provides relatively frictionless movements while bearing great weight under severe pressure. (Note: the "secondary" cartilage of the mandibular condyle involves a specialized growth system. See Chapters 3 and 4.)
5. Because cartilage has an interstitial as well as a membrane-dependent appositional mode of growth, it can thereby still grow in those pressure areas where an enclosing connective tissue membrane is absent. This includes articular surfaces, synchondroses, and epiphyseal plates.

FIGURE 18–5. The dual nature of cartilage growth is illustrated in this series of growth changes. In a center of chondrogenesis *(a)* some of the mesenchymal cells differentiate into a mass of precartilage *(2)*, and other stem cells become involved in the formation of the enclosing perichondrium *(1)*. Note that the perichondrium has blood vessels, but the cartilage itself is avascular. Then, in stage *c*, some of the chondroblasts present in the inner part of the perichondrium of stage *b* secrete matrix. When fully enclosed within its own secretory material (fibers and ground substance), each cell becomes a chondrocyte *(4)*. Note that the new, young chondrocytes are still elongate; the biomechanical forces acting on the perichondrium have caused its cells, and the other cells in the cartilage derived from them, to be elliptical in shape. The laying down of new cartilage by the perichondrium is **appositional** growth. The cartilage cells already existing also enlarge (lipids and glycogen accumulate in the cytoplasm in large amounts), and the cells become rounded *(3)*. The chondrocytes become separated from each other by the increasing amounts of intercellular matrix. These changes, combined with repeated divisions of the chondrocytes, constitute **interstitial** growth *(5)*. The inset in stage c shows the fibrous matrix as it would appear if the hyalinizing ground substance were removed. It is a feltwork of fine, densely arranged collagenous fibers. (From Enlow, D.H.: *The Human Face*. New York, Harper & Row, 1968, p.4.)

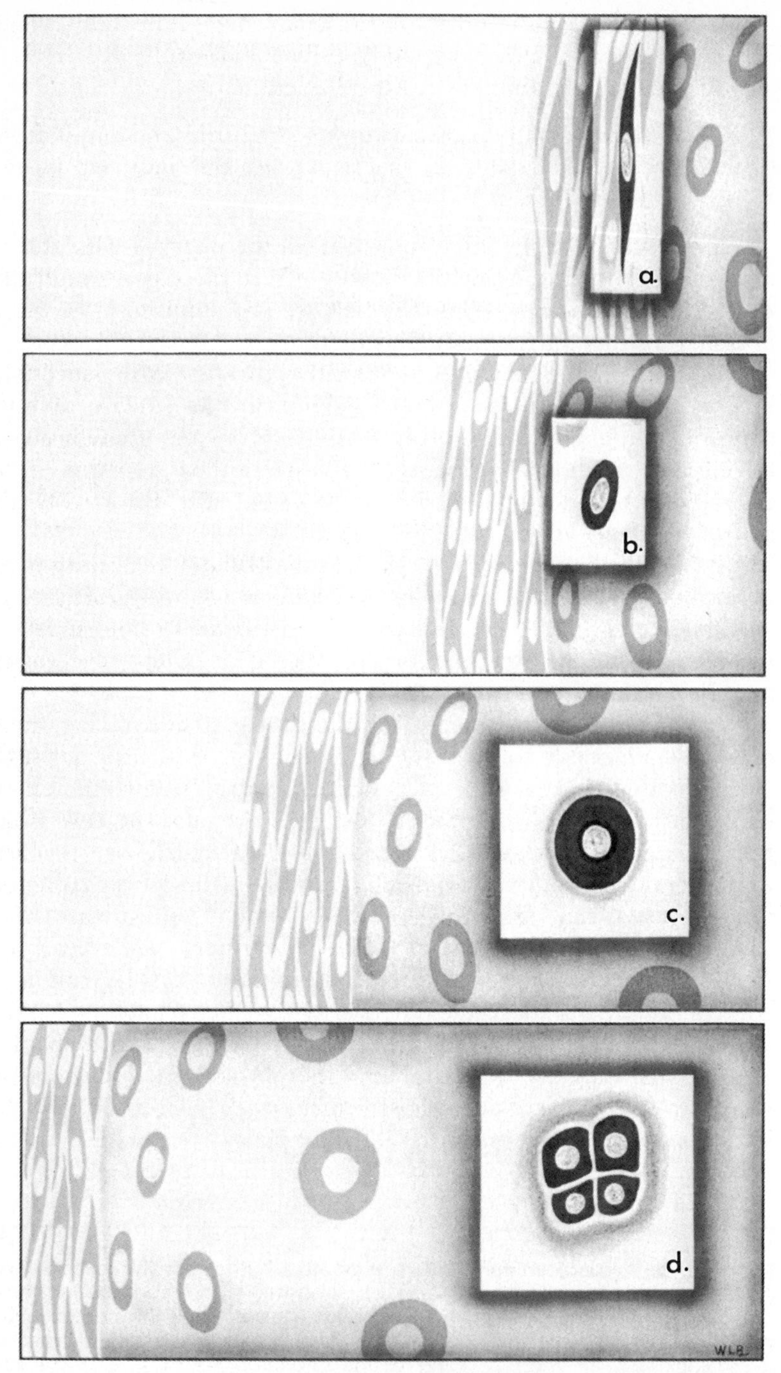

FIGURE 18–6. The sequence of changes involved in cartilage growth is schematized in this illustration. An undifferentiated connective tissue cell in stage *a* comes activated into a chondroblast. When its own matrix secretion completely encloses it, this cell is then a chondrocyte *(b)*. This is **appositional** growth, a process that requires a connective tissue membrane. As new cartilage continues to be formed, the chondrocyte becomes buried more and more deeply within the enlarging cartilage mass. The cell increases in size and becomes rounded. Intercellular matrix is produced by all the cells, and this pushes them apart *(c)*. The chondrocytes, soon after inclusion, undergo mitotic cell division. These are **interstitial** growth changes. When growth slows and finally ceases, matrix production is curtailed, and the cells are not thereby further separated. They remain clustered in isogenous groups *(d)*. In many **growth cartilages**, these isogenous cell groups have a special linear arrangement that coincides with the longitudinal direction of growth of the bone. The mandibular condyle is an exception. (From Enlow, D.H.: *The Human Face*. New York, Harper & Row, 1968, p. 6.)

6. Because the ordinary cartilage matrix is not calcified, cell divisions can take place—which is not the case in bone—thus providing interstitial growth.

One can readily see how each and all of the above features interrelate and are directly interdependent. To perform the functions of a **growth cartilage**, no one of these features could work without all the others. Cartilage exists as a separate, special tissue type because of these special features, and its functions could not be carried out by **any** of the other soft or hard tissues.

BONE

Bone provides the specialized feature of **hardness**. Because of this, it has several unique developmental characteristics. A bone, of course, cannot enlarge by interstitial growth because its cells are locked into a nonexpandable matrix. It is thus dependent upon a covering vascular membrane providing the osteogenic capacity for an appositional system of growth. This is also why bone is necessarily a **traction** (tension)-adapted kind of tissue and why "bone" is said to be "pressure-sensitive." The covering soft tissue membrane is sensitive to direct compression, because any undue amount would occlude blood vessels and interfere with osteoblastic deposition of new bone. Actually, it is the **membrane** and not the hard part of the bone itself that is pressure-sensitive, as described later.

The degree of vascular flow is affected by the amount and type of mechanical force acting on a soft tissue, and this is directly involved in initiating either chondrogenesis or osteogenesis. More extreme levels of hypoxia caused by higher levels of pressure are known to stimulate formation of chondroblasts, rather than osteoblasts, from undifferentiated connective tissue cells. For all of these various reasons, two basic modes of bone growth exist, one adapted to a localized environment of tension (or at least levels of pressure less than that of capillary pressure) and the other to more extreme compressive forces.

In osteogenic areas where **membrane** tension* exists or where the level of any pressure present is minimal, the intramembranous mode of bone growth and development takes place (Fig. 18–7). Thus, bone formed by the periosteum and by sutures occurs in locations where severe membrane compression is not involved. In specific places where such compressive forces do occur, however, the endochondral mode of osteogenesis occurs (Figs. 18–8 and 18–9). Chondroblasts, rather than osteoblasts, develop from stem cells, since, presumably, the level of hypoxia is increased because of decreased vascular capacity.

Soft tissues, in general, grow (1) by increasing the **number** of cells (as in epithelia); (2) by increasing the **size** of cells (as in skeletal muscle); or (3) by

*The whole subject of pressure and tension is greatly oversimplified in most routine descriptions, as it is also in the present discussion. The actual nature of the forces that act on a bone is quite complex and can seldom be designated purely as either tension or pressure. A bundle of periosteal fibers, for example, can exert tension on a bone surface, but that same surface can also be under a compressive influence from other sources, such as a flexure pressure effect on a concave curvature. Pressure effects can also be exerted by intercellular fluids in a region otherwise classified as under tension. An osteogenic cell located between collagenous fibers under tension actually receives a direct pressure effect. Moreover, compressive effects on the hard part of the bone itself can have an osteoblastic trigger effect, while compressive effects on blood vessels can have an osteoclastic effect. The old, greatly oversimplified, and inaccurate concept that one-to-one tension-deposition and pressure-resorption relationships exist is no longer acceptable, as described in Chapter 8. The overall control system is much more complex, and it is incompletely understood at present.

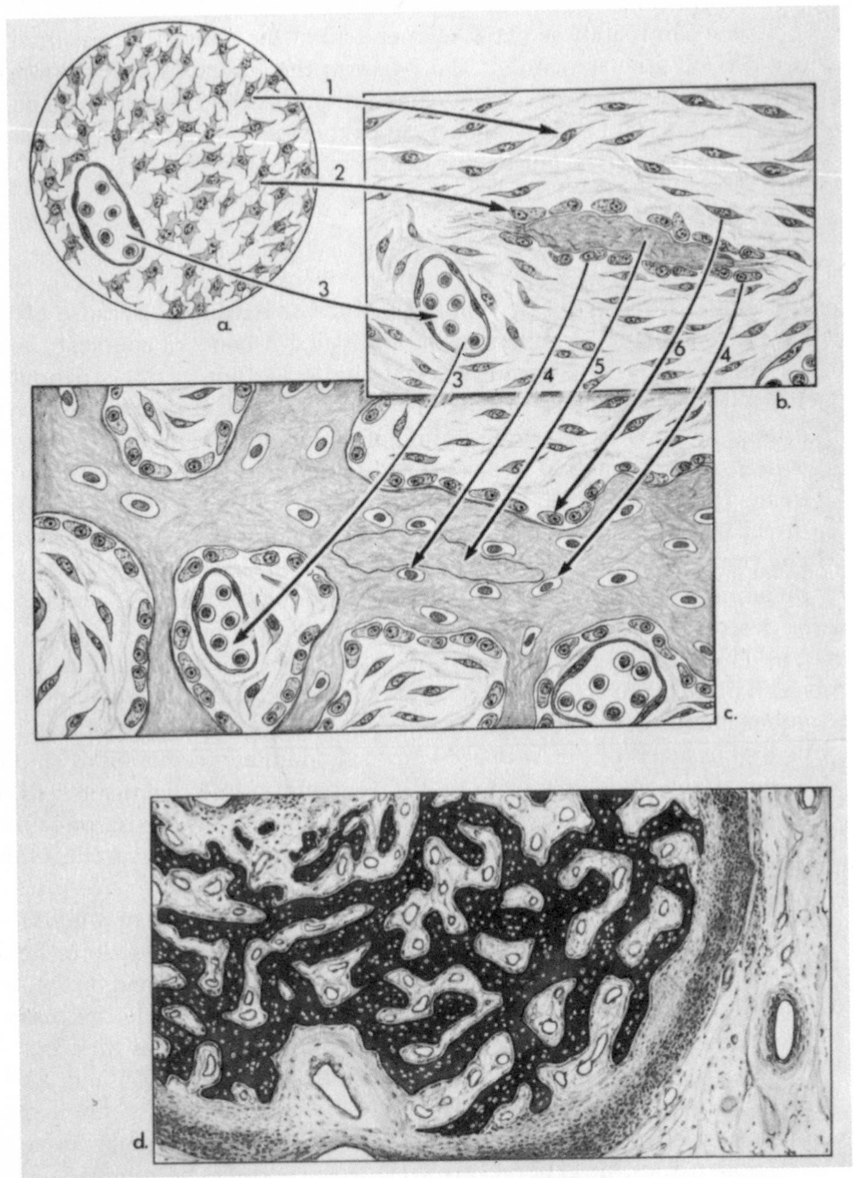

FIGURE 18–7. Intramembranous bone formation. In a center of ossification *(a)*, the cells and matrix of the undifferentiated connective tissue (late mesenchyme) undergo a series of changes that produce small spicules of bone. Some cells *(1)* remain relatively undifferentiated, but others *(2)* develop into osteoblasts that lay down the first fibrous bone matrix (osteoid), which subsequently becomes mineralized (stage *b*). Blood vessels are retained within the spaces among the formative bony trabeculae *(3)*. As bone deposition by osteoblasts continues, some of these cells are enclosed by their own deposits and thus become osteocytes *(4)*. Some undifferentiated cells develop into new osteoblasts *(6)*, and the remaining preosteoblasts undergo cell division to accommodate enlargement of the trabeculae. The outline of an early bone spicule *(5)* is shown in the enlarged trabeculae for reference. The spaces contain a scattering of fibers, undifferentiated connective tissue cells, and osteoblasts. At lower magnification, the characteristic fine cancellous nature of the developing cortex is seen. This bone tissue type is widely distributed in the prenatal as well as young postnatal skeleton. It is a particularly fast-growing variety of bone tissue. Note that the periosteum (also formed from undifferentiated cells in the ossification center) has become arranged into inner (cellular) and outer (fibrous) layers. (From Enlow, D.H.: *The Human Face*. New York, Harper & Row, 1968, p. 44.)

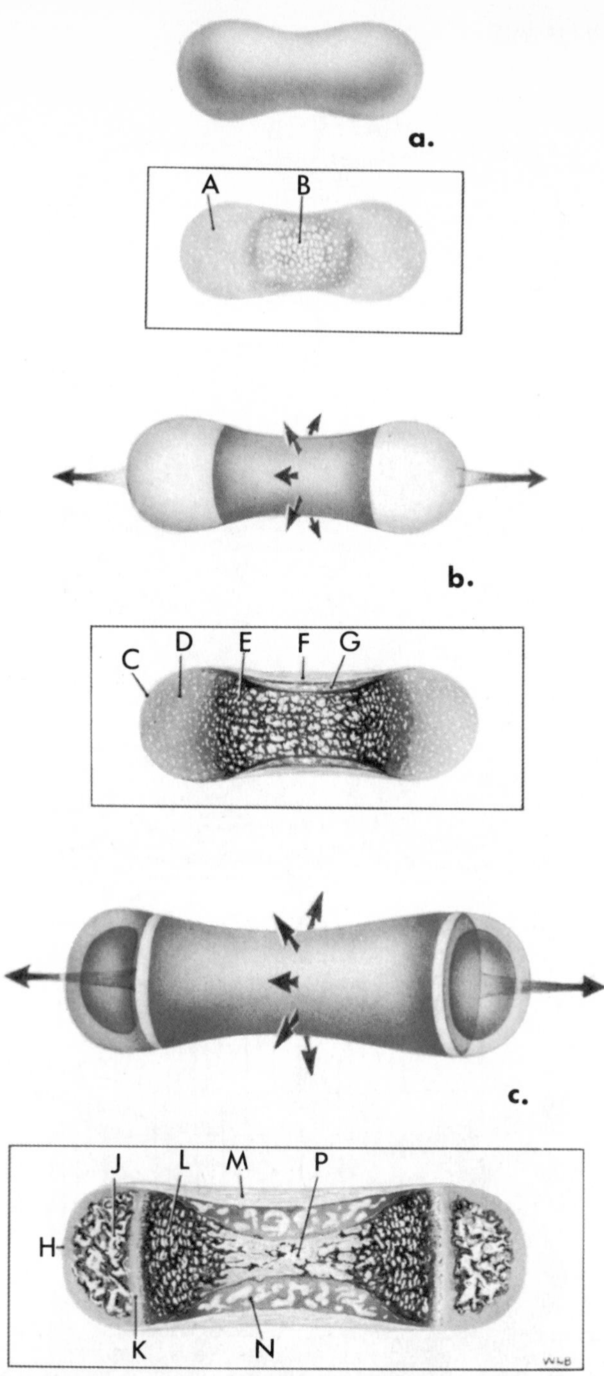

FIGURE 18–8. Endochondral ossification. In the cartilaginous primordium of a bone *(a)*, a primary center of ossification first appears *(B)*; the remainder of the element is still cartilaginous *(A)*. The "bone" then grows in two directions, longitudinally and circumferentially (arrows in stage *b*). The epiphyses are still composed entirely of cartilage *(C)*, but the articular surfaces lack a perichondrium. The underlying cartilage mass continues to undergo rapid interstitial proliferation *(D)*. As it grows in a linear course, it is successively replaced by endochondral bone, which thus also lengthens in a linear direction in the medullary region *(E)*. The former perichondrium is now functioning as a periosteum *(F)*, and it lays down the subperiosteal (intramembranous) bone of the enclosing bony cortex *(G)*. With continued growth in both longitudinal and diametrical directions (stage *c*, arrows), secondary centers of ossification *(J)* form the bony epiphyses. The epiphyseal plate of "growth cartilage" remains between the primary and secondary parts *(K)*. A special kind of "secondary" cartilage develops into the articular cartilage *(H)*. As the epiphyseal plate proliferates by interstitial growth in a direction toward its own end of the bone, successive replacement by endochondral bone takes place *(L)*. At the same time, cancellous bone previously occupying the center of the bone undergoes removal *(P)*. The periosteum *(M)* continues to lay down bone, and the **cortex** increases in thickness as well as length *(N)*. (From Enlow, D.H.: *The Human Face*. New York, Harper & Row, 1968, p. 46.)

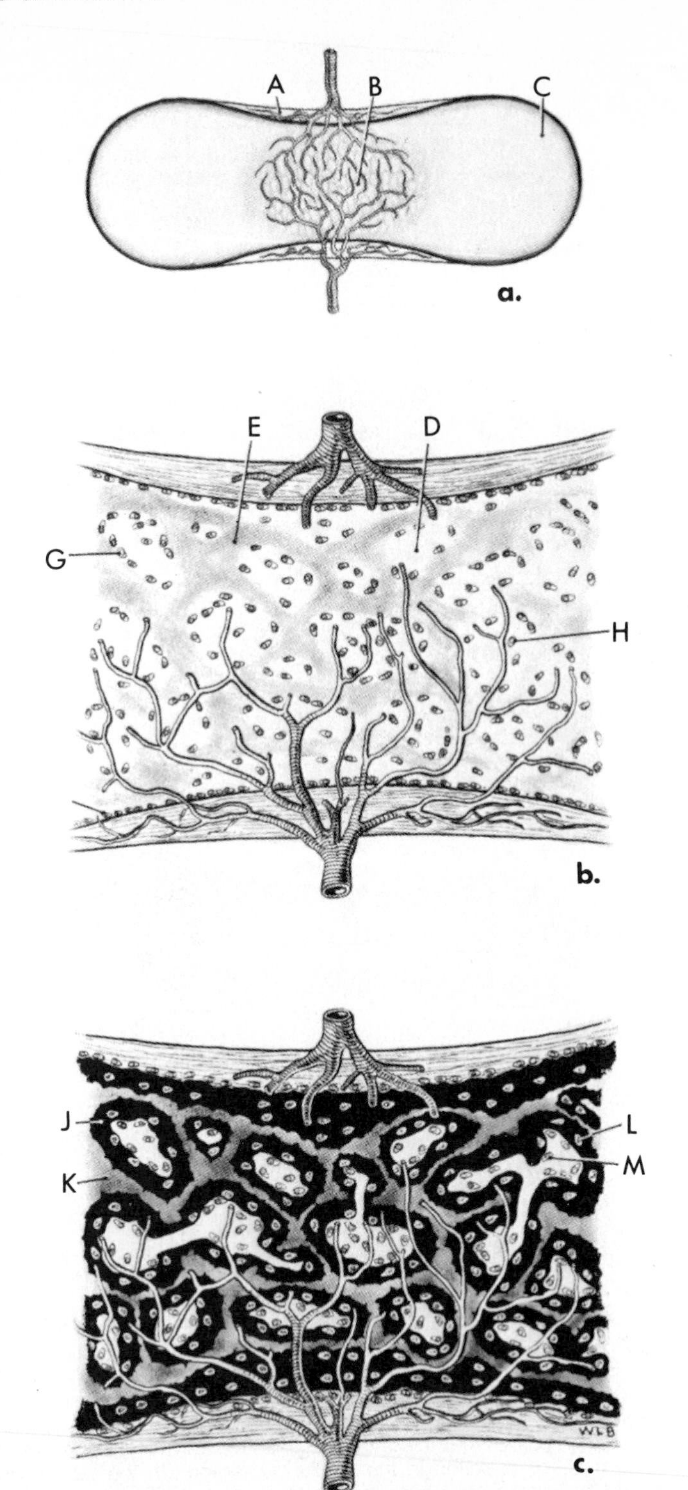

FIGURE 18–9. Endochondral ossification. The cartilaginous prototype of the bone *(C)* has a primary center of ossification *(B)* that involves chondrocyte hypertrophy, matrix calcification, and invasion by vascular buds from the periosteum *(A)* bringing in undifferentiated connective tissue cells. In stage *b*, the calcified matrix *(E)* has become permeated by anastomosing erosion tunnels, and each space *(D)* contains vessels and undifferentiated cells *(H)*. Osteoblasts develop from these cells; and in stage *c*, a thin shell of bone *(J)* has been laid down on the remnants of the calcified cartilage matrix *(K)*. Endochondral bone, as a type, can easily be recognized in sections by the presence of these identifying spicules. Some of the osteoblasts *(M)* have been incorporated in the fine cancellous trabeculae as osteocytes *(L)*. (From Enlow, D. H.: *The Human Face*. New York, Harper & Row, 1968, p. 48.)

increasing the amount of **matrix** between cells (as in loose connective tissue). Many tissue types combine two or all three of these different modes of growth (as in cartilage). All are interstitial systems of growth because they involve expansive changes of tissue components already present. Bone, because it is hard, must necessarily grow by a process of adding new cells and new matrix onto the existing **surfaces** of previously formed generations of bone tissue. It cannot, of course, expand interstitially by division and proliferation of osteocytes because the cells have no place to divide to; they and their genes are locked into their calcified, nonexpanding matrix. Bone **must**, therefore, grow in relationship to a covering or lining membrane or some other soft tissue such as cartilage or tendon. It is the bone surface, either periosteal or endosteal, that is the site of growth activity. This type of growth is termed "appositional," in contrast to "interstitial" expansion. All bone that is fully calcified grows in this manner regardless of its mode of osteogenesis (endochondral or intramembranous).

Where **compression** is involved that exceeds the connective tissue membrane's level of capillary pressure, as mentioned above, the intramembranous growth mechanism (which is dependent upon vascular membranes, as the name indicates) does not have the capacity to function. Thus, the articular ends of a bone and the epipyseal plates are composed of cartilage, which **can** grow and function in a pressure environment. The cartilage grows **toward** the site of compression. Epiphyseal plates, synchondroses, and other "growth cartilages" provide for the linear enlargement of bones that have pressure contacts at their ends. The cartilage grows interstitially on one side, and as it does so, the older part of the cartilage on the other side is removed and replaced by bone. The cartilage functions essentially as a kind of advance ram that protects the sensitive endosteal bone membrane beneath it and, importantly, **also provides for the growth elongation of the bone at the same time.** The **other** areas of the bone grow intramembranously.

During active growth, a bone surface is constantly changing because of new additions; any given point within the compact substance of bone **used** to be an actual exposed surface, either periosteal or endosteal. The point in stage *a* of Figure 18–10 is on the periosteal surface, but as new deposits are laid down by the osteogenic layer of the periosteum, it is "relocated" in stage *a′*. The point itself has not moved, but its relative position has become shifted because of the surface additions that cover it. If a **metallic implant marker** or **vital dye** (such as alizarin or the procion dyes) is used on a living bone, growth changes that occur after the marker is implanted or the dye is administered can be determined. Such markers and the lines formed by the vital dyes become covered over wherever subsequent surface growth occurs, just as point *a* is covered. (Note: Vital dyes stain only that bone actively being laid down during the period in which the dye is in the bloodstream. Thus, one injection forms a thin, colored line on the surface of the bone. Subsequently formed bone deposits are not colored.) Metallic implants (tiny tantalum pins) can be injected into the cortex of a growing bone by means of a special "gun." These radiopaque markers can then be seen in x-ray films taken days or weeks later in order to determine where and how much a bone has been **remodeled** by deposition and resorption relative to the markers. **Displacement** movements of whole bones can also be determined by noting the directions and amounts of separation among implants previously inserted into two or more separate bones.

Bone **deposition** is only part of the overall process of bone enlargement; it is one phase of a multiphase growth system. **Resorption** is another part, and this is just as important and necessary as deposition (Fig. 18–11). Ordinary

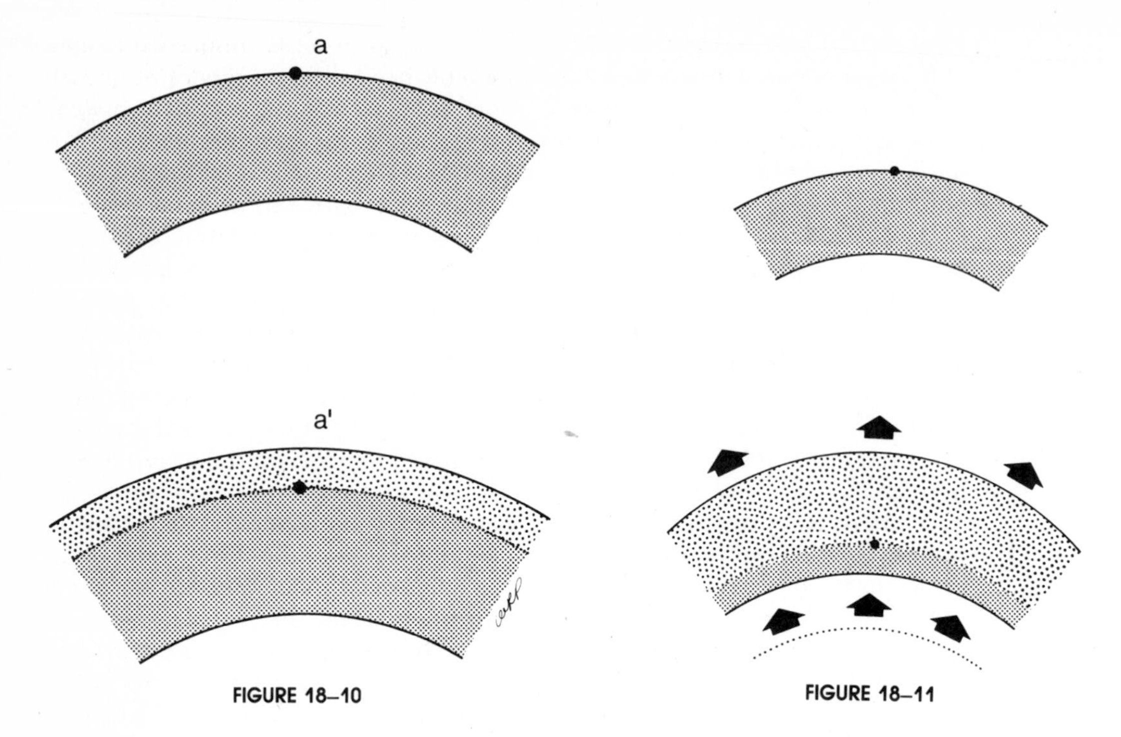

FIGURE 18–10

FIGURE 18–11

resorption associated with growth is not "pathologic," although beginning students sometimes mistakenly view it as evil because resorption is a destructive process or because it can also be involved in some diseases. Resorption **must** accompany deposition, and deposition on one side of a cortical plate (or cancellous trabecula) with resorption from the other brings about the growth movement of the part. Thus, deposition on the periosteal side with resorption on the endosteal surface of a cortex moves the whole cortex outward and at the same time proportionately increases its thickness. As pointed out in Chapter 2, however, **remodeling** also takes place throughout the entire bone because the bone does not simply expand by generalized external deposition and internal resorption.

Remodeling (Fig. 18–12) involves various combinations of deposition *(1, 3)* and resorption *(2, 4)* on different periosteal and endosteal surfaces in order to **move** a part of a growing bone into a new location (see Chapter 2). The process of displacement also occurs as a basic part of the growth mechanism.

Except at contact surfaces involving direct, high-level pressure, the bone grows by the surface depository activity of the osteoblasts present as the innermost cellular layer of the thick periosteum or the very thin endosteum (the latter is only about one cell thick and lacks a thick, fibrous layer because muscles, tendons, and other such force-adapted tissues are not attached to it). Vessels enclosed within endosteal circumferential layers of bone characteristically enter at a right angle because the endosteum is not under tension and its vessels therefore are not drawn out toward the long axis of the bone. Also, periosteal "slippage" on the bone is involved in the periosteum in relation to the forces acting on it as it grows, and this causes the periosteal vessels to become enclosed at much more acute angles. The old concept of perpendicular "Volksmann's" canals entering the bone from the periosteum needs to be forgotten.

As new bone is progressively laid down, the covering periosteal membrane **moves** outward. The membrane moves inward if the periosteal surface is resorptive. Thus, in Figure 18–13, periosteum *a* moves in an outward direction to *a′* as its osteoblasts lay down the new bone in "space" *x*. The endosteum *b* moves to *b′* as its osteoclasts remove bone. (Reverse the sequence if the bone is growing inward by endosteal deposition and periosteal resorption.) However, these membranes do not simply "back off" as they bring about the cortical movement. They are not merely pushed or pulled into their new positions. Rather, the membranes each **grow** from the one location to the other. It is the growth movement of the membranes that leads to growth movement of the bony cortex located between the membranes. The membrane has its own internal, interstitial growth process. Just as the bone **remodels** during growth, the periosteum also undergoes its own internal remodeling process. Remember that the membrane itself paces the bone changes and that the "fields" of growth activity (described on page 45) reside in this membrane and the other soft tissues, rather than in the bone.

As collagenous fibers and ground substance are laid down by the osteoblasts (*g* in Figure 18–14), this layer of **osteoid** almost simultaneously undergoes calcification to become a new layer of bone tissue (*x*). Some of the osteoblasts are enclosed to become new osteocytes, and some of the periosteal blood vessels (*f*) contiguous with the bone surface are also incorporated. These anastomosing vessels then lie within a network of **vascular canals** as bone is formed around them. Note that the fibers of attachment (Sharpey's fibers) become more deeply embedded as new bone is formed around the collagenous fibers in the innermost layer of the periosteum (*d*). The periosteal fibers now **grow** outward. Fiber

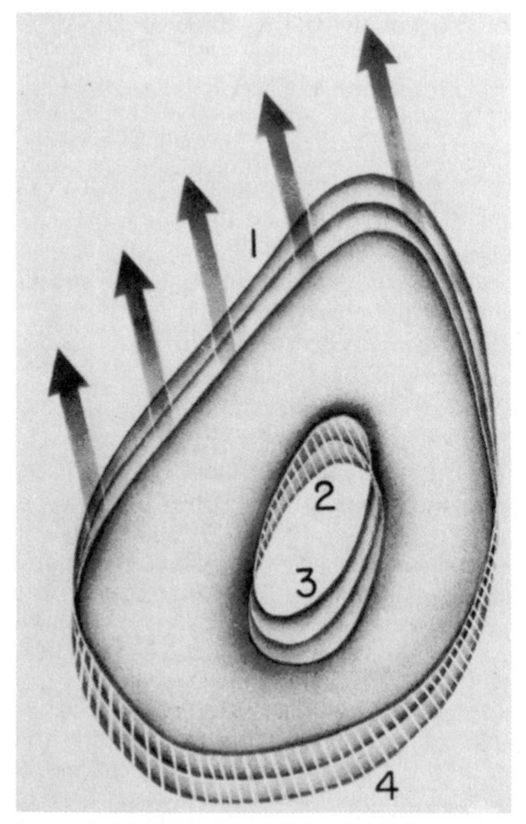

FIGURE 18–12. (From Enlow, D. H.: *Principles of Bone Remodeling.* Springfield, Ill., Charles C Thomas, 1963.)

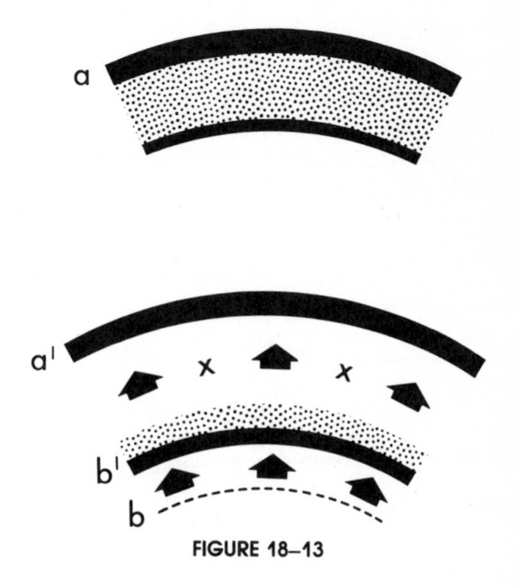

FIGURE 18–13

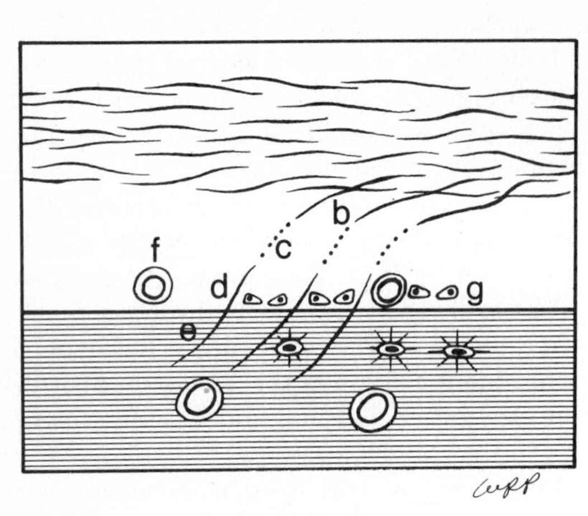

FIGURE 18–14

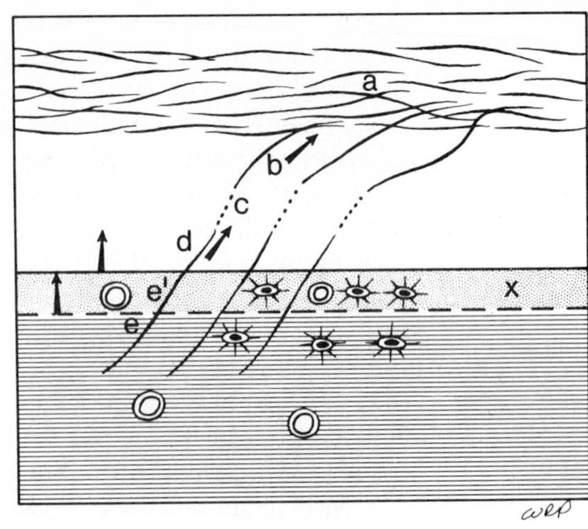

FIGURE 18–15

segment *d* lengthens on the outside while it is being enclosed by new bone on the inside (*e'* in Figure 18–15). It does this by a **conversion** of segment *c* into a new addition onto *d*, which elongates by this process. Segment *c* is a special **precollagenous** fibril. It is a very thin fibril that requires special staining methods (see Kraw and Enlow, 1967). Many such fibrils (that is, "linkage" fibrils) form a distinctive zone within the intermediate part of the periosteal membrane. Under the control of the rich population of fibroblasts, these slender fibrils become remodeled into the thick collagenous fibers that form the lengthening segments of *d*. This is done by binding many of the fibrils together with ground substance (proteoglycans).

Segment *c* now lengthens in a direction away from the bone surface. It is not presently known whether this is done by a remodeling conversion from segment *b* through enzymatic removal of the binding ground substance to release its numerous slender precollagenous fibrils, or whether new precollagenous lengths are added directly to *c* by the fibroblasts in this zone. As these changes occur, however, new *b* segments are being formed by fibroblastic activity as they join the expanding outer "fibrous" layer of the periosteum (*a*). The entire periosteum thus **drifts** outward as the bone surface correspondingly drifts in the same direction. If the periosteal surface is **resorptive**, rather than depository, the sequence of operations is the same, but the direction is reversed. That is, the periosteum and its fibers grow **toward**, rather than away from, the bone surface, which is moving inward, rather than outward.

How does a muscle or tendon maintain continuous attachment onto a bone surface that is **resorptive**? Moreover, how does a muscle **migrate** over a bone

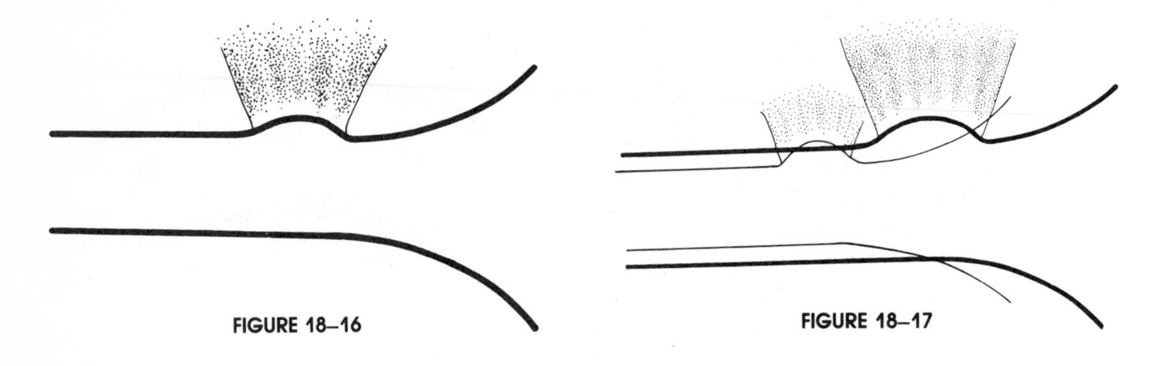

FIGURE 18–16

FIGURE 18–17

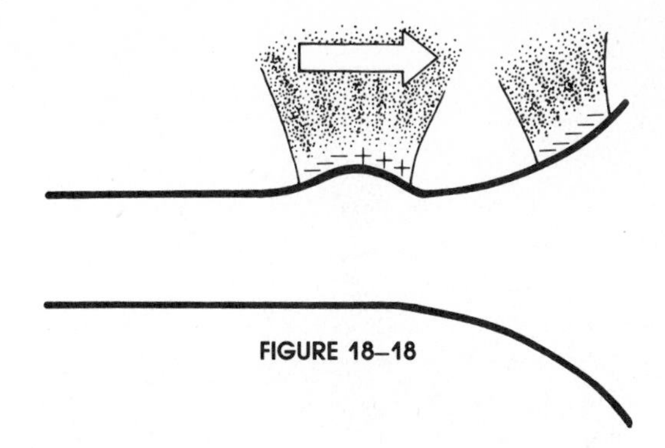

FIGURE 18–18

surface (whether resorptive or depository) as the bone lengthens? For example, the muscle shown in Figures 18–16 to 18–18 **moves** its insertion. It must also sustain constant attachment on a surface, part of which is undergoing progressive resorption, thus seemingly destroying its fibrous anchorage. The other muscle shown in Figure 18–18 attaches entirely onto a resorptive surface. Bone resorption is customarily regarded as a process that results in total destruction of the bone tissue, including its fibers of attachment. In many non-stress-bearing locations, this is true. In some (not all) areas involving muscle and tendon attachments, however, the proces of fiber destruction is **not** complete. Importantly, some of the fibers in the ordinary bone matrix are not removed by the resorptive process (Figs. 18–19 and 18–20). These fibers become uncovered as the remainder of the bone matrix around them is resorbed, and they are then freed to function as fibers of the **periosteum** while retaining continuity with and attachment to the fibers in the bone of which they were once a part. Thus, surface layer *m* is resorbed (Fig. 18–21). This releases *b* fibers, which are continuous with the periosteal *a* fibers on one side and the *c* bone fibers on the other side. The *b* fibers thereby become transferred from bone fibers to periosteal fibers. A second commonly seen histogenetic mechanism also provides for connective tissue attachments onto resorptive surfaces

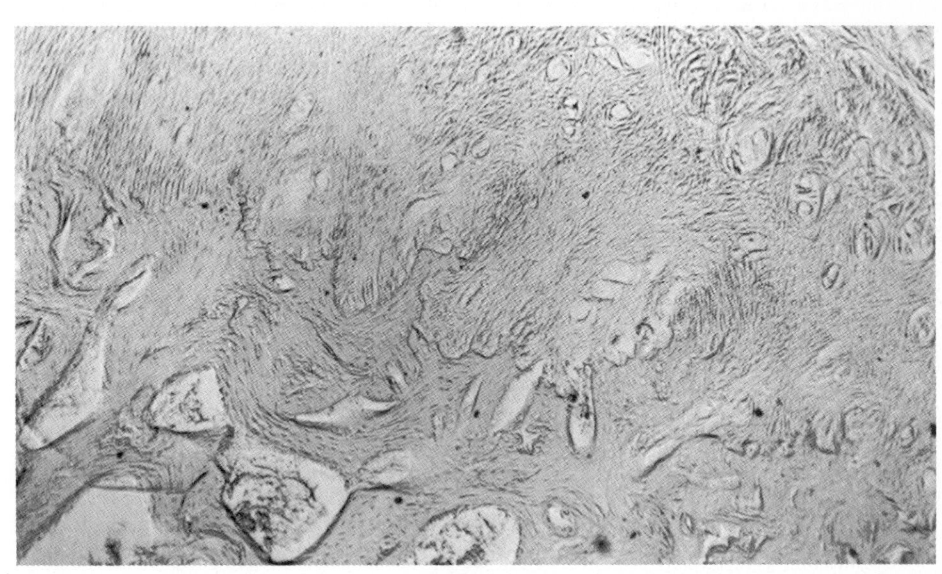

FIGURE 18–19

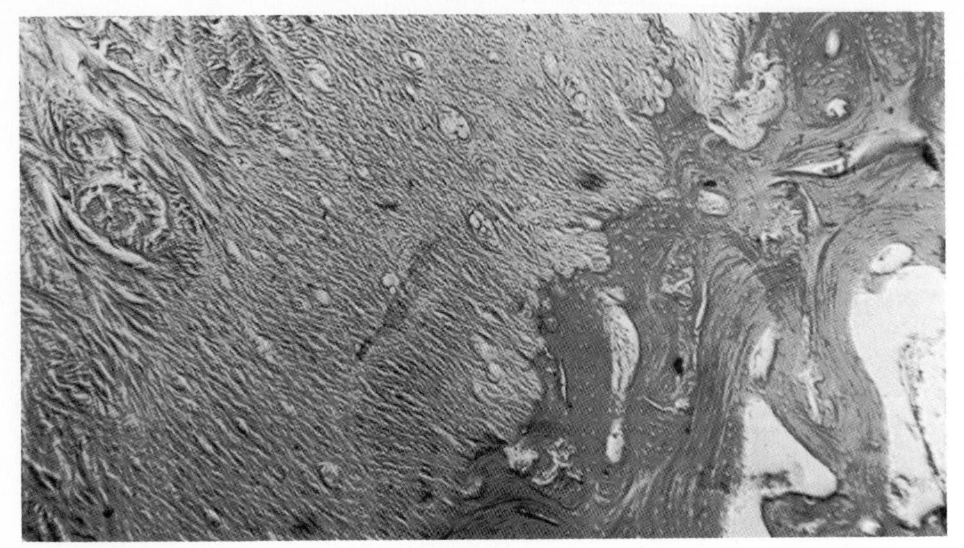

FIGURE 18–20

of bone (Kurihara and Enlow, 1980b and c). This involves an "adhesive" mode of attachment in which certain fibroblast-like cells secrete proteoglycans onto the naked surface, previously resorptive in nature. New precollagenous fibrils are then formed as the adhesive secretion of proteoglycans continues. The new fibrils then link to the more mature fibers in the periosteum, the proteoglycans presumably serving as the binding agent. If new bone is subsequently deposited by the periosteum, the calcified adhesive interface then becomes a "reversal line."

In other stress-bearing locations, another separate mechanism is seen that provides for continuous fibrous attachments on resorptive and remodeling bone surfaces. This involves a reconstruction of the bone **deep** to the resorptive surface in order to provide fibrous anchorage. "Undermining" resorption takes place, in which resorption canals are formed well below the surface of the inward-advancing resorptive periosteal front. New bone is then laid down in these spaces, and it is **this** attachment that provides additional fibrous anchorage while the outside bone surface itself is undergoing resorptive removal. Thus,

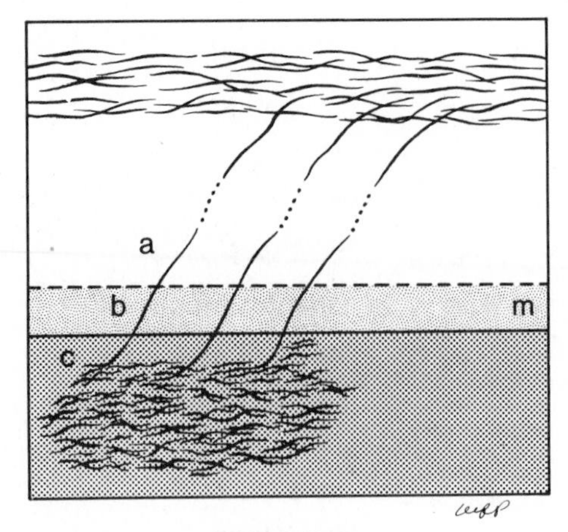

FIGURE 18–21

as periosteal surface *a* is resorbed, a large number of resorption spaces are formed at *b* in Figure 18–22 (in thin sections, many appear as cut-off canals). These spaces anastomose with each other. The fibers in the new bone subsequently deposited in them are continuous with the fibers of the periosteum, and anchorage is thereby maintained (Fig. 18–23). The structural result is a generation of haversian systems (secondary osteons) **deep** to the surface. The fibrous matrix of each osteon and its connection by labile linkage fibrils to the inward-moving periosteum are protected from resorption until the resorptive front reaches them. However, new waves of haversian systems are constantly being formed in advance of the resorptive front, so that new deeper osteons replace, in turn, those that become exposed as they reach the resorptive periosteal surface. It is believed that the combination of this process in some areas with the process of bone matrix fiber release and/or adhesive reattachments in other areas maintains continuous muscle attachment during the period of active skeletal remodeling. Moreover, a muscle is believed to move and migrate along a bone surface by this same process of haversian formation, as well as by lateral reconnections of the labile linkage fibrils (*x*) in the intermediate part of the periosteum (or equivalent areas for direct tendon insertions). Thus, the precollagenous linkage fibrils connecting with fiber *a* in the outer part of the periosteum (Figs. 18–24 and 18–25) become recombined with the precollagenous fibrils of fiber *b′* in the inner part of the periosteum, and so on. This progressively moves the entire muscle across the bone surface to keep pace with the elongation of the entire bone. Separations of fiber bundles by enzymatic removal of the ground substance binding, together with fibril regroupings by new ground substance formation, are believed to carry this out.

Sutures have an osteogenic process comparable to periosteal bone growth. The suture is an inward reflection of the periosteal membrane; and the various fibrous, linkage, and osteoblastic zones are directly continuous from one to the other (Figs. 18–26, 18–27A, and 18–28). As a new layer of bone is added (*x*), inner collagenous fibers *d* become embedded to form new attachment fibers (*e′*) in the bone matrix. The *d* fibers, however, lengthen by conversion from the labile linkage fibers (*c*) (just as previously described for the periosteum), and *d* fibers converted, in turn, into lengthening *c* fibers (or fibroblastic activity

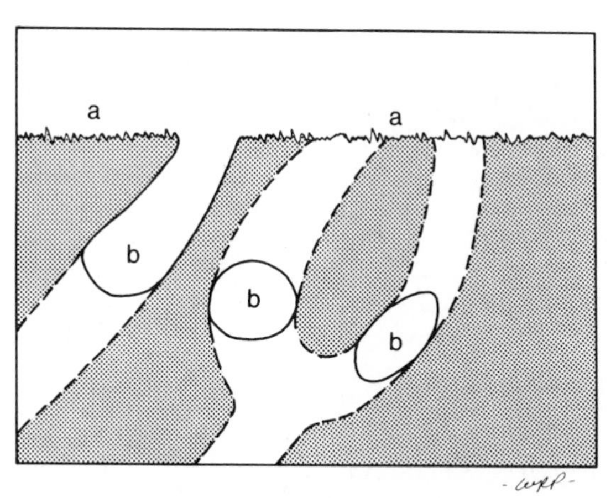

FIGURE 18–22

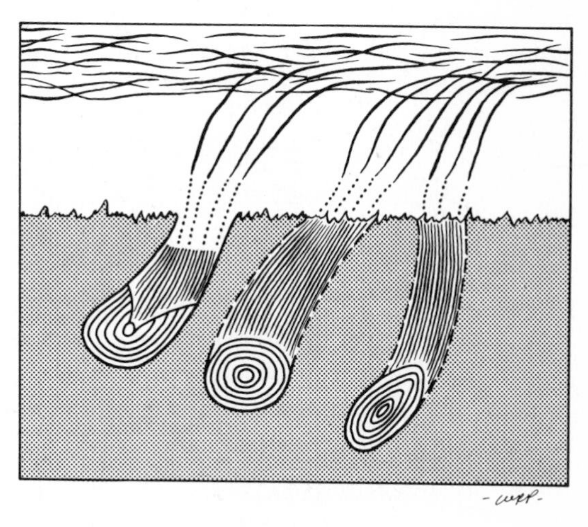

FIGURE 18–23

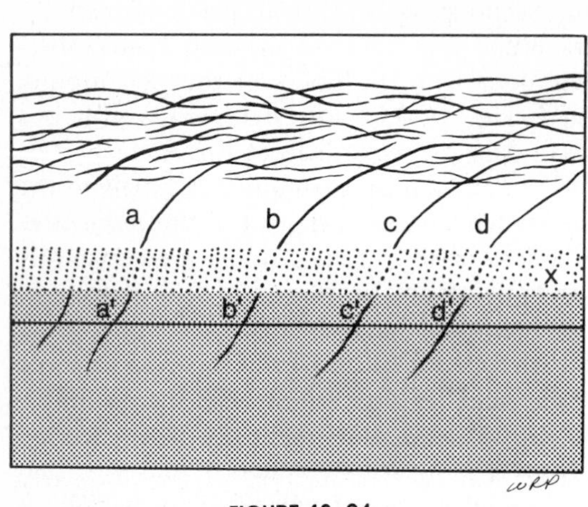

FIGURE 18–24

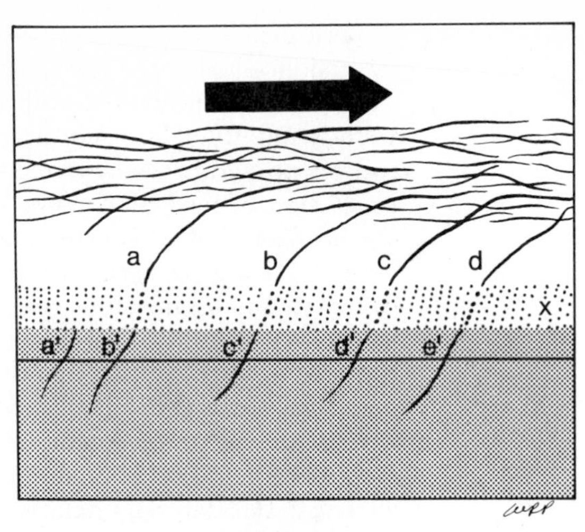

FIGURE 18–25

may bring about direct lengthening of *c*, as in the periosteum). As new bone is added to the sutural contact surfaces, the bones are simultaneously displaced away from one another (see Chapter 3 for a discussion of the physical forces that cause this displacement). Many sutures have three basic layers (on each side), as indicated here. Some sutural types, however, have another layer of loosely arranged fibers located within the center of the dense-fibered capsular layer dividing the two sides. The basic plan of growth and remodeling, however, is the same. When the process of growth ceases, the suture becomes essentially a mature ligament, and the precollagenous linkage fibrils are no longer present.

Importantly, the source of the propulsive force that produces the "downward and forward" displacement movement of the nasomaxillary complex at its various sutures has long been a subject of controversy. It has recently been suggested that the abundant population of actively contractile fibroblasts ("myofibroblasts," *m* cells in Figure 18–27B) within the linkage zone of the sutural membrane provide, at least in part, a contractile force that exerts tension on the fibrous framework. This in turn presumably pulls one bone along its sutural interface with another bone or at least contributes to the moving, relinking placements of the fibers as some other propulsive force causes the bone movements. The bone thus "slides" along the suture as new bone tissue, at the same time, is laid down on the sutural edges. The midface is thereby **pulled**

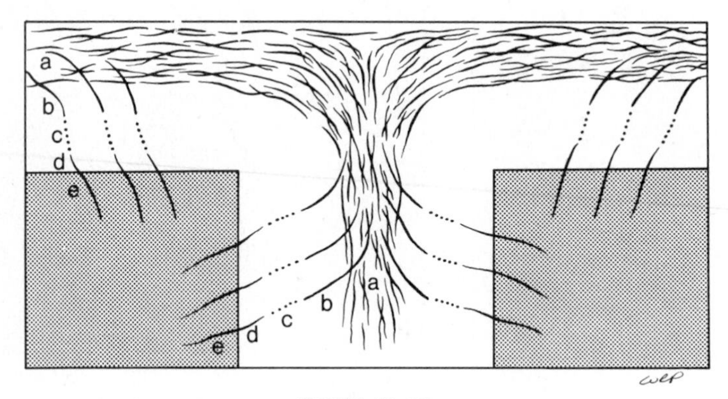

FIGURE 18–26

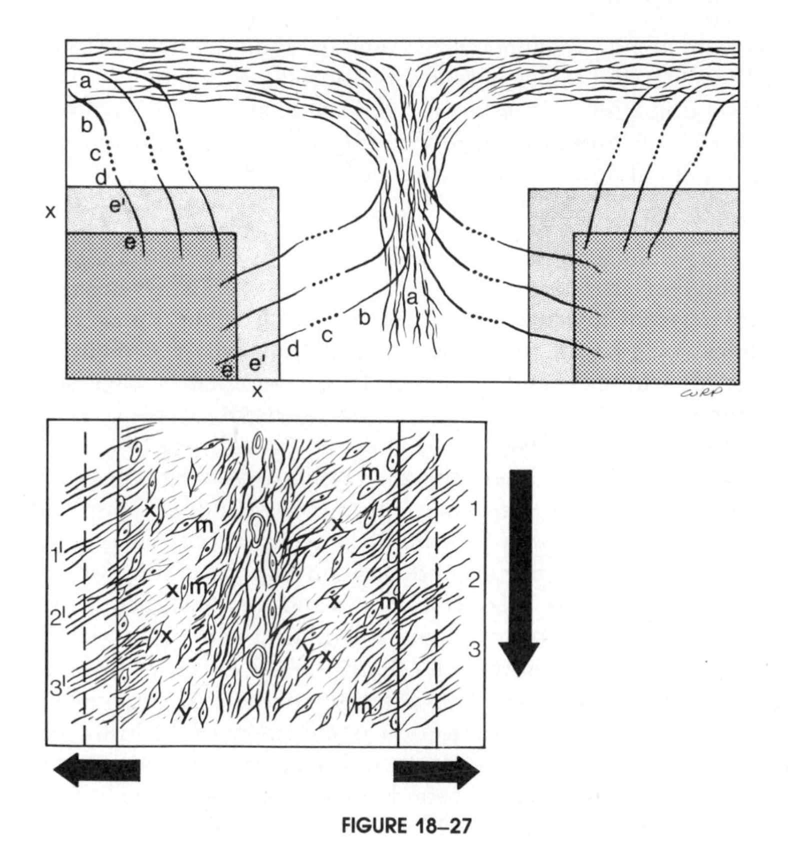

FIGURE 18–27

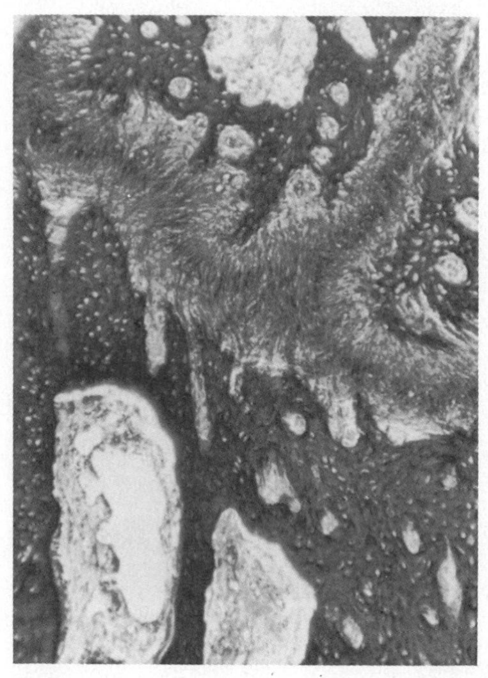

FIGURE 18–28. (From Enlow, D.H.: *The Human Face*. New York, Harper & Row, 1968, p. 96.)

forward and downward along its zygomaxillary, among others, sutural surfaces. Special collagen-degrading and collagen-producing fibroblasts (x and y) provide the fiber remodeling and ground substance relinkage changes also involved. Fibers at level *1*, which were formerly linked with fibers at level *1'*, have become relinked with fibers at *2'*, and so on. (See Azuma et al., 1975.)

BONE TISSUE TYPES

Many different varieties of bone tissue exist, and each of these relates to **specific** circumstances of growth, physiology, and pathology. At present, at least most histology texts imply that bone generally is composed of haversian systems. The **secondary osteon** is described as the universal "unit" of bone tissue structure. This is not true. Like the 14 clubs in the golf bag, there are special kinds of bone for different circumstances. Very fast-growing types exist wherever rapid bone growth must keep pace with a correspondingly fast rate of soft tissue expansion. There are slow-growing bone types as well. Some kinds of bone tissue are adapted to large masses of deposition; others, to thin cortical deposits. Some bone types are associated with muscle and tendon attachments; others, with processes of cortical reconstruction; still others, with alveolar tooth attachments. Some bone types grow outward; others, inward. Some are densely vascularized; others, sparsely vascular. The processes of normal remodeling produce almost limitless variations in the localized arrangement of the histologic components of bone. The cortex becomes layered during growth and remodeling, and the pattern of stratification and the bone tissue types in each layer differ in the different parts of a bone (Fig. 18–29). A section made at one level of a bone is always different from sections made elsewhere in that same bone

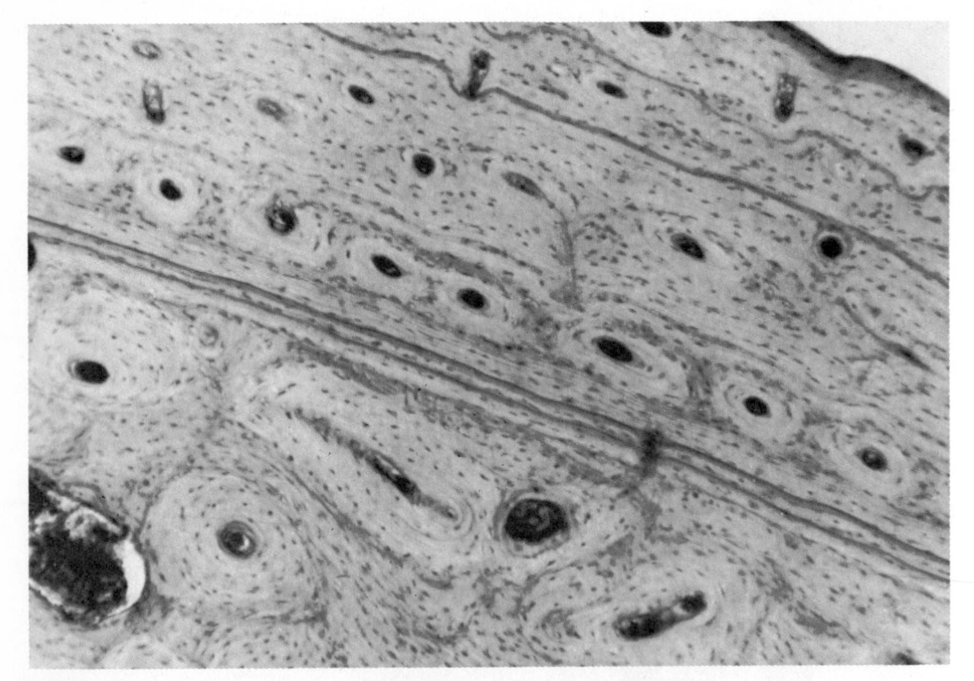

FIGURE 18–29. This cortical section illustrates the stratified nature of the cortex and also shows the difference in structure between secondary osteons (haversian systems) located in the lower region and the smaller primary osteons in the center of the field. See text for details. (From Enlow, D. H.: *Principles of Bone Remodeling*. Springfield, Ill. Charles C Thomas, 1963.)

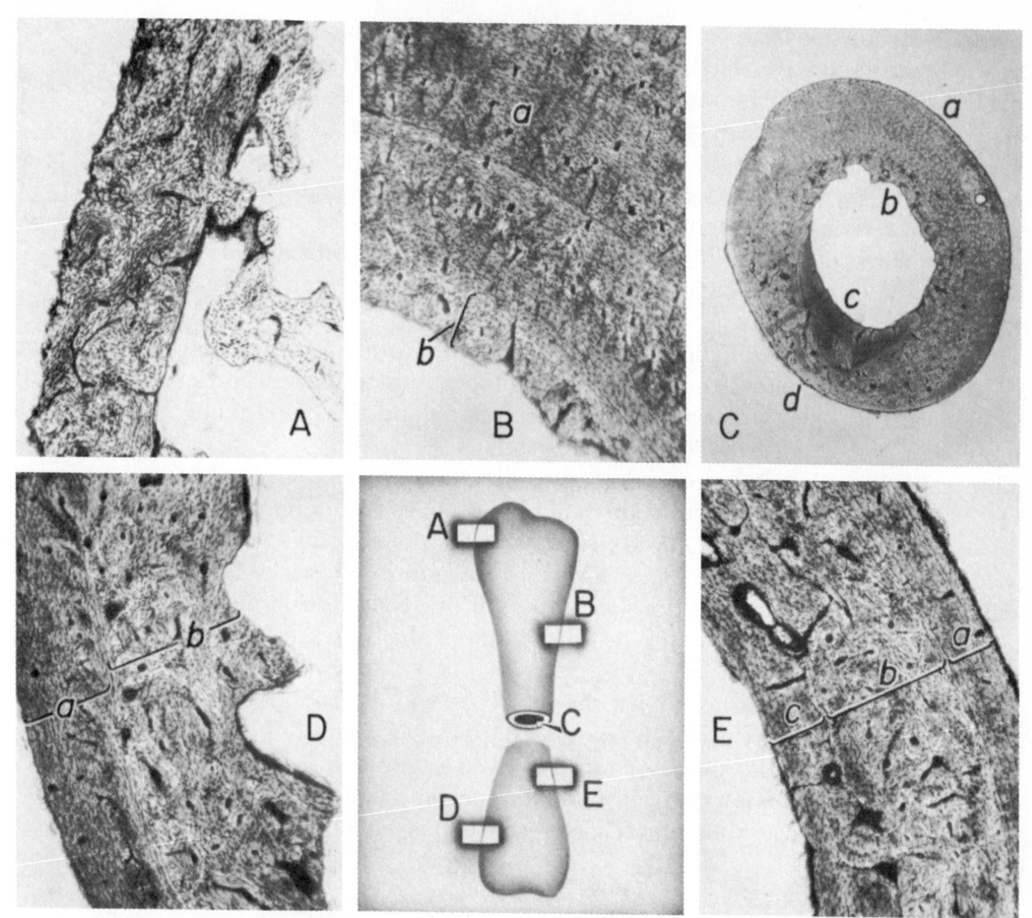

FIGURE 18–30. Sections taken from different parts of a bone always show marked differences in histologic pattern owing to the differences in **regional** growth and remodeling. The cortex taken from location *A* is composed entirely of endosteal bone. The periosteal surface is resorptive, and the endosteal surface is depository. The bone tissue is of the "compacted cancellous" type. Section *B* shows a wide periosteal zone *(a)* composed of primary vascular bone and a remnant of an endosteal zone *(b)* that now has a resorptive inner surface. In section *C*, the cortex is drifting in a northeast direction to form the lateral curvature of the bone. Surface *a* is depository, *b* is resorptive, *c* is depository, and *d* is resorptive. Section *D* shows a zone of periosteal bone *(a)* separated from a layer of endosteal bone *(b)* by a reversal line. The endosteal surface, formerly depository, is now resorptive following outward reversal, Section *E* is composed of a layer of periosteal bone *(a)*, a middle zone of compacted cancellous bone *(b)*, and an inner layer of circumferential bone *(c)*. Compare with Figure 18–42.

because of the characteristic nature of localized remodeling changes in each of the different parts of any individual bone (Fig. 18–30). Just as a geologist can reconstruct the history of sedimentation and erosion in large rock outcroppings, the histologist can reconstruct the depository and resorptive remodeling history of a bone recorded in the rocklike substance of the cortex.

The bone tissue of a newborn is quite different histologically from the bone of an older child. This is different from the bone tissue of a mature adult, which, in turn, differs from aged bone. The reason is that the **circumstances** (rate of growth, amount, and so forth) are different for each age. The pattern of bone structure in a human being differs from that in a rat. The rat is tiny, the cortex of any given bone is much thinner, the metabolic rate is different, the rates of growth differ, the mechanical forces are different, and the rat lives only a year or two. Research investigators are continually "discovering" that the bone of a 6-month-old rat weighing a few ounces is considerably different histologically from that of a 50-**year**-old, 200-**pound**-man. Surprise is always expressed because no haversian systems (the legendary universal units of bone) are observed in the rat. The secondary osteon in no way is a "unit" of bone construction. It is not present at all in most vertebrates. In those species in which it indeed occurs, haversian systems are mainly a feature of larger animals having a longer life span and are found in abundance only at older age levels. When present, haversian systems form in specific locations of a bone. The orthodontist, when manipulating the bone tissue of a child, is **not** dealing with the haversian type of bone. What, then, is the structure of bone in the human child?

Outlined below is a brief summary of some major varieties of bone tissue. There are many more. However, these are the types most often encountered by the clinical researcher concerned with human bone and the bone tissues of common laboratory animals. (See Enlow, 1968b.)

Fine cancellous bone tissue is the fastest-growing bone type of all (Fig. 18–31). It is characteristically found throughout all bones in the fetal skeleton and in some parts of the postnatal skeleton as well. It is also involved in the callus formation of a healing fracture. Osteophytes and exostoses, commonly present in many pathologic conditions, are layers of fine cancellous bone added to the outer or inner surfaces of the compact bone of the cortex (see also Fig. 18–3).

Fine cancellous bone forms almost the entire cortex of fetal bones, and it is found in some cortical areas of the **fast**-growing parts of a child's bone (as in the posterior border of the growing ramus). Even though this bone type is located in the cortex, it is not "compact" bone; it is porous. The spaces, however, are not nearly as large as those in "coarse" cancellous bone (another major

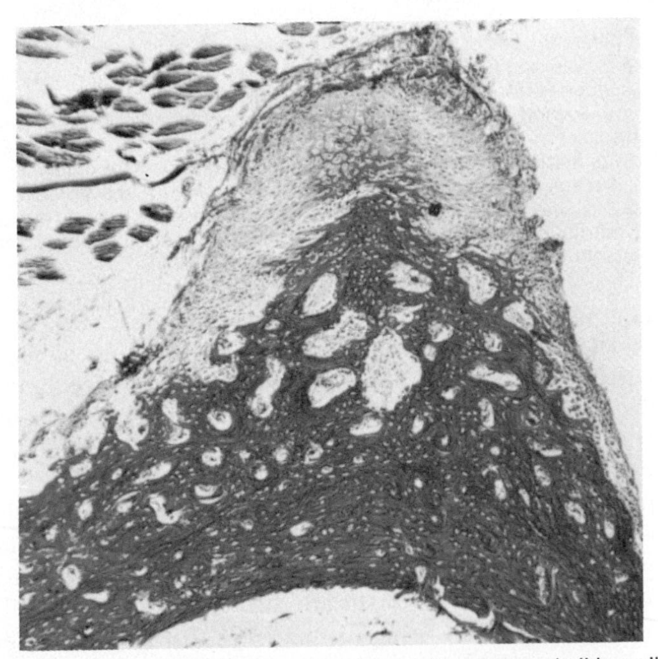

FIGURE 18–31. At the apex of many tubercles and tuberosities, as seen in this section, chondroid bone is formed. This unique bone type is also found on alveolar crests in the mandible and maxilla during the active growth period. Note the cartilage-like appearance of the large cells. Fine cancellous, nonlamellar bone occurs deep to the chondroid apex of the tubercle. (From Enlow, D. H.: A study of the postnatal growth and remodeling of bone. Am. J. Anat., 110:79, 1962.)

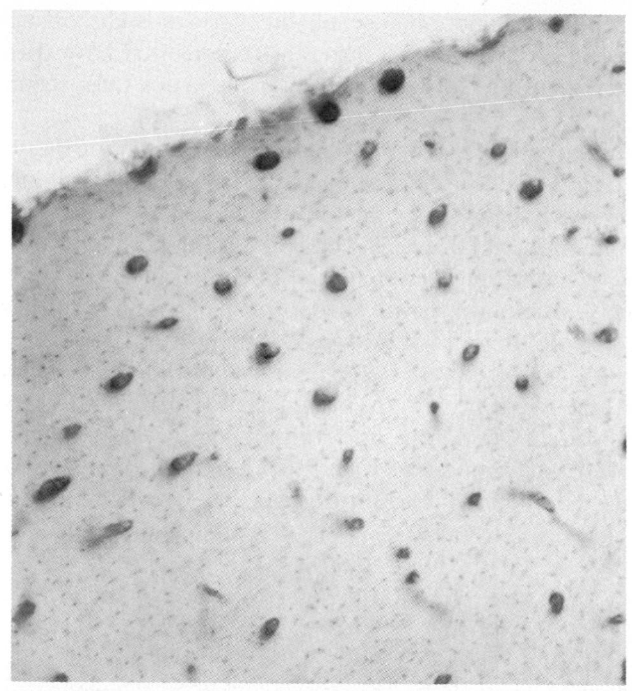

FIGURE 18–32. Areas of fast-forming bone tend to have a richer concentration of vascular canals. This kind of bone is also more resistant to the normal process of necrocytosis. Compare with Figure 18–35. (After Enlow, D. H.: Functions of the haversian system. Am. J. Anat., 110:269, 1962.)

type, described below). Immature connective tissue, not marrow, is located within the spaces. Fine cancellous bone can be formed by both the periosteum and the endosteum.

Coarse cancellous bone has irregularly large spaces containing red or yellow marrow. This is the familiar "spongy bone" composed of trabeculae or thin bony plates. It is a feature of the medulla and is particularly abundant in epiphyses and between the cortices of the flat bones of the skull (where it is termed "diploë"). Coarse cancellous bone is always produced by the endosteum; it never normally forms on the outside of a bone (see Fig. 18–2).

Nonlamellar bone is the type that makes up fine cancellous bone tissue (see Fig. 18–31). It has a matrix in which fiber orientation does not produce the distinct stratification that characterizes the more familiar lamellar bone tissue. One common type of nonlamellar bone is sometimes termed "woven" bone tissue because of the interwoven nature of the fibers throughout the matrix. The osteocytes have a characteristically irregular, jumbled arrangement with no alignment into rows (Fig. 18–32). This is in contrast to **lamellar** bone, in which the collagenous fibers all have a common alignment in each layer (lamella), and the cells are arranged so that their long axes are parallel. Because the fiber directions differ in adjacent layers, a stratified appearance is produced. This fiber arrangement gives a plywood-like strength to the bone substance.

While composing the fine cancellous type of bone, nonlamellar bone tissue may also be formed as compact masses, and this occurs when the rate of deposition is somewhat less rapid. In the child, thus, nonlamellar bone is present wherever growth involves a moderate to rapid rate of bone deposition to accompany surrounding soft tissue expansion. Lamellar bone is slower growing. In any given bone of a young child, mixed layers and areas of lamellar and nonlamellar bone always occur. This is because all parts of a bone do not grow at the same rate and because all the various areas of a bone become shifted (as a result of "relocation") into new regions that have different growth rates. Layers of these and other different bone types thus accumulate in the cortex.

Primary vascular bone is the most common type of bone tissue of periosteal origin in the child's skeleton, including all facial and cranial bones. This is **the** standard periosteal type of bone in the growing skeleton and it may also be formed in some endosteal areas as well (Figs. 18–33, 18–34; also see Fig. 18–45). Curiously, it has never been included in elementary histology texts. There are two general classes of bone tissue: primary and secondary. All bone deposited directly on the outside and inside surfaces of the cortex is "primary," and blood vessels enclosed in these deposits (as the bone matrix is formed around them) are correspondingly designated as primary in type. In certain circumstances, primary bone can undergo a process of cortical reconstruction. The result is "secondary" bone, which includes the haversian system. Primary vascular canals do not have concentric rings of lamellae surrounding them;

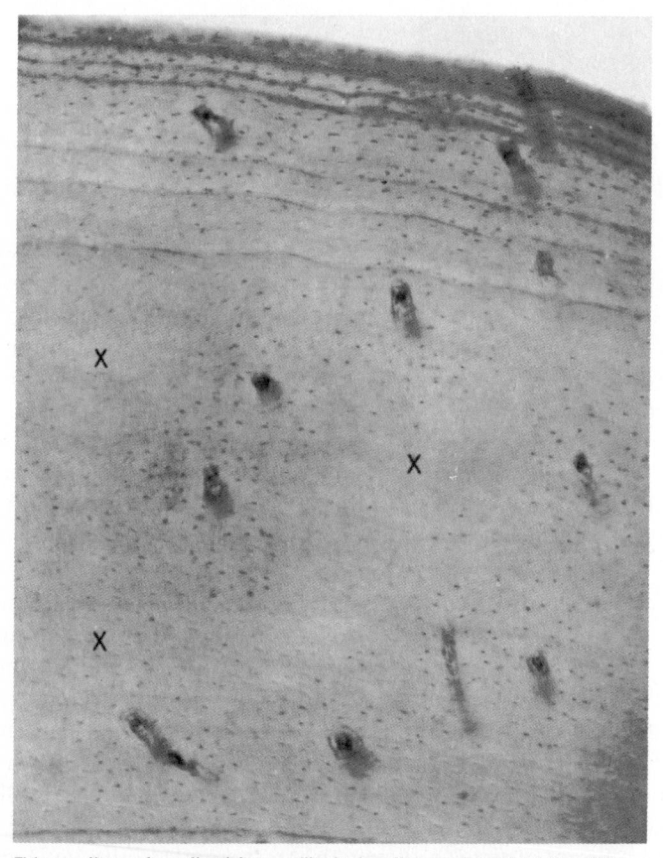

FIGURE 18–33. This section of cortical bone illustrates the concept of the "tissue cylinder." The capillary in each canal supplies a cylindrical region of cells surrounding it. When the spread of capillaries is relatively scant, the intervascular areas outside these physiologic cylinders of effective capillary supply eventually undergo necrosis (note the empty lacunae in areas *X*). In time, such necrotic areas spread as the canaliculi and canals become filled with minerals, and the bone tissue is replaced by the process of haversian reconstruction. The resorption canals and secondary osteons that then develop represent the radius of effective supply from the blood vessel to the osteocytes. The haversian system is a graphic example of the "tissue cylinder"; such physiologic cylinders exist for all vascular soft tissues as well. Compare with Figure 18–38. (After Enlow, D. H.: Functions of the haversian system. Am. J. Anat., 110:269, 1962.)

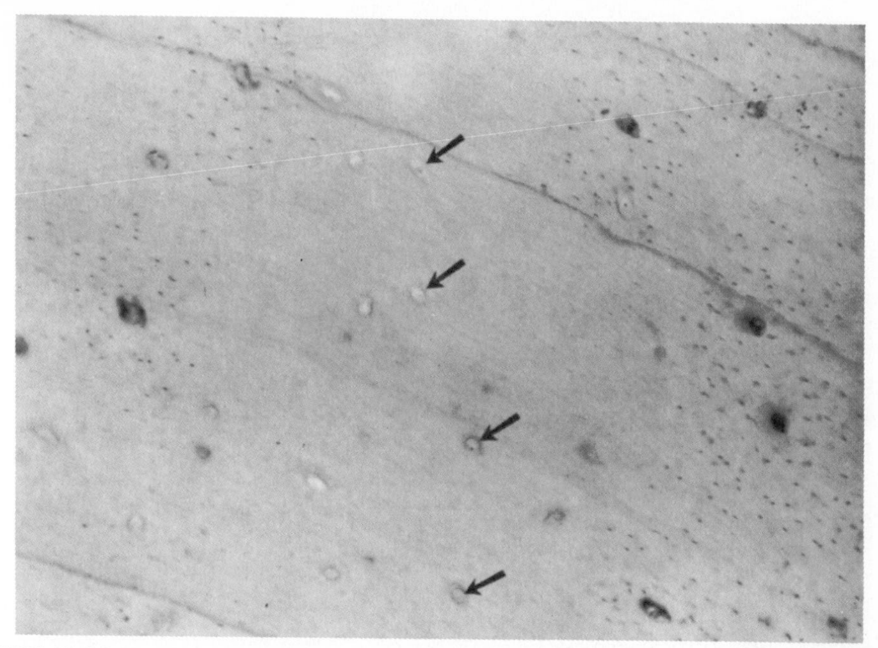

FIGURE 18–34. Empty vascular canals or canals that have become plugged with mineral (arrows) lead to the early necrosis (note empty lacunae) of the "tissue cylinders" surrounding them.

they are simply canals located within a matrix that is composed largely of circumferential lamellae. Primary vascular bone is also the standard form of periosteal bone tissue among most other animals. Only in a relatively few groups is this basic kind of bone replaced by secondary bone tissue. Although primary vascular bone is the dominant type in the human child and young adult, many primary vascular canals are still found in the aged skeleton.

Nonvascular bone is found as the sole type of bone in some species. It is a very **slow**-growing variety. In the human skeleton, it forms in specific locations where the cortex forms leisurely over a period of time. In **all** bone tissue types, in general, the extent of vascularity is an index of the rate of bone deposition. Abundant vascular canals are characteristic in any cortical region that is rapidly formed (see Fig. 18–32). The spectrum of canal density then grades down to totally nonvascular areas that are the slowest forming of all (as in the zone seen in Figure 18–35).

Bone undergoes a normal process of **necrosis**. Osteocytes, like most connective tissue cells, have a limited life span. In bone, the cells live for about 7 years or so, but the length of time is quite variable; the degree of vascularization is a basic factor determining this life span. Less abundant vascular canals that are farther apart favor earlier bone cell death. There is a cylinder-like area of physiologic tissue dependence surrounding each vascular canal. (This is a general rule, also, for the various kinds of tissues and the capillaries within them.) The vessel in each canal supplies the particular cylindrical field around it. When vascular canals are abundantly distributed, all the osteocytes have an adequate vascular supply. However, when the canals are widely separated, as in the slower-growing areas, the outlying regions between the canals have a much lower level of vascular supply because the transport distance from the capillary through the canalicular system of the lacunae is greater. In time, these areas farther removed from the blood vessels begin to undergo necrocytosis (Figs. 18–33 and 18–35). It is a common and, indeed, normal circumstance.

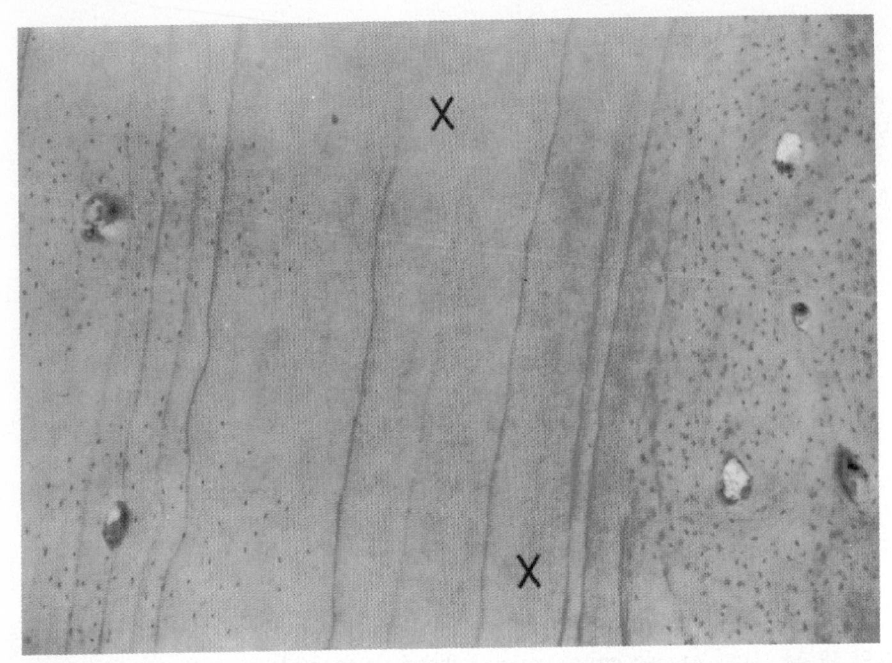

FIGURE 18–35. Zones of nonvascular bone *(X)* often occur in thick, slow-forming cortical areas. Note that this layer has become necrotic (empty lacunae). The bone cells surrounding the vascular canals, however, still survive. (From Enlow, D. H.: Functions of the haversian system. Am. J. Anat., 110:269, 1962.)

Moreover, the canaliculi and even the lacunae and vascular canals often become plugged with calcium deposits. This causes the death of all the bone cells located downstream (see Fig. 18–34). In Figure 18–32, the close proximity of the vascular canals is such that the effective physiologic **tissue cylinder** of each is sufficient to reach all osteocytes. In Figure 18–33, however, the spread of canals is such that the intervascular areas lie beyond the physiologic limits that the canals can effectively supply over a long period. The osteocytes in these areas undergo necrosis in time. Localized necrotic regions can be recognized because the lacunae are empty, and they are often filled with calcium deposits (the clear areas in Figure 18–36; this circumstance is comparable to that found in **sclerotic dentin**). Such necrotic spots are more widespread in adult bone because nonvascular areas or regions having less dense vascularization are correspondingly more widespread, as noted below.

Haversian bone tissue is "secondary" because it replaces the original primary vascular bone. This involves a process of cortical reconstruction. When it is said that "bone remodels throughout life," this is what is meant. There are different kinds of remodeling, as noted in Chapter 2. **Biochemical remodeling** involves the exchanges of minerals and other ions between the bone substance and blood in order to maintain calcium and other levels. Growth remodeling carries out the relocation of a bone's various parts as the bone increases in size, and the remodeling changes are produced by various combinations of resorption and deposition on the periosteal and endosteal surfaces. **Haversian remodeling** is a process of internal reconstruction **within** the cortex, not changes on the outer and inner cortical surfaces. There are several functional reasons why it occurs.

During the period of childhood growth, the constant replacement of bone takes place because of **growth** remodeling. That is, new bone is constantly

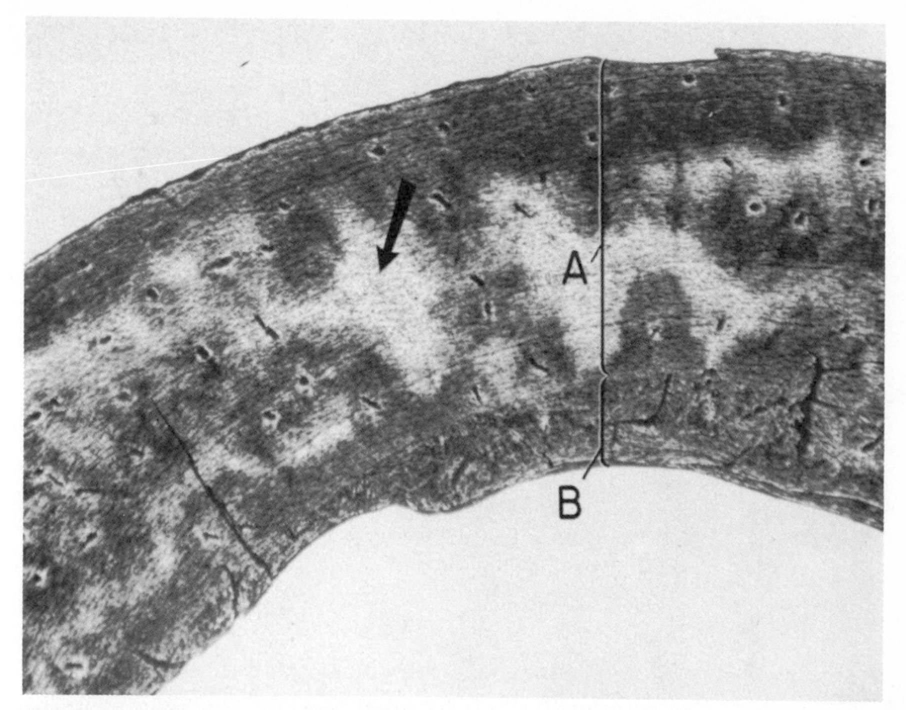

FIGURE 18–36. The outer zone of this cortex *(A)* is composed of primary vascular bone tissue. Zone *B* is composed of irregular endosteal bone. Note the intervascular areas of necrosis (arrow). In a ground section, such as this, necrosis can be recognized by the cleared appearance of the matrix (as seen also in sclerotic dentin). (From Enlow, D.H.: Functions of the haversian system. Am. J. Anat., 110:269, 1962.)

being formed, and previously deposited bone is removed. The bone is not present long enough for any marked extent of osteocyte necrosis to develop. Moreover, the childhood types of bone tissue are more highly vascular because of their more rapid rates of formation. This favors a longer life span for the bone cells. As the child matures, however, slower-growing types of bone are then laid down, and these tend to be less vascular or even nonvascular in many areas. The decreasing abundance of vascular canals in the adult skeleton leads to earlier cell death. Because the bone is no longer being replaced, owing to the turnover by the process of growth remodeling, it remains in the skeleton until much more widespread cell necrosis has time to occur. **Haversian** remodeling is a process that replaces this older dead or dying bone with progressively new generations of living bone tissue. Secondary osteons thus develop by a process of reconstructing the primary bone. This occurs by the same resorptive and depository mechanism utilized in growth remodeling. **Resorption canals** develop by enlarging the original primary canals through osteoclastic activity (Fig. 18–37). New resorption canals can be formed in previously nonvascular regions by vascular invasion. The resorption canal represents the removal of the primary physiologic "tissue cylinder," which is that specific territory supplied by the vessels of each canal. The radius of this tissue cylinder, and the resorption canal that replaces it, is determined by the effective distance metabolites can diffuse to and from the centrally located capillary. Lamellar bone is then laid down from outside to inside within each resorption space. The result is a secondary osteon, and it is structurally (as well as functionally) a true cylinder. **All** vascular tissues, soft as well as hard, have "tissue cylinders"; the haversian system in bone is a graphic example that can actually be seen (Fig. 18–38). The

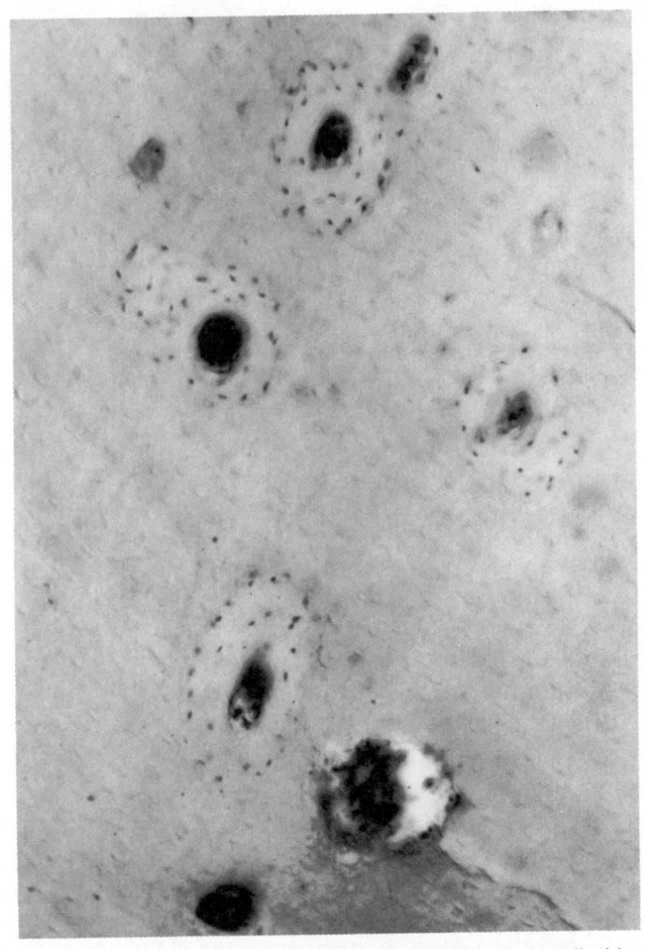

FIGURE 18–37. Note the resorptive canal and the replacement secondary osteons that have developed within an area of cortical bone that is necrotic (empty lacunae). (From Enlow, D.H.: Functions of the haversian system. Am. J. Anat., 110:269, 1962.)

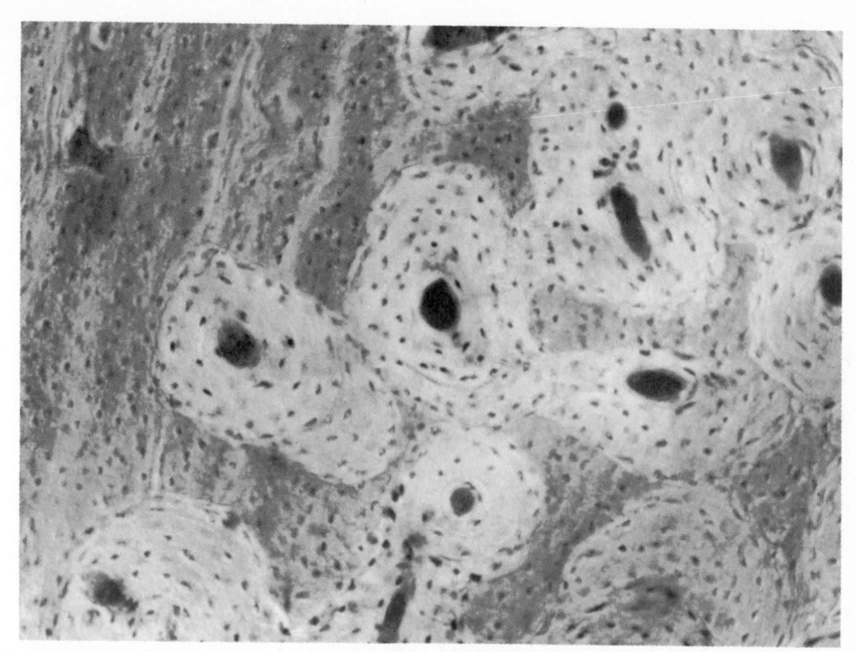

FIGURE 18–38. Secondary osteons. Each haversian system is formed by a process of secondary reconstruction of existing primary or secondary bone. A resorption canal first develops (see Figs. 18–4 and 18–37). Concentric lamellae are then laid down within the enlarged tubular space until the resulting central canal has the diameter of the original primary vascular canal. Second **generations** of osteons have developed in this section (osteons superimposed over one another). Note remnants of the original primary lamellar bone between the haversian systems. Interestingly, haversian **systems** were not known to Clopton Havers, an English physician and anatomist. Osteons as such were not described until well into the following century. Havers, however, believed that he was the first to observe vascular **canals** in bone (during the 1680s), but Leeuwenhoek had actually published on their existence a decade earlier. Thus they should properly be called Leeuwenhoekian canals, but this is not likely. Since capillaries could not be seen at the time, Havers logically presumed that the smaller canals functioned to transport "medullary oil" to the joints for lubrication. (For an account of historical landmarks in bone biology, see Enlow, 1963.) (From Enlow, D.H.: *The Human Face.* New York, Harper & Row, 1968, p. 34.)

haversian reconstruction process also contributes to **mineral homeostasis** in the older individual. Aged bone tends to be less labile in surface ion exchange, and the process of secondary reconstruction provides calcium turnover. Furthermore, the calcified matrix of aging bone has been shown to undergo a kind of structural fatigue in which microfractures develop over a period of time. The process of secondary reconstruction is believed to accommodate this situation as it simultaneously provides new bone replacement for the other, mutually interrelated reasons already outlined.

Haversian systems are thus a feature of the **older** skeleton. They begin to appear during early maturity, and the distribution then accumulates over the years. A rat (as well as many other species) lacks haversian systems because (1) it does not live long enough, and (2) its cortical bone tends to be highly vascular in type or so thin that vessels in the periosteum and endosteum can effectively supply it.

The haversian system provides another key function. It is involved in the attachments and progressive reattachments of muscles on growing, remodeling surfaces of bones (see Fig. 18–4). Even in the young child, haversian systems are formed specifically in some sites of muscle insertion. Here they function to secure the muscle and tendon on resorptive surfaces by anchoring the tendon into the bone deep to the resorptive surface (see page 476). This has nothing to do with reconstruction of necrotic bone, which does not become a major process until much later in life. While other mechanisms are also operative in maintaining muscle attachments, the development of secondary osteons represents one process that does this (in larger mammals) wherever shifting tendons are inserted on a remodeling tubercle, tuberosity, or other localized resorptive surface (page 461). This type of haversian bone is not nearly as widespread as the haversian system formation that takes place in adult bone.

"Primary" osteons are a feature of the young, fast-growing skeleton. They are formed by the deposition of new bone in cortical fine cancellous spaces (Fig. 18–39). The resultant structures resemble secondary osteons (haversian systems), but they are not formed by the process of secondary reconstruction involving the prior enlargement of canals by resorption. The bone between the primary osteons is usually of the nonlamellar type, because it was fine cancellous in type (see above). Primary osteons are much smaller than haversian systems, and they are not surrounded by a "cementing line" (reversal line) because the formation of a resorption canal is not involved. Primary osteons often occur as narrow layers (see Fig. 18–29) within the cortex, but large areas may also exist that are composed entirely of these structures. Primary osteons function to convert fine cancellous into compact bone (see also Fig. 18–32).

Compacted coarse cancellous bone is the most common of all bone types (Figs. 18–40 and 18–41). It is found in almost all mammalian species and in almost all bones. It constitutes a major part of the compact, cortical bone in the growing child as well as in the adult. Incredibly, this major, ubiquitous type of bone tissue is not (at least at present) described in basic histology textbooks. Compacted coarse cancellous bone (which has no formal name but is sometimes called, simply, "convoluted" bone tissue) is formed **only** by the endosteum during periods or at locations involving inward directions of cortical growth. This involves half to two thirds of the skeletal formation process throughout the body (see page 38). Convoluted bone forms by lamellar deposition within

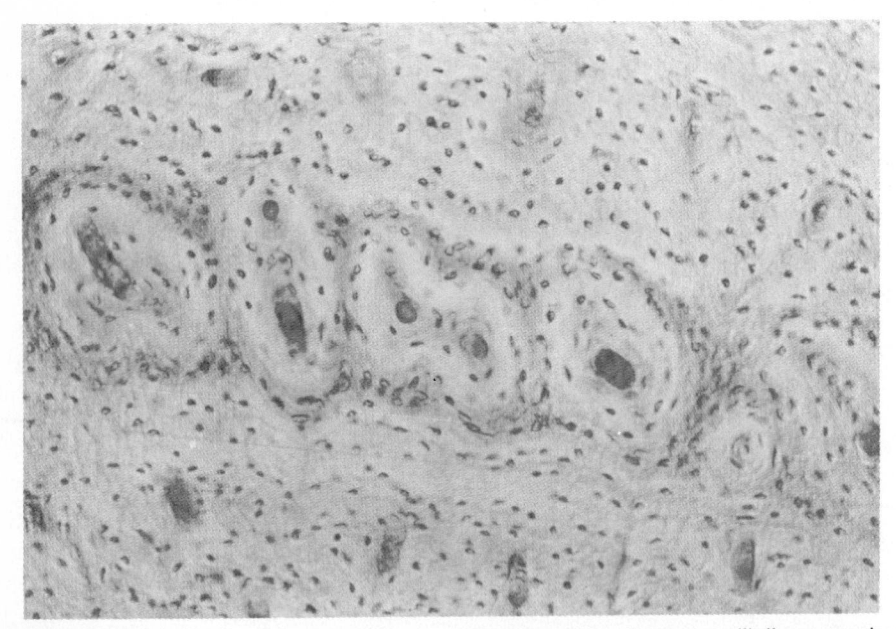

FIGURE 18–39. A row of small primary osteons is seen in this section. Compare with the secondary osteons in Figure 18–38. (From Enlow, D.H.: *The Human Face*. New York, Harper & Row, 1968, p. 23.)

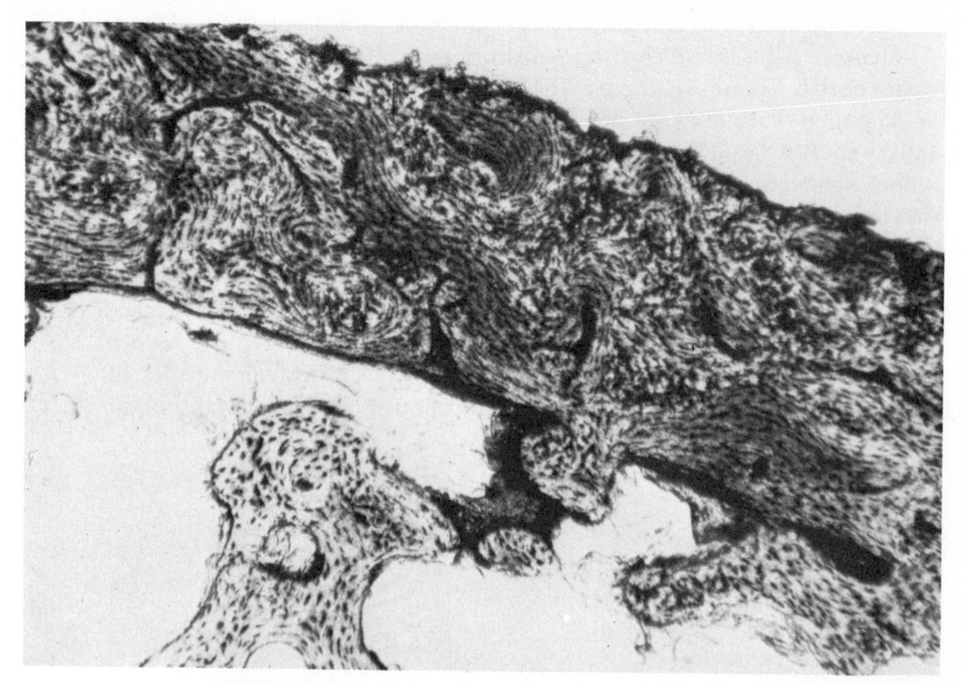

FIGURE 18–40. This is endosteal cortical bone tissue that was formed by cancellous compaction. The periosteal surface is resorptive, and the endosteal side is depository. (From Enlow, D. H.: *Principles of Bone Remodeling*. Springfield, Ill. Charles C Thomas, 1963.)

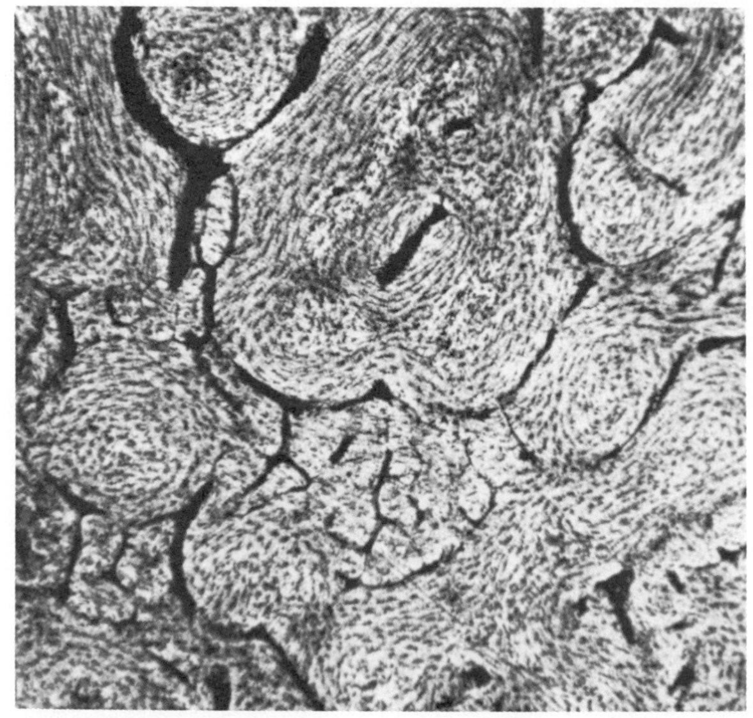

FIGURE 18–41. Compacted cancellous bone. This was originally coarse cancellous bone in the medulla, but the endosteal mode of growth resulted in its conversion into cortical compact bone. (From Enlow, D.H.: *The Human Face*. New York, Harper & Row, 1968, p. 32.)

the large spaces between the convoluted trabeculae of cancellous bone. This is a process of compaction in which the **medullary cancellous** bone is converted into **compact cortical** bone. In Figure 18–40, the cortex is growing **inward**. It moves into the medullary area already occupied by cancellous bone. To change this medullary spongy bone into the compact bone of the inward-moving cortex, the cancellous spaces are filled in until each resulting lumen is reduced to the dimension of an ordinary cortical vascular canal. Because the cancellous trabeculae are the templates upon which the new bone is laid down, the microscopic appearance is one of irregular whorls and convolutions of compact bone, shown in Figure 18–42. When an outward reversal later takes place because this area becomes relocated to a new level, as seen in Figure 18–43, this endosteal bone then becomes covered by one of the periosteal varieties of bone (Fig. 18–44). In some medullary areas, cancellous bone may be absent, as

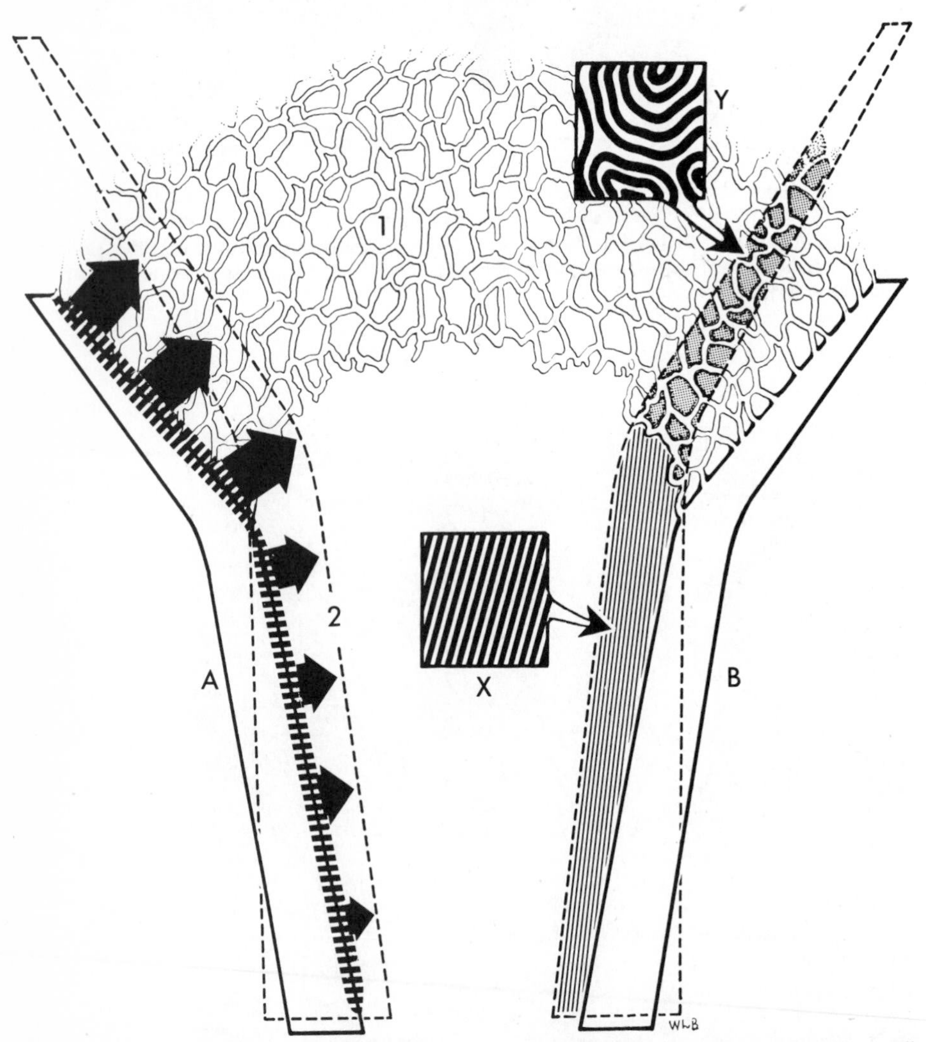

FIGURE 18–42. In those areas of a bone that involve an **endosteal** direction of cortical growth (together with periosteal resorption), a process of direct conversion from medullary cancellous *(1)* to cortical compact bone *(Y)* takes place by filling in the cancellous spaces until they are "vascular canal" in diameter. In those places *(A* and *B)* where cancellous trabeculae are not present *(2)*, inner circumferential lamellar or nonlamellar bone is laid down *(X)*. See Figures 18–30 and 18–40. (From Enlow, D. H.: *Principles of Bone Remodeling.* Springfield, Ill. Charles C Thomas, 1963.)

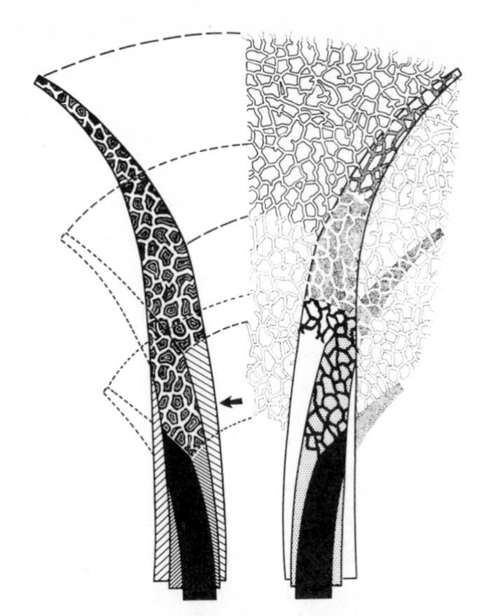

FIGURE 18–43. The wide end of a bone grows in a longitudinal direction by deposition on the endosteal side and resorption from the periosteal side. This is because the **inside** surface actually faces toward the direction of growth. Medullary bone is converted into cortical bone by cancellous compaction. In areas where cancellous trabeculae are no longer present, however, inward growth involves deposition of inner circumferential bone (arrow). The inward mode of growth also serves to reduce the wide part into the more narrow part (diaphysis) as the whole bone lengthens. In the diaphysis, the direction of growth reverses, and periosteal bone is laid down. (From Enlow, D.H.: *Principles of Bone Remodeling.* Springfield, Ill. Charles C Thomas, 1963.)

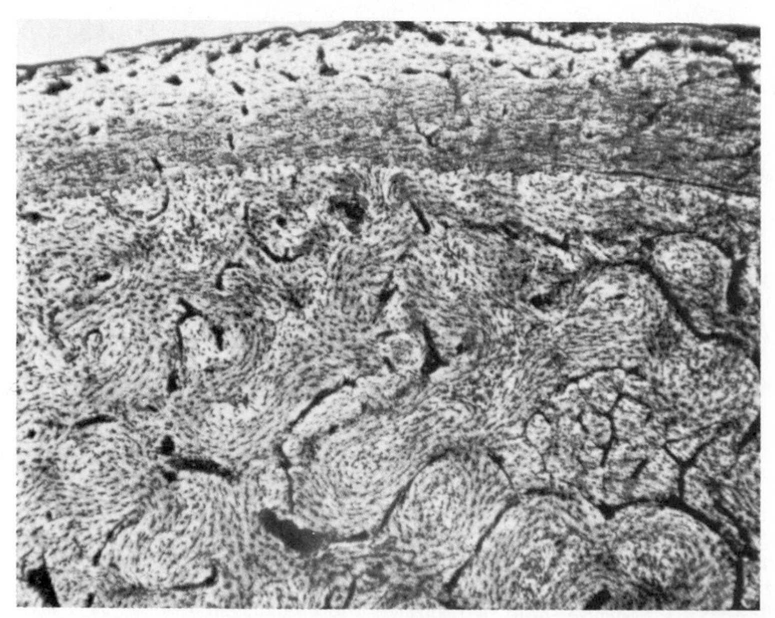

FIGURE 18–44. The inner layer in this transverse section of cortical bone was produced during a period of inward (endosteal) growth and was formed by the process of cancellous compaction. After outward reversal, a periosteal layer of primary vascular bone was subsequently laid down. Note the reversal front between these two zones. (From Enlow, D. H.: A study of the postnatal growth and remodeling of bone. Am. J. Anat., 110:79, 1962.)

in the mid diaphysis of a long bone. Here uniform layers of **inner circumferential** lamellae are laid down, rather than the convoluted type (see Fig. 18–42). In other areas of the diaphysis, the endosteal surface becomes resorptive as the whole cortex grows outward, as seen in Figure 18–45. Whether the inner surface is depository or resorptive depends on the regional growth circumstances, as illustrated in Figure 14–30.

Plexiform bone is a type in which massive cortical deposits take place in relatively short periods of time (Fig. 18–46). It is most common in medium- and large-sized animals. In man, it is sometimes found in fast-forming areas such as the maxillary tuberosity. This bone type is composed of a symmetrical, three-dimensional plexus of primary vascular canals. The canals extend in longitudinal, radial, and circumferential directions, giving the cortex a brick wall-like appearance.

Bundle bone is so called because it contains massive bundles of parallel collagenous fibers attaching the periodontal membrane to the bony alveolar wall (see Fig. 3–124). This type of bone is formed only on the depository ("tension") side of the socket. A reversal in drift direction may occur, however, and the surface of the bundle bone may then become resorptive on the new "pressure" side. Bundle bone is always laid down in a direction **toward** the

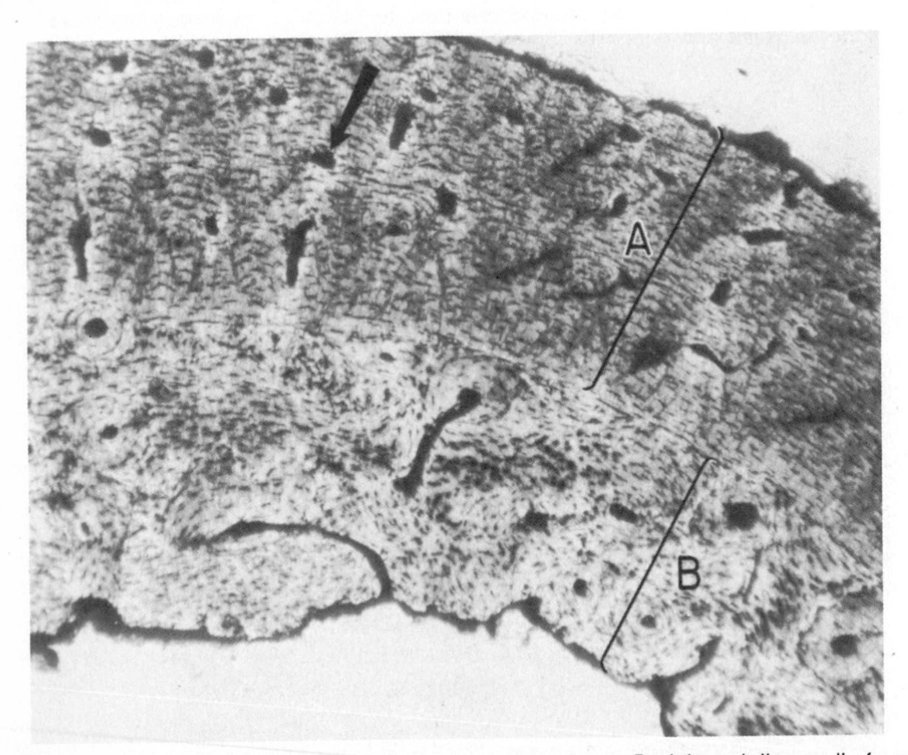

FIGURE 18–45. The cortex is often composed of two or more layers. Each layer is the result of past growth-remodeling changes, and the different types reflect the circumstances at the time of their formation. This cortical section shows an inner zone *(B)* that was produced during a former growth period involving an inward direction of growth and resultant endosteal bone formation. It is composed of compacted cancellous bone with superimposed secondary osteons that once functioned in muscle attachment on what was then a resorptive periosteal surface. A reversal then took place when this part of the bone became "relocated." The endosteal surface became resorptive, and a layer of periosteal bone was deposited *(A)*. This outer layer is composed of primary vascular, lamellar bone tissue. (After Enlow, D. H.: *Principles of Bone Remodeling.* Springfield, Ill., Charles C Thomas, 1963.)

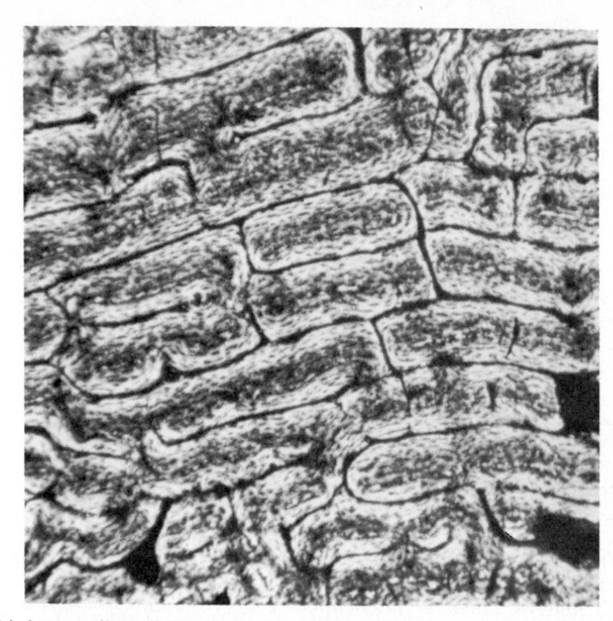

FIGURE 18–46. This is a section of plexiform bone tissue. It is characterized by a three-dimensional plexus of vascular canals, and the symmetrical arrangement of the bony laminae gives this bone type a characteristic "brick wall" appearance. Plexiform bone is a fast-growing type that provides large amounts of deposits for thick cortical areas. (From Enlow, D. H.: *The Human Face*. New York, Harper & Row, 1968, p. 26.)

moving tooth root, and as it forms, the outer periodontal fibers become enclosed within its matrix.

The bone on the pressure side of the socket is often the "convoluted" type. This is because the bony alveolar plate drifts into the cancellous bone deep to it, and compaction of the spaces converts the cancellous trabeculae into the compact bone of the alveolar cortical lining. If the alveolar wall is very thin, however, with no cancellous bone separating adjacent sockets and with only a single bony plate between them, bone apposition proceeds in one socket, and that bone is then removed by resorption on its other surface in the other socket as the teeth drift (see Fig. 3–126). Thus, the alveolar bone in the socket on the resorptive side was actually produced in the adjacent socket prior to translocation over to the resorptive side.

Chondroid bone is found on the rapidly growing crests of alveolar ridges (see Fig. 18–31). It forms as a cap on the apex of the crest. Chondroid bone may also develop elsewhere in the skeleton, such as the apex of a growing tubercle to which a muscle is attached. It has a distinctive, cartilage-like appearance with large, closely arranged cells and a nonlamellar, often basophilic, matrix. Chrondroid bone undergoes direct metaplasia into ordinary nonlamellar bone, suggesting that this is one type that can actually undergo a kind of interstitial growth or remodeling.

References

Abraham, S., C. L. Johnson, and M. F. Najjar: Weight and height of adults 18–74 years of age: United States 1971–74. DHEW Publ. No. (PHS) 79–1659. Vital Health Stat. (11), No. 211, 1979.

Acheson, R. M., and M. Archer: Radiological studies of the growth of the pituitary fossa in man. J. Anat., 93:52, 1959.

Ackerman, J. L., Y. Tagaki, W. R. Proffit, and M. J. Baer: Craniofacial growth and development in cebocephalia. Oral Surg., 19:543, 1965.

Ackerman, J. L., J. Cohen, and M. I. Cohen: The effects of quantified pressures on bone. Am. J. Orthod., 52:34, 1966.

Adams, C. D., M. C. Meikle, K. W. Norwick, and D. L. Turpin: Dentofacial remodelling produced by intermaxillary forces in *Macaca mulatta.* Arch. Oral Biol., 17:1519, 1972.

Adams, D., and M. Harkness: Histological and radiographic study of the spheno-occipital synchondrosis in cynomolgus monkeys, *Macaque irus.* Anat. Rec., 172:127, 1972.

Aduss, H.: Form, function, growth, and craniofacial surgery. Otolaryngol. Clin. N. Am., 14:939, 1981.

Agnoli, L., and G. Hildebrandt: Computer-tomographic investigations in malformations of the occipito-cervical junction. Neurosurg. Rev., 6:177, 1983.

Agronin, K. J., and V. G. Kokich: Displacement of glenoid fossa: A cephalometric evaluation of growth during treatment. Am. J. Orthod. Dentofacial Orthop., 91:42, 1987.

Alexander, T. L., and H. P. Hitchcock: Cephalometric standards for American Negro children. Am. J. Orthod., 74:298, 1978.

Altuna, G., and D. G. Woodside: Response of the midface to treatment with increased vertical occlusal forces: Treatment and posttreatment effects in monkeys. Angle Orthod., 55:251, 1985.

Amprino, R., and G. Marotti: Topographic quantitative study of bone formation and reconstruction. In: *Proceedings of the First European Symposium on Bone and Tooth.* Ed. by H. H. J. Blackwood. New York, Macmillan, 1964.

Anderson, C. E.: The structure and function of cartilage. J. Bone Joint Surg., 44:777, 1962.

Anderson, C. E.: The mechanisms of cartilage growth and replacement in endochondral ossification. In: *Cranio-facial Growth in Man.* Ed. by R. E. Moyers and W. M. Krogman. Oxford, Pergamon Press, 1971.

Anderson, D. L., and F. Popovich: Lower cranial height vs craniofacial dimensions in Angle Class II malocclusion. Angle Orthod., 53:253, 1983.

Anderson, D. L., G. W. Thompson, and F. Popovich: Tooth, chin, bone and body size correlations. Am. J. Phys. Anthropol., 46:7, 1977.

Anderson, J. H., L. Furstman, and S. Bernick: The postnatal development of the rat palate. J. Dent. Res., 46:366, 1967.

Andrew, R.: *The Anatomy of Aging in Man and Animals.* New York, Grune & Stratton, 1971.

Angle, E. H.: *Treatment of Malocclusion of the Teeth.* Vol. 7. Philadelphia, White Dental Manufacturing Company, 1907, p. 132.

Angle, E. H.: Bone growing. Dent. Cosmos., 52:261, 1910.

Anson, B. J.: Development of the incus of the human ear. Illustrated in Atlas Series. Quart. Bull. Northwest. Univ. Med. Sch., 33:110, 1959a.

Anson, B. J.: Development of the stapes of the human ear. Illustrated in Atlas Series. Quart. Bull. Northwest. Univ. Med. Sch., 33:44, 1959b.

Anson, B. J., and T. H. Bast: The development of the auditory ossicles and associated structures in man. Ann. Otol., 55:467, 1946.

Anson, B. J., T. H. Bast, and S. F. Richany: The fetal and early postnatal development of the tympanic ring and related structures in man. Ann. Otol., 64:802, 1955a.

Anson, B. J., T. H. Bast, and S. F. Richany: The fetal development of the tympanic ring, and related structures in man. Quart. Bull. Northwest. Univ. Med. Sch., 29:21, 1955b.

Appleton, J.: The ultrastructure of the articular tissue of the mandibular condyle in the rat. Arch. Oral Biol., 20:823, 1975.

Appleton, J.: The fine structure of a surface layer over the fibrous articular tissue of the rat mandibular condyle. Arch. Oral Biol., 23:719, 1978.

Araki, N. G., and C. T. Araki: Head angulation and variations in the maxillo-mandibular relationship. Part I. The effects on the vertical dimension of occlusion. J. Prosthet. Dent., 58:96, 1987.

Araki, N. G., and C. T. Araki: Head angulation and variations in the maxillo-mandibular relationship. Part II. The effects on facial contour and lip support. J. Prosthet. Dent., 58:218, 1987.

Archer, S. Y., and P. S. Vig: Effects of head position on intraoral pressures in Class I and Class II adults. Am. J. Orthod., 87:311, 1985.

Arena, S. A., and A. A. Gianelly: Resection of the mandibular body and its effect on mandibular growth. Am. J. Orthod., 76:218, 1979.

Arridge, S., J. P. Moss, A. D. Linney, and D. R. James: Three dimensional digitization of the face and skull. J. Maxillofac. Surg., 13:136, 1985.

Arvystas, M. G., P. Antonellis, and A. F. Justin: Progressive facial asymmetry as a result of early closure of the left coronal suture. Am. J. Orthod., 87:240, 1985.

Ashley-Montagu, M. F.: Form and dimensions of the palate in the newborn. Int. J. Orthod., 20:694, 1934.

Ashton, E. H.: Age changes in the basicranial axis of primates. J. Anat., 91:601, 1957a.

Ashton, E. H.: Age changes in the axis of the anthropoidea. Proc. Zool. Soc. (Lond.), 129:61, 1957b.

Asling, C. W., and H. R. Frank: Roentgen cephalometric studies on skull development in rats. I. Normal and hypophysectomized females. Am. J. Phys. Anthropol., 21:527, 1963.

Atkinson, P. J.: Structural aspects of ageing bone. Gerontologia (Basel), 15:171, 1969.

Atkinson, P. J., and C. Woodhead: Changes in human mandibular structure with age. Arch. Oral Biol., 13:1453, 1968.

Atkinson, P. J., K. Powell, and C. Woodhead: Cortical structure of the pig mandible after insertion of metallic implants into alveolar bone. Arch. Oral Biol., 22:383, 1977.

Atkinson, P. J., J. A. Spence, G. Aitchison, and A. R. Sykes: Mandibular bone in ageing sheep. J. Comp. Pathol., 92:67, 1982.

Atwood, D. A.: Reduction of residual ridges: A major oral disease entity. J. Prosthet. Dent., 26:266, 1971.

Avery, J. K.: Children with cleft lips and cleft palate: Embryological basis for defects of the face and palate. In: *Handicapped Children—Problems, Programs, Services in Michigan.* University of Michigan Educational Series, No. 93. Ann Arbor, University of Michigan, 1961.

Avery, J. K., and R. K. Devine: The development of the ossification centers in the face and palate of normal and cleft palate human embryos. Cleft Palate Bull., 9:25, 1959.

Avis, V.: The relation of the temporal muscle to the form of the coronoid process. Am. J. Phys. Anthropol., 17:99, 1959.

Azuma, M.: Study of histologic changes of periodontal membrane incident to experimental tooth movement. Tokyo Med. Dent. Univ., 17:149, 1970.

Azuma, M., and D. H. Enlow: Fine structure of fibroblasts in the periodontal membrane and their possible role in tooth drift and eruption. Jpn. J. Orthod., 36:1, 1977.

Azuma, M., D. H. Enlow, R. G. Frederickson, and L. G. Gaston: A myofibroblastic basis for the physical forces that produce tooth drift and eruption, skeletal displacement at sutures, and periosteal migration. In: *Determinants of Mandibular Form and Growth.* Ed. by J. A. McNamara, Jr. Ann Arbor, University of Michigan, Center for Human Growth and Development, 1975.

Babineau, T. A., and J. H. Kronman: A cephalometric evaluation of the cranial base in microcephaly. Angle Orthod., 39:57, 1969.

Babula, W. J., G. R. Smiley, and A. D. Dixon: The role of the cartilaginous nasal septum in midfacial growth. Am. J. Orthod., 58:250, 1970.

Baer, M. J.: Patterns of growth of the skull as revealed by vital staining. Hum. Biol., 26:80, 1954.

Baer, M. J.: *Growth and Maturation: An Introduction to Physical Development.* Cambridge, Mass., Doyle Publishing Company, 1973.

Baer, M. J., and J. A. Gavan: Symposium on bone growth as revealed by *in vitro* markers. Am. J. Phys. Anthropol., 29:155, 1968.

Baer, M. J., and J. E. Harris: A commentary on the growth of the human brain and skull. Am. J. Phys. Anthropol. 30:39, 1969.

Bahreman, A. A., and J. E. Gilda: Differential cranial growth in rhesus monkeys revealed by several bone markers. Am. J. Orthod., 53:703, 1967.

Bakker, V. M., and L. E. Johnston, Jr.: The effect of Class II elastic forces on craniofacial growth in rats. J. Dent. Res., 64:44, 1985.

Bakwin, M., and J. D. McLaughlin: Secular increase in height: Is the end in sight? Lancet, 2:1195, 1964.

Balbach, D. R.: The cephalometric relationship between the morphology of the mandible and its future occlusal position. Angle Orthod., 39:29, 1969.

Bang, S., and D. H. Enlow: Postnatal growth of the rabbit mandible. Arch. Oral Biol., 12:993, 1967.

Barber, C.: Effects of the physical consistency of diet on the condylar growth of the rat mandible. J. Dent. Res., 42:848, 1963.

Barber, T. K., and S. Pruzansky: An evaluation of the oblique cephalometric film. J. Dent. Child., 28:94, 1961.

Barbosa, J., and D. Martins: Longitudinal study of anterior facial development as related to overbite in Brazilian Caucasian teenagers. Ortodontia, 13:86, 1980.

Barnett, D. P.: Variations in the soft tissue profile and their relevance to the clinical assessment of skeletal pattern. Br. J. Orthod., 2:235, 1975.

Bass, N. M.: Orthopedic coordination of dentofacial development in skeletal Class II malocclusion in conjunction with edgewise therapy. Part II. Am. J. Orthod., 84:466, 1983.

Bassett, C. A. L.: Current concepts of bone formation. J. Bone Joint Surg., 44:1217, 1962.

Bassett, C. A. L.: Electrical effects in bone. Sci. Am., 213:18, 1965.

Bassett, C. A. L.: Electro-mechanical factors regulating bone architecture. In: *Proceedings of the Third European Symposium on Calcified Tissues.* Davos, Switzerland. New York, Springer-Verlag, 1966.

Bassett, C. A. L.: Biologic significance of piezoelectricity. Calcif. Tissue Res., 1:252, 1968.

Bassett, C. A. L.: A biological approach to craniofacial morphogenesis. Acta Morphol. Neerl. Scand., 10:71, 1972.
Baughan, B., and A. Demirjian: Sexual dimorphism in the growth of the cranium. Am. J. Phys. Anthropol., 49:383, 1978.
Baume, L. J.: Physiological tooth migration and its significance for the development of occlusion. II. The biogenesis of accessional dentition. J. Dent. Res., 29:331, 1950.
Baume, L. J.: A biologist looks at the sella point. Trans. Eur. Orthod. Soc., 1957:150, 1958.
Baume, L. J.: Principles of cephalofacial development revealed by experimental biology. Am. J. Orthod., 47:881, 1961a.
Baume, L. J.: The postnatal growth activity of the nasal cartilage septum. Helv. Odont. Acta, 5:9, 1961b.
Baume, L. J.: Ontogenesis of the human temporomandibular joint. I. Development of the condyles. J. Dent. Res., 41:1327, 1962.
Baume, L. J.: Patterns of cephalofacial growth and development: A comparative study of the basicranial growth centers in rat and man. Int. Dent. J., 18:489, 1968.
Baume, L. J.: Cephalofacial growth patterns and the functional adaptation of the temporomandibular joint structures. Eur. Orthod. Soc. Trans., 1969:79, 1970.
Baume, L. J., and H. Becks: The development of the dentition of the *Macaca mulatta:* Its difference from the human pattern. Am. J. Orthod., 36:723, 1950.
Baume, L. J., and H. Derichsweiler: Is the condylar growth center responsive to orthodontic treatment? Oral Surg., 14:347, 1961a.
Baume, L. J., and H. Derichsweiler: Response of condylar growth cartilage to induced stresses. Science, 134:53, 1961b.
Baume, R. M., P. H. Buschang, S. Weinstein: Stature, head height, and growth of the vertical face. Am. J. Orthod., 83:477, 1983.
Baumhammers, A., and R. E. Stallard: S^{35}-sulphate utilization and turnover by the connective tissues of the periodontium. J. Periodont. Res., 3:187, 1968.
Baumhammers, A., R. E. Stallard, and H. A. Zander: Remodeling of alveolar bone. J. Periodontol., 36:439, 1965.
Baumrind, S.: Reconsideration of the propriety of the "pressure-tension" hypothesis. Am. J. Orthod., 55:12, 1969.
Baumrind, S.: Mapping the skull in 3-D. J. Calif. Dent. Assoc., 48:21, 1972.
Baumrind, S., and R. C. Frantz: The reliability of head film measurements (1). Am. J. Orthod., 60:111, 1971a.
Baumrind, S., and R. C. Frantz: The reliability of head film measurements (2). Am. J. Orthod., 60:505, 1971b.
Baumrind, S., and D. Miller: Computer-aided minimization of landmark location errors of head films. J. Dent. Res., 52 (Suppl.):211, 1973.
Baumrind, S., and D. M. Miller: Computer-aided headfilm analysis: The University of California, San Francisco Method. Am. J. Orthod., 78:192, 1980.
Baumrind, S., F. H. Moffett, and S. Curry: Three-dimensional x-ray stereometry from paired coplanar images: A progress report. Am. J. Orthod., 84:292, 1983a.
Baumrind, S., F. H. Moffett, and S. Curry: The geometry of three-dimensional measurement from paired coplanar x-ray images. Am. J. Orthod., 84:313, 1983b.
Baumrind, S., E. L. Korn, R. J. Isaacson, E. E. West, and R. Molthen: Superimpositional assessment of treatment-associated changes in the temporomandibular joint and the mandibular symphysis. Am. J. Orthod., 84:443, 1983c.
Baumrind, S., E. L. Korn, and E. E. West: Prediction of mandibular rotation: An empirical test of clinician performance. Am. J. Orthod., 86:371, 1984.
Baumrind, S., E. L. Korn, Y. Ben-Basset, and E. E. West: Quantitation of maxillary remodeling. 2. Masking of remodeling effects when an "anatomical" method of superimposition is used in the absence of metallic implants. Am. J. Orthod. Dentofacial Orthop., 91:463, 1987.
Bayley, N.: Growth curves of height and weight by age for boys and girls, scaled according to physical maturity. J. Pediatr., 48, 187, 1956.
Beardsley, T.: Developmental dialectics: An intricate hierarchy of genes seems to organize the organism. Sc. Am., 259:40, 1988.
Becker, R. O.: The bioelectric factors in amphibian limb regeneration. J. Bone Joint Surg., 43:643, 1961.
Becker, R. O.: The direct current field: A primitive control and communication system related to growth processes. In: *Proceedings of the SVI International Congress of Zoology.* Vol. 3. Washington, D. C., 1963.
Becker, R. O., and D. G. Murray: A method for producing cellular dedifferentiation by means of very small electrical currents. Trans. NY. Acad. Sci., 29:606, 1967.
Becker, R. O., C. A. L. Bassett, and C. H. Bachman: The bioelectric factors controlling bone structure. In: *Bone Biodynamics.* Boston, Little, Brown, 1964.
Beecher, R. M.: Function and fusion at the mandibular symphyses. Am. J. Phys. Anthropol., 47:325, 1977.
Beecher, R. M., and R. S. Corruccini: Effects of dietary consistency on craniofacial and occlusal development in the rat. Angle Orthod., 51:61, 1981.

Beer, G. R., de: *The Development of the Vertebrate Skull.* Oxford, Clarendon Press, 1937.
Beertsen, W.: Migration of fibroblasts in the periodontal ligament of the mouse incisor as revealed by autoradiography. Arch. Oral Biol., 20:659, 1975.
Beertsen, W.: Remodeling of collagen fibers in the periodontal ligament and the supra-alveolar region. Angle Orthod., 49:218, 1979.
Behrents, R. G.: Déjà vu: Neurotropism and the regulation of craniofacial growth. In: *Factors Affecting the Growth of the Midface.* Ed. by J. A. McNamara, Jr. Ann Arbor, University of Michigan, Center for Human Growth and Development, 1979.
Behrents, R. G.: The continuity of mandibular form in mandibulofacial dysostosis. J. Dent. Res., 61:1240, 1982.
Behrents, R. G.: *An Atlas of Growth in the Aging Craniofacial Skeleton.* Craniofacial Growth Series. Ed. by J. A. McNamara and K. A. Ribbens. Ann Arbor, University of Michigan, Center for Human Growth and Development, 1983.
Behrents, R. G.: The Biological Basis for Understanding Craniofacial Growth During Adulthood. In: *Normal and Abnormal Bone Growth: Basic and Clinical Research.* Ed. by A. G. Dixon, and B. G. Sarnat. New York, Alan R. Liss, 1985a.
Behrents, R. G.: The biological basis for understanding craniofacial growth during adulthood. Prog. Clin. Biol. Res., 187:307, 1985b.
Behrents, R. G.: *Growth in the Aging Craniofacial Skeleton.* Monograph 17. Craniofacial Growth Series. Ed. by D. S. Carlson and K. A. Ribbens. Ann Arbor, University of Michigan. Center for Human Growth and Development, 1985c.
Behrents, R. G.: Adult craniofacial growth. JCO interview. J. Clin. Orthod., 20:842, 1986a.
Behrents, R. G.: *Atlas of Growth in the Aging Craniofacial Skeleton.* Monograph 18. Craniofacial Growth Series. Ed. by D. S. Carlson and K. A. Ribbens. Ann Arbor, University of Michigan, Center for Human Growth and Development, 1986b.
Behrents, R. G.: The consequences of adult craniofacial growth. In: *Orthodontics in an Aging Society.* Monograph 22. Craniofacial Growth Series. Ed. by D. Carlson and A. Ferrara. Ann Arbor, University of Michigan, Center for Human Growth and Development, 1989.
Behrents, R. G., and E. F. Harris: Normal length changes in hand bones during adulthood: Exploratory study of ray two. Ann. Hum. Biol., 14:277, 1987.
Behrents, R. G., and L. E. Johnston: The influence of the trigeminal nerve on facial growth and development. Am. J. Orthod., 85:199, 1984.
Behrents, R. G., D. S. Carlson, and T. Abdelnour: *In vivo* analysis of bone strain about the sagittal suture in *Macaca mulatta* during masticatory movements. J. Dent. Res., 57:904, 1978.
Behrents, R. G., J. A. McNamara, and J. K. Avery: A case of mandibulofacial dysostosis in utero. In: *Symposium on Diagnosis and Treatment of Craniofacial Anomalies.* Ed. by J. M. Converse, J. G. McCarthy, and D. Wood-Smith. St. Louis, C. V. Mosby, 1979b.
Bell, W. H., W. Gonyea, R. A. Finn, K. A. Storum, C. Johnston, and G. S. Throckmorton: Muscular rehabilitation after orthognathic surgery. Oral Surg., 56:229, 1983.
Benkert, P.: Experimental animal studies on the effect of the size and position of the tongue on mandibular growth. Zahn. Mund. Kieferheilkd., 72:818, 1984.
Beresford, W. A.: Schemes of zonation in the mandibular condyle. Am. J. Orthod., 68:189, 1975.
Beresford, W. A.: *Chondroid Bone, Secondary Cartilage and Metaplasia.* Baltimore, Urban & Schwarzenberg, 1981.
Bergen, R., J. Hallenberg, and O. Malmgren: Computerized cephalometrics. Acta Odont. Scand., 36:349, 1978.
Bergersen, E. O.: Enlargement and distortion in cephalometric radiography: Compensation tables for linear measurements. Angle Orthod., 50:230, 1980.
Bergsma, D. (Ed.): *Birth Defects Atlas and Compendium.* Baltimore, Williams & Wilkins, 1973.
Berkowitz, S.: State of the art in cleft palate orofacial growth and dentistry: A historical perspective. Am. J. Orthod., 74:564, 1978.
Berkowitz, S.: Some questions, a few answers in maxillomandibular surgery: The role of muscle and growth. Clin. Plast. Surg., 9:603, 1982.
Bernabei, R. L.: The effect of bovine somatotrophic hormone on the *in situ* growth of isolated mandibular segments. In: *Factors Affecting the Growth of the Midface.* Ed. by J. A. McNamara, Jr. Ann Arbor, University of Michigan, Center for Human Growth and Development, 1976.
Bevis, R. R., et al.: Facial growth response to human growth hormone in hypopituitary dwarfs. Angle Orthod., 47:193, 1977.
Bhaskar, S. N., J. P. Weinmann, and I. Schour: Role of Meckel's cartilage in the development and growth of the rat mandible. J. Dent. Res., 32:398, 1953.
Bhat, M., and D. H. Enlow: Facial variations related to headform type. Angle Orthod., 55:269, 1985.
Bhatia, S. N., G. W. Wright, and B. C. Leighton: A proposed multivariate model for prediction of facial growth. Am. J. Orthod., 75:264, 1979.
Bibby, R. E.: A cephalometric study of sexual dimorphism. Am. J. Orthod., 76:256, 1979.
Biggerstaff, R. H.: Cusp size, sexual dimorphism, and heritability of cusp size in twins. Am. J. Phys. Anthropol., 42:127, 1975.
Biggerstaff, R. H., R. C. Allen, O. C. Tuncay, and J. Berkowitz: A vertical cephalometric analysis of the human craniofacial complex. Am. J. Orthod., 72:397, 1977.

Billewicz, W. Z., H. M. Fellowes, and A. M. Thomson: Pubertal changes in boys and girls in Newcastle-upon-Tyne. Ann. Hum. Biol., 8:211, 1981.
Bimler, H. P.: Stomatopedics in theory and practice. Int. J. Orthod., 3:5, 1965.
Bimler, H. P.: Physiologic and pathologic variants of the mandible in form, position and size. Fortschr. Kieferorthop., 46:261, 1985.
Birch, R. H.: Foetal retrognathia and the cranial base. Angle Orthod., 38:231, 1968.
Birkby, W. H.: An evaluation of race and sex identification from cranial measurements. Am. J. Phys. Anthropol., 24:21, 1966.
Bishara, S. E.: Longitudinal cephalometric standards from 5 years to age adulthood. Am. J. Orthod., 79:35, 1981.
Bishara, S. E., and R. M. Thorp: Effects of Von Langenbeck palatoplasty on facial growth. Angle Orthod., 47:34, 1977.
Bjork, A.: The face in profile. Sven. Tandlak. Tidskr., 40:56, 1947.
Bjork, A.: Cranial base development. Am. J. Orthod., 41:198, 1955a.
Bjork, A.: Facial growth in man, studied with the aid of metallic implants. Acta Odont. Scand., 13:9, 1955b.
Bjork, A.: Variations in the growth pattern of the human mandible: Longitudinal radiographic studies by the implant method. J. Dent. Res., 42:400, 1963.
Bjork, A.: Sutural growth of the upper face studied by the metallic implant method. Acta Odont. Scand., 24:109, 1966.
Bjork, A.: The use of metallic implants in the study of facial growth in children: Method and application. Am. J. Phys. Anthropol., 29:243, 1968.
Bjork, A.: Prediction of mandibular growth rotation. Am. J. Orthod., 55:535, 1969.
Bjork, A.: The role of genetic and local environmental factors in normal and abnormal morphogenesis. Acta Morphol. Neerl. Scand., 10:49, 1972.
Bjork, A., and T. Kudora: Congenital bilateral hypoplasia of the mandibular condyles. Am. J. Orthod., 54:584, 1968.
Bjork, A., and V. Skieller: Facial development and tooth eruption. Am. J. Orthod., 62:339, 1972.
Bjork, A., and V. Skieller: Growth of the maxilla in three dimensions as revealed radiographically by the implant method. Br. J. Orthod., 4:53, 1977.
Bjork, A., and V. Skieller: Normal and abnormal growth of the mandible. A synthesis of longitudinal cephalometric implant studies over a period of 25 years. Eur. J. Orthod., 5:1, 1983.
Blackwood, H. J.: Vascularization of the condylar cartilage. J. Dent. Res., 37:753, 1958.
Blackwood, H. J.: Vascularization of the condylar cartilage of the human mandible. J. Anat., 99:551, 1965a.
Blackwood, H. J.: Cell differentiation in the mandibular condyle of the rat and man. Calcif. Tissue Res., 1964:23, 1965b.
Blackwood, H. J.: Growth of the mandibular condyle of the rat studied with tritiated thymidine. Arch. Oral Biol., 11:493, 1966.
Blechschmidt, M.: Biokinetics of the developing basicranium. In: *Development of the Basicranium.* Ed. by J. F. Bosma. DHEW Pub. 76:989, NIH, Bethesda, Md., 1976.
Bloore, J. A., L. Furstman, and S. Bernick: Postnatal development of the cat palate. Am. J. Orthod., 56:505, 1969.
Bluestone, C. D.: The role of tonsils and adenoids in the obstruction of respiration. In: *Nasorespiratory Function and Craniofacial Growth.* Ed. by J. A. McNamara, Jr. Ann Arbor, University of Michigan, Center for Human Growth and Development, 1979.
Bollobas, E.: The body and processes of the fetal maxilla. Acta Morphol. Hung., 32:217, 1984.
Bookstein, F. L.: Looking at mandibular growth: Some new geometric methods. In: *Craniofacial Biology.* Ed. by D. S. Carlson. Ann Arbor, University of Michigan, Center for Human Growth and Development, 1981.
Bookstein, F. L.: On the cephalometrics of skeletal change. Am. J. Orthod., 82:177, 1982.
Bookstein, F. L., and R. E. Moyers: Do horizontal growers grow horizontally? Swed. Dent. J., 15:179, 1982.
Bosma, J. F.: Maturation of function of the oral and pharyngeal region. Am. J. Orthod., 49:94, 1963.
Bosma, J. F.: Form and function in the infant's mouth and pharynx. In: *Third Symposium on Oral Sensation and Perception.* Ed. by J. Bosma. Springfield, Ill., Charles C Thomas, 1972.
Bosma, J. F.: Form and function in the mouth and pharynx of the human infant. In: *Control Mechanisms in Craniofacial Growth.* Ed. by J. A. McNamara, Jr. Ann Arbor, University of Michigan, Center for Human Growth and Development, 1975.
Bowden, C. M., and M. W. Kohn: Mandibular deformity associated with unilateral absence of the condyle. J. Oral Surg., 31:469, 1973.
Brader, A. C.: Dental arch form related with intraoral forces: PR = C. Am. J. Orthod., 61:541, 1972.
Brain, D. J., and W. P. Rock: The influence of nasal trauma during childhood on growth of the facial skeleton. Laryngol. Otol., 97:917, 1983.
Brash, J. C.: *The Growth of the Jaws and Palate.* London, Dental Board of the United Kingdom, 1924.

Brash, J. C.: The growth of the alveolar bone and its relation to the movements of the teeth, including eruption. Dent. Rec., 46:641, 1926.
Brash, J. C.: The growth of the alveolar bone and its relation to the movements of the teeth, including eruption. Dent. Rec., 47:1, 1927.
Brash, J. C.: The growth of the alveolar bone and its relation to the movements of the teeth, including eruption. Int. J. Orthod., 14:196, 1928.
Brash, J. C.: Some problems in the growth and developmental mechanics of bone. Edinb. Med. J., 41:305, 1934.
Brash, J. C., H. T. A. McKeag, and J. H. Scott: The aetiology of irregularity and malocclusion of the teeth. London, Dental Board of the United Kingdom, 1956.
Bremers, L. M. H.: De condylus mandibulae *in vitro.* Thesis, Katholieke Univ. te Nijmegen, 1973.
Bresolin, D., G. G. Shapiro, P. A. Shapiro, S. W. Dassel, C. T. Furukawa, W. W. Pierson, M. Chapko, and C. W. Bierman: Facial characteristics of children who breathe through the mouth. Pediatrics, 73:622, 1984.
Brieden, C. M., V. Pangrazio-Kulbersh, and R. Kulbersh: Maxillary skeletal and dental change with Frankel appliance therapy: An implant study. Angle Orthod., 54:226, 1984.
Brin, I., D. Hom, and D. Enlow: Correlation between nasal width and maxillary incisal alveolar width in postnatal facial development. Europ. J. Orthod. In press.
Brin, I., M. B. Kelley, J. L. Ackerman, and P. A. Green: Molar occlusion and mandibular rotation: A longitudinal study. Am. J. Orthod., 8:397, 1982.
Broadbent, B. H.: A new x-ray technique and its application to orthodontia. Angle Orthod., 1:45, 1931.
Broadbent, B. H.: The face of the normal child. Angle Orthod., 7:183, 1937.
Broadbent, B. H., B. H. Broadbent, Jr., and W. H. Golden: *Bolton Standards of Dentofacial Developmental Growth.* St. Louis, C. V. Mosby, 1975.
Broch, J., O. Slagsvold, and M. Rosler: Error in landmark identification in lateral radiographic headplates. Europ. J. Orthod., 3:9, 1981.
Brodie, A. G.: Present status of knowledge concerning movement of the tooth germ through the jaw. J.A.D.A., 21:1830, 1934.
Brodie, A. G.: Behavior of normal and abnormal facial growth patterns. Am. J. Orthod. Oral Surg., 27:633, 1941a.
Brodie, A. G.: On the growth pattern of the human head. Am. J. Anat., 68:209, 1941b.
Brodie, A. G.: Facial patterns: A theme on variation. Angle Orthod., 16:75, 1946.
Brodie, A. G.: The growth of the jaws and the eruption of the teeth. Oral Surg., 1:334, 1948.
Brodie, A. G.: Cephalometric roentgenology: History, technics and uses. J. Oral Surg., 7:185, 1949.
Brodie, A. G.: Late growth changes in the human face. Angle Orthod., 23:146, 1953.
Brodie, A. G.: The behavior of the cranial base and its components as revealed by serial cephalometric roentgenograms. Angle Orthod., 25:148, 1955.
Brodie, A. G.: Craniometry and cephalometry as applied to the living child. In: *Pediatric Dentistry.* Ed. by M. M. Cohen. St. Louis, C. V. Mosby, 1961.
Brodie, A. G.: The apical base: Zone of interaction between the intestinal and skeletal systems. Angle Orthod., 36:136, 1966.
Bromage, T. G.: Mapping remodeling reversals with the aid of the scanning electron microscope. Am. J. Phys. Anthropol., 81:314, 1982.
Bromage, T. G.: Interpretation of scanning electon microscopic images of abraded forming bone surfaces. Am. J. Phys. Anthropol., 64:161, 1984a.
Bromage, T. G.: Surface remodelling studies on fossil bone. J. Dent. Res., 63:491, 1984b.
Bromage, T. G.: Taung facial remodeling: A growth and development study. In: *Hominid Evolution: Past, Present, and Future.* Ed. by P. V. Tobias. New York, Alan R. Liss, 1985.
Bromage, T. G., and A. Boyde: Microscopic criteria for the determination of directionality of cutmarks on bone. Am. J. Phys. Anthropol., 65:359, 1984.
Bruce, R. A., and J. R. Hayward: Condylar hyperplasia and mandibular asymmetry. J. Oral Surg., 26:281, 1968.
Brundtland, G. H., and L. Walløe: Menarchal age in Norway: Halt in the trend towards earlier maturation. Nature, 241:478, 1973.
Brundtland, G. H., K. Liestol, and L. Walløe: Height, weight and menarcheal age of Oslo schoolchildren during the last 60 years. Ann. Hum. Biol., 7:307, 1980.
Buchi, E. C.: Anderung der Korperform bein erwachsenen Menschen, eine Untersuchung nach der Individual-Methode. Anthropol. Forsch. (Anthropol. Gesellsch. Wien), 1:1, 1950.
Buchner, R. J.: Induced growth of the mandibular condyle in the rat. Oral Rehabil., 9:7, 1982.
Buckler, J. M. H.: *A Reference Manual of Growth and Development.* Blackwell Scientific Publications, 1979.
Burdi, A. R.: Sagittal growth of the naso-maxillary complex during the second trimester of human prenatal development. J. Dent. Res., 44:112, 1965.
Burdi, A. R.: Catenary analysis of the maxillary dental arch during human embryogenesis. Anat. Rec., 154:13, 1966.
Burdi, A. R.: Morphogenesis of mandibular dental arch shape in human embryos. J. Dent. Res., 47:50, 1968.
Burdi, A. R.: Cephalometric growth analyses of the human upper face region during the last two trimesters of gestation. J. Anat., 125:113, 1969.

Burdi, A. R.: The premaxillary-vomerine junction: an anatomic viewpoint. Cleft Palate J., 8:364, 1971.
Burdi, A. R.: Early development of the human basicranium: its morphogenic controls, growth patterns and relations. In: *Development of the Basicranium.* Ed. by J. F. Bosma. DHEW Pub. 76:989, NIH, Bethesda, Md., 1976.
Burdi, A. R.: Biological forces which shape the human midface before birth. In: *Factors Affecting the Growth of the Midface.* Ed. by J. A. McNamara, Jr. Ann Arbor, University of Michigan, Center for Human Growth and Development, 1976.
Burdi, A. R., and K. Faist: Morphogenesis of the palate in normal human embryos with special emphasis of the mechanisms involved. Am. J. Anat., 120:149, 1967.
Burdi, A. R., and R. G. Silvey: Sexual differences in closure of the human palatal shelves. Cleft Palate J., 6:1, 1969.
Burdi, A. R., and M. N. Spyropoulos: Prenatal growth patterns of the human mandible and masseter muscle complex. Am. J. Orthod., 74:380, 1978.
Burke, P. H.: Growth of the soft tissues of middle third of the face between 9 and 16 years. Eur. J. Orthod., 1:1, 1979.
Burres, S. A.: Facial biomechanics: The standards of normal. Laryngoscope, 95:708, 1985.
Burstein, S., and R. L. Rosenfield: Pubertal data for growth velocity charts (letter). J. Pediatr., 109:564, 1986.
Burstone, C. J.: Integumental profile. Am. J. Orthod., 44:1, 1958.
Burstone, C. J.: Integumental contour and extension patterns. Angle Orthod., 29:93, 1959.
Burstone, C. J.: Biomechanics of tooth movement. In: *Vistas in Orthodontics.* Ed. by B. T. Kraus and R. A. Riedel. Philadelphia, Lea & Febiger, 1962.
Burstone, C. J.: Lip posture and its significance in treatment planning. Am. J. Orthod., 53:262, 1967.
Buschang, P. H., R. Baume, and G. G. Nass: A craniofacial growth maturity gradient for males and females between 4 and 16 years of age. Am. J. Phys. Anthropol., 61:373, 1983.
Bushey, R. S.: Adenoid obstruction of the nasopharynx. In: *Naso-respiratory Function and Craniofacial Growth.* Ed. by J. A. McNamara, Jr. Ann Arbor, University of Michigan, Center for Human Growth and Development, 1979.
Cachel, S. M.: A functional analysis of the primate masticatory system and the origin of the anthropoid post-orbital septum. Am. J. Phys. Anthropol., 50:1, 1979.
Cameron, N.: The methods of auxological anthropometry. In *Human Growth.* Vol. 3, 2nd Ed. Edited by F. Falkner and J. M. Tanner. New York, Plenum Press, 1986.
Cameron, N., J. M. Tanner, and R. H. Whitehouse: A longitudinal analysis of the growth of limb segments in adolescence. Ann. Hum. Biol., 9:211, 1982.
Campbell, P. M.: The dilemma of Class III treatment: Early or late? Angle Orthod., 53:175, 1983.
Cannon, J.: Craniofacial height and depth increments in normal children. Angle Orthod., 40:212, 1970.
Carlson, B. M.: Relationship between the tissue and epimorphic regeneration of muscles. Am. Zool., 10:175, 1970.
Carlson, D. S.: Patterns of morphological variation in the human midface and upper face. In: *Factors Affecting the Growth of the Midface.* Ed. by J. A. McNamara, Jr. Ann Arbor, University of Michigan, Center for Human Growth and Development, 1976.
Carlson, D. S.: Condylar translation and the function of the superficial masseter muscle in the Rhesus monkey (*M. mulatta*). Am. J. Phys. Anthropol., 47:53, 1977.
Carlson, D. S.: Growth of the masseter muscle in rhesus monkeys (*Macaca mulatta*). Am. J. Phys. Anthropol., 60:401, 1983.
Carlson, D. S., and E. D. Schneiderman: Cephalometric analysis of adaptations after lengthening of the masseter muscle in adult rhesus monkeys, *Macaca mulatta.* Arch. Oral Biol., 28:627, 1983.
Carlson, D. S., and D. P. Van Gerven: Masticatory function and post-Pleistocene evolution in Nubia. Am. J. Anthropol., 46:495, 1977.
Carlson, D. S., J. A. McNamara, Jr., and D. H. Jaul: Histological analysis of the mandibular condyle in the Rhesus monkey (*Macaca mulatta*). Am. J. Anat., 151:103, 1978.
Carlson, D. S., E. E. Ellis III, and P. C. Dechow: Adaptation of the suprahyoid muscle complex to mandibular advancement surgery. Am. J. Orthod. Dentofacial Orthop., 92:134, 1987.
Carlsson, G. E., and G. Persson: Morphological changes of the mandible after extraction and wearing of dentures: A longitudinal, clinical and x-ray cephalometric study covering 5 years. Odontol. Rev., 18:27, 1967.
Castelli, W. A., P. C. Ramirez, and A. R. Burdi: Effect of experimental surgery on mandibular growth in Syrian hamsters. J. Dent. Res., 50:356, 1971.
Cederquist, R., and A. Dahlberg: Age changes in facial morphology of an Alaskan Eskimo population. Int. J. Skeletal Res., 6:39, 1979.
Chaconas, S. J.: A statistical evaluation of nasal growth. Am. J. Orthod., 56:403, 1969.
Chaconas, S. J., and J. D. Bartroff: Prediction of normal soft tissue facial changes. Angle Orthod., 45:12, 1975.
Chaconas, S. J., and A. A. Caputo: Observation of orthopedic force distribution produced by maxillary orthodontic appliances. Am. J. Orthod., 82:492, 1982.
Chaconas, S. J., A. A. Caputo, and J. C. Davis: The effect of orthopedic forces on the craniofacial complex utilizing cervical and headgear appliances. Am. J. Orthod., 69:527, 1976.

Chalanset, C.: Utilisation de la méthode des ellipses équiprobables dans l'étude de la variabilité de points cranio-faciaux, chez l'enfant de race blanche, au cours de la croissance: Orientation vestibulaire sur des projections sagittales. Thèse, Acad. Paris, Univ. Paris VI, 1973.
Charles-Severe, J.: Le coefficient (r') d'allongemenet elliptique. Son application a l'étude des nuages de points craniométriques d'une population de jeunes adultes. Thèse, Faculté, Libre de Médecine de Lille, 1972.
Charlier, J. P.: Les facteurs mécaniques dans la croissance de l'arc basal mandibulaire à la lumiere de l'analyse des caractères structuraux et des propriétés biologiques du cartilage condylien. Orthod. Fr., 38:1, 1967.
Charlier, J. P., and A. Petrovic: Recherches sur la mandibule de rat en culture d'organes: Le cartilage condylien a-t-il un potentiel de croissance indépendant? Orthod. Fr., 38:165, 1967.
Charlier, J. P., A. Petrovic, and J. Herrmann: Déterminisme de la croissance mandibulaire: Effets de l'hyperpulsion et de l'hormone somatotrope sur la croissance condylienne de jeunes rats. Orthod. Fr., 39:567, 1968.
Charlier, J. P., A. Petrovic, and J. Herrmann-Stutzmann: Effects of mandibular hyperpulsion on the prechondroblastic zone of young rat condyle. Am. J. Orthod., 55:71, 1969a.
Charlier, J. P., A. Petrovic, and G. Linck: La fronde mentonnière et son action sur la croissance mandibulairé. Recherches experimentales chez la rat. Orthod. Fr., 40:99, 1969b.
Chase, S. W.: The early development of the human premaxilla. J.A.D.A., 29:1991, 1942.
Cheng, M., D. Enlow, M. Papsidero, H. Broadbent, Jr., O. Oyen, and M. Sabat: Developmental effects of impaired breathing in the face of the growing child. Angle Orthod., 58:309, 1988.
Chierici, G., and A. J. Miller: Experimental study of muscle reattachment following surgical detachment. J. Oral Maxillofac. Surg., 42:485, 1984.
Christiansen, R. L., and C. J. Burstone: Centers of rotation within the periodontal space. Am. J. Orthod., 55:353, 1969.
Christie, T. E.: Cephalometric patterns of adults with normal occlusion. Angle Orthod., 47:128, 1977.
Chung, C. S., and J. D. Niswander: Genetic and epidemiologic studies of oral characteristics in Hawaii's schoolchildren. V. sibling correlations in occlusion traits. J. Dent. Res., 54:324, 1975.
Cicmanec, J. L., D. H. Enlow, and B. J. Cohen: Polyostotic osteophytosis in a rhesus monkey. Lab. Anim. Sci., 22:2, 1972.
Cleall, J. F.: Bone marking agents for the longitudinal study of growth in animals. Arch. Oral Biol., 9:627, 1964.
Cleall, J. F.: Normal craniofacial skeletal growth of the rat. Am. J. Phys. Anthropol., 29:225, 1968.
Cleall, J. F.: Growth of the craniofacial complex in the rat. Am. J. Orthod., 60:368, 1971.
Cleall, J. F.: Growth of the palate and maxillary dental arch. J. Dent. Res., 53:1226, 1974.
Cleall, J. F., G. W. Wilson, and D. S. Garnett: Normal craniofacial skeletal growth of the rat. Am. J. Phys. Anthropol., 29:225, 1968.
Cleall, J. F., E. A. BeGale, and F. S. Chebib: Craniofacial morphology: A principal component analysis. Am. J. Orthod., 75:650, 1979.
Clements, B. S.: Nasal imbalance in the orthodontic patient. Am. J. Orthod., 55:244, 1969.
Coben, S. E.: The integration of facial skeletal variants. Am. J. Orthod., 41:407, 1955.
Coben, S. E.: Growth concepts. Angle Orthod., 31:194, 1961.
Coben, S. E.: Growth and Class II treatment. Am. J. Orthod., 52:5, 1966.
Coben, S. E.: The biology of Class II treatment. Am. J. Orthod., 59:470, 1971.
Cochran, G. V. B., R. J. Pawluk, and C. A. L. Bassett: Stress generated electric potentials in the mandible and teeth. Arch. Oral Biol., 12:917, 1967.
Cohen, A. M.: Reorientation of the mandible and tongue during growth. Br. J. Orthod., 4:175, 1977.
Cohen, A. M.: Skeletal changes during the treatment of Class II/I malocclusions. Br. J. Orthod., 10:147, 1983.
Cohen, A. M.: Uncertainty in cephalometrics. Br. J. Orthod., 11:44, 1984.
Cohen, A. M., and P. S. Vig: A serial growth study of the tongue and intermaxillary space. Angle Orthod., 46:332, 1976.
Cohen, M. M., Jr.: Dysmorphic syndromes with craniofacial manifestations. In: *Oral Facial Genetics.* Edited by R. E. Stewart and G. H. Prescott. St. Louis, C. V. Mosby, 1976.
Cohen, M. M., Jr.: A critical review of cephalometric studies of dysmorphic syndromes. Proc. Finn. Dent. Soc., 77:17, 1981.
Cohen, M. M., Jr.: *The Child with Multiple Birth Defects.* New York, Raven Press, 1982.
Cohen, M. M., Jr.: Dysmorphic growth and development and the study of craniofacial syndromes. J. Craniofac. Genet. Dev. Biol., 1:251, 1985.
Cohen, M. M., Jr.: Children, birth defects and multiple birth defects. Ann. R. Coll. Phys. Surg. Can. Part 1, 19:375, 1986; Part 2, 19:465, 1986.
Cohen, M. M., Jr., and T. H. Hohl: Etiologic heterogeneity in holoprosencephaly and facial dysmorphia with comments on the facial bones and cranial base. In: *Development of the Basicranium.* Ed. by J. F. Bosma. DHEW Pub. 76:989, NIH, Bethesda, Md., 1976.
Cole, P., and J. S. Haigt: Posture and nasal patency. Am. Rev. Respir. Dis., 129:351, 1984.
Conklin, J. L., D. H. Enlow, and S. Bang: Methods for the demonstration of lipid as applied to compact bone. Stain Technology, 40:183, 1965.

Converse, J. M., J. G. McCarthy, and D. Wood-Smith: Clinical aspects of craniofacial synostosis. In: *Symposium on Diagnosis and Treatment of Craniofacial Anomalies.* Ed. by J. M. Converse, J. G. McCarthy, and D. Wood-Smith. St. Louis, C. V. Mosby, 1979.

Copray, J. C., and H. S. Duterloo: A comparative study on the growth of craniofacial cartilages *in vitro.* Eur. J. Orthod., 8:157, 1986.

Copray, J. C., and H. W. B. Jansen: Cyclic nucleotides and growth regulation of the mandibular condylar cartilage of the rat *in vitro.* Arch. Oral Biol., 30:749, 1985.

Copray, J. C., H. W. B. Jansen, and H. Duterloo: Growth of the mandibular condylar cartilage of the rat in serum-free organ culture. Arch. Oral Biol., 28:967, 1983.

Copray, J. C., H. W. B. Jansen, and H. S. Duterloo: Growth and growth pressure of mandibular condylar and some primary cartilages of the rat *in vitro.* Am. J. Orthod., 90:19, 1986.

Copray, J. C., J. M. H. Dibbets, and T. Kantomaa: The role of condylar cartilage in the development of the temporomandibular joint. Angle Orthod., 58:369, 1988.

Corruccini, R. S., and R. H. Y. Potter: Genetic analysis of occlusal variation in twins. Am. J. Orthod., 78:140, 1980.

Corruccini, R. S., L. D. Whitley, S. Kaul, L. Flander, and C. Morrow: Facial height and breadth relative to dietary consistency and oral breathing in two populations (North India and U.S.). Hum. Biol., 57:151, 1985.

Costaras-Volarich, M., and S. Pruzansky: Is the mandible intrinsically different in Apert and Crouzon syndromes? Am. J. Orthod., 85:475, 1984.

Couly, G.: Growth of the head in humans. Arch. Fr. Pediatr., 41:375, 1984.

Cousin, R. P., and R. Fenart: La rotation globale de la mandibule infantile envisagée dans sa variabilité. Étude en orientation vestibulaire. Orthod. Fr., 42:225, 1971.

Cousin, R. P., J. Dardenne, and R. Fenart: Apports de l'orientation vestibulaire a l'étude des corrélations cranio-faciales chez l'adulte. Report of the 44th Congress of Monaco. Trans. Eur. Orthod. Soc., 44:209, 1968.

Cox, N. H.: Morfogenese en vascularisatie van het secundaire palatum van de rat. Thesis, Katholieke Univ. te Nijmegen, 1973.

Craven, A. H.: Growth in width of the head of the Macaca Rhesus monkey as revealed by vital staining. Am. J. Orthod., 42:341, 1956.

Crelin, E. S., and W. E. Koch: An autoradiographic study of chondrocyte transformation into chondroclasts and osteocytes during bone formation in vitro. Anat. Rec., 158:473, 1967.

Cross, J. J.: Facial growth: Before, during, and following orthodontic treatment. Am. J. Orthod., 71:68, 1977.

Dahl, E.: Craniofacial morphology in congenital clefts of the lip and palate. An x-ray cephalometric study of young adult males. Acta Odont. Scand., 28(Suppl. 57):11, 1970.

Dahlberg, A. A.: Concepts of occlusion in physical anthropology and comparative anatomy. J.A.D.A., 46:530, 1953.

Dahlberg, A. A.: Evolutionary background of dental and facial growth. J. Dent. Res., 44(Suppl.):151, 1965.

Dale, J. G., A. M. Hunt, G. Pudy, and D. Wagner: Autoradiographic study of the developing temporomandibular joint. Can. Dent. Assoc. J., 29:27, 1963.

Damon, A.: Secular trend in height and weight within old American families at Harvard, 1870–1965. Am. J. Phys. Anthropol., 29:45, 1968.

Damon, A.: Larger body size and earlier menarche: The end may be in sight. Soc. Biol. 21:8, 1974.

Dardenne, J.: Etude comparative des principaux paramètres sagittaux de la face et du crane, chez l'Homme et les chimpanzes par la méthode vestibulaire d'orientation. Thèse, Univ. de Lille, 1970.

Das, A., J. Meyer, and H. Sicher: X-ray and alizarin studies on the effect of bilateral condylectomy in the rat. Angle Orthod., 35:138, 1965.

Dass, R., and S. S. Makhni: Ossification of ear ossicles: The stapes. Arch. Otolaryngol., 84:88, 1966.

Davenport, C. B.: Postnatal development of the head. Proc. Am. Phil. Soc., 83:1, 1940.

Davidovitch, Z., J. L. Shanfeld, and P. J. Batastini: Increased production of cyclic AMP in mechanically stressed alveolar bone in cats. Eur. Orthod. Soc. Trans. p. 477, 1972.

Davidovitch, Z., M. D. Finkelson, S. Steigman, J. L. Shanfeld, P. C. Montgomery, and E. Korostaff: Electric currents, bone remodeling, and orthodontic tooth movement. Part I. The effect of electric currents on periodontal cyclic nucleotides. Am. J. Orthod., 77:14, 1980a.

Davidovitch, Z., M. D. Finkelson, S. Steigman, J. L. Shanfeld, P. C. Montgomery, and E. Korostaff: Electric currents, bone remodeling, and orthodontic tooth movement. Part II. Increase in rate of tooth movement and periodontal cyclic nucleotide levels by combining force and electric current. Am. J. Orthod., 77:33, 1980b.

Davidovitch, Z., S. Steigman, M. D. Finkelson, R. W. Yost, P. C. Montgomery, J. L. Shanfeld, and E. Korostoff: Immunohistochemical evidence that electric currents increase periosteal cell cyclic nucleotide levels in feline alveolar bone *in vivo.* Arch. Oral Biol., 25:321, 1980c.

De Angelis, V.: Autoradiographic investigation of calvarial growth in the rat. Am. J. Anat., 123:359, 1968.

De Angelis, V.: Observations on the response of alveolar bone to orthodontic force. Am. J. Orthod., 58:284, 1970.

DeBeer, G. R.: *The Development of the Vertebrate Skull.* London, Oxford, 1937.

de Bont, L. G.: Temporomandibular joint articular cartilage structure and function. Thesis, Univ. of Groningen, 1985.
de Bont, L. G., R. S. B. Liem, and G. Boering: Ultrastructure of the articular cartilage of the mandibular condyle: Aging and degeneration. Oral Surg. Oral Med. Oral Pathol., 60:631, 1985.
DeCoster, L.: Une méthode d'analyse des malformations maxillo-faciales. La Province Dentaire, 5:269, 1931.
DeCoster, L.: Une nouvelle ligne de référence pour l'analyse des télé-radiographics sagittales en orthodontie. Rev. Stomatol., 11:937, 1951.
Delaire, J.: La croissance des os de la voute du crane: Principes generaux. Rev. Stomatol., 62:518, 1961.
Delaire, J.: Malformations faciales et asymétrie de la base du crane. Rev. Stomatol., 66:379, 1965.
Delaire, J.: Considérations sur la croissance faciale (en particulier du maxillaire supérieur). Déductions thérapeutiques. Rev. Stomatol., 72:57, 1971.
Delaire, J.: The potential role of facial muscles in monitoring maxillary growth and morphogenesis. In: *Muscle Adaptation in the Craniofacial Region.* Ed. by D. S. Carlson and J. A. McNamara, Jr. Ann Arbor, University of Michigan, Center for Human Growth and Development, 1976.
Delaire, J., and J. Billet: Considérations sur la croissance de la région zygomato-malaire et ses anomalies morphologiques. Rev. Stomatol., 66:205, 1965.
Delattre, A., and R. Fenart: L'hominisation du crane. Éditions du Centre National de la Recherche Scientifique, Paris, 1960.
Demoge, P. H.: La recherche en orthopédie dento-faciale. Orthod. Fr., 39:1, 1968.
Dempster, W. T.: Patterns of vascular channels in the cortex of the human mandible. Anat. Rec., 135:189, 1959.
Dempster, W. T., and D. H. Enlow: Osteone organization and the demonstration of vascular canals in the compacta of the human mandible. Anat. Rec., 133:268, 1959.
Dempster, W. T., W. J. Adams, and R. A. Duddles: Arrangement in the jaws of the roots of the teeth. J.A.D.A., 67:779, 1963.
DeMyer, W.: Median facial malformations and their implications for brain malformations. In: *Morphogenesis and Malformation of Face and Brain.* Ed. by D. Bergsma, J. Langman, and N. W. Paul. National Foundation—March of Dimes. Birth Defects Original Article Series, Vol. 11, No. 7. New York, Alan Liss, 1975.
Denoix, J. M.: Comparative anatomy of the mandible: Functional aspects. Bull. Assoc. Anat., 67:395, 1983.
Dermaut, L. R., and M. I. T. O'Reilly: Changes in anterior facial height in girls during puberty. Angle Orthod., 48:163, 1978.
Diamond, M.: Posterior growth of the maxilla. Am. J. Orthod., 32:359, 1946.
Dibbets, J. M. H.: Juvenile temporomandibular joint dysfunction and craniofacial growth. Thesis, Dept. Orthod., Univ. of Groningen, 1977.
Dibbets, J. M. H., and L. van der Weele: Orthodontic treatment in relation to symptoms attributed to dysfunction of the temporomandibular joint: A 10-year report on the University of Groningen Study. Am. J. Orthod., 91:193, 1987.
Dibbets, J. M. H., L. van der Weele, and G. Boering: Craniofacial morphology and temporomandibular joint dysfunction in children. In: *Developmental Aspects of Temporomandibular Joint Disorders.* Edited by D. S. Carlson, J. A. McNamara, and K. A. Ribbens. Ann Arbor, University of Michigan, Center for Human Growth and Development, 1985a.
Dibbets, J. M. H., L. van der Weele, and A. Uildriks: Symptoms of TMJ dysfunction: Indicators of growth patterns. J. Pedodont., 9:265, 1985b.
Dibbets, J. M. H., R. de Bruin, and L. van der Weele: Shape change in the mandible during adolescence. In: *Craniofacial Growth during Adolescence.* Ed. by D. S. Carlson, and K. A. Ribbens. Ann Arbor, University of Michigan, Center for Human Growth and Development, 1987.
Diewart, V. M.: A quantitative coronal plane evaluation of craniofacial growth and spatial relations during secondary palate development in the rat. Arch. Oral Biol., 23:607, 1978.
Diewert, V. M.: Experimental induction of premature movement of rat palatal shelves *in vivo.* J. Anat., 129:597, 1979.
Diewert, V. M.: Differential changes in cartilage cell proliferation and cell density in the rat craniofacial complex during secondary palate development. Anat. Rec., 198:219, 1980.
Diewert, V. M.: Contributions of differential growth of cartilages to changes in craniofacial morphology. Prog. Clin. Biol. Res., 101:229, 1982.
Di Paolo, R. J.: An individualized approach to locating the occlusal plane. Am. J. Orthod. Dentofacial Orthop., 92:41, 1987.
Dixon, A. D.: The early development of the maxilla. Dent. Pract., 33:331, 1953.
Dixon, A. D.: The development of the jaws. Dent. Pract., 9:10, 1958.
Dixon, A. D., and D. A. N. Hoyte: A comparison of autoradiographic and alizarin techniques in the study of bone growth. Anat. Rec., 145:101, 1963.
Dolan, K.: Cranial suture closure in two species of South American monkeys. Am. J. Phys. Anthropol., 35:109, 1971.
Dorenbos, J.: Craniale Synchondroses. Doctoral thesis, Central Drukkerij n.v. Nijmegen, University of Nijmegen, 1971.
Dorenbos, J.: In vivo cerebral implantation of the anterior and posterior halves of the spheno-occipital synchondrosis in rats. Arch. Oral Biol., 17:1067, 1972.

Dorenbos, J.: Morphogenesis of the spheno-occipital and the presphenoidal synchondrosis in the cranial base of the fetal Wistar rat. Acta Morphol. Neerl. Scand., 11:63, 1973.
Dorst, J. P., D. B. Crawford, and R. E. Ensor: The cranium in achondroplasia. In: *Development of the Basicranium.* Ed. by J. F. Bosma. DHEW Pub. 76:989, NIH, Bethesda, Md., 1976.
Downs, W. B.: Variations in facial relations: Their significance in treatment and prognosis. Am. J. Orthod., 34:812, 1948.
Downs, W. B.: The role of cephalometrics in orthodontic case analysis and diagnosis. Am. J. Orthod., 38:162, 1952.
Downs, W. B.: Analysis of the dento-facial profile. Angle Orthod., 26:191, 1956.
Drachman, D. B. (Ed.): Trophic functions of the neuron. Ann. N.Y. Acad. Sci., 228:1, 1974.
Droel, R., and R. J. Isaacson: Some relationships between the glenoid fossa position and various skeletal discrepancies. Am. J. Orthod., 61:64, 1972.
Droschl, H.: The effect of heavy orthopedic forces on the sutures of the facial bones. Angle Orthod., 45:26, 1975.
Drummond, R. A.: A determination of cephalometric norms for the Negro race. Am. J. Orthod., 54:670, 1968.
Du Brul, E. L.: Early Hominid feeding mechanisms. Am. J. Phys. Anthropol., 47:305, 1977.
Du Brul, E. L., and D. M. Laskin: Preadaptive potentiality of the mammalian skull. Anat. Rec., 138:345, 1960.
Du Brul, E. L., and D. M. Laskin: Preadaptive potentialities of the mammalian skull: An experiment in growth and form. Am. J. Anat., 109:117, 1961.
Du Brul, E. L., and H. Sicher: *The Adaptive Chin.* Springfield, Ill., Charles C Thomas, 1954.
Dudas, M., and V. Sassouni: The hereditary components of mandibular growth: A longitudinal twin study. Angle Orthod., 43:314, 1973.
Dufresnoy, P.: Recherches des différences sexuelles du neurocrane sagittal par la méthode des "droites frontières." Thèse, Univ. de Nancy, 1973.
Dullemeijer, P.: Some methodology problems in a holistic approach to functional morphology. Acta Biotheor., 18:203, 1968.
Dullemeijer, P.: Comparative ontogeny and cranio-facial growth. In: *Cranio-facial Growth in Man.* Ed. by R. E. Moyers and W. M. Krogman. Oxford, Pergamon Press, 1971.
Dunn, P. M.: Congenital postural deformities. Br. Med. Bull., 32:71, 1976.
Durkin, J. R.: Secondary cartilage: A misnomer? Am. J. Orthod., 62:15, 1972.
Durkin, J. F., J. T. Irving, and J. D. Heeley: A comparison of the circulatory and calcification patterns in the mandibular condyle in the guinea pig with those found in the tibial epiphyseal and articular cartilages. Arch. Oral Biol., 14:1365, 1969.
Durkin, J. F., J. D. Heeley, and J. T. Irving: The cartilage of the mandibular condyle. Oral Sci. Rev., 2:29, 1973.
Duterloo, H. S.: In vivo implantation of the mandibular condyle of the rat. Doctoral Dissertation, Univ. of Nijmegen, 1967.
Duterloo, H. S., and M. Bierman: Morphological changes in alveolar bone during the development of the dentition in man. In: *Craniofacial Biology.* Ed. by J. A. McNamara, Jr. Ann Arbor, University of Michigan, Center for Human Growth and Development, 1977.
Duterloo, H. S., and D. H. Enlow: A comparative study of cranial growth in *Homo* and *Macaca.* Am. J. Anat., 127:357, 1970.
Duterloo, H. S., and H. W. B. Jansen: Chondrogenesis and osteogenesis in the mandibular condylar blastema. Eur. Orthod. Soc. Trans., 1969:109, 1970.
Duterloo, H. S., and H. W. B. Jansen: Potentials of *in vivo* transplantation as a method in craniofacial growth. Proc. Finn. Dent. Soc., 77:27, 1981.
Duterloo, H. S., and H. Vilmann: Translative and transformative growth of the rat mandible. Acta Odont. Scand., 36:25, 1978.
Duterloo, H. S., and J. M. Wolters: Experiments of the significance of articular function as a stimulating chondrogenic factor for the growth of secondary cartilages of the rat mandible. Eur. Orthod. Soc. Trans., 1971:103, 1972.
Duterloo, H. S., G. Kragt, and A. M. Algra: Holographic and cephalometric study of the relationship between craniofacial morphology and the initial reactions to high-pull headgear traction. Am. J. Orthod., 88:297, 1985.
Eccles, J. D.: Studies on the development of the periodontal membrane: The principal fibers of the molar teeth. Dent. Pract., 10:31, 1959.
Edwards, L. F.: The edentulous mandible. J. Prosthet. Dent., 4:222, 1954.
Ehrlich, J., A. Yaffe, J. L. Shanfeld, P. C. Montgomery, and Z. Davidovitch: Immunohistochemical localization and distribution of cyclic nucleotides in the rat mandibular condyle in response to an induced occlusal change. Arch. Oral Biol., 22:545, 1980.
Eisenfeld, J., et al.: Soft-hard tissue correlations and computer drawings for the frontal view. Angle Orthod., 45:267, 1975.
Elgoyhen, J. C., R. E. Moyers, J. A. McNamara, Jr., and M. L. Riolo: Craniofacial adaptation of protrusive function in young rhesus monkeys. Am. J. Orthod., 62:469, 1972a.
Elgoyhen, J. C., M. L. Riolo, L. W. Graber, R. E. Moyers, and J. A. McNamara, Jr.: Craniofacial growth in juvenile *Macaca mulatta:* A cephalometric study. Am. J. Phys. Anthropol., 36:369, 1972b.

El-Najjar, M. Y., and G. L. Dawson: The effect of artificial cranial deformation on the incidence of wormian bones in the lambdoidal suture. Am. J. Phys. Anthropol., 46:155, 1977.
Engel, M. B.: Lability of bone. Angle Orthod., 22:116, 1952.
Engelsma, S. O., H. W. B. Jansen, and H. S. Duterloo: An *in vivo* transplantation study of growth of the mandibular condyle in a functional position in the rat. Arch. Oral Biol., 25:305, 1980.
Engle, M. B., and A. G. Brodie: Condylar growth and mandibular deformities. Surgery, 22:975, 1947.
Engle, M. B., J. B. Richmond, and A. G. Brodie: Mandibular growth disturbance in rheumatoid arthritis of childhood. Am. J. Dis. Child., 78:728, 1949.
Engstrom, C., S. Killiardis, and B. Thilander: The relationship between masticatory function and craniofacial morphology. II. A histological study in the growing rat fed a soft diet. Swed. Dent. J., 36:1, 1986.
Enlow, D. H.: A plastic-seal method for mounting sections of ground bone. Stain Technol., 29:21, 1954.
Enlow, D. H.: Decalcification and staining of ground thin sections of bone. Stain Technol., 36:250, 1961.
Enlow, D. H.: Functions of the haversian system. Am. J. Anat., 110:269, 1962a.
Enlow, D. H.: A study of the postnatal growth and remodeling of bone. Am. J. Anat., 110:79, 1962b.
Enlow, D. H.: *Principles of Bone Remodeling.* Springfield, Ill., Charles C Thomas, Publisher, 1963.
Enlow, D. H.: Direct medullary-to-periosteal transition and the occurrence of subperiosteal haematopoietic islands. Arch. Oral Biol., 10:545, 1965a.
Enlow, D. H.: Mesial drift as a function of growth. Symposium on Growth, University of the West Indies. West Indian Med. J., 14:124, 1965b.
Enlow, D. H.: A comparative study of facial growth in *Homo* and *Macaca.* Am. J. Phys. Anthropol., 24:293, 1966a.
Enlow, D. H.: An evaluation of the use of bone histology in forensic medicine and anthropology. In: *Studies on the Anatomy and Function of Bone and Joints.* Ed. by F. G. Evans. New York, Springer-Verlag, 1966b.
Enlow, D. H.: A morphogenetic analysis of facial growth. Am. J. Orthod., 52:283, 1966c.
Enlow, D. H.: Osteocyte necrosis in normal bone. J. Dent. Res., 45:213, 1966d.
Enlow, D. H.: Morphogenic interpretation of cephalometric data. J. Dent. Res., 46:1209, 1967.
Enlow, D. H.: The bone of reptiles. In: *Biology of the Reptilia.* New York, Academic Press, 1968a.
Enlow, D. H.: *The Human Face: An Account of the Postnatal Growth and Development of the Craniofacial Skeleton.* New York, Harper & Row, 1968b.
Enlow, D. H.: Wolff's law and the factor of architectonic circumstance. Am. J. Orthod., 54:803, 1968c.
Enlow, D. H.: Postnatal facial growth. In: *Cleft Lip and Palate.* Ed. by W. Grabb, S. W. Rosenstein, and K. R. Bzoch. Boston, Little, Brown, 1971.
Enlow, D. H.: Facial growth and development. In: *Handbook of Orthodontics.* Ed. by R. E. Moyers, Chicago, Year Book Medical Publishers, 1973a.
Enlow, D. H.: Alveolar bone. In: *International Workshop on Complete Denture Occlusion.* Ann Arbor, University of Michigan, School of Dentistry. Report on the Workshop, 1973b.
Enlow, D. H.: Growth and the problem of the local control mechanism. Editorial, Am. J. Anat., 178:2, 1973c.
Enlow, D. H.: Croissance et architecture de la face. Pédod. Fr., 6:122, 1974a.
Enlow, D. H.: The PM boundary: A natural cephalometric plane. Anat. Rec., 178, 1974b.
Enlow, D. H.: Postnatal growth and development of the face and cranium. In: *Scientific Foundations of Dentistry.* Ed. by B. Cohen and I. R. H. Kramer. London, Heinemann, 1975a.
Enlow, D. H.: Mandibular rotations during growth. In: *Determinants of Mandibular Form and Growth.* Ed. by J. A. McNamara, Jr. University of Michigan, Center for Human Growth and Development, 1975b.
Enlow, D. H.: The West Virginia Anatomical Board Rules and Regulations, 1975c.
Enlow, D. H.: The growth of the human basicranium. In: *The Basicranium.* Ed. by J. Bosma. USPHS publication, 1976.
Enlow, D. H.: The remodeling of bone. Am. J. Phys. Anthropol. Yearbook, 1977a.
Enlow, D. H.: Faces. Dental Horizons, 1:4, 1977b.
Enlow, D. H.: Normal maxillo-facial growth. In: *Reconstruction of Jaw Deformities.* Ed. by L. Whitaker. St. Louis, C. V. Mosby, 1978.
Enlow, D. H.: Facial growth and development. Int. J. Oral Myology, 5:7, 1979.
Enlow, D. H.: Morphologic factors involved in the biology of relapse. Journal of the Charles Tweed Foundation, 8:16, 1980a.
Enlow, D. H.: Postnatal facial growth and development. In: *Diagnosis and Treatment of Craniofacial Anomalies.* Ed. by J. Converse, J. McCarthy, and D. Wood-Smith, St. Louis, C. V. Mosby, 1980b.
Enlow, D. H.: Growth of the face after birth. In: *Advances in Oral Surgery.* By W. Irby. St. Louis, C. V. Mosby, 1980c.
Enlow, D. H.: Childhood facial growth. In: *Oral Histology: Structure and Function* by R. Ten Cate. St. Louis, C. V. Mosby, 1980d.
Enlow, D. H.: Mechanisms of craniofacial growth. In: *Orthodontics: The State of the Art.* Ed. by H. Barrier. Philadelphia, University of Pennsylvania Press, 1980e.

Enlow, D. H.: The mandibular condyle. In: *The Temporomandibular Joint: Biologic Basis for Clinical Practice,* 3rd Ed. By B. G. Sarnat and D. M. Laskin. Springfield, Ill., Charles C Thomas, 1980f.
Enlow, D. H.: The growth of the face. In: *Pediatric Dental Medicine.* By D. J. Forrester. Philadelphia, Lea & Febiger, 1981a.
Enlow, D. H.: The dynamics of skeletal growth and remodeling. In: *Scientific Foundations of Orthopedic Surgery.* London, Heinemann, 1981b.
Enlow, D. H.: Craniofacial growth mechanisms: Normal and disturbed. In: *Effect of Surgical Intervention on Craniofacial Growth.* Ed. by J. McNamara. Ann Arbor, University of Michigan, Center for Human Growth and Development, 1982a.
Enlow, D. H.: *Handbook of Facial Growth.* Philadelphia, W. B. Saunders, 1982b. Japanese translation by F. Miura, T. Kuroda, and M. Azuma, 1980. Spanish translation, 1981, 1984; Italian translation, 1986; German translation, 1989.
Enlow, D. H.: Role of the TMJ in Facial Growth and Development. In: *President's Conference on Examination, Diagnosis, and Management of Temporomandibular Disorders.* Ed. by D. Laskin et al. American Dental Association, Chicago, 1983a.
Enlow, D. H.: Editor's Interview. J. Clin. Orthod., 17:669, 1983b.
Enlow, D. H.: Biological targets in the control process of facial growth. In: *Physiologic Principles of Functional Appliances.* Ed. by T. M. Graber. St. Louis, C. V. Mosby, 1985.
Enlow, D. H.: Normal Craniofacial Growth. In *Craniosynostosis; Diagnosis, Evaluation, and Management.* Ed. by M. Cohen. New York, Raven Press, 1986.
Enlow, D. H.: Structural and functional "balance" during craniofacial growth. In: *Orthodontics, State of the Art, Essence of the Science.* Ed. by L. W. Graber. St. Louis, C. V. Mosby, 1986.
Enlow, D. H.: A Review of the Facial Growth Process. In *Human Growth: A Multidisciplinary Review.* Ed. by A. Dmirjian. London, Taylor and Francis, 1986.
Enlow, D. H.: Normal and abnormal patterns of craniofacial growth. In: *Scientific Foundations and Surgical Treatment of Craniosynostosis.* Ed. by J. A. Persing, M. T. Edgerton, and J. A. Jane. Baltimore, Williams & Wilkins, 1989.
Enlow, D. H., and M. Azuma: Functional growth boundaries in the human and mammalian face. In: *Morphogenesis and Malformation of the Face and Brain.* Ed. by D. Bergsma, J. Langman, and N. W. Paul. National Foundation—March of Dimes. Birth Defects Original Article Series, Vol. 11, No. 7. New York, Alan R. Liss, 1975.
Enlow, D. H., and S. Bang: Growth and remodeling of the human maxilla. Am. J. Orthod., 51:446, 1965.
Enlow, D. H., and M. Bhat: Facial morphology associated with headform variations. Journal of the Charles Tweed Foundation, 12:21, 1984.
Enlow, D. H., and S. O. Brown: A comparative histological study of fossil and recent bone tissues. Part I. Introduction, methods, fish and amphibian bone tissues. Tex. J. Sci., 7:405, 1956.
Enlow, D. H., and S. O. Brown: A comparative histological study of fossil and recent bone tissues. Part II. Reptilian and bird bone tissues. Tex. J. Sci., 9:186, 1957.
Enlow, D. H., and S. O. Brown: A comparative histological study of fossil and recent bone tissues. Part III. Mammalian bone tissues. General discussion. Tex. J. Sci., 10:187, 1958.
Enlow, D. H., and S. Comet-Epstein: A comparative population-distribution study of dental specialties in Ohio. Ohio Dent. J., 53:34, 1979.
Enlow, D. H., and D. B. Harris: A study of the postnatal growth of the human mandible. Am. J. Orthod., 50:25, 1964.
Enlow, D. H., and W. S. Hunter: A differential analysis of sutural and remodeling growth in the human face. Am. J. Orthod., 52:823, 1966.
Enlow, D. H., and W. S. Hunter: Growth of the face in relation to the cranial base. European Orthodontic Society, 44:321, 1968.
Enlow, D. H., and J. McNamara: The neurocranial basis for facial form and pattern. Angle Orthod., 43:256, 1973a.
Enlow, D. H., and J. McNamara: Varieties of *in vivo* tooth movements. Angle Orthod., 43:216, 1973b.
Enlow, D. H., and R. E. Moyers: Growth and architecture of the face. J.A.D.A., 82:763, 1971.
Enlow, D. H., J. L. Conklin, and S. Bang: Observations on the occurrence and the distribution of lipids in compact bone. Clin. Orthop., 38:157, 1965.
Enlow, D. H., R. E. Moyers, W. S. Hunter, and J. A. McNamara, Jr.: A procedure for the analysis of intrinsic facial form and growth. Am. J. Orthod., 56:6, 1969a.
Enlow, D. H., P. Williams, and K. Williams: An instrument for the analysis of facial growth. Angle Orthod., 39:316, 1969b.
Enlow, D. H., T. Kuroda, and A. B. Lewis: The morphological and morphogenetic basis for craniofacial form and pattern. Angle Orthod., 41:161, 1971a.
Enlow, D. H., T. Kuroda, and A. B. Lewis: Intrinsic craniofacial compensations. Angle Orthod., 41:161, 1971b.
Enlow, D. H., H. Bianco, and S. Eklund: The remodeling of the edentulous mandible. J. Prosthet. Dent., 36:685, 1976.
Enlow, D. H., E. Harvold, R. Latham, B. Moffett, R. Christiansen, and H. G. Hausch: Research on control of craniofacial morphogenesis: An NIDR Workshop. Am. J. Orthod., 71:509, 1977.
Enlow, D. H., C. Pfister, and E. Richardson: An analysis of Black and Caucasian craniofacial patterns. Angle Orthod., 52:279, 1982.

Enlow, D. H., D. DiGangi, J. A. McNamara, and M. Mina: Morphogenic effects of the functional regulator as revealed by the counterpart analysis. Eur. J. Orthod., 10:192, 1988.
Epker, B. N., and L. C. Fish: The surgical-orthodontic correction of mandibular deficiency. Part I. Am. J. Orthod., 84:408, 1983.
Epker, B. N., and H. M. Frost: Correlation of bone resorption and formation with the physical behavior of loaded bone. J. Dent. Res., 44:33, 1965.
Epker, B. N., F. A. Henny, and H. M. Forst: Biomechanical control of bone modeling and architecture. J. Bone Joint Surg., 50:1261, 1958.
Erickson, L. P., and W. S. Hunter: Class II, division 2 treatment and mandibular growth. Angle Orthod., 55:215, 1985.
Erskine, R. B.: A comparison of xeroradiographs with conventional lateral skull radiographs. Br. J. Orthod., 5:193, 1978.
Evans, C. A., and R. L. Christiansen: Facial growth associated with a cranial base defect: A case report. Angle Orthod., 49:44, 1979.
Evans, F. G.: *Stress and Strain in Bones.* Springfield, Ill., Charles C Thomas, 1957.
Eveleth, P. B.: Differences between populations in body shape of children and adolescents. Am. J. Phys. Anthropol., 49:373, 1978.
Eveleth, P. B., E. J. Bowers, and J. I. Schall: Secular change in growth of Philadelphia Black adolescents. Hum. Biol., 51:213, 1979.
Farkas, L. G., and G. Cheung: Facial asymmetry in healthy North American Caucasians. An anthropometrical study. Angle Orthod., 51:70, 1981.
Fastlicht, J.: Crowding of mandibular incisors. Am. J. Orthod., 58:156, 1970.
Fastlicht, J.: *The Universal Orthodontic Technique.* Philadelphia, W. B. Saunders, 1972.
Faulkner, J. A., L. C. Maxwell, and T. P. White: Adaptations in skeletal muscle. In: *Muscle Adaptation in the Craniofacial Region.* Ed. by D. S. Carlson and J. A. McNamara, Jr. Ann Arbor, University of Michigan, Center for Human Growth and Development, 1978.
Fawcett, E.: The development of the bones around the mouth. Five lectures on the growth of the jaws, normal and abnormal in health and disease. London, Dental Board of the United Kingdom, 1924.
Fell, H. B.: Chondrogenesis in cultures of endosteum. Proc. R. Soc. Lond., 112:417, 1933.
Felts, W. J. L.: Transplantation studies in skeletal organogenesis. I. The subcutaneously implanted immature long-bone of the rat and mouse. Am. J. Phys. Anthropol., 17:201, 1959.
Felts, W. J. L.: In vivo implantation as a technique in skeletal biology. Int. Rev. Cytol., 12:243, 1961.
Fenart, R.: Influence des modifications: experimentales et teratologiques de la station et de la locomotion, sur la morphologie cephalique des mammiferes quadrupedes: Étude par la méthode vestibulaire. Arch. Anat. Histol. Embryol. (Strasb.), 69:5, 1966a.
Fenart, R.: Changements morphologiques de l'encephale, chez la rat amputé des membres antérieurs. J. Hirnforsch., 8:493, 1966b.
Fenart, R.: L'hominisation du crane. Bull. Acad. Dent. (Paris), 14:33, 1970.
Ferguson, M. W. J.: Palatal shelf elevation in the Wistar rat fetus. J. Anat., 125:555, 1978.
Ferre, J.-C.: Contribution a l'étude du "syndrome asymétique cranio-facial." Thèse, Univ. de Nantes, 1973.
Ferre, J. C., J. Barbin, J. Helary, and J. Lumineau: A physicomathematical approach to the structure of the mandible. Anat. Clin. 6:45, 1984.
Finch, C. E., and L. Hayflick (Eds.): *Handbook of the Biology of Aging.* Van Nostrand Reinhold, New York, 1977.
Finlay, L. M.: Craniometry and cephalometry: A history prior to the advent of radiography. Angle Orthod., 50:312, 1980.
Fisher, R. A.: The correlations between relatives on the supposition of mendelian inheritance. Trans. R. Soc. Edinb., 52:399, 1918.
Fishman, L. S.: Chronological versus skeletal age, an evaluation of craniofacial growth. Angle Orthod., 49:181, 1979.
Fisk, R. O.: When malocclusion concerns the public. J. Can. Dent. Assoc., 26:397, 1960.
Fjellvang, H., and B. Solow: Craniocervical postural relations and craniofacial morphology in 30 blind subjects. Am. J. Orthod. Dentofacial Orthop., 90:327, 1986.
Fonseca, R. J., and W. D. Klein: A cephalometric evaluation of American Negro women. Am. J. Orthod., 73:152, 1978.
Ford, E. H.: Growth of the foetal skull. J. Anat., 90:63, 1956.
Ford, E. H.: Growth of the human cranial base. Am. J. Orthod., 44:498, 1958.
Forrester, D. J., N. K. Carstens, and D. B. Shurteff: Craniofacial configuration of hydrocephalic children. J.A.D.A., 72:1399, 1966.
Forsberg, C. M.: Facial morphology and aging: A longitudinal cephalometric investigation of young adults. Eur. J. Orthod., 1:15, 1979.
Fotis, V., B. Melson, S. Williams, and H. Droschi: Posttreatment changes of skeletal morphology following treatment aimed at restriction of maxillary growth. Am. J. Orthod., 88:288, 1985.
Frake, S. E., and D. H. Goose: A comparison between mediaeval and modern British mandibles. Arch. Oral Biol., 22:55, 1977.
Frankel, R.: The applicability of the occipital reference base in cephalometrics. Am. J. Orthod., 77:379, 1980.

Frankel, R.: Biomechanical aspects of the form/function relationship in craniofacial morphogenesis: A clinician's approach. In: *Clinical Alteration of the Growing Face.* Ed. by J. A. McNamara, Jr., K. A. Ribbens, and R. P. Howe. Monograph 14. Craniofacial Growth Series. Ann Arbor, University of Michigan, Center for Human Growth and Development, 1983.

Frankel, R.: Concerning recent articles on Frankel appliance therapy. Am. J. Orthod., 85:441, 1984.

Frankel, R., and C. Frankel: A functional approach to treatment of skeletal open bite. Am. J. Orthod., 84:54, 1983.

Fraser, F. C.: Experimental teratogenesis in relation to congenital malformations in man. In: *Proceedings of the Second International Congress on Congenital Malformations.* New York, International Medical Congress, 1964.

Fraser, F. C.: Gene-environment interactions in the production of cleft palate. In: *Methods for Teratological Studies in Experimental Animals and Man.* Ed. by H. Nishimura and J. R. Miller, Tokyo, Igaku Shoin, 1969.

Fraser, F. C.: The genetics of cleft lip and cleft palate. Am. J. Hum. Genet., 22:336, 1970.

Fraser, F. C.: Etiology of cleft lip and palate. In: *Cleft Lip and Palate.* Ed. by W. C. Grabb, S. W. Rosenstein, and K. R. Bzoch. Boston, Little, Brown, 1971.

Fraser, F. C., and H. Pashayan: Relation of face shape to susceptibility to congenital cleft lip. J. Med. Genet., 7:112, 1970.

Fraser, F. C., B. E. Walker, and D. G. Trasler: Experimental production of congenital cleft palate: Genetic and environmental factors. Pediatrics, 19:782, 1957.

Freeman, E., and A. R. Ten Cate: Development of the periodontium: An electron microscopic study. J. Periodont., 42:387, 1971.

Freng, A., and E. Kvam: Facial sagittal growth following partial basal resection of the nasal septum: A retrospective study in man. Europ. J. Orthod., 1:89, 1979.

Friede, H.: A histological and enzyme-histochemical study of growth sites of the premaxilla in human foetuses and neonates. Arch. Oral Biol., 20:809, 1975.

Friedenberg, Z. B., R. H. Dyer, Jr., and C. T. Brighton: Electro-osteograms of long bones of immature rabbits. J. Dent. Res., 50:635, 1971.

Friedi, H., B. Johanson, J. Ahlgren, and B. Thilander: Metallic implants as growth markers in infants with craniofacial anomalies. Acta Odont. Scand., 35:265, 1977.

Frommer, J.: Prenatal development of the mandibular joint in mice. Anat. Rec. 150:449, 1964.

Frommer, J., and M. R. Margolis: Contribution of Meckel's cartilage to ossification of the mandible in mice. J. Dent. Res., 50:1250, 1971.

Frommer, J., C. W. Monroe, J. R. Morehead, and W. D. Belt: Autoradiographic study of cellular proliferation during early development of the mandibular condyle in mice. J. Dent. Res., 47:816, 1968.

Frost, H. M.: In vivo osteocyte death. J. Bone Joint Surg., 42:138, 1960a.

Frost, H. M.: Micropetrosis. J. Bone Joint Surg., 42:144, 1960b.

Frost, H. M.: Tetracycline bone labeling in anatomy. Am. J. Phys. Anthropol., 29:183, 1968.

Fukada, E., and I. Yasuda: On the piezoelectric effect of bone. J. Physiol. Soc. Jpn., 12:1158, 1957.

Furstman, L.: The early development of the human mandibular joint. Am. J. Orthod., 49:672, 1963.

Gans, G. J., and B. G. Sarnat: Sutural facial growth of the Macaca rhesus monkey: A gross and serial roentgenographic study by means of metallic implants. Am. J. Orthod., 37:827, 1951.

Gans, C.: Three considerations in evaluating factors affecting the growth of the midface. In: *Factors Affecting the Growth of the Midface.* Ed. J. A. McNamara, Jr. University of Michigan, Center for Human Growth and Development, 1976.

Gans, C., and G. C. Gorniak: Concepts of muscle: An introduction to the intact animal. In: *Muscle Adaptation in the Craniofacial Region.* Ed. by D. S. Carlson and J. A. McNamara, Jr. University of Michigan, Center for Human Growth and Development, 1978.

Garcia, C. J.: Cephalometric evaluations of Mexican Americans using the Downs and Steiner analyses. Am. J. Orthod., 68:67, 1975.

Garn, S. M.: Inheritance of symphyseal size during growth. Angle Orthod., 33:222, 1963.

Garn, S. M.: *The Earlier Gain and the Later Loss of Cortical Bone in Nutritional Perspective.* Springfield, Ill., Charles C Thomas, 1970.

Garn, S. M.: The course of bone gain and the phases of bone loss. Orthop. Clin. North Am., 3:503, 1972.

Garn, S. M.: Genetics of dental development. In: *Craniofacial Biology.* Ed. by J. A. McNamara, Jr., Ann Arbor, University of Michigan, Center for Human Growth and Development, 1977.

Garn, S. M.: Living together as a factor in family-line resemblances. Hum. Biol., 51:565, 1979.

Garn, S. M.: Contributions of the radiographic image to our knowledge of human growth. Am. J. Roentgenol., 137:231, 1981.

Garn, S. M.: The secular trend in size and maturational timing and its implications for nutritional assessment. J. Nutr., 117:17, 1987.

Garn, S. M., and C. G. Rohmann: Communalities of the ossification centers of the hand and wrist. Am. J. Phys. Anthropol., 17:319, 1959.

Garn, S. M., and B. Wagner: The adolescent growth of the skeletal mass and its implications to mineral requirements. In: *Adolescent Nutrition and Growth.* Ed. by F. P. Held. New York, Appleton-Century-Crofts, 1969.

Garn, S. M., A. B. Lewis, and J. H. Vicinus: Third molar polymorphism and its significance to dental genetics. J. Dent. Res., 42(Suppl. 6):1344, 1963.
Garn, S. M., C. G. Rohmann, and P. Nolan, Jr.: Developmental nature of bone changes during aging. In: *Relations of development and aging: A symposium presented before the Gerontological Society.* 15th Annual Meeting, Miami Beach, Florida. Ed. by J. E. Birren. Springfield, Ill., Charles C Thomas, 1964.
Garn, S. M., A. B. Lewis, and R. M. Blizzard: Endocrine factors in dental development. J. Dent. Res., 44:243, 1965.
Garn, S. M., C. G. Rohmann, B. Wagner, and W. Ascoli: Continuing bone growth throughout life: A general phenomenon. Am. J. Phys. Anthropol., 26:313, 1967a.
Garn, S. M., C. G. Rohmann, and B. Wagner: Bone loss as a general phenomenon in man. Fed. Proc., 26:1729, 1967b.
Garn, S. M., C. G. Rohmann, and F. N. Silverman: Radiographic standards for postnatal ossification and tooth calcification. Med. Radiogr. Photogr., 43:45, 1967c.
Garn, S. M., C. G. Rohman, R. Wagner, and W. Ascoli: Continuing bone growth throughout life: a general phenomenon. Am. J. Phys. Anthropol., 26:313, 1967d.
Garn, S. M., A. B. Lewis, and A. Walenga: Evidence for a secular trend in tooth size over two generations. J. Dent. Res., 47:503, 1968.
Garn, S. M., A. K. Poznanski, and J. M. Nagy: The operational meaning of maturity critera. Am. J. Phys. Anthropol., 35:319, 1971.
Garn, S. M., B. H. Smith, and R. E. Moyers: Structured (patterned) dimensional and developmental asymmetry. Proc. Finn. Dent. Soc., 77:33, 1981.
Garn, S. M., B. H. Smith, and M. LaVelle: Applications of patterns profile analysis to malformations of the head and face. Radiology, 150:683, 1984.
Garner, L. D., and Yu, P. L.: Is partial anodontia a syndrome of Black Americans? Angle Orthod., 48:85, 1978.
Gasser, R. F.: Early formation of the basicranium in man. In: *Development of the Basicranium.* Ed. by J. F. Bosma. DHEW Pub. 76:989, NIH, Bethesda, Md., 1976.
Gasson, N., and J. Lavergne: Maxillary rotation during human growth: Annual variation and correlations with mandibular rotation: A metal implant study. Acta Odont. Scand., 35:13, 1977.
Gasson, N., and J. Lavergne: The maxillary rotation: Its relation to the cranial base and the mandibular corpus. An implant study. Acta Odont. Scand., 35:89, 1977.
Gasson, M. N., and M. A. Petrovic: Méchanismes et regulation de la croissance antéro-postérieure du maxillaire supérieur: Recherches expérimentales, chez le jeune rat, sur le rôle de l'hormone somatotrope et du cartilage de la cloison nasale. Orthod. Fr., 43:255, 1972.
George, S. L.: A longitudinal and cross-sectional analysis of the growth of the postnatal cranial base angle. Am. J. Phys. Anthropol., 49:171, 1978.
Gerling, J. A., P. M. Sinclair, and R. L. Roa: The effect of pulsating electromagnetic field on condylar growth in guinea pigs. Am. J. Orthod., 87:211, 1985.
Ghafari, J., and C. Degroote: Condylar cartilage response to continuous mandibular displacement in the rat. Angle Orthod., 56:49, 1986.
Ghafari, J., and J. D. Heeley: Condylar adaptation to muscle alteration in the rat. Angle Orthod., 52:26, 1982.
Gianelly, A. A.: Force-induced changes in the vascularity of the periodontal ligament. Am. J. Orthod., 55:5, 1969.
Gianelly, A. A., and H. M. Goldman: *Biologic Basis of Orthodontics.* Philadelphia, Lea & Febiger, 1971.
Gianelly, A. A., and C. F. A. Moorrees: Condylectomy in the rat. Arch. Oral Biol., 10:101, 1965.
Gianelly, A. A., P. Brosnan, M. Martignoni, and L. Bernstein: Mandibular growth, condyle position and Frankel appliance therapy. Angle Orthod., 53:131, 1983.
Giles, W. B., C. L. Phillips, and D. R. Joondeph: Growth in the basicranial synchondroses of adolescent *Macaca mulatta.* Anat. Rec., 199:259, 1981.
Gillooly, C. J., Jr., R. T. Hosley, J. R. Mathews, and D. L. Jewett: Electric potentials recorded from mandibular alveolar bone as a result of forces applied to the tooth. Am. J. Orthod., 54:649, 1968.
Gingerich, P. D.: The human mandible: Lever, link or both? Am. J. Phys. Anthropol., 51:135, 1979.
Girgis, F. G., and J. J. Pritchard: Experimental production of cartilages during the repair of fractures of the skull vault in rats. J. Bone Joint Surg., 403:274, 1958.
Glasstone, S.: Differentiation of the mouse embryonic mandible and squamo-mandibular joint in organ culture. Arch. Oral Biol., 16:723, 1971.
Godard, H.: Les zones de croissance de la mandible. C. R. Assoc. Anat., 43:357, 1957.
Goland, P. P., and N. G. Grand: Chloro-s-triazines as markers and fixatives for the study of growth in teeth and bones. Am. J. Phys. Anthropol., 29:201, 1968.
Goldberg, G., and D. H. Enlow: Some anatomical characteristics of the craniofacial skeleton in several syndromes of the head as revealed by the counterpart analysis. J. Oral Surg., 39:489, 1981.
Goldhaber, P.: The effect of hyperoxia on bone resorption in tissue culture. Arch. Pathol., 66:635, 1958.

Goldhaber, P.: Oxygen dependent bone resorption in tissue culture. In: *Parathyroids.* Ed. by R. O. Greep and R. V. Talmage. Springfield, Ill., Charles C Thomas, 1961.
Goldhaber, P.: Some chemical factors influencing bone resorption in tissue culture. In: *Mechanisms of Hard Tissue Destruction: A Symposium.* Ed. by R. F. Sognnaes. A. A. A. S. Symposium, 1963.
Goldstein, M. S.: Changes in dimension and form of the face and head with age. Am. J. Phys. Anthropol., 22:37, 1936.
Goldstein, M. S.: Development of the head in the same individuals. Hum. Biol., 11:197, 1939.
Goldstein, D. F., S. L. Kraus, W. B. Williams, and M. Glasheen-Wray: Influence of cervical posture on mandibular movement. J. Prosthet. Dent., 52:421, 1984.
Goodman, H. O.: Genetic parameters of dentofacial development. J. Dent. Res., 44:174, 1965.
Gordon, H. J.: Human cranial base development during the late embryonic and the fetal periods. Master's Thesis, Univ. of Illinois (Orthodont.), 1955.
Goret-Nicaise, M., and A. Dhem: Presence of chondroid tissue in the symphyseal region of the growing human mandible. Acta Anat. (Basel), 113:189, 1982.
Goret-Nicaise, M., and D. Pilet: A few observations about Meckel's cartilage in the human. Anat. Embryol. (Berl.), 167:365, 1983.
Gorlin, R. J., and J. J. Pindborg: *Syndromes of the Head and Neck.* New York, McGraw-Hill, 1964.
Gorlin, R. J., J. Cervenka, and S. Pruzansky: Facial clefting and its syndromes. Birth Defects, 7:3, 1971.
Gorlin, R. J., M. M. Cohen, Jr., and L. S. Levin: *Syndromes of the Head and Neck,* 3rd Ed., Oxford University Press, 1990.
Gorniak, G. C.: Correlation between histochemistry and muscle activity of jaw muscles in cats. J. Appl. Physiol., 60:1393, 1986.
Goyne, T. E.: Reconstructing the face from the skull as a means of identification. Med. Leg. Bull., 31:1, 1982.
Grabb, W. C., S. W. Rosenstein, and K. R. Bzoch: *Cleft Lip and Palate.* Boston, Little, Brown, 1971.
Graber, L. W.: Congenital absence of teeth: A review with emphasis on inheritance patterns. J.A.D.A., 96:266, 1978.
Graber, T. M.: Implementation of the roentgenographic cephalometric technique. Am. J. Orthod., 44:906, 1958.
Graber, T. M.: Clinical cephalometric analysis. In: *Vistas of Orthodontics.* Ed. by B. S. Kraus and R. A. Reidel. Philadelphia, Lea & Febiger, 1962.
Graber, T. M.: A study of cranio-facial growth and development in the cleft palate child from birth to six years of age. In: *Early Treatment of Cleft Lip and Palate.* Ed. by R. Hotz. Berne, Switzerland, Hans Huber, 1964.
Graber, T. M.: *Orthodontics: Principles and Practice.* Philadelphia, W. B. Saunders, 1966.
Graber, T. M.: Extrinsic control factors influencing craniofacial growth. In: *Control Mechanisms in Craniofacial Growth.* Ed. by J. A. McNamara, Jr., Ann Arbor, University of Michigan, Center for Human Growth and Development, 1975.
Grahnen, H.: Hereditary factors in relation to dental caries and congenitally missing teeth. *Genetics and Dental Health.* Ed. by W. Witkop. McGraw-Hill, 1962.
Grant, D., and S. Bernick: Formation of the periodontal ligament. J. Periodontol., 43:17, 1972.
Grave, K. C., and T. Brown: Skeletal ossification and the adolescent growth spurt. Am. J. Orthod., 69:611, 1976.
Grayson, B. H., J. G. McCarthy, and F. Bookstein: Analysis of craniofacial asymmetry by multiplane cephalometry. Am. J. Orthod., 84:217, 1983.
Grayson, B. H., S. Boral, S. Eisig, A. Kolber, and J. G. McCarthy: Unilateral craniofacial microsomia. Part I. Mandibular analysis. Am. J. Orthod., 84:225, 1983.
Grayson, B. H., C. B. Cutting, and C. R. Dufresne: Three-dimensional computer simulation of craniofacial anatomy. N.Y. State Dent. J., 52:29, 1986.
Greenberg, A.: Life cycle of the human mandible. N.Y. Dent. J., 31:98, 1965.
Greenspan, J. S., and H. J. Blackwood: Histochemical studies of chondrocyte function in the cartilage of the mandibular condyle of the rat. J. Anat., 100:615, 1966.
Gregory, W. K.: *Our Face from Fish to Man.* New York, Putnam, 1929.
Gregory, W. K.: Certain critical stages in the evolution of the vertebrate jaws. Int. J. Orthod., 17:1138, 1931.
Greulich, W. W., and S. I. Pyle: *Radiographic Atlas of Skeletal Development of the Hand and Wrist,* 2nd Ed. Stanford, Stanford University Press, 1959.
Griffiths, D. L., L. Furstman, and S. Bernick: Postnatal development of the mouse palate. Am. J. Orthod., 53:757, 1967.
Gross, S. J., J. M. Oehler, and C. O. Eckerman: Head growth and development outcome in very low-birth-weight infants. Pediatrics, 71:70, 1983.
Gugino, C. F.: *An Orthodontic Philosophy,* 6th Ed. RM/Communicators Division of Rocky Mountain/Associates International Inc., Denver, Colo., 1971.
Guileminault, C., R. Riley, and N. Powell: Obstructive sleep apnea and abnormal cephalometric measurements: Implication for treatment. Chest, 86:793, 1984.
Gussen, R.: Articular and internal remodeling in the human otic capsule. Am. J. Anat., 122:397, 1968b.
Gustafsson, M., and J. Ahlgren: Mentalis and orbicularis oris activity in children with incompetent lips. An electromyographic and cephalometric study. Acta Odont. Scand., 33:355, 1975.

Guth, L.: Regulation of metabolic and functional properties of muscle. In: *Regulation of Organ and Tissue Growth.* Ed. by R. J. Goss. New York, Academic Press, 1973.

Guyer, E. C., E. E. Ellis, J. A. McNamara, and R. G. Behrents: Components of Class III malocclusion in juveniles and adolescents. Angle Orthod., 1:7, 1986.

Haas, A. J.: A biological approach to diagnosis, mechanics and treatment of vertical dysplasia. Angle Orthod., 50:279, 1980a.

Haas, A. J.: Long-term post-treatment evaluation of rapid palatal expansion. Angle Orthod., 50:189, 1980b.

Hägg, W., and J. Taranger: Menarche and voice change as indicators of the pubertal growth spurt. Acta Odontol. Scand., 38:179, 1980.

Hahn, F. J., W. Chu, and J. Cheung: CT measurements of cranial growth: alternative measurement method. A.J.N.R., 6:537, 1985.

Hall, B. K.: *In vitro* studies on the mechanical evocation of adventitious cartilage in the chick. J. Exp. Zool., 168:238, 1968.

Hall, B. K.: Differentiation of cartilage and bone from common germinal cells. J. Exp. Zool., 173:383, 1970.

Hall, B. K.: The origin and fate of osteoclasts. Anat. Rec., 183:1975.

Hall, B. K.: Initiation of osteogenesis by mandibular mesenchyme of the embryonic chick in response to mandibular and non-mandibular epithelia. Arch. Oral Biol., 23:1157, 1978a.

Hall, B. K.: *Development and Cellular Skeletal Biology.* New York, Academic Press, 1978b.

Hall, B. K.: How is mandibular growth controlled during development and evolution? J. Craniofac. Genet. Dev. Biol., 2:45, 1982a.

Hall, B. K.: Mandibular morphogenesis and craniofacial malformations. J. Craniofac. Genet. Dev. Biol., 2:309, 1982b.

Hall-Craggs, E. C. B.: Influence of epiphyses on the regulation of bone growth. Nature 221:1245, 1969.

Hall-Craggs, E. C. B., and C. A. Lawrence: The effect of epiphysial stapling on growth in length of the rabbit's tibia and femur. J. Bone Joint Surg., 513:359, 1969.

Hamill, P. V. V., F. E. Johnston, and W. Grams: Height and weight of children, United States. DHEW Publ. No. (PHS)1000, Vital Health Stat. (11), No. 104, 1970.

Hamill, P. V. V., T. A. Drizd, C. L. Johnston, R. B. Reed, and A. F. Roche: NCHA growth curves for children, birth–18 years, United States. DHEW Publ. No. (PHS) 78-1650. Vital Health Stat. (11), No. 165, 1977.

Handelman, C. S., and G. Osborne: Growth of the nasopharynx and adenoid development from one to eighteen years. Angle Orthod., 46:243, 1976.

Hanson, J., and D. W. Smith: U-shaped palatal defect in the Robin anomalad: Developmental and clinical relevance. J. Pediatr. 87:30, 1975.

Hanson, J. R., and B. J. Anson: Development of the malleus of the human ear. Illustrated in Atlas Series. Quart. Bull. Northwest. Univ. Med. Sch., 36:119, 1962.

Hanson, J. R., B. J. Anson, and T. H. Bast: The early embryology of the auditory ossicles in man. Illustrated in Atlas Series. Quart. Bull. Northwest. Univ. Med. Sch., 33:358, 1959.

Haralabakis, H. N., and E. G. Dagalakis: Effect of prednisolone and human growth hormone on growth of cranial bones and cranial base synchondroses in rats. Eur. J. Orthod., 2:239, 1980.

Harkness, E. M., and W. D. Trotter: Growth spurt in rat cranial bases transplanted into adult hosts. J. Anat., 131:39, 1980.

Harper, R. P., H. deBruin, and J. Burcea: Lateral pterygoid muscle activity in mandibular retrognathism and response to mandibular advancement surgery. Am. J. Orthod. Dentofacial Orthop., 9:70, 1987.

Harris, E. F., and R. J. Smith: A study of occlusion and arch widths in families. Am. J. Orthod., 78:155, 1980.

Harris, J. E.: A cephalometric analysis of mandibular growth rate. Am. J. Orthod., 48:161, 1962.

Harris, J. E., and C. J. Kowalski: All in the family: Use of familial information in orthodontic diagnosis, case assessment, and treatment planning. Am. J. Orthod., 69:493, 1976.

Harris, J. E., C. J. Kowalski, and S. S. Watnick: Genetic factors in the shape of the cranio-facial complex. Angle Orthod., 43:107, 1973.

Harris, J. E., C. J. Kowalski, and S. J. Walker: Dentofacial differences between "normal" sibs of Cl II and Cl III patients. Angle Orthod., 45:103, 1975.

Hart, J. C., G. R. Smiley, and A. D. Dixon: Sagittal growth of the craniofacial complex in normal embryonic mice. Arch. Oral Biol., 14:995, 1969.

Harvold, E. P.: The asymmetries of the upper facial skeleton and their morphological significance. Eur. Orthod. Soc. Trans., 1951.

Harvold, E. P.: The role of function in the etiology and treatment of malocclusion. Am. J. Orthod., 54:883, 1968.

Harvold, E. P.: Skeletal and dental irregularities in relation to neuromuscular dysfunctions. In: *Patterns of Orofacial Growth and Development.* Report 6. Washington, D.C., American Speech and Hearing Association, 1971.

Harvold, E. P.: Neuromuscular and morphological adaptations in experimentally induced oral respiration. In: *Naso-respiratory Function and Craniofacial Growth.* Ed. by J. A. McNamara, Jr. Ann Arbor, University of Michigan, Center for Human Growth and Development, 1979.

Harvold, E. P.: Bone remodeling and orthodontics. Eur. J. Orthod., 7:217, 1985.

Harvold, E. P., and K. Vargervik: Morphogenic response to activator treatment. Am. J. Orthod., 60:478, 1970.
Harvold, E. P., G. Chierici, and K. Vargervik: Experiments on the development of dental malocclusions. Am. J. Orthod., 61:38, 1972.
Harvold, E. P., K. Vargervik, and G. Chierici: Primate experiments on oral sensation and dental malocclusion. Am. J. Orthod., 63:494, 1973.
Haskell, B. S.: The human chin and its relationship to mandibular morphology. Angle Orthod., 49:153, 1979.
Hasund, A., and R. Sivertsen: Dental arch space and facial type. Angle Orthod., 41:140, 1971.
Healy, M. J. R.: Statistics of growth standards. In: *Human Growth.* Vol. 3, 2nd Ed. Ed. by F. Falkner and J. M. Tanner. New York, Plenum Press, 1986.
Heine, H.: Unfolding of the brain and secondary mandibular joint and face development in mammals: A synergetic and synchronous event. Gegenbaurs Morphol. Jahrb., 129:699, 1983.
Hellman, M.: A preliminary study in development as it affects the human face. Dent. Cosmos., 71:250, 1927a.
Hellman, M.: Changes in the human face brought about by development. Int. J. Orthod. Oral Surg., 13:475, 1927b.
Hellman, M.: An introduction of growth of the human face from infancy to adulthood. Int. J. Orthod. Oral Surg. Radiol., 18:777, 1932.
Hellman, M.: The face in its developmental career. Dent. Cosmos., 77:685, 1935.
Hellsing, G.: Functional adaptation to changes in vertical dimension. J. Prosthet. Dent., 52:862, 1984.
Hellsing, E., C. M. Forsberg, S. Linder-Aronson, and A. Sheibholeslam: Changes in postural EMG activity in the neck and masticatory muscles following obstruction of the nasal airways. Eur. J. Orthod., 8:247, 1986.
Henry, H. L.: Experimental study of external force application of the maxillary complex. In: *Factors Affecting the Growth of the Midface.* Ed. by J. A. McNamara, Jr. University of Michigan, Center for Human Growth and Development, 1976.
Herovici, C.: A polychrome stain for differentiating precollagen from collagen. Stain Technol., 38:204, 1963.
Herring, S. W.: Sutures: A tool in functional cranial analysis. Acta Anat., 83:222, 1972.
Herring, S. W., and T. C. Lakars: Cranibiolofacial development in the absence of muscle contraction. J. Craniofac. Genet. Dev. Biol., 1:341, 1982.
Hertzberg, S. R., Z. F. Muhl, and E. A. Begole: Muscle sarcomere length following passive jaw opening in the rabbit. Anat. Rec., 197:435, 1980.
Hewitt, A. B.: A radiographic study of facial asymmetry. Br. J. Orthod., 2:37, 1975.
Hildyard, L. T., W. J. Moore, and M. E. Corbett: Logarithmic growth of the hominoid mandible. Anat. Rec., 186:405, 1976.
Himes, J. H., and A. F. Roche: Clinical longitudinal growth charts for stature of American children (letter). J. Pediatr. 108:487, 1986.
Hinrichsen, G. J., and E. Storey: The effect of force on bone and bones. Angle Orthod., 38:155, 1963.
Hinton, R. J.: Changes in articular eminence morphology with dental function. Am. J. Phys. Anthropol., 54:439, 1981.
Hinton, R. J., and D. S. Carlson: Temporal changes in human temporomandibular joints' size and shape. Am. J. Phys. Anthropol., 50:325, 1979.
Hinton, W. L.: Form and function in the temporomandibular joint. In: *Craniofacial Biology.* Ed. by D. S. Carlson. Ann Arbor, University of Michigan, Center for Human Growth and Development, 1981.
Hirabayashi, S., K. Harii, A. Sakurai, E. K. Takaki, and O. Fukuda: An experimental study of craniofacial growth in a heterotopic rat head transplant. Plastic Reconstr. Surg., 82:236, 1988.
Hirschfeld, W. J., and R. E. Moyers: Prediction of craniofacial growth: the state of the art. Am. J. Orthod., 60:435, 1971.
Hirschfeld, W. J., R. E. Moyers, and D. H. Enlow: A method of deriving subgroups of a population: A study of craniofacial taxonomy. Am. J. Phys. Anthropol., 39:279, 1973.
Hixon, E. H.: Prediction of facial growth. Eur. Orthod. Soc. Rep. Congr., 44:127, 1968.
Hixon, E. H.: Cephalometrics: A perspective. Angle Orthod., 42:200, 1972.
Hixon, E. H., and S. L. Horowitz: *Nature of Orthodontic Diagnosis.* St. Louis, C. V. Mosby Co., 1966.
Hoffer, F., Jr., and R. D. Walters: Adaptive changes in the face of the *Macaca mulatta* monkey following orthopedic opening of the midpalatal suture. Angle Orthod., 45:282, 1975.
Hohl, T. H., and W. Tucek: Measurement of condylar loading forces by instrumented prosthesis in the baboon. J. Maxillofac. Surg., 10:1, 1982.
Holdaway, R. A.: Changes in relationship of points A and B during orthodontic treatment. Am. J. Orthod., 42:176, 1956.
Holdaway, R. A.: A soft-tissue cephalometric analysis and its use in orthodontic treatment planning. Part I. Am. J. Orthod., 84:1, 1983.
Holdaway, R. A.: A soft-tissue cephalometric analysis and its use in orthodontic treatment planning. Part II. Am. J. Orthod., 85:279, 1984.
Holt, S. B.: Quantitative genetics of finger-print patterns. Br. Med. Bull., 17:247, 1961.
Hooton, E. P.: *Up From the Ape,* 3rd Ed. New York, Macmillan, 1946.

Hopkin, G. B.: The cranial base as an aetiological factor in malocclusion. Angle Orthod., 38:250, 1968.
Horowitz, S. L.: Modifications of mandibular architecture following removal of the temporalis muscle in the rat. J. Dent. Res., 30:276, 1951.
Horowitz, S. L.: The role of genetic and local environmental factors in normal and abnormal morphogenesis. Acta Morphol. Neerl. Scand., 10:59, 1972.
Horowitz, S. L.: Variability of the maxillary complex in common dental malformations. Presented at the Third International Orthodontic Congress, London, 1975.
Horowitz, S. L., and R. Osborne: The genetic aspects of cranio-facial growth. In: *Cranio-facial Growth in Man.* Ed. by R. E. Moyers and W. M. Krogman. Oxford, Pergamon Press, 1971.
Horowitz, S. L., and R. H. Thompson: Variations of the craniofacial skeleton in post-adolescent males and females with special references to the chin. Angle Orthod., 34:97, 1964.
Horowitz, S. L., R. H. Osborne, and F. V. DeGeorge: Hereditary factors in tooth dimensions; a study of the anterior teeth of twins. Angle Orthod., 28:86, 1958.
Horowitz, S. L., R. H. Osborne, and F. V. DeGeorge: A cephalometric study of craniofacial variation in adult twins. Angle Orthod., 30:1, 1960.
Horowitz, S. L., L. J. Gerstman, and J. M. Converse: Craniofacial relationships in mandibular prognathion. Arch. Oral Biol., 14:121, 1969.
Houghton, P.: Rocker jaws. Am. J. Phys. Anthropol., 47:365, 1977.
Houghton, P.: Polynesian mandibles. J. Anat., 127:251, 1978.
Houpt, M. I.: Growth of the craniofacial complex of the human fetus. Am. J. Orthod., 58:373, 1970.
Houston, W. J. B.: The current status of facial growth prediction: A review. Br. J. Orthod., 6:11, 1978.
Houston, W. J.: Relationship between skeletal maturity estimated from hand-wrist radiographs and the timing of the adolescent growth spurt. Eur. J. Orthod., 2:81, 1980.
Houston, W. J.: A comparison of the reliability of measurement of cephalometric radiographs by tracings and direct digitization. Swed. Dent. J., 15:99, 1982.
Houston, W. J.: The analysis of errors in orthodontic measurements. Am. J. Orthod., 83:382, 1983.
Houston, W. J., and W. A. B. Brown: Family likeness as a basis for facial growth prediction. Eur. J. Orthod., 2:13, 1980.
Houston, W. J., and R. T. Lee: Accuracy of different methods of radiographic superimposition on cranial base structures. Eur. J. Orthod., 7:127, 1985.
Houston, W. J. B., J. C. Miller, and J. M. Tanner: Prediction of the timing of the adolescent growth spurt from ossification events in hand-wrist films. Br. J. Orthod., 6:142, 1979.
Howe, R. P., J. A. McNamara, and K. A. O'Connor: An examination of dental crowding and its relationship to tooth size and arch dimension. Am. J. Orthod., 83:363, 1983.
Howells, W. W.: Body measurements in the light of familial influences. Am. J. Phys. Anthropol., 7:101, 1949.
Howells, W.: *Mankind So Far.* Garden City, New York, Doubleday, 1949.
Hoyte, D. A. N.: The relative contribution of sutural and ectocranial deposition of bone to cranial growth in rodents. J. Anat., 92:654, 1958.
Hoyte, D. A. N.: Resorption and skull expansion in rats. Anat. Rec., 139:307, 1961a.
Hoyte, D. A. N.: The postnatal growth of the ear capsule in the rabbit. Am. J. Anat., 108:1, 1961b.
Hoyte, D. A. N.: Facts and fallacies of alizarin staining: The exterior of the infant pig skull. J. Dent. Res., 43:814, 1964a.
Hoyte, D. A. N.: Facts and fallacies of alizarin staining: The interior of the infant pig skull. Anat. Rec., 148:292, 1964b.
Hoyte, D. A. N.: The role of the soft tissues in skull growth in rabbits. West Indian Med. J., 14:125, 1965a.
Hoyte, D. A. N.: The sphenoidal complex in the first three months of its postnatal growth in rabbits: An alizarin study. Anat. Rec., 151:364, 1965b.
Hoyte, D. A. N.: Experimental investigations of skull morphology and growth. Int. Rev. Gen. Exp. Zool., 2:345, 1966.
Hoyte, D. A. N.: Alizarin red in the study of the apposition and resorption of bone. Am. J. Phys. Anthropol., 29:157, 1968.
Hoyte, D. A. N.: Mechanisms of growth in the cranial vault and base. J. Dent. Res., 50:1447, 1971a.
Hoyte, D. A. N.: The modes of growth of the neurocranium: The growth of the sphenoid bone in animals. In: *Cranio-facial Growth in Man.* Ed. by R. E. Moyers and W. M. Krogman. Oxford, Pergamon Press, 1971b.
Hoyte, D. A. N.: Basicranial elongation. 2. Is there differential growth within a synchondrosis? Anat. Rec., 1975:347, 1973a.
Hoyte, D. A. N.: Basicranial elongation. 3. Differential growth between synchondroses and basion. Proceedings of the Third European Anatomy Congress, Manchester, England, 231, 1973b.
Hoyte, D. A. N.: A critical analysis of the growth in length of the cranial base. In: *Morphogenesis and Malformation of Face and Brain.* Ed. by D. Bergsma, J. Langman, and N. W. Paul. National Foundation—March of Dimes. Birth Defects Original Article Series, Vol. 11, No. 7. Alan R. Liss, New York, 1975.
Hoyte, D. A. N.: Contributions of the spheno-ethmoidal complex to basicranial growth in the

rabbit. In: *Development of the Basicranium.* Ed. by J. F. Bosma. DHEW Pub. 76:989, NIH, Bethesda, Md., 1976.
Hoyte, D. A. N., and D. H. Enlow: Wolff's law and the problem of muscle attachment on resorptive surfaces of bone. Am. J. Phys. Anthropol., 24:205, 1966.
Hughes, P. C. R., J. M. Tanner, and J. P. G. Williams: A longitudinal radiographic study of the growth of the rat skull. J. Anat., 127:83, 1978.
Hulse, F. S.: Exogamie et heterosis. Arch. Suisse Anthropol. Gen., 22:103, 1957.
Humphrey, T.: Reflex activity in the oral and facial area of the human fetus. In: *Second Symposium on Oral Sensation and Perception.* Ed. by James Bosma. Springfield, Ill., Charles C Thomas, 1970.
Humphrey, T.: The development of oral and facial motor mechanisms in human fetuses and their relation to craniofacial growth. J. Dent. Res., 50:1428, 1971a.
Humphrey, T.: Human prenatal activity sequences in the facial region and their relationship to postnatal development. In: *Patterns of Orofacial Growth and Development.* Report 6. Washington, D.C., American Speech and Hearing Association, 1971b.
Humphrey, T.: Human prenatal activity sequences in the facial region and their relationship to postnatal development. In: *Proceedings of the Conference on Patterns of Orofacial Growth and Development,* Ann Arbor, Michigan, March 4–6, 1970. ASHA Report #6. American Speech and Hearing Association, Washington, D.C., March 1971.
Humphry, G.: On the growth of the jaws. Cambridge, Trans. Cambridge Phil. Soc., 1864.
Hunter, W. S.: The inheritance of mesiodistal tooth diameter in twins. Ph.D. Dissertation, Univ. of Michigan, 1959.
Hunter, W. S.: A study of the inheritance of the craniofacial characteristics, as seen in lateral cephalograms of 72 like-sexed twins. Trans. Eur. Orthod. Soc., 59:70, 1965.
Hunter, W. S.: The dynamics of mandibular arch perimeter change from mixed to permanent dentitions. In: *Craniofacial Biology.* Ed. by J. A. McNamara, Jr. Ann Arbor, University of Michigan, Center for Human Growth and Development, 1977.
Hunter, W. S.: Review: The Michigan cleft palate study. J. Craniofac. Genet. Dev. Biol., 1:235, 1981.
Hunter, W. S., and D. J. Dijkman: The timing of height and weight deficits in twins discordant for cleft of the lip and/or palate. Cleft Palate J., 14:158, 1977.
Hunter, W. S., and S. Garn: Evidence for a secular trend in face size. Angle Orthod., 39:320, 1969.
Hunter, W. S., and S. M. Garn: Differential secular increase in the progeny of small faced parents. J. Dent. Res., 52:212(abstr. 613), 1973.
Hunter, W. S., D. R. Balbach, and D. E. Lamphiear: The heritability of attained growth in the human face. Am. J. Orthod., 58:128, 1970.
Hurst, R. V. V., B. Schwaninger, R. Shaye, and J. M. Chadha: Landmark identification accuracy in xeroradiographic cephalometry. Am. J. Orthod., 73:568, 1978.
Hylander, W. L.: The human mandible: Lever or link? Am. J. Phys. Anthropol., 43:227, 1975.
Hylander, W. L.: *In vivo* bone strain in the mandible of *Galago crassicaudatus.* Am. J. Phys. Anthropol., 46:309, 1977a.
Hylander, W. L.: Bone strain in the mandibular symphysis of *Macaca fascicularis.* J. Dent. Res., Special Issue B, 46:344, 1977b.
Hylander, W. L.: *In vivo* bone strain in the mandible of *Galago crassicaudatus.* Am. J. Phys. Anthropol., 46:309, 1977c.
Hylander, W. L.: Incisal bite force direction in humans and the functional significance of mammalian mandibular translation. Am. J. Phys. Anthropol., 48:1, 1978.
Hylander, W. L.: An experimental analysis of temporomandibular joint reaction forces in macaques. Am. J. Phys. Anthropol., 51:433, 1979.
Hylander, W. L.: Mandibular function in *Gagalgo crassicadatus* and *Macaca fascicularis:* An *in vivo* approach to stress analysis of the mandible. J. Morphol., 159:253, 1979.
Hylander, W. L.: Functional anatomy. In: *The Temporomandibular Joint,* 3rd Ed. Ed. by B. G. Sarnat and D. M. Laskin. Springfield, Ill., Charles C Thomas, 1980.
Hylander, W. L.: Patterns of stress and strain in the macaque mandible. In: *Craniofacial Biology.* Ed. by D. S. Carlson. Ann Arbor, University of Michigan, Center for Human Growth and Development, 1981.
Hylander, W. L.: Stress and strain in the mandibular symphysis of primates: A test of competing hypotheses. Am. J. Phys. Anthrop., 64:1, 1984.
Infante, P. F.: Malocclusion in the deciduous dentition in White, Black and Apache Indian children. Angle Orthod., 45:213, 1975.
Ingerslev, C. H., and B. Solow: Sex differences in craniofacial morphology. Acta Odont. Scand., 33:85, 1975.
Ingervall, B., and E. Helkimo: Masticatory muscle force and facial morphology in man. Arch. Oral Biol., 23:203, 1978.
Ingervall, B., and B. Thilander: The human spheno-occipital synchondrosis. I. The time of closure appraised macroscopically. Acta Odont. Scand., 30:349, 1972.
Ingervall, B., E. Carlsson, and B. Thilander: Postnatal development of the human temporomandibular joint. II. A microradiographic study. Acta Odont. Scand., 34:133, 1976.
Inoue, N.: A study of the developmental changes in the dentofacial complex during the fetal period by means of roentgenographic cephalometrics. Tokyo Med. Dent. Univ. Bull., 8:205, 1961.

Isaacson, J. R., R. J. Isaacson, T. M. Speidel, and F. W. Worms: Extreme variation in vertical facial growth and associated variation in skeletal and dental relations. Angle Orthod., 41:219, 1971.

Isaacson, R. J., R. J. Zapfel, F. W. Worms, and A. G. Erdman: Effect of rotational jaw growth on the occlusion and profile. Am. J. Orthod., 72:276, 1977.

Isaacson, R. J., A. G. Erdman, and B. W. Hultgren: Facial and dental effects of mandibular rotation. In: *Craniofacial Biology*. Ed. by D. S. Carlson. Ann Arbor, University of Michigan, Center for Human Growth and Development, 1981.

Isaacson, R. J., et al.: Some effects of mandibular growth on the dental occlusion and profile. Angle Orthod., 47:97, 1977.

Isotupa, K.: Alizarin trajectories in bone growth studies. Proc. Finn. Dent. Soc., 77:63, 1981.

Isotupa, K., K. Koski, and L. Makinen: Changing architecture of growing cranial bones at sutures as revealed by vital staining with alizarin red S in the rabbit. Am. J. Phys. Anthropol., 23:19, 1965.

Israel, H.: Loss of bone and remodeling-redistribution in the craniofacial skeleton with age. Fed. Proc., 26:1723, 1967.

Israel, H.: Continuing growth in the human cranial skeleton. Arch. Oral Biol., 13:133, 1968.

Israel, H.: Age factor and the pattern of change in craniofacial structures. Am. J. Phys. Anthropol., 39:111, 1973a.

Israel, H.: Recent knowledge concerning craniofacial aging. Angle Orthod., 43:176, 1973b.

Israel, H.: Age factor and the pattern of change in craniofacial structures. Am. J. Phys. Anthrop., 39:111, 1973c.

Israel, H.: The dichotomous pattern of craniofacial expansion during aging. Am. J. Phys. Anthropol., 47:47, 1977.

Jacobs, B., and J. H. Kronman: The zygomatic arch and its possible influence on craniofacial growth and development. Angle Orthod., 47:136, 1977.

Jacobson, A.: The craniofacial skeletal pattern of the South African Negro. Am. J. Orthod., 73:681, 1978.

Jamison, J. E., S. E. Bishara, L. C. Peterson, W. H. DeKock, and C. R. Kremenak: Longitudinal changes in the maxilla and the maxillary-mandibular relationship between 8 and 17 years of age. Am. J. Orthod., 82:217, 1982.

Jane, J. A.: Radical reconstruction of complex cranioorbitofacial abnormalities. In: *Morphogenesis and Malformation of Face and Brain.* Ed. by D. Bergsma, J. Langman, and N. W. Paul. National Foundation—March of Dimes. Birth Defects Original Article Series, Vol. 11, No. 7. New York, Alan R. Liss, 1975.

Janzen, E. K., and J. A. Bluher: The cephalometric, anatomic and histologic changes in *Macaca mulatta* after application of a continuous action retraction force on the mandible. Am. J. Orthod., 51:823, 1965.

Jarabak, J. R., and J. R. Thompson: Cephalometric appraisal of the cranium and mandible of the rat following condylar resection. J. Dent. Res., 28:655, 1949.

Jarvinen, S.: Variability of linear and angular cephalometric variables. Proc. Finn. Dent. Soc., 79:13, 1983.

Jaskoll, T., and M. Melnick: The effects of long term fetal constraint *in vitro* on the cranial base and other skeletal components. Am. J. Med. Genet., 12:289, 1982.

Johnson, P. A., P. J. Atkinson, and W. J. Moore: The development and structure of the chimpanzee mandible. J. Anat., 122:467, 1976.

Johnston, L. E.: A statistical evaluation of cephalometric prediction. Master's Thesis, University of Michigan, Ann Arbor, 1964.

Johnston, L. E.: A statistical evaluation of cephalometric prediction. Angle Orthod., 38:284, 1968.

Johnston, L. E.: A simplified approach to prediction. Am. J. Orthod., 67:253, 1975.

Johnston, L. E.: The functional matrix hypothesis: Reflections in a jaundiced eye. In: *Factors Affecting the Growth of the Midface.* Ed. by J. A. McNamara, Jr. Ann Arbor, University of Michigan, Center for Human Growth and Development, 1976.

Johnston, L. E., and R. G. Behrents: A system for graphic representation during stereotaxic procedures. Brain Res. Bull., 13:335, 1984.

Johnston, M. C.: The neural crest in abnormalities of the face and brain.In: *Morphogenesis and Malformation of Face and Brain.* Ed. by D. Bergsma, J. Langman, and N. W. Paul. National Foundation—March of Dimes. Birth Defects Original Article Series, Vol. 11, No. 7. New York, Alan R. Liss, 1975.

Johnston, M. C., et al.: An expanded role of the neural crest in oral and pharyngeal development. In: *Oral Sensation and Perception: Development in the Fetus and Infant.* Ed. by J. Bosma. Washington, D.C., DHEW Pub. No. 73-546, 1973.

Joho, J. P.: Changes in form and size of the mandible in the orthopaedically treated *Macacus irus* (an experimental study). Eur. Orthod. Soc. Trans. 1968:161, 1969.

Jones, B. H., and H. V. Meredith: Vertical changes in osseous and odontic portions of human face height between the ages of 5 and 15 years. Am. J. Orthod., 52:902, 1966.

Jones, K. L., and M. C. Higginbottom: Dysmorphology: An approach to diagnosing children with structural defects of the head and neck. Head Neck Surg., 1:35, 1978.

Jonsson, G., and B. Thilander: Occlusion, arch dimensions, and craniofacial morphology after palatal surgery in a group of children with clefts in the secondary palate. Am. J. Orthod., 76:243, 1979.

Joondeph, D. R., and L. E. Wragg: Facial growth during the secondary palate closure in the rat. Am. J. Orthod., 6:88, 1966.

Jovanocic, S., N. Jelicic, and A. Kargovska-Klisarova: Postnatal development and anatomical relationship of the maxillary sinus. Acta Anat., 118:122, 1984.

Justus, R., and J. H. Luft: A mechanochemical hypothesis for bone remodeling induced by mechanical stress. Calcif. Tissue Res., 5:222, 1970.

Kaban, L. B., G. T. Cisneros, S. Heyman, and S. Treves: Assessment of mandibular growth by skeletal scintigraphy. J. Oral. Maxillofac. Surg., 40:18, 1982.

Kanouse, M. C., S. P. Ramfjord, and C. E. Nasjleti: Condylar growth in rhesus monkeys. J. Dent. Res., 48:1171, 1969.

Kantomaa, T.: Role of the mandibular condyle in facial growth. Proc. Fin. Dent. Soc., 81:111, 1985.

Kantomaa, T.: Reactions of the condylar tissues to attempts to increase mandibular growth. Scand. J. Dent. Res., 95:335, 1987.

Kardos, T. B., and L. O. Simpson: A new periodontal membrane biology based upon thixotropic concepts. Am. J. Orthod., 77:508, 1980.

Katz, M., and S. Kvinnsland: Matrix formation in the mandibular condyle of the rat: (^{35}S)-Sulfate incorporation studies. Acta Odont. Scand., 37:137, 1979.

Kawamura, Y.: *Physiology of Mastication.* Basel, S. Karger, 1974.

Kean, M. R., and P. Houghton: The role of function in the development of human craniofacial form: A perspective. Anat. Rec., 218:107, 1987.

Keith, A.: Contribution to the mechanism of growth of the human face. Internatl. J. Orthod., 8:607, 1922.

Keith, B. S., and J. D. Decker: The prenatal inter-relationships of the maxilla and premaxilla in the facial development of man. Acta Anat., 40:278, 1960.

Keizer, S., and D. B. Tuinzing: Spontaneous regeneration of a unilaterally absent mandibular condyle. J. Oral Maxillofac. Surg., 43:130, 1985.

Kelling, A. L., P. Zipse, and S. A. Miller: A biochemical comparison of development of various facial bones in neonatal rats. Arch. Oral Biol., 24:719, 1979.

Kerr, W. J. S.: A method of superimposing serial lateral cephalometric films for the purpose of comparison: A preliminary report. Br. J. Orthod., 5:51, 1978.

Kerr, W. J. S.: A longitudinal cephalometric study of dento-facial growth from five to fifteen years. Br. J. Orthod., 6:115, 1979.

Kerr, W. J.: The nasopharynx, face height, and overbite. Angle Orthod., 55:31, 1985.

Kerr, W. J., and C. P. Adams: Cranial base and jaw relationship. Am. J. Phys. Anthropol., 77:213, 1988.

Kerr, W. J., and D. Hirst: Craniofacial characateristics of subjects with normal and postnormal occlusion: A longitudinal study. Am. J. Orthod. Dentofacial Orthop., 92:207, 1987.

Kier, E. L.: The infantile sella turcica: New radiologic and anatomic concepts based on a developmental study of the sphenoid bone. Am. J. Roentgenol., 102:747, 1968.

Kier, E. L.: Phylogenetic and ontogenetic changes of the brain relevant to the evolution of the skull. In: *Development of the Basicranium.* Ed. by J. F. Bosma. DHEW Pub. 76:989, NIH, Bethesda, Md., 1976.

Kier, E. L., and S. L. G. Rothman: Radiologically significant anatomic variations of the developing sphenoid in humans. In: *Development of the Basicranium.* Ed. by J. F. Bosma. DHEW Publ. No. 76:989, NIH, Bethesda, Md., 1976.

Kiliaridis, S.: Masticatory muscle function and craniofacial morphology. An experimental study in the growing rat fed a soft diet. Swed. Dent. J., 36:1, 1986.

Kiliaridis, S.: Masticatory muscle function and craniofacial morphology. Swed. Dent. J., 36(Suppl.):7, 1986.

Kimura, K.: A consideration of the secular trend in Japanese for height and weight by the graphic method. Am. J. Phys. Anthropol. 27:89, 1967.

Kisling, E.: *Cranial Morphology in Down's Syndrome,* Vol. 58. Copenhagen, Munksgaard, 1966, p. 106.

Klami, O., and S. L. Horowitz: An analysis of the relationship between posterior dental cross-bite and vertical palatal asymmetry. Am. J. Orthod., 76:51, 1979.

Klink-Heckmann, U., T. Dahl, R. Grabowski, B. Moller, and J. Towe: Comparative study of the craniofacial structures in different occlusal anomalies. Zahn. Mund. Kieferheilkd., 71:230, 1983.

Knott, V. B.: Ontogenetic change of four cranial base segments in girls. Growth, 33:123, 1969.

Knott, V. B.: Change in cranial base measures of human males and females from age 6 years to early adulthood. Growth, 35:145, 1971.

Kodama, G.: Developmental studies of the presphenoid of the human sphenoid bone. In: *Development of the Basicranium.* Ed. by J. F. Bosma. DHEW Pub. 76:989, NIH, Bethesda, Md., 1976.

Kohn, R. R.: *Principles of Mammalian Aging.* 2nd Ed. Englewood Cliffs, Prentice-Hall, 1978.

Kokich, V. G.: Age changes in the human frontozygomatic sutures from 20 to 95 years. Am. J. Orthod., 69:411, 1976.

Konjevich, N.: Origin and maturation of the spheno-occipital synchondrosis, M.S. (Orthodont.) Thesis, Univ. of Illinois, 1963.

Koski, K.: Some aspects of the growth of the cranial base and the upper face. Odontol. Tidskr., 68:344, 1960.
Koski, K.: Cranial growth centers: Facts or fallacies? Am. J. Orthod., 54:566, 1968.
Koski, K.: Some characteristics of cranio-facial growth cartilages. In: *Cranio-facial Growth in Man.* Ed. by R. E. Moyers and W. M. Krogman. Oxford, Pergamon Press, 1971.
Koski, K.: Variability of the cranio-facial skeleton: An exercise in roentgencephalometry. Am. J. Orthod., 64:188, 1973.
Koski, K.: Cartilage in the face. In: *Morphogenesis and Malformation of Face and Brain.* Ed. by D. Bergsma, J. Langman, and N. W. Paul. National Foundation—March of Dimes. Birth Defects Original Article Series, Vol. 11, No. 7. New York, Alan R. Liss, 1975.
Koski, K.: Reflections on craniofacial growth research. Acta Morphol. Neerl. Scand., 23:357, 1985.
Koski, K., and L. Makinen: Growth potential of transplanted components of the mandibular ramus of the rat. I. Suom. Hammaslaak. Toim., 59:296, 1963.
Koski, K., and K. E. Mason: Growth potential of transplanted components of the mandibular ramus of the rat. II. Suom. Hammaslaak. Toim., 60:209, 1964.
Koski, K., and O. Rönning: Growth potential of transplanted components of the mandibular ramus of the rat. III. Suom. Hammaslaak. Toim., 61:292, 1965.
Koski, K., and O. Rönning: Pitkan luun rustoisen paan siirrannaisen kasvupotentiaalista rotalla. Suom. Hammaslaak. Toim., 62:165, 1966.
Koski, K., and O. Rönning: Growth potential of subcutaneously transplanted cranial base synchondroses of the rat. Acta Odont. Scand., 27:343, 1969.
Koski, K., and Rönning: Growth potential of intracerebrally transplanted cranial base synchondroses in the rat. Arch. Oral Biol., 15:1107, 1970.
Koskinen, L., and K. Koski: Regeneration in transplanted epiphysectomized humeri of rats. Am. J. Phys. Anthropol., 27:33, 1967.
Koskinen, L., K. Isotupa, and K. Koski: A note on craniofacial sutural growth. Am. J. Phys. Anthropol., 45:511, 1976.
Koskinen-Moffett, L., and B. Moffett: Influence of prenatal jaw functions on human facial development. Birth Defects, 20:47, 1984.
Koskinen-Moffett, L., R. McMinn, K. Isotupa, and B. Moffett: Migration of craniofacial periosteum in rabbits. Proc. Finn. Dent. Soc., 77:83, 1981.
Kowalski, C. J., C. E. Nasjleti, and G. F. Walker: Differential diagnosis of adult male black and white populations. Angle Orthod., 44:346, 1974.
Kowalski, C. J., J. E. Harris, and S. J. Walker: The craniofacial morphology of Nubian school children. Angle Orthod., 45:180, 1975.
Kowalski, C. J., C. E. Nasjleti, and G. F. Walker: Dentofacial variations within and between four groups of adult American males. Angle Orthod., 45:146, 1975.
Kragt, G., and H. S. Duterloo: The initial effects of orthopedic forces: A study of alterations in the craniofacial complex of a macerated human skull owing to high-pull headgear traction. Am. J. Orthod., 8:57, 1982.
Kraus, B. S.: Prenatal growth and morphology of the human bony palate. J. Dent. Res., 39:1177, 1960a.
Kraus, B. S.: The prenatal inter-relationships of the maxilla and premaxilla in the facial development of man. Acta Anat., 40:278, 1960b.
Kraus, B. S.: *Human Dentition before Birth.* Philadelphia, Lea & Febiger, 1965.
Kraus, B. S., H. Kitamura, and R. A. Latham: *Atlas of Developmental Anatomy of the Face.* New York, Harper & Row, 1966.
Kraw, A. G., and D. H. Enlow: Continuous attachment of the periodontal membrane. Am. J. Anat., 120:133, 1967.
Kreiborg, S., B. L. Jensen, E. Moller, and A. Bjork: Craniofacial growth in a case of congenital muscular dystrophy. A roentgencephalometric and electromyographic investigation. Am. J. Orthod., 74:207, 1978.
Kremenak, C. R., Jr.: Circumstances limiting the development of a complete explanation of craniofacial growth. Acta Morphol. Neerl. Scand., 10:127, 1972.
Kremenak, C. R., D. F. Hartshorn, and S. E. Demjen: The role of the cartilaginous nasal septum in maxillofacial growth: Experimental septum removal in beagle pups. J. Dent. Res., 48: Abstr. 32, 1969.
Krogman, W. M.: Studies in growth changes in the skull and face of anthropoids. Am. J. Anat., 46:315, 1930.
Krogman, W. M.: Principles of human growth. Ciba Found. Symp., 5:1458, 1943.
Krogman, W. M.: The growth periods from birth to adulthood. Syllabus. Third Annual Midwestern Seminar of Dental Medicine, University of Illinois, 1950.
Krogman, W. M.: Craniometry and cephalometry as research tools in growth of head and face. Am. J. Orthod., 37:406, 1951a.
Krogman, W. M.: The problem of 'timing' in facial growth, with special reference to the period of the changing dentition. Am. J. Orthod., 37:253, 1951b.
Krogman, W. M.: *The Human Skeleton in Forensic Medicine.* Springfield, Ill., Charles C Thomas, 1962.
Krogman, W. M.: Role of genetic factors in the human face, jaws and teeth: A review. Eugenics Rev., 59:165, 1967.

Krogman, W. M.: Growth of head, face, trunk and limbs in Philadelphia White and Negro children of elementary and high school age. Monogr. Soc. Res. Child Develop., Serial No. 136, 35:1, 1970.
Krogman, W. M.: Craniofacial growth and development: An appraisal. Yearbook Phys. Anthropol., 18:31, 1974.
Krogman, W. M.: Physical anthropology and the dental and medical specialties. Am. J. Phys. Anthropol., 45:531, 1976.
Krogman, W. M., and D. D. B. Chung: The craniofacial skeleton at the age of one month. Angle Orthod., 25:305, 1965.
Krogman, W. M., and V. Sassouni: *Syllabus in Roentgenographic Cephalometry.* Philadelphia, Philadelphia Center for Research in Child Growth, 1957.
Krompecher, S., and L. Toth: Die Konzeption von Kompression, Hypoxie und konsekutiver Mucopolysaccharidbildung in der kausalen Analyse der Chondrogenese. Z. Anat. Entwicklungesch., 124:268, 1964.
Kummer, B.: Anatomy and biomechanics of the mandible. Fortschr. Kieferorthop., 46:335, 1985.
Kurihara, S., and D. H. Enlow: Remodeling reversals in anterior parts of the human mandible and maxilla. Angle Orthod., 50:98, 1980a.
Kurihara, S., and D. H. Enlow: An electron microscopic study of attachments between periodontal fibers and bone during alveolar remodeling. Am. J. Orthod., 77:516, 1980b.
Kurihara, S., and D. H. Enlow: A histochemical and electron microscopic study of an adhesive type of collagen attachment on resorptive surfaces of alveolar bone. Am. J. Orthod., 77:532, 1980c.
Kurisu, K., J. D. Niswander, M. C. Johnston, and M. Mazaheri: Facial morphology as an indicator of genetic predisposition to cleft lip and palate. Am. J. Hum. Gen., 26:702, 1974.
Kuroda, T.: A longitudinal cephalometric study on the craniofacial development in Japanese children. Presented at the Annual Meeting of the Int. Assoc. Dent. Res., New York, 1970, Abstr. 32.
Kuroda, T., F. Miura, T. Nakamura, and K. Noguchi: Cellular kinetics of synchondroseal cartilage in organ culture. Proc. Finn. Dent. Soc., 77:89, 1981.
Kvam, E., B. Ostball, and O. Slagsvold: Growth in width of the frontal bones after fusion of the metopic suture. Acta Odont. Scand., 33:227, 1975.
Kvinnsland, S.: Observation on the early ossification of the upper jaw. Acta Odont. Scand., 27:649, 1969.
Kvinnsland, S.: The sagittal growth of the upper face during foetal life. Acta Odont. Scand., 29:717, 1971a.
Kvinnsland, S.: The sagittal growth of the lower face during foetal life. Acta Odont. Scand., 29:733, 1971b.
Kvinnsland, S.: The sagittal growth of the foetal cranial base. Acta Odont. Scand., 29:699, 1971c.
Kvinnsland, S., and S. Kvinnsland. Transplantation studies of the nasal septal cartilage in rats. In: *Factors Affecting the Growth of the Midface.* Ed. by J. A. McNamara, Jr. Ann Arbor, University of Michigan, Center for Human Growth and Development, 1976.
Kvinnsland, S., and G. Rakstang: Proliferation of osteogenic cells in the facial bones of the rat. Eur. J. Orthod., 2:95, 1980.
Kylamarkula, S., and J. Huggari: Head posture and the morphology of the first cervical vertebra. Eur. J. Orthod., 7:151, 1985.
Kylamarkula, S., and O. Rönning: Transplantation of a basicranial sychrondrosis to a sutural area in the isogeneic rat. Eur. J. Orthod., 1:145, 1979.
LaBanc, J. P., and B. N. Epker: Changes of the hyoid bone and tongue following advancement of the mandible. Oral Surg., 57:351, 1984.
Laitman, J. T., and E. S. Crelin: Postnatal development of the basicranium and vocal tract region in man. In: *Development of the Basicranium.* Ed. by J. F. Bosma. DHEW Pub. 76:989, NIH, Bethesda, Md., 1976.
Lande, M. S.: Growth behaviour of the human bony profile as revealed by serial cephalometric statistical appraisal of dentofacial growth. Angle Orthod., 32:205, 1962.
Langman, J., P. Rodier, and W. Webster: Interference with proliferative activity in the CNS and its relation to facial abnormalities. In: *Morphogenesis and Malformation of Face and Brain.* Ed. by D. Bergsma, J. Langman, and N. W. Paul. National Foundation—March of Dimes. Birth Defects Original Article Series, Vol. 11, No. 7. New York, Alan R. Liss, 1975.
Largo, M. A.: Visusveranderungen im Verlaufe der Pubertat und Augendominanz im Alter von 10 Jahren. Helv. Paediatr. Acta, 32:59, 1978.
Largo, M. A. and A. Prader: Pubertal development in Swiss boys. Helf. Paediatr. Acta, 38:211, 1983a.
Largo, M. A., and A. Prader: Pubertal development in Swiss girls. Helv. Paediatr. Acta, 38:229, 1983b.
Larsson, A.: Light microscopic and ultrastructural observations of the calcifying zone of the mandibular condyle in the rat. Anat. Rec., 185:171, 1976.
Larsson, E.: The effect of finger-sucking on the occlusion: A review. Eur. J. Orthod., 9:279, 1987.
Laskin, D. M., W. A. Ryan, and C. S. Greene: Incidence of temporomandibular symptoms in patients with major skeletal malocclusions: A survey of oral and maxillofacial surgery training programs. Oral Surg. Oral Med. Oral Pathol., 61:537, 1986.
Latham, R. A.: Skull growth in the Rhesus monkey (*Macaca mulatta*). J. Anat., 92:654, 1958.

Latham, R. A.: Observations on the growth of the cranial base in the human skull. J. Anat., 100:435, 1966.

Latham, R. A.: The sliding of cranial bodies at sutural surfaces during growth. J. Anat., 102:593, 1968.

Latham, R. A.: The structure and development of the intermaxillary suture. J. Anat., 106:167, 1970a.

Latham, R. A.: Maxillary development and growth: The septopremaxillary ligament. J. Anat., 107:471, 1970b.

Latham, R. A.: Mechanism of maxillary growth in the human cyclops. J. Dent. Res., 50:929, 1971a.

Latham, R. A.: The development, structure, and growth pattern of the human mid-palatal suture. J. Anat., 108:1, 31–41, 1971b.

Latham, R. A.: The sella point and postnatal growth of the human cranial base. Am. J. Orthod., 61:156, 1972a.

Latham, R. A.: The different relationship of the sella point to growth sites of the cranial base in fetal life. J. Dent. Res., 51:1646, 1972b.

Latham, R. A.: An appraisal of the early maxillary growth mechanism. In: *Factors Affecting the Growth of the Midface.* Ed. by J. A. McNamara, Jr. Ann Arbor University of Michigan, Center for Human Growth and Development, 1976a.

Latham, R. A.: Malformations of the cranial base in human fetuses. In: *Development of the Basicranium.* Ed. by J. F. Bosma. DHEW Pub. 76:989, NIH, Bethesda, Md., 1976b.

Latham, R. A., and W. R. Burston: The postnatal pattern of growth at the sutures of the human skull. Dent. Pract., 17:61, 1966.

Latham, R. A., and J. H. Scott: A newly postulated factor in the early growth of the human middle face and the theory of multiple assurance. Arch. Oral Biol., 15:1097, 1970.

Lau, S. P. and K. P. Fung: Secular trend of growth in Hong Kong children. Hong Kong J. Pediatr., 4:33, 1987. Quoted in: A. K. Leung: Contemporary growth charts (letter). Lancet, 2:974, 1987.

Lauritzen, C., J. Lilja, and J. Jaristedt: Airway obstruction and sleep apnea in children with craniofacial anomalies. Plastic Reconstr. Surg., 77:1, 1986.

Lavelle, C. L. B.: An analysis of foetal craniofacial growth. Ann. Hum. Biol., 1:3, 269, 1974.

Lavelle, C. L. B.: Dental and other bodily dimensions in different orthodontic categories. Angle Orthod., 45:65, 1975.

Lavelle, C. L. B.: A study of dental arch and body growth. Angle Orthod., 46:361, 1976.

Lavelle, C. L. B.: A study of the taxonomic significance of the dental arch. Am. J. Phys. Anthropol., 46:415, 1977a.

Lavelle, C. L. B.: An analysis of the craniofacial complex in different occlusal categories. Am. J. Orthod., 71:574, 1977b.

Lavelle, C. L. B.: A study of the craniofacial skeleton. Angle Orthod., 48:227, 1978.

Lavelle, C. L. B.: A study of craniofacial form. Angle Orthod., 49:65, 1979.

Lavelle, C. L.: A study of mandibular shape. Br. J. Orthod., 11:69, 1984.

Lavelle, C. L.: A preliminary study of mandibular shape. J. Craniofac. Genet. Dev. Biol., 5:159, 1985a.

Lavelle, C. L.: An analysis of craniofacial form; the need for new analytic techniques. Anat. Anz., 160:157, 1985b.

Lavelle, C.: Change in facial shape in micro-, macro-, and normocephalics. Acta Anat. (Basel), 126:248, 1986.

Lavelle, C.: An analysis of basicranial axis form. Anat. Anz., 164:169, 1987.

Lavelle, C. L. B., R. P. Shellis, and D. F. G. Poole: *Evolutionary Changes to the Primate Skull and Dentition,* Springfield, Ill., Charles C Thomas, 1977.

Lavergne, J., and N. Gasson: A metal implant study of mandibular rotation. Angle Orthod., 46:144, 1976.

Lavergne, J., and N. Gasson: The influence of jaw rotation on the morphogenesis of malocclusion. Am. J. Orthod., 73:658, 1978.

Lebret, L.: Growth changes of the palate. J. Dent. Res., 41:1391, 1962.

Lecerf, J.-P.: Rapports ontogéniques et phylogéniques entre quelques plans manducatoires et le foramen magnum. Thèse, Univ. de Lille, 1974.

Le Diascorn, H.: Anatomie et physiologie des sutures de la face. Thèse, Univ. de Nantes, 1971.

Leonard, M. S. and G. F. Walker: A cephalometric study of the relationship between the malar bones and the maxilla in White American females. Angle Orthod., 47:42, 1977.

Leung, A. K., T. O. Siu, P. C. Lai, S. Gabos, and W. L. Robson: Physical growth parameters of Chinese children in Calgary. Can. Fam. Physician, 33:396, 1987.

Levihn, W. C.: A cephalometric roentgenographic cross-sectional study of the craniofacial complex in fetuses from 12 weeks to birth. Am. J. Orthod., 53:822, 1967.

Lewis, A. B., and A. F. Roche: Elongation of the cranial base in girls during pubescence. Angle Orthod., 42:358, 1972.

Lewis, A. B., and A. F. Roche: The saddle angle: Constancy or change? Angle Orthod., 47:46, 1977.

Lewis, A. B., and A. F. Roche: Late growth changes in the craniofacial skeleton. Angle Orthod., 58:127, 1988.

Lewontin, R. C.: The analysis of variance and the analysis of causes. Am. J. Hum. Gen., 26:400, 1974.

Liebgott, B.: Factors of human skeletal craniofacial morphology. Angle Orthod., 47:222, 1977.
Liggett, J.: *The Human Face.* New York, Stein and Day, Publishers, 1974.
Limwongse, V., and M. DeSantis: Cell body locations and axonal pathways of neurons innervating muscles of mastication in the rat. Am. J. Anat., 149:477, 1977.
Linder-Aronson, S.: Adenoids: Their effect on mode of breathing and nasal airflow and their relationship to characteristics of the facial skeleton and the dentition. Acta Otolaryngol. 265 (Suppl.):3, 1970.
Linder-Aronson, S.: Respiratory function in relation to facial morphology and the dentition. Br. J. Orthod., 6:59, 1979.
Linder-Aronson, S.: Naso-respiratory function and craniofacial growth. In: *Naso-respiratory Function and Craniofacial Growth.* Ed. by J. A. McNamara, Jr. Ann Arbor, University of Michigan, Center for Human Growth and Development, 1979.
Linder-Aronson, S.: The relation between nasorespiratory function and dentofacial morphology. Am. J. Orthod., 83:443, 1983.
Linder-Aronson, S.: Nasorespiratory considerations in orthodontics. In: *Orthodontics State of the Art Essence of the Science.* Ed. by L. W. Graber, 116, 1986.
Linder-Aronson, S., and J. Lindgien: The skeletal and dental effects of rapid maxillary expansion. Br. J. Orthod., 6:25, 1979.
Linder-Aronson, S., and D. G. Woodside: The growth in the sagittal depth of the bony nasopharynx in relation to some other facial variables. In: *Naso-respiratory Function and Craniofacial Growth.* Ed. by J. A. McNamara, Jr. University of Michigan, Center for Human Growth and Development, 1979.
Lindgren, G.: Height, weight and menarche in Swedish urban school children in relation to socio-economic and regional factors. Ann. Hum. Biol., 3:501, 1976.
Lindgren, G.: Growth of school children with early, average, and late ages of peak height velocity. Ann. Hum. Biol., 5:253, 1978.
Lindsay, K. N.: An autoradiographic study of cellular proliferation of the mandibular condyle after induced dental malocclusion in the mature rat. Arch. Oral Biol., 22:711, 1977.
Linge, L.: Tissue reactions in facial sutures subsequent to external mechanical influences. In: *Factors Affecting the Growth of the Midface.* Ed. by J. A. McNamara, Jr. Ann Arbor, University of Michigan, Center for Human Growth and Development, 1976.
Liskova, M., and J. Hert: Reaction of bone to mechanical stimuli. Folia Morphol. (Warsz.), 19:301, 1971.
Litt, R. A.: Relapse after total mandibular advancement: A possible solution. Angle Orthod., 48:262, 1978.
Ljung, B. O., A. Bergsten-Brucefors, and G. Lundgren: The secular trend in physical growth in Swed. Ann. Hum. Biol., 1:245, 1974.
Long, R., R. C. Greulich, and B. G. Sarnat: Regional variations in chondrocyte proliferation in the cartilaginous nasal septum of the growing rabbit. J. Dent. Res., 47:137, 1968.
Long, S. Y.: Delay in the straightening of the basicranium associated with cleft palate in the mouse. In: *Development of the Basicranium.* Ed. by J. F. Bosma. DHEW Pub. 76:989, NIH, Bethesda, Md., 1976.
Longacre, J. J.: *Craniofacial Anomalies: Pathogenesis and Repair.* Philadelphia, J. B. Lippincott, 1968.
Low, W. D., L. S. Kung, J. Leong, L. Hsu, D. Fang, A. Yau, and F. Lisowski: The secular trend in the growth of southern Chinese girls in Hong Kong. Z. Morphol. Anthropol., 72:77, 1981.
Lowe, A. A.: Correlations between orofacial muscle activity and craniofacial morphology in a sample of control and anterior open-bite subjects. Am. J. Orthod., 78:89, 1980.
Lowe, A. A., and K. Takada: Associations between anterior temporal masseter and orbicularis oris muscle activity and craniofacial morphology in children. Am. J. Orthod., 86:319, 1984.
Lowe, A. A., K. Takada, and L. M. Taylor: Muscle activity during function and its correlation with craniofacial morphology in a sample of subjects with Class II, Division 1 malocclusions. Am. J. Orthod., 84:204, 1983.
Lowe, A. A., K. Takada, Y. Yamagata, and M. Sakuda: Dentoskeletal and tongue soft-tissue correlates: a cephalometric analysis of rest position. Am. J. Orthod., 88:333, 1985.
Lubsen, C. C., R. Nanda, T. Sakamoto, and S. Nakamura: Histomorphometric analysis of cartilage and subchondral bone in mandibular condyles of young human adults at autopsy. Arch. Oral Biol., 30:129, 1985.
Luder, H. U.: Mandibular condylar structure and growth in *Macaca fascicularis* are related to skeletal rather than dentitional development. J. Biol. Buccale, 13:133, 1985.
Luder, H. U.: Facial pattern and anterior apical base. Angle Orthod., 56:58, 1986.
Lulla, P., and A. A. Gianelly: The mandibular plane and mandibular rotation. Am. J. Orthod., 70:567, 1976.
Lundstrom, A.: Importance of genetic and non-genetic factors in the facial skeleton studied in 100 pairs of twins. Eur. Orthod. Soc. Trans., 1954.
Lundstrom, A.: The clinical significance of profile x-ray analysis. Eur. Orthod. Soc. Trans., 31:190, 1955a.
Lundstrom, A.: The significance of genetic and non-genetic factors in the profile of the facial skeleton. Am. J. Orthod., 12:910, 1955b.
Lundstrom, A.: A twin study of postnormal occlusion. Trans. Europ. Orthod. Soc., 39:43, 1963.

Lundstrom, A.: The biology of occlusal development: Discussion. In: *Craniofacial Biology.* Ed. by J. A. McNamara, Jr. Ann Arbor, University of Michigan, Center for Human Growth and Development, 1977.

Lundstrom, A., and J. McWilliam: Comparison of some cephalometric distances and corresponding facial proportions with regard to heritability. Eur. J. Orthod., 10:27, 1988.

Lundstrom, A., and D. G. Woodside: Longitudinal changes in facial type in cases with vertical and horizontal mandibular growth directions. Eur. J. Orthod., 5:259, 1983.

Macho, G. A.: Cephalometric and craniometric age changes in adult humans. Ann. Hum. Biol., 13:49, 1986.

MacMahon, B.: Age at menarche: United States. DHEW Publ. No. (HRA) 74–1615. Vital Health Stat. (11), No. 133, 1973.

Mahan, P. E., T. M. Wilkinson, C. H. Gibbs, A. Mauderli, and L. S. Brannon: Superior and inferior bellies of the lateral pterygoid muscle EMG activity at basic jaw positions. J. Prosthet. Dent., 50:710, 1983.

Maier, A.: Occurrence and distribution of muscle spindles in masticatory and suprahyoid muscles of the rat. Am. J. Anat., 155:483, 1979.

Maj, G., and C. Luzi: Analysis of mandibular growth on 28 normal children followed from 9 to 13 years of age. Eur. Orthod. Soc. Trans., 1962.

Maj, G., F. Alleva, and F. P. Lucchese: Changes in length and width of the mandibular arch from the mixed dentition to completion of the permanent dentition. Eur. J. Orthod., 1:259, 1979.

Malina, R. M.: Comparison of the increase in body size between 1899 and 1970 in a specially selected group with that in the general population. Am. J. Phys. Anthropol., 37:135, 1972.

Malina, R. M.: Secular changes in size and maturity: Causes and effects. In: *Secular trends in human growth, maturation, and development.* Ed. by A. F. Roche., Monogr. Soc. Res. Child Dev., 44:59, 1979.

Mann, A. W., III: Craniofacial morphological variations in an adult sample: A radiographic cephalometric study. Br. J. Orthod., 6:95, 1979.

Manson, J. D.: *A Comparative Study of the Postnatal Growth of the Mandible.* London, Henry Kimpton, 1968.

Manson, J. D., and R. B. Lucas: A microradiographic study of age changes in the human mandible. Arch. Oral Biol., 7:761, 1962.

Marden, P. M., D. W. Smith, and M. J. McDonald: Congenital anomalies in the newborn infant, including minor variations. J. Pediatr., 64:358, 1964.

Maresh, M. M.: Linear growth of long bones of extremities from infancy through adolescence. Am. J. Dis. Child., 89:725, 1955.

Margolis, H. I.: A basic facial pattern and its application in clinical orthodontics. Am. J. Orthod., 39:425, 1953.

Marshall, D.: Interpretation of the posteroanterior skull radiograph: Assembly of disarticulated bones. Dent. Radiogr. Photogr., 42:27, 1969.

Marshall, W. A.: Growth and secondary sexual development and related abnormalities. Clin. Obstet. Gynecol., 1(3):593, 1974a.

Marshall, W. A.: Interrelationships of skeletal maturation, sexual development and somatic growth in man. Ann. Hum. Biol., 1:29, 1974b.

Marshall, W. A.: The relation of variation in children's growth rates to seasonal climatic variations. Ann. Hum. Biol., 2:243, 1975.

Marshall, W. A.: *Human Growth and Its Disorders.* London, Academic Press, 1977.

Marshall, W. A., and J. M. Tanner: Variations in pattern of pubertal changes in girls. Arch. Dis. Child., 44:291, 1969.

Marshall, W. A., and J. M. Tanner: Variations in pattern of pubertal changes in boys. Arch. Dis. Child., 45:13, 1970.

Marshall, W. A. and J. M. Tanner: Puberty. In *Human Growth.* Vol. 2. 2nd Ed., Ed. by F. Falkner and J. M. Tanner. New York, Plenum Press, 1986.

Mashouf, K., and M. B. Engel: Maturation of periodontal connective tissue in new-born rat incisor. Arch. Oral Biol., 20:161, 1975.

Massler, M., and J. M. Frankel: Prevalence of malocclusion in children aged 14 to 18 years. Am. J. Orthod., 37:751, 1951.

Mathews, J. R., and W. H. Ware: Longitudinal mandibular growth in children with tantalum implants. Am. J. Orthod., 74:633, 1978.

Matsumoto, K.: Secular acceleration of growth in height in Japanese and its social background. Ann. Hum. Biol., 9:399, 1982.

Mauser, C., D. H. Enlow, D. O. Overman, and R. McCafferty: Growth and remodeling of the human fetal face and cranium. In: *Determinants of Mandibular Form and Growth.* Ed. by J. A. McNamara. Monograph 5. Craniofacial Growth Series. Ann Arbor, University of Michigan, Center for Human Growth and Development, 1975.

Maxwell, L. C., D. S. Carlson, J. A. McNamara, Jr., and J. A. Faulkner: Histochemical characteristics of the masseter and temporalis muscles of the Rhesus monkey (*Macaca mulatta*). Anat. Rec., 193:389, 1979.

Maxwell, L. C., J. A. McNamara, Jr., D. S. Carlson, and J. A. Faulkner: Histochemistry of fibers of masseter and temporalis muscles of edentulous monkeys (*Macaca mulatta*). Arch. Oral Biol., 25:87, 1980.

Maxwell, L. C., D. S. Carlson, J. A. McNamara, Jr., and J. A. Faulkner: Effect of shortening or lengthening of the mandible upon the characteristics of masticatory muscle fibers in Rhesus monkeys. In: *Craniofacial Biology.* Ed. by D. S. Carlson. Ann Arbor, University of Michigan, Center for Human Growth and Development, 1981.
McDonald, F.: The effect of articular function on the mandibular condyle of the rat. Eur. J. Orthod., 9:87, 1987.
McDougall, P. D., J. A. McNamara, and J. M. Dierkes: Arch width development in Class II patients treated with the Frankel appliance. Am. J. Orthod., 82:10, 1982.
McDowell, A., A. Engel, J. T. Massey, and K. Maurer: Plan and operation of the second National Health and Nutrition Examination Survey, 1976–1980. DHEW Publ. No. (PHS)81–1317, Vital Health Stat. (1), No. 15, 1981.
McLain, J. B., and W. R. Proffit: Oral health status in the United States: Prevalence of malocclusion. J. Dent. Ed., 49:386, 1985.
McKeown, M.: The allometric growth of the skull. Am. J. Orthod., 67:412, 1975.
McKeown, M.: The influence of environment on the growth of the craniofacial complex: A study on domestication. Angle Orthod., 45:137, 1975.
McKeown, M., and A. Richardson: The nature of cranial vault variation and its relation to facial height. Angle Orthod., 41:15, 1971.
McNamara, J. A., Jr.: *Neuromuscular and Skeletal Adaptations to Altered Orofacial Function.* Monograph 1. Craniofacial Growth Series. Ann Arbor, University of Michigan, Center for Human Growth and Development, 1972.
McNamara, J. A., Jr.: Neuromuscular and skeletal adaptation to altered function in the orofacial region. Am. J. Orthod., 64:578, 1973a.
McNamara, J. A., Jr.: Increasing vertical dimension in the growing face: An experimental study. Am. J. Orthod., 64:364, 1973b.
McNamara, J. A., Jr.: Procion dyes as vital markers in rhesus monkeys. J. Dent. Res. 52:634, 1973c.
McNamara, J. A., Jr.: The role of muscle and bone interaction in craniofacial growth. In: *Control Mechanisms in Craniofacial Growth.* Ed. by J. A. McNamara, Jr. Ann Arbor, University of Michigan, Center for Human Growth and Development, 1975.
McNamara, J. A., Jr.: An experimental study of increased vertical dimension in the growing face. Am. J. Orthod., 71:382, 1977.
McNamara, J. A., Jr.: Functional determinants of craniofacial size and shape. Eur. J. Orthod., 2:131, 1980.
McNamara, J. A., Jr.: Functional determinants of craniofacial size and shape. In: *Craniofacial Biology.* Ed. by D. S. Carlson. Ann Arbor, University of Michigan, Center for Human Growth and Development, 1981.
McNamara, J. A., Jr.: Influence of respiratory pattern on craniofacial growth. Angle Orthod., 51:269, 1981.
McNamara, J. A., Jr.: JCO interviews Dr. James A. McNamara, Jr. on the Frankel appliance. Part I. Biological basis and appliance design. J. Clin. Orthod. 16:320, 1982a.
McNamara, J. A., Jr.: JCO interviews Dr. James A. McNamara, Jr. on the Frankel appliance. Part II. Clinical management. J. Clin. Orthod., 16:390, 1982b.
McNamara, J. A., Jr.: Dentofacial adaptations in adult patients following functional regulator therapy. Am. J. Orthod. 85:57, 1984.
McNamara, J. A., Jr., and F. A. Bryan: Long-term mandibular adaptation to protrusive function: an experimental study in Macaca mulatta. Am. J. Orthod. Dentofacial Orthop., 92:98, 1987.
McNamara, J. A., Jr., and D. S. Carlson: Quantitative analysis of temporomandibular joint adaptations to protrusive function. Am. J. Orthod., 76:593, 1979.
McNamara, J. A., Jr., and L. W. Graber: Mandibular growth in the rhesus monkey (*Macaca mulatta*). Am. J. Phys. Anthropol., 42:15, 1975.
McNamara, J. A., Jr., M. L. Riolo, and D. H. Enlow: Growth of the maxillary complex in the rhesus monkey (*Macaca mulatta*). Am. J. Phys. Anthropol., 44:15, 1976.
McNamara, J. A., Jr., D. S. Carlson, G. M. Yellich, and R. P. Hendricksen: Musculoskeletal adaptation following orthognathic surgery. In: *Muscle Adaptation in the Craniofacial Region.* Ed. by D. S. Carlson and J. A. McNamara, Jr. Ann Arbor, University of Michigan, Center for Human Growth and Development, 1978.
McNamara, J. A., Jr., T. G. Bookstein, and T. G. Shaughnessy: Skeletal and dental changes following functional regulator therapy on Class II patients. Am. J. Orthod., 88:91, 1985.
McWilliam, J., and S. Linder-Aronson: Hypoplasia of the middle third of the face: A morphological study. Angle Orthod., 46:260, 1976.
Mednick, L. W., and S. L. Washburn: The role of the sutures in the growth of the braincase of the infant pig. Am. J. Phys. Anthropol., 14:175, 1956.
Meikle, M. C.: The role of the condyle in the postnatal growth of the mandible. Am. J. Orthod., 64:50, 1973.
Meikle, M. C.: In vivo transplantation of the mandibular joint of the rat: An autoradiographic investigation into cellular changes at the condyle. Arch. Oral Biol., 18:1011, 1973.
Meikle, M. C.: The distribution and function of lysosomes in condylar cartilage. J. Anat., 119:85, 1975.
Meikle, M. C.: Remodeling. In: *The Temporomandibular Joint,* 3rd Ed. Ed. by B. G. Sarnat and D. M. Laskin. Springfield, Ill., Charles C Thomas, 1980.

Melcher, A. M.: Behaviour of cells and condylar cartilage of foetal mouse mandible maintained *in vitro.* Arch Oral Biol., 16:1379, 1971.
Melsen, B.: Time of closure of the spheno-occipital synchondrosis determined on dry skulls: A radiographic craniometric study. Acta Odont. Scand., 27:73, 1969.
Melsen, B.: Computerized comparison of histological methods for the evaluation of craniofacial growth. Acta Odont. Scand., 29:295, 1971a.
Melsen, B.: The postnatal growth of the cranial base in *Macaca rhesus* analyzed by the implant method. Tandlaegebladet, 75:1320, 1971b.
Melsen, B.: Time and mode of closure of the spheno-occipital synchondrosis determined on human autopsy material. Acta Anat., 83:112, 1972.
Melsen, B.: The cranial base. Acta Odont. Scand., 32 (Suppl. 62), 1974.
Melsen, B.: Histological analysis of the postnatal development of the nasal septum. Angle Orthod., 47:83, 1977.
Melsen, B., F. Melsen, and M. L. Moss: Postnatal development of the nasal septum studied on human autopsy material. In: *Craniofacial Biology.* Ed. by D. S. Carlson. Ann Arbor, University of Michigan, Center for Human Growth and Development, 1981.
Melsen, B., J. Bjerregaard, and M. Bundgaard: The effect of treatment with functional appliance on a pathologic growth pattern of the condyle. Am. J. Orthod. Dentofacial Orthop., 90:503, 1986.
Melsen, B., L. Attina, M. Santuari, and A. Atting: Relationships between swallowing pattern, mode of respiration and development of malocclusion. Angle Orthod., 57:113, 1987.
Meng, H., J. M. H. Dibbets, L. van der Weele, and G. Boering: Symptoms of temporomandibular joint dysfunction and predisposing factors. J. Prosthet. Dent., 57:215, 1987.
Menius, J. A., M. D. Largent, and C. J. Vincent: Skeletal development of cleft palate children as determined by hand-wrist roentgenographs: A preliminary study. Cleft Palate J., 3:67, 1966.
Meredith, H. V.: Serial study of change in a mandibular dimension during childhood and adolescence. Growth, 25:229, 1961.
Meredith, H. V.: Childhood interrelations of anatomic growth rates. Growth, 26:23, 1962.
Meredith, H. V.: Change in the stature and body weight of North American boys during the last 80 years. Adv. Child Dev. Behav., 1:69, 1963.
Meredith, H. V.: Findings from Asia, Australia, Europe, and North America on secular change in mean height of children, youths, and young adults. Am. J. Phys. Anthropol., 44:315, 1976.
Merow, W. W.: A cephalometric statistical appraisal of dento-facial growth. Angle Orthod., 32:205, 1962.
Mestre, J. C.: A cephalometric appraisal of cranial and facial relationships at various stages of human fetal development. Am. J. Orthod., 45:473, 1959.
Mew, J. R.: Factors influencing mandibular growth. Angle Orthod., 56:31, 1986.
Michejda, M.: Ontogenic changes of the cranial base in *Macaca mulatta:* Histologic study. Proc. 3rd Internatl. Congr. Primat., Zurich, 1:215, 1970.
Michejda, M.: The role of the basicranial synchondroses in flexure processes and ontogenetic development of the skull base. Am. J. Phys. Anthropol., 37:143, 1972a.
Michejda, M.: Significance of basiocranial synchondroses in nonhuman primates and man. *Medical Primatology.* Proc. 3rd Conf. Exp. Med. Surg. Primates, Lyon, Vol. 1. Basel, S. Karger, 1972b.
Michejda, M., and D. Lamey: Flexion and metric age changes of the cranial base in the *Macaca mulatta.* I. Infant and juveniles. Folia Primatol. (Basel), 14:84, 1971.
Midy, J.: Trajet et inclinaison des germes dentaires et des dents temporaires et permanentes dans les axes vestibulaires d'orientation au cours de l'ontogenèse humaine. Thèse, Acad. Paris, Univ. Paris VI, 1973.
Miller, A. J.: Electromyography of craniofacial musculature during oral respiration in the rhesus monkey (*Macaca mulatta*). Arch. Oral Biol., 23:145, 1978.
Miller, A. J., and G. Chierici: Concepts related to adaptation of neuromuscular function and craniofacial morphology. Birth Defects, 18:21, 1982.
Miller, A. J., and K. Vargervik: Neuromuscular changes during long-term adaptation of the rhesus monkey to oral respiration. In: *Naso-Respiratory Function and Craniofacial Growth.* Ed. by J. A. McNamara, Jr. Craniofacial Growth Series. Ann Arbor, University of Michigan, Center for Human Growth and Development, 1979.
Miller, A. J., K. Vargervik, and G. Chierici: Experimentally induced neuromuscular changes during and after nasal airway obstruction. Am. J. Orthod., 85:385, 1984.
Miller, S. C., J. G. Shupe, E. H. Redd, M. A. Miller, and T. H. Omura: Changes in bone mineral and bone formation rates during pregnancy and lactation in rats. Bone, 7:283, 1986.
Mills, J. R. E.: The effect of orthodontic treatment on the skeletal pattern. Br. J. Orthod., 5:133, 1978.
Mills, J. R.: A clinician looks at facial growth. Br. J. Orthod., 10:58, 1983.
Mitani, H.: Behavior of the maxillary first molar in three planes with emphasis on its role of providing room for the second and third molars during growth. Angle Orthod., 45:159, 1975.
Mitani, H.: Unilateral mandibular hyperplasia associated with a lateral tongue thrust. Angle Orthod., 46:268, 1976.
Miura, F., N. Inoue, and S. Kazuo: The standards of Steiner's analysis for Japanese. Bull. Tokyo Med. Dent. Univ., 10:387, 1963.

Miura, F., N. Inoue, M. Azuma, and G. Ito: Development and organization of periodontal membrane and physiologic tooth movements. Bull. Tokyo Med. Dent. Univ., 17:123, 1970.

Moffett, B., and L. Koskinen-Moffett: A biologic look at mandibular growth rotation. In: *Craniofacial Biology*. Ed. by D. S. Carlson. Ann Arbor, University of Michigan, Center for Human Growth and Development, 1981.

Moffett, B. C., Jr.: The prenatal development of the human temporomandibular joint. Contrib. Embryol. Carneg. Inst., 36:19, 1957.

Moffett, B. C., Jr.: A research perspective on craniofacial morphogenesis. Acta Morphol. Neerl. Scand., 10:99, 1972.

Moffett, B. C., Jr., L. C. Johnson, J. B. McCabe, and H. C. Askew: Articular remodeling in the adult human temporomandibular joint. Am. J. Anat., 115:119, 1964.

Mongini, F., G. Preti, P. M. Calderale, and G. Barberi: Experimental strain analysis on the mandibular condyle under various conditions. Med. Biol. Eng. Comput., 19:521, 1981.

Monteith, B. D.: A cephalometric method to determine the angulation of the occlusal plane in edentulous patients. J. Prosthet. Dent., 54:81, 1985.

Moore, A. W.: Head growth of the macaque monkey as revealed by vital staining, embedding, and undecalcified sectioning. Am. J. Orthod., 35:654, 1949.

Moore, A. W.: Observations on facial growth and its clinical significance. Am. J. Orthod., 45:399, 1959.

Moore, A. W.: Cephalometrics as a diagnostic tool. J.A.D.A. 82:775, 1971.

Moore, R. N.: A cephalometric and histologic study of the cranial base in foetal monkeys, *Macaca nemestrina*. Arch. Oral Biol., 23:57, 1978.

Moore, R. N., and C. Phillips: Sagittal craniofacial growth in the fetal Macaque monkey *Macaca nemestrina*. Arch. Oral Biol., 25:19, 1980.

Moore, W. J.: Masticatory function and skull growth. J. Zool., 146:123, 1965.

Moore, W. J.: Skull growth in the albino rat (*Rattus norvegicus*). J. Zool. (Lond.), 149:137, 1966.

Moore, W. J.: The influence of muscular function on the growth of the skull. Scientia, 103:333, 1968.

Moore, W. J.: Associations in the hominoid facial skeleton. J. Anat., 123:111, 1977.

Moore, W. J.: The use of shape measures in the study of skeletal growth and development. Prog. Clin. Biol. Res., 187:495, 1985.

Moore, W. J., and C. L. B. Lavelle: *Growth of the Facial Skeleton in the Hominoidea*. New York, Academic Press, 1974.

Moore, W. J., and T. F. Spence: Age changes in the cranial base of the rabbit (*Cryctolagus cuniculus*). Anat. Rec., 165:355, 1969.

Moore, W. J., P. A. Johnson, and P. J. Atkinson: The influence of remodelling on the structure of the chimpanzee mandible. J. Anat., 122:198, 1976.

Moorrees, C. F. A.: Normal variation and its bearing on the use of cephalometric radiographs in orthodontic diagnosis. Am. J. Orthod., 39:942, 1953.

Moorrees, C. F. A.: Natural head position, a basic consideration in the interpretation of cephalometric radiographs. Am. J. Phys. Anthropol., 16:213, 1958.

Moorrees, C. F. A.: *Dentition of the Growing Child, a Longitudinal Study of Dental Development Between 3 and 18 Years of Age*. Cambridge, Harvard University Press, 1959.

Moorrees, C. F. A.: Register of longitudinal studies of facial and dental development. International Society of Craniofacial Biology, Washington, D.C., 1967.

Moorrees, C. F. A.: Patterns of dental maturation. In: *Craniofacial Biology*. Ed. by J. A. McNamara, Jr. Ann Arbor, University of Michigan, Center for Human Growth and Development, 1977.

Moorrees, C. F. A., and R. B. Reed: Correlations among crown diameters of human teeth. Arch. Oral Biol., 9:685, 1964.

Moorrees, C. F. A., E. A. Fanning, and E. E. Hunt, Jr.: Age variation of formation stages for ten permanent teeth. J. Dent. Res., 42:1490, 1963.

Moorrees, C. F. A., M. E. van Venrooij, L. M. L. Lebret, C. B. Glatky, R. L. Kent, Jr., and R. B. Reed: New norms for the mesh diagram analysis. Am. J. Orthod., 69:57, 1976.

Morax, S. J.: Occulomotor disorders in craniofacial malformations. J. Maxillofac. Surg., 12:1, 1984.

Moss, J. P., and D. R. James: An investigation of a group of 35 consecutive patients with a first arch syndrome. Br. J. Oral Maxillofac. Surg., 22:157, 1984.

Moss, M. L.: Genetics, epigenetics and causation. Am. J. Orthod., 36:481, 1950.

Moss, M. L.: Postnatal growth on the human skull base. Angle Orthod., 25:77, 1955a.

Moss, M. L.: Correlation of cranial base angulation with cephalic malformations and growth disharmonies of dental interest. N.Y. Dent. J., 21:452, 1955b.

Moss, M. L.: Rotations of the cranial components in the growing rat and their experimental alteration. Acta Anat., 32:65, 1958a.

Moss, M. L.: Fusion of the frontal suture in the rat. Am. J. Anat., 102:141, 1958b.

Moss, M. L.: Embryology, growth and malformations of the temporomandibular joint. In: *Disorders of the Temporomandibular Joint*. Ed. by L. Schwartz. Philadelphia, W. B. Saunders, 1959.

Moss, M. L.: Functional analysis of human mandibular growth. J. Prosthet. Dent., 10:1149, 1960a.

Moss, M. L.: A functional approach to craniology. Am. J. Phys. Anthropol., 18:281, 1960b.

Moss, M. L.: The functional matrix. In: *Vistas of Orthodontics*. Ed. by B. S. Kraus and R. A. Riedel. Philadelphia, Lea & Febiger, 1962.

Moss, M. L.: Functional cranial analysis of mammalian mandibular ramal morphology. Acta Anat., 71:423, 1968a.
Moss, M. L.: The primacy of functional matrices in orofacial growth. Dent. Pract., 19:65, 1968b.
Moss, M. L.: The primary role of functional matrices in facial growth. Am. J. Orthod., 55:566, 1969.
Moss, M. L.: Functional cranial analysis and functional matrix. Am. Speech Hear. Assoc. Rep., 6:5, 1971a.
Moss, M. L.: Neurotropic processes in orofacial growth. J. Dent. Res., 50:1492, 1971b.
Moss, M. L.: Ontogenic aspects of cranio-facial growth. In: *Cranio-facial Growth in Man*. Ed. by R. E. Moyers and W. M. Krogman. Oxford, Pergamon Press, 1971c.
Moss, M. L.: The regulation of skeletal growth. In: *Regulation of Organ and Tissue Growth*. Ed. by R. J. Goss. New York, Academic Press, 1972a.
Moss, M. L.: An introduction to the neurobiology of orofacial growth. Acta. Biotheor., 22:236, 1972b.
Moss, M. L.: Twenty years of functional cranial analysis. Am. J. Orthod., 61:479, 1972c.
Moss, M. L.: New studies of cranial growth. In: *Morphogenesis and Malformation of Face and Brain*. Ed. by D. Bergsma, J. Langman, and N. W. Paul. National Foundation—March of Dimes, New York, Alan R. Liss, 11:7, 1975a.
Moss, M. L.: Neurotropic regulation of craniofacial growth. In: *Control Mechanisms in Craniofacial Growth*. Ed. by J. A. McNamara, Jr. Ann Arbor, University of Michigan, Center for Human Growth and Development, 1975b.
Moss, M. L.: The effect of rhombencephalic hypoplasia on posterior cranial base elongation in rodents. Arch. Oral Biol., 20:489, 1975c.
Moss, M. L.: Experimental alteration of basi-synchondrosal cartilage growth in rat and mouse. In: *Development of the Basicranium*. Ed. by J. F. Bosma. DHEW Pub. 76:989, NIH, Bethesda, Md., 1976a.
Moss, M. L.: The role of the nasal septal cartilage in midfacial growth. In: *Factors Affecting the Growth of the Midface*. Ed. by J. A. McNamara, Jr. Ann Arbor, University of Michigan, Center for Human Growth and Development, 1976b.
Moss, M. L.: Beyond roentgenographic cephalometry—what? Am. J. Orthod., 84:77, 1983.
Moss, M. L.: The application of the finite element method to the description of craniofacial skeletal growth and form comparison. In *Human Growth: a multidisciplinary review*. Ed. by A. Demirjian and M. Dubuc. London and Philadelphia, Taylor and Francis, 1986.
Moss, M. L., and S. N. Greenberg: Functional cranial analysis of the human maxillary bone. Angle Orthod., 37:151, 1967.
Moss, M. L., and M. A. Meehan: Functional cranial analysis of the coronoid process in the rat. Acta Anat., 77:11, 1970.
Moss, M. L., and L. Moss-Salentijn: The muscle-bone interface: An analysis of a morphological boundary. In: *Muscle Adaptation in the Craniofacial Region*. Ed. by D. S. Carlson and J. A. McNamara, Jr. Ann Arbor, University of Michigan, Center for Human Growth and Development, 1978.
Moss, M. L., and R. M. Rankow: The role of the functional matrix in mandibular growth. Angle Orthod., 38:95, 1968.
Moss, M. L., and L. Salentijn: The primary role of functional matrices in facial growth. Am. J. Orthod., 55:566, 1969a.
Moss, M. L., and L. Salentijn: The capsular matrix. Am. J. Orthod., 56:474, 1969b.
Moss, M. L., and L. Salentijn: The logarithmic growth of the human mandible. Acta Anat., 77:341, 1970.
Moss, M. L., and L. Salentijn: Differences between the functional matrices in anterior open and deep overbite. Am. J. Orthod., 60:264, 1971a.
Moss, M. L., and L. Salentijn: The unitary logarithmic curve descriptive of human mandibular growth. Acta Anat., 78:532, 1971b.
Moss, M. L., C. R. Noback, and G. G. Robertson: Growth of certain human fetal cranial bones. Am. J. Anat., 98:191, 1956.
Moss, M. L., B. E. Bromberg, I. C. Song, and G. Eisenman: The passive role of nasal septal cartilage in mid-facial growth. Plast. Reconstr. Surg., 41:536, 1968.
Moss, M. L., M. A. Meehan, and L. Salentijn: Transformative and translative growth processes in neurocranial development of the rat. Acta Anat., 81:161, 1972.
Moss, M. L., R. Skalak, G. Dasgupta, and H. Vilmann: Space, time and space-time in craniofacial growth. Am. J. Orthod., 77:591, 1980.
Moss, M. L., H. Vilmann, G. Dasgupta, and R. Skalak: Craniofacial growth in space-time. In: *Craniofacial Biology*. Ed. by D. S. Carlson. University of Michigan, Center for Human Growth and Development, 1981.
Moss, M. L., R. Skalak, L. Moss-Salentijn, G. M. Dasgupta, H. Vilmann, and P. Mehta: The allometric center. The biological basis of an analytical model of craniofacial growth. Proc. Finn. Dent. Soc., 77:119, 1981.
Moss, M. L., R. Skalak, M. Shinozuka, H. Patel, L. Moss-Salentijn, H. Vilmann, and P. Mehta: The testing of an allometric center model of craniofacial growth. Z. Morphol. Anthropol., 74:295, 1984.

Moss-Salentijn, L.: The human fetus. In: *Development of the Basicranium.* Ed. by J. F. Bosma. DHEW Pub. 76:989, NIH, Bethesda, Md., 1976.
Moss-Salentijn, L.: Changes in the morphology and vasculature of a rabbit growth plate following experimental growth rate acceleration. In: *Muscle Adaptation in the Craniofacial Region.* Ed. by D. S. Carlson and J. A. McNamara, Jr. Ann Arbor, University of Michigan, Center for Human Growth and Development, 1978.
Moss-Salentijn, L.: Experimental retardation of early endochondral ossification in the phalangeal epiphyses of rat. Proc. Finn. Dent. Soc., 77:129, 1981a.
Moss-Salentijn, L.: Reattachment of the mammalian mylohyoid muscle during late embryonic development. In: *Craniofacial Biology.* Ed. by D. S. Carlson. Ann Arbor, University of Michigan, Center for Human Growth and Development, 1981b.
Motohashi, N., and K. Kuroda: Morphological analysis of congenital craniofacial malformations. Kokubyo Gakkai Zasshi, 49:698, 1982.
Moyers, R. E.: Temporomandibular muscle contraction patterns in Angle Class II, Division 1 malocclusions: An electromyographic analysis. Am. J. Orthod., 35:837, 1949.
Moyers, R. E.: Periodontal membrane in orthodontics. J.A.D.A., 40:22, 1950.
Moyers, R. E.: Some recent electromyographic findings in the oro-facial muscles. Eur. Orthod. Soc. Trans., 32:225, 1956.
Moyers, R. E.: Role of musculature in malocclusion. Eur. Orthod. Soc. Trans., 37:40, 1961.
Moyers, R. E.: The infantile swallow. Trans. Eur. Orthod. Soc., 40:180, 1964.
Moyers, R. E.: Development of occlusion. Dent. Clin. North Am., 13:523, 1969.
Moyers, R. E.: Some comments about the nature of orthodontic relapse. Orthodoncia, 54:215, 1970.
Moyers, R. E.: Postnatal development of the orofacial musculature. In: *Patterns of Orofacial Growth and Development.* Report 6. Washington, D.C., American Speech and Hearing Association, 1971.
Moyers, R. E.: Skeletal contributions to occlusal development. In: *Craniofacial Biology.* Ed. by J. A. McNamara, Jr. Ann Arbor, University of Michigan, Center for Human Growth and Development, 1977.
Moyers, R. E.: *Handbook of Orthodontics,* 4th Ed. Chicago, Year Book Medical Publishers, 1988.
Moyers, R. E., and F. L. Bookstein: The inappropriateness of conventional cephalometrics. Am. J. Orthod., 75:599, 1979.
Moyers, R. E., and F. Muira: The use of serial cephalograms to study racial differences in development. I. and II. Trans. VIII Congress of Anthrop. and Ethnol. Sci., Tokyo, 284, 1968.
Moyers, R. E., J. Elgoyhen, M. Riolo, J. McNamara, and T. Kuroda: Experimental production of Class III in rhesus monkeys. Eur. Orthod. Soc. Trans., 46:61, 1970.
Moyers, R. E., F. Bookstein, and K. E. Guire: The concept of pattern in craniofacial growth. Am. J. Orthod., 76:136, 1979.
Mugnier, A., and M. Schouker-Jolly: Physio-pathologic des malocclusions dento-maxillaires moyens prophylactiques et thérapeutiques précoces. Pédod. Fr., 5:101, 1973.
Muhl, Z. F., J. H. Netwon: Change of digastric muscle length in feeding rabbits. J. Morphol., 17:15, 1982.
Muhleman, H. R.: Genetic viewpoints regarding caries and periodontal disease in man. Oral Res. Abstr., 5236, 8, 1973.
Muzj, E.: *Oro-facial Anthropometrics.* Hempstead, Index Publishers, Inc., 1970.
Nahoum, H. I., S. L. Horowitz, and E. A. Benedicto: Varieties of anterior open bite. Am. J. Orthod., 61:486, 1972.
Najjar, M. F., and M. Rowland: Anthropometric reference data and prevalence of overweight, United States, 1976–1980. DHHS publ. No. (PHS) 87–1688. Vital Health Stat. (11), No. 238, 1987.
Nakasima A., M. Ichinose, and S. Nakata: Genetic and environmental factors in the development of so-called pseudo- and true mesiocclusions. Am. J. Orthod. Dentofac. Orthoped., 90:106, 1986.
Nakata, M.: Twin studies in craniofacial genetics: A review. Acta Genet. Med. Gemellol. (Roma), 34:1, 1985.
Nakata, M., P. L. Yu, B. Davis, and W. E. Nance: The use of genetic data in the prediction of craniofacial dimensions. Am. J. Orthod., 63:471, 1973.
Nanda, R. S.: Rates of growth of several facial components measured from serial cephalometric roentgenograms. Am. J. Orthod., 41:658, 1955.
Nanda, R., and B. Goldin: Biomechanical approaches to the study of alterations in facial morphology. Am. J. Orthod., 78:213, 1980.
Nanda, R. S., and R. C. Taneja: Growth of the face during the transitional period. Angle Orthod., 42:165, 1972.
Nemeth, R. B., and R. J. Isaacson: Vertical anterior relapse. Am. J. Orthod., 65:565, 1974.
Nickel, J. C., K. R. McLachlan, and D. M. Smith: Eminence development of the postnatal human temporomandibular joint. J. Dent. Res., 67:896, 1988.
Nielsen, I. L.: Facial growth during treatment with the function regulator appliance. Am. J. Orthod., 85:401, 1984.
Nishemura, H., R. Semba, T. Tanimura, and O. Tanacha: *Prenatal Development of the Human Skeleton with Special Reference to Craniofacial Structures: An Atlas.* Washington, D.C., U.S. Government Printing Office, 1977.

Noback, C. R.: Developmental anatomy of the human osseous skeleton during embryonic fetal and circumnatal periods. Anat. Rec., 88:91, 1944.
Noback, C. R., and G. G. Robertson: Sequences of appearance of ossification centers in the human skeleton during the first five prenatal months. Am. J. Anat., 89:1, 1951.
Noden, D. M.: Patterns and organization of craniofacial skeletogenic and myogenic mesenchyme: A perspective. Prog. Clin. Biol. Res., 101:167, 1982.
Norton, L. A.: Implications of bioelectric growth control in orthodontics and dentistry. Angle Orthod., 45:34, 1975.
Oberg, T.: Morphology, growth and matrix formation in the mandibular joint of the guinea pig. Trans. Roy. Schools of Dent., Stockholm & Umea, No. 10, 1964.
Odegaard, J.: Mandibular rotation studied with the aid of metal implants. Am. J. Orthod., 58:448, 1970.
Odegaard, J., and A. G. Brodie: On the growth of the human head from birth to the third month of life. Anat. Rec., 103:311, 1949.
Ohtsuki, F.: Developmental changes of the cranial bone thickness in the human fetal period. Am. J. Phys. Anthropol., 46:141, 1977.
Ohtsuki, F.: Areal growth in the human fetal parietal bone. Am. J. Phys. Anthropol., 53:5, 1980.
Opdebeeck, H., and W. H. Bell: The short face syndrome. Am. J. Orthod., 73:499, 1978.
Opitz, J. M., J. Herrmann, and H. Dieker: The study of malformation syndromes in man. Birth Defects, 5(2):1, 1969.
O'Reilly, M. T.: A longitudinal growth study: Maxillary length at puberty in females. Angle Orthod., 49:234, 1979.
Orlowski, W. A.: Biochemical study of collagen turnover in rat incisor periodontal ligament. Arch. Oral Biol., 23:1163, 1978.
O'Ryan, F., and B. N. Epker: Temporomandibular joint function and morphology: Observations on the spectra of normalcy. Oral Surg. Oral Med. Oral Pathol., 58:272, 1984.
Osborn, J. W.: Relationship between the mandibular condyle and the occlusal plane during hominid evolution: Some of its effects on jaw mechanics. Am. J. Phys. Anthrop., 73:193, 1987.
Osborn, R. H., and F. V. DeGeorge: *Genetic Basis of Morphological Variation.* Harvard University Press, 1959.
Ossenberg, N. S.: Within and between race distances in population studies based on discrete traits of the human skull. Am. J. Phys. Anthropol., 45:701, 1976.
Oudet, C., and A. G. Petrovic: Variations in the number of sarcomeres in series in the lateral pterygoid muscle as a function of the longitudinal deviation of the mandibular position produced by the postural hyperpropulsor. In: *Muscle Adaptation in the Craniofacial Region.* Ed. by D. S. Carlson and J. A. McNamara, Jr. Ann Arbor, University of Michigan, Center for Human Growth and Development, 1978.
Ousterhout, D. K., K. Vargervik, and A. Miller: Nasal airway function as it relates to the timing of mid and lower facial osteotomies. Ann. Plast. Surg., 11:175, 1983.
Owen, A. H., 3rd: Morphologic changes in the transverse dimension using the Frankel appliance. Am. J. Orthod., 83:200, 1983.
Oyen, O. J.: Masticatory function and histogenesis of the middle and upper face in chimpanzees (*P. troglodytes*). In: *Factors and Mechanisms Influencing Bone Growth.* Ed. by A. R. Dixon and B. G. Sarnat. Prog. Clin. Biol. Res. Series (101), New York, Alan R. Liss, 1982.
Oyen, O. J.: Palatal growth in baboons. Primates (Japan), 25:337, 1984.
Oyen, O. J.: Bone strain in the orbital region in growing monkeys. Am. J. Phys. Anthropol., 72:239, 1987.
Oyen, O. J.: Analyses of postnatal bone growth in long bones versus the craniofacial skeleton and the formation of explanatory models of growth. Am. J. Phys. Anthropol., 75:256, 1988.
Oyen, O. J.: Biomechanical analysis of craniofacial form and function. In: *Biostereometrics '88, Fifth International Meeting.* Ed. by J. U. Baumann and R. E. Herron. Proc. SPIE, 1030, 118, 1989.
Oyen, O. J., and D. H. Enlow: Structural-functional relationships between masticatory biomechanics, skeletal biology, and craniofacial development in primates. Anthrop. Contemp., 3(2):251, 1980.
Oyen, O. J., and P. E. Lestrel: A stereometric analysis of skull growth in *P. cynocephalus anubis.* Biostereometrics '82, 361:330, 1982.
Oyen, O. J., and R. W. Rice: Supraorbital development in chimpanzees (Pan), macaques (*Macaca*), and baboons (*Papio*). J. Med. Primatol. (Lond.), 9:161, 1980.
Oyen, O. J., and M. D. Russell: Histogenesis of the craniofacial skeleton and models of facial growth. In: *The Effect of Surgical Intervention on Craniofacial Growth.* Ed. by J. A. McNamara, Jr., D. S. Carlson, and K. A. Ribbens. Craniofacial Growth Series. Ann Arbor, University of Michigan, Center for Human Growth and Development, 1982.
Oyen, O. J., and A. Walker: Stereometric craniometry. Am. J. Phys. Anthropol., 46:177, 1977.
Oyen, O. J., A. C. Walker, and R. W. Rice: Craniofacial growth in the olive baboon (*Papio cynocephalus anubis*): Browridge formation. Growth, 43:174, 1979.
Oyen, O. J., R. W. Rice, and M. S. Cannon: Browridge structure and function in Neanderthals and extant primates. Am. J. Phys. Anthropol., 51:83, 1979.
Oyen, O. J., R. W. Rice, and D. H. Enlow: Cortical surface patterns in human and non-human primates. Am. J. Phys. Anthropol., 54:415, 1981.
Oyen, O. J., and T. P. Tsay: A biomechanical analysis of craniofacial form and bite force. Am. J. Orthod. Dentofac. Orthoped. In press.

Pancherz, H.: A cephalometric analysis of skeletal and dental changes contributing to Class II correction in activator treatment. Am. J. Orthod., 85:125, 1984.
Pancherz, H., A. Winnberg, and P. Westesson: Masticatory muscle activity and hyoid bone behavior during cyclic jaw movements in man. A synchronized electromyographic and videofluorographic study. Am. J. Orthod., 89:122, 1986.
Parker, W. S.: The comparative anatomy of the internal and external pterygoid muscles: Functional variations among species. Angle Orthod., 53:9, 1983.
Patten, B. M.: Embryology of the palate and maxillofacial region. In: *Cleft Lip and Palate.* Ed. by W. C. Grabb, S. W. Rosenstein, and K. R. Bzoch. Boston, Little, Brown, 1971.
Penrose, L. S.: *The Biology of Mental Defects.* London, Sidgwick & Jackson, 1949.
Perry, H. T.: The temporomandibular joint. Am. J. Orthod., 52:399, 1966.
Perry, H. T.: Relation of occlusion to temporomandibular joint dysfunction: The orthodontic viewpoint. J.A.D.A., 79:137, 1969.
Perry, H. T.: Temporomandibular joint and occlusion. Angle Orthod., 46:284, 1976.
Persson, M.: Mandibular asymmetry of hereditary origin. Am. J. Orthod., 63:1, 1973a.
Persson, M.: Structure and growth of facial sutures. Odontol. Rev., 24:26, 1973b.
Persson, M., and W. Roy: Suture development and bony fusion in the fetal rabbit palate. Arch. Oral Biol., 24:283, 1979.
Persson, M., B. C. Magnusson, and B. Thilander: Sutural closure in rabbit and man: A morphological and histochemical study. J. Anat., 125:313, 1978.
Petit-Maire, N.: Morphogenèse du crane de primates. L'Anthropologie, 75:85, 1971.
Petrovic, A.: Recherches sur les mécanismes histophysiologiques de la croissance osseuse craniofaciale. Ann. Biol., 9:63, 1970.
Petrovic, A.: Mechanisms and regulation of condylar growth. Acta Morphol. Neerl. Scand., 10:25, 1972.
Petrovic, A. G.: Experimental and cybernetic approaches to the mechanisms of action of functional appliances on mandibular growth. Ed. by J. A. McNamara, Jr., and K. A. Ribbens. In: *Malocclusion and the Periodontium.* Monograph 15. Craniofacial Growth Series. Ann Arbor, University of Michigan, Center for Human Growth and Development, 1984.
Petrovic, A., and J. P. Charlier: La synchondrose spheno-occipitale de jeune rat en culture d'organes: mise en evidence d'un potential de croissance indépendent. C. R. Acad. Sci. (D), (Paris) 265:1511, 1967.
Petrovic, A., and J. Stutzmann: Le muscle pterygoidien externe et la croissance du condyle mandibulaire. Recherches expérimentales chez le jeune rat. Orthod. Fr., 43:271, 1972.
Petrovic, A., J. P. Charlier, and J. Herrman: Les mécanismes de croissance du crane. Recherches sur le cartilage de la cloison nasale et sur les sutures craniennes et faciales de jeunes rats en culture d'organes. Bull. Assoc. Anat. (Nancy), 143:1376, 1968.
Petrovic, A., C. Oudet, and N. Gasson: Effects des appareils de propulsion et de rétropulsion mandibulaire sur le nombre des sarcomeres en série du muscle pterygoidien externe et sur la croissance du cartilage condylien de jeune rat. Orthod. Fr., 44:191, 1973a.
Petrovic, A., J. Stutzmann, and C. Oudet: Effects de l'hormone somatotrope sur la croissance du cartilage condylien mandibulaire et de la synchondrose sphéno-occipitale de jeune rat, en culture organotypique. C. R. Acad. Sci. (D) (Paris) 276:3053, 1973b.
Petrovic, A., J. Stutzmann, and C. Oudet: Condylectomy and mandibular growth in young rats. A quantitative study. Proc. Finn. Dent. Soc., 77:139, 1981a.
Petrovic, A. G., J. J. Stutzmann, and N. Gasson: The final length of the mandible: Is it genetically predetermined? In: *Craniofacial Biology.* Ed. by D. S. Carlson. Ann Arbor, University of Michigan, Center for Human Growth and Development, 1981b.
Petrovic, A. G., J. Stutzmann, and C. Oudet: Orthopedic appliances modulate the bone formation in the mandible as a whole. Swed. Dent. J., 15:197, 1982.
Pettersen, J. C.: An anatomical study of two cases of cebocephaly. In: *Development of the Basicranium.* Ed. by J. F. Bosma. DHEW Pub. 76:989, NIH, Bethesda, Md., 1976.
Pfeiffer, J. P.: Should orthopaedic treatment of severe Class II malocclusions be related to growth? Eur. J. Orthod., 2:249, 1980.
Phelps, A. E.: A comparison of lower face changes. Angle Orthod., 48:283, 1978.
Pia, H. W.: Craniocervical malformations. Neurosurg. Rev., 6:169, 1983.
Pierce, R. M., M. W. Mainen, and J. F. Bosma: The cranium of the newborn: An atlas of tomography and anatomical sections. Am. J. Orthod., 75:693, 1978.
Pike, J. B.: A cephalometric investigation of facial profile changes in high angle nongrowing cases. Angle Orthod., 45:115, 1975.
Pimenidis, M. Z., and A. A. Gianelly: The effect of early postnatal condylectomy on the growth of the mandible. Am. J. Orthod., 62:42, 1972.
Pimenidis, M. Z., and A. A. Gianelly: Class III malocclusion produced by oral facial sensory deprivation in the rat. Am. J. Orthod., 71:94, 1977.
Poole, A. E., I. Greene, and P. Buschang: The effect of growth hormone therapy on longitudinal growth of the oral facial structures in children. Prog. Clin. Biol. Res., 101:499, 1982.
Popli, I. K.: Cephalometric appraisal of dentoskeletal pattern in mentally retarded children. Angle Orthod., 47:123, 1977.
Popovich, F., and G. W. Thompson: Craniofacial templates for orthodontic case analysis. Am. J. Orthod., 71:406, 1977.

Popovich, F., and G. W. Thompson: The maxillary interincisal diastema and its relationship to the superior labial frenum and intermaxillary suture. Angle Orthod., 47:265, 1977.
Popovich, F., G. W. Thompson, and S. Saunders: Craniofacial measurements in siblings of the Burlington Growth Center sample. J. Dent. Res., 56:A113, 1977.
Poswillo, D. E.: The late effects of mandibular condylectomy. Oral Surg., 33:500, 1972.
Poswillo, D. E.: Orofacial malformations. Proc. R. Soc. Med., 67:343, 1974.
Poswillo, D.: Hemorrhage in development of the face. In: *Morphogenesis and Malformation of Face and Brain.* Ed. by D. Bergsma, J. Langman, and N. W. Paul. National Foundation—March of Dimes. Birth Defects Original Article Series, Vol. 11, No. 7. New York, Alan R. Liss, 1975.
Poswillo, D. E.: Etiology and pathogenesis of first and second branchial arch defects: the contribution of animal studies. In: *Symposium on Diagnosis and Treatment of Craniofacial Anomalies.* Ed. by J. M. Converse, J. G. McCarthy, and D. Wood-Smith. St. Louis, C. V. Mosby Co., 1979.
Poswillo, D. E.: Congenital malformations: Prenatal experimental studies. In: *The Temporomandibular Joint,* 3rd Ed. Ed. by B. G. Sarnat and D. M. Laskin. Springfield, Ill., Charles C Thomas, 1980.
Poulton, D. R.: Influence of extraoral traction. Am. J. Orthod., 53:8, 1967.
Powell, S. J., and R. K. Rayson: The profile in facial aesthetics. Br. J. Orthod., 3:207, 1976.
Powell, T. W., and A. G. Brodie: Laminagraphic study of the spheno-occipital synchondrosis. Anat. Rec., 147:15, 1963.
Precious, D., and J. Delaire: Balanced facial growth: a schematic interpretation. Oral. Surg. Oral Med. Oral Pathol., 63:637, 1987.
Precious, D., and D. A. Miles: The lateral craniofacial cephalometric radiography. J. Oral Maxillofac. Surg., 45:737, 1987.
Prescott, G. H., D. F. Mitchell, and H. Fahmy: Procion dyes as matrix markers in growing bone and teeth. Am. J. Phys. Anthropol., 29:219, 1968.
Preston, C. B.: Pituitary fossa size and facial type. Am. J. Orthod., 75:259, 1979.
Preston, C. B., and W. G. Evans: The cephalometric analysis of *Cercopithecus aethiops.* Am. J. Phys. Anthropol., 44:105, 1976.
Pritchard, J. J.: The control or trigger mechanism induced by mechanical forces which cause responses of mesenchymal cells in general and bone application and resorption in particular. Acta. Morphol. Neerl. Scand., 10:63, 1972.
Pritchard, J. J., J. H. Scott, and F. G. Girgis: The structure and development of cranial and facial sutures. J. Anat., 90:73, 1956.
Proffit, W. R.: The facial musculature in its relation to the dental occlusion. In: *Muscle Adaptation in the Craniofacial Region.* Ed. by D. S. Carlson and J. A. McNamara, Jr. University of Michigan, Center for Human Growth and Development, 1978.
Proffit, W. R., and J. L. Ackerman: Rating the characteristics of malocclusion: A systematic approach for planning treatment. Am. J. Orthod., 64:258, 1973.
Proffit, W. R., and H. W. Fields: Occlusal forces in normal- and long-face children. J. Dent. Res., 62:571, 1983.
Proffit, W. R., H. W. Fields, and W. L. Nixon: Occlusal forces in normal- and long-face adults. J. Dent. Res., 62:566, 1983.
Pruzansky, S.: Congenital anomalies of the face and associated structures. Springfield, Ill., Charles C Thomas, 1961.
Pruzansky, S.: Anomalies of face and brain. In: *Morphogenesis and Malformation of Face and Brain.* Ed. by D. Bergsma, J. Langman, and N. W. Paul. National Foundation, 11:7, 1975.
Pruzansky, S.: Radiocephalometric studies of the basicranium in craniofacial malformation. In: *Development of the Basicranium.* Ed. by J. F. Bosma. DHEW Pub. 76:989, NIH, Bethesda, Md., 1976.
Pruzansky, S., and E. F. Lis: Cephalometric roentgenography of infants: Sedation, instrumentation, and research. Am. J. Orthod., 44:159, 1958.
Pruzansky, S., and J. B. Richmond: Growth of the mandible in infants with micrognathia. Am. J. Dis. Child., 88:29, 1954.
Pucciarelli, H. M.: The influence of experimental deformation of craniofacial development in rats. Am. J. Phys. Anthropol., 48:455, 1978.
Pucciarelli, H. M.: The effect of race, sex, and nutrition on craniofacial differentiation in rats: A multivariate analysis. Am. J. Phys. Anthropol., 53:359, 1980.
Pullinger, A., and L. Hollender: Variation in condyle-fossa relationships according to different methods of evaluation in tomograms. Oral Sug. Oral. Med. Oral Pathol., 62:719, 1986.
Rabine, M.: The role of uninhibited occlusal development. Am. J. Orthod., 74:51, 1978.
Rahme, J., A. el-Danaf, J. Chassagne, P. Maxant, M. Stricker, and F. Flot: The influence of the masticatory muscles on craniofacial growth: A microsurgical study in the rat. Rev. Stom. Chir. Maxillofac., 88:108, 1987.
Rak, Y.: Australopithecine taxonomy and phylogeny in light of facial morphology. Am. J. Phys. Anthropol., 66:281, 1985.
Rallison, M. L.: *Growth disorders in infants, children, and adolescents.* John Wiley and Sons, New York, 1986.
Ramadan, M. F.: Effect of experimental nasal obstruction on growth of alveolar arch. Arch. Otolaryngol., 110:566, 1984.

Ramfjord, S. P., and R. D. Enlow: Anterior displacement of the mandible in adult rhesus monkeys: Long-term observations. J. Prosthet. Dent., 26:517, 1971.

Rangel, R. D., O. Oyen, and M. Russell: Changes in masticatory biomechanics and stress magnitude that affect growth and development of the facial skeleton. Prog. Clin. Biol. Res., 187:281, 1985.

Rees, L. A.: The structure and function of the temporomandibular joint. Br. Dent. J., 96:125, 1954.

Reitan, K.: Tissue behavior during orthodontic tooth movement. Am. J. Orthod., 46:881, 1960.

Reitan, K.: Bone formation and resorption during reversed tooth movement. In: *Vistas in Orthodontics.* Ed. by B. T. Kraus and R. A. Riedel. Philadelphia, Lea & Febiger, 1962.

Reitan, K.: Biomechanical principles and reactions. In: *Current Orthodontic Concepts and Techniques.* Ed. by T. M. Graber. Philadelphia, W. B. Saunders, 1969.

Remmer, K. R., A. H. Mamandras, W. S. Hunter, and D. C. Way: Cephalometric changes associated with treatment using the activator, the Frankel appliance, and the fixed appliance. Am. J. Orthod., 88:363, 1985.

Richardson, A.: A comparison of traditional and computerized methods of cephalometric analysis. Europ. J. Orthod., 3:15, 1981.

Richardson, A., and A. V. Krayachich: The prediction of facial growth. Angle Orthod., 50:135, 1980.

Richardson, E. R.: Racial differences in dimensional traits of the human face. Angle Orthod., 50:301, 1980.

Richardson, M. E.: Late lower arch crowding: Facial growth or forward drift? Europ. J. Orthod., 1:219, 1979.

Ricketts, R. M.: Planning treatment on the basis of the facial pattern and an estimate of its growth. Angle Orthod., 27:14, 1957.

Ricketts, R. M.: Cephalometric synthesis. Am. J. Orthod., 46:647, 1960a.

Ricketts, R. M.: The influence of orthodontic treatment on facial growth and development. Angle Orthod., 30:103, 1960b.

Ricketts, R. M.: The evolution of diagnosis to computerized cephalometrics. Am. J. Orthod., 55:795, 1969.

Ricketts, R. M.: A principle of arcial growth of the mandible. Angle Orthod., 42:368, 1972a.

Ricketts, R. M.: The value of cephalometrics and computerized technology. Angle Orthod., 42:368, 1972b.

Ricketts, R. M.: New perspectives on orientation and their benefits to clinical orthodontics: Part I. Angle Orthod., 45:238, 1975a.

Ricketts, R. M.: A four-step method to distinguish orthodontic changes from natural growth. J. Clin. Orthod., 9:208, 1975b.

Ricketts, R. M.: New perspectives on orientation and their benefits to clinical orthodontics: Part II. Angle Orthod., 46:26, 1976.

Ricketts, R. M.: The interdependence of the nasal and oral capsules. In: *Naso-respiratory Function and Craniofacial Growth.* Ed. by J. A. McNamara, Jr. Ann Arbor, University of Michigan, Center for Human Growth and Development, 1979.

Ricketts, R. M.: Perspectives in the clinical application of orthodontics. The first fifty years. Angle Orthod., 51:115, 1981.

Ricketts, R. M., R. W. Bench, J. J. Hilgers, and R. Schulhof: An overview of computerized cephalometrics. Am. J. Orthod., 61:1, 1972.

Ricketts, R. M., R. J. Schulhof, and L. Bagha: Orientation: Sella-nasion or Frankfort horizontal. Am. J. Orthod., 69:648, 1976.

Riedel, R.: The relation of maxillary structures to cranium in malocclusion and in normal occlusion. Angle Orthod., 22:142, 1952.

Riedel, R.: An analysis of dentofacial relationships. Am. J. Orthod., 43:103, 1957.

Riedel, R.: A review of the retention problem. Angle Orthod., 30:179, 1960.

Righellis, E. G.: Treatment effects of Frankel, activator and extraoral traction appliances. Angle Orthod., 53:107, 1983.

Ringqvist, M.: Histochemical studies of the human masseter muscle. In: *Muscle Adaptation in the Craniofacial Region.* Ed. by D. S. Carlson and J. A. McNamara, Jr. Ann Arbor, University of Michigan, Center for Human Growth and Development, 1978.

Riolo, M. L.: Growth and remodeling of the cranial floor: A multiple microfluoroscopic analysis with serial cephalometrics. M. S. Thesis, Georgetown University, Washington, D.C., 1970.

Riolo, M. L., and J. A. McNamara, Jr.: Cranial base growth in the rhesus monkey from infancy to adulthood. J. Dent. Res., 52:249, 1973.

Riolo, M. L., R. E. Moyers, J. A. McNamara, and W. S. Hunter: *An Atlas of Craniofacial Growth: Cephalometric Standards from the University School Growth Study, The University of Michigan.* Monograph 2. Craniofacial Growth Series. Ann Arbor, University of Michigan, Center for Human Growth and Development, 1974.

Ritucci, R., and R. Nanda: The effect of chin cup therapy on the growth and development of the cranial base and midface. Am. J. Orthod. Dentofacial Orthop., 90:475, 1986.

Roberts, G. J., and J. J. Blackwood: Growth of the cartilages of the mid-line cranial base: a radiographic and histological study. J. Anat., 36:307, 1983.

Robertson, N. R.: Facial form of patients with cleft lip and palate. The long-term influence of presurgical oral orthopaedics. Br. Dent. J., 23:155, 1983.

Robinson, I. B., and B. G. Sarnat: Growth pattern of the pig mandible: A serial roentgenographic study using metallic implants. Am. J. Anat., 96:37, 1955.
Roche, A. F.: Increase in cranial thickness during growth. Hum. Biol., 25:81, 1953.
Roche, A. F.: Secular trends in stature, weight, and maturation. In: *Secular Trends in Human Growth, Maturation and Development.* Ed. by A. F. Roche. Monogr. Soc. Res. Child Dev., 44:3, 1979.
Roche, A. F.: The measurement of skeletal maturation. In: *Human Physical Growth and Maturation.* Ed. by F. E. Johnston, A. F. Roche, and C. Susanne. New York, Plenum Press, 1980.
Roche, A. F.: Bone growth and maturation. In: *Human Growth.* Vol 2. 2nd Ed. Ed. by F. Falkner and J. M. Tanner. New York, Plenum Press, 1986.
Roche, A. F., and A. B. Lewis: Late growth changes in the cranial base. In: *Development of the Basicranium.* Ed. by J. F. Bosma. DHEW Pub. 76:989, NIH, Bethesda, Md., 1976.
Roche, A. F., K. Manuel, and F. S. Seward: Unusual patterns of growth in the frontal and parietal bones. Anat. Rec., 152:459, 1965.
Roche, A. F., C. Rohmann, N. French, and G. Davila: Effect of training on replicability of assessments of skeletal maturity (Greulich-Pyle). Am. J. Roentgenol. Radium Ther. Nucl. Med., 108:511, 1970.
Roche, A. F., J. Roberts, and P. Hamill: Skeletal maturity of children 6–11 years, United States. DHEW Publ. No. (HRA) 75–1622. Vital Health Stat. (11), No. 140, 1974.
Roche, A. F., J. Roberts, and P. Hamill: Skeletal maturity of children 6–11 years: racial, geographic area of residence, and socioeconomic differentials, United States. DHEW Publ. No. (HRA) 76–1631. Vital Health Stat. (11), No. 149, 1975.
Roche, A. F., W. C. Chumlea, D. Thissen *Assessing the Skeletal Maturity of the Hand-Wrist: Fels Method.* Springfield, Ill., Charles C Thomas, 1988.
Roger-Thooris, M. O.: Relations entre: croissance et variabilité sur le profil cranien. Thèse, Univ. de Nancy, 1973.
Romeo, D. A., and R. Nanda: Generation cycle of mesenchymal cells of palatal shelves of rat foetuses. Arch. Oral Biol., 23:123, 1978.
Rönning, O.: Observations on the intracerebral transplantation of the mandibular condyle. Acta Odont. Scand., 24:443, 1966.
Rönning, O., and K. Koski: The effect of the articular disc on the growth of condylar cartilage transplants. Eur. Orthod. Soc. Trans., 1969:99, 1970.
Rönning, O., and K. Koski: The effect of periostomy on the growth of the condylar process in the rat. Proc. Finn. Dent. Soc., 70:28, 1974.
Rönning, O., K. Paunio, and K. Koski: Observations on the histology, histochemistry and biochemistry of growth cartilages in young rats. Suom. Hammaslaak. Toim., 63:187, 1967.
Rosa, R. A., and M. G. Arvystas: An epidemiologic survey of malocclusions among American Negros and American Hispanics. Am. J. Orthod., 73:258, 1978.
Rose, J. S.: A survey of congenitally missing teeth, excluding third molars, in 6000 orthodontic patients. Dent. Pract., 17:107, 1966.
Rosenstein, S. W.: Pathological and congenital disturbances: The orthodontic viewpoint. J.A.D.A., 82:763, 1971.
Ross, R. B.: Lateral facial dysplasia (first and second branchial arch syndrome, hemifacial microsomia). In: *Morphogenesis and Malformation of Face and Brain.* Ed. by D. Bergsma, J. Langman, and N. W. Paul. National Foundation—March of Dimes, New York, Alan R. Liss, 11:7, 1975.
Ross, R. B.: Treatment variables affecting facial growth in complete unilateral cleft lip and palate. Part 1. treatment affecting growth. Cleft Palate J., 24:5, 1987.
Ross, R. B., and M. C. Johnston: *Cleft Lip and Palate.* Baltimore, Williams & Wilkins, 1972.
Rubin, R. M.: Facial deformity: A preventable disease? Angle Orthod., 49:98, 1979.
Rubin, R. M.: Mode of respiration and facial growth. Am. J. Orthod., 78:504, 1980.
Ruff, R.: Orthodontic treatment in the mixed dentition. Am. J. Orthod., 52:502, 1970a.
Ruff, R.: Orthodontic treatment in the permanent dentition. Am. J. Orthod., 58:597, 1970b.
Sakuda, M., and K. Wada: Principal component analysis and craniofacial morphology and individual growth pattern. J. Dent. Res., 58:45, 698, 1979.
Salentijn, L., and M. L. Moss: Morphological attributes of the logarithmic growth of the human face: gnomic growth. Acta Anat., 78:185, 1971.
Salyer, K. E., I. R. Munro, L. A. Whitaker, and I. Jackson: Difficulties and problems to be solved in the approach to craniofacial malformations. In: *Morphogenesis and Malformation of Face and Brain.* Ed. by D. Bergsma, J. Langman, and N. W. Paul. National Foundation—March of Dimes, New York, Alan R. Liss, 11:7, 1975.
Salzmann, J. A.: The research workshop on cephalometrics. Am. J. Orthod., 46:834, 1960a.
Salzmann, J. A.: *Roentgenographic Cephalometrics. Proceedings of the second research workshop.* Philadelphia, J. B. Lippincott Company, 1960b.
Salzmann, J. A.: *Practice of Orthodontics.* Philadelphia, J. B. Lippincott, 1966.
Sandham, A., and R. Duball: Natural head posture in cephalometric radiography. Radiography, 51:287, 1985.
Sarnas, C., and B. Solow: Early adult changes in the skeletal and soft-tissue profile. Eur. J. Orthod., 2:1, 1980.
Sarnat, B. G.: Facial and neurocranial growth after removal of the mandibular condyle in the Macaca rhesus monkey. Am. J. Surg., 94:19, 1957.

Sarnat, B. G.: Postnatal growth of the upper face: Some experimental considerations. Angle Orthod., 33:139, 1963.
Sarnat, B. G.: *The Temporomandibular Joint,* 2nd Ed. Springfield, Ill., Charles C Thomas, 1964.
Sarnat, B. G.: The face and jaws after surgical experimentation with the septovomeral region in growing and adult rabbits. Acta Otolaryngol. Suppl., 268:1970.
Sarnat, B. G.: Clinical and experimental considerations in facial bone biology: Growth, remodeling, and repair. J.A.D.A., 82:876, 1971a.
Sarnat, B. G.: Surgical experimentation and gross postnatal growth of the face and jaws. J. Dent. Res., 50:1462, 1971b.
Sarnat, B. G.: The postnatal maxillary-nasal-orbital complex: Experimental surgery. In: *Factors Affecting the Growth of the Midface.* Ed. by J. A. McNamara, Jr. Ann Arbor, University of Michigan, Center for Human Growth and Development, 1976.
Sarnat, B. G.: Growth pattern of the mandible: some reflections. Am. J. Orthod. Dentofacial Orthop., 90:221, 1986.
Sarnat, B. G., and M. B. Engel: A serial study of mandibular growth after the removal of the condyle of the Rhesus monkey. Plast. Reconstr. Surg., 7:364, 1951.
Sarnat, B. G., and D. Laskin: *Temporomandibular Joint: Biologic Basis for Clinical Practice.* 3rd Ed. Springfield, Ill., Charles C Thomas, 1980.
Sarnat, B. G., and H. Muchnic: Facial skeletal changes after mandibular condylectomy in the adult monkey. J. Anat., 108:338, 1971a.
Sarnat, B. G., and H. Muchnic: Facial skeletal changes after mandibular condylectomy in growing and adult monkeys. Am. J. Orthod., 60:33, 1971b.
Sarnat, B. G., and A. J. Selman: Growth pattern of the rabbit snout dorsum: A serial cephalometric radiographic study with radio-opaque implants. J. Anat., 124:469, 1977.
Sarnat, B. G., and P. D. Shanedling: Postnatal growth of the orbit and upper face in rabbits. Arch. Ophthalmol., 73:829, 1965.
Sarnat, B. G., and M. R. Wexler: Growth of the face and jaws after resection of the septal cartilage in the rabbit. Am. J. Anat., 118:755, 1966.
Sarnat, B. G.: J. A. Feigenbaum, and W. M. Krogman: Adult monkey coronoid process after resection of trigeminal nerve motor root. Am. J. Anat., 150:129, 1977.
Sassouni, V.: Roentgenographic cephalometric analysis of cephalo-facio-dental relationships. Am. J. Orthod., 41:734, 1955.
Sassouni, V.: *The Face in Five Dimensions.* Philadelphia, Philadelphia Center for Research in Child Growth, 1960.
Sassouni, V.: *Heredity and Growth of the Human Face.* Pittsburgh, University of Pittsburgh, 1965.
Sassouni, V., G. Friday, H. Shnorhokian, Q. Beery, T. Ullo, D. Miller, S. Murphey, and R. Landay: The influence of perennial allergic rhinitis on facial type and a pilot study of the effect of allergy management on facial growth patterns. Ann. Allergy, 54:493, 1985.
Saunders, S. R.: Surface and cross-sectional comparisons of bone growth remodeling. Growth, 49:105, 1985.
Saunders, S. R., F. Popovich, and G. W. Thompson: A family study of craniofacial dimensions in the Burlington Growth Centre sample. Am. J. Orthod., 78:394, 1980.
Savara, B. S., and I. J. Singh: Norms of size and annual increments of seven anatomical measures of maxillae in boys from three to sixteen years of age. Angle Orthod., 38:104, 1968.
Sawin, P. B., M. Ranlett, and D. D. Crary: Morphogenetic studies of the rabbit. XXV. The spheno-occipital synchondrosis of the dachs (chondrodystrophy) rabbit. Am. J. Anat., 105:257, 1959.
Scammon, R. E.: The measurements of the body in childhood. In: *The Measurement of Man.* Ed. by J. A. Harris, C. M. Jackson, D. G. Paterson, and R. E. Scammon. Minneapolis, University of Minnesota Press, 1930.
Scheiman-Tagger, E., and A. G. Brodie: Lead acetate as a marker of growing calcified tissues. Anat. Rec., 150:435, 1964.
Schouker-Jolly, M.: Utilisation d'appareillages extra-oraux récents dans le prognathisme mandibulaire, associe à une hypoplasie maxillaire. Méd. Infant., 6:479, 1972.
Schudy, F. F.: Cant of the occlusal plane and axial inclinations of teeth. Angle Orthod., 33:69, 1963.
Schudy, F. F.: Vertical growth vs. anteroposterior growth as related to function and treatment. Angle Orthod., 34:75, 1964.
Schudy, F. F.: The rotation of the mandible resulting from growth: Its implications in orthodontic treatment. Angle Orthod., 35:36, 1965.
Schudy, F. F.: The control of vertical overbite in clinical orthodontics. Angle Orthod., 38:19, 1968.
Schulhof, R. J., and L. Bagha: A statistical evaluation of the Ricketts and Johnston growth-forecasting method. Am. J. Orthod., 67:258, 1975.
Schulter, F. P.: Studies of the basicranial axis: A brief review. Am. J. Phys. Anthropol., 45:545, 1976.
Schumacher, G. H.: Factors influencing craniofacial growth. Prog. Clin. Biol. Res., 187:3, 1985.
Schumacher, G. H., D. Ivankievicz, E. Ehler, and H. Pfau: Comparative study of the leverage of the mandible in man and mammals. Fogorv. Sz., 76:48, 1983.

Scott, J. H.: The cartilage of the nasal septum. Br. Dent. J., 95:37, 1953.
Scott, J. H.: Growth and function of the muscles of mastication in relation to the development of the facial skeleton and of the dentition. Am. J. Orthod., 40:429, 1954a.
Scott, J. H.: The growth of the human face. Proc. R. Soc. Med., 47:91, 1954b.
Scott, J. H.: Craniofacial regions: Contribution to the study of facial growth. Dent. Pract., 5:208, 1955.
Scott, J. H.: Growth at facial sutures. Am. J. Orthod., 42:381, 1956.
Scott, J. H.: The growth in width of the facial skeleton. Am. J. Orthod., 43:366, 1957.
Scott, J. H.: The cranial base. Am. J. Phys. Anthropol., 16:319, 1958a.
Scott, J. H.: The analysis of facial growth. Part I. The anteroposterior and vertical dimensions. Am. J. Orthod., 44:507, 1958b.
Scott, J. H.: The analysis of facial growth. Part II. The horizontal and vertical dimensions. Am. J. Orthod., 44:585, 1958c.
Scott, J. H.: The face in foetal life. Eur. Orthod. Soc. Trans., 37:168, 1961.
Scott, J. H., and A. D. Dixon: *Anatomy for Students of Dentistry,* 3rd Ed. Edinburgh and London, Churchill Livingstone, 1972.
Searls, J. C.: A radioautographic study of chondrocytic proliferation in nasal septal cartilage of the newborn rat. Am. J. Anat., 150:559, 1977.
Searls, J. C.: A comparative radioautographic study of chondrocytic proliferation in nasal septal cartilage of the 5-day-old rat, rabbit, guinea pig, and beagle. Am. J. Anat., 154:437, 1979.
Sekiguchi, T., and B. Savara: Variability of cephalometric landmarks used for face growth studies. Am. J. Orthod., 61:603, 1972.
Sellke, T. A., and B. J. Schneider: The effects of reduced attrition on craniofacial and dentoalveolar development in the rat. Angle Orthod., 47:313, 1977.
Servoss, J. M.: An in vivo and in vitro autoradiographic investigation of growth in synchondrosal cartilages. Am. J. Anat., 136:479, 1973.
Seward, S.: Relation of basion to articulare. Angle Orthod., 51:151, 1981.
Shah, S. M., and M. R. Joshi: An assessment of asymmetry in the normal craniofacial complex. Angle Orthod., 48:141, 1978.
Shapiro, P. A.: Responses of the nonhuman maxillary complex to mechanical forces. In: *Factors Affecting the Growth of the Midface.* Ed. by J. A. McNamara, Jr. University of Michigan, Center for Human Growth and Development, 1976.
Shapiro, G. G., and P. Shapiro: Nasal airway obstruction and facial development. Clin. Rev. Allergy, 2:225, 1984.
Shaw, W. C.: Problems of accuracy and reliability in cephalometric studies with implants in infants with cleft lip and palate. Br. J. Orthod., 4:93, 1977.
Shaw, R. E., L. Mark, D. Jenkins, and E. Mingolla: A dynamic geometry for predicting growth of gross craniofacial morphology. Prog. Clin. Biol. Res., 101:423, 1982.
Shea, B. T.: Eskimo craniofacial morphology: Cold stress and the maxillary sinus. Am. J. Phys. Anthropol., 47:289, 1977.
Sheppard, S. M.: Asymptomatic Morphologic variations in the mandibular condyle-ramus region. J. Prosthet. Dent., 47:539, 1982.
Sheridan, J. D.: Cell coupling and cell communication during embryogenesis. In: *Cell Surface in Animal Embryogenesis and Development.* Cell Surf. Rev. 1:409, 1977.
Sherman, M. S.: The nerves of bone. J. Bone Joint Surg., 45:522, 1963.
Shore, R. C., and B. K. B. Berkovitz: An ultrastructural study of periodontal ligament fibroblasts in relation to their possible role in tooth eruption and intracellular collagen degradation in the rat. Am. J. Orthod., 24:155, 1979.
Sicher, H.: The growth of the mandible. Am. J. Orthod., 33:30, 1947.
Sicher, H.: Some aspects of the anatomy and pathology of the temporomandibular articulation. The Bur, 48:14, 1948.
Sicher, H., and J. Tandler: *Anatomie für Zahnärzte.* Vienna and Berlin, 1928.
Sicher, H., and J. P. Weinmann: Bone growth and physiologic tooth movement. Am. J. Orthod. Oral Surg., 30:109, 1944.
Siegel, M. I.: The facial and dental consequences of nasal septum resections in baboons. Med. Primat., 1972:204, 1972.
Silbermann, M., and J. Frommer: Further evidence for the vitality of chondrocytes in the mandibular condyle as revealed by ^{35}S-sulfate autoradiography. Anat. Rec., 174:503, 1972a.
Silbermann, M., and J. Frommer: Vitality of chondrocytes in the mandibular condyle as revealed by collagen formation. An autoradiographic study with ^{3}H-proline. Am. J. Anat., 135:359, 1972b.
Silbermann, M., and J. Frommer: The nature of endochondral ossification in the mandibular condyle of the mouse. Anat. Rec., 172:659, 1972c.
Silbermann, M., and D. Lewinson: An electron microscopic study of the premineralizing zone of the condylar cartilage of the mouse mandible. J. Anat., 125:55, 1978.
Silbermann, M., and E. Livne: Age-related degenerative changes in the mouse mandibular joint. J. Anat., 129:507, 1979.

Silbermann, M., A. Weiss, and E. Raz: Studies on hormonal regulation of the growth of the craniofacial skeleton. II. Effects of a glucocorticoid hormone on sulfate incorporation by neonatal condylar cartilage. J. Craniofac. Genet. Dev. Biol., 2:299, 1982.
Simpson, M. M.: Lip incompetence and its relationship to skeletal and dental morphology. Br. J. Orthod., 3:177, 1976.
Simpson, M.: An electromyographic investigation of the perioral musculature in Class II Division 1 malocclusion. Br. J. Orthod., 4:17, 1977.
Sinclair, D.: *Human Growth after Birth,* 3rd Ed. Oxford University Press, New York, 1978.
Singer, J.: Physiologic timing of orthodontic treatment. Angle Orthod., 50:322, 1980.
Sirianni, J. E., and L. Newell-Morris: Craniofacial growth of fetal *Macaca nemestrina:* A cephalometric roentgenographic study. Am. J. Phys. Anthropol., 53:407, 1980.
Sirianni, J. E., and A. L. Van Ness: Postnatal growth of the cranial base in *Macaca nemestrina.* Am. J. Phys. Anthropol., 49:329, 1978.
Sirianni, J. E., A. L. Van Ness, and D. R. Swindler: Growth of the mandible in adolescent pigtailed macaques (*Macaca nemestrina*). Hum. Biol., 54:31, 1982.
Skieller, V., A. Bjork, and T. Linde-Hansen: Prediction of mandibular growth rotation evaluated from a longitudinal implant sample. Am. J. Orthod., 86:359, 1984.
Skuzuki, S.: Histomorphometric study on growing condyle of rat. Bull. Tokyo Med. Dent. Univ., 33:23, 1986.
Slavkin, H. C.: Overview of research on craniofacial malformations: gene regulations. In: *Symposium on Diagnosis and Treatment of Craniofacial Anomalies.* Ed. by J. M. Converse, J. G. McCarthy, and D. Wood-Smith. St. Louis, C. V. Mosby, 1979a.
Slavkin, H. C.: *Developmental Craniofacial Biology.* Philadelphia, Lea & Febiger, 1979b.
Slavkin, H. C.: Gene regulation in the development of oral tissues. J. Dent. Res., 67:1142, 1988.
Smiley, G. R.: A histological study of the formation and development of the soft palate in mice and man. Arch. Oral Biol., 20:297, 1975.
Smiley, G. R., and W. E. Koch: A comparison of secondary palate development with different *in vitro* techniques. Anat. Rec., 181:711, 1975.
Smith, B. H., S. M. Garn, and W. S. Hunter: Secular trends in face size. Angle Orthod., 56:196, 1986.
Smith, R. J.: Mandibular biomechanics and temporomandibular joint functions in primates. Am. J. Phys. Anthropol., 49:341, 1978.
Smith, R. J.: Development of occlusion and malocclusion. Pediatr. Clin. North Am., 29:475, 1982.
Smith, R. J.: Functions of condylar translation in human mandibular movement. Am. J. Orthod., 88:191, 1985.
Smith, R. J., and H. L. Bailit: Problems and methods in research on the genetics of dental occlusion. Angle Orthod., 47:65, 1977.
Smith, R. J., and J. Frommer: Condylar growth gradients: Possible mechanism for spiral or arcial growth of the mandible. Angle Orthod., 50:274, 1980.
Sofaer, J. A., and C. J. MacLean: Heredity and morphological variation in early and late developing human teeth of the same morphological class. Arch. Oral Biol., 17:811, 1972.
Solomon, W. R.: Allergic responses in the upper respiratory system. In: *Naso-respiratory Function and Craniofacial Growth.* Ed. by J. A. McNamara, Jr. Ann Arbor, University of Michigan, Center for Human Growth and Development, 1979.
Solow, B.: The pattern of craniofacial associations: A morphological and methodological correlation and factor analysis study on young male adults. Acta Odontol. Scand., 24 (Suppl. 46): 1966.
Solow, B.: Automatic processing of growth data. Angle Orthod., 39:186, 1969.
Solow, B.: Factor analysis of cranio-facial variables. In: *Cranio-facial Growth in Man.* Ed. by R. E. Moyers and W. M. Krogman. Oxford, Pergamon Press, 1971.
Solow, B., and E. Greve: Craniocervical angulation and nasal respiratory resistance. In: *Naso-respiratory Function and Craniofacial Growth.* Ed. by J. A. McNamara, Jr. Ann Arbor, University of Michigan, Center for Human Growth and Development, 1979.
Solow, B., and A. Tallgren: Head posture and craniofacial morphology. Am. J. Phys. Anthropol., 44:417, 1976.
Solow, B., and A. Tallgren: Dentofacial morphology in relation to craniocervical posture. Angle Orthod., 47:157, 1977.
Souyris, F., V. Moncarz, P. Rey: Facial asymmetry of developmental etiology: A report of nineteen cases. Oral Surg., 56:113, 1983.
Speidel, T. M., R. J. Isaacson, and F. W. Worms: Tongue-thrust therapy and anterior dental open bite. Am. J. Orthod., 62:287, 1972.
Spranger, J. W., K. Benirschke, J. G. Hall, and W. Lenz: Errors of morphogenesis: Concepts and terms. J. Ped., 100:160, 1982.
Spyropoulos, M. N.: The morphogenetic relationship of the temporal muscle to the coronoid process in human embryos and fetuses. Am. J. Anat., 150:395, 1977.
Srivastava, H. C., and D. C. Vyas: Postnatal development of rat soft palate. J. Anat., 128:97, 1979.
Stanley, R. B., and R. A. Latham: The regression pattern of Meckel's cartilage in normal mandibular development. I.A.D.R. Abstr., 1973:216, 1973.

Stein, K. F., T. J. Kelly, and E. W. Wood: Influence of heredity in the etiology of malocclusion. Am. J. Orthod., 42:125, 1956.
Steiner, C. C.: Cephalometrics for you and me. Am. J. Orthod., 39:729, 1953.
Steiner, C. C.: Cephalometrics in clinical practice. Angle Orthod., 29:8, 1959.
Stenstrom, S. J., and B. L. Thilander: Effects of nasal septal cartilage resections on young guinea pigs. Plast. Reconstr. Surg., 45:160, 1970.
Steuer, L.: The cranial base for superimposition of lateral cephalometric radiographs. Am. J. Orthod., 16:493, 1972.
Stockli, P. W., and H. G. Willert: Tissue reactions in the temporomandibular joint resulting from anterior displacement of the mandible in the monkey. Am. J. Orthod., 60:142, 1971.
Stohler, C. S.: A comparative electromyographic and kinesiographic study of deliberate and habital mastication in man. Arch. Oral Biol., 31:669, 1986.
Storey, A. T.: Physiology of a changing vertical dimension. J. Prosthet. Dent., 1:912, 1962.
Storksen, K., S. Aukland, and S. Kvinnsland: Matrix formation in craniofacial cartilages of the rat. Acta Odont. Scand., 37:29, 1979.
Stoudt, H. W., A. Damon, R. McFarland, and J. Roberts: Weight, height, and selected body dimensions of adults: United States, 1960–1962. Vital Health Stat. (11), No. 8, 1965.
Stramrud, L.: External and internal cranial base: A cross-sectional study of growth and of association in form. Acta Odont. Scand., 17:239, 1959.
Strickland, A. L., and R. B. Shearin: Diurnal height variation in children. J. Pediatr., 80:1023, 1972.
Stutzmann, J., and A. Petrovic: Particularités de croissance de la suture palatine sagittale de jeune rat. Bull. Assoc. Anat. (Nancy), 148:552, 1970.
Stutzmann, J. J., and A. G. Petrovic: Experimental analysis of general and local extrinsic mechanisms controlling upper jaw growth. In: *Factors Affecting the Growth of the Midface.* Ed. by J. A. McNamara, Jr. Ann Arbor, University of Michigan, Center for Human Growth and Development, 1976.
Stutzmann, J., and A. Petrovic: Intrinsic regulation of the condylar cartilage growth rate. Eur. J. Orthod., 1:41, 1979.
Subtelny, J. D.: Longitudinal study of soft tissue facial structures and their profile characteristics, defined in relation to underlying skeletal structures. Am. J. Orthod., 45:481, 1959.
Subtelny, J. D.: The soft tissue profile, growth and treatment changes. Angle Orthod., 31:105, 1961.
Subtelny, J. D.: Oral respiration: Facial maldevelopment and corrective dentofacial orthopedics. Angle Orthod., 50:147, 1980.
Subtelny, J. D.: The degenerative, regenerative mandibular condyle: Facial asymmetry. J. Craniofac. Genet. Dev. Biol., 1:227, 1985.
Subtelny, J. D., and M. Sakuda: Muscle function, oral malformation and growth changes. Am. J. Orthod., 52:495, 1966.
Subtelny, J. D., and J. Subtelny: Oral habits: Studies in form, function, and therapy. Angle Orthod., 43:347, 1973.
Susanne, C.: Genetic and environmental influences on morphological characteristics. Ann. Hum. Biol., 2:279, 1975.
Susanne, C.: Ageing, continuous changes in adulthood. In: *Human Physical Growth and Maturation.* Ed. by F. E. Johnston, A. F. Roche, and C. Susanne. New York, Plenum Press, 1978.
Swindler, D. R., J. E. Sirianni, and L. H. Tarrant: A longitudinal study of cephalofacial growth in *Papio cynocephalus* and *Macaca nemestrina* from three months to three years. IVth International Congress of Primatology, Vol. 3, *Craniofacial Biology of Primates.* Basel, S. Karger, 1973.
Symons, N. B. B.: Studies on the growth and form of the mandible. Dent. Rec., 71:41, 1951.
Symons, N. B. B.: The development of the human mandibular joint. J. Anat., 86:326, 1952.
Symons, N. B. B.: A histochemical study of the secondary cartilage of the mandibular condyle in the rat. Arch. Oral Biol., 10:579, 1965.
Takagi, Y.: Human postnatal growth of vomer in relation to base of cranium. Ann. Otol., 73:238, 1964.
Takahashi, E.: Growth and environmental factors in Japan. Hum. Biol. 38:112, 1966.
Takahashi, R: *The Formation of the Nasal Septum and the Etiology of Septal Deformity: The Concept of Evolutionary Paradox..* Acta Otolaryng. Suppl. 443, 1987.
Takeuchi, Y., B. S. Savara, and R. J. Shadel: Biennial size norms of eight measures of the temporal bone from four to twenty years of age. Angle Orthod., 50:107, 1980.
Tallgren, A.: Neurocranial morphology and ageing: A longitudinal roentgen cephalometric study of adult Finnish women. Am. J. Phys. Anthropol., 41:285, 1974.
Tallgren, A., and B. Solow: Hyoid bone position, facial morphology and head posture in adults. Eur. J. Orthod., 9:1, 1987.
Tanner, J. M.: *Growth and Adolescence.* Oxford, Blackwell, 1955.
Tanner, J. M.: *Growth at Adolescence,* 2nd Ed. Oxford, Blackwell, 1962.
Tanner, J. M.: Earlier maturation in man. Sci. Am., 218:21, 1968.
Tanner, J. M.: Trend towards earlier menarche in London, Oslo, Copenhagen, the Netherlands and Hungary. Nature, 243:95, 1973.

Tanner, J. M.: Physical growth and development. In: *Textbook of Paediatrics*. Vol. 1, 2nd Ed. Ed. by J. O. Forfar and G. C. Arneil. Edinburgh, Churchill Livingstone, 1978a.

Tanner, J. M.: *Foetus into Man: Physical Growth from Conception to Maturity*. Cambridge, Harvard University Press, 1978b.

Tanner, J. M.: Normal growth and techniques of growth assessment. Clin. Endocrinol. Metab., 15:411, 1986a.

Tanner, J. M.: Use and abuse of growth standards. In *Human Growth*. Vol. 3, 2nd Ed. Ed. by F. Falkner and J. M. Tanner. Plenum Press, New York, 1986b.

Tanner, J. M.: Issues and advances in adolescent growth and development. J. Adolesc. Health Care, 8:470, 1987.

Tanner, J. M., and P. Davies: Clinical longitudinal standards for height and height velocity for North American children. J. Pediatr., 107:317, 1985.

Tanner, J. M., and R. H. Whitehouse: Revised standards for triceps and subscapular skinfolds in British children. Arch. Dis. Child., 50:142, 1975.

Tanner, J. M., and R. H. Whitehouse: Clinical longitudinal standards for height, weight, height velocity, and stages of puberty. Arch. Dis. Child., 51:170, 1976.

Tanner, J. M., R. H. Whitehouse, and M. Takaishi: Standards from birth and maturity for height, weight, height velocity, and weight velocity: British children, 1965. Arch. Dis. Child., 41:454, 613, 1966.

Tanner, J. M., R. H. Whitehouse, and B. S. Carter: Prediction of adult height from height, bone age, age and occurrence of menarche at ages 4 to 16 with allowance for mid parent height. Arch. Dis. Child., 50:14, 1975a.

Tanner, J. M., R. H. Whitehouse, W. A. Marshall, M. Healy, and H. Goldstein: *Assessment of Skeletal Maturity and Prediction of Adult Height (TW2 Method)*. London, Academic Press, 1975b.

Tanner, J. M., T. Hayashi, M. A. Preece, and N. Cameron: Increase in length of leg relative to trunk in Japanese children and adults from 1957–1977: Comparison with British and with Japanese Americans. Ann. Hum. Biol., 91:411, 1982.

Tanner, J. M., K. W. Landt, N. Cameron, B. S. Carter, and J. Patel: Prediction of adult height from height and bone age in childhood. Arch. Dis. Child., 58:767, 1983a.

Tanner, J. M., R. H. Whitehouse, M. Cameron, W. A. Marshall, M. Healy, and H. Goldstein: *Assessment of Skeletal Maturity and Prediction of Adult Height (TW2 Method)*, 2nd Ed. London, Academic Press, 1983b.

Tanuma, K.: Changes with advance of age of the human maxillomandibularis, zygomaticomandibularis and superficial temporalis. Okajimas. Folia. Anat. Jpn., 6:1, 1984.

Taranger, J., and U. Hägg: The timing and duration of adolescent growth. Acta Odontol. Scand., 38:57, 1980.

Taranger, J., I. Engstrom, H. Lichtenstein, and I. Svennberg-Redegren: VI Somatic pubertal development. Acta Paediatr. Scand. [Suppl] 258:121, 1976a.

Taranger, J., B. Bruning, I. Claesson, P. Karlberg, T. Landstrom, and B. Lindstrom: A new method for the assessment of skeletal maturity: the MAT-method (mean appearance time of bone stages). Acta Paediatr. Scand. [Suppl] 258:109, 1976b.

Taranger, J., B. Bruning, I. Claesson, P. Karlberg, T. Landstrom, and B. Lindstrom: Skeletal development from birth to 7 years. Acta Paediatr. Scand. [Suppl] 258:98, 1976c.

Taranger, J., T. Lewin, and P. Karlberg: Continuing secular trend of height of Swedish conscripts. Ann. Hum. Biol., 5203, 1978.

Taranger, J., P. Karlberg, B. Bruning, and I. Engstrom: Standard deviation score charts of skeletal maturity and its velocity in Swedish children assessed by the Tanner-Whitehouse method (TW2-20). Ann. Hum. Biol., 14:357, 1987.

Taylor, R. G.: Craniofacial growth during closure of the secondary palate in the hamster. J. Anat., 125:361, 1978.

Taylor, R. M.: Nonlever action of the mandible. Am. J. Phys. Anthropol., 70:417, 1986.

Ten Cate, A. R.: Development of the periodontium. In: *Biology of the Periodontium*. Ed. by A. H. Melcher and W. H. Bowen. New York, Academic Press, 1969.

Ten Cate, A. R.: Formation of supporting bone in association with periodontal ligament organization in the mouse. Arch. Oral Biol., 20:137, 1975.

Ten Cate, A. R., and D. A. DePorter: The degradative role of the fibroblast in the remodeling and turnover of collagen in soft connective tissue. Anat. Rec., 182:1, 1975.

Ten Cate, A. R., C. Mills, and G. Solomon: The development of the periodontium. A transplantation and autoradiographic study. Anat. Rec., 170:365, 1971.

Ten Cate, A. R., E. Freeman, and J. B. Dicker: Sutural development: Structure and its response to rapid expansion. Am. J. Orthod., 71:622, 1977.

Terk, B.: Modifications apportées chez le rat, à l'évolution des dents, par l'hydrocéphalie expérimentale étude en orientation vestibulaire. Thèse, Acad. Paris, Univ. Paris VI, 1973.

Tessier, P. J.: Ostéotomies totales de la face: syndrome de Crouzon, syndrome d'Apert., oxycéphalies, scaphocéphalies, turriecéphalies. Ann. Chir. Plast., 12:273, 1967.

Tessier, P., J. Delaire, J. Billet, and H. Landais: Considérations sur le développement de l'orbite: Ses incidences sur la croissance faciale. Rev. Stomatol., 63:1–2, 27–39, 1964.

Teuscher, U.: A growth-related concept for skeletal Class II treatment. Am. J. Orthod., 74:258, 1978.
Thesleff, I.: Use of organ culture techniques in craniofacial developmental biology. Proc. Finn. Dent. Soc., 77:159, 1981.
Theunissen, J. J. W.: Het fibreuze periosteum. Thesis, Katholieke Univ. te Nymegen, 1973.
Thilander, B.: Innervation of the temporomandibular joint capsule in man. Trans. R. Sch. Dent. Stockholm, 7:9, 1961.
Thilander, B.: The structure of the collagen of the temporomandibular joint disc in man. Acta Odont. Scand., 22:135, 1964.
Thilander, B., and B. Ingervall: The human spheno-occipital synchondrosis. II. A histological and microradiographic study of its growth. Acta Odont. Scand., 31:323, 1973.
Thilander, B., G. E. Carlsson, and B. Ingervall: Postnatal development of the human temporomandibular joint. I. A histological study. Acta Odont. Scand., 34:117, 1976.
Thompson, J. L., and G. S. Kendrick: Changes in the vertical dimensions of the human male skull during the third and fourth decades of life. Anat. Rec., 150:209, 1964.
Throckmorton, G. S., R. A. Finn, and W. H. Bell: Biomechanics of differences in lower facial height. Am. J. Orthod., 77:410, 1980.
Thurow, R. C.: Cephalometric methods in research and private practice. Angle Orthod., 21:104, 1951.
Thurow, R. C.: *Atlas of Orthodontic Principles*. St. Louis, C. V. Mosby, 1970.
Thurow, R. C.: Fifty years of cephalometric radiology. Editorial. Angle Orthod., 51:89, 1981.
Todd, T. W.: Prognathism; a study in development of the face. J.A.D.A., 19:2172, 1932.
Tomarin, A., and A. Boyde: Facial and visceral arch development in the mouse embryo: A study by scanning electron microscopy. J. Anat., 124:563, 1977.
Tomer, B. S., and E. P. Harvold: Primate experiments on mandibular growth direction. Am. J. Orthod., 82:114, 1982.
Tommasone, D., R. Rangel, S. Kurihara, and D. Enlow: Remodeling patterns in the facial and cranial skeleton of the human cleft palate fetus. Kalevi Koski Festschrift, Proc. Finnish Dent. Soc. (Special Issue), 77:171, 1981.
Tonge, E. A., J. H. Heath, and M. C. Meikle: Anterior mandibular displacement and condylar growth: An experimental study in the rat. Am. J. Orthod., 82:277, 1982.
Tonna, E. A.: Topographic labelling method using (^{3}H)-proline autoradiography in assessment of ageing paradental bone in the mouse. Arch. Oral Biol., 21:729, 1976.
Tracy, W. E., B. S. Savara, and J. W. A. Brant: Relation of height, width, and depth of the mandible. Angle Orthod., 35:269, 1965.
Tradowsky, M., and J. B. Dworkin: Determination of the physiologic equilibrium point of the mandible by electronic means. J. Prosthet. Dent., 48:89, 1982.
Trenouth, M. J.: Shape changes during human fetal craniofacial growth. J. Anat., 139:639, 1984.
Trenouth, M. J.: Changes in the jaw relationships during human foetal cranio-facial growth. Br. J. Orthod., 12:33, 1985.
Trenouth, M. J.: Asymmetry of the human skull during fetal growth. Anat. Res., 211:205, 1985.
Treuenfels, H.: Head position, atlas position and breathing in open bite. Fortschr. Kieferothop., 45:111, 1984.
Trouten, J. C., D. H. Enlow, M. Rabine, A. E. Phelps, and D. Swedlow: Morphologic factors in open bite and deep bite. Angle Orthod., 53:192, 1983.
Turley, P. K., P. A. Shapiro, and B. C. Moffett: The loading of bioglass-coated aluminum oxide implants to produce sutural expansion of the maxillary complex in the pigtail monkey *(Macaca nemestrina)*. Arch. Oral Biol., 25:459, 1980.
Turpin, D. L.: Growth and remodeling of the mandible in the *Macaca mulatta* monkey. Am. J. Orthod., 54:251, 1968.
Tweed, C. H.: The Frankfort-mandibular plane angle in orthodontic diagnosis, classification, treatment planning, and prognosis. Am. J. Orthod. Oral Surg., 32:175, 1946.
Tweed, C. H.: The Frankfort-mandibular incisor angle (FMIA) in orthodontic diagnosis, treatment planning and prognosis. Angle Orthod., 24:121, 1954.
Tyler, M. S.: Epithelial influences on membrane bone formation in the maxilla of the embryonic chick. Anat. Rec., 192:225, 1978.
Tyler, M. S., and B. K. Hall: Epithelia influences on skeletogenesis in the mandible of the embryonic chick. Anat. Rec., 188:229, 1977.
Tyler, M. S., and W. E. Koch: *In vitro* development of palatal tissues from embryonic mice: Differentiation of the secondary palate from 12-day mouse embryos. Anat. Rec., 182:297, 1975.
Urban, J. P., and J. F. McMullin: Swelling pressure of the lumbar intervertebral discs: influence of age, spinal level, composition, and degeneration. Spine, 13:179, 1987.
Urist, M. R., and B. S. Strates: Bone morphogenetic protein. J. Dent. Res., 50:1392, 1971.
Urist, M. R., B. Silverman, K. Buring, F. Dubuc, and J. Rosenberg: The bone induction principle. Clin. Orthop., 53:243, 1967.
Utley, R. K.: The activity of alveolar bone incident to orthodontic tooth movement as studied by oxytetracyline-induced fluorescence. Am. J. Orthod., 54:167, 1968.

van Beek, H.: The transfer of mesial drift potential along the dental arch in *Macaca irus:* An experimental study of tooth migration rate related to the horizontal vectors of occlusal forces. Eur. J. Orthod., 1:125, 1979.
van der Klaauw, C. J.: Cerebral skull and facial skull: A contribution to the knowledge of skull structure. Arch. Neerl. Zool., 9:16, 1946.
van der Klaauw, C. J.: Size and position of the functional components of the skull: A contribution to the knowledge of the architecture of the skull, based on data in the literature. Arch. Neerl. Zool., 9:176, 1948.
van der Klaauw, C. J.: Size and position of the functional components of the skull (continuation). Arch. Neerl. Zool., 9:177, 1951.
van der Klaauw, C. J.: Size and position of the functional components of the skull (conclusion). Arch. Neerl. Zool., 9:369, 1952.
van der Linden, F. P. G. M.: Interrelated factors in the morphogenesis of teeth, the development of the dentition and craniofacial growth. Schweiz. Monatsschr. Zahnheilkd., 80:518, 1970.
van der Linden, F. P. G. M.: A study of roentgenocephalometric bony landmarks. Am. J. Orthod., 59:111, 1971.
van der Linden, F. P. G. M.: Changes in the position of posterior teeth in relation to ruga points. Am. J. Orthod., 74:142, 1978.
van der Linden, F. P. G. M.: Control mechanisms regulating the development of the dentition. In: *Control Mechanisms in Craniofacial Growth.* Ed. by J. A. McNamara, Jr. Ann Arbor, University of Michigan, Center for Human Growth and Development, 1979.
van der Linden, F. P. G. M.: Changes in the dentofacial complex during and after orthodontic treatment. Eur. J. Orthod., 1:97, 1979.
van der Linden, F. P.: Bone morphology and growth potential: a perspective of postnatal normal bone growth. Prog. Clin. Biol. Res., 187:181, 1985.
van der Linden, F., and H. S. Duterloo: *Development of the Human Dentition.* Hagerstown, Md., Harper & Row, 1976.
van der Linden, F. P. G. M., and D. H. Enlow: A study of the anterior cranial base. Angle Orthod., 41:119, 1971.
Van der Weele, L. Th., and J. M. H. Dibbets: Conceptual models for analyzing symptoms of temporomandibular joint dysfunction. J. Craniomandib. Practice, 4:357, 1986.
Van der Weele, L. Th., and J. M. H. Dibbets: Helkimo's index: A scale or just a set of symptoms? J. Oral Rehabil., 14:229, 1987.
van Limborgh, J.: The regulation of the embryonic development of the skull. Acta Morphol. Neerl. Scand., 7:101, 1968.
van Limborgh, J.: A new view on the control of the morphogenesis of the skull. Acta Morphol. Neerl. Scand., 8:143, 1970.
van Limborgh, J.: The role of genetic and local environmental factors in the control of postnatal craniofacial morphogenesis. Acta Morphol. Neerl. Scand., 10:37, 1972.
van Limborgh, J., and H. L. Verwoerd-Verhoef: Effects of artifical unilateral facial clefts on growth of the skull in young rabbits. J. Dent. Res., 47:1013, 1968.
Van Ness, A. L.: Implantation of cranial base metallic markers in nonhuman primates. Am. J. Phys. Anthropol., 49:85, 1978.
Van Ness, A. L., O. M. Merrill, and J. R. Hansel: Cephalometric roentgenography for nonhuman primates utilizing a surgically implanted head positioner. Am. J. Phys. Anthropol., 43:141, 1975.
Van Vlierberghe, M. V., L. Dermant, and H. Jansen: Influence of the digastric muscle and occlusion on the sagittal growth of the mandible: An experimental investigation in minipigs. Eur. J. Orthod., 8:1, 1986.
van Wieringen, J. C.: Secular growth changes. In: *Human Growth.* Vol. 3, 2nd Ed. Ed. by F. Falkner and J. M. Tanner. New York, Plenum Press, 1986.
Van Wyk, J. J., L. E. Underwood, R. N. Marshall, and R. C. Lister: The somatomedins: a new class of growth-regulating hormones. (With an introduction by Stanley M. Garn.) In: *Control Mechanisms in Craniofacial Growth.* Ed. by J. A. McNamara, Jr. Ann Arbor, University of Michigan, Center for Human Growth and Development, 1975.
Vargervik, K.: Morphological evidence of muscle influence on dental arch width. Am. J. Orthod., 76:21, 1979.
Vargervik, K., and E. Harvold: Experiments on the interaction between orofacial function and morphology. Ear Nose Throat J., 66:201, 1987.
Vargervik, K., and A. J. Miller: Observations on the temporal muscle in craniosynostosis. Birth Defects, 18:45, 1982.
Varjanne, I., and K. Koski: Cranial base, sagittal jaw relationship and occlusion. A radiological-craniometric appraisal. Proc. Finn. Dent. Soc., 78:179, 1982.
Vermeij-Keers, C.: Transformation in the facial region of the human embryo. Adv. Anat. Embryol. Cell Biol., 46:5, 1972.
Vidic, B.: The morphogenesis of the lateral nasal wall in the early prenatal life of man. Am. J. Anat., 130:121, 1971.
Vig, P. S.: Respiratory mode and morphological types: Some thoughts and preliminary conclusions.

In: *Naso-Respiratory Function and Craniofacial Growth.* Ed. by J. A. McNamara, Jr. Ann Arbor, University of Michigan, Center for Human Growth and Development, 1979.
Vig, P. S., and A. B. Hewitt: Asymmetry of the human facial skeleton. Angle Orthod., 45:125, 1975.
Vig, P. S., D. M. Sarver, D. J. Hall, and D. W. Warren: Quantitative evaluation of nasal airflow in relation to facial morphology. Am. J. Orthod., 79:263, 1981.
Vig, P. S., J. F. Rink, K. J. Showfety: Adaptation of head posture in response to relocating the center of mass: A pilot study. Am. J. Orthod., 83:138, 1983.
Vignery, A., and R. Baron: Dynamic histomorphometry of alveolar bone remodeling in the adult rat. Anat. Rec., 196:191, 1980.
Vilmann, H.: The growth of the cranial base in the albino rat revealed by roentgencephalometry. J. Zool. (Lond.), 159:283, 1969.
Vilmann, H.: The growth of the cranial base in the Wistar albino rat studied by vital staining with alizarin red S. Acta Odont. Scand., 29 (Suppl. 59), 1971.
Vilmann, H.: Osteogenesis in the basioccipital bone of the Wistar albino rat. Scand. J. Dent. Res., 80:410, 1972.
Vilmann, H.: Growth of the cranial base in the rat. In: *Development of the Basicranium.* Ed. by J. F. Bosma. DHEW Pub. 76:989, NIH, Bethesda, Md., 1976.
Vinkka, H., L. Odent, and K. Koski: Variability of the craniofacial skeleton. II. Comparison between two age groups. Am. J. Orthod., 67:34, 1975.
Vinkka, H., L. Odent, D. Odent, K. Koski, and J. A. McNamara: Variability of the craniofacial skeleton. III. Radiographic cephalometry of juvenile *Macaca mulatta.* Am. J. Orthod., 68:1, 1975.
Virapongse, C., R. Shapiro, M. Sarwar, S. Bhlmani, and E. Crelin: Computed tomography in the study of the development of the skull base. 1. Normal morphology. J. Comput. Assist. Tomogr., 9:85, 1985.
Vlastovsky, V. G.: The secular trend in the growth and development of children and young persons in the Soviet Union. Hum. Biol., 38:219, 1966.
von Truenfels, H., and D. von Torklus: Relation between atlas position, prognathism and prognathous jaw anomalies. Z. Orthop., 121:657, 1983.
Voorhies, J. W., and J. W. Adams: Polygonic interpretations of cephalometric findings. Angle Orthod., 21:194, 1951.
Walker, A. C.: Functional anatomy of oral tissues: mastication and deglutition. In: *Textbook of Oral Biology.* Ed. by J. H. Shaw, E. A. Sweeney, C. C. Cappuccino, and S. M. Meller. Philadelphia, W. B. Saunders, 1978.
Walker, G.: The composite vectorgram: A key to the analysis and synthesis of craniofacial growth. J. Dent. Res., 50:1508, 1971.
Walker, G.: A new approach to the analysis of craniofacial morphology and growth. Am. J. Orthod., 61:221, 1972.
Walker, G., and C. J. Kowalski: A two-dimensional coordinate model for the quantification, description, analysis, prediction and simulation of craniofacial growth. Growth, 35:119, 1971.
Walker, G., and C. J. Kowalski: On the growth of the mandible. Am. J. Phys. Anthropol., 36:111, 1972a.
Walker, G., and C. J. Kowalski: Use of angular measurements in cephalometric analysis. J. Dent. Res., 51:1015, 1972b.
Walker, G. F., and C. J. Kowalski: Early diagnosis and growth estimation of prognathism. J. Dent. Res., 56:B193, 1977.
Walters, R. D.: Facial changes in the *Macaca mulatta* monkey by orthopedic opening of the midpalatal suture. Angle Orthod., 45:169, 1975.
Warren, D. W.: Aerodynamic studies of upper airway: Implications for growth, breathing and speech. In: *Naso-respiratory Function and Craniofacial Growth.* Ed. by J. A. McNamara, Jr. Ann Arbor, University of Michigan, Center for Human Growth and Development, 1979.
Washburn, S. L.: The effect of facial paralysis on the growth of the skull of rat and rabbit. Anat. Rec., 94:163, 1946a.
Washburn, S. L.: The effect of removal of the zygomatic arch in the rat. J. Mammal., 27:169, 1946b.
Washburn, S. L.: The relation of the temporal muscle to the form of the skull. Anat. Rec., 99:239, 1947.
Waterlow, J. C., R. Buzina, W. Keller, J. Lane, M. Nichaman, and J. Tanner: The presentation and use of height and weight data for comparing the nutritional status of groups of children under the age of 10 years. Bull. WHO, 55:489, 1977.
Wedden, S. E., and C. Tickle: Facial morphogenesis and pattern formation. Prog. Clin. Biol. Res., 217A:335, 1986.
Weidenreich, F.: The special form of the human skull in adaptation to the upright gait. Z. Morphol. Anthropol., 24:157, 1924.
Weidenreich, F.: The brain and its role in the phylogenetic transformation of the human skull. Trans. Am. Phil. Soc., 31:321, 1941.
Weidenreich, F.: Generic, specific, and subspecific characters in human evolution. Am. J. Phys. Anthropol., 31:413, 1946b.

Weijs, W. A., and B. Hillen: Relationships between masticatory muscle cross-section and skull shape. J. Dent. Res., 638:154, 1984.
Weijs, W. A., and B. Hillen: Correlations between the cross-sectional area of the jaw muscles and craniofacial size and shape. Am. J. Phys. Anthropol., 70:423, 1986.
Weinmann, J. P.: Adaptation of the periodontal membrane to physiologic and pathologic changes. Oral Surg., 8:977, 1955.
Weinmann, J. P., and H. Sicher: *Bone and Bones,* 2nd Ed. St. Louis, C. V. Mosby, 1955.
Weinstein, S., D. C. Haack, L. Y. Morris, B. B. Snyder, and H. E. Attaway: On an equilibrium theory of tooth position. Angle Orthod., 33:1, 1963.
Wendell, P. D., R. Nanda, T. Sakamoto, and S. Nakamura: The effects of chin cup therapy on the mandible: A longitudinal study. Am. J. Orthod., 87:265, 1985.
West, E. E.: Facial patterns in malocclusion. J. Dent. Res., 31:464, 1952.
West, E. E.: Analysis of early Class I, Division 1 treatment. Am. J. Orthod., 43:769, 1955.
Wexler, M. R., and B. G. Sarnat: Rabbit snout growth after dislocation of nasal septum. Arch. Otolaryngol., 81:68, 1965.
Whetten, L. L., and L. E. Johnston: The control of condylar growth: An experimental evaluation of the role of the lateral pterygoid muscle. Am. J. Orthod., 8:181, 1985.
Whitaker, L. A., and J. A. Katowitz: Nasolacrimal apparatus in craniofacial deformity. In: *Symposium on Diagnosis and Treatment of Craniofacial Anomalies.* Ed. by J. M. Converse, J. G. McCarthy, and D. Wood-Smith. St. Louis, C. V. Mosby, 1979.
Widmalm, S. E., J. H. Lillie, and M. M. Ash, Jr.: Anatomical and electromyographic studies of the lateral pterygoid muscle. J. Oral Rehabil., 14:429, 1987.
Widman, D. J.: Functional and morphologic considerations of the articular eminence. Angle Orthod., 58:221, 1988.
Williams, R. E., and R. F. Ceen: Craniofacial growth and the dentition. Pediatr. Clin. North Am., 29:503, 1982.
Williams, S., and B. Melsen: Condylar development and mandibular rotation and displacement during activator treatment. An implant study. Am. J. Orthod., 8:322, 1982a.
Williams, S., and B. Melsen: The interplay between sagittal and vertical growth factors: An implant study of activator treatment. Am. J. Orthod., 8:327, 1982b.
Winnberg, A., and H. Pancherz: Head posture and masticatory muscle function: An EMG investigation. Eur. J. Orthod., 5:209, 1983.
Winnberg, A.: Suprahyoid biomechanics and head posture: An electromyographic, videofluorographic and dynamographic study of hyo-mandibular function in man. Swed. Dent. J., 46:1, 1987.
Wisth, P. J.: Nose morphology in individuals with angle Class I, II, or III occlusions. Acta Odont. Scand., 33:53, 1975.
Wisth, P. J.: Mandibular function and dysfunction in patients with mandibular prognathism. Am. J. Orthod., 85:193, 1984.
Woo, J. K.: On the asymmetry of the human skull. Biometrika, 22:324, 1931.
Woo, J. K.: Ossification and growth of the human maxilla, premaxilla and palate bones. Anat. Rec., 105:737, 1949.
Wood, W. W.: A functional comparison of the deep and superficial parts of the human anterior temporal muscle. J. Dent. Res., 65:924, 1986.
Wood, N. D., L. E. Wragg, O. G. Stuteville, and R. G. Oglesby: Osteogenesis of the human upper jaw: Proof of the nonexistence of a separate premaxillary centre. Arch. Oral Biol., 14:1331, 1969.
Woodside, D. G.: Distance, velocity and relative growth rate standards for mandibular growth for Canadian males and females age three to twenty years. Toronto, Canada, American Board of Orthodontics, Thesis, 1969.
Woodside, D. G., and S. Linder-Aronson: The channelization of upper and lower anterior face heights compared to population standard in males between ages 6 to 20 years. Eur. J. Orthod., 1:25, 1979.
Woodside, D. G., A. Metaxas, and G. Altuna: The influence of functional appliance therapy on glenoid fossa remodeling. Am. J. Orthod. Dentofacial Orthop., 92:181, 1987.
World Health Organization: A growth chart for international use in maternal and child health care. Geneva: Office Publications, WHO, 1978.
World Health Organization: Measuring change in nutritional status. Geneva: Office of Publications, WHO, 1983.
World Health Organization: The growth chart: A tool for use in infant and child health care. Geneva: Office of Publications, WHO, 1986.
Worms, F. W., L. H. Meskin, and R. J. Isaacson: Open-bite. Am. J. Orthod., 59:589, 1971.
Wright, D. M., and B. C. Moffett: The postnatal development of the human temporomandibular joint. Am. J. Anat., 141:235, 1974.
Wright, H. V., I. Kjaer, and C. W. Asling: Roentgen cephalometric studies on skull development in rats. II. Normal and hypophysectomized males; sex differences. Am. J. Phys. Anthropol., 25:103, 1966.
Wylie, W. L.: Assessment of antero-posterior dysplasia. Angle Orthod., 17:97, 1947.

Wylie, W. L., and E. L. Johnson: Rapid evaluation of facial dysplasia in the vertical plane. Angle Orthod., 22:164, 1952.
Wyshak, G.: Secular changes in age of menarche in a sample of US women. Ann. Hum. Biol., 10:75, 1983.
Young, R. W.: The influence of cranial contents on postnatal growth of the skull in the rat. Am. J. Anat., 105:383, 1959.
Young, W. F.: The influence of the growth of the teeth and nasal septum on growth of the face. In: *Report of 36th Congress.* Ed. by G. E. M. Hallett. Eur. Orthod. Soc., London, 1959, p. 385.
Youssef, E. H.: The development of the skull in a 34 mm human embryo. Acta Anat., 57:72, 1964.
Youssef, E. H.: The chondrocranium in the albino rat. Acta Anat., 64:586, 1966.
Youssef, E. H.: Development of the membrane bones and ossification of the chondrocranium in the albino rat. Acta Anat., 72:603, 1969.
Youdelis, R. A.: The morphogenesis of the human temporomandibular joint and its associated structures. J. Dent. Res., 45:182, 1966.
Zacharias, L. and W. Rand: Adolescent growth in height and its relation to menarche in contemporary American girls. Ann. Hum. Biol., 10:209, 1983.
Zacharias, L., W. Rand, and R. Wurtman: A prospective study of sexual development and growth in American girls: The statistics of menarche. Obstet. Gynecol. Surv., 31:325, 1976.
Zengo, A. N., C. A. L. Bassett, R. J. Pawluk, and G. Prountzos: In vivo bioelectric potentials in the dentoalveolar complex. Am. J. Orthod., 66:130, 1974.
Zins, J., J. Kusiak, L. Whitaker, and D. H. Enlow: Influence of recipient site on bone grafts to the face. J. Plast. Reconstr. Surg., 73:371, 1984.
Zuckerman, S.: Age changes in the basicranial axis of the human skull. Am. J. Phys. Anthropol., 13:521, 1955.
Zwarych, P. D., and M. B. Quigley: The intermediate plexus of the periodontal ligament: History and further investigations. J. Dent. Res., 44:383, 1965.

Index

Note: Numbers in *italics* refer to illustrations; numbers followed by (t) indicate tables.